TRANSLATIONAL NEUROIMMUNOLOGY IN MULTIPLE SCLEROSIS

TRANSLATIONAL NEUROIMMUNOLOGY IN MULTIPLE SCLEROSIS

FROM DISEASE MECHANISMS TO CLINICAL APPLICATIONS

Edited by

RUTH ARNON
Weizmann Institute of Science, Rehovot, Israel

ARIEL MILLER
Technion-Israel Institute of Technology; Carmel Medical Center, Haifa, Israel

AMSTERDAM • BOSTON • HEIDELBERG • LONDON
NEW YORK • OXFORD • PARIS • SAN DIEGO
SAN FRANCISCO • SINGAPORE • SYDNEY • TOKYO
Academic Press is an imprint of Elsevier

Academic Press is an imprint of Elsevier
125 London Wall, London EC2Y 5AS, United Kingdom
525 B Street, Suite 1800, San Diego, CA 92101-4495, United States
50 Hampshire Street, 5th Floor, Cambridge, MA 02139, United States
The Boulevard, Langford Lane, Kidlington, Oxford OX5 1GB, United Kingdom

Library of Congress Cataloging-in-Publication Data
A catalog record for this book is available from the Library of Congress

British Library Cataloguing-in-Publication Data
A catalogue record for this book is available from the British Library

ISBN: 978-0-12-801914-6

For information on all Academic Press publications
visit our website at https://www.elsevier.com/

ELSEVIER
Book Aid
International
Working together
to grow libraries in
developing countries
www.elsevier.com • www.bookaid.org

Typeset by TNQ Books and Journals

Contents

18. Currently Approved Injectable Disease-Modifying Drugs: Cytotoxic Immunosuppressive Drugs

G. EDAN

V

CURRENTLY APPROVED DMDs-ORAL

19. Fingolimod (Gilenya®)

S. SCHIPPLING

20. Oral Dimethyl Fumarate (BG-12; Tecfidera®) for Multiple Sclerosis

A. SALMEN, R.A. LINKER, AND R. GOLD

21. Emerging Therapies for Multiple Sclerosis

M.S. FREEDMAN

VI

SYMPTOMATIC & COMPLEMENTARY TREATMENTS

22. Treatment of Acute Relapses in Multiple Sclerosis

R. BERKOVICH

23. Shedding Light on Vitamin D and Multiple Sclerosis

J. SMOLDERS, R. HUPPERTS, AND J. DAMOISEAUX

24. Symptomatic and Complementary Treatments

P.S. SORENSEN, K. SCHREIBER, AND A.K. ANDREASEN

25. Cognitive Impairment in Multiple Sclerosis

M.P. AMATO AND B. GORETTI

VII

NOVEL & EMERGING STRATEGIES

26. Personalized Medicine and Theranostics: Applications to Multiple Sclerosis

T. PAPERNA, E. STAUN-RAM, N. AVIDAN, I. LEJBKOWICZ, AND A. MILLER

27. Stem Cell-Based Therapies, Remyelination, and Repair Promotion in the Treatment of Multiple Sclerosis

A. MERLINI, D. DE FEO, M. RADAELLI, C. LATERZA, F. CICERI, AND G. MARTINO

28. Stem Cells in Multiple Sclerosis

I. KASSIS, P. PETROU, AND D. KARUSSIS

29. T-Cell Vaccination: An Insight Into T-Cell Regulation

I.R. COHEN, N. FRIEDMAN, AND F.J. QUINTANA

30. Reversal of Misfortune: Therapeutic Strategies on the Horizon

L. STEINMAN

List of Contributors

R. Aharoni Weizmann Institute of Science, Rehovot, Israel

M.P. Amato University of Florence, Florence, Italy

A.K. Andreasen University Hospital Rigshospitalet, Copenhagen, Denmark

J. Antel McGill University, Montreal, QC, Canada

R. Arnon Weizmann Institute of Science, Rehovot, Israel

N. Avidan Technion-Israel Institute of Technology, Haifa, Israel

C. Avolio University of Foggia, Foggia, Italy

A. Bar-Or McGill University, Montreal, QC, Canada

K. Baruch Weizmann Institute of Science, Rehovot, Israel

R. Berkovich University of Southern California, Los Angeles, CA, United States

Z.N. Berneman University of Antwerp, Antwerp, Belgium

F. Ciceri San Raffaele Scientific Institute, Milan, Italy

I.R. Cohen The Weizmann Institute of Science, Rehovot, Israel

A. Coles University of Cambridge, Cambridge, United Kingdom

G. Comi Vita-Salute San Raffaele University, Milan, Italy

A. Compston University of Cambridge, Cambridge, United Kingdom

N. Cools University of Antwerp, Antwerp, Belgium

J. Damoiseaux Maastricht University Medical Center, Maastricht, The Netherlands

D. De Feo San Raffaele Scientific Institute, Milan, Italy

N. Deckx University of Antwerp, Antwerp, Belgium

G. Edan University Hospital, Rennes, France; West Neurosciences Network of Excellence (WENNE), Rennes, France

G. Fadda McGill University, Montreal, QC, Canada

M. Filippi Vita-Salute San Raffaele University, Milan, Italy

M.S. Freedman University of Ottawa, Ottawa, ON, Canada

N. Friedman The Weizmann Institute of Science, Rehovot, Israel

S. Giatti Department of Pharmacological and Biomolecular Sciences, University of Milan, Milan, Italy

R. Gold Ruhr-University Bochum, Bochum, Germany

S.M. Gold University Medical Center Hamburg-Eppendorf, Hamburg, Germany

B. Goretti University of Florence, Florence, Italy

E. Grazioli UPMC Northshore Neurology, Erie, PA, United States

R.H. Gross Icahn School of Medicine at Mount Sinai, New York, NY, United States

L.M. Healy McGill University, Montreal, QC, Canada

R. Hohlfeld Institute of Clinical Neuroimmunology, Ludwig-Maximilian University, Munich, Germany

R. Hupperts Zuyderland Medical Center, Sittard, The Netherlands; Maastricht University Medical Center, Maastricht, The Netherlands

J. Jones University of Cambridge, Cambridge, United Kingdom

D. Karussis Hebrew University, Ein–Karem, Jerusalem, Israel

I. Kassis Hebrew University, Ein–Karem, Jerusalem, Israel

B.C. Kieseier Heinrich-Heine University, Düsseldorf, Germany

C. Kolb SUNY University at Buffalo, UBMD Neurology, Buffalo, NY, United States

H. Lassmann Medical University of Vienna, Vienna, Austria

C. Laterza San Raffaele Scientific Institute, Milan, Italy

I. Lejbkowicz Technion-Israel Institute of Technology, Haifa, Israel

V.I. Leussink Heinrich-Heine University, Düsseldorf, Germany

R.A. Linker Friedrich Alexander University Erlangen-Nürnberg, Erlangen, Germany

F. Lublin Icahn School of Medicine at Mount Sinai, New York, NY, United States

G. Martino San Raffaele Scientific Institute, Milan, Italy

J.L. McCauley University of Miami, Miami, FL, United States

R.C. Melcangi Department of Pharmacological and Biomolecular Sciences, University of Milan, Milan, Italy

A. Merlini San Raffaele Scientific Institute, Milan, Italy

M.A. Michell-Robinson McGill University, Montreal, QC, Canada

A. Miller Technion-Israel Institute of Technology, Haifa, Israel; Carmel Medical Center, Haifa, Israel

J.R. Oksenberg University of California San Francisco, San Francisco, CA, United States

T. Paperna Technion-Israel Institute of Technology, Haifa, Israel

P. Petrou Hebrew University, Ein–Karem, Jerusalem, Israel

F.J. Quintana Harvard Medical School, Boston, MA, United States

M. Radaelli San Raffaele Scientific Institute, Milan, Italy

V.T.S. Rao McGill University, Montreal, QC, Canada

K. Regev Harvard Medical School, Boston, MA, United States

M.A. Rocca Vita-Salute San Raffaele University, Milan, Italy

A. Salmen Ruhr-University Bochum, Bochum, Germany

S. Schippling Neuroimmunology and Multiple Sclerosis Research, Department of Neurology, University Hospital Zurich and University of Zurich, Zurich, Switzerland

K. Schreiber University Hospital Rigshospitalet, Copenhagen, Denmark

M. Schwartz Weizmann Institute of Science, Rehovot, Israel

J. Smolders Canisius Wilhelmina Hospital, Nijmegen, The Netherlands; Zuyderland Medical Center, Sittard, The Netherlands

P.S. Sorensen University Hospital Rigshospitalet, Copenhagen, Denmark

E. Staun-Ram Technion-Israel Institute of Technology, Haifa, Israel

L. Steinman Stanford University, Stanford, CA, United States

H. Touil McGill University, Montreal, QC, Canada

M. Trojano University of Bari, Bari, Italy

C. Warnke Heinrich-Heine University, Düsseldorf, Germany

H.L. Weiner Harvard Medical School, Boston, MA, United States

B. Weinstock-Guttman SUNY University at Buffalo, Buffalo, NY, United States

H. Wekerle Max-Planck-Institute of Neurobiology, Martinsried, Munich, Germany

R.O. Weller University of Southampton, Southampton, United Kingdom

Preface

Science becomes productive when accompanied by creative thinking, and the results found are fruitful if they evoke new ideas and improve our understanding. The beauty of translational research is that it harnesses such creative thinking and new ideas for the development of remedies and therapeutic means for the benefit of mankind. Research on multiple sclerosis (MS), culminating in translational neuroimmunological approaches, is certainly a case in point.

Traditionally, MS has been considered an autoimmune disease, in which the immune system reacts against the body's own constituents, namely the myelin in the central nervous system. However, novel technologies including imaging systems have greatly progressed our understanding of the disease, leading to the recognition of the paramount role of the neurodegenerative processes in MS. Consequently, the various therapeutic modalities, both those that are currently in use and potential therapies in the pipeline, are all based on our understanding of its neuropathological characteristics.

This book presents the various aspects related to this issue, as addressed by the most illustrious researchers in the field of MS. We are aware of the complexities of preparing a book attempting to provide up-to-date information, and certain that there are issues that have not been covered; however, we hope that most of the areas in the field have been addressed, and will facilitate cross-fertilization among basic, translational, and clinical research practitioners. Divided in several clusters, the book starts with the pathology of MS and the various underlying mechanisms, including genetic and environmental factors, that help us understand its complexity. A separate section is devoted to magnetic resonance imaging, as that single technology has made the most significant advancement both in MS research and toward the development of therapies. This leads directly to the clinical applications, namely the various therapeutic modalities in the field of MS.

It should be borne in mind that although the disease was characterized more than 50 years ago, approved therapies have been in use only in the last 20 years. First to be introduced were immunosuppressive drugs, followed by immunomodulatory agents intended to alleviate the symptoms that result from the neuroimmune and neurodegenerative processes. Better understanding of the mechanisms of MS led to the design of more sophisticated and more specific medications. Three clusters in the book are devoted to various treatment modalities that are currently in use, mainly the injectable, oral, and complementary treatments, respectively. Finally, the last section is devoted to novel and emerging strategies, including neuroprotective and neurorepair-promoting strategies such as stem cell–based approaches, which are considered as key components of the future of MS treatment. Additionally, in view of the process of shifting from treating the disease to treating the patient, the book also includes a spotlight into the world of predictive and personalized medicine being applied to MS. As is well known, currently available MS treatment modalities are mainly disease-modifying agents and symptomatic treatments. Hopefully, with further understanding of the nature of MS and its underlying mechanisms, translational research will hopefully lead to its future cure.

Ruth Arnon
Ariel Miller

SECTION I

MS PATHOLOGY AND MECHANISMS

CHAPTER

1

Multiple Sclerosis Pathology: Inflammation Versus Neurodegeneration

H. Lassmann
Medical University of Vienna, Vienna, Austria

OUTLINE

1.1 INTRODUCTION

The key features of multiple sclerosis (MS) pathology have already been defined during the late 19th century. The cardinal alteration seen in MS patients is the presence of large confluent demyelinated plaques within the white matter of the brain and spinal cord, which can already be depicted by the naked eye[13,17] (Fig. 1.1). These demyelinated lesions are invariably centered by a vein, which shows perivascular accumulation of small round cells, mainly representing lymphocytes.[15,18,59] In contrast to lesions seen in other brain diseases, myelin loss is defined by the term primary demyelination, which implies that axons largely remain intact.[4] Active destruction of myelin sheaths occurs in close contact with macrophages, which

Translational Neuroimmunology in Multiple Sclerosis
http://dx.doi.org/10.1016/B978-0-12-801914-6.00001-5

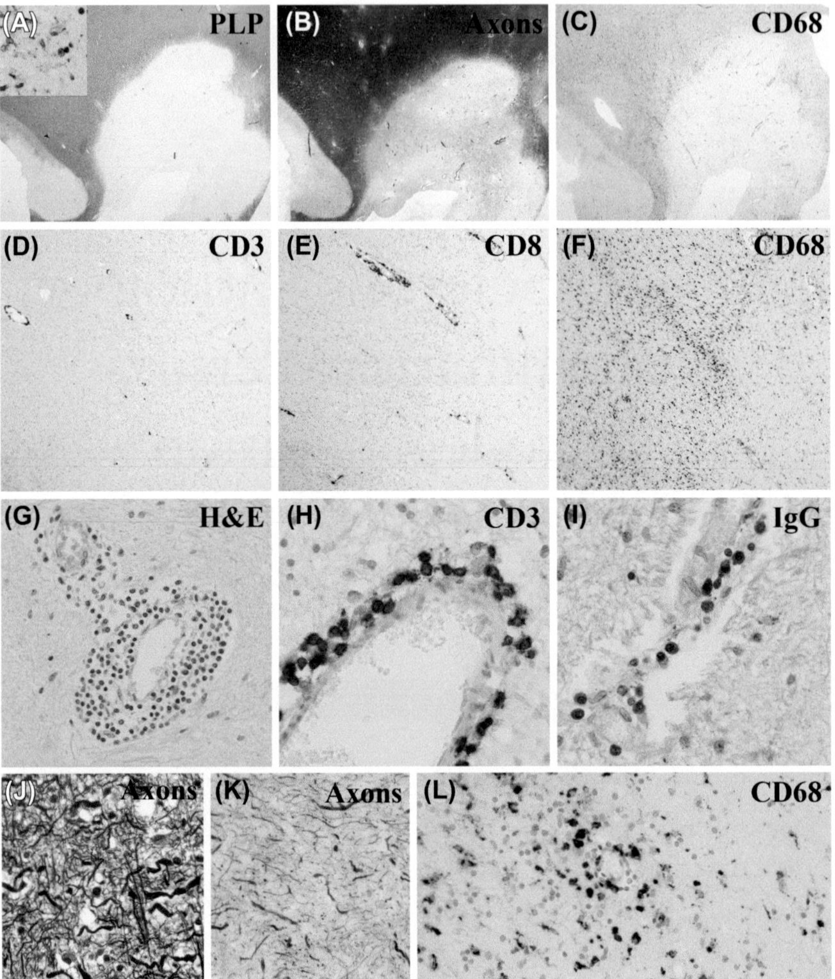

FIGURE 1.1 **Slowly expanding focal white matter lesion in progressive multiple sclerosis (MS): 56-year-old male patient with secondary-progressive MS with a disease duration of 372 months.** Periventricular demyelinated plaque in sections stained for PLP (A), Bielschowsky silver impregnation for axons (B), and the macrophage marker CD68 (C). In the myelin stain the plaque appears sharply demarcated from the surrounding normal-appearing white matter. However, at high magnification (insert, A) some macrophages with early myelin degradation products are visible. In the staining for axons (B) a profound reduction of axonal density can be seen in the plaque center, amounting to a more than 70% loss (K) in comparison to the periplaque white matter (J). Staining for the macrophage marker CD68 shows reduced macrophage/microglia density within the inactive demyelinated plaque center, but increased macrophage numbers at the active plaque edge. Immunocytochemistry for leukocyte markers shows the presence of inflammatory cells within such an active lesion, consisting of CD3 (D, H) and CD8 (E) positive T-cells. Note that there are very similar numbers of CD3 and CD8 positive cells in the lesions, suggesting that the vast majority of T-cells are CD8 positive cells. At the plaque edge, at the site of active demyelination, profound infiltration of CD68 positive macrophages is seen (F, L); extensive perivascular infiltrates are also seen in sections stained for hematoxylin and eosin (G). In addition to T-cells, high numbers of IgG-containing plasma cells are seen in the inflammatory infiltrates (I).

take up fragments of the disrupted myelin sheaths.[4] However in some cases—in particular, in those with very rapid disease evolution—dissolution of myelin sheaths may also occur in the absence of macrophages, suggesting a humoral factor to be involved in the destructive process.[48] Demyelination in the white matter is associated with profound astrocytic scar formation.[15] Signs suggestive of remyelination have been observed in MS lesions, defined as areas with reduced myelin density due to unusually thin myelin sheaths.[48] When present in the entire plaque such lesions have been named *Markschattenherde* (shadow plaques) by Schlesinger.[61]

Although there was widespread agreement in the early literature on the disease-specific nature of primary demyelination in MS, it was already clear in the earliest pathological descriptions of the disease that axons are injured and lost in the course of the disease process.[15] The respective studies described that acute axonal injury is mainly seen in fresh and active lesions, and associated with macrophage infiltration and characterized by focal swelling of damaged axons (axon spheroids[20]). Its extent is variable between cases and brain regions, and even attempts of axonal regeneration such as the formation of thin axonal sprouts can be seen (for review see Kornek and Lassmann[37]). Thus in the early decades of the 20th century most of the basic features of the pathology of MS were well defined. However, the pathological view of the disease was mainly centered on focal lesions in the white matter.

1.2 THE NEUROPATHOLOGY OF MULTIPLE SCLEROSIS AS SEEN IN 2015

The plaque-centered view on MS pathology served as the basis for detailed studies on inflammation, immune-mediated tissue injury, and remyelination. However, it became apparent that the MS brain is not only affected by focal demyelinated lesions in the white matter, but that there are also characteristic and severe changes in the gray matter and even in the white matter, which on gross inspection of the brain appears unaffected. Thus, MS is a disease that not only leads to focal brain lesions, but affects the entire central nervous system.

1.2.1 Gray Matter Demyelination

A major advance in the understanding of the disease process in MS, which came from neuropathological studies, was the discovery of the importance of gray matter lesions, and in particular of cortical lesions (Fig. 1.2). Although it has been recognized in early studies on MS pathology that demyelination may also affect the cortex, the brain stem nuclei, and the gray matter of the spinal cord,[11,34] technical limitations in myelin staining did not allow researchers to grasp the global extent of these lesions. This changed with the introduction of very sensitive immunocytochemical methods, which allowed reproducible detection of even the very thin myelin sheaths, which normally are present in gray matter areas.[7,52] Studies based on this new technology showed that demyelination is extensive in the gray matter and these studies were also instrumental in defining the shape and distribution of such lesions in the MS brain.[40] In the cerebral cortex three different lesion types were identified:

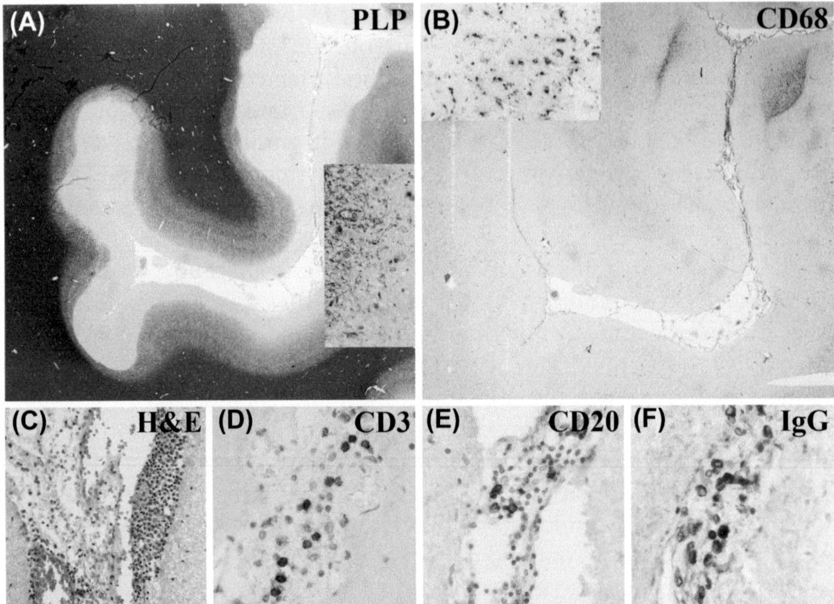

FIGURE 1.2 **Active subpial demyelinating lesion in the cortex of a multiple sclerosis (MS) patient with secondary-progressive MS (SPMS): 41-year-old male patient with SPMS with a disease duration of 137 months.** Extensive band-like subpial demyelination within a cortical sulcus; most areas of this lesion are inactive, but there is a small zone with active demyelination, characterized by focal accumulation of macrophages at the border of the cortical lesion ((B) and insert to (B)); active demyelination is also documented by the presence of macrophages with early myelin degradation products (insert to (A)). The active part of the cortical lesion is present in an area of adjacent profound meningeal inflammation, while little inflammation is seen in the meninges overlying the inactive demyelinated cortical areas (B). The meningeal inflammatory infiltrates show aggregates of inflammatory cells in part resembling lymph follicle-like structures (C), and are composed of CD3 positive T-cells (D), some CD20 positive B-cells (E), and large numbers of IgG-containing plasma cells (F).

the cortico-subcortical lesions, affecting both the gray and white matter; the small focal intracortical lesions; and the subpial lesions, which have their base on the outer surface of the cortex, and may extend over several adjacent gyri and sulci.[52] The latter expand from the outer toward the inner cortical layers and in extreme examples may reach the cortico-subcortical border (Fig. 1.2). Interestingly, subpial lesions only rarely pass the cortico-subcortical border and extend into the white matter. Focal cortico-subcortical lesions and small intracortical lesions are seen at all stages of the disease, being present even in patients with acute or relapsing-remitting MS (RRMS).[40] Also subpial lesions are sometimes seen in early disease stages, in particular in biopsy tissue from patients with large tumefactive lesions in the white matter.[43] However, their incidence and size are small in early disease stages, but such lesions become very prominent in patients with secondary or primary progressive disease. In the progressive stage an average of 20–30% of the forebrain cortex is demyelinated and cortical demyelination may reach levels of up to 90% in extreme cases.[40] Besides the cerebral cortex, demyelination is also prominent in the cerebellar cortex,[39] in the basal ganglia, in the hippocampus[26] and in the spinal cord gray matter.[27] In the latter, gray matter demyelination frequently exceeds demyelination in the white matter.

Subpial cortical demyelination in the forebrain cortex is most abundant in the cortical sulci and in the deep invaginations of the brain surface, such as the insular cortex, the cingulate cortex, and the temporal lobe including the hippocampus.[40] A potential reason for this distribution is that subpial demyelination is related to meningeal inflammation and the degree of active demyelination and neurodegeneration correlates well with the severity of the meningeal inflammatory process[45] (Fig. 1.2). Meningeal inflammatory infiltrates are composed of T- and B-lymphocytes and plasma cells, but these cells are retained in the meninges and only rarely infiltrate the cortical parenchyme.[8] Active demyelination is associated with a rim of activated microglia between the demyelinated lesion and the adjacent normal-appearing gray matter. Such a pattern of lesion formation can best be explained by the action of a soluble factor, produced by meningeal inflammatory cells, which diffuses into the cortex and induces demyelination either directly or indirectly through microglia activation. It has to be emphasized that subpial cortical lesions are the most specific neuropathological changes that can be found in the MS brain. It is not seen in any other inflammatory, metabolic, or neurodegenerative disease.[23,49] This indicates that demyelination and neurodegeneration in MS is not simply a bystander reaction in response to the inflammatory process, but is mediated by a mechanism that is specific for MS. This view is supported by experimental data. Such cortical lesions do not occur in models of experimental autoimmune encephalomyelitis (EAE), which are purely driven by encephalitogenic T-cells. Similar lesions, however, can be seen in EAE animals with a demyelinating anti-MOG (myelin oligodendrocyte glycoprotein) autoantibody response.[53,64] However, despite extensive efforts, no evidence has been found so far that demyelinating antibodies may play a role in the induction of cortical lesions in MS patients, and in contrast to cortical lesions in experimental models, active demyelination is not associated with complement activation.[10]

1.2.2 Diffuse Injury of the Normal-Appearing White and Gray Matter

Another feature of MS pathology, which has been defined during the 2000s, is the diffuse injury of the normal-appearing white and gray matter. It has been shown decades ago that in MS patients the normal-appearing white matter (NAWM) is frequently abnormal.[2] Changes consist of a low-grade diffuse inflammation with perivascular accumulation and parenchymal infiltration of lymphocytes, diffuse microglia activation, diffuse astrocytic gliosis, some small perivascular areas of demyelination, and a diffuse neuroaxonal loss and injury.[40] A characteristic alteration in areas of diffuse white matter injury is the appearance of multiple small microglia nodules, which originally had been regarded as a specific feature of MS pathology.[54] However, similar microglia nodules can also be seen in the white matter adjacent to stroke lesions and may represent a tissue reaction to retrograde or anterograde axonal degeneration.[60] Diffuse white matter changes are mild or absent in acute MS and early RRMS, but become very prominent in patients with primary-progressive MS (PPMS) or secondary-progressive MS (SPMS).[40] They may in part be explained as a secondary reaction to axonal degeneration in focal demyelinated plaques. However, their extent does not correlate with white matter lesion load,[19,21] but shows a partial correlation with the extent of demyelination in the cerebral cortex and thus may be in part secondary to neuronal and axonal degeneration in the demyelinated cortex.[40] In addition, however, they may occur independently from focal lesions, since there is also a correlation between diffuse changes in the NAWM of the spinal cord with the degree of meningeal inflammation.[3]

1.2.3 Remyelination

Another area of MS pathology where major progress has been made during the last decade is related to spontaneous remyelination of demyelinated lesions. A widely held view on MS pathophysiology is that in contrast to the situation in experimental animals there is a general failure of remyelination within the lesions and that stimulation of remyelination is a suitable therapeutic option in the patients.[24] This is based on the observation that the majority of plaques in the white matter in the majority of patients are chronic demyelinated lesions. However, neuropathological evidence suggests that this view may be an oversimplification.

The analysis of classical active plaques, which dominate in the early stages of acute and relapsing MS, has shown that in the majority of the lesions early stages of remyelination are abundant.[55] This is reflected by the presence of numerous, apparently newly differentiated oligodendrocytes and the formation of thin myelin sheaths with shortened internodes around the demyelinated axons in active lesions, which still are infiltrated by numerous macrophages that contain myelin degradation products.[57] Thus the primary regenerative response to demyelination in such active MS lesions is similar to that seen in experimental models. Although such changes of early remyelination are seen very frequently, this is not the case in all patients. In fact, different patterns of oligodendrocyte loss have been described, when active lesions of different patients were compared with each other.[44]

Despite the quite extensive early remyelination within many active lesions, shadow plaques, reflecting stable and durable remyelination, are rare even in early stages of MS.[9] Thus, in these lesions failure of stable remyelination may not be due to lack of remyelination, but may be related to a problem of stability of the newly formed myelin. In a systematic study on remyelination in MS, Prineas et al.[56] have shown that remyelinated areas are targets for recurrent demyelination. Furthermore, the likelihood of new active demyelination in the MS brain is higher in remyelinated areas than in the NAWM.[9] These data indicate that remyelination in early MS stages is quite effective, but unstable as long as the inflammatory disease process is active.

A different situation is encountered in slowly expanding lesions (Fig. 1.1), which are the dominant type of active lesions in patients with PPMS or SPMS.[9] In such lesions very little remyelination is seen in the inactive lesion center.[28] This is in part due to massive reduction of axons within these areas. However, even the preserved axons in the plaques in their vast majority remain demyelinated. In contrast to classical active lesions of early MS, macrophage infiltration is sparse in slowly expanding lesions and the number of microglia cells within the lesions is reduced in comparison to that in the adjacent NAWM.[41] Thus, these lesions lack a cell population, which may be a source for remyelination-stimulating cytokines or growth factors. NG2 positive progenitor cells are present within such lesions, although in variable numbers, and differentiation into premyelinating oligodendrocytes is sparse or absent.[14,38] In part oligodendrocyte progenitor cell differentiation and remyelination may be inhibited by the astrocytic scar and by remyelination-inhibiting factors present in the extracellular matrix.[62] Alternatively, remyelination failure may be due to lack of proper axonal signals or a nonpermissive environment, lacking the respective cytokines and growth factors. Despite the very low extent of active early remyelination in slowly expanding lesions of progressive MS, extensive stable remyelination, reflected by the presence of shadow plaques, can be seen in some patients, and in such patients up to 80% of their entire plaques can be remyelinated.[50,51]

Overall the incidence of shadow plaques is higher in patients with PPMS compared to patients with SPMS.[9] In addition, incidence and extent of remyelination depends upon the location of the lesions. Periventricular lesions show very little remyelination, while a much higher incidence of shadow plaques is seen in subcortical lesions.[51] Furthermore, gray matter lesions tend to remyelinate more effectively compared to white matter plaques.[1] The discrepancy between the very low extent of early remyelination in slowly expanding lesions and the variable, but sometimes very abundant number of shadow plaques in patients with progressive MS is currently not well understood. Shadow plaques may represent lesions with stable and permanent remyelination, which have originally arisen during the early disease stages. Alternatively, gradual and slow myelin repair could occur in chronically demyelinated lesions when active inflammatory demyelination has subsided.

1.3 THE RELATION BETWEEN INFLAMMATION WITH DEMYELINATION AND NEURODEGENERATION

There is good agreement that inflammation plays an important role in the pathogenesis of MS. As described earlier, pathological studies have defined inflammation as a hallmark of the disease process. Genome-wide association studies revealed multiple candidate gene or gene regions to be associated with MS susceptibility and the vast majority of the putative target genes code for proteins, involved in T-cell mediated inflammation.[33] Furthermore clinical investigations identified an MS-associated immunological signature, suggestive of a major role of T-cells in the initiation of the disease.[58]

However, based on divergent clinical, imaging, and neuropathological evidence there is currently an open debate, whether inflammation is the driving force of tissue injury in all stages of MS (outside-in hypothesis); whether inflammation starts an injury cascade, which in the progressive stage of the disease becomes independent from immune mechanisms; or whether neurodegeneration is the primary event in MS, which is modified and amplified by inflammatory mechanisms (inside-out hypothesis).[65] Several arguments have been put forward supporting the inside-out hypothesis, but experience from pathology offers alternative explanations.

It is now well established that current antiinflammatory treatments are effective in the early relapsing stage of the disease, but fail when patients have entered the progressive stage.[67] A widely held view is that the disease starts with inflammation, but triggers a cascade of events, which in the progressive stage lead to progression of tissue injury independent from the inflammatory process. However, pathological studies show that active demyelination and neurodegeneration in all stages of MS, including the progressive stage, is associated with inflammatory infiltrates, composed by T-cells, B-cells, plasma cells, and activated macrophages and microglia[25] (Figs. 1.1 and 1.2). Whether this inflammatory reaction drives tissue injury or represents recruitment of bystander inflammation to sites of tissue damage cannot be decided from current pathological evidence, in particular since little is known so far regarding the exact phenotype and activation status of inflammatory cells in different lesion or disease stages. Nevertheless, in patients at late stages of progressive MS inflammation may decline to levels seen in age-matched controls. In these patients no active demyelination occurs and also acute axonal neurodegeneration declines to levels seen in age-matched controls.[25] The lack of efficacy of antiinflammatory treatments in patients with progressive MS can also be explained by the observation that at this stage of the disease the inflammatory reaction is at least in part hidden behind a closed or repaired blood–brain barrier.[31]

Another argument for the inside-out hypothesis comes from the observation that in active lesions in the cortex as well as in the white matter, ongoing demyelination is associated with activated microglia, while lymphocytes are mainly seen in perivascular spaces and in the meninges.[5,36,45] As discussed earlier, this pattern of tissue damage can be well explained by the presence of a soluble demyelinating factor, which induces demyelination either directly or indirectly through microglia activation.

Several studies have provided convincing evidence that active cortical demyelination and neurodegeneration is associated with meningeal inflammation.[16,32,40] This, however, was not seen in other studies.[8,35] The difference between these studies is that the association between meningeal inflammation and subpial cortical demyelination is mainly seen in patients at earlier stages of the progressive phase of the disease and in patients with more rapid and severe clinical deterioration. Furthermore, in those reports, which could not find an association between meningeal inflammation and cortical demyelination, no clear evidence for demyelinating activity defined by stringent criteria is provided. It may not be surprising that an association between inflammation and tissue injury in the cortex is absent when inactive demyelinated lesions are analyzed (see also Fig. 1.2).

Thus, overall, the currently available data suggest that inflammation is indeed the driving force of demyelination and tissue injury in all stages of the disease, although final proof for this view is lacking. Despite these data there is some other evidence that clinically relevant neurodegeneration may occur in MS patients, which is truly independent from inflammation. As mentioned, inflammation in certain MS patients may be present at only very low levels, which are similar to those seen in age-matched controls. In these patients also active axonal neurodegeneration is very low, but it is still present to an extent, which is similar to that seen in age-matched controls.[25] However, due to the preexisting brain damage, functional reserve capacity may be exhausted[6] and thus, even a low grade of progressive age-related neurodegeneration, which may be clinically insignificant in aged controls, may result in disease progression in MS patients. Recent data indicate that a dominant mechanism of demyelination and neurodegeneration in MS patients is oxidative injury and mitochondrial damage. Evidence for this view comes from several observations. Active tissue injury in MS lesions is associated with profound microglia activation and microglia cells express enzymes involved in the production of reactive oxygen species, such as nicotinamide adenin dinucleotide phosphatase (NADPH) oxidase or myeloperoxidase.[22,29] This is associated with profound accumulation of oxidized lipids and DNA in oligodendrocytes, neurons, and axons, where these markers for oxidative stress label cells in the process of degeneration.[23,30] Mitochondria are particularly vulnerable to oxidative injury and this is reflected by profound alterations in mitochondrial function, by the reduced expression of proteins of the respiratory chain,[47] and by expansion of mitochondria with gene deletions.[12] Mitochondrial injury results in energy deficiency, which is particularly important in MS due to the increased energy demand of demyelinated axons. These mechanisms have been summarized in detail in several reviews.[42,46] Interestingly, similar mechanisms of tissue injury have also been suggested to operate in normal brain aging and in age-related neurodegenerative diseases.[66]

1.4 SUMMARY AND CONCLUSIONS

Neuropathology has provided major insights into the pathophysiology of MS. It is now clear that MS is not simply a disease with focal lesions in the white matter, but that the disease process affects the entire brain and spinal cord. The pathological hallmark of MS is the presence of lesions with primary demyelination. Although primary demyelination is also seen in some other diseases (eg, those with virus infection of oligodendrocytes), widespread subpial cortical demyelination, associated with meningeal inflammation, is unique to MS. MS is a chronic inflammatory disease of the central nervous system and most of the data suggest that tissue injury in the brain of the patients is indeed driven by the inflammatory process. However, demyelination and neurodegeneration are not simply a bystander reaction to chronic inflammation, since they are not present in comparable form in any other inflammatory human brain disease. Oxidative injury and mitochondrial damage, leading to a state of histotoxic hypoxia, seems to be a dominant pathway of cell and tissue injury in MS, but it is currently unresolved why these changes are so prominent in MS in comparison to other inflammatory or neurodegenerative diseases in humans. Overall the neuropathological changes seen in MS are only incompletely reproduced in models of EAE and this is particularly the case for the commonly used mouse models, induced by sensitization with the 35–55 peptide of myelin oligodendrocyte glycoprotein.[63]

References

1. Albert M, Antel J, Brück W, Stadelmann C. Extensive cortical remyelination in patients with chronic multiple sclerosis. *Brain Pathol*. 2007;17:129–138.
2. Allen IV, McQuid S, Miradkhur M, Nevin G. Pathological abnormalities in the normal-appearing white matter in multiple sclerosis. *Neurol Sci*. 2001;22:141–144.
3. Androdias G, Reynolds R, Chanal M, Ritleng C, Confavreux C, Nataf S. Meningeal T cells associate with diffuse axonal loss in multiple sclerosis spinal cords. *Ann Neurol*. 2010;68:465–476.
4. Babinski J. Recherches sur l'anatomie pathologique de la sclerose en plaque et etude comparative des diverses varietes de la scleroses de la moelle. *Arch Physiol (Paris)*. 1885;5-6:186–207.
5. Barnett MH, Prineas JW. Relapsing and remitting multiple sclerosis: pathology of the newly forming lesion. *Ann Neurol*. 2004;55:458–468.
6. Bjartmar C, Wujek JR, Trapp BD. Axonal loss in the pathology of MS: consequences for understanding the progressive phase of the disease. *J Neurol Sci*. 2003;206(2):165–171.
7. Bo L, Vedeler CA, Nyland HI, Trapp BD, Mork SJ. Subpial demyelination in the cerebral cortex of multiple sclerosis patients. *J Neuropath Exp Neurol*. 2003;62:723–732.
8. Bo L, Vedeler CA, Nyland H, Trapp BD, Mork SJ. Intracortical multiple sclerosis lesions are not associated with increased lymphocyte infiltration. *Mult Scler*. 2003;9:323–331.
9. Bramow S, Frischer JM, Lassmann H, et al. Demyelination versus remyelination in progressive multiple sclerosis. *Brain*. 2010;133:2983–2998.
10. Brink BP, Veerhuis R, Breij EC, van der Valk P, Dijkstra CD, Bo L. The pathology of multiple sclerosis is location dependent: no significant complement activation is detected in purely cortical lesions. *J Neuropath Exp Neurol*. 2005;64:147–155.
11. Brownell B, Hughes JT. The distribution of plaques in the cerebrum in multiple sclerosis. *J Neurol Neurosurg Psychiatry*. 1962;25:315–320.
12. Campbell GR, Ziabreva I, Reeve AK, et al. Mitochondrial DNA deletions and neurodegeneration in multiple sclerosis. *Ann Neurol*. 2011;69:481–492.
13. Carswell R. *Pathological Anatomy: Illustrations on Elementary Forms of Disease*. London: Longman; 1838.

14. Chang A, Tourtellotte WW, Rudick R, Trapp BD. Premyelinating oligodendrocytes in chronic lesions of multiple sclerosis. *New Engl J Med*. 2002;346:165–173.
15. Charcot JM. *Lecons sur les maladies du systeme nerveux faites a la Salpetriere*. 4th ed. Vol. 1. Paris: Tome; 1880.
16. Choi S, Howell OW, Carassiti D, et al. Meningeal inflammation plays a role in the pathology of primary progressive multiple sclerosis. *Brain*. 2012;135:2925–2937.
17. Cruveilhier J. *Anatomie pathologique du corps humain*. Paris: J.B. Bailliere; 1842.
18. Dawson JW. The histology of disseminated sclerosis. *Trans R Soc*. 1916;50:517–540.
19. DeLuca GC, Williams K, Evangelou N, Ebers GC, Esiri MM. The contribution of demyelination to axonal loss in multiple sclerosis. *Brain*. 2006;129:1507–1516.
20. Doinikow B. Über De- und Regenerationserscheinungen an Achsenzylindern bei der multiplen Sklerose. *Z Ges Neurol Psych*. 1915;27:151–178.
21. Evangelou N, DeLuca GC, Owens T, Esiri MM. Pathological study of spinal cord atrophy in multiple sclerosis suggests limited role of focal lesions. *Brain*. 2005;128:29–34.
22. Fischer MT, Sharma R, Lim J, et al. NADPH oxidase expression in active multiple sclerosis lesions in relation to oxidative tissue damage and mitochondrial injury. *Brain*. 2012;135:886–899.
23. Fischer MT, Wimmer I, Hoftberger R, et al. Disease-specific molecular events in cortical multiple sclerosis lesions. *Brain*. 2013;136(Pt 6):1799–1815.
24. Franklin RJ, Ffrench-Constant C. Remyelination in the CNS: from biology to therapy. *Nat Rev Neurosci*. 2008;9(11):839–855.
25. Frischer JM, Bramow S, Dal Bianco A, et al. The relation between inflammation and neurodegeneration in multiple sclerosis brains. *Brain*. 2009;132:1175–1189.
26. Geurts JJ, Bo L, Roosendaal SD, et al. Extensive hippocampal demyelination in multiple sclerosis. *J Neuropath Exp Neurol*. 2007;66:819–827.
27. Gilmore CP, Donaldson I, Bo L, Owens T, Lowe J, Evangelou N. Regional variations in the extent and pattern of grey matter demyelination in multiple sclerosis: a comparison between the cerebral cortex, cerebellar cortex, deep grey matter nuclei and the spinal cord. *J Neurol Neurosurg Psychiatry*. 2009;80:182–187.
28. Goldschmidt T, Antel J, König FB, Brück W, Kuhlmann T. Remyelination capacity of the MS brain decreases with disease chronicity. *Neurology*. 2009;27:1914–1921.
29. Gray E, Thomas TL, Betmouni S, Scolding N, Love S. Elevated activity of microglial expression of myeloperoxidase in demyelinated cerebral cortex in multiple sclerosis. *Brain Pathol*. 2008;18:86–95.
30. Haider L, Fischer MT, Frischer JM, et al. Oxidative damage and neurodegeneration in multiple sclerosis lesions. *Brain*. 2011;134:1914–1924.
31. Hochmeister S, Grundtner R, Bauer J, et al. Dysferlin is a new marker for leaky brain blood vessels in multiple sclerosis. *J Neuropathol Exp Neurol*. 2006;65:855–865.
32. Howell OW, Reeves CA, Nicholas R, et al. Meningeal inflammation is widespread and linked to cortical pathology in multiple sclerosis. *Brain*. 2011;134:2755–2771.
33. International Multiple Sclerosis Genetics Consortium (IMSGC). Analysis of immune-related loci identifies 48 new susceptibility variants for multiple sclerosis. *Nat Genet*. 2013;45:1353–1360.
34. Kidd T, Barkhof F, McConnell R, Algra PR, Allen IV, Revesz T. Cortical lesions in multiple sclerosis. *Brain*. 1999;122:17–26.
35. Kooi EJ, Geurts JJ, van Horssen J, Bo L, van der Valk P. Meningeal inflammation is not associated with cortical demyelination in chronic multiple sclerosis. *J Neuropathol Exp Neurol*. 2009;68:1021–1028.
36. Kooi EJ, Strijbis EM, van der Valk P, Geurts JJ. Herterogeneity of cortical lesions in multiple sclerosis: clinical and pathologic implications. *Neurology*. 2012;79:1369–1376.
37. Kornek B, Lassmann H. Axonal pathology in multiple sclerosis: a historical note. *Brain Pathol*. 1999;9:651–656.
38. Kuhlmann T, Miron V, Cui Q, Wegner C, Antel J, Brück W. Differentiation block of oligodendroglial progenitor cells as a cause for remyelination failure in chronic multiple sclerosis. *Brain*. 2009;131:1749–1758.
39. Kutzelnigg A, Faber-Rod JC, Bauer J, et al. Widespread demyelination in the cerebellar cortex in multiple sclerosis. *Brain Pathol*. 2007;17:38–44.
40. Kutzelnigg A, Lucchinetti CF, Stadelmann C, et al. Cortical demyelination and diffuse white matter injury in multiple sclerosis. *Brain*. 2005;128:2705–2712.
41. Lassmann H. The architecture of active multiple sclerosis lesions. *Neuropathol Appl Neurobiol*. 2011;37:698–710.
42. Lassmann H, van Horssen J, Mahad D. Progressive multiple sclerosis: pathology and pathogenesis. *Nat Rev Neurol*. 2012;8:647–656.

43. Lucchinetti CF, Popescu BFG, Bunyan RF, et al. Inflammatory cortical demyelination in early multiple sclerosis. *N Engl J Med*. 2011;365:2188–2197.
44. Lucchinetti C, Brück W, Parisi J, Scheithauer B, Rodriguez M, Lassmann H. A quantitative analysis of oligodendrocytes in multiple sclerosis lesions. A study of 117 cases. *Brain*. 1999;122:2279–2295.
45. Magliozzi R, Howell O, Vora A, et al. Meningeal B-cell follicles in secondary progressive multiple sclerosis associate with early onset of disease and severe cortical pathology. *Brain*. 2007;130:1089–1104.
46. Mahad D, Trapp B, Lassmann H. Pathological mechanisms in progressive multiple sclerosis. *Lancet Neurol*. 2015;14:183–193.
47. Mahad D, Ziabreva I, Lassmann H, Turnbull D. Mitochondrial defects in acute multiple sclerosis lesions. *Brain*. 2008;131:1722–1735.
48. Marburg O. Die sogenannte "akute multiple Sklerose". *Jahrb Psychiatr*. 1906;27:211–312.
49. Moll NM, Rietsch AM, Ransohoff AJ, et al. Cortical demyelination in PML and MS: similarities and differences. *Neurology*. 2008;70:336–343.
50. Patani R, Balaratnam M, Vora A, Reynolds R. Remyelination can be extensive in multiple sclerosis despite a long disease course. *Neuropathol Appl Neurobiol*. 2007;33:277–287.
51. Patrikios P, Stadelmann C, Kutzelnigg A, et al. Remyelination is extensive in a subset of multiple sclerosis patients. *Brain*. 2006;129:3165–3172.
52. Peterson JW, Bo L, Mork S, Chang A, Trapp BD. Transected neurites, apoptotic neurons and reduced inflammation in cortical multiple sclerosis lesions. *Ann Neurol*. 2001;50:389–400.
53. Pomeroy IM, Matthews PM, Frank JA, Jordan EK, Esiri MM. Demyelinated neocortical lesions in marmoset autoimmune encephalomyelitis mimic those in multiple sclerosis. *Brain*. 2005;128:2713–2721.
54. Prineas JW, Kwon EE, Cho ES, et al. Immunopathology of secondary-progressive multiple sclerosis. *Ann Neurol*. 2001;50(5):646–657.
55. Prineas JW, Barnard RO, Kwon EE, Sharer LR, Cho ES. Multiple sclerosis: remyelination of nascent lesions. *Ann Neurol*. 1993;33:137–151.
56. Prineas JW, Barnard RO, Revesz T, Kwon EE, Sharer L, Cho ES. Multiple sclerosis. Pathology of recurrent lesions. *Brain*. 1993;116:681–693.
57. Prineas JW, Kwon EE, Goldenberg PZ, et al. Multiple sclerosis. Oligodendrocyte proliferation and differentiation in fresh lesions. *Lab Invest*. 1989;61:489–503.
58. Raj T, Rothamel K, Mostafavi S, et al. Polarization of the effects of autoimmune and neurodegenerative risk alleles in leukocytes. *Science*. 2014;344:519–523.
59. Rindfleisch E. Histologisches Detail zur grauen Degeneration von Gehirn und Rückenmark. *Arch Pathol Anat Physiol Klin Med (Virchow)*. 1863;26:474–483.
60. Singh S, Metz I, Amor S, van der Valk P, Stadelmann C, Brück W. Microglia nodules in early multiple sclerosis white matter are associated with degenerating axons. *Acta Neuropathol*. 2013;125:595–608.
61. Schlesinger H. Zur Frage der akuten multiplen Sklerose und der encephalomyelitis disseminata im Kindesalter. *Arb Neurol Inst (Wien)*. 1909;17:410–432.
62. Sloane JA, Batt C, Ma Y, Harris ZM, Trapp B, Vartanian T. Hyaluronan blocks oligodendrocyte progenior maturation and remyelination through TLR2. *Proc Natl Acad Sci USA*. 2010;107:11555–11560.
63. Schuh C, Wimmer I, Hametner S, et al. Oxidative tissue injury in multiple sclerosis is only partly reflected in experimental disease models. *Acta Neuropathol*. 2014;128:247–266.
64. Storch MK, Bauer J, Linington C, Olsson T, Weissert R, Lassmann H. Cortical demyelination can be modeled in specific rat models of autoimmune encephalomyelitis and is major histocompatability complex (MHC) haplotype-related. *J Neuropathol Exp Neurol*. 2006;65:1137–1142.
65. Trapp BD, Nave KA. Multiple sclerosis: an immune or neurodegenerative disorder? *Annu Rev Neurosci*. 2008;31:247–269.
66. Wang X, Wang W, Li L, Perry G, Lee HG, Zhu X. Oxidative stress and mitochondrial dysfunction in Alzheimer's disease. *Biochim Biophys Acta*. 2014;1842:1240–1247.
67. Wiendl H, Hohlfeld R. Multiple sclerosis therapeutics: unexpected outcomes clouding undisputed successes. *Neurology*. 2009;72(11):1008–1015.

CHAPTER

2

Immune Dysregulation in Multiple Sclerosis

K. Regev, H.L. Weiner

Harvard Medical School, Boston, MA, United States

OUTLINE

2.1 INTRODUCTION

Multiple sclerosis (MS) is considered to be a chronic inflammatory, demyelinating disease of the central nervous system (CNS) of an autoimmune nature. A complex interplay between genetic susceptibility and environmental factors leads to disease evolution. For many years MS has primarily been considered an inflammatory T-cell-mediated disease, a hypothesis based on observations made from the animal model, experimental autoimmune encephalomyelitis (EAE). As more has been learned about both MS and the complex immune networks that drive autoimmunity, a pivotal role for the innate immune system, B cells, endothelial function, and different T-cell subpopulations has been identified as key components in MS immune dysregulation.

Translational Neuroimmunology in Multiple Sclerosis
http://dx.doi.org/10.1016/B978-0-12-801914-6.00002-7

2.2 INNATE IMMUNITY

2.2.1 Dendritic Cells

Dendritic cells (DCs) are professional antigen-presenting cells (APCs) that play a key role in promoting activation and differentiation of naïve T cells. DCs sense environmental, microbial, and endogenous antigens and, after migrating to the lymphoid organs, present processed antigens to naïve T cells and induce either antigen-specific immunity or immune tolerance. The interaction of DCs with T-cells is crucial in determining T-cell differentiation into either effector T cells (Th1, Th2, and Th17 cells) or regulatory T cells (Tregs), thus shaping the adaptive immune response.[1] Several observations point to a DC role in the immune pathogenesis of MS including the presence of DCs in vascular-rich regions of the healthy CNS and accumulation of DCs in the CNS parenchyma during a wide range of inflammatory processes.[2–4] Patients with MS have increased numbers of blood-derived myeloid DCs, which express the maturation marker CD80 and secrete interleukin (IL)-12 and tumor necrosis factor (TNF) as compared to healthy controls.[5] Furthermore, DCs differentiated in vitro from blood monocytes taken from patients with MS secrete higher levels of proinflammatory cytokines IFN-γ, TNF-α, and IL-6 as compared to cells taken from healthy individuals.[6] DCs are also important for the activation of $CD8^+$ T cells and natural-killer (NK) cells, as they can induce either cytotoxic or regulatory NK cells.[7]

In EAE pathogenesis, several studies have reported the involvement of DCs, showing accumulation of these cells in CNS during inflammation.[8,9] In addition, transfer of DCs pulsed in vitro with antigen[10,11] activates encephalitogenic T cells and result in either induction of disease or tolerance,[12] depending upon the activation state of DCs and mechanism of antigen uptake.[13] DCs isolated from the CNS of EAE mice are the most potent stimulators of naïve T cells in the presence or absence of endogenous peptide, suggesting the contribution of DCs in epitope spreading in the CNS during the disease.[8,14] Thus DCs can both initiate adaptive autoimmunity in MS and contribute to disease progression at later stages of disease.

2.2.2 Natural-Killer Cells

NK cells are cells of the innate immune system that are involved in the early defense against various forms of stress such as microbial infection or tumor transformation.

NK cells contribute to both effector and regulatory functions of innate immunity via their cytotoxic activity and their ability to secrete pro- and antiinflammatory cytokines and growth factors. NK subsets are defined based on surface marker expression. CD56dim$CD16^+$ NK cells are a cytolytic subset and CD56bright$CD16^+$ NK cells lack perforin and secrete abundant quantities of antiinflammatory cytokines. In EAE, some studies suggest that NK cells contribute to disease progression, while others demonstrate a protective role and even a role in repair through the production of neurotrophic factors such as brain-derived neurotrophic factor (BDNF).[15]

NK cells are present in demyelinating lesions of patients with MS[16] and decreased cytotoxic activity of circulating NK cells has been described in MS during clinical relapses.[17] Studies have suggested a beneficial effect of the CD56bright NK cell subset in MS patients as their number is increased by immunomodulatory and immunosuppressant therapy and correlates with a treatment response.[18,19] In a study of daclizumab, a humanized antibody against the IL-2 receptor chain, an expansion of CD56bright immunoregulatory NK cells was observed and was correlated with a decreased number of Gad$^+$ lesions on MRI, whereas only marginal effects on CD4$^+$ T cells were reported.[18]

2.2.3 Microglia, Macrophages, and Monocytes

Microglial cells are the resident macrophage population of the CNS and regulate local innate and adaptive immune responses in CNS tissue. Microglia constitute up to 10–15% of the total glial population in the CNS. Microglial activation consists of the release of proinflammatory cytokines, reactive oxygen intermediates, proteinases (eg, matrix metalloproteinases (MMPs)), and complement proteins. In addition, microglial cells can recruit a variety of leukocytes via chemokine secretion as well as performing phagocytosis of debris, apoptotic cells, and pathogens, with consequent antigen presentation to T cells. In addition, microglia cells may also play a nourishing and antiinflammatory role by secreting growth factors and antiinflammatory cytokines.[20] In MS, activated phagocytes can be found in white matter lesions (early and late) and in gray matter subpial lesions,[21–24] though it is not clear whether these are resident microglia or infiltrating monocytes. In EAE studies, a reduction in disease severity was observed when activated microglia/monocytes were killed either by ganciclovir administration to EAE induced in CD11b-HSV-TK mice or by using clodronate liposomes.[25]

In both MS and EAE, microglia/monocytes were shown to be involved in demyelination and phagocytosis of degraded myelin and neuronal debris.[26] They also have a role in antigen presentation to T cells, and play an important role in epitope spreading observed in Theiler's virus infection and EAE progression in SJL mice (though they are not as efficient as DCs).[14] Microglia and macrophages can also support remyelination, promote neural survival, and suppress the adaptive immune response in the CNS. For example, depletion of microglia and macrophages leads to impaired remyelination in toxin-induced demyelinating animal models.[27] Microglia can also secrete growth factors such as nerve growth factor and BDNF, thus supporting neural survival. Consistent with this, BDNF was found to be expressed by microglia/macrophages in MS lesions.[28] Microglia/macrophages can also decrease the adaptive immune response in the CNS by secreting IL-10, TGFβ, and nitric oxide, or by expressing inhibitory molecules such as PD-L1 (B7-h1).[29,30] In addition, microglia upregulate indoleamine 2,3-dioxygenase, an enzyme that modulates tryptophan metabolism and thus generates an antiinflammatory microenvironment.[31] We have identified a unique microglial signature that will allow a better understanding of microglia and monocytes in MS[32] and have applied this work to the investigation of EAE.[33]

2.3 ADAPTIVE IMMUNITY

2.3.1 CD4 T Cells

Based on the findings from animal experiments and human studies, MS is presumed to be a T-cell-driven autoimmune disease of the CNS characterized by the formation of inflammatory lesions within the CNS white matter consisting of activated T and B lymphocytes, macrophages, and microglia. The autoimmune T-cell hypothesis of MS was suggested by the work of Kabat and colleagues,[34] who demonstrated that immunization of monkeys with myelin antigens dissolved in an adjuvant resulted in spinal cord and brain inflammation similar to the neuropathology of MS. This model disease, initially designated experimental allergic encephalomyelitis (EAE), was then established in a variety of species, including rodents. A milestone was achieved when an adoptive transfer model was developed by Paterson[35] demonstrating that the disease could be transferred from an immunized rat to a naïve rat by injection of lymph node cells. Later, this model of adoptive transfer (or "passive") EAE was further refined when intravenous inoculation of T cells reactive only against myelin basic protein led to the development of clinical paralysis in syngeneic rats.[36] The EAE model is now termed experimental autoimmune encephalomyelitis and serves as the primary model for the development of new therapeutic strategies in MS.

Other evidence supporting the pivotal role of CD4 T cells in MS includes their presence within CNS lesions and the cerebrospinal fluid (CSF) of MS patients. MS is associated with HLA-DR and HLA-DQ molecules, molecules responsible for presenting processed antigen to the T-cell receptors on CD4 T cells. Mice transgenic for human HLA-DR or HLA-DQ molecules are susceptible to EAE.[37,38] Furthermore, mice that express both MS-associated HLA-DR and MS patient-derived T-cell receptors develop spontaneous EAE. Major histocompatibility complex (MHC) class I molecules, on the other hand, which are recognized by CD8 cytotoxic T lymphocytes, show only weaker genetic associations.[39]

Evidence that myelin-reactive T cells are able to induce inflammatory demyelination in humans derives from a clinical trial[40] in which MS patients were treated with an altered peptide ligand of myelin basic protein (MBP) to shift the MBP T-cell repertoire into a more tolerogenic state. Although there were some positive effects, two patients developed exacerbations associated with increased contrast-enhancing lesions and a 2000-fold expansion of MBP-specific T cells with an increased in avidity and cross-reactivity.

The chronology of the developing MS lesion as currently understood begins with the activation of autoreactive CD4 T cells in the periphery, which then transmigrate through the blood–brain barrier into the CNS. Once in the CNS they are locally reactivated by APCs and recruit additional T cells and macrophages to establish the inflammatory lesion. These cells mediate myelin, oligodendrocyte, and axon damage, leading to neurologic dysfunction (Fig. 2.1). In parallel, immunomodulatory elements are activated, and begin to limit inflammation and initiate repair, which often results in at least partial remyelination and clinical remission.

The thymus negatively selects high-avidity autoreactive T cells. This process prevents such cells from populating the peripheral immune repertoire and plays a crucial role in preventing autoimmunity (central tolerance). However, low-avidity autoreactive T cells that cross-react with foreign antigens in a higher avidity regularly escape this selection process.[41,42] T cells reactive to myelin antigens are found in the peripheral blood of virtually every healthy individual;[43] nevertheless, myelin-specific T cells from MS patients have a higher avidity, state of activation, and potency to induce inflammation than myelin-specific T cells from healthy individuals.[44]

Apart from a priori defects in thymic selection and maturation,[45] two possible tolerance-breaking mechanisms have been proposed. The molecular mimicry hypothesis has been coined following the observation that similarities between peptides from pathogens and

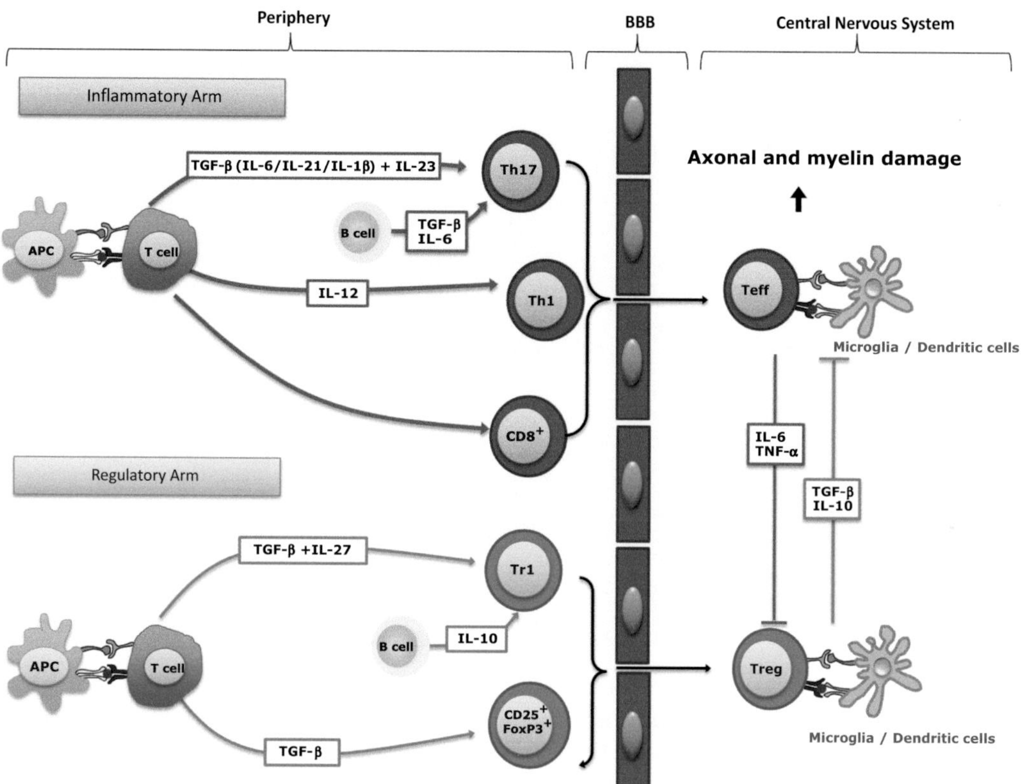

FIGURE 2.1 **Immune mechanisms in multiple sclerosis.**

autoantigens can result in cross-activation of autoreactive T cells.[46] It is less clear if it primarily serves a role in maintaining the T-cell repertoire (ie, is entirely physiologic) or is also involved in initiating autoimmune diseases. A second possible mechanism is that autoreactive T cells are activated by a nonspecific inflammatory event. This mechanism is termed bystander activation and could partly explain increased relapse rates after bacterial viral and parasite infections (Fig. 2.2).[47–49]

Once naïve CD4 T cells have been primed by APCs in peripheral lymphoid organs and differentiated into effector CD4 T cells, they reenter the circulation. A local reactivating signal, such as a repeated stimulation by antigen, MHC class II complexes, or a highly inflammatory cytokine environment, would enable them to exert their effector function fully. Activated CD4 T cells secrete proinflammatory cytokines such as IFN-γ and TNF-α, which cause upregulation of several endothelial adhesion molecules, including ICAM-1 and VCAM-1. Corresponding to the increase in endothelial adhesion molecules, CD4 T cells express higher amounts of their integrin counterparts, LFA-1 and VLA-4.[50]

After CD4 T cells have bound to the endothelium, they can transmigrate via either a transcellular or a paracellular route, during which time tight junction proteins need to be modulated.[51] In MS, there is evidence that activated autoreactive CD4 T cells and activated

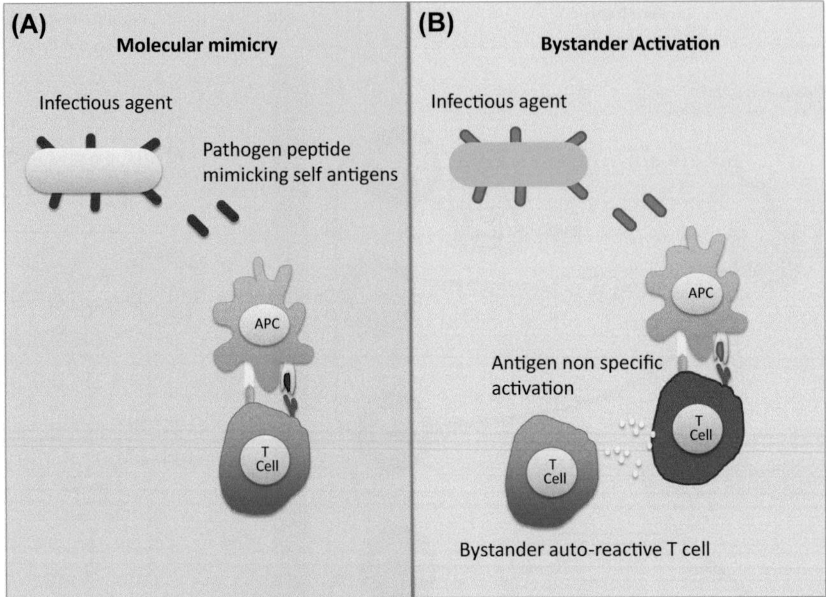

FIGURE 2.2 **Initiation of autoimmunity.**

monocytes secrete elevated amounts of proteolytic enzymes, such as MMPs, and apart from mediating direct CNS tissue damage, these proteases can degrade tight junction proteins and thereby cause blood–brain barrier disruption and tissue edema.[52] Correspondingly, expression of MMP-2, -3, -7, and -9 is increased in EAE and MS lesions and higher levels of MMP-9 are found in the CSF during relapses.[53]

Upon cell contact with an activated and antigen-loaded perivascular APC, CD4 autoreactive T cells are strongly reactivated, as seen by an acute upregulation of proinflammatory cytokines (ie, IFN-γ, IL-17, and TNF-α), proteases (MMP-9), and chemokines (CCL5). This reactivation enables them to infiltrate the parenchyma.[54]

A growing body of evidence now suggests that a subset of T cells perform regulatory function, Tregs, and they serve to control and modulate autoimmune reactions.[55] Several subtypes of Tregs have been described. Forkhead box protein 3 (FoxP3)-expressing natural Tregs and IL-10-producing, T-regulatory type 1 cells (Tr1) are the best-studied types of $CD4^+$ Tregs in humans and experimental animal models. It was shown that they play a crucial role during autoimmune neuroinflammation. Both cell types seem to be particularly important for MS. Depletion of $CD4^+CD25^+$ Tregs results in the onset of systemic autoimmune diseases in mice.[56] Furthermore, cotransfer of Tregs with $CD4^+CD25^-$ cells prevents the development of experimentally induced autoimmune diseases.[57] A population of $CD4^+$CD25hi Tregs in human peripheral blood and thymus[58,59] are anergic to in vitro antigenic stimulation and strongly suppress the proliferation of responder T cells upon coculture.

Similar to $FoxP3^+$ Tregs, evidence that Tr1 cells might play a role in MS came first from studies of experimental animal models. Numerous studies have demonstrated the importance of IL-10 in murine EAE and the induction of Tr1-like cells by dexamethasone and vitamin D3 was beneficial in the treatment of EAE.[60]

Regarding the encephalitogenic T-cell population, IL-12-induced proinflammatory CD4 Th1 cells were believed to be the main T-effector cells; however, EAE studies revealed that the inhibition or knockout of the Th1 cytokine IFN-γ results in a more exacerbated disease course.[61–64]

In 2005, Th17 cells were described[65,66] and have been shown to be prominently involved in a number of autoimmune diseases, including psoriasis, inflammatory bowel disease, rheumatoid arthritis, lupus, and MS.[67] Several pieces of evidence suggest that Th17 cells are important in both EAE and MS.[68] However, similar to Th1-lineage cytokines, modulation of Th17-related cytokines, such as IL-17A, IL-17F, and IL-22, yielded conflicting results in EAE.[69–71] In MS lesions, IL-17 is upregulated[72] and expressed not only by CD4 T cells but also by CD8 T cells and astrocytes.[73] A different migration pattern has been found for Th17-type immune cells, probably based on the impact of the bradykinin system.[74] The damage process itself is mediated by the local upregulation and release of specific harmful cytokines, including IFN-γ and members of the TNF family such as TNF-related apoptosis-inducing ligand or CD95 (Fas/APO-1) ligand and the activation of complement.[75,76] As a consequence, not only are oligodendrocytes damaged but neurons undergo cell death, presumably via apoptosis.

Adoptive transfer EAE experiments showed that both Th1 and Th17 cells are independently able to induce CNS autoimmunity[77–79] with distinct patterns of EAE histology, CNS chemokine profile, and response to selective cytokine depending on cell subset.

A subpopulation of CD4 T cells in MS patients was found to produce both IFN-γ and IL-17, hence exhibiting characteristics of both Th1 and Th17 cells (Th117 cells).[80] Th117 cells, like pure Th17 cells, are elevated during relapses compared to states of remission, migrate more efficiently through human brain endothelium than IL-17 or IFN-γ single expressors, and are also found to be involved in EAE.[81,82] The importance of Th17 in MS has been shown in initial clinical trials in which anti-IL17 antibody has been shown to have a positive effect on MS relapses.[83]

2.3.2 CD8 T Cells

$CD8^+$ T cells in MS have not been investigated to the extent of $CD4^+$ T cells. However, a striking feature of MS lesions is that $CD8^+$ T cells are more abundant than $CD4^+$ T cells; one group found that perivascular cuffs at the edges of active demyelinating plaques contained up to 50 times more $CD8^+$ than $CD4^+$ T cells[84] contrasting the $CD4^+$:$CD8^+$ ratio of about 3:1 to 6:1 in the CSF and 2:1 in the peripheral blood.[85] $CD8^+$ T cells are found to be oligoclonally expanded within lesions, CSF, and peripheral blood, suggesting involvement of antigen-specific responses.[86] Although the associations of HLA class I alleles with MS are weaker, HLA-A*0301 was found to double the risk for MS in an HLA-DR2-independent fashion.[87]

In addition to CD8 effector cells, there is evidence for the role of regulatory/suppressor $CD8^+$ T cells, as this subset was decreased in both CSF and blood in patients with exacerbations versus patients in remission or controls.[88] Genetic evidence supporting this observation is that the MHC class I gene HLA-A2 (A*0201) confers protection against disease, suggesting that some $CD8^+$ T cells are beneficial, whereas others are pathogenic.

Finally, IL-17-secreting $CD8^+$ T cells are also observed within active lesions, along with equal amounts of IL-17-expressing $CD4^+$ T cells.[73]

2.3.3 B Cells

Multiple lines of evidence support a role for B cells and antibodies in MS. A classic hallmark of MS is the presence of oligoclonal IgG appearing as oligoclonal bands (OCBs) in isoelectric focusing in the CSF. OCBs are present in around 95% of MS patients and arise from clonally expanded Ig-secreting cells.[89] Furthermore, clonally expanded B cells are found in brain parenchyma, meninges, and CSF of MS patients.[90] Although there have been many studies of autoantibodies in MS, no clear autoantibody has been identified.[91]

The most important piece of evidence suggesting that B cells play an important role in MS is the observation that rituximab, a depleting anti-CD20 mAb dramatically reduces clinical relapses and gadolinium-enhancing lesions in patients with relapsing-remitting MS.[92] While rituximab is not being further developed in MS, another anti-CD20 mAb, ocrelizumab, has showed similar clinical efficacy.[93] Long-lived plasma cells, which normally reside in the bone marrow and in MS patients also in the CNS, are not eliminated by this treatment and there is no immediate change in antibody levels. The clinical and MRI effects of anti-B-cell therapy occurs within a few months of treatment, suggesting that the effect is related to the

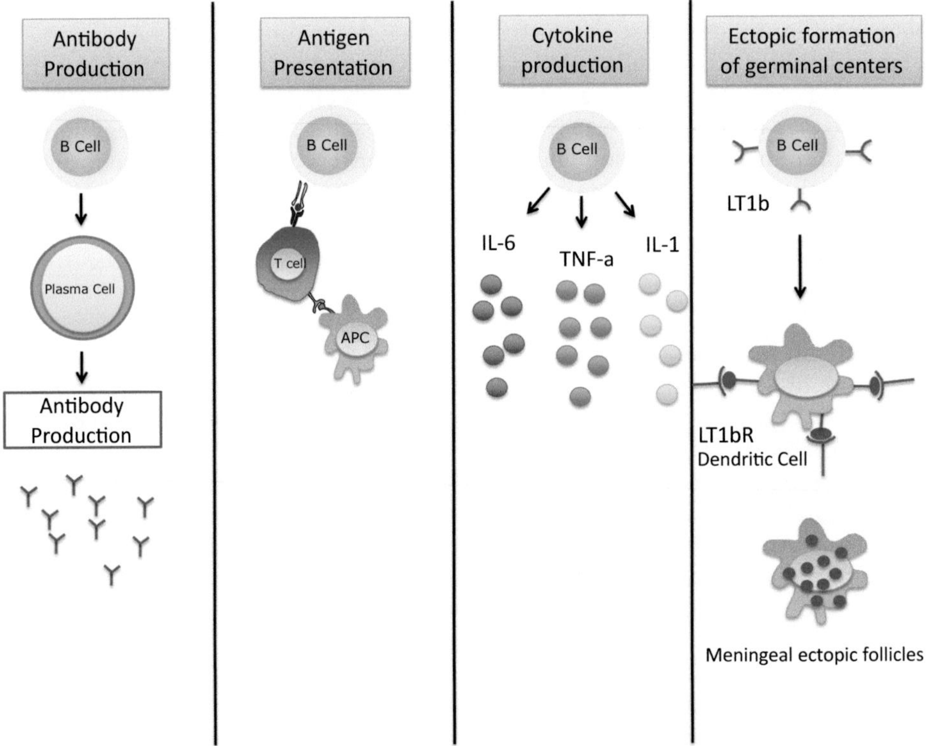

FIGURE 2.3 **The diverse functional roles of B cells.**

antigen-presenting properties of B cells and their effect on T cells. Indeed there are different subsets of B cells and it appears that the effect of anti-B-cell therapy is to remove a B-cell subset that drives proinflammatory T-cell responses.[94]

Further evidence for the role of B cells in MS is provided by genetic studies. In a recent report, the cell types in which the known MS genetic risk alleles are active were identified combining data on MS-associated single nucleotide polymorphisms and DNase I hypersensitive sites. In addition to Th1, Th17, $CD8^+$ cytotoxic T cells, and $CD56^+$ NK cells, $CD19^+$ B cells showed significant enrichment of genomic regions associated with MS.[95] The associated risk alleles include genes particularly important for B-cell function and homeostasis such as CD40, CXCR4, and CXCR5. Finally, investigators have identified ectopic lymphoid follicles within the meninges in two-thirds of secondary-progressive MS brain specimens.[96] These structures harbor B cells and plasma cells lie in close proximity to subpial lesions, and correlate with early onset of disease.[97] Fig. 2.3 summarizes B-cell function of potential relevance in MS.

2.4 CONCLUSION

The immune system is dysregulated at a number of levels in MS. Proof of the central role of the immune system in MS is that the large number of FDA-approved therapies for MS act by affecting the dysregulated immune system, either by suppressing proinflammatory responses or enhancing regulatory networks. This review has focused on relapsing forms of MS in which peripheral immune dysregulation plays a crucial role. Immune mechanisms associated with progressive forms of MS are less well understood and appear to involve the local innate immune response in the CNS and neurodegeneration triggered by the initial inflammatory response.[98]

References

1. Liu YJ, Kanzler H, Soumelis V, Gilliet M. Dendritic cell lineage, plasticity and cross-regulation. *Nat Immunol.* 2001;2(7):585–589.
2. Pashenkov M, Link H. Dendritic cells and immune responses in the central nervous system. *Trends Immunol.* 2002;23:69–70.
3. Serafini B, Rosicarelli B, Magliozzi R, et al. Dendritic cells in multiple sclerosis lesions: maturation stage, myelin uptake, and interaction with proliferating T cells. *J Neuropathol Exp Neurol.* February 2006;65(2):124–141.
4. Lande R, Gafa V, Serafini B, et al. Plasmacytoid dendritic cells in multiple sclerosis: intracerebral recruitment and impaired maturation in response to interferon-beta. *J Neuropathol Exp Neurol.* May 2008;67(5):388–401.
5. Karni A, Abraham M, Monsonego A, et al. Innate immunity in multiple sclerosis: myeloid dendritic cells in secondary progressive multiple sclerosis are activated and drive a proinflammatory immune response. *J Immunol.* September 15, 2006;177(6):4196–4202.
6. Huang YM, Xiao BG, Ozenci V, et al. Multiple sclerosis is associated with high levels of circulating dendritic cells secreting pro-inflammatory cytokines. *J Neuroimmunol.* September 1, 1999;99(1):82–90.
7. Barreira da Silva R, Münz C. Natural killer cell activation by dendritic cells: balancing inhibitory and activating signals. *Cell Mol Life Sci.* 2011;68:3505–3518.
8. Miller SD, McMahon EJ, Schreiner B, Bailey SL. Antigen presentation in the CNS by myeloid dendritic cells drives progression of relapsing experimental autoimmune encephalomyelitis. *Ann NY Acad Sci.* April 2007;1103:179–191.
9. Serafini B, Columba-Cabezas S, Di Rosa F, Aloisi F. Intracerebral recruitment and maturation of dendritic cells in the onset and progression of experimental autoimmune encephalomyelitis. *Am J Pathol.* December 2000;157(6):1991–2002.

10. Dittel BN, Visintin I, Merchant RM, Janeway Jr CA. Presentation of the self antigen myelin basic protein by dendritic cells leads to experimental autoimmune encephalomyelitis. *J Immunol*. July 1, 1999;163(1):32–39.
11. Weir CR, Nicolson K, Bäckström BT. Experimental autoimmune encephalomyelitis induction in naive mice by dendritic cells presenting a self-peptide. *Immunol Cell Biol*. February 2002;80(1):14–20.
12. Khoury SJ, Gallon L, Chen W, et al. Mechanisms of acquired thymic tolerance in experimental autoimmune encephalomyelitis: thymic dendritic-enriched cells induce specific peripheral T cell unresponsiveness in vivo. *J Exp Med*. August 1, 1995;182(2):357–366.
13. El Behi M, Dubucquoi S, Lefranc D, et al. New insights into cell responses involved in experimental autoimmune encephalomyelitis and multiple sclerosis. *Immunol Lett*. January 15, 2005;96(1):11–26.
14. McMahon EJ, Bailey SL, Castenada CV, Waldner H, Miller SD. Epitope spreading initiates in the CNS in two mouse models of multiple sclerosis. *Nat Med*. March 2005;11(3):335–339.
15. Hammarberg H, Lidman O, Lundberg C, et al. Neuroprotection by encephalomyelitis: rescue of mechanically injured neurons and neurotrophin production by CNS-infiltrating T and natural killer cells. *J Neurosci*. July 15, 2000;20(14):5283–5291.
16. Traugott U. Characterization and distribution of lymphocyte subpopulations in multiple sclerosis plaques versus autoimmune demyelinating lesions. *Springer Semin Immunopathol*. 1985;8(1–2):71–95.
17. Kastrukoff LF, Lau A, Wee R, Zecchini D, White R, Paty DW. Clinical relapses of multiple sclerosis are associated with 'novel' valleys in natural killer cell functional activity. *J Neuroimmunol*. December 2003;145(1–2):103–114.
18. Bielekova B, Catalfamo M, Reichert-Scrivner S, et al. Regulatory CD56(bright) natural killer cells mediate immunomodulatory effects of IL-2Ralpha-targeted therapy (daclizumab) in multiple sclerosis. *Proc Natl Acad Sci USA*. April 11, 2006;103(15):5941–5946.
19. Saraste M, Irjala H, Airas L. Expansion of CD56Bright natural killer cells in the peripheral blood of multiple sclerosis patients treated with interferon-beta. *Neurol Sci*. June 2007;28(3):121–126.
20. Kettenmann H, Hanisch U-K, Noda M, Verkhratsky A. Physiology of microglia. *Physiol Rev*. 2011;91:461–553.
21. Trebst C, et al. $CCR1^+/CCR5^+$ mono-nuclear phagocytes accumulate in the central nervous system of patients with multiple sclerosis. *Am J Pathol*. 2001;159:1701–1710.
22. Peterson JW, Bo L, Mörk S, Chang A, Trapp BD. Transected neurites, apoptotic neurons, and reduced inflammation in cortical multiple sclerosis lesions. *Ann Neurol*. 2001;50:389–400.
23. Moll NM, Rietsch AM, Ransohoff AJ, et al. Cortical demyelination in PML and MS: similarities and differences. *Neurology*. January 29, 2008;70(5):336–343.
24. Kutzelnigg A, et al. Cortical demyelination and diffuse white matter injury in multiple sclerosis. *Brain*. 2005;128:2705–2712.
25. Tran EH, Hoekstra K, van Rooijen N, Di-jkstra CD, Owens T. Immune invasion of the central nervous system parenchyma and experimental allergic encephalomyelitis, but not leukocyte extravasation from blood, are prevented in macrophage-depleted mice. *J Immunol*. 1998;161:3767–3775.
26. Bauer J, Sminia T, Wouterlood FG, Dijkstra CD. Phagocytic activity of macrophages and microglial cells during the course of acute and chronic relapsing experimental autoimmune encephalomyelitis. *J Neurosci Res*. 1994;38:365–375.
27. Kotter MR, Zhao C, van Rooijen N, Franklin RJM. Macrophage-depletion induced impairment of experimental CNS remyelination is associated with a reduced oligo-dendrocyte progenitor cell response and altered growth factor expression. *Neurobiol Dis*. 2005;18:166–175.
28. Stadelmann C, Kerschensteiner M, Misgeld T, Brück W, Hohlfeld R, Lassmann H. BDNF and gp145trkB in multiple sclerosis brain lesions: neuroprotective interactions between immune and neuronal cells? *Brain*. 2002;125:75–85.
29. Magnus T, Schreiner B, Korn T, et al. Microglial expression of the B7 family member B7 homolog 1 confers strong immune inhibition: implications for immune responses and autoimmunity in the CNS. *J Neurosci*. March 9, 2005;25(10):2537–2546.
30. Duncan DS, Miller SD. CNS expression of B7-H1 regulates pro-inflammatory cytokine production and alters severity of Theiler's virus-induced demyelinating disease. *PLoS One*. 2011;6:e18548.
31. Kwidzinski E, Bunse J, Aktas O, et al. Indolamine 2,3-dioxygenase is expressed in the CNS and down-regulates autoimmune inflammation. *FASEB J*. August 2005;19(10):1347–1349.
32. Butovsky O, Jedrychowski MP, Moore CS, et al. Identification of a unique TGF-β-dependent molecular and functional signature in microglia. *Nat Neurosci*. January 2014;17(1):131–143.
33. Yamasaki R, Lu H, Butovsky O, et al. Differential roles of microglia and monocytes in the inflamed central nervous system. *J Exp Med*. July 28, 2014;211(8).

34. Kabat EA, Wolf A, Bezer AE. The rapid production of acute disseminated encephalomyelitis in rhesus monkeys by injection of heterologous and homologous brain tissue with adjuvants. *J Exp Med*. January 1, 1947;85(1):117–130.
35. Paterson PY. Transfer of allergic encephalomyelitis in rats by means of lymph node cells. *J Exp Med*. January 1, 1960;111:119–136.
36. Ben-Nun A, Wekerle H, Cohen IR. The rapid isolation of clonable antigen-specific T lymphocyte lines capable of mediating autoimmune encephalomyelitis. *Eur J Immunol*. March 1981;11(3):195–199.
37. Das P, Drescher KM, Geluk A, Bradley DS, Rodriguez M, David CS. Complementation between specific HLA-DR and HLA-DQ genes in transgenic mice determines susceptibility to experimental autoimmune encephalomyelitis. *Hum Immunol*. March 2000;61(3):279–289.
38. Kawamura K, et al. Hla-DR2-restricted responses to proteolipid protein 95-116 peptide cause autoimmune encephalitis in transgenic mice. *J Clin Invest*. 2000;105:977–984.
39. Sospedra M, Martin R. Immunology of multiple sclerosis. *Annu Rev Immunol*. 2005;23:683–747. [Review].
40. Bielekova B, Goodwin B, Richert N, et al. Encephalitogenic potential of the myelin basic protein peptide (amino acids 83-99) in multiple sclerosis: results of a phase II clinical trial with an altered peptide ligand. *Nat Med*. October 2000;6(10):1167–1175. Erratum in: *Nat Med*. December 2000;6(12):1412.
41. Jones DE, Diamond AG. The basis of autoimmunity: an overview. *Baillieres Clin Endocrinol Metab*. January 1995;9(1):1–24. [Review].
42. Steinman L. Multiple sclerosis: a coordinated immunological attack against myelin in the central nervous system. *Cell*. May 3, 1996;85(3):299–302. [Review].
43. Hellings N, Barée M, Verhoeven C, et al. T-cell reactivity to multiple myelin antigens in multiple sclerosis patients and healthy controls. *J Neurosci Res*. February 1, 2001;63(3):290–302.
44. Bielekova B, Sung MH, Kadom N, Simon R, McFarland H, Martin R. Expansion and functional relevance of high-avidity myelin-specific $CD4^+$ T cells in multiple sclerosis. *J Immunol*. March 15, 2004;172(6):3893–3904.
45. Hug A, Korporal M, Schröder I, et al. Thymic export function and T cell homeostasis in patients with relapsing remitting multiple sclerosis. *J Immunol*. July 1, 2003;171(1):432–437.
46. Libbey JE, McCoy LL, Fujinami RS. Molecular mimicry in multiple sclerosis. *Int Rev Neurobiol*. 2007;79:127–147. [Review].
47. Rapp NS, Gilroy J, Lerner AM. Role of bacterial infection in exacerbation of multiple sclerosis. *Am J Phys Med Rehabil*. November–December 1995;74(6):415–418.
48. Andersen O, Lygner PE, Bergström T, Andersson M, Vahlne A. Viral infections trigger multiple sclerosis relapses: a prospective seroepidemiological study. *J Neurol*. July 1993;240(7):417–422.
49. Correale J, Farez M. Association between parasite infection and immune responses in multiple sclerosis. *Ann Neurol*. February 2007;61(2):97–108.
50. Baron JL, Madri JA, Ruddle NH, Hashim G, Janeway Jr CA. Surface expression of alpha 4 integrin by CD4 T cells is required for their entry into brain parenchyma. *J Exp Med*. January 1, 1993;177(1):57–68.
51. Wolburg H, Wolburg-Buchholz K, Engelhardt B. Involvement of tight junctions during transendothelial migration of mononuclear cells in experimental autoimmune encephalomyelitis. *Ernst Schering Res Found Workshop*. 2004;47:17–38. [Review].
52. Yang Y, Estrada EY, Thompson JF, Liu W, Rosenberg GA. Matrix metalloproteinase-mediated disruption of tight junction proteins in cerebral vessels is reversed by synthetic matrix metalloproteinase inhibitor in focal ischemia in rat. *J Cereb Blood Flow Metab*. April 2007;27(4):697–709.
53. Leppert D, Ford J, Stabler G, et al. Matrix metalloproteinase-9 (gelatinase B) is selectively elevated in CSF during relapses and stable phases of multiple sclerosis. *Brain*. December 1998;121(Pt 12):2327–2334.
54. Bartholomäus I, Kawakami N, Odoardi F, et al. Effector T cell interactions with meningeal vascular structures in nascent autoimmune CNS lesions. *Nature*. November 5, 2009;462(7269):94–98.
55. Lopez-Diego RS, Weiner HL. Novel therapeutic strategies for multiple sclerosis–a multifaceted adversary. *Nat Rev Drug Discov*. November 2008;7(11):909–925.
56. Sakaguchi S, et al. Organ-specific autoimmune diseases induced in mice by elimination of T cell subset. I. Evidence for the active participation of T cells in natural self-tolerance; deficit of a T cell subset as a possible cause of autoimmune disease. *J Exp Med*. 1985;161:72–87.
57. Read S, Malmstrom V, Powrie F. Cytotoxic T lymphocyte-associated antigen 4 plays an essential role in the function of CD25(+) CD4(+) regulatory cells that control intestinal inflammation. *J Exp Med*. 2000;192:295–302.
58. Baecher-Allan C, et al. $CD4^+CD25$high regulatory cells in human peripheral blood. *J Immunol*. 2001;167:1245–1253.

59. Dieckmann D, et al. Ex vivo isolation and characterization of CD4(+) CD25(+) T cells with regulatory properties from human blood. *J Exp Med.* 2001;193:1303–1310.
60. Kleinewietfeld M, Hafler DA. Regulatory T cells in autoimmune neuroinflammation. *Immunol Rev.* May 2014;259(1):231–244.
61. Kalsi JK, et al. Functional and modelling studies binding human monoclonal anti-DNA antibodies DNA. *Mol Immunol.* 1996;33:471–483.
62. Willenborg DO, Fordham S, Bernard CC, Cowden WB, Ramshaw IA. IFN-gamma plays a critical down-regulatory role in the induction and effector phase of myelin oligodendrocyte glycoprotein-induced autoimmune encephalomyelitis. *J Immunol.* October 15, 1996;157(8):3223–3227.
63. Becher B, Durell BG, Noelle RJ. Experimental autoimmune encephalitis and inflammation in the absence of interleukin-12. *J Clin Invest.* August 2002;110(4):493–497.
64. Gran B, Zhang GX, Yu S, et al. IL-12p35-deficient mice are susceptible to experimental autoimmune encephalomyelitis: evidence for redundancy in the IL-12 system in the induction of central nervous system autoimmune demyelination. *J Immunol.* December 15, 2002;169(12):7104–7110.
65. Harrington LE, Hatton RD, Mangan PR, et al. Interleukin 17-producing $CD4^+$ effector T cells develop via a lineage distinct from the T helper type 1 and 2 lineages. *Nat Immunol.* November 2005;6(11):1123–1132.
66. Park H, Li Z, Yang XO, et al. A distinct lineage of CD4 T cells regulates tissue inflammation by producing interleukin 17. *Nat Immunol.* November 2005;6(11):1133–1141.
67. Korn T, Bettelli E, Oukka M, Kuchroo VK. IL-17 and Th17 cells. *Annu Rev Immunol.* 2009;27:485–517.
68. Brucklacher-Waldert V, Stuerner K, Kolster M, Wolthausen J, Tolosa E. Phenotypical and functional characterization of T helper 17 cells in multiple sclerosis. *Brain.* December 2009;132(Pt 12):3329–3341.
69. Komiyama Y, Nakae S, Matsuki T, et al. IL-17 plays an important role in the development of experimental autoimmune encephalomyelitis. *J Immunol.* July 1, 2006;177(1):566–573.
70. Kreymborg K, Etzensperger R, Dumoutier L, et al. IL-22 is expressed by Th17 cells in an IL-23-dependent fashion, but not required for the development of autoimmune encephalomyelitis. *J Immunol.* December 15, 2007;179(12):8098–8104.
71. Haak S, Croxford AL, Kreymborg K, et al. IL-17A and IL-17F do not contribute vitally to autoimmune neuroinflammation in mice. *J Clin Invest.* January 2009;119(1):61–69. http://dx.doi.org/10.1172/JCI35997. Epub December 15, 2008.
72. Lock C, Hermans G, Pedotti R, et al. Gene-microarray analysis of multiple sclerosis lesions yields new targets validated in autoimmune encephalomyelitis. *Nat Med.* May 2002;8(5):500–508.
73. Tzartos JS, Friese MA, Craner MJ, et al. Interleukin-17 production in central nervous system-infiltrating T cells and glial cells is associated with active disease in multiple sclerosis. *Am J Pathol.* January 2008;172(1):146–155.
74. Schulze-Topphoff U, Prat A, Prozorovski T, et al. Activation of kinin receptor B1 limits encephalitogenic T lymphocyte recruitment to the central nervous system. *Nat Med.* July 2009;15(7):788–793.
75. Aktas O, Schulze-Topphoff U, Zipp F. The role of TRAIL/TRAIL receptors in central nervous system pathology. *Front Biosci.* May 1, 2007;12:2912–2921. [Review].
76. Hemmer B, Archelos JJ, Hartung HP. New concepts in the immunopathogenesis of multiple sclerosis. *Nat Rev Neurosci.* April 2002;3(4):291–301. [Review].
77. Kroenke MA, Carlson TJ, Andjelkovic AV, Segal BM. IL-12- and IL-23-modulated T cells induce distinct types of EAE based on histology, CNS chemokine profile, and response to cytokine inhibition. *J Exp Med.* July 7, 2008;205(7):1535–1541.
78. Luger D, Silver PB, Tang J, et al. Either a Th17 or a Th1 effector response can drive autoimmunity: conditions of disease induction affect dominant effector category. *J Exp Med.* April 14, 2008;205(4):799–810.
79. Steinman L. A rush to judgment on Th17. *J Exp Med.* July 7, 2008;205(7):1517–1522.
80. Kebir H, Ifergan I, Alvarez JI, et al. Preferential recruitment of interferon-gamma-expressing TH17 cells in multiple sclerosis. *Ann Neurol.* September 2009;66(3):390–402.
81. Lee Y, Awasthi A, Yosef N, et al. Induction and molecular signature of pathogenic TH17 cells. *Nat Immunol.* October 2012;13(10):991–999.
82. Murphy AC, Lalor SJ, Lynch MA, Mills KH. Infiltration of Th1 and Th17 cells and activation of microglia in the CNS during the course of experimental autoimmune encephalomyelitis. *Brain Behav Immun.* May 2010;24(4):641–651.
83. Secukinumab Shines in Small MS Trial. http://www.medpagetoday.com/MeetingCoverage/ECTRIMS/35300.

84. Hauser SL, Bhan AK, Gilles F, Kemp M, Kerr C, Weiner HL. Immunohistochemical analysis of the cellular infiltrate in multiple sclerosis lesions. *Ann Neurol*. June 1986;19(6):578–587.
85. Friese MA, Fugger L. Autoreactive CD8$^+$ T cells in multiple sclerosis: a new target for therapy? *Brain*. August 2005;128(Pt 8):1747–1763.
86. Jacobsen M, Cepok S, Quak E, et al. Oligoclonal expansion of memory CD8$^+$ T cells in cerebrospinal fluid from multiple sclerosis patients. *Brain*. March 2002;125(Pt 3):538–550.
87. Fogdell-Hahn A, Ligers A, Grønning M, Hillert J, Olerup O. Multiple sclerosis: a modifying influence of HLA class I genes in an HLA class II associated autoimmune disease. *Tissue Antigens*. February 2000;55(2):140–148.
88. Correale J, Villa A. Isolation and characterization of CD8$^+$ regulatory T cells in multiple sclerosis. *J Neuroimmunol*. March 2008;195(1–2):121–134.
89. Kabat EA, Moore DH, Landow H. An electrophoretic study of the protein components in cerebrospinal fluid and their relationship to the serum proteins. *J Clin Invest*. September 1942;21(5):571–577.
90. Lovato L, Willis SN, Rodig SJ, et al. Related B cell clones populate the meninges and parenchyma of patients with multiple sclerosis. *Brain*. February 2011;134(Pt 2):534–541.
91. Schirmer L, Srivastava R, Hemmer B. To look for a needle in a haystack: the search for autoantibodies in multiple sclerosis. *Mult Scler*. March 2014;20(3):271–279.
92. Bourdette D, Yadav V. B-cell depletion with rituximab in relapsing-remitting multiple sclerosis. *Curr Neurol Neurosci Rep*. September 2008;8(5):417–418.
93. Chaudhuri A. Ocrelizumab in multiple sclerosis: risks and benefits. *Lancet*. March 31, 2012;379(9822):1196–1197. author reply 1197.
94. Barun B, Bar-Or A. Treatment of multiple sclerosis with anti-CD20 antibodies. *Clin Immunol*. January 2012;142(1):31–37.
95. Disanto G, Sandve GK, Berlanga-Taylor AJ, et al. Genomic regions associated with multiple sclerosis are active in B cells. *PLoS One*. 2012;7(3):e32281.
96. Serafini B, Rosicarelli B, Magliozzi R, Stigliano E, Aloisi F. Detection of ectopic B-cell follicles with germinal centers in the meninges of patients with secondary progressive multiple sclerosis. *Brain Pathol*. April 2004;14(2):164–174.
97. Magliozzi R, Howell O, Vora A, et al. Meningeal B-cell follicles in secondary progressive multiple sclerosis associate with early onset of disease and severe cortical pathology. *Brain*. April 2007;130(Pt 4):1089–1104.
98. Mayo L, Quintana FJ, Weiner HL. The innate immune system in demyelinating disease. *Immunol Rev*. July 2012;248(1):170–187. http://dx.doi.org/10.1111/j.1600-065X.2012.01135.x. [Review].

CHAPTER

3

Central Nervous System Immune Inflammation

L.M. Healy, H. Touil, V.T.S. Rao, M.A. Michell-Robinson, J. Antel

McGill University, Montreal, QC, Canada

R.O. Weller

University of Southampton, Southampton, United Kingdom

OUTLINE

Translational Neuroimmunology in Multiple Sclerosis
http://dx.doi.org/10.1016/B978-0-12-801914-6.00003-9

3.1 INTRODUCTION

In this chapter on regulation of inflammation within the central nervous system (CNS), we consider both those mechanisms that regulate access of constituents of the systemic immune system into the CNS and those that regulate the activities of the cells that have accessed the CNS. The latter will focus on the role of endogenous glial cells, microglia and astrocytes, that can directly mediate effects linked to the immune system. The properties of these glial cells are themselves responsive to signals derived from the CNS environment and/or from the infiltrating exogenous immune cells and products. It is now well recognized that immune surveillance is ongoing within the CNS under physiologic conditions. In context of multiple sclerosis (MS), we ask what changes in immune regulation at the level of the CNS contribute to the acute and chronic inflammatory activity that characterizes the disease. Attention will be drawn to therapeutic agents and approaches that have successfully or unsuccessfully been used to modulate these processes in MS or its experimental models. Keep in mind that modulating properties of both infiltrating immune cells (adaptive and innate) and endogenous glial cells will also impact their neural cell protection and repair processes.

3.2 REGULATION OF IMMUNE TRAFFICKING TO AND FROM THE CENTRAL NERVOUS SYSTEM

Understanding the dynamic processes whereby cells of the immune system access the CNS under physiologic conditions and during the course of CNS-directed inflammatory disease has greatly advanced with development of techniques to label systemic immune cells and follow their trafficking patterns by live imaging techniques.[1–7] Such studies allow evaluation of trafficking when selected molecules associated with either immune, endothelial, or glial cells are genetically deleted or blocked by administration of specific antibodies or pharmacologic agents. The dominant experimental animal approach used to investigate cell trafficking and used as a model of MS is experimental autoimmune encephalomyelitis (EAE), which involves active immunization with neural autoantigens and/or adoptive systemic transfer of autoreactive lymphocytes considered to play a distinct role in inducing disease.

3.2.1 Blood–Brain Barrier

A traditional view has been that trafficking of immune cells from the systemic to the CNS compartment during the course of MS or in EAE requires passage of such cells across the blood–brain barrier (BBB) at the level of postcapillary venules.[1–7] Contributing to the BBB are specialized endothelial cells that interact with each other; such interaction is dependent on an array of proteins, belonging to the tight junction (claudins, occludin, and junctional adhesion molecules) and adherens families of molecules (reviewed in Ref. 5). The endothelial cells are separated from the parenchyma by two basement membranes—vascular and parenchymal (glia)—between which is a potential perivascular space. The endothelial barrier-regulating molecular interactions are responsive to molecules provided via astrocytes whose end-feet contact and cover the parenchymal basement membrane of the BBB. Astrocyte-derived molecules including sonic hedgehog (SHH) and angiotensin II induce signaling cascades that regulate endothelial cell–cell interactions promoting a competent BBB.[8,9] As an example, SHH is shown to stimulate netrin1 expression on human BBB endothelial cells, resulting in increased tight junction protein expression.[10] This regulatory pathway is responsive to inflammatory conditions reflecting an attempt to restore homeostasis. Conversely, production of proinflammatory molecules by activated astrocyte molecules would enhance permeability of the barrier.

Immune cell transmigration across most endothelial barriers is dependent on a sequence of molecular steps that begin with selectin-dependent rolling (see later comment regarding cerebral microvessels) leading to firm adhesion that results from interaction with integrins; the latter are activated by chemokines produced by the cellular constituents (glia and endothelial cells) forming the BBB (reviewed in Refs 1,4). Such transmigrating cells enter the perivascular space formed between the vascular and parenchymal basement membranes; chemokine-related signals, specifically CXCL12, which is constitutively expressed by BBB endothelial cells, promote their retention in this space.[1] Transmigration of cells across the parenchymal barrier and into the actual tissue involves actions of matrix metalloproteinases on this barrier.[11] As will be discussed, these molecular sequences have been and continue to be targets of therapeutic interventions.

3.2.2 Choroid Plexus/Cerebrospinal Fluid/Draining Vein Trafficking

Cerebrospinal fluid (CSF) in the subarachnoid space (SAS) is largely separate from interstitial fluid (ISF) in the brain parenchyma. Pia mater separates the SAS from the brain and from vessels entering and leaving the surface of the brain.[12,13] The SAS does not exhibit the same immunological privilege as the brain parenchyma (as discussed later); CSF drains to cervical lymph nodes from the SAS predominantly through channels that traverse the cribriform plate of the ethmoid bone and join nasal lymphatics[14–16] (see Fig. 3.1). Traffic of antigen-presenting cells has been observed along this route of lymphatic drainage[17] (Fig. 3.1). Other routes for lymphatic drainage of CSF via dural lymphatics and cranial nerve roots have been described.[18]

There are no traditional lymphatic vessels in the brain. Lymphatic drainage of ISF from the brain is along very narrow 100-nm-wide basement membranes in the walls of cerebral capillaries and between smooth muscle cells in the tunica media of cerebral arteries[16,19] (see Fig. 3.1). Although this route allows lymphatic drainage of fluid and solutes, it is too narrow for the trafficking of antigen-presenting cells to regional lymph nodes[19]; this feature may contribute to immunological privilege in the brain.[20,21]

As arteries enter the cerebral cortex at the surface of the brain, there is no clear perivascular space between the artery wall and brain tissue.[13] However, CSF does enter the brain from the SAS and tracers pass along the outer aspect of cerebral arteries into the brain parenchyma,[22,23] where mixing of CSF with ISF is mediated by aquaporin 4.[22] This route has been termed convective influx[23] or the glymphatic system.[22] A mixture of CSF and ISF with brain metabolites passes out of the brain alongside veins to the CSF.[22] It is not yet clear whether the route alongside veins allows traffic of antigen-presenting cells from the brain parenchyma into the CSF. The convective influx/glymphatic system is separate from the rapid periarterial lymphatic drainage system that is a direct link between ISF and cervical lymph nodes[16,24] (see Fig. 3.1).

An emerging concept regarding access of immune cells under physiologic conditions (immune surveillance) and at the initiation of an autoreactive CNS-directed immune response implicates trafficking across a blood/CSF barrier.[2,3] This is referred to as a two-wave model.

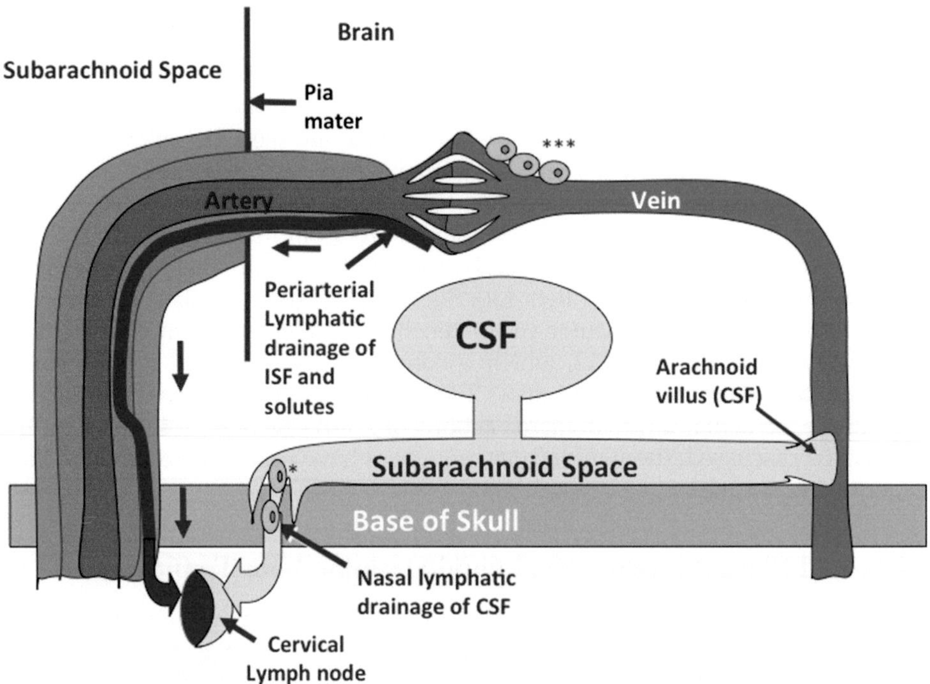

FIGURE 3.1 **Diagram showing the largely separate pathways of lymphatic drainage of interstitial fluid (ISF) and cerebrospinal fluid (CSF) from the brain.** (1) Blood flows into the brain along branches of the carotid (and vertebral) arteries (red) and forms a network of capillaries (red) that are the sites of the blood–brain barrier through which nutrients enter brain tissue. ISF and solutes drain from the brain along perivascular pathways (*curved blue line*) in the walls of capillaries and in the tunica media (brown) and adventitia (light blue) of arteries to cervical lymph nodes. (2) Blood flows into postcapillary venules (blue) from which lymphocytes and monocytes (***) enter the brain by receptor-mediated mechanisms, gather in perivenular and perivenous spaces, and migrate into brain tissue. (3) CSF is formed in the ventricles (yellow), flows into the subarachnoid space from which it drains via nasal lymphatics to cervical lymph nodes or into venous sinuses through arachnoid villi or granulations. Antigen-presenting cells (**) traffic with CSF to cervical lymph nodes by this route. The convective influx/glymphatic system is not shown in this diagram. *Modified from Fig. 3.1 in Weller RO, Galea I, Carare RO, Minagar A. Pathophysiology of the lymphatic drainage of the central nervous system: implications for pathogenesis and therapy of multiple sclerosis. Pathophysiology. 2010;17:295–306. doi:10.1016/j.pathophys.2009.10.007 and reproduced with permission.*

Trafficking begins at the level of the choroid plexus, where the endothelial cells lack the full molecular complement of junctional proteins found with parenchymal endothelial cells.[5] This allows immune cell movement into the CSF. The SAS contains competent antigen-presenting cells including dendritic cells.[25] Studies of MS cases illustrate that potential disease-relevant antigens such as myelin components can be identified in these meningeal spaces, providing a means to induce and sustain a potentially disease-relevant immune response.[26] A sustained meningeal response in MS is postulated to result in proinflammatory molecules penetrating into the underlying gray matter and result in subpial demyelination; the latter is not observed in other meningeal inflammatory conditions including chronic meningitis and lymphomas.[27] As mentioned, there is no perivascular space around penetrating arteries, raising the issue of how cells trafficking via the choroid/SAS pathway would reach the perivascular space around small venules that are the site of penetration of the BBB. Immune cells when present in the brain under physiologic conditions are usually found between the basement membranes of glia limitans and walls of venules and veins.

3.2.3 Cell Trafficking-Related Therapeutic Interventions

Optimal therapeutic interventions directed at immune cell trafficking would aim to target molecules that enhance trafficking of putative cell types mediating a pathologic response while sparing migration of cells most involved in physiologic surveillance, which may themselves serve to regulate immune activity. Selection of therapeutic strategies aimed at specific members of the families of molecules that regulate trafficking will need to consider both the individual subpopulations of immune cells and the different migration sites (choroid/meninges vs BBB) being targeted. In the EAE model, blocking E- and P-selectins that are expressed in leptomeningeal and choroid plexus vessels but not in the parenchymal microvessels does not inhibit lesion formation. This suggests that CSF accumulation of immune cells is not necessary for lesion development.[5]

Preclinical studies in the EAE model of monoclonal antibodies and small molecules directed at adhesion molecules have often produced inconsistent results, likely reflecting the complexity of the migration process and the limitations of single-antigen/single-strain models.[28–31] Natalizumab, which recognizes VLA-4, whose partner ligands are vascular cell adhesion molecule (CAM) and fibronectin, is now an approved therapy for relapsing MS although its use is limited by the risk of development of progressive multifocal leukoencephopathy (PML).[32] The small molecule firategrast that also targets VLA-4 was found to reduce disease activity in phase II clinical trials, albeit to a lesser extent than observed in the natalizumab trials.[33] Concerns regarding development of PML have limited evaluation of efalizumab, a monoclonal antibody recognizing LFA-1, whose partner ligand is intercellular adhesion molecule.[34] An expanded array of adhesion molecules participating in the cell trafficking process continue to be identified including activated leukocyte CAM, melanoma CAM, and Ninjurin-1 (reviewed in Ref. 5) with evidence of selectivity for regulating migration of specific leukocyte subtypes to the CNS as exemplified by mucosal vascular addressin cell adhesion molecule regulation of Th17-cell–endothelial-cell interaction.[35–37] Further to be considered is that the overall state of activation of the immune system may be impacted by multiple variables that may be operative under different pathologic and experimental conditions. Examples in the EAE model include use of adjuvants or pertussis that acts on the BBB or whether animals are housed under clean or dirty conditions. Variables in the clinical situation include intercurrent viral and bacterial infections and the status of the microbiome. These issues call attention to the role of innate immune cells including neutrophils and innate lymphocytes in initiating CNS inflammatory responses.[38–40]

3.2.4 Immune Trafficking at Specialized Central Nervous System Sites

Cell passage into the CNS can occur relatively more directly at distinct sites where the BBB is less developed such as the area postrema.[41,42] Studies of myeloid cells in chimeric models indicate macrophage entry through these less protected areas with subsequent migration through the parenchyma.[42] Access of cytokines, chemokines, and other products of systemic inflammation such as endotoxins, to the CNS from the systemic compartment also will occur more readily through sites less protected by the BBB such as the circumventricular organs; as reviewed by Sankowsi et al., such inflammatory mediators can interfere with neuronal and glial cell well-being, leading to a break of balance in brain homeostasis.[43] Cognitive and behavioral manifestations commonly resulting from such occurrences during acute infections including anorexia, malaise, depression, and decreased physical activity are collectively referred to as sickness behavior.

3.2.5 Antibody/Immunoglobulin Access to the Central Nervous System

In context of MS and its therapies, a specific issue to consider is the capacity of antibodies to pass from the systemic compartment into the CNS. The role of antibodies in MS lesion pathogenesis continues to be defined including the relative contribution of blood-derived versus intrathecally produced immunoglobulin (Ig). In neuromyelitis optica, most of the target-specific antibody (antiaquaporin A) would seem to be derived from a systemic source.[44] High titers of anti-interferon (IFN)β antibodies can develop in some patients receiving this therapy, raising the question as to whether these could access the CNS in sufficient amounts to impact endogenous production of IFNβ by astrocytes and disrupt IFNβ-dependent signaling pathways.[45] Of further concern is the use of systemic delivery routes for monoclonal antibodies aimed at modulating inflammation or promoting repair within the CNS. Similar consideration may be given to therapy with whole Ig molecule preparations (I-V Ig) and Fc receptor fractions.[46]

3.3 IMMUNE REGULATION WITHIN THE CENTRAL NERVOUS SYSTEM

In this chapter, the focus is on the role of endogenous CNS cells, microglia and astrocytes, although acknowledging the role of T and B lymphocytes within the meningeal, perivascular, and parenchymal compartments. In the case of MS, where blood-derived macrophages access the CNS, their relative role compared to microglia as innate immune cells mediating regulatory and effector functions also needs to be considered; similar recognition should be given to dendritic cells. Further to be recognized is the bidirectional effect these cells have on each other and with other endogenous neural cells, neurons, and oligodendrocytes (see Fig. 3.2).

3.3.1 Microglia

Microglia are demonstrated to be a mesodermal cell lineage distinct from blood-derived monocytes/macrophages, arising during primitive hematopoiesis in yolk-sac blood islands.[47–51] Microglia are estimated to make up 6–18% of neocortical cells in the human brain.[52,53] They are long-lived cells that are not replaced by blood-derived cells under usual conditions.[54–57] If depleted using a small molecule inhibitor of colony-stimulating factor-1 (CSF-1) receptor, microglia cells will be repopulated, implicating presence of progenitor populations.[58] Presence of microglia may also inhibit access of blood-derived myeloid cells to the

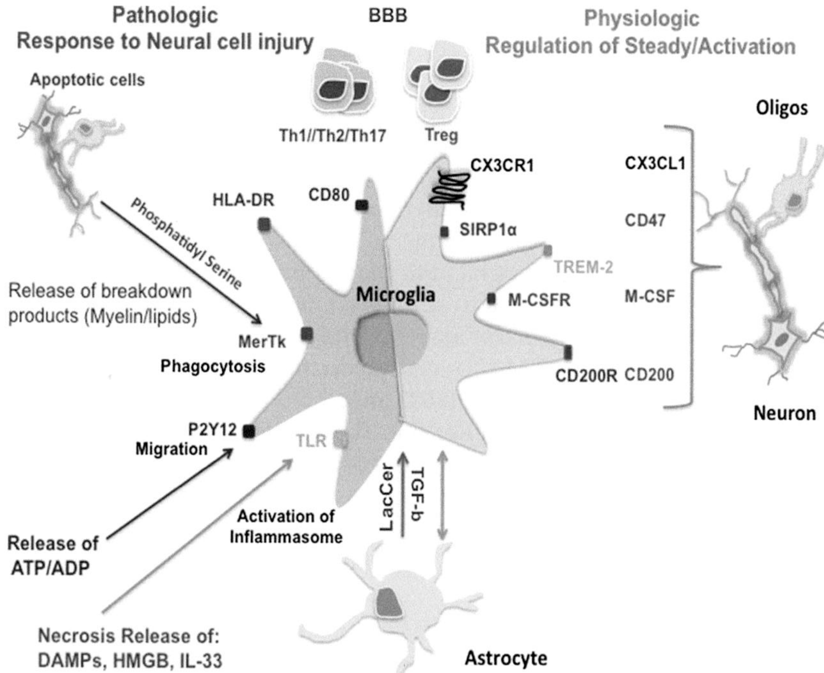

FIGURE 3.2 **Cellular interactions regulating microglia functional properties within the central nervous system under physiologic and pathologic conditions.** Physiologic conditions: Maintenance of quiescent state of microglia by interaction of specific surface molecules with regulatory signals derived from adjacent cells (neurons, astrocytes) and extracellular matrix. Interactions include CX3CR/CX3CL1, SIRP1a/CD47, M-CSFR/M-CSF or IL-34, CD200R/CD200, and TREM-2. Pathologic conditions: Response of microglia to signals derived from injured and dying cells (neurons, oligodendrocytes), astrocytes, and infiltrating immune cells or their products. Responses to dying/injured cells include MerTK-dependent phagocytosis of apoptotic cells expressing phosphatidyl serine on their surface and of myelin breakdown products; P2Y12-dependent migration in response to release of ATP/ADP by injured cells, and TLR-dependent inflammasome activation by molecules (danger-associated molecules, HMGB, and IL-33) released by cells undergoing necrosis. Astrocytes and infiltrating immune cells released molecules that can modulate activation/polarization state of microglia.

CNS parenchyma as shown by irradiated chimeric rodent studies where blood-derived macrophages populate the brain only after ablation of microglia.[59] At least in the human system, microglia unlike macrophages do not appear to be dependent on signals delivered by CSF-1 for survival through its protein tyrosine kinase receptor, CSF-1R.[60]

Gene sequencing studies have defined a unique molecular signature of microglia compared to blood-derived myeloid cells under homeostatic conditions.[58] For mice, the in situ homeostatic signature is best defined in microglia maintained in transforming growth factor (TGF)β. Human microglia present a similar distinct profile compared to blood monocytes as assessed using cells isolated from surgical samples of noninflamed adult human brain and maintained under basal culture conditions for 2–4 days.[61] Analysis of microglia immediately upon isolation from normal human brain will be needed to define their physiologic homeostatic profile and what culture conditions best model these. As discussed later, the molecular signature of microglia in situ and in vitro, as well as that of blood-derived macrophages, is responsive to external stimuli. This property would be analogous to T cells that have distinct genetically determined lineage markers, CD4 and CD8 T cells, but where an array of their functional properties is dependent on exposure to specific external signals, for example, selective cytokines that induce Th1, Th2, Th17/Treg phenotypes. The gene sequencing studies also identify distinct profiles of expression of microRNAs that would support activation of inflammatory responses (eg, Mir-155) or favor a state of quiescence (eg, Mir-124) that acts by restricting Sp1 (PU.1) and CSF-1 expression.[62]

The functional phenotypes of microglia (and macrophages) continue to be defined based on exposing these cells in vitro or in vivo to defined combinations of molecules. A continuum of phenotypes may exist, especially under actual complex in vivo conditions as described in a later section. Classically activated (GM-CSF, endotoxin, proinflammatory cytokines) myeloid cells have been characterized by their rounded amoeboid morphology, which is consistent with a state of hyperactivity (reviewed in Ref. 58). These cells, often referred to as being of the M1 phenotype, express surface markers such as costimulatory molecules CD80 and CD86, the chemokine receptor CCR7, and increase the expression of HLA-DR. In addition, they produce reactive nitrogen species and proinflammatory cytokines such as IL-12,[60] and express the microRNA-155 that is a master regulator of proinflammatory genes via its effects on suppressor of cytokines secretion −1.

In contrast, cells generated by activation with macrophage colony-stimulating factor (M-CSF) and anti-inflammatory cytokines (IL-4 and IL-13) and conveniently referred to as M2, possess a typical elongated morphology, express cell surface markers including the scavenger receptor CD163, C-type lectins CD206 and CD209, and produce anti-inflammatory cytokines such as TGFβ1 and IL-10.[63–66] M2 cells are noted to be more efficient than M1 cells with regard to phagocytosis of opsonized targets.[67] We observe M2-polarized human microglia to be more efficient at phagocytosing myelin compared to either M1 microglia or any polarized macrophage phenotype.[68] When comparing functional immune-related properties of M1- and M2-polarized human microglia, we found that the M1 cells are more capable of supporting proliferation of naïve T cells in a mixed lymphocyte reaction[69] and in producing effector molecules, especially tumor necrosis factor, which will induce injury of oligodendrocyte progenitor cells.[70]

The M2 phenotype has been further subclassified according to specific activation stimuli, with M2 cells generated as described earlier (IL-4 and IL-13) being referred to as M2a, M2b are generated by using immune complexes and lipopolysaccharide, and M2c using IL-10 and TGFβ.[71] M2b cells are described as continuing to express M1-associated costimulatory molecules and produce nitric oxide, but also secreting anti-inflammatory cytokines such as IL-10. M2c macrophages, originally termed deactivated, are associated with immune suppression and tissue remodeling. Miron et al. ascribed the remyelination promoting effects of microglia following lysolecithin-induced demyelination in mice to production of activin A by cells having acquired an M2 phenotype.[72]

The signals that innate immune cells are likely actually to encounter within the CNS can be considered under the concept of "stranger" and "danger" signals. The former refer to molecules such as endotoxins produced, expressed, and released by exogenous agents, specifically bacterial infections. There is now recognition that comparable molecules, referred to as alarmins or danger-associated molecules, can be released by endogenous cells whose surface membranes are being ruptured (cell necrosis) under inflammatory or injury conditions (see Fig. 3.2).[73] Mitochondria are a particular source of such products, reflecting their ancestral bacterial origin.[73] These stimuli would favor generation of a proinflammatory phenotype in myeloid cells. Toll-like receptors would serve as a common receptor family for these molecules. Danger signals arise as a consequence of cell injury or death and include surface molecules such as phophatidyl serine expressed by cells undergoing programmed cell death and from cell-type-specific constituents such as myelin and iron that would be released from damaged oligodendrocytes in the case of MS. Iron is observed to promote a rapid switch from the M2 to M1 phenotype.[74]

In specific context of the inflamed environment of MS lesions, we need to consider signals derived from the immune constituents present in the lesions. Previously mentioned was the potential for specific crystallizable fragment (Fc) components of Ig to interact with the corresponding Fc receptors expressed on the myeloid cells. We have previously shown that supernatants derived from Th1- and Th2-polarized cells will respectively induce an M1- or M2-biased phenotype on human microglia.[69,75] The Th2 cells were induced using glatiramer acetate, an agent that does not access the CNS, providing an example of how systemic immune-modulatory therapy may have an indirect effect on innate immunity within the CNS.

Adding further to the complexity in concluding how the array of interactive environmental signals will produce a net effect on microglia/macrophage properties in MS is the status of specific receptors on the myeloid cells and their ligands expressed by endogenous neural cells, both neurons and astrocytes, whose interactions regulate the functional properties of the myeloid cells (see Fig. 3.2). These include triggering receptor expressed on myeloid cells-2 (TREM2), whose ligand remains uncertain and that modulates phagocytosis[76]; SIRP-1 (previously known as CD172a), which interacts with its ligand CD47 that is expressed by neurons and astrocytes, to induce a "do not eat me" signal[77,78]; CD200R, which interacts with CD200 that can be soluble or expressed on the neuron surface; and CX3CR1 (known as fractalkine receptor), which binds to its soluble ligand CX3CL1. Both CD200R and CX3CR1 once activated mediate a resting phenotype of the microglia. M-CSFR response to M-CSF or IL-34 enhances macrophage although not necessarily human microglia survival.[60]

The molecular profiling studies of microglia referred to previously include identification or confirmation of a number of surface molecules that are linked to functional responses of the microglia and thus could be potential therapeutic targets.[79,80] In this regard, however, we need to consider how their expression is modulated, depending on the polarization state of the cells and whether the distinct differences with macrophages are maintained under all such polarization conditions. In animal model studies, we can bypass the challenge of distinguishing microglia and macrophages by use of genetic labeling techniques and by tracking each of the populations under all the selected test conditions. Such studies have delineated differences in participation of macrophages and microglia in their contributions to myelin destruction and myelin clearance, respectively.[81] To date no unique natural lineage marker exists to distinguish microglia and macrophages with relative expression of CD45 and the fractalkine receptor being the most commonly used.[82,83]

Our studies of expression of the purinergic receptor P2Y12 on human microglia and macrophages illustrate this complexity. This receptor is responsive to ATP/ADP that we and others have shown guide migratory responses of microglia, whereas macrophages are more responsive to chemokine signals.[84] P2Y12 is more highly expressed on microglia compared to monocytes and macrophages under basal culture conditions and further increases under M2 conditions in vitro and in situ as illustrated by its expression in parasite-infected brain samples.[84] In contrast, P2Y12 is downregulated under M1 conditions including the regions of active MS lesions.[84] We also observe that expression is upregulated on macrophages under M2 conditions. Thus application as an in situ marker would be complicated, depending on the polarization state of the cells as would be any prediction regarding targeting this receptor for therapeutic purposes (ie, regulating migration of myeloid cells).

Microglia/Microglia-Directed Therapy

Because these cells are the dominant cell type in both acute and chronic active MS lesions and are contributors to the changes in the normal-appearing white matter, these cells are recognized as potential targets to reduce tissue injury and enhance repair. As mentioned, an indirect approach would be via modulation of properties of T cells that infiltrate the CNS and interact with the myeloid populations present in the meningeal or parenchymal compartments of the CNS.[85] Using ganciclovir treatment of CD11b-HSVTK transgenic mice to ablate either microglia or blood-derived macrophages, Heppner et al. concluded that microglial paralysis inhibits the development and maintenance of inflammatory CNS lesions in EAE[86]; establishing potential effects in progressive aspects of MS is limited by lack of animal models. Current clinical trials using parasite-associated molecules to deviate the immune response in an M2 direction are not CNS selective.[87] The identification of the molecular signature of microglia and macrophages and linking them with functional properties raises the potential to modulate functions via targeting specific molecules. These would include migration (P2Y12, chemokine receptors), phagocytosis (MerTK, TREM2), production of inflammatory molecules (mir-155), and production of trophic molecules (activin A).[82]

3.3.2 Astrocytes

These cells can also be considered as contributing to enhancing or inhibiting the injury and repair processes ongoing in MS at all stages of disease evolution. Mayo et al. found that the depletion of reactive astrocytes during the acute phase of disease resulted in a significant worsening of EAE, whereas astrocyte depletion during the progressive phase led to a significant amelioration of disease.[88] To be noted is that there is significant topographic heterogeneity in the distribution of astrocytes in the CNS.[89] We observe relative differences in microRNA expression between astrocytes laser dissected from gray and white matter of noninflamed adult human tissue sections, a potential contributor to the differences in inflammatory reactivity seen in white and gray matter MS lesions.[90]

Astrocytes can contribute to multiple steps in the cascade of events related to immune cell entry into the CNS and formation of a new lesion in MS, with much direct evidence coming from studies using the EAE model. Already mentioned is their contribution to maintaining the integrity of the BBB. Astrocytes can express molecules such as major histocompatibility complex (MHC) class II required for participation in antigen presentation although the consensus would be that they are less competent than myeloid cells.[91] Astrocytes are also sources of chemoattractants for immune cells.[92]

Astrocytes can also play a direct and indirect role in actual tissue injury. We have observed that soluble products released by human fetal brain-derived astrocytes when exposed to supernatants derived from Th1-activated T cells will be cytotoxic to oligodendrocyte progenitor cells, an effect that could be inhibited using an anti-CXCL10 antibody.[70] An example of indirect effects is provided by the observation that IL-15 released by astrocytes can induce expression of NGKG2 molecules (NKG2D and C) on T cells resulting in enhanced cytotoxic capacity of CD8 T cells and allowing CD4 cells to acquire promiscuous (non-MHC-restricted) cytotoxic capability including those of oligodendrocytes.[93]

The functional contribution of astrocytes in the EAE model is further illustrated by studies showing that preventing activation of astrocytes by sphingosine-1-phosphate (S1P) by genetically depleting S1P-receptors on astrocytes results in amelioration of EAE.[94] Using a gene array screening approach in the progressive NOD EAE, Mayo et al. identified overexpression of B4GALT6, which codes for a LacCer synthase in the progressive phase of the disease.[88] *B4GALT6* expression and LacCer levels are increased in MS CNS lesions. Inhibiting B4GALT6 suppressed local CNS innate immunity and neurodegeneration in this model. The suppression linked genes comprised of interferon-sensitive response elements and NF-κB response elements. Conversely, astrocytes can play a protective role by serving as scavengers of reactive oxygen species and removing excess glutamate from the microenvironment (reviewed in Ref. 95).

3.3.3 Central Nervous System Compartment-Directed Immunomodulatory Therapies

The therapeutic era in MS evolved from use of therapies that targeted constituents of the systemic immune system. Continuing therapeutic objectives include more selectively intervening in immune trafficking to the CNS and directly modulating immune reactivity within this compartment. Initial success with anti-VLA4-directed antibody therapy provided proof of principle that targeting lymphocyte/monocyte trafficking would interrupt new lesion formation in MS. As the molecular mechanisms are defined that regulate trafficking through the now-recognized two-wave model and that may distinguish molecules (adhesion, chemoattractant) used for trafficking by distinct immune cell subsets, more disease-selective therapies will be developed.

The CNS compartment-directed approach would have relevance to reducing the extent of initial tissue injury, inhibiting ongoing injury, and promoting repair. Such therapies would seem to require access of the agent to this compartment. Initial emphasis has been on development of conventional drugs that can achieve such access. Two approved agents for treatment of relapsing forms of MS can access the CNS with in vitro and EAE-related data suggesting effects on both microglia and astrocytes. Fingolimod is shown to inhibit the S1P-induced proinflammatory responses of astrocytes to S1P while concurrently inducing signaling pathways that result in inhibition of intracellular calcium release and nitric oxide production.[94,96,97] However, clinical trials up to 2015 have shown that this agent impacts on the relapsing but not progressive phases of the disease as measured by clinical criteria and magnetic resonance imaging indices of tissue loss. Dimethyl fumarate (DMF) is demonstrated to have both anti-inflammatory- and antioxidant-inducing effects on astrocytes in vitro and in experimental models.[98] These effects can also be reproduced in vitro using adult human CNS-derived microglia.[99] However, the in vitro effects are reproduced using DMF rather than by monomethyl fumarate (MMF) even though the MMF receptor HCAR-2 is present on these cells.[99] Oral DMF is rapidly metabolized to MMF, leaving open the question of its mechanism of action within the CNS. The receptors on microglia that regulate their functional responses, including migration, phagocytosis, and production of immune regulatory and effector molecules, provide potential drug-able targets as do the receptor–ligand interactions that regulate overall cell activation. For astrocytes, potential exists to inhibit production of inflammatory mediators and upregulating protective processes as exemplified by the studies cited regarding effects of fingolimod and inhibition of lac ser production.

Ongoing efforts are aimed at developing additional therapies that will effectively access the CNS compartment. Successful experimental approaches include use of regulatory microRNAs, specifically mir-155 to reduce inflammation.[80] Gene therapy approaches using myeloid cells are already in use for enzyme replacement purposes in inherited disorders (leukodystrophies).[100]

3.4 CONCLUSION

This chapter emphasizes the need to understand the precise pathogenic mechanisms underlying MS at each stage of disease evolution including both the neurobiologic and immunologic aspects so that novel therapies or combinations thereof can be developed and optimally utilized.

References

1. Sallusto F, Impellizzieri D, Basso C, et al. T-cell trafficking in the central nervous system. *Immunol Rev.* 2012;248(1):216–227.
2. Engelhardt B, Ransohoff RM. Capture, crawl, cross: the T cell code to breach the blood–brain barriers. *Trends Immunol.* 2012;33(12):579–589.
3. Ransohoff RM, Engelhardt B. The anatomical and cellular basis of immune surveillance in the central nervous system. *Nat Rev Immunol.* 2012;12(9):623–635.
4. Ousman SS, Kubes P. Immune surveillance in the central nervous system. *Nat Neurosci.* 2012;15(8):1096–1101.
5. Larochelle C, Alvarez JI, Prat A. How do immune cells overcome the blood–brain barrier in multiple sclerosis? *FEBS Lett.* 2011;585(23):3770–3780.
6. Saunders NR, Dreifuss JJ, Dziegielewska KM, et al. The rights and wrongs of blood–brain barrier permeability studies: a walk through 100 years of history. *Front Neurosci.* 2014;8:404.
7. Hussain RZ, Hayardeny L, Cravens PC, et al. Immune surveillance of the central nervous system in multiple sclerosis – relevance for therapy and experimental models. *J Neuroimmunol.* 2014;276(1–2):9–17.
8. Wosik K, Cayrol R, Dodelet-Devillers A, et al. Angiotensin II controls occludin function and is required for blood–brain barrier maintenance: relevance to multiple sclerosis. *J Neurosci.* 2007;27(34):9032–9042.
9. Alvarez JI, Dodelet-Devillers A, Kebir H, et al. The Hedgehog pathway promotes blood–brain barrier integrity and CNS immune quiescence. *Science.* 2011;334(6063):1727–1731.
10. Podjaski C, Alvarez JI, Bourbonniere L, et al. Netrin 1 regulates blood–brain barrier function and neuroinflammation. *Brain.* 2015;138(Pt 6):1598–1612.
11. Kaushik DK, Hahn JN, Yong VW. EMMPRIN, an upstream regulator of MMPs, in CNS biology. *Matrix Biol.* 2015;44–46C:138–146.
12. Hutchings M, Weller RO. Anatomical relationships of the pia mater to cerebral blood vessels in man. *J Neurosurg.* 1986;65:316–325.
13. Zhang ET, Inman CB, Weller RO. Interrelationships of the pia mater and the perivascular (Virchow-Robin) spaces in the human cerebrum. *J Anat.* 1990;170:111–123.

14. Johnston M, Zakharov A, Papaiconomou C, Salmasi G, Armstrong D. Evidence of connections between cerebrospinal fluid and nasal lymphatic vessels in humans, non-human primates and other mammalian species. *Cerebrospinal Fluid Res*. 2004;1:2–15.
15. Kida S, Pantazis A, Weller RO. CSF drains directly from the subarachnoid space into nasal lymphatics in the rat. Anatomy, histology and immunological significance. *Neuropathol Appl Neurobiol*. 1993;19:480–488.
16. Weller RO, Galea I, Carare RO, Minagar A. Pathophysiology of the lymphatic drainage of the central nervous system: implications for pathogenesis and therapy of multiple sclerosis. *Pathophysiology*. 2010;17:295–306. http://dx.doi.org/10.1016/j.pathophys.2009.10.007.
17. Kaminski M, Bechmann I, Pohland M, Kiwit J, Nitsch R, Glumm J. Migration of monocytes after intracerebral injection at entorhinal cortex lesion site. *J Leukoc Biol*. 2012;92:31–39. http://dx.doi.org/10.1189/jlb.0511241.
18. Louveau A, Smirnov I, Keyes TJ, et al. Gg structural and functional features of central nervous system lymphatic vessels. *Nature*. June 1, 2015;523(7560):337–341. http://dx.doi.org/10.1038/nature14432.
19. Carare RO, Bernardes-Silva M, Newman TA, et al. Solutes, but not cells, drain from the brain parenchyma along basement membranes of capillaries and arteries. Significance for cerebral amyloid angiopathy and neuroimmunology. *Neuropathol Appl Neurobiol*. 2008;34:131–144.
20. Laman JD, Weller RO. Drainage of cells and soluble antigen from the CNS to regional lymph nodes. *J Neuroimmune Pharmacol*. 2013;8:840–856. http://dx.doi.org/10.1007/s11481-013-9470-8.
21. Weller RO, Djuanda E, Yow HY, Carare RO. Lymphatic drainage of the brain and the pathophysiology of neurological disease. *Acta Neuropathol*. 2009;117:1–14. http://dx.doi.org/10.1007/s00401-008-0457-0.
22. Iliff JJ, Wang M, Liao Y, et al. A paravascular pathway facilitates CSF flow through the brain parenchyma and the clearance of interstitial solutes, including amyloid beta. *Sci Transl Med*. 2012;4:147ra111. http://dx.doi.org/10.1126/scitranslmed.3003748.
23. Rennels ML, Gregory TF, Blaumanis OR, Fujimoto K, Grady PA. Evidence for a 'paravascular' fluid circulation in the mammalian central nervous system, provided by the rapid distribution of tracer protein throughout the brain from the subarachnoid space. *Brain Res*. 1985;326:47–63.
24. Szentistvanyi I, Patlak CS, Ellis RA, Cserr HF. Drainage of interstitial fluid from different regions of rat brain. *Am J Physiol*. 1984;246:F835–F844.
25. Kooi EJ, van Horssen J, Witte ME, et al. Abundant extracellular myelin in the meninges of patients with multiple sclerosis. *Neuropathol Appl Neurobiol*. 2009;35(3):283–295.
26. Lassmann H. Multiple sclerosis: lessons from molecular neuropathology. *Exp Neurol*. 2014;262(Pt A):2–7.
27. Laman JD, Weller RO. Drainage of cells and soluble antigen from the CNS to regional lymph nodes. *J Neuroimmune Pharmacol*. 2013;8:840–856.
28. Yednock TA, Cannon C, Fritz LC, Sanchez-Madrid F, Steinman L, Karin N. Prevention of experimental autoimmune encephalomyelitis by antibodies against alpha 4 beta 1 integrin. *Nature*. March 5, 1992;356(6364):63–66.
29. Theien BE, Vanderlugt CL, Eagar TN, et al. Discordant effects of anti-VLA-4 treatment before and after onset of relapsing experimental autoimmune encephalomyelitis. *J Clin Invest*. April 2001;107(8):995–1006.
30. Cannella B, Gaupp S, Tilton RG, Raine CS. Differential efficacy of a synthetic antagonist of VLA-4 during the course of chronic relapsing experimental autoimmune encephalomyelitis. *J Neurosci Res*. February 1, 2003;71(3):407–416.
31. Bullard DC, Hu X, Schoeb TR, Collins RG, Beaudet AL, Barnum SR. Intercellular adhesion molecule-1 expression is required on multiple cell types for the development of experimental autoimmune encephalomyelitis. *J Immunol*. 2007;178:851–857.
32. Romme Christensen J, Ratzer R, Börnsen L, et al. Natalizumab in progressive MS: results of an open-label, phase 2A, proof-of-concept trial. *Neurology*. 2014;82(17):1499–1507.
33. Miller DH, Weber T, Grove R, et al. Firategrast for relapsing remitting multiple sclerosis: a phase 2, randomised, double-blind, placebo-controlled trial. *Lancet Neurol*. February 2012;11(2):131–139.
34. Schwab N, Ulzheimer JC, Fox RJ, et al. Fatal PML associated with efalizumab therapy: insights into integrin $\alpha L\beta 2$ in JC virus control. *Neurology*. February 14, 2012;78(7):458–467.
35. Kebir H, Kreymborg K, Ifergan I, et al. Human TH17 lymphocytes promote blood–brain barrier disruption and central nervous system inflammation. *Nat Med*. 2007;13(10):1173–1175.
36. Cayrol R, Wosik K, Berard JL, et al. Activated leukocyte cell adhesion molecule promotes leukocyte trafficking into the central nervous system. *Nat Immunol*. 2008;9(2):137–145.

37. Greenwood J, Heasman SJ, Alvarez JI, Prat A, Lyck R, Engelhardt B. Review: leucocyte-endothelial cell crosstalk at the blood–brain barrier: a prerequisite for successful immune cell entry to the brain. *Neuropathol Appl Neurobiol*. 2011;37(1):24–39.
38. Rumble JM, Huber AK, Krishnamoorthy G, et al. Neutrophil-related factors as biomarkers in EAE and MS. *J Exp Med*. 2015;212(1):23–35.
39. Russi AE, Walker-Caulfield ME, Ebel ME, Brown MA. Cutting edge: c-Kit signaling differentially regulates type 2 innate lymphoid cell accumulation and susceptibility to central nervous system demyelination in male and female SJL mice. *J Immunol*. 2015;194(12):5609–5613.
40. Sayed BA, Walker ME, Brown MA. Cutting edge: mast cells regulate disease severity in a relapsing-remitting model of multiple sclerosis. *J Immunol*. 2011;186:3294–3298.
41. Loeffler C, Dietz K, Schleich A, et al. Immune surveillance of the normal human CNS takes place in dependence of the locoregional blood–brain barrier configuration and is mainly performed by $CD3^+/CD8^+$ lymphocytes. *Neuropathology*. 2011;31:230–238.
42. Schwartz M, Kipnis J, Rivest S, Prat A. How do immune cells support and shape the brain in health, disease, and aging? *J Neurosci*. 2013;33(45):17587–17596.
43. Sankowski R, Mader S, Valdés-Ferrer SI. Systemic inflammation and the brain: novel roles of genetic, molecular, and environmental cues as drivers of neurodegeneration. *Front Cell Neurosci*. 2015;9:28.
44. Lucchinetti CF, Guo Y, Popescu BF, Fujihara K, Itoyama Y, Misu T. The pathology of an autoimmune astrocytopathy: lessons learned from neuromyelitis optica. *Brain Pathol*. 2014;24(1):83–97.
45. Shapiro AM, Jack CS, Lapierre Y, Arbour N, Bar-Or A, Antel JP. Potential for interferon beta-induced serum antibodies in multiple sclerosis to inhibit endogenous interferon-regulated chemokine/cytokine responses within the central nervous system. *Arch Neurol*. 2006;63(9):1296–1299.
46. Jensen MA, Arnason BG, White DM. A novel Fc gamma receptor ligand augments humoral responses by targeting antigen to Fc gamma receptors. *Eur J Immunol*. 2007;37(4):1139–1148.
47. Benarroch EE. Microglia: multiple roles in surveillance, circuit shaping, and response to injury. *Neurology*. 2013;81(12):1079–1088.
48. Michell-Robinson MA, Touil H, Healy LM, et al. Roles of microglia in brain development, tissue maintenance and repair. *Brain*. 2015;138(Pt 5):1138–1159.
49. Cuadros MA, Martin C, Coltey P, Almendros A, Navascues J. First appearance, distribution, and origin of macrophages in the early development of the avian central nervous system. *J Comp Neurol*. 1993;330:113–129.
50. Ginhoux F, Greter M, Leboeuf M, et al. Fate mapping analysis reveals that adult microglia derive from primitive macrophages. *Science*. 2010;330:841–845.
51. Kierdorf K, Erny D, Goldmann T, et al. Microglia emerge from erythromyeloid precursors via Pu.1- and Irf8-dependent pathways. *Nat Neurosci*. 2013;16:273–280.
52. Pelvig DP, Pakkenberg H, Stark AK, Pakkenberg B. Neocortical glial cell numbers in human brains. *Neurobiol Aging*. 2008;29:1754–1762.
53. Lyck L, Santamaria ID, Pakkenberg B, et al. An empirical analysis of the precision of estimating the numbers of neurons and glia in human neocortex using a fractionator-design with sub-sampling. *J Neurosci Methods*. 2009;182:143–156.
54. Ajami B, Bennett JL, Krieger C, Tetzlaff W, Rossi FM. Local self-renewal can sustain CNS microglia maintenance andfunction throughout adult life. *Nat Neurosci*. 2007;10:1538–1543.
55. Mildner A, Schmidt H, Nitsche M, et al. Microglia in the adult brain arise from Ly-6ChiCCR2$^+$ monocytes only under defined host conditions. *Nat Neurosci*. 2007;10:1544–1553.
56. Sieweke MH, Allen JE. Beyond stem cells: self-renewal of differentiated macrophages. *Science*. 2013;342(6161):1242974.
57. Jenkins SJ, Hume DA. Homeostasis in themononuclear phagocyte system. *Trends Immunol*. 2014;35:358–367.
58. Elmore MR, Najafi AR, Koike MA, et al. Colony-stimulating factor 1 receptor signaling is necessary for microglia viability, unmasking a microglia progenitor cell in the adult brain. *Neuron*. 2014;82:380–397.
59. Simard AR, Rivest S. Bone marrow stem cells have the ability to populate the entire central nervous system into fully differentiated parenchymal microglia. *FASEB J*. 2004;18(9):998–1000.
60. Durafourt BA, Moore CS, Blain M, Antel JP. Isolating, culturing, and polarizing primary human adult and fetal microglia. *Methods Mol Biol*. 2013;1041:199–211.
61. Butovsky O, Jedrychowski MP, Moore CS, et al. Identification of a unique TGF-beta-dependent molecular and functional signature in microglia. *Nat Neurosci*. 2014;17:131–143.

62. Ponomarev ED, Veremeyko T, Barteneva N, Krichevsky AM, Weiner HL. MicroRNA-124 promotes microglia quiescence and suppresses EAE by deactivating macrophages via the C/EBP-alpha- PU.1 pathway. *Nat Med.* 2011;17:64–70.
63. Van Ginderachter JA, Movahedi K, Hassanzadeh Ghassabeh G, et al. Classical and alternative activation of mononuclear phagocytes: picking the best of both worlds for tumor promotion. *Immunobiology.* 2006;211:487–501.
64. Fairweather D, Cihakova D. Alternatively activated macrophages in infection and autoimmunity. *J Autoimmun.* 2009;33:222–230.
65. Mantovani A, Sica A, Sozzani S, Allavena P, Vecchi A, Locati M. The chemokine system in diverse forms of macrophage activation and polarization. *Trends Immunol.* 2004;25:677–686.
66. Martinez FO, Gordon S, Locati M, Mantovani A. Transcriptional profiling of the human monocyte-to-macrophage differentiation and polarization: new molecules and patterns of gene expression. *J Immunol.* 2006;177:7303–7311.
67. Leidi M, Gotti E, Bologna L, et al. M2 macrophages phagocytose rituximab-opsonized leukemic targets more efficiently than m1 cells in vitro. *J Immunol.* 2009;182:4415–4422.
68. Durafourt BA, Moore CS, Zammit DA, et al. Comparison of polarization properties of human adult microglia and blood-derived macrophages. *Glia.* 2012;60:717–727.
69. Kim HJ, Ifergan I, Antel JP, et al. Type 2 monocyte and microglia differentiation mediated by glatiramer acetate therapy in patients with multiple sclerosis. *J Immunol.* 2004;172:7144–7153.
70. Moore CS, Cui Q, Warsi NM, et al. Direct and indirect effects of immune and central nervous system-resident cells on human oligodendrocyte progenitor cell differentiation. *J Immunol.* 2015;194(2):761–772.
71. Edwards JP, Zhang X, Frauwirth KA, Mosser DM. Biochemical and functional characterization of three activated macrophage populations. *J Leukoc Biol.* 2006;80(6):1298–1307.
72. Miron VE, Boyd A, Zhao JW, et al. M2 microglia and macrophages drive oligodendrocyte differentiation during CNS remyelination. *Nat Neurosci.* 2013;16:1211–1218.
73. Gadani SP, Walsh JT, Lukens JR, Kipnis J. Dealing with danger in the CNS: the response of the immune system to injury. *Neuron.* 2015;87(1):47–62.
74. Krysko DV, Agostinis P, Krysko O, et al. Emerging role of damage-associated molecular patterns derived from mitochondria in inflammation. *Trends Immunol.* 2011;32(4):157–164.
75. Kroner A, Greenhalgh AD, Zarruk JG, Passos Dos Santos R, Gaestel M, David S. TNF and increased intracellular iron alter macrophage polarization to a detrimental M1 phenotype in the injured spinal cord. *Neuron.* 2014;83:1098–1116.
76. Seguin R, Biernacki K, Prat A, et al. Differential effects of Th1 and Th2 lymphocyte supernatants upon human microglia. *Glia.* 2003;42:36–45.
77. Klesney-Tait J, Turnbull IR, Colonna M. The TREM receptor family and signal integration. *Nat Immunol.* 2006;7:1266–1273.
78. van Beek EM1, Cochrane F, Barclay AN, van den Berg TK. Signal regulatory proteins in the immune system. *J Immunol.* 2005;175(12):7781–7787.
79. Biber K, Neumann H, Inoue K, Boddeke HW. Neuronal 'On' and 'Off' signals control microglia. *Trends Neurosci.* 2007;30:596–602.
80. Butovsky O, Siddiqui S, Gabriely G, et al. Modulating inflammatory monocytes with a unique microRNA gene signature ameliorates murine ALS. *J Clin Invest.* 2012;122:3063–3087.
81. Butovsky O, Ziv Y, Schwartz A, et al. Microglia activated by IL-4 or IFN-gamma differentially induce neurogenesis and oligodendrogenesis from adult stem/progenitor cells. *Mol Cell Neurosci.* 2006;31:149–160.
82. Yamasaki R, Lu H, Butovsky O, et al. Differential roles of microglia and monocytes in the inflamed central nervous system. *J Exp Med.* 2014;211(8):1533–1549.
83. Wlodarczyk A, Løbner M, Cédile O, Owens T. Comparison of microglia and infiltrating CD11c$^+$cells as antigen presenting cells for T cell proliferation and cytokine response. *J Neuroinflammation.* 2014;11:57.
84. Moore CS, Ase A, Kinsara A, et al. P2Y12 expression and function in alternatively activated human microglia. *Neurol Neuroimmunol Neuroinflamm.* 2015;2(2):e80.
85. Healy L, Michell-Robinson M, Antel JP. Regulation of human glia by multiple sclerosis disease modifying therapies. *Semin Immunopathol.* 2014;37(6):639–649.
86. Heppner FL, Greter M, Marino D, et al. Experimental autoimmune encephalomyelitis repressed by microglial paralysis. *Nat Med.* 2005;11(2):146–152.
87. Fleming JO. Helminths and multiple sclerosis: will old friends give us new treatments for MS? *J Neuroimmunol.* 2011;233(1–2):3–5.

88. Mayo L, Trauger SA, Blain M, et al. Regulation of astrocyte activation by glycolipids drives chronic CNS inflammation. *Nat Med*. 2014;20:1147–1156.
89. Oberheim NA, Takano T, Han X, et al. Uniquely hominid features of adult human astrocytes. *J Neurosci*. 2009;29:3276–3287.
90. Rao VTS, Ludwin SK, Fuh SC, et al. MicroRNA expression patterns in human astrocytes in relation to anatomical location and age. *J Neuropathol Exp Neurol*. 2016. [in press].
91. Constantinescu CS, Tani M, Ransohoff RM, et al. Astrocytes as antigen-presenting cells: expression of IL-12/IL-23. *J Neurochem*. 2005;95:331–340.
92. Kim RY, Hoffman AS, Itoh N, et al. Astrocyte CCL2 sustained immune cell infiltration in chronic experimental autoimmune encephalomyelitis. *J Neuroimmunol*. 2014;274:53–61.
93. Saikali P, Antel JP, Pittet CL, Newcombe J, Arbour N. Contribution of astrocyte-derived IL-15 to CD8 T cell effector functions in multiple sclerosis. *J Immunol*. 2010;185:5693–5703.
94. Choi JW, Gardell SE, Herr DR, et al. FTY720 (fingolimod) efficacy in an animal model of multiple sclerosis requires astrocyte sphingosine 1-phosphate receptor 1 (S1P1) modulation. *Proc Natl Acad Sci USA*. 2011;108:751–756.
95. Hamby ME, Sofroniew MV. Reactive astrocytes as therapeutic targets for CNS disorders. *Neurotherapeutics*. 2010;7:494–506.
96. Antel J. Mechanisms of action of fingolimod in multiple sclerosis [Review]. *Clin Exp Neuroimmunol*. 2014;5:49–54.
97. Groves A, Kihara Y, Chun J. Fingolimod: direct CNS effects of sphingosine 1-phosphate (S1P) receptor modulation and implications in multiple sclerosis therapy. *J Neurol Sci*. 2013;328(1–2):9–18.
98. Salmen A, Gold R. Mode of action and clinical studies with fumarates in multiple sclerosis. *Exp Neurol*. 2014;262(Pt A):52–56.
99. Michell-Robinson MA, Moore CS, Healy LM, et al. Effects of fumarates on circulating and CNS myeloid cells in multiple sclerosis. *Ann Clin Transl Neurol*. 2015;3:27–41.
100. Krägeloh-Mann I, Groeschel S, Kehrer C, et al. Juvenile metachromatic leukodystrophy 10 years post-transplant compared with a non-transplanted cohort. *Bone Marrow Transpl*. 2013;48(3):369–375.

CHAPTER

4

Genetics of Multiple Sclerosis

J.R. Oksenberg
University of California San Francisco, San Francisco, CA, United States

J.L. McCauley
University of Miami, Miami, FL, United States

OUTLINE

4.1 INTRODUCTION

The convergence of epidemiological observations and empirical laboratory data firmly established the role of DNA variation as an important determinant of risk in multiple sclerosis (MS). Concurrently, a broad consensus has emerged bracketing MS with the so-called complex genetic disorders, a group of relatively frequent diseases characterized by multifaceted gene–environment interactions and a genomic risk-signature composed of many allelic DNA variants, each rather common in the population and exerting modest effects on the total risk. The multifactorial, polygenic model of MS genetics provided a useful conceptual framework for the systematic agnostic analysis of genetic susceptibility by means of large association studies, which transformed our understanding of MS pathogenesis.

Translational Neuroimmunology in Multiple Sclerosis
http://dx.doi.org/10.1016/B978-0-12-801914-6.00004-0

4.2 EPIDEMIOLOGICAL DATA DRIVES THE RATIONALE FOR GENETIC RESEARCH

A genetic component in MS is implied by the familial aggregation of cases and the relative high incidence in some ancestral groups irrespective of geographic location. High frequency rates of MS (~1–2 in 1000) are found in North America and Europe.[1–5] Notwithstanding challenges in surveillance, the disease is uncommon among African Blacks, Asians, and native populations of the Americas, New Zealand, and Australia.[2] Early estimates suggested that MS is also significantly less prevalent in African Americans than in European Americans (relative risk of 0.64).[6] Contemporary studies, however, are challenging the widely held belief that African Americans are at a reduced risk for developing MS.[7,8] Relatively lower frequencies are still observed among Hispanics living in the United States. Interestingly, three ranks in disease prevalence were recorded in Israel. As expected, the highest MS rates were in Israeli-born Jews followed by Jewish immigrants from Europe and the United States. On the other hand, Jewish immigrants from African or Asian countries and Christian Arabs had intermediate MS rates, whereas Moslem Arabs, Druze, and Bedouins had the lowest rates of MS.[9] According to most observers, the distinctive global geoprevalence distribution of MS reflects precipitating environmental triggers, such as pathogens and/or the disappearance of protective environmental factors.[10–12] Alternatively, this uneven geographical distribution might also be explained, at least in part, by ancestral differences in the frequency of genetic risk factors, reflecting past migrations, admixture, and other population-level events.[13,14]

The incidence of MS seems to have increased considerably over the last century, and this increase may have occurred primarily in women.[15] It has been argued that since it is unlikely that the distribution and frequency of genetic risk factors have changed over such a period of time, the explanation must relate to the increasing (or diminishing) exposure to some environmental factor or factors. Interestingly, the surge in incidence overlaps with the extraordinary population growth that occurred in Europeans in the last 200 years (over 11%/generation in eight generations). Even if we assume a conservative rate of 10^{-8}–10^{-9} mutations per site/per generation, given the population growth, the accumulation of rare and new mutations in recent generations is high,[16] but much will be concentrated within recent genealogy branches. Thus, these new rare variants will be private to small groups of individuals or families, and could, in principle, affect disease risk and incidence in modern Europeans and their descendants. Whether multiple rare variants are a component of the genetic risk burden in MS, effectively challenging the exclusivity of the broadly accepted common variant hypothesis for complex diseases, remains to be addressed empirically.[17–19]

Family-based studies provided fundamental insights to our understanding of MS risk and heritability. Monozygotic twins have a higher concordance rate (20–30%) compared to dizygotic twins of the same sex (2–5%).[20–22] After adjusting for age, nontwin siblings of an affected individual are 10–15 times more likely to develop MS than the general population,[23,24] and second- and third-degree relatives are also at an increased risk.[23,25] These statistics and the observation that there are few reported multicase families with affected individuals across three or more generations are consistent with the polygenic, low-penetrant, common allelic variants model of MS heritability. Studies in Canadian half-siblings,[26] spouses,[27] and adoptees[28] further support the conclusion that sharing genetics is the primary driver of familial clustering. Decoding the MS genome represents a worthy scientific goal, as the demonstration of even a modest functional effect of a known gene or group of genes on the course of MS is likely to elucidate fundamental disease mechanisms and yield new therapeutic opportunities. Interestingly, the extent of disease concordance among monozygotic twins seems to correlate with latitude,[29] an observation that further highlights the complexity of the underlying multifactorial interactions associated with MS susceptibility.

4.3 GENOME-WIDE ASSOCIATION STUDIES

Early genetic association studies sought to identify allelic variants in single or a handful of specific genes with unevenly distributed frequencies in a group of unrelated individual carriers of a quantifiable trait compared to a group of unrelated controls. Technological advances and overall assay miniaturization (DNA chips or microarrays) has enabled these screens to now rapidly interrogate large numbers of individuals for a vast and increasing number of markers located throughout the genome. The key assumption in these aptly named genome-wide association studies (GWASs) is that every region in the genome is equally likely to be associated with the trait under study. The arrays typically contain probes for genetic polymorphisms selected on the basis of frequency in the population, technical reproducibility of allelic calls, and linkage disequilibrium (LD) parameters to efficiently capture large portions of common variation across the majority of the genome. LD describes and quantifies the condition in which the frequency of a particular haplotype (combination of alleles at different genes on the same chromosome that are transmitted together as a block) in the population is significantly different from that expected if the different polymorphisms or genes were assorting independently. Thus, these selected single nucleotide polymorphisms (SNPs) act as surrogate markers or tags for putative causal variants located in the same broad genomic loci, which may be gene-dense regions and, in some instances even include multiple disease-causing candidates. Given the very large number of simultaneous tests in each study ($>10^6$ with current arrays), large datasets are required and stringent *P*-value thresholds are established a priori to assess the statistical significance of the associations. However, the gold standard for defining a true association is replication in multiple independent studies. Replication is important because even highly statistically significant associations can occur by chance alone, and within a single study it is possible that factors unrelated to the trait (such as genotyping errors) can create spurious associations.

Twelve GWASs and two metaanalyses of published GWASs have been reported to date for MS susceptibility, all in populations of European descent, including scans of high-risk population isolates and progressive phenotypes (summarized in Ref. 18). As expected, the information content of each study correlated with the size of the dataset. Up until 2013, this decade-long effort driven primarily by the International Multiple Sclerosis Genetics Consortium had identified just over 50 susceptibility alleles distributed across all autosomes (sex chromosomes were not included in the majority of these analyses). Genome-wide, the strongest susceptibility signal was mapped to the human leukocyte antigen (HLA) class II region of the major histocompatibility complex (MHC) in chromosome 6p21, explaining up to 10.5% of the genetic variance underlying risk. The HLA association with MS, which was first described several decades ago,[30,31] is consistent with the idea that MS is, at its core, an antigen-specific autoimmune disease. Moreover, this hypothesis is supported by the observation that the non-MHC-associated variants appear to locate predominantly in or near genes influencing the function of the adaptive immune system.[32]

Interestingly, some of the non-MHC allelic variants associated with MS have also emerged in GWASs of other autoimmune diseases,[32–34] suggesting that common underlying risk mechanisms likely exist across multiple immune-related conditions. To better describe this overlap and refine the regions of interest in susceptibility loci, a megaconsortium was established to conduct cost-effective association studies of the risk loci emerging from the autoimmunity GWASs across multiple immune-mediated diseases using a common, high-coverage SNPs array known as the ImmunoChip.[35] This array was designed in 2010, with a total of 196,524 variants passing the manufacturing quality control. Typing the ImmunoChip in European and US MS datasets identified an additional 48 novel susceptibility variants with genome-wide significance.[36]

To date, 110 polymorphisms in 103 discrete loci outside the MHC have been firmly associated with susceptibility through these screens, all of which have modest individual effects as indicated by odds ratios (ORs) between 1.09 and 1.34.[32,36] It is also noteworthy that significant variability exists across MS patients in the number of disease risk variants each individual carries. In aggregate, the proportion of the genetic variance accounting for disease risk explained by these polymorphisms, including the MHC, is roughly 30%, but the mapping of additional risk variants is likely to be immediate through ongoing multicenter initiatives utilizing specialized arrays, very large sample collections, and meta-analysis statistical methods. Taken together, the data from GWASs appear to support the long-held view that MS susceptibility rests in large part on the action of common allelic variants (that is, risk alleles with a population frequency of >1%) in multiple genes that may act independently or as part of functional networks.[37,38] It is important to note, however, that the associated SNPs represented in these works do not necessarily represent the causative variants. The use of highly saturated and redundant SNP arrays or direct sequencing of entire coding and noncoding regions of interest, coupled with the study of diverse population groups characterized by different LD patterns, is likely to further narrow the associated loci and facilitate the final assignment of candidate causative variants.[14] It is expected, however, that in vitro and in vivo functional studies will be needed to unambiguously determine the disease causative variants and their potential mechanisms of action.[39]

4.4 THE HUMAN LEUKOCYTE ANTIGEN GENE CLUSTER IN MULTIPLE SCLEROSIS

The association of the *HLA* locus with MS risk has been observed across all populations studied, and in both primary progressive and relapsing-remitting patients. The primary signal within the MHC maps to the *HLA-DRB1* gene, or more specifically to the *DRB1*15:01* allele, in the class II segment of this locus. In general, individuals who are *HLA-DRB1*15:01* homozygotes carry a high-risk genotype with ORs exceeding 7.0, compared to a range between 3.5 and 5.0 for heterozygotes *HLA-DRB1*15:01/X*. Complex allelic hierarchical lineages, cis/trans haplotypic effects, and independent protective signals in the class I region of the locus have been described as well (Fig. 4.1). Using SNP data from GWASs (5091 cases/9595 controls), a 2013 study identified 11 statistically independent effects: six *HLA-DRB1* alleles and one

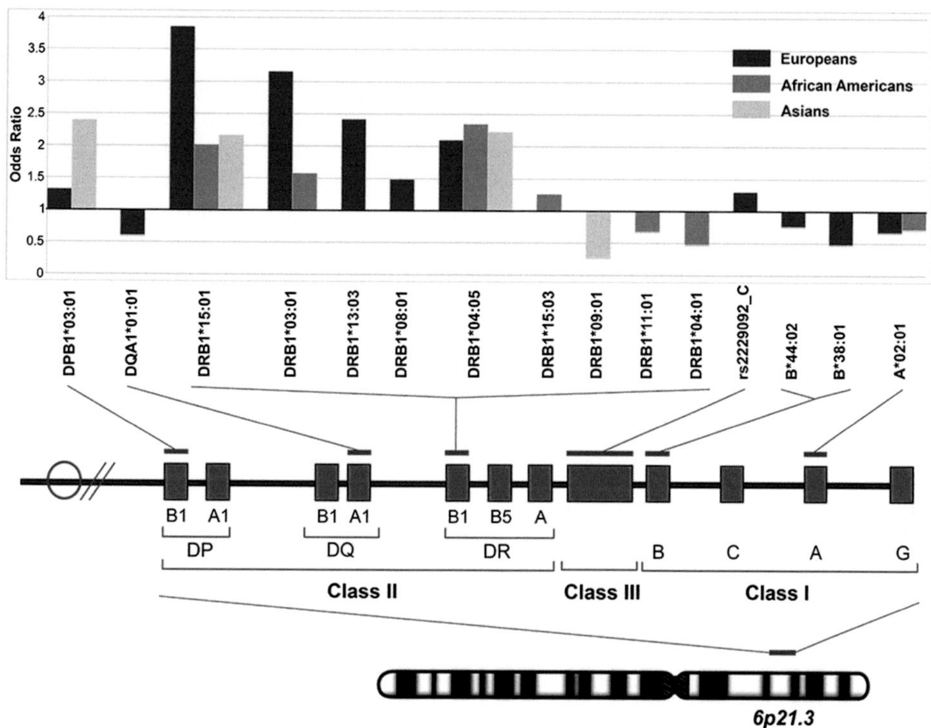

FIGURE 4.1 **The HLA system in multiple sclerosis.** The human leukocyte antigen (*HLA*) gene complex is located on the short arm of chromosome 6 at p21.3, spanning almost 4000 kb of DNA. The full sequence of the region was completed and reported in 1999. From 224 identified loci, 128 were predicted to be expressed and about 40% to have immune-response functions.[47] There are two major classes of HLA-encoding genes involved in antigen presentation. The telomeric stretch contains the *class I* genes, whereas the centromere proximal region encodes *HLA-class II* genes. The *HLA-class II* gene *DRB5* (in brackets) is only present in the DR51 haplotypic group (*DRB1*15* and *16 alleles). *HLA class I* and *class II* encoded molecules are cell-surface glycoproteins, whose primary role in an immune response is to display and present short antigenic peptide fragments to peptide/MHC-specific T cells, which can then become activated by a second stimulatory signal and initiate an immune response. In addition, HLA molecules are present on stromal cells on the thymus during development, helping to determine the specificity of the mature T-cell repertoire. A third group of genes collectively known as *class III*, cluster between the *class I* and *II* regions and include genes coding for complement proteins, 21a-hydroxylase, tumor necrosis factor, and heat shock proteins. This super-locus contains the strongest susceptibility signal for MS genome-wide, the *HLA-DRB1* gene. *HLA-DRB1* allelic heterogeneity and copy number effects have been described. In addition to the signal from the *HLA-DRB1* gene, complex hierarchical cis and trans effects can be found across the locus, including protective effect of *DQA1*01:01* only in the presence of *DRB1*15:01*,[48] and independent protective effects conferred by the telomeric class I region.[49,50] The figure displays OR values for different alleles in different ancestral groups reported to confer susceptibility or resistance to MS.

HLA-DPB1 allele in class II, one *HLA-A* allele and two *HLA-B* alleles in class I, and one signal in a region spanning from *MICB* to *LST1*.[40]

Additionally, more complex genetic interactions within the MHC region have been identified, with perhaps the most interesting being related to *HLA-DRB5*, a class II gene immediately telomeric of *HLA-DRB1*.[41] HLA-DRβ5DRα heterodimers appear to be effective myelin antigen-presenting molecules, with elegant experiments using triple *DRB1-DRB5-hTCR* transgenic mice supporting functional epistasis between *HLA-DRB1* and *HLA-DRB5* genes, whereby DRβ5 modifies the T-cell response driven by DRβ1 through activation-induced cell death, resulting in a milder and relapsing form of autoimmune demyelinating experimental disease.[42] Similarly, in African Americans with MS, carriers of the *HLA-DRB1*15:01* with *HLA-DRB5*null* haplotypes have a more severe disease course than *HLA-DRB5* carriers.[41] Although based on a small number of individuals with the rare *DRB5*null* mutation, the convergence of findings obtained from HLA-humanized experimental autoimmune encephalomyelitis mice with human MS genetic data supports a modulatory role of *HLA-DRB5* gene products on the progression of autoimmune demyelinating disease. Despite the remarkable molecular dissection of the *HLA* region in MS in recent years, it is evident that further studies are needed to determine the mechanism or mechanisms by which *HLA* genes fully contribute to MS susceptibility.

4.5 FUNCTIONAL GENOMICS

As suggested earlier, an important issue that remains to be addressed in order to maximize the value of genetic research in MS is the need to advance beyond gene discovery and explain the physiological effect of each risk variant on gene (or pathway) function and ultimately disease mechanism. Although each risk variant identified in GWAS explains statistically only a small proportion of overall disease risk, this does not necessarily mean that their molecular effects are weak or modest in nature. Some progress has already been made with the functional analysis of a restricted number of associated variants, particularly those located within the boundaries of a coding gene, demonstrating they may affect pathogenesis through a variety of mechanisms, including immune deviation and altering the ratio of membrane bound versus soluble immune receptors (Table 4.1).

Many established risk variants map outside the coding segments of annotated genes, posing a challenge to the design of molecular studies to understand function. Leveraging reference transcriptional and epigenomic data to inform on the putative role of the disease-associated variation is an area of active investigation.[43–45] It is becoming apparent that a significant number of noncoding risk variants lie within enhancers, many of which gain histone acetylation and transcribe enhancer-associated RNA upon immune stimulation.[46] Understanding the DNA-encoded regulatory mechanisms associated with susceptibility could have profound therapeutic implications.

4.6 CONCLUSIONS

A wealth of data confirms that genetic variation is an important determinant of MS risk. Population, family, and molecular studies provide strong empirical support for a polygenic model of inheritance, driven primarily by allelic variants relatively common in the general population. However, SNP arrays are not well suited to discover rare variants and their role cannot be forcefully dismissed. The MHC region of chromosome 6p21.3 represents by far the strongest MS susceptibility locus genome-wide. The primary signal arises from the *HLA-DRB1* gene in the class II segment of the locus, with hierarchical allelic and haplotypic effects.

TABLE 4.1 From SNP Identification to Function

Candidate Gene	Variant	Putative Mechanism
IL7R	rs6897932 (exon 6)	Low-level skipping of exon 6, changes in the soluble/membrane-bound ratio with higher sIL7R.[51]
IL2RA	rs2104286 (intronic)	Changes in the soluble/membrane-bound ratio with higher sIL2RA.[52]
TNFRSF1A	rs1800693 (intronic)	Skipping of exon 6, changes in the soluble/membrane-bound ratio with higher sTNFR1.[53] Enhanced NFkB response to TNFα and increase inflammatory cytokines.[54]
CD58	rs2300747 (intronic)	Higher membrane expression of CD58 and correction of $CD4^+$ regulatory cell function.[55]
IRF8	rs17445836 (intronic)	Widespread effect on the type I interferon transcriptional responses.[56]
TYK2	rs34536443 (exon 21)	Decrease TYK2 kinase activity and cytokine shifting toward Th2.[57]
CD6	rs1782933 (intronic)	Decreased expression of full-length CD6 in $CD4^+$ cells leading to altered proliferation.[58]
EVI5	rs11810217 (intronic)	Regulation of adjacent gene *GFI1*.[59]
EVI5	rs11808092 (exon)	Changes in the EVI5 interactome.[60]
CYP27B1	rs12368653 (intronic)	Underexpression in tolerogenic dendritic cells (DC2).[61]
PRKCA	GGTG ins/del (exon 3)	Risk associated with lower PRKCA expression.[62]
CBLB	rs12487066 (intronic)	Increased binding of the transcription factor C/EBPβ, diminished CBL-B expression and altered Type I IFN function in $CD4^+$ cells.[63]
NKkB1	rs228614 (intergenic)	Increase NFkB signaling and negative regulation of the NFkB pathway.[54]
CD40	rs1883832 (-1 bp of the transcription start site)	Changes in the full length/truncated CD40 mRNA ratio. Reduced cell-surface expression in B and dendritic cells.[64]
CLEC16	rs12927355 (intronic)	Expression quantitative locus for CLEC16A and SOCS1 in $CD4^+$ T cells.[65]
SOCS1	rs423674 (intergenic)	Expression quantitative locus for SOCS1 in dendritic cells.[66]
SP140	rs288445040 (exon 7)	Exon 7 skipping and protein expression levels of SP140.[67]
CD226	rs727088 (3′UTR) and rs763361 (exon 7)	Reduced expression of memory T cells and impaired regulatory T cell function.[68]

Independent protective signals in the telomeric class I region of the locus have been described as well. From 2005 to 2015, the collation of large multicenter DNA collections have thrived, providing the means for the pursuit of comprehensive GWASs to identify the non-MHC genetic component of MS. These studies have provided unambiguous evidence for the association of over 100 non-MHC loci with disease susceptibility. Predominantly, MS-associated variants appear to influence the function of the adoptive and innate immune system rather than the nervous system, but this may be due to our incomplete knowledge about any possible CNS roles of many genes initially identified to have some immune function; some identified genes, such as *GALC* on chromosome 14, coding for the lysosomal enzyme galactosylceramidase, clearly have CNS functions that could be relevant to MS.

Despite the expanding roster of established risk variants, our understanding of MS genetics remains incomplete. Furthermore, the potential for the discovery of additive risk variance extractable from large DNA screens will be quickly exhausted and collectively, the screen-based data will fail to fully explain the heritability of MS. Similarly to other complex genetic diseases, in addition to unidentified rare variants affecting susceptibility, multiple explanations for the missing heritability have been proposed including gene-by-gene and gene-by-environment interactions; cis/trans regulators of allelic expression, population and/or disease heterogeneity; genes of interest in technically problematic genomic regions to assay, neglecting the analysis of sex chromosomes; and hidden epigenetic effects. In addition, because the large genetic studies up to 2015 have focused on susceptibility and not clinical expression or course of MS, genetic contributors to progression remain to be revealed. The modest familial recurrence risk of the disease and the fact that most individuals in the general population carry common MS risk alleles and never develop the disease, make establishing an SNP-based diagnostic test that can reliably identify individuals at a high risk for developing MS simply not feasible at the present time.

The goals and aspirations of genetic research remain grand. The ultimate objective of MS genetic studies is to provide a better understanding of MS in order to improve patient care in the clinic. The convergence of genetics and functional genomics with next generation neuroimaging and informatics is expected to generate a genetic road map to guide the discovery of new drugs for MS treatment.

References

1. Rosati G, Granieri E, Carreras M, Tola R. Multiple sclerosis in southern Europe. A prevalence study in the sociosanitary district of Copparo, northern Italy. *Acta Neurol Scand*. 1980;62:244–249.
2. Pugliatti M, Sotgiu S, Rosati G. The worldwide prevalence of multiple sclerosis. *Clin Neurol Neurosurg*. 2002;104:182–191.
3. Pugliatti M, Rosati G, Carton H, et al. The epidemiology of multiple sclerosis in Europe. *Eur J Neurol*. 2006;13:700–722.
4. Kingwell E, Marriott JJ, Jette N, et al. Incidence and prevalence of multiple sclerosis in Europe: a systematic review. *BMC Neurol*. 2013;13:128.
5. Evans C, Beland SG, Kulaga S, et al. Incidence and prevalence of multiple sclerosis in the Americas: a systematic review. *Neuroepidemiology*. 2013;40:195–210.
6. Wallin MT, Page WF, Kurtzke JF. Multiple sclerosis in US veterans of the Vietnam era and later military service: race, sex, and geography. *Ann Neurol*. 2004;55:65–71.
7. Wallin MT, Culpepper WJ, Coffman P, et al. The Gulf War era multiple sclerosis cohort: age and incidence rates by race, sex and service. *Brain*. 2012;135:1778–1785.
8. Langer-Gould A, Brara SM, Beaber BE, Zhang JL. Incidence of multiple sclerosis in multiple racial and ethnic groups. *Neurology*. 2013;80:1734–1739.
9. Alter M, Kahana E, Zilber N, Miller A. Multiple sclerosis frequency in Israel's diverse populations. *Neurology*. 2006;66:1061–1066.
10. Karni A, Kahana E, Zilber N, Abramsky O, Alter M, Karussis D. The frequency of multiple sclerosis in Jewish and Arab populations in greater Jerusalem. *Neuroepidemiology*. 2003;22:82–86.
11. Cabre P. Migration and multiple sclerosis: the French West Indies experience. *J Neurol Sci*. 2007;262:117–121.
12. Ebers GC. Environmental factors and multiple sclerosis. *Lancet Neurol*. 2008;7:268–277.
13. Reich D, Patterson N, De Jager PL, et al. A whole-genome admixture scan finds a candidate locus for multiple sclerosis susceptibility. *Nat Genet*. 2005;37:1113–1118.

14. Isobe N, Gourraud PA, Harbo HF, et al. Genetic risk variants in African Americans with multiple sclerosis. *Neurology*. 2013;81:219–227.
15. Orton SM, Herrera BM, Yee IM, et al. Sex ratio of multiple sclerosis in Canada: a longitudinal study. *Lancet Neurol*. 2006;5:932–936.
16. Nachman MW, Crowell SL. Estimate of the mutation rate per nucleotide in humans. *Genetics*. 2000;156:297–304.
17. Gibson G. Rare and common variants: twenty arguments. *Nat Rev Genet*. 2011;13:135–145.
18. Sawcer S, Franklin RJ, Ban M. Multiple sclerosis genetics. *Lancet Neurol*. 2014;13:700–709.
19. Kemppinen AK, Baker A, Liao W, et al. Exome sequencing in single cells from the cerebrospinal fluid in multiple sclerosis. *Mult Scler*. 2014;20:1564–1568.
20. Willer CJ, Dyment DA, Risch NJ, Sadovnick AD, Ebers GC. Twin concordance and sibling recurrence rates in multiple sclerosis. *Proc Natl Acad Sci USA*. 2003;100:12877–12882.
21. Hansen T, Skytthe A, Stenager E, Petersen HC, Bronnum-Hansen H, Kyvik KO. Concordance for multiple sclerosis in Danish twins: an update of a nationwide study. *Mult Scler*. 2005;11:504–510.
22. Hawkes CH, Macgregor AJ. Twin studies and the heritability of MS: a conclusion. *Mult Scler*. 2009;15:661–667.
23. Robertson NP, Fraser M, Deans J, Clayton D, Walker N, Compston DA. Age-adjusted recurrence risks for relatives of patients with multiple sclerosis. *Brain*. 1996;119:449–455.
24. Hemminki K, Li X, Sundquist J, Hillert J, Sundquist K. Risk for multiple sclerosis in relatives and spouses of patients diagnosed with autoimmune and related conditions. *Neurogenetics*. 2009;10:5–11.
25. Carton H, Vlietinck R, Debruyne J, et al. Risks of multiple sclerosis in relatives of patients in Flanders, Belgium. *J Neurol Neurosurg Psychiatry*. 1997;62:329–333.
26. Sadovnick AD, Ebers GC, Dyment DA, Risch NJ, Group. CCS. Evidence for genetic basis of multiple sclerosis. *Lancet*. 1996;347:1728–1730.
27. Ebers GC, Yee IM, Sadovnick AD, Duquette P. Conjugal multiple sclerosis: population-based prevalence and recurrence risks in offspring. *Ann Neurol*. 2000;48:927–931.
28. Ebers GC, Sadovnick AD, Risch NJ. A genetic basis for familial aggregation in multiple sclerosis. *Nature*. 1995;377:150–151.
29. Islam T, Gauderman WJ, Cozen W, Hamilton AS, Burnett ME, Mack TM. Differential twin concordance for multiple sclerosis by latitude of birthplace. *Ann Neurol*. 2006;60:56–64.
30. Bertrams J, Kuwert E, Liedtke U. HLA antigens and multiple sclerosis. *Tissue Antigens*. 1972;2:405–408.
31. Maito S, Manerow N, Mickey MR, Terasaki PI. Multiple sclerosis: association with HL-A3. *Tissue Antigens*. 1972;2:1–4.
32. The International Multiple Sclerosis Genetics Consortium; Wellcome Trust Case Control Consortium 2. Genetic risk and a primary role for cell-mediated immune mechanisms in multiple sclerosis. *Nature*. 2011;476:214–219.
33. Cotsapas C, Voight BF, Rossin E, et al. Pervasive sharing of genetic effects in autoimmune disease. *PLoS Genet*. 2011;7:e1002254.
34. Solovieff N, Cotsapas C, Lee PH, Purcell SM, Smoller JW. Pleiotropy in complex traits: challenges and strategies. *Nat Rev Genet*. 2013;14:483–495.
35. Cortes A, Brown MA. Promise and pitfalls of the immunochip. *Arthritis Res Ther*. 2011;13:101.
36. The International Multiple Sclerosis Genetics Consortium. Analysis of immune-related loci identifies 48 new susceptibility variants for multiple sclerosis. *Nat Genet*. 2013;45:1353–1360.
37. Baranzini SE, Galwey NW, Wang J, et al. Pathway and network-based analysis of genome-wide association studies in multiple sclerosis. *Hum Mol Genet*. 2009;18:2078–2090.
38. The International Multiple Sclerosis Genetics Consortium. Network-based multiple sclerosis pathway analysis with GWAS data from 15,000 cases and 30,000 controls. *Am J Hum Genet*. 2013;92:854–865.
39. Attfield KE, Dendrou CA, Fugger L. Bridging the gap from genetic association to functional understanding: the next generation of mouse models of multiple sclerosis. *Immunol Rev*. 2012;248:10–22.
40. Patsopoulos NA, Barcellos LF, Hintzen RQ, et al. Fine-mapping the genetic association of the major histocompatibility complex in multiple Sclerosis: HLA and non-HLA effects. *PLoS Genet*. 2013;9:e1003926.
41. Caillier SJ, Briggs F, Cree BA, et al. Uncoupling the roles of HLA-DRB1 and HLA-DRB5 genes in multiple sclerosis. *J Immunol*. 2008;181:5473–5480.
42. Gregersen JW, Kranc KR, Ke X, et al. Functional epistasis on a common MHC haplotype associated with multiple sclerosis. *Nature*. 2006;443:574–577.
43. Maurano MT, Humbert R, Rynes E, et al. Systematic localization of common disease-associated variation in regulatory DNA. *Science*. 2012;337:1190–1195.

44. Trynka G, Sandor C, Han B, et al. Chromatin marks identify critical cell types for fine mapping complex trait variants. *Nat Genet*. 2013;45:124–130.
45. Ziller MJ, Gu H, Muller F, et al. Charting a dynamic DNA methylation landscape of the human genome. *Nature*. 2013;500:477–481.
46. Farh KK, Marson A, Zhu J, et al. Genetic and epigenetic fine mapping of causal autoimmune disease variants. *Nature*. 2014;518:337–343.
47. The MHC Sequencing Consortium. Complete sequence and gene map of a human major hisocompatibility complex. *Nature*. 1999;401:921–923.
48. Lincoln MR, Ramagopalan SV, Chao MJ, et al. Epistasis among HLA-DRB1, HLA-DQA1, and HLA-DQB1 loci determines multiple sclerosis susceptibility. *Proc Natl Acad Sci USA*. 2009;106:7542–7547.
49. Brynedal B, Duvefelt K, Jonasdottir G, et al. HLA-A confers an HLA-DRB1 independent influence on the risk of multiple sclerosis. *PLoS One*. 2007;2:e664.
50. The IMAGEN Consortium. Mapping of multiple susceptibility variants within the MHC region for senven immune-mediated diseases. *Proc Natl Acad Sci USA*. 2009;106:18680–18685.
51. Gregory SG, Schmidt S, Seth P, et al. Interleukin 7 receptor alpha chain (IL7R) shows allelic and functional association with multiple sclerosis. *Nat Genet*. 2007;39:1083–1091.
52. Maier LM, Lowe CE, Cooper J, et al. IL2RA genetic heterogeneity in multiple sclerosis and type 1 diabetes susceptibility and soluble interleukin-2 receptor production. *PLoS Genet*. 2009;5:e1000322.
53. Gregory AP, Dendrou CA, Attfield KE, et al. TNF receptor 1 genetic risk mirrors outcome of anti-TNF therapy in multiple sclerosis. *Nature*. 2012;488(7412):508–511.
54. Housley WJ, Fernandez SD, Vera K, et al. Genetic variants associated with autoimmunity drive NFkB signaling and responses to inflammatory stimuli. *Sci Transl Med*. 2015;7:291ra93.
55. De Jager PL, Baecher-Allan C, Maier LM, et al. The role of the CD58 locus in multiple sclerosis. *Proc Natl Acad Sci USA*. 2009;106:5264–5269.
56. De Jager PL, Jia X, Wang J, et al. Meta-analysis of genome scans and replication identify CD6, IRF8 and TNFRSF1A as new multiple sclerosis susceptibility loci. *Nat Genet*. 2009;41(7):776–782.
57. Couturier N, Bucciarelli F, Nurtdinov RN, et al. Tyrosine kinase 2 variant influences T lymphocyte polarization and multiple sclerosis susceptibility. *Brain*. 2011;134:693–703.
58. Kofler DM, Severson CA, Mousissian N, De Jager PL, Hafler DA. The CD6 multiple sclerosis susceptibility allele is associated with alterations in $CD4^+$ T cell proliferation. *J Immunol*. 2011;187:3286–3291.
59. Martin D, Pantoja C, Fernandez Minan A, et al. Genome-wide CTCF distribution in vertebrates defines equivalent sites that aid the identification of disease-associated genes. *Nat Struct Mol Biol*. 2011;18:708–714.
60. Didona A, Isobe N, Caillier SJ, et al. A non-synonymous single nucleotide polymorphism associated with multiple sclerosis risk affects the EVI5 interactome. *Hum Mol Genet*. 2015;24:7151–7158.
61. Shahijanian F, Parnell GP, McKay FC, et al. The CYP27B1 variant associated with an increased risk of autoimmune disease is underexpressed in tolerizing dendritic cells. *Hum Mol Genet*. 2014;23:1425–1434.
62. Paraboschi EM, Rimoldi V, Solda G, et al. Functional variations modulating PRKCA expression and alternative splicing predispose to multiple sclerosis. *Hum Mol Genet*. 2014;23:6746–6761.
63. Sturner KH, Borgmeyer U, Schulze C, Pless O, Martin R. A multiple sclerosis-associated variant of CBLB links genetic risk with type I IFN function. *J Immunol*. 2014;193:4439–4447.
64. Field J, Shahijanian F, Schibeci S, et al. The MS risk allele of CD40 is associated with reduced cell membrane bound expression in antigen presenting cells: implications for gene function. *PLoS One*. 2015;10(6):e0127080.
65. Leikfoss IS, Keshari PK, Gustavsen MW, et al. Multiple sclerosis risk allele in CLEC16A acts as an expression quantitative trait locus for CLEC16A and SOCS1 in $CD4^+$ T cells. *PLoS One*. 2015;10:e0132957.
66. Lopez de Lapuente A, Pinto-Medel MJ, Astobiza I, et al. Cell-specific effects in different immune subsets associated with SOCS1 genotypes in multiple sclerosis. *Mult Scler*. 2015;21:1498–1512.
67. Matesanz F, Potenciano V, Fedetz M, et al. A functional variant that affects exon-skipping and protein expression of SP140 as genetic mechanism predisposing to multiple sclerosis. *Hum Mole Genet*. 2015;24:5619–5627.
68. Piedavent-Salomon M, Willing A, Engler JB, et al. Multiple sclerosis associated genetic variants of CD226 impair regulatory T cell function. *Brain*. 2015;138:3263–3274.

CHAPTER

5

Multiple Sclerosis Subtypes: How the Natural History of Multiple Sclerosis Was Challenged due to Treatment

B. Weinstock-Guttman
SUNY University at Buffalo, Buffalo, NY, United States

E. Grazioli
UPMC Northshore Neurology, Erie, PA, United States

C. Kolb
SUNY University at Buffalo, UBMD Neurology, Buffalo, NY, United States

OUTLINE

Multiple sclerosis (MS) is an immune-mediated disease of the central nervous system (CNS) with a heterogeneous course spanning decades. The clinical course varies between individuals with MS as well as over time in the same individual and is marked by relapses, progression, or a combination of these processes. Since the first approved disease-modifying

Translational Neuroimmunology in Multiple Sclerosis
http://dx.doi.org/10.1016/B978-0-12-801914-6.00005-2

therapy (DMT) for MS, interferon beta-1b (IFN b-1b) in 1993, there has been a rapid expansion of treatment options aimed at reducing relapses and disability. Currently approved DMTs for MS include IFN b-1b (Betaseron; Bayer HealthCare and Extavia; Novartis), interferon beta 1a (IFN b-1a) (Avonex®; Biogen & Idec and Rebif®; EMD Serono), peginterferon beta 1a (Plegridy®; Biogen Idec), glatiramer acetate (GA) (Copaxone®; TEVA Neuroscience), natalizumab (Tysabri®; Biogen Idec), fingolimod (Gilenya®; Novartis), teriflunomide (Aubagio®; Genzyme & Sanofi), dimethyl fumarate (Tecfidera®; Biogen Idec), and alemtuzumab (Lemtrada®; Genzyme & Sanofi). The use of mitoxantrone (Novantrone®; EMD Serono), approved for secondary progressive and rapidly worsening relapsing MS, decreased significantly, secondary to its toxicity profile (primarily cardiotoxicity and risk for leukemia).

5.1 DEFINING THE CLINICAL COURSE OF MULTIPLE SCLEROSIS

The initial clinical presentation of MS (in up to 80% of patients) consists of an acute neurological deficit (relapses or exacerbations) followed by complete or incomplete recovery, known as relapsing-remitting MS (RRMS). Less frequently, patients present from onset with a relentless, continuous progressive clinical deterioration known as primary-progressive MS (PPMS). Varied manifestations of the disease course may be seen in the same patient at different time points such that a patient who had initially experienced acute relapsing symptoms followed by recovery, develops later disease progression with continuous decline, at different rates, with possible superimposed acute relapses known as secondary-progressive MS (SPMS). The disease activity and severity can change sometimes with unpredictable temporal measures. Criteria for MS diagnosis and disease course have been constantly updated for the last 15 years, providing the basis for an earlier more accurate classification and conceivable a more appropriate therapeutic intervention. Former criteria were based solely on clinical features, whereas more recent classifications include MRI established outcomes.

In 1983, Poser published new diagnostic criteria for MS with the primary intent to be applied in clinical trials, although immediately they were adopted in practice as well. MS was defined as definite or probable. These criteria could be met clinically or could be laboratory supported with evidence from spinal fluid markers of oligoclonal bands or increased IgG production. Clinically definite MS required the occurrence of at least two clinical attacks and evidence of two separate lesions. Paraclinical evidence from evoked potentials could be used to define one of the lesions. In laboratory-supported definite MS, the spinal fluid data could substitute for one of the attacks or one of the lesions. Probable MS criteria were met with two attacks and evidence of one lesion (clinically or paraclinically) or one attack and evidence of two lesions.[53]

In 2001, the International Panel on the Diagnosis of Multiple Sclerosis presented the first version of the McDonald Criteria for MS classification. These revised diagnostic criteria integrated MRI data with the clinical and paraclinical data used by Poser. MRI criteria could substitute for clinical criteria to prove either dissemination in space or time. MRI criteria for dissemination in space required three of the following four:

1. One gadolinium-enhancing lesion or nine T2 hyperintense lesions
2. At least one infratentorial lesion

3. At least one juxtacortical lesion
4. At least three periventricular lesions

Dissemination in time required that if a first scan occurs 3 months or more after the onset of the clinical event, the presence of a gadolinium-enhancing lesion was sufficient to demonstrate dissemination in time, provided that it was not at the site implicated in the original clinical event. If there was no enhancing lesion, a follow-up scan was required, typically in 3 months. A new T2 lesion or gadolinium-enhancing lesion at that time fulfilled the criteria for dissemination in time. If the first scan was performed less than 3 months after the onset of the clinical event, a second scan done 3 or more months after the clinical event fulfilled evidence for dissemination in time. If no enhancing lesion was seen at this second scan, another scan that shows a new T2 lesion in no less than 3 months would fulfill criteria.[35]

The 2005 revisions to the McDonald Criteria[41] attempted to further incorporate spinal cord imaging, shorten MRI interval needed for dissemination in time, and simplify criteria for PPMS. In the 2005 revision, dissemination in time could be demonstrated by detection of a gadolinium-enhancing lesion at least 3 months after the clinical event if not at the site corresponding to the initial event or by the detection of a new T2 lesion if it appears at any time compared with a reference scan after the onset of the clinical event. For the 2005 criteria for dissemination in space, a spinal cord lesion could now be considered equivalent to a brain infratentorial lesion, an enhancing spinal cord lesion considered equivalent to an enhancing brain lesion, and an individual spinal cord lesion could count toward the total number of T2 lesions. MS with progression from onset could be retrospectively or prospectively determined. Two of the following criteria for progressive MS also needed to be met:

1. Positive brain MRI (nine T2 lesions or four or more T2 lesions with positive pattern reversal visual evoked potential).
2. Positive spinal cord MRI (two focal T2 lesions).
3. Positive cerebrospinal fluid (oligoclonal bands or increased IgG index).

The version of the McDonald Criteria from 2010 again attempted to simplify MRI criteria. Dissemination in space can be demonstrated by one or more T2 lesions in at least two or four areas of the CNS (periventricular, juxtacortical, infratentorial, spinal cord). Dissemination in time can be demonstrated by a new T2 or gadolinium-enhancing lesion on follow-up MRI, with reference to baseline scan, irrespective of timing of the baseline MRI or the simultaneous present of asymptomatic gadolinium-enhancing and nonenhancing lesions at any time.[43] The MRI parameters in the 2010 revisions to the McDonald Criteria have led to earlier diagnosis of MS then what was provided with the earlier versions of diagnostic criteria.

Clinically isolated syndrome (CIS) has been defined as a first attack involving the CNS coupled with MRI findings in a pattern consistent with a demyelinating disease. This classically presents as a spinal cord syndrome (incomplete transverse myelitis), optic neuritis, or brain stem dysfunction.[13] Without intervention, most CIS of patients will develop a second event over months or years, indicating conversion to MS.[5,40]

Radiologically isolated syndrome (RIS) is a condition in which MRI findings suggestive of MS are present, but no clinical manifestations of MS have been noted.[39] Many such patients ultimately develop neurological symptoms such that RIS likely represents very early MS. Approximately two-thirds of patients with RIS develop new MRI lesions and one-third develop neurological symptoms consistent with MS over a 5-year time frame.[17]

In 1996, the National Multiple Sclerosis Society (NMSS) Advisory Committee published a report that described the pattern of disease based on clinical criteria that defined four MS disease courses:[33]

1. RRMS: "clearly defined relapses with full recovery or with sequelae and residual deficits upon recovery; the periods between disease relapses are characterized by a lack of disease progression."
2. SPMS: "initial RR disease course followed by progression with or without occasional relapses, minor remissions, and plateaus."
3. PPMS: "disease progression from onset with occasional plateaus and temporary minor improvements allowed."
4. Progressive-relapsing MS: "progressive disease from the onset, with clear acute relapses, with or without full recovery; periods between relapses characterized by continuing progression."

There are also two consensus definitions based on clinical severity:

1. Benign MS: "disease in which the patient remains fully functional in all neurologic systems 15 years after disease onset." This diagnosis is based on a retrospective analysis.
2. Malignant MS: "disease with a rapid progressive course, leading to significant disability in multiple neurologic systems or death in a relatively short time after disease onset."

In 2013, the NMSS committee joined with the European Committee for Treatment Research in MS to update clinical course definitions such that disease activity and progression may be based on clinical or MRI information. Active disease is defined as clinical relapse followed by full or partial recovery or as MRI occurrence of contrast-enhancing or new or unequivocally enlarging T2 lesions. If CIS fulfills criteria for active disease, the classification becomes RRMS. Progressive disease is clinically defined as steadily increasing objectively documented neurological dysfunction/disability without unequivocal recovery. Criteria for MRI-defined progression is not yet standardized but could include an increasing number and volume of T1 lesions, increased brain atrophy measures, and changes in diffusion tensor or magnetic transfer imaging. PPMS is defined as progressive accumulation of disability from onset, whereas SPMS is defined as progressive accumulation of disability after initial relapsing course. Progressive disease (PPMS and SPMS) may also be further classified as (1) active (presence of relapses or MRI activity) with progression or active without progression or (2) not active with progression or without progression (stable disease).[34]

The pathology of MS has both inflammatory and neurodegenerative components. While both processes are present throughout the disease course, in the relapsing stage, inflammatory pathology appears to be primarily responsible for the clinical presentation while in progressive phases of the disease manifestations related to neurodegeneration predominate.[32] The available DMTs are primarily known to control the inflammatory aspect of the disease, with less distinct evidence for neuroprotection or repair properties.[4]

Most observational studies providing data on MS disease course were acquired prior to the advent of widespread use of DMTs. The natural historical data on the disease course suggest that after 11–15 years from disease onset, 58% of RRMS converted to SPMS, while after 16–25 years duration, 66% had progressive MS.[52] This natural history pattern is supported by Runmarker and Andersen, with 80% of their MS population becoming progressive by 25 years.[47]

More than 20 years passed since the approval of first DMTs for treatment of MS. However, long-term impact data on disease progression following the introduction of DMTs are limited and controversial. The efficacy of a drug evaluated usually under well-controlled but short (1–3 years) randomized studies cannot be fully translated into a real-world setting. The long-term effect of a drug can be assessed from data obtained from extension studies following the initial pivotal randomized clinical trials or using observational studies from registries or large single-center databases. Either method has its limitations, as the observational studies are usually biased by the nonrandomized treatment assignment, while the long-term extension studies are usually able to compare only between early versus later treatment initiation cohorts, often with concerns on the available retention cohort. Similarly, different statistical methods can only partly address the observational studies' limitations (ie, propensity scores methods, regression models adjusting for unbalanced baseline).[49]

5.2 DATA FROM RANDOMIZED CONTROLLED CLINICAL TRIALS

The first DMT approved for MS was IFN b-1b (Betaseron) in 1993. Since that time, multiple studies have examined the impact of DMT on relapse rate, progression, and conversion of CIS to MS. IFN b-1b, IFN b-1a IM (Avonex), GA, and teriflunomide are approved for use in CIS. The Controlled High-Risk Subjects Avonex Multiple Sclerosis Prevention Study (CHAMPS) and the Controlled High-Risk Avonex Multiple Sclerosis Prevention Study in Ongoing Neurological Surveillance (CHAMPIONS) evaluated the effectiveness of Avonex on transition to clinically definite MS (CDMS) in subjects diagnosed with CIS and presenting with optic neuritis, partial transverse myelitis, or a brain stem syndrome. Over a 3-year period in CHAMPS, treatment with IFN showed a relative risk reduction for CDMS of 33%.[22] CHAMPIONS was a 10-year follow-up of the CHAMPS subjects and compared immediate treatment (IT) to delayed treatment (DT) of CIS patients with Avonex. The DT group started IFN b-1a IM q week at a median of 30 months from the beginning of the CHAMPS study. At 10 years, the percentage of patients who developed CDMS was lower in the IT group than in the DT group (38% vs 53%). Ninety-two percent of patients in CHAMPIONS were classified as having relapsing MS at the 10-year mark with 81% having an Expanded Disability Status Scale (EDSS) of less than 3. However, it should be noted that no significant difference in long-term disability outcome was observed between the IT and DT groups.[31] Similarly in the BENEFIT trial, IFN b-1b 250 μg every other day significantly reduced the risk for development of MS over a 2-year period, with 28% conversion to CDMS in the treatment group versus 45% in the placebo group.[25,26] An extension study of Betaseron in CIS favored IT over DT with risk of conversion to CDMS of 37% in IT and 51% in DT groups as well as a 40% reduced risk of disability in the IT group.[27] GA also reduced conversion to MS in CIS with a monosymptomatic onset by 45% versus placebo in the PreCISe study.[11] Teriflunomide is a once-daily oral DMT. The TOPIC trial demonstrated a 43% reduction for a 14-mg dose of teriflunomide and a 37% reduction for a 7-mg dose of teriflunomide in risk of conversion to CDMS over a 2-year study period.[36]

The randomized-controlled pivotal DMT trials for MS have consistently demonstrated reductions in number of relapses and often a beneficial effect on preventing disease progression as well. IFNs are considered to have an antiinflammatory mechanism with decreased

T-cell migration through the blood–brain barrier and downregulation of proinflammatory cytokines.[1] Over a 2–3-year period, IFN b-1b subcutaneously every other day reduced relapse rate by approximately 30%.[19] This relapse rate reduction was maintained over a 5-year period in the original cohort. Fewer patients in the IFN b-1b treatment group also had EDSS progression; however, this difference in disability was not statistically significant.[20] Over 2 years, IFN b-1a 30 μg administered weekly intramuscularly (MSCRG study) slowed disability progression sustained for at least 6 months and reduced the relapse rate. IFN b-1a, IM weekly demonstrated a 32% relapse rate reduction versus placebo. The rate of EDSS progression of at least one point was 37% less in the IFN b-1a treatment group compared to the placebo group.[21] Also fewer patients in the IFN b-1a-treated group reached an EDSS level of 4 or 6.[46] The PRISMS study examined three times weekly subcutaneous IFN b-1a at doses of 22 and 44 μg over 2 years. Relapse rates were reduced by approximately 30%. Time to sustained disability was delayed with IFN treatment at both of these doses; however, in patients with a higher baseline EDSS (greater than 3.5), a dose effect was noted such that a significant delay in disability progression was noted only in the 44-μg group.[44] A blinded extension study of PRISMS-4 compared treatment and crossover groups (placebo group randomized to 22 or 44 μg of IFN b-1a). Patients treated for 4 years had greater relapse rate reductions than those in the crossover groups with 2-year treatment duration and time to confirmed disability was prolonged in the 44 μg treatment group compared to crossover groups.[45] This again suggests benefit for early versus delayed treatment. Peginterferon b-1a is a pegylated formulation of IFN b-1a dosed 125 mg given subcutaneously every other week. Relapse rate reduction of 36% and decreased proportion of patients with disability progression of 38% were seen in treated group compared to placebo after 1 year.[7]

GA is a synthetic mixture of L-alanine, L-glutamate, L-lysine and L-Ent tyrosine thought to induce an antiinflammatory Th1 to Th2 shift. Development of GA-specific cells with a Th2 behavior have the potential of entering into the CNS. GA has been demonstrated to influence the development of antiinflammatory antigen-presenting cells of M2 phenotypes[30] that may be involved in the regenerative processes of repair or neuroprotection of the CNS (antiinflammatory).[37] Subjects receiving 20 mg of GA subcutaneously daily had a 29% reduction of relapse rate compared to placebo over 2 years. Patients in the placebo arm were more likely to have an increased EDSS than those in the treatment arm.[23] In the GALA study utilizing 40 mg GA three times weekly, at 24 months, the proportion of relapse-free patients significantly favored early initiation over delaying start of GA by 12 months (67.5% vs 56.6%).[24]

Natalizumab is a monoclonal antibody that binds to the alpha 4 integrin on T lymphocytes and prevents migration through the vascular endothelium. In the AFFIRM study, natalizumab 300 mg intravenously every 4 weeks for up to 28 months compared to placebo reduced annualized relapse rate by 59% at year 2. EDSS progression was reduced by 42% (sustained for 3 months) and 54% (sustained for 6 months) over 2 years as well.[42]

Teriflunomide is an oral pyrimidine synthesis inhibitor. Teriflunomide reduced the annualized relapse rate in MS by approximately 31% at both the 7- and 14-mg dosage levels and decreased in the proportion of patients with confirmed disability progression (over 2 years) by approximately 30% only in the 14-mg dose.[38]

Dimethyl fumarate has been shown to activate the nuclear factor-like 2 pathway that is involved in the cellular response to oxidative stress. Twice daily oral dimethyl fumarate relatively reduced relapse rate by 53% and rate of confirmed disability progression by 38% compared to placebo over 2 years.[15]

Fingolimod binds to sphingosine-1-phosphate receptors, blocking egress of lymphocytes from the lymph nodes. Fingolimod 0.5 mg orally daily reduced relapse rate by 54% over 2 years. The probability of disability progression on 0.5 mg daily of fingolimod was reduced to 17.7% compared with 24.1% for placebo.[28] In the TRANSFORMS trial, fingolimod showed superior efficacy to IFN b-1a in terms of annualized relapse rate (ARR = 0.16 vs IFN b-1a ARR = 0.33).[8]

Alemtuzumab is an anti-CD52 monoclonal antibody. The CAREMS trials (1 and 2) were rater-blinded trials that compared alemtuzumab dosed at 12 mg daily intravenously for 5 days at baseline and for 3 days in the second year with the active comparator IFN b-1a 44 μg subcutaneously 3 times weekly.[9,10] There was an approximately 50% reduction in relapse rate in both studies for the alemtuzumab group compared to the IFN group. The CAREMS 2 population had previously relapsed on platform DMT, while in CAREMS 1, patients were naïve to previous DMT. In CAREMS 2, a 42% reduction in sustained disability in the alemtuzumab group versus the active comparator IFN was observed.

5.3 LONG-TERM FOLLOW-UP EXTENSION TRIALS

The pivotal MS trials, although high-quality, typically randomized placebo-controlled studies, provide information on the MS disease course during a relatively short window, often 2–3 years, which may not be representative of the full MS disease course typically spanning decades. Long-term extension studies of the initial randomized controlled trials are available primarily for the older first-line injectable agents. No long-term data is yet available for the newer second- and third-line agents. Fifteen-year follow-up data from the MSCRG on intramuscular IFN b-1a based on patient-reported outcomes revealed that fewer patients had progressed to EDSS milestones of 6 or 7 in the early IFN-treated group than those who were treated later.[2] In the 8-year follow-up data of SQ IFN b-1a, TIW (PRISMS) early treatment showed benefit also in preventing long-term irreversible progression versus the delayed therapy group.[26] Twenty-one-year follow-up data on IFN b-1b MS study-group patients showed a significant reduction in all-cause mortality in early treated patients versus the delayed group.[16]

Long-term data also exists for the GA pivotal study. After 15 years, 65% of patients retained on GA study had not transitioned to SPMS.[12] These data may suggest improved outcomes in terms of disease progression with treatment as compared to the natural history data that report 58% of patients having a progressive course after 15 years.[52] Nevertheless, a direct comparison with the historical cohort is biased as the retention in the long-term GA cohort was approximately 50% from the initial enrolled cohort.

5.3.1 Observational Studies on the Long-Term Effect of Disease-Modifying Therapy

A retrospective study on a prospective followed cohort of RRMS patients included in the British Columbia MS database treated with IFN-b was compared with a contemporary untreated cohort as well as with an historical untreated cohort available from the same database.

After adjusting for baseline characteristics (EDSS, age, disease duration, sex), no significant difference was noted in the risk of reaching an EDSS of 6 between the treated or untreated cohorts after a follow-up of 4–10.8 years.[48] Bias related to the control group appeared to influence the results as treatment with IFN b appeared to increase the rate of progression to EDSS of 6 when treated patients were compared to the contemporary untreated group (with more benign disease) while it appeared to reduce it when compared to the historical group. In contrast, another large Italian observational study did show beneficial effect in slowing the disease progression in IFN-treated patients such that times to reach SPMS, EDSS of 4, and EDSS of 6 were reduced by 3.8 years, 1.7 years, and 2.2 years, respectively, in favor of IFN-treated patients.[50] Two additional studies from Canada comparing the disease progression in large prospective cohorts before and after introduction of IFN therapy showed a significant reduction on yearly EDSS worsening after treatment initiation.[6,51] The risk of converting from RRMS to SPMS was also shown in an Italian study to be significantly reduced by treatment with IFN as well as with GA.[3]

5.4 DISEASE ACTIVITY-FREE STATUS

Another concept to assess effectiveness of DMT on the basis of a drug's ability to achieve disease-free activity status in the MS patient is being discussed and applied in clinical-trial design. The concept of disease activity-free status (DAFS) or "no discernible disease activity" is borrowed from the treat-to-target approach in the field of rheumatology. DAFS in MS is commonly defined as the absence of relapses, no accumulation of disability, and no new MRI lesions (new or enlarging T2 lesions or gadolinium-enhancing lesions). For example, post hoc analysis of data from the trial of natalizumab, fingolimod, and dimethyl fumarate has found a greater proportion of patients who obtained DAFS with DMT compared to placebo.[14,18,29]

There are, however, important differences between the disease pathology in rheumatologic diseases and MS that may limit its application. While the rheumatologic diseases such as rheumatoid arthritis are primarily inflammatory, MS again has both significant inflammatory and neurodegenerative components. Current measures of DAFS are limited by EDSS-defined disability measures that may underestimate certain aspects of disease such as cognitive dysfunction and fatigue. Patient-reported measures of quality of life, which also consider side-effect profiles of DMT and employment, can also broaden measures of efficacy.

5.5 CONCLUSION

Evaluating the long-term efficacy and safety profile of DMTs in chronic diseases such as MS is of significant importance. Based on current evidence, early DMT initiation and good adherence are key elements for success. Recognizing that the newer DMTs are also associated with an increased side-effect profile, a well-defined risk–benefit assessment at the time of initiation of a specific DMT is required. Combining long-term follow-up data from randomized controlled trials as well as from large observational studies after introduction of the more efficacious, second-line DMTs will be important for understanding the real impact of DMTs on MS disease course. For long-term disability prevention, efficient intervention with neuroprotective and/or repair agents in addition to the antiinflammatory products available today will be necessary.

References

1. Bermel R, Rudick R. Interferon-B treatment for multiple sclerosis. *Neurotherapeutics*. 2007;4:633–646.
2. Bermel R, Weinstock-Guttman B, Bourdette D, Foulds P, You X, Rudick R. Intramuscular interferon beta-1a therapy in patients with relapsing-remitting multiple sclerosis: a 15-year follow-up study. *Mult Scler*. 2010;16(5):588–596.
3. Bergamaschi R, Quaglini S, Trojano M, et al. Early prediction of the long term evolution of multiple sclerosis: the Bayesian Risk Estimate for Multiple Sclerosis (BREMS) score. *JNNP*. 2007;78:757–759.
4. Bitsch A, Schuchardt J, Bunkowski S, Kuhlmann T, Bruck W. Acute axonal injury in multiple sclerosis. Correlation with demyelination and inflammation. *Brain*. 2000;123:1174–1183.
5. Brex PA, Ciccarelli O, O'Riordan JI, Sailer M, Thompson AJ, Miller DH. A longitudinal study of abnormalities on MRI and disability from multiple sclerosis. *N Engl J Med*. 2002;346:158–164.
6. Brown M, Kirby S, Skedgel C, et al. How effective are disease-modifying drugs in delaying progression in relapsing-onset MS? *Neurology*. 2007;69:1498–1507.
7. Calabresi P, Kiesseier B, Arnold D, et al. Pegylated interferon beta-1a for relapsing remitting multiple sclerosis (ADVANCE): a randomised, phase 3, double-blind study. *Lancet Neurol*. 2014;13(7):657–665.
8. Cohen J, Barkhof T, Comi G, et al. Oral fingolimod or intramuscular interferon for relapsing multiple sclerosis. *NEJM*. 2010;362:402–415.
9. Cohen J, Coles A, Arnold D, et al. Alemtuzumab versus interferon beta 1a as first-line treatment for patients with relapsing-remitting multiple sclerosis: a randomised controlled phase 3 trial. *Lancet*. 2012;380:1819–1828.
10. Coles A, Twyman C, Arnold D, et al. Alemtuzumab for patients with relapsing multiple sclerosis after disease modifying therapy: a randomised controlled phase 3 trial. *Lancet*. 2012;380:1829–1839.
11. Comi G, Martinelli V, Rodegher M, et al. Effect of glatiramer acetate on conversion to clinically definite multiple sclerosis in patients with clinically isolated syndrome (PreCISe study): a randomised, double-blind, placebo-controlled trial. *Lancet*. 2009;374:1503–1511.
12. Ford C, Goodman A, Johnson K, et al. Continuous long-term immunomodulatory therapy in relapsing multiple sclerosis: results from the 15 year analysis of the US prospective open-label study of glatiramer acetate. *Mult Scler*. 2010;16:342–350.
13. Frohman EM. Multiple sclerosis. *Med Clin North Am*. 2003;87(4):867–897.
14. Giovanni G, Gold R, Kappos L, et al. Analysis of clinical and radiological disease activity-free status in patients with relapsing-remitting multiple sclerosis treated with BG-12: findings from the DEFINE study. *J Neurol*. 2012;259:S106.
15. Gold R, Kappos L, Arnold D, et al. Placebo-controlled phase 3 study of oral BG-12 for relapsing multiple sclerosis. *NEJM*. 2012;367:1098–1107.
16. Goodin D, Reder A, Ebers G, et al. Survival in MS. A randomized cohort study 21 years after the start of the pivotal IFNB-1b trial. *Neurology*. 2012;78:1315–1322.
17. Granberg T, Martola J, Kristoffersen-Wilberg M, et al. Radiologically isolated syndrome-incidental magnetic resonance imaging findings suggestive of multiple sclerosis, a systematic review. *Mult Scler*. 2013;19(3):271–280.
18. Hardova E, Galetta S, Hutchinson M, et al. Effect of natalizumab on clinical and radiological disease activity in multiple sclerosis: a retrospective analysis of the natalizumab safety and efficacy in relapsing remitting multiple sclerosis (AFFIRM) study. *Lancet Neurol*. 2009;8:254–260.
19. IFNB Multiple Sclerosis Study Group. Interferon beta-1b is effective in relapsing-remitting multiple sclerosis. Clinical results of multicenter, randomized, double blind, placebo-controlled trial. *Neurology*. 1993;43:655–661.
20. IFNB Multiple Sclerosis Study Group. Interferon beta-1b in the treatment of multiple sclerosis: final outcomes of the randomized controlled trial. *Neurology*. 1995;45:1277–1285.
21. Jacobs LD, Cookfair DI, Rudick RA, et al. Intramuscluar interferon beta-1a for disease progression in multiple sclerosis. *Ann Neurol*. 1996;39:285–294.
22. Jacobs LD, Beck RW, Simon JH, et al. Intramuscular interferon beta-1a therapy initiated during a first demyelinating event in multiple sclerosis. *NEJM*. 2000;343(13):898–904.
23. Johnson K, Brooks B, Cohen J, et al. Copolymer 1 reduces relapse rate and improves disability in relapsing-remitting multiple sclerosis: results of a phase II multicenter, double-blind, placebo-controlled trial. *Neurology*. 1995;45:1268–1278.

24. Kahn O, Rieckmann P, Boyko A, et al. 24-month efficacy and safety of glatiramer acetate 40 mg/1 mL 3-times weekly: open-label extension study of the GALA trial in subjects with relapsing-remitting multiple sclerosis. *Neurology*. 2014;83:S31.003.
25. Kappos L, Polman C, Freedman M, et al. Treatment with interferon beta-1b delays conversion to clinically definite and McDonald MS in patients with clinically isolated syndromes. *Neurology.* 2006;67(7):1242–1249.
26. Kappos L, Traboulsee A, Constantinescu C, et al. Long-term subcutaneous interferon beta-1a therapy in patients with relapsing-remitting MS. *Neurology*. 2006;67:944–953.
27. Kappos L, Freedman M, Polman C, et al. Effect of early versus delayed interferon beta-1b treatment on disability after a first clinical event suggestive of multiple sclerosis: a 3-year follow-up analysis of the BENEFIT study. *Lancet*. 2007;370:389–397.
28. Kappos L, Ernst-Wilhelm R, O'Connor P, et al. A placebo-controlled trial of oral fingolimod in relapsing multiple sclerosis. *NEJM*. 2010;362:387–401.
29. Kappos L, O'Connor P, Amato M, et al. Fingolimod treatment increased the proportion of patients who are free from disease activity in multiple sclerosis; results from a phase 3, placebo-controlled study (FREEDOMS). In: *Paper Presented at the 63rd Annual Meeting or the American Academy of Neurology. Honolulu, HI.* ; April 9, 2011.
30. Kim H, Ifergan I, Antel J, et al. Type 2 monocyte and microglia differentiation mediated by glatiramer acetate therapy in patients with multiple sclerosis. *J Immunol*. 2004;172(11):7144–7153.
31. Kinkel RP, Dontchev M, Kollman C, et al. Association between immediate initiation of intermuscular interferon beta 1a at the time of clinically isolated syndrome and long-term outcomes. *Arch Neurol*. 2012;69(2):183–190.
32. Lassmann H, Bruck W, Lucchinetti C. The immunopathology of multiple sclerosis: an overview. *Brain Pathol*. 2007;17(2):210–218.
33. Lublin FD, Reingold SC. Defining the clinical course of multiple sclerosis: results of an international survey. National Multiple Sclerosis Society (USA) Advisory Committee on Clinical Trials of New Agents in Multiple Sclerosis. *Neurology*. 1996;46:907–911.
34. Lublin FD, Reingold SC, Cohen JA, et al. Defining the clinical course of multiple sclerosis. The 2013 revisions. *Neurology*. 2014;83:1–9.
35. McDonald W, Compston A, Edan G, et al. Recommended diagnostic criteria for multiple sclerosis: guidelines from the International Panel on the diagnosis of multiple sclerosis. *Ann Neurol*. 2001;50:121–127.
36. Miller A, Wolinsky J, Kappos L, et al. Oral teriflunomide for patients with a first clinical episode suggestive of multiple sclerosis (TOPIC): a randomised, double-blind, placebo-controlled, phase 3 trial. *Lancet Neurol*. 2014;13:977–986.
37. Miron V, Boyd A, Zhao J, et al. M2 microglia and macrophages drive oligodendrocyte differentiation during CNS remyelination. *Nat Neurosci*. 2013;16:1211–1218.
38. O'Connor P, Wolinsky J, Confavreux C, et al. Randomized trial of oral teriflunimide for relapsing multiple sclerosis. *NEJM*. 2011;365:1293–1303.
39. Okuda D, Srinivasan R, Oksenberg J, et al. Genotype-phenotype correlations in multiple sclerosis: HLA genes influence disease severity inferred by 1HMR spectroscopy and MRI measures. *Brain*. 2009;132:250–259.
40. Optic Neuritis Study Group. Multiple sclerosis risk after optic neuritis. *Arch Neurol*. 2008;65:727–732.
41. Polman C, Reingold S, Edan G, et al. Diagnostic criteria for multiple sclerosis: 2005 revisions to the McDonald Criteria. *Ann Neurol*. 2005;58(6):840–846.
42. Polman C, O'Connor P, Hardova E, et al. A randomized placebo-controlled trial of natalizumab for relapsing multiple sclerosis. *N Engl J Med*. March 2, 2006;354(9):899–910.
43. Polman C, Reingold SC, Banwell B, et al. Diagnostic criteria for multiple sclerosis: 2010 revisions to the McDonald Criteria. *Ann Neurol*. 2011;69:292–302.
44. PRISMS Study Group. Randomised double-blind placebo-controlled study of interferon beta-1a in relapsing remitting multiple sclerosis. *Lancet*. 1998;352:1498–1504.
45. PRISMS Study Group and the University of British Columbia MS/MRI Analysis Group. PRISMS-4: long term efficacy of interferon B-1a in relapsing MS. *Neurology*. 2001;56:1628–1636.
46. Rudick RA, Goodkin DE, Jacobs LD, et al. Impact of interferon beta-1a on neurological disability in relapsing multiple sclerosis. *Neurology*. 1997;49:358–363.
47. Runmarker B, Andersen O. Prognostic factors in a multiple sclerosis incidence cohort with twenty-five years of follow-up. *Brain*. 1993;116(pt 1):117–134.
48. Shirani A, Yinshan Z, Karim M, et al. Association between use of interferon beta and progression of disability in patients with relapsing remitting multiple sclerosis. *JAMA*. 2012;308:247–256.

49. Sormani MP, Bruzzi P. Can we measure long-term treatment effects in multiple sclerosis? *Nat Rev Neurol.* 2015;11(3):176–182.
50. Trojano M, Pellegrini F, Fuiani A. New natural history of interferon beta treated multiple sclerosis. *Ann Neurol.* 2007;61(4):300–306.
51. Veugelers P, Fisk J, Brown M, et al. Disease progression among multiple sclerosis patients before and during a disease modifying drug program: a longitudinal population-based evaluation. *Mult Scler.* 2009;15:1286–1294.
52. Weinshenker B, Bass B, Rice G, et al. The natural history of multiple sclerosis: a geographically based study. *Brain.* 1989;112:1419–1428.
53. Poser CM. The pathogenesis of multiple sclerosis. Additional considerations. *J Neurol Sci.* 1993;115:S3–S15.

CHAPTER

6

Pediatric-Onset Multiple Sclerosis as a Window Into Early Disease Targets and Mechanisms

G. Fadda, A. Bar-Or
McGill University, Montreal, QC, Canada

OUTLINE

6.1 INTRODUCTION

Pediatric-onset multiple sclerosis (MS) represents an important unmet clinical need as well as a unique opportunity to study early disease mechanisms. While some features (including degree of focal–inflammatory activity, apparent recovery from acute relapse) differ between pediatric- and adult-onset MS, shared genetic and environmental risk factors across the age spectrum suggest that mechanisms underlying MS development in children are in essence the same as in adults, and that any differences may be more likely to reflect manifestation of

Translational Neuroimmunology in Multiple Sclerosis
http://dx.doi.org/10.1016/B978-0-12-801914-6.00006-4

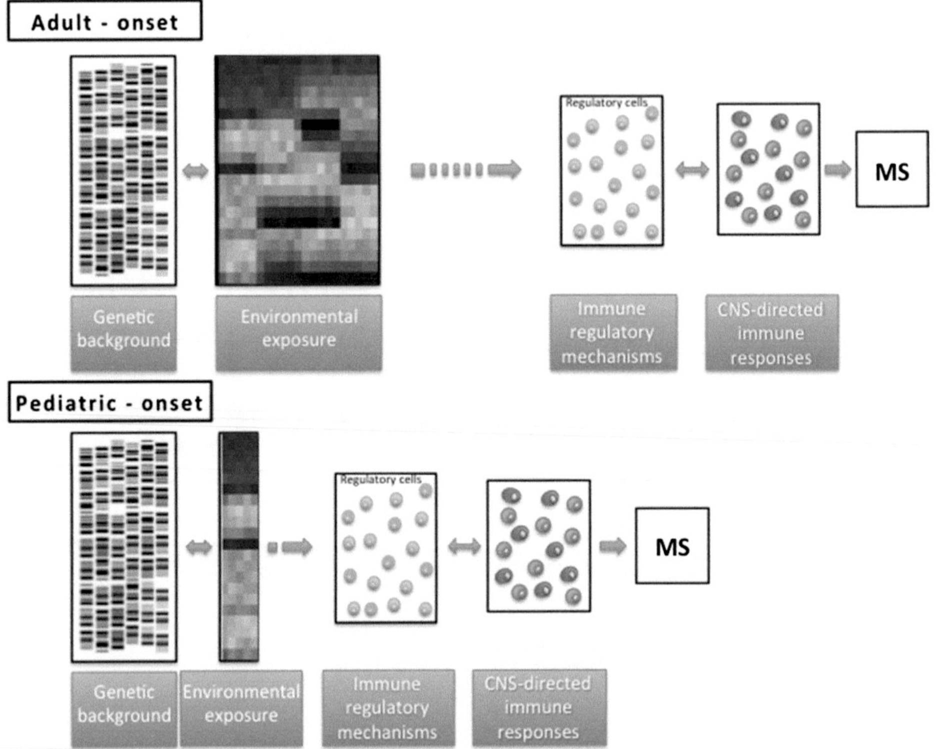

FIGURE 6.1 Development of Pediatric- and Adult-Onset Multiple Sclerosis (MS). Genetic and environmental factors interact over time to confer risk of developing MS. Patients with pediatric-onset MS come to clinical attention at a time that is closer on average to biological disease onset. This would be associated with a narrower range of environmental exposures acting on genetic susceptibility, providing a unique window to investigate early targets and mechanisms involved in the regulation of CNS-directed immune responses.

the same disease in younger and still maturing immune and nervous systems. The shorter interval between biological and clinical onset in children with MS limits the number of irrelevant exposures or biological abnormalities that may emerge as a consequence (rather than a cause) of disease (Fig. 6.1). Here we review genetic and environmental risk factors implicated in the pediatric MS population and summarize emerging insights into early humoral and cellular immune targets and mechanisms involved in disease pathogenesis.

6.2 GENETIC SUSCEPTIBILITY AND PEDIATRIC-ONSET MULTIPLE SCLEROSIS

While the number of pediatric MS genetic studies remains relatively small, it appears as though the same genetic risk factors implicated in adult-onset MS are also implicated in the risk of developing MS in the pediatric age group (Table 6.1). An increased frequency of the HLA-DR2 haplotype was first reported in a Russian pediatric MS cohort compared to controls.[1] Subsequently, HLA DRB1*1501 (the most strongly implicated risk gene in adult-onset MS)

TABLE 6.1 Common Genetic and Environmental Risk Factors in Adult- and Pediatric-Onset Multiple Sclerosis

Risk Factor	Adult Onset	Pediatric Onset
≥1 HLA-DRB1 alleles (%MS vs control)	28–33% vs 9–15% (OR 3.08)[108]	37–43% vs 17–24% (OR 3.2, 95% CI 1.7–6.1)[1,2]
GWAS implicated genes	57 non-HLA SNPs conferring increased risk of MS[a]	Implication of the 57 SNPs identified in adult-MS in pediatric-MS[6]
Vitamin D	50 nmol/L increase in serum level OR 0.59 (95% CI 0.36–0.97)[14]	10 nmol/L increase in serum level; HR 0.9 (95% CI 0.8–1.0)[9]
Prior EBV infection (%MS vs control)	99.5% vs 94%[11]	83–99% vs 42–72%[20–22] OR 3.72 (95% CI 1.48–8.85)[23]
Cigarette smoke exposure	RR 1.6 (95% CI 1.2–2.1)[b]	RR 2.12 (95% CI 1.43–3.15)[31]
Sex	F:M = 3.2:1[c]	F:M = 1.1:1 (<10 years) F:M = 1.8:1 (>10 years)[22]
Obesity	BMI > 27 kg/m² OR 2.2 (95% CI 1.5–3.0)[37]	BMI > 30 kg/m² OR 1.78 (95% CI 0.70–4.49)[39]

MS: multiple sclerosis; SNP: single nucleotide polymorphism; OR: odds ratio; CI: confidence interval; EBV: Epstein–Barr virus; HR: hazard ratio; BMI: body mass index; RR: relative risk.

[a] *International Multiple Sclerosis Genetics C, Wellcome Trust Case Control C, Sawcer S, et al. Genetic risk and a primary role for cell-mediated immune mechanisms in multiple sclerosis.* Nature. *2011;476(7359):214–219.*

[b] *Hernan MA, Olek MJ, Ascherio A. Cigarette smoking and incidence of multiple sclerosis.* Am J Epidemiol. *2001;154(1):69–74.*

[c] *Orton SM, Herrera BM, Yee IM, et al. Sex ratio of multiple sclerosis in Canada: a longitudinal study.* Lancet Neurol. *2006;5(11):932–936.*

was found to be overrepresented (OR 2.7) in children with MS compared to children with monophasic acquired demyelinating syndrome (ADS), also indicating that this genetic risk is specific to MS rather than to central nervous system (CNS) inflammatory diseases more generally.[2] A number of studies over the years have focused on other molecules implicated in disease susceptibility. One study sequencing the promotor, coding regions, and exon–intron boundaries of the putative MS target-antigen myelin-oligodendrocyte glycoprotein (MOG) did not identify particular polymorphisms distinguishing children with MS ($n = 75$) from controls ($n = 100$).[3] Another study examining particular polymorphisms of tumor necrosis factor-alpha (TNF-α), a cytokine implicated in MS pathogenesis, reported no differences between children with MS ($n = 24$) and healthy control subjects from the same population ($n = 93$).[4] An exploratory study of mitochondrial DNA (mtDNA) variants in 213 children with ADS and 166 healthy controls matched for age and sex analyzed a total of 31 single nucleotide polymorphisms (SNPs) including haplogroup-defining SNPs and mtDNA variants previously reported to be associated with adult MS.[5] While no single SNP was implicated, haplogroup H was associated with an increased risk of MS (OR 2.60; 95% CI 1.21–5.55) raising the possibility of an association between mtDNA variants and the risk of MS in children.

Although genome-wide association studies (GWASs) have not been carried out in pediatric-onset MS because of limited numbers of subjects available for study, a compound genetic score based on the 57 MS-risk SNPs that had been implicated at the time in adult-onset GWASs, was found to be significantly higher in children with MS compared to children with monophasic ADS, and was similar to the score previously observed in the adult-onset MS patients.[6] The odds ratios of individual GWAS-implicated risk polymorphisms in pediatric MS remain unknown because of the relative rarity of this condition. Genetic analyses in pediatric MS populations are also hampered by the heterogeneous distribution of ethnicities in the available pediatric MS cohorts. Nonetheless, studies as of 2015 have indicated that the genetic susceptibility identified in adults with MS appears also relevant in children, lending support to the view that MS in children shares genetic mechanisms of susceptibility with the adult-onset MS spectrum. It remains unknown whether the earlier-life disease onset of MS in children reflects a greater genetic burden than in those presenting in adulthood. The availability of now over 170 candidate risk polymorphisms derived from adult MS GWASs presents opportunities to test these loci in children, regarding copy number variation, epigenetic modification, effects on cellular functions, and gene–environment interaction.

6.3 ENVIRONMENTAL EXPOSURES IN PEDIATRIC MULTIPLE SCLEROSIS

Epidemiological studies in migrants developing MS have indicated that the age of emigration influences whether risk is carried over from the region of origin, or acquired at the new destination.[7,8] These findings suggest that particular environmental factors experienced during early life importantly contribute to MS risk in genetically susceptible individuals. As with genetic risk, environmental risk factors implicated in adult-onset MS including low levels of vitamin D, previous infection with Epstein–Barr virus (EBV), and smoking exposure, appear to be relevant also for pediatric-onset MS[9] (see Table 6.1).

6.3.1 Vitamin D

Vitamin D can be obtained from the diet or synthesized in the skin after ultraviolet B radiation exposure, the latter traditionally thought to account for the biggest contribution of vitamin D supply.[10] Epidemiological studies revealing higher MS incidence and prevalence in regions of higher latitude with more temperate climates and limited sun exposure have long-since implicated low levels of vitamin D in MS risk.[11–13] Similar to the results obtained from studies in adult-onset MS,[14] higher serum levels of 25-hydroxyvitamin D in children presenting with incident acute demyelinating syndromes were correlated with a reduced risk of MS diagnosis[9] and increased levels of circulating vitamin D have been correlated with a reduction in relapse rates in pediatric MS.[15]

A number of mechanisms have been proposed to explain the relation of vitamin D to MS pathophysiology (reviewed in Refs 16,17). These include the demonstrated ability of vitamin D (mostly based on in vitro experiments) to (1) limit differentiation, maturation, and activation of dendritic cells; (2) reduce T-cell proinflammatory (Th1, Th17) responses while enhancing Th2 responses; (3) induce regulatory T-cell function; (4) inhibit B-cell and plasma-cell responses; and (5) downregulate the expression of MMP9, otherwise implicated in disrupting blood–brain barrier integrity. The association of particular polymorphisms in the vitamin D response gene with MS risk, and identification of a vitamin D response element in the promoter region of the HLA-DRB1 gene,[18] provide possible links between low vitamin D levels and genetic susceptibility of MS, and it is of interest to speculate that low levels of vitamin

D during fetal development or early life might influence thymic tolerance, favoring escape of proinflammatory CNS autoreactive T cells. The contribution of vitamin D levels to future MS risk may become increasingly relevant as vitamin D deficiency is increasingly common in early life given social influences including sun-avoidance behavior.

6.3.2 Infectious Exposures

The implication of early EBV exposure with development of MS in adults[19] has now been supported by several studies in children that have documented EBV seropositivity in 83–99% of pediatric MS patients compared to 42–72% in controls.[20–23] Serological evidence of remote EBV infection was associated with a threefold increased risk of pediatric MS, and EBNA1 titers were significantly higher in children with MS compared to controls.[22] Furthermore, serial analysis of saliva samples has demonstrated that children with MS shed EBV (but not other viruses) more frequently than controls,[24] suggesting a selective failure of control of EBV and/or more frequent EBV reactivation in children with MS. Mechanisms proposed to explain this association have included (1) chronic activation of EBV-infected memory B cells (eg, through viral expression of LMP1 that mimics immune activation mediated by CD40),[25] resulting in exaggerated activation of pathogenic T cells; (2) molecular mimicry based on shared peptide sequences of EBV and putative CNS target antigens such as myelin basic protein (MBP); and (3) expression by EBV of an IL-10-like molecule that might interfere with immune-regulatory functions of human IL-10.[26]

In contrast to EBV, remote infection with cytomegalovirus appears to be associated with a decreased risk of developing MS.[23] Of interest, exposure to certain infections may also be associated with decreased risk of developing neuromyelitis optica (NMO).[27] Whether a particular sequence of exposures to different infections influences risk of (or protection from) developing MS or NMO is unknown.

Synergy may exist between genetic and environmental risk factors for MS onset in children. A multivariable analysis in children with early MS and controls indicated that previous EBV infection, low vitamin D levels, and harboring HLA-DRB1*15 could contribute as independent risk factors for developing disease.[9] While the presence of all three risk factors was not sufficient for disease development, the absence of all three was associated with significantly decreased disease risk. Particular interactions have been described between HLA DRB1 and vitamin D,[18] as well as between HLA DRB15 and both EBV and herpes simplex virus-associated MS risk.[23,28] In addition, interactions between both HLA DRB15 status[29] and vitamin D levels[30] with host immune responses to pediatric MS-implicated viruses have been suggested.

6.3.3 Cigarette Smoking

Passive cigarette smoking was found to be associated with an increased risk of developing MS in children of parents smoking at home, and a positive correlation was noted with cumulative exposure,[31] in keeping with previous implication of this association in adults with MS. Mechanisms by which smoking may influence development of MS are not fully understood; suggestions have included induction of proinflammatory immune responses including priming in the lungs; alteration of blood–brain barrier function, including through impacting the renin-angiotensin system; and direct toxic effects on the CNS.[32–35]

6.3.4 Obesity and Sex Hormones

Presence of obesity at age of 20 was associated with doubling of MS risk in both females and males.[36,37] Moderate/severe obesity in adolescent girls was also associated with increased MS risk.[38,39] Several adipokines (secreted in part by white adipose tissue) have immunomodulatory properties. Altered levels of leptin, resistin, visfatin, and adiponectin, observed in both obese individuals and MS patients, are associated with higher levels of proinflammatory cytokines (eg, TNF-α, IL-1β) and with reduced number of regulatory T cells.[40] A high body mass index was found to be associated with decreased plasma levels of vitamin D in both adults[41] and children[42,43] pointing to the potential interaction between these MS risk factors. The MS female preponderance is observed postpubertally, whereas MS incidence is comparable in females and males prior to puberty.[44–47] The incidence increases substantially in older children where the female:male ratio ranges around 2:1, similar to the proportion observed in adult-onset MS. A possible explanation may relate to protective effects of the increased testosterone levels that follow puberty in males.[48] The roles of ovarian hormones appear complex.[48,49] Early onset of menarche has been associated with an increased risk of MS,[49,50] and an earlier age at clinical onset.[51] Indeed, female mice have elevated T-cell responses to myelin antigens after puberty compared to age-matched mice with prepubertal ovariectomy.[49]

6.4 PATHOLOGY IN EARLY MULTIPLE SCLEROSIS

Relatively little is known about early MS pathology, since most studies have been carried out using postmortem material, which typically comprises tissue from patients with long-standing progressive MS (refer to Chapter 1: MS Pathology: Inflammation versus Neurodegeneration). Biopsy studies that generally involve atypical lesions requiring tissue to establish the diagnosis may nonetheless offer valuable insights into early processes including the demonstration of MS lesion heterogeneity,[52,53] greater acute axonal damage in younger aged individuals and in those with shorter disease duration,[54] and some evidence of more extensive remyelination in early disease stages.[55] The first detailed histological study in pediatric MS was reported in 2015.[56]

General pathologic features were similar to those classically described in adult MS, with white matter lesions characterized by perivenous inflammatory demyelination, accumulation of myelin debris in macrophage and axonal injury, in association with a high number of infiltrating macrophages and lymphocytes, mostly CD8 and CD4 T cells, and relatively few B cells. However, a substantially higher degree of acute axonal damage was noted in white matter lesions of pediatric patients compared to adults, especially in the youngest children.[56] There was also a clear relationship between age and the profile and extent of inflammation and axonal injury within the lesions. The number of activated microglia and macrophage appeared greatest in lesions of the youngest children and decreases with age (from prepubertal, to postpubertal, to adults), while numbers of infiltrating T cells did not appreciably differ across age groups. The axonal damage within pediatric MS lesions positively correlated with the extent of macrophage infiltration. These observations are noteworthy as they indicate that the well-described capacity of children to recover quite well (motorically) from relapses is not due to a lesser destructive injury. While the tendency of children to recover from a given motor deficit better than adults may reflect enhanced repair potential, an alternate explanation is that children have greater functional reserve than adults with MS whose biological disease duration has likely been longer (including greater loss of reserve from subclinical disease activity).

MRI studies provide a noninvasive means to observe the pathological events taking place in the inflamed CNS. In comparison to adult MS patients, children have higher lesion volume[57,61] and increased frequency of brain stem lesions. Pediatric-onset MS is associated with gray matter atrophy and, in particular, with early reduction of thalamic volume.[58,59] Brain atrophy is moderately correlated with white matter lesion volume and with disease duration, suggesting a contribution of both focal inflammatory and more diffuse (potentially neurodegenerative) processes.[59] Altered diffusion tensor imaging values in normal-appearing white matter (NAWM) suggest a diffuse pathological involvement in pediatric MS.[60] Additionally, children with MS reportedly exhibit lower magnetization transfer ratio (MTR) values within the brain lesions, NAWM, and normal-appearing gray matter, suggesting greater tissue damage and lower degree of remyelination compared to adult-onset MS.[61] The MTR pattern does not suggest increased remyelinating capacity in children, supporting the notion that the improved motor recovery seen in the pediatric population is more likely to reflect greater preservation of functional reserve rather than improved repair per se.[62,63] While relatively little is known about cortical pathology, clear failure of age-expected growth is disconcerting[64] and reflects the serious impact this illness may have on the young and still-developing CNS.

6.5 CEREBROSPINAL FLUID IN PEDIATRIC-ONSET MULTIPLE SCLEROSIS

As in adults with MS, increased cerebrospinal fluid (CSF) immunoglobulin (Ig) levels and synthesis rates, as well as CSF-restricted Ig oligoclonal bands (OCBs) and immune cell pleocytosis, are all relatively common (though nonspecific) features of pediatric-onset MS. It is noteworthy that CSF studies in pediatric MS populations have yielded considerable variability in estimated frequencies of patients with OCBs (8–92%) or pleocytosis (33–73%).[65] This in part is likely to reflect age-related changes in normal CSF composition (particularly around puberty) that develop independently of MS.[66] For example, the amount of intrathecal IgG can double during normal puberty due to changes in the albumin quotient. Failure to take physiologic age-associated changes in CSF composition into account can result in underestimating intrathecal IgG synthesis, thus contributing to false-negative results.[66,67] Even with correction for age, CSF abnormalities may be absent in early pediatric MS, particularly in the youngest children.[67–69] They nonetheless may be helpful in predicting outcome following initial presentations of ADS.[70] Prepubertal children with MS reportedly have slightly greater CSF pleocytosis, and increased proportion of polymorphonuclear cells compared to older children.[67] It is not known whether such differences reflect the developing immune system or changes in MS-specific immune responses that evolve with age or biological disease duration. As also reported in adults, a subset of children with MS may harbor CSF IgM OCB though no clear association has been established between their presence and clinical MS features.[71,72] Children with MS are more likely to exhibit intrathecal synthesis of Igs against measles, rubella, and varicella zoster virus, compared to children with other CNS inflammatory demyelinating syndromes, including acute disseminated encephalomyelitis (ADEM).[66] In addition to the utility of CSF analyses to aid in early diagnosis, CSF from children with MS may also provide unique insights into early disease targets and mechanisms. An example is a proteomic analysis comparing CSF obtained from children presenting with ADS who were subsequently diagnosed with MS or with monophasic ADS.[73] This study set out to assess early disease targets associated specifically with MS injury. Surprisingly, levels of compact myelin proteins (traditionally implicated in MS) were not different in the CSF of MS children and controls. Rather, several differences were noted in molecules that comprise the axo-glial apparatus, a small physiologic structure at the edge of the node-of-Ranvier that represents the region of contact between the myelinating oligodendrocyte and the underlying axon. Such early implication of the axo-glial apparatus is attractive in a disease where injury to both the myelinating structures and underlying axons is now well established.

6.6 CENTRAL NERVOUS SYSTEM-DIRECTED ANTIBODIES

Over the years, antibodies directed toward a range of CNS antigens have been reported in serum, CSF, and within the brain lesions of MS patients. The contribution of these autoantibodies to disease initiation and propagation continues to be discussed. It is plausible that the pathogenic mechanisms differ across individuals and that autoantibodies may play different roles in distinct subsets of patients.[74] An additional challenge, particularly later in disease, is that longstanding inflammation may result in antibody responses to an expanded range of targets (including antibodies that emerge as epiphenomena of disease rather than contributors to disease). This may confound the ability to identify disease-relevant autoantibodies. It has also become clear that the technique used to identify and quantify CNS-directed antibodies can be crucial. Unlike solid-phase assays (including Western blot and ELISA), cell-based assays can better assess antibody binding to conformationally true structures[75] as they are expressed on the surface of cells in vivo. These cell-based assays have proven most reliable in assessing autoantibodies that bind the extracellular domains of proteins such as aquaporin-4, diagnostic of NMO-spectrum disorders (NMO-SD),[76] and MOG,[77] which have been investigated in recent years for their potential to distinguish other non-MS patient subgroups. While MOG has represented a particularly appealing antibody target in MS (given its exclusive CNS expression on the outermost surface of the myelin sheath[78]; its high encephalitogenic potential for experimental autoimmune encephalomyelitis (EAE)[79]; and the identification within, and elution from, MS lesions of antibodies with affinity to MOG[80,81]), it is noteworthy that the great majority of adult patients with MS do not appear to harbor anti-MOG antibodies.

Studies of CNS-directed antibodies in pediatric demyelinating disease cohorts have also been pursued both to further characterize potentially distinct disease subgroups as well as in an effort to elucidate early disease targets. Unlike NMO-SD in which anti-AQP4 antibody is thought to be directly involved in mediating tissue injury, it remains unknown whether anti-CNS antibodies are primary contributors to pediatric MS immunopathogenesis. Anti-MOG antibodies have been described in a substantial minority of children presenting with ADS, where they may be found at higher titers compared to adults, particularly in very young patients.[82–84] Indeed, the detection of these autoantibodies in ADS children best correlates with an early age of disease onset rather than with a specific clinical presentation.[82,83] Serial serum measurements in a longitudinal pediatric study indicated that anti-MOG reactivity may persist in children with a relapsing course, while it may diminish in children with monophasic disease phenotypes.[85] However, such persistence of anti-MOG antibodies may be a better indicator of non-MS relapsing syndromes than of MS; in keeping with this, children with anti-MOG antibodies are less likely to present with CSF OCB positivity and show a pattern of demyelinating lesions on MRI atypical for MS.[86] As in adults, studies are ongoing to evaluate whether presence of anti-MOG antibodies may define non-MS disease phenotypes in the pediatric age group.[77,87,88] Studies in adults with MS have implicated autoantibodies to axonal antigens such as neurofascin and contactin 2[89,90] though a study in pediatric MS did not detect anti Contactin 2 antibodies.[91]

Following a seminal study describing the presence of serum antibodies to KIR4.1 (the ATP-sensitive inward-rectifying potassium channel expressed principally on glial cells) in approximately half of adult patients with MS,[92] the same group described antibodies to KIR4.1 in a similar subset of children with MS.[93] As in the adult cohort, presence of this antibody was not associated with a particular clinical phenotype in the children with MS. To date, other studies in adults have not replicated the original findings,[94,95] highlighting technical challenges associated with the assay, as well as the importance of direct assay comparisons as previously carried out for anti-AQP4.[74]

In spite of the challenges ascribing functional significance to anti-CNS antibodies in MS, it remains likely that at least some antibodies are pathogenic in and/or have the capacity to modulate MS disease expression. An example of a potential modulatory effect of CNS-directed antibodies on clinical phenotype was provided in a study of antibodies against MBP in children with pediatric-onset MS and controls[96]: While similar frequencies and binding affinities of serum anti-MBP antibodies were found across groups (indicating no causal implication of such antibodies in MS), those children with MS who harbored such antibodies were more likely to exhibit widespread CNS involvement than MS children who did not harbor such antibodies.[96] Such a disease-modulatory effect of antimyelin antibodies would be consistent with classical observations in animal studies in which passive transfer of antimyelin antibodies alone could not trigger EAE; however, the same antibody transfer during T cell-induced EAE resulted in a more aggressive phenotype of CNS inflammation.[97]

Serum antibody profiling may also prove useful diagnostically in children presenting with ADS as well as provide insights into fundamental disease mechanisms. In one study, a prediction algorithm, using myelin antibody array platforms, was proposed to distinguish ADEM and MS cases on the basis of distinct autoantibodies targets and Ig isotypes in serum.[98] In another study using an antigen array, antibody reactivities to a range of CNS and other antigens were assessed in serum samples collected from a cohort of children at the time of incident presentations with ADS and three months later.[99] Children were followed prospectively and ascertained as having either MS (based on clinical or imaging evidence of further disease activity) versus remaining with monophasic disease. While the range of anti-CNS antibody reactivities constricted over the three-month interval between samples in the ADS children who remained monophasic, a substantial expansion of such reactivities was seen in ADS children ascertained as having MS, consistent with both intra- and intermolecular epitope spreading of antibodies to CNS targets. This evolution of the CNS-directed humoral response in children with MS reflects an ongoing presumably insufficiently regulated disease process, in contrast to the transient nature of monophasic syndromes.[99]

6.7 CELLULAR IMMUNE RESPONSES IN PEDIATRIC MULTIPLE SCLEROSIS

Current concepts of MS immune pathophysiology are largely based on studies in adult patients and complementary animal modeling (refer to Chapter 2: Immune Dysregulation in Multiple Sclerosis). The basic immune model of MS[100] has considered that particular subsets of CNS self-reactive T cells are abnormally activated in the periphery (outside the CNS), possibly involving interactions with other cell types including myeloid cells and B cells. These abnormally activated immune cells then traffic into the CNS where they are thought to become reactivated and mediate local inflammatory injury. Immune responses to additional CNS targets may evolve with ongoing disease, reflecting a process called epitope spreading.

It is noteworthy that T cells and B cells with the potential to recognize many self-antigens (including CNS self-antigens) are present as part of the normal adaptive immune repertoire in both children and adults. Pathogenic immune responses targeting the CNS in patients with MS can therefore not be solely attributed to the presence (versus absence) of such autoreactive cells, but more likely occur when responses of such cells become overly aggressive or incompletely regulated. While abnormal effector T cell responses and reduced suppressive capacity by regulatory T cells (Treg) have both been described in adult MS patients,[101,102] a number of key questions remain unanswered. These include the identity of initiating and propagating antigenic targets in MS, the particular subsets of CD4 and CD8 T cells involved, the roles of additional cell types of the immune system, and the molecular mechanisms by which disease-relevant immune-cell subsets interact with one another and with cells of the CNS. A number of these questions could be conceptually best addressed early in the disease course, though technical challenges such as access to sufficient biological samples from well-characterized patients and appropriate controls pose substantial practical limitations.

Regarding the initial antigenic targets in MS, the commonly invoked molecular mimicry hypothesis postulates that one or more non-CNS antigens (potentially of infectious origin) trigger responses of peripheral T cells whose T-cell receptors are cross-reactive with one or more CNS antigens. Compact myelin antigens such as MBP and proteolipid protein, as well as the outer-membrane layer MOG, have been traditionally considered leading candidate autoantigens in MS, largely based on the experience from the commonly used animal model (EAE). Relatively few studies to date have examined CNS-reactivities of circulating T cells in pediatric demyelinating diseases and controls.[103–105] Similar to studies in adults, myelin-reactive T cells have been variably implicated in pediatric-onset MS, with some studies identifying heightened proliferative and/or inflammatory cytokine (including Th17) responses of myelin-reactive T cells in children with MS and CIS, and others noting no differences between MS children and controls. One study documented abnormalities in T-cell proliferative responses to myelin antigens in children with early MS compared to both healthy controls and children with Type 1 diabetes (T1D; representing another inflammatory disease control population); however, the same abnormalities were seen in children with remote head injury.[104] This underscores the challenge, even in children, to distinguish cellular immune abnormalities that contribute to injury from those that may emerge as a consequence of prior injury and persist even in the absence of ongoing CNS inflammation.

It is certainly possible that the magnitude of the proliferative response of T cells to CNS antigen is not as important as their (pro- or antiinflammatory) response profile.[105] Further elucidation of the profile of CNS-reactivity to traditional myelin antigens may require considerably more elaborate approaches including preexpansion of immune subsets of particular interest as shown in adults with MS.[102] It is also possible that relevant targets of the cellular immune response extend beyond the traditionally implicated myelin antigens. As mentioned earlier, an exploratory proteomic analysis of CSF obtained from MS children at the time of ADS implicated the axoglial apparatus, and not the traditionally implicated myelin antigens, as a potential target of early injury in pediatric MS.[73] More studies are required to ascertain whether this proteomic implication of novel antigenic targets in MS can indeed be supported with demonstration of corresponding abnormalities in antigen-specific cellular (eg, CD4, CD8 T cells) and/or humoral (autoreactive antibody) immune responses.

Comparison of T-cell reactivities between young children diagnosed with MS or T1D revealed increased T-cell reactivities in both groups toward several dietary antigens, as well as increased reactivities to a range of self-antigens beyond the disease-specific target-organ self-antigens.[104] These findings, on one hand, raise the emerging specter of how the gut microbiome may contribute in early disease, but also raise speculation that some degree of nontarget organ reactivity may reflect a broader early abnormality of immune regulation in pediatric autoimmune disease, or reflect a more general characteristic of the less mature immune system. Direct comparisons of such immune responses across several pediatric- and adult-onset autoimmune diseases and healthy controls may help elucidate which of these interesting possibilities is more likely.

With respect to the regulatory limb of the immune system, Balint et al. reported decreased regulatory T-cell function in a study of pediatric MS patients compared to controls,[106] consistent with prior reports in adults with MS (see Chapter 2: Immune Dysregulation in Multiple Sclerosis). The same study implicated abnormalities in T-cell homeostasis in the MS children, who appeared to exhibit reduced frequencies of naïve T cells emerging from the thymus (recent thymic emigrants; RTE), including regulatory RTE T cells (RTE-Treg), while the same children exhibited higher frequencies of circulating memory T cells. These abnormalities resulted in decreased ratios between naïve and memory T cells that were similar to ratios expected in considerably older individuals, raising the interesting possibility that early in the MS disease process, T cells (including regulatory T cells) undergo premature senescence. Such a process could conceivably predispose to the emergence of autoimmune disease through compensatory extra-thymic proliferation that might favor the expansion of autoreactive effector (but not regulatory) T cells.[107]

6.8 CONCLUDING COMMENTS

Pediatric-onset MS represents not only an important unmet clinical need but also a unique opportunity to investigate early disease targets and mechanisms. The observations that genetic and environmental risk factors are shared with adult-onset MS patients indicates that pediatric-onset MS is biologically essentially the same illness as in adults, though issues related to relative immaturity of both the immune and nervous systems especially in younger children may influence expression of disease, and also have important therapeutic implications. While studies in pediatric MS pose particular challenges, exciting opportunities exist to further define disease-relevant abnormalities in the T-cell repertoire, investigate the potential contributions of other immune cell subsets that are increasingly implicated in disease (including regulatory and effector B cells; cells of the innate immune system), and to elucidate how the expanding range of SNPs implicated in genetic risk, and the established environmental risk factors (including low levels of vitamin D, early exposure to EBV, and smoking) interact to ultimately translate into the insufficiently regulated CNS-directed effector immune response that contributes to MS pathogenesis. Beyond the contributions of traditional cellular and humoral immune responses, a more complete understanding of pediatric MS disease mechanisms will require thorough investigation of the neurobiology of early-life MS including how the immature CNS responds to and participates in ongoing injury and repair. While it is rare to clinically observe progressive MS in the pediatric age group, it is disconcerting to realize that children presenting with the initial clinical presentation of MS already exhibit significant brain atrophy (more correctly referred to in this age group as failure of normal brain growth) involving both white and gray matter structures.[64] Moreover, the skull sizes of children with MS at initial clinical attack already appear smaller on average than skull sizes of children with monophasic ADS,[59] which begs the question of how early injury mechanisms are initiated in the MS disease spectrum.

References

1. Boiko AN, Guseva ME, Guseva MR, et al. Clinico-immunogenetic characteristics of multiple sclerosis with optic neuritis in children. *J Neurovirol*. 2000;6(suppl 2):S152–S155.
2. Disanto G, Magalhaes S, Handel AE, et al. HLA-DRB1 confers increased risk of pediatric-onset MS in children with acquired demyelination. *Neurology*. 2011;76(9):781–786.
3. Ohlenbusch A, Pohl D, Hanefeld F. Myelin oligodendrocyte gene polymorphisms and childhood multiple sclerosis. *Pediatr Res*. 2002;52(2):175–179.
4. Anlar B, Alikasifoglu M, Kose G, Guven A, Gurer Y, Yakut A. Tumor necrosis factor-alpha gene polymorphisms in children with multiple sclerosis. *Neuropediatrics*. 2001;32(4):214–216.
5. Venkateswaran S, Zheng K, Sacchetti M, et al. Mitochondrial DNA haplogroups and mutations in children with acquired central demyelination. *Neurology*. 2011;76(9):774–780.
6. van Pelt ED, Mescheriakova JY, Makhani N, et al. Risk genes associated with pediatric-onset MS but not with monophasic acquired CNS demyelination. *Neurology*. 2013;81(23):1996–2001.
7. Gale CR, Martyn CN. Migrant studies in multiple sclerosis. *Prog Neurobiol*. 1995;47(4–5):425–448.
8. Kennedy J, O'Connor P, Sadovnick AD, Perara M, Yee I, Banwell B. Age at onset of multiple sclerosis may be influenced by place of residence during childhood rather than ancestry. *Neuroepidemiology*. 2006;26(3):162–167.
9. Banwell B, Bar-Or A, Arnold DL, et al. Clinical, environmental, and genetic determinants of multiple sclerosis in children with acute demyelination: a prospective national cohort study. *Lancet Neurol*. 2011;10(5):436–445.
10. Holick MF. Vitamin D: a millenium perspective. *J Cell Biochem*. 2003;88(2):296–307.
11. Ascherio A, Munger KL. Environmental risk factors for multiple sclerosis. Part II: Noninfectious factors. *Ann Neurol*. 2007;61(6):504–513.
12. Ascherio A, Munger KL, Simon KC. Vitamin D and multiple sclerosis. *Lancet Neurol*. 2010;9(6):599–612.
13. Pugliatti M, Harbo HF, Holmoy T, et al. Environmental risk factors in multiple sclerosis. *Acta Neurol Scand Suppl*. 2008;188:34–40.
14. Munger KL, Levin LI, Hollis BW, Howard NS, Ascherio A. Serum 25-hydroxyvitamin D levels and risk of multiple sclerosis. *JAMA*. 2006;296(23):2832–2838.
15. Mowry EM, Krupp LB, Milazzo M, et al. Vitamin D status is associated with relapse rate in pediatric-onset multiple sclerosis. *Ann Neurol*. 2010;67(5):618–624.
16. Correale J, Ysrraelit MC, Gaitan MI. Immunomodulatory effects of vitamin D in multiple sclerosis. *Brain*. 2009;132(Pt 5):1146–1160.

17. Wobke TK, Sorg BL, Steinhilber D. Vitamin D in inflammatory diseases. *Front Physiol*. 2014;5:244.
18. Ramagopalan SV, Maugeri NJ, Handunnetthi L, et al. Expression of the multiple sclerosis-associated MHC class II Allele HLA-DRB1*1501 is regulated by vitamin D. *PLoS Genet*. 2009;5(2):e1000369.
19. Levin LI, Munger KL, Rubertone MV, et al. Temporal relationship between elevation of epstein-barr virus antibody titers and initial onset of neurological symptoms in multiple sclerosis. *JAMA*. 2005;293(20):2496–2500.
20. Alotaibi S, Kennedy J, Tellier R, Stephens D, Banwell B. Epstein-Barr virus in pediatric multiple sclerosis. *JAMA*. 2004;291(15):1875–1879.
21. Pohl D, Krone B, Rostasy K, et al. High seroprevalence of Epstein-Barr virus in children with multiple sclerosis. *Neurology*. 2006;67(11):2063–2065.
22. Banwell B, Krupp L, Kennedy J, et al. Clinical features and viral serologies in children with multiple sclerosis: a multinational observational study. *Lancet Neurol*. 2007;6(9):773–781.
23. Waubant E, Mowry EM, Krupp L, et al. Common viruses associated with lower pediatric multiple sclerosis risk. *Neurology*. 2011;76(23):1989–1995.
24. Yea C, Tellier R, Chong P, et al. Epstein-Barr virus in oral shedding of children with multiple sclerosis. *Neurology*. 2013;81(16):1392–1399.
25. Graham JP, Arcipowski KM, Bishop GA. Differential B-lymphocyte regulation by CD40 and its viral mimic, latent membrane protein 1. *Immunol Rev*. 2010;237(1):226–248.
26. Yoon SI, Jones BC, Logsdon NJ, Harris BD, Kuruganti S, Walter MR. Epstein-Barr virus IL-10 engages IL-10R1 by a two-step mechanism leading to altered signaling properties. *J Biol Chem*. 2012;287(32):26586–26595.
27. Graves J, Grandhe S, Weinfurtner K, et al. Protective environmental factors for neuromyelitis optica. *Neurology*. 2014;83(21):1923–1929.
28. Mechelli R, Umeton R, Policano C, et al. A "candidate-interactome" aggregate analysis of genome-wide association data in multiple sclerosis. *PLoS One*. 2013;8(5):e63300.
29. Waubant E, Mowry EM, Krupp L, et al. Antibody response to common viruses and human leukocyte antigen-DRB1 in pediatric multiple sclerosis. *Mult Scler*. 2013;19(7):891–895.
30. Mowry EM, James JA, Krupp LB, Waubant E. Vitamin D status and antibody levels to common viruses in pediatric-onset multiple sclerosis. *Mult Scler*. 2011;17(6):666–671.
31. Mikaeloff Y, Caridade G, Tardieu M, Suissa S, Group KS. Parental smoking at home and the risk of childhood-onset multiple sclerosis in children. *Brain*. 2007;130(Pt 10):2589–2595.
32. Naik P, Fofaria N, Prasad S, et al. Oxidative and pro-inflammatory impact of regular and denicotinized cigarettes on blood–brain barrier endothelial cells: is smoking reduced or nicotine-free products really safe? *BMC Neurosci*. 2014;15:51.
33. Hedstrom AK, Bomfim IL, Barcellos LF, et al. Interaction between passive smoking and two HLA genes with regard to multiple sclerosis risk. *Int J Epidemiol*. 2014;43(6):1791–1798.
34. O'Gorman C, Bukhari W, Todd A, Freeman S, Broadley SA. Smoking increases the risk of multiple sclerosis in Queensland, Australia. *J Clin Neurosci*. 2014;21(10):1730–1733.
35. Correale J, Farez MF. Smoking worsens multiple sclerosis prognosis: two different pathways are involved. *J Neuroimmunol*. 2015;281:23–34.
36. Munger KL, Chitnis T, Ascherio A. Body size and risk of MS in two cohorts of US women. *Neurology*. 2009;73(19):1543–1550.
37. Hedstrom AK, Olsson T, Alfredsson L. High body mass index before age 20 is associated with increased risk for multiple sclerosis in both men and women. *Mult Scler*. 2012;18(9):1334–1336.
38. Munger KL. Childhood obesity is a risk factor for multiple sclerosis. *Mult Scler*. 2013;19(13):1800.
39. Langer-Gould A, Brara SM, Beaber BE, Koebnick C. Childhood obesity and risk of pediatric multiple sclerosis and clinically isolated syndrome. *Neurology*. 2013;80(6):548–552.
40. Versini M, Jeandel PY, Rosenthal E, Shoenfeld Y. Obesity in autoimmune diseases: not a passive bystander. *Autoimmun Rev*. 2014;13(9):981–1000.
41. Wortsman J, Matsuoka LY, Chen TC, Lu Z, Holick MF. Decreased bioavailability of vitamin D in obesity. *Am J Clin Nutr*. 2000;72(3):690–693.
42. Rodriguez-Rodriguez E, Navia-Lomban B, Lopez-Sobaler AM, Ortega RM. Associations between abdominal fat and body mass index on vitamin D status in a group of Spanish schoolchildren. *Eur J Clin Nutr*. 2010;64(5):461–467.
43. Sioen I, Mouratidou T, Kaufman JM, et al. Determinants of vitamin D status in young children: results from the Belgian arm of the IDEFICS (Identification and Prevention of Dietary- and Lifestyle-Induced Health Effects in Children and Infants) Study. *Public Health Nutr*. 2012;15(6):1093–1099.

44. Ghezzi A, Deplano V, Faroni J, et al. Multiple sclerosis in childhood: clinical features of 149 cases. *Mult Scler*. 1997;3(1):43–46.
45. Haliloglu G, Anlar B, Aysun S, et al. Gender prevalence in childhood multiple sclerosis and myasthenia gravis. *J Child Neurol*. 2002;17(5):390–392.
46. Ruggieri M, Iannetti P, Polizzi A, Pavone L, Grimaldi LM. Italian Society of Paediatric Neurology Study Group on Childhood Multiple S. Multiple sclerosis in children under 10 years of age. *Neurol Sci*. 2004;25(suppl 4):S326–S335.
47. Banwell B, Kennedy J, Sadovnick D, et al. Incidence of acquired demyelination of the CNS in Canadian children. *Neurology*. 2009;72(3):232–239.
48. Voskuhl RR, Palaszynski K. Sex hormones in experimental autoimmune encephalomyelitis: implications for multiple sclerosis. *Neuroscientist*. 2001;7(3):258–270.
49. Ahn JJ, O'Mahony J, Moshkova M, et al. Puberty in females enhances the risk of an outcome of multiple sclerosis in children and the development of central nervous system autoimmunity in mice. *Mult Scler*. 2015;21(6):735–748.
50. Ramagopalan SV, Valdar W, Criscuoli M, et al. Age of puberty and the risk of multiple sclerosis: a population based study. *Eur J Neurol*. 2009;16(3):342–347.
51. Sloka JS, Pryse-Phillips WE, Stefanelli M. The relation between menarche and the age of first symptoms in a multiple sclerosis cohort. *Mult Scler*. 2006;12(3):333–339.
52. Lucchinetti C, Bruck W, Parisi J, Scheithauer B, Rodriguez M, Lassmann H. Heterogeneity of multiple sclerosis lesions: implications for the pathogenesis of demyelination. *Ann Neurol*. 2000;47(6):707–717.
53. Bruck W, Popescu B, Lucchinetti CF, et al. Neuromyelitis optica lesions may inform multiple sclerosis heterogeneity debate. *Ann Neurol*. 2012;72(3):385–394.
54. Kuhlmann T, Lingfeld G, Bitsch A, Schuchardt J, Bruck W. Acute axonal damage in multiple sclerosis is most extensive in early disease stages and decreases over time. *Brain*. 2002;125(Pt 10):2202–2212.
55. Goldschmidt T, Antel J, Konig FB, Bruck W, Kuhlmann T. Remyelination capacity of the MS brain decreases with disease chronicity. *Neurology*. 2009;72(22):1914–1921.
56. Pfeifenbring S, Bunyan RF, Metz I, et al. Extensive acute axonal damage in pediatric multiple sclerosis lesions. *Ann Neurol*. 2015;77(4):655–667.
57. Ghassemi R, Narayanan S, Banwell B, et al. Quantitative determination of regional lesion volume and distribution in children and adults with relapsing-remitting multiple sclerosis. *PLoS One*. 2014;9(2):e85741.
58. Mesaros S, Rocca MA, Absinta M, et al. Evidence of thalamic gray matter loss in pediatric multiple sclerosis. *Neurology*. 2008;70(13 Pt 2):1107–1112.
59. Kerbrat A, Aubert-Broche B, Fonov V, et al. Reduced head and brain size for age and disproportionately smaller thalami in child-onset MS. *Neurology*. 2012;78(3):194–201.
60. Vishwas MS, Chitnis T, Pienaar R, Healy BC, Grant PE. Tract-based analysis of callosal, projection, and association pathways in pediatric patients with multiple sclerosis: a preliminary study. *AJNR*. 2010;31(1):121–128.
61. Yeh EA, Weinstock-Guttman B, Ramanathan M, et al. Magnetic resonance imaging characteristics of children and adults with paediatric-onset multiple sclerosis. *Brain*. 2009;132(Pt 12):3392–3400.
62. Rocca MA, Absinta M, Ghezzi A, Moiola L, Comi G, Filippi M. Is a preserved functional reserve a mechanism limiting clinical impairment in pediatric MS patients? *Hum Brain Mapp*. 2009;30(9):2844–2851.
63. Rocca MA, Absinta M, Moiola L, et al. Functional and structural connectivity of the motor network in pediatric and adult-onset relapsing-remitting multiple sclerosis. *Radiology*. 2010;254(2):541–550.
64. Aubert-Broche B, Fonov V, Narayanan S, et al. Onset of multiple sclerosis before adulthood leads to failure of age-expected brain growth. *Neurology*. 2014;83(23):2140–2146.
65. Vargas-Lowy D, Chitnis T. Pathogenesis of pediatric multiple sclerosis. *J Child Neurol*. 2012;27(11):1394–1407.
66. Reiber H, Teut M, Pohl D, Rostasy KM, Hanefeld F. Paediatric and adult multiple sclerosis: age-related differences and time course of the neuroimmunological response in cerebrospinal fluid. *Mult Scler*. 2009;15(12):1466–1480.
67. Chabas D, Ness J, Belman A, et al. Younger children with MS have a distinct CSF inflammatory profile at disease onset. *Neurology*. 2010;74(5):399–405.
68. Pohl D, Rostasy K, Reiber H, Hanefeld F. CSF characteristics in early-onset multiple sclerosis. *Neurology*. 2004;63(10):1966–1967.
69. Sinclair AJ, Wienholt L, Tantsis E, Brilot F, Dale RC. Clinical association of intrathecal and mirrored oligoclonal bands in paediatric neurology. *Dev Med Child Neurol*. 2013;55(1):71–75.
70. Heussinger N, Kontopantelis E, Gburek-Augustat J, et al. Oligoclonal bands predict multiple sclerosis in children with optic neuritis. *Ann Neurol*. 2015;77(6):1076–1082.

71. Villar LM, Sadaba MC, Roldan E, et al. Intrathecal synthesis of oligoclonal IgM against myelin lipids predicts an aggressive disease course in MS. *J Clin Invest*. 2005;115(1):187–194.
72. Stauch C, Reiber H, Rauchenzauner M, et al. Intrathecal IgM synthesis in pediatric MS is not a negative prognostic marker of disease progression: quantitative versus qualitative IgM analysis. *Mult Scler*. 2011;17(3):327–334.
73. Dhaunchak AS, Becker C, Schulman H, et al. Implication of perturbed axoglial apparatus in early pediatric multiple sclerosis. *Ann Neurol*. 2012;71(5):601–613.
74. Meinl E, Derfuss T, Krumbholz M, Probstel AK, Hohlfeld R. Humoral autoimmunity in multiple sclerosis. *J Neurol Sci*. 2011;306(1–2):180–182.
75. Waters PJ, McKeon A, Leite MI, et al. Serologic diagnosis of NMO: a multicenter comparison of aquaporin-4-IgG assays. *Neurology*. 2012;78(9):665–671. discussion 669.
76. Lennon VA, Wingerchuk DM, Kryzer TJ, et al. A serum autoantibody marker of neuromyelitis optica: distinction from multiple sclerosis. *Lancet*. 2004;364(9451):2106–2112.
77. Waters P, Woodhall M, O'Connor KC, et al. MOG cell-based assay detects non-MS patients with inflammatory neurologic disease. *Neurol Neuroimmunol Neuroinflamm*. 2015;2(3):e89.
78. Iglesias A, Bauer J, Litzenburger T, Schubart A, Linington C. T- and B-cell responses to myelin oligodendrocyte glycoprotein in experimental autoimmune encephalomyelitis and multiple sclerosis. *Glia*. 2001;36(2):220–234.
79. Schluesener HJ, Sobel RA, Linington C, Weiner HL. A monoclonal antibody against a myelin oligodendrocyte glycoprotein induces relapses and demyelination in central nervous system autoimmune disease. *J Immunol*. 1987;139(12):4016–4021.
80. Genain CP, Cannella B, Hauser SL, Raine CS. Identification of autoantibodies associated with myelin damage in multiple sclerosis. *Nat Med*. 1999;5(2):170–175.
81. O'Connor KC, Appel H, Bregoli L, et al. Antibodies from inflamed central nervous system tissue recognize myelin oligodendrocyte glycoprotein. *J Immunol*. 2005;175(3):1974–1982.
82. McLaughlin KA, Chitnis T, Newcombe J, et al. Age-dependent B cell autoimmunity to a myelin surface antigen in pediatric multiple sclerosis. *J Immunol*. 2009;183(6):4067–4076.
83. Brilot F, Dale RC, Selter RC, et al. Antibodies to native myelin oligodendrocyte glycoprotein in children with inflammatory demyelinating central nervous system disease. *Ann Neurol*. 2009;66(6):833–842.
84. Di Pauli F, Mader S, Rostasy K, et al. Temporal dynamics of anti-MOG antibodies in CNS demyelinating diseases. *Clin Immunol*. 2011;138(3):247–254.
85. Probstel AK, Dornmair K, Bittner R, et al. Antibodies to MOG are transient in childhood acute disseminated encephalomyelitis. *Neurology*. 2011;77(6):580–588.
86. Hacohen Y, Absoud M, Deiva K, et al. Myelin oligodendrocyte glycoprotein antibodies are associated with a non-MS course in children. *Neurol Neuroimmunol Neuroinflamm*. 2015;2(2):e81.
87. Reindl M, Di Pauli F, Rostasy K, Berger T. The spectrum of MOG autoantibody-associated demyelinating diseases. *Nat Rev Neurol*. 2013;9(8):455–461.
88. Ketelslegers IA, Van Pelt DE, Bryde S, et al. Anti-MOG antibodies plead against MS diagnosis in an acquired demyelinating syndromes cohort. *Mult Scler*. 2015;21(12):1513–1520.
89. Derfuss T, Parikh K, Velhin S, et al. Contactin-2/TAG-1-directed autoimmunity is identified in multiple sclerosis patients and mediates gray matter pathology in animals. *Proc Natl Acad Sci USA*. 2009;106(20):8302–8307.
90. Mathey EK, Derfuss T, Storch MK, et al. Neurofascin as a novel target for autoantibody-mediated axonal injury. *J Exp Med*. 2007;204(10):2363–2372.
91. Hacohen Y, Absoud M, Woodhall M, et al. Autoantibody biomarkers in childhood-acquired demyelinating syndromes: results from a national surveillance cohort. *J Neurol Neurosurg Psychiatry*. 2014;85(4):456–461.
92. Srivastava R, Aslam M, Kalluri SR, et al. Potassium channel KIR4.1 as an immune target in multiple sclerosis. *N Engl J Med*. 2012;367(2):115–123.
93. Kraus V, Srivastava R, Kalluri SR, et al. Potassium channel KIR4.1-specific antibodies in children with acquired demyelinating CNS disease. *Neurology*. 2014;82(6):470–473.
94. Nerrant E, Salsac C, Charif M, et al. Lack of confirmation of anti-inward rectifying potassium channel 4.1 antibodies as reliable markers of multiple sclerosis. *Mult Scler*. 2014;20(13):1699–1703.
95. Brickshawana A, Hinson SR, Romero MF, et al. Investigation of the KIR4.1 potassium channel as a putative antigen in patients with multiple sclerosis: a comparative study. *Lancet Neurol*. 2014;13(8):795–806.
96. O'Connor KC, Lopez-Amaya C, Gagne D, et al. Anti-myelin antibodies modulate clinical expression of childhood multiple sclerosis. *J Neuroimmunol*. 2010;223(1–2):92–99.

97. Lassmann H, Brunner C, Bradl M, Linington C. Experimental allergic encephalomyelitis: the balance between encephalitogenic T lymphocytes and demyelinating antibodies determines size and structure of demyelinated lesions. *Acta Neuropathol*. 1988;75(6):566–576.
98. Van Haren K, Tomooka BH, Kidd BA, et al. Serum autoantibodies to myelin peptides distinguish acute disseminated encephalomyelitis from relapsing-remitting multiple sclerosis. *Mult Scler*. 2013;19(13):1726–1733.
99. Quintana FJ, Patel B, Yeste A, et al. Epitope spreading as an early pathogenic event in pediatric multiple sclerosis. *Neurology*. 2014;83(24):2219–2226.
100. Bar-Or A. The immunology of multiple sclerosis. *Semin Neurol*. 2008;28(1):29–45.
101. Zozulya AL, Wiendl H. The role of regulatory T cells in multiple sclerosis. *Nat Clin Pract Neurol*. 2008;4(7):384–398.
102. Cao Y, Goods BA, Raddassi K, et al. Functional inflammatory profiles distinguish myelin-reactive T cells from patients with multiple sclerosis. *Sci Transl Med*. 2015;7(287):287ra274.
103. Correale J, Tenembaum SN. Myelin basic protein and myelin oligodendrocyte glycoprotein T-cell repertoire in childhood and juvenile multiple sclerosis. *Mult Scler*. 2006;12(4):412–420.
104. Banwell B, Bar-Or A, Cheung R, et al. Abnormal T-cell reactivities in childhood inflammatory demyelinating disease and type 1 diabetes. *Ann Neurol*. 2008;63(1):98–111.
105. Vargas-Lowy D, Kivisakk P, Gandhi R, et al. Increased Th17 response to myelin peptides in pediatric MS. *Clin Immunol*. 2013;146(3):176–184.
106. Balint B, Haas J, Schwarz A, et al. T-cell homeostasis in pediatric multiple sclerosis: old cells in young patients. *Neurology*. 2013;81(9):784–792.
107. Bar-Or A, Muraro PA. Premature immune senescence in children with MS: too young to go steady. *Neurology*. 2013;81(9):778–779.
108. Gourraud PA, Harbo HF, Hauser SL, Baranzini SE. The genetics of multiple sclerosis: an up-to-date review. *Immunol Rev*. 2012;248(1):87–103.

CHAPTER

7

Sex-Related Factors in Multiple Sclerosis

S.M. Gold
University Medical Center Hamburg-Eppendorf, Hamburg, Germany

OUTLINE

Translational Neuroimmunology in Multiple Sclerosis
http://dx.doi.org/10.1016/B978-0-12-801914-6.00007-6

7.1 SEX DIFFERENCES

7.1.1 The Effects of Sex on Disease Incidence

Clinical Evidence

It is well known that women are more frequently affected by a number of autoimmune diseases.[1] In multiple sclerosis (MS), the sex ratio is approximately 1:3 for the relapsing-remitting form of the disease, making sex one of the top risk factors for developing MS.[2] Intriguingly, the progressive onset form of the disease affects men and women equally.[3] In addition, within the relapsing-remitting form, the sex ratio quite considerably varies across different ages at onset, with a sharp incline of female preponderance around puberty (reviewed in Ref. 4). Moreover, age at onset of puberty is associated with MS risk in girls but not in boys.[5] One intriguing observation coming from a comparison of epidemiological studies over several decades is that disease incidence of MS appears to be rising in women but not in men, leading to an increase in sex ratio over time.[6]

Taken together, there is very strong epidemiological evidence that sex has a major impact on MS risk, which, however, does not appear to be uniform across different age groups and may change over time. Thus the mechanisms underlying sex differences could include modifiable and nonmodifiable biological as well as environmental or even behavioral factors.

7.1.2 The Effects of Sex on Disease Activity and Progression

Clinical Evidence

Within the major subtypes of relapsing-remitting MS (RRMS) and secondary-progressive MS (SPMS), early predictors of future disability have included sex, age of onset, and degree of recovery from first episode.[7,8] Some studies indicated that women with MS have more contrast-enhancing lesions compared to male patients,[9,10] although this was not confirmed in larger studies controlling for age of onset and type of MS.[11–13] Having said that, there are clear sex differences in immune responses. Responses in females to various immune challenges are stronger than in males.[14] This is also true when comparing autoantigen-specific responses in women versus men with MS.[15–17] Overall, evidence seems to suggest that inflammatory processes are more pronounced in females compared to males.

Whether or not there are sex differences in progression of disease with respect to accumulation of irreversible disability is controversial. A review of placebo-treated subjects in large clinical trials revealed no apparent sex effect on progression.[18] However, a natural history study of untreated MS patients observed that male RRMS patients had a shorter time to conversion to SPMS and reach the progressive stage at an earlier age.[19] A few studies have investigated imaging correlates of neurodegeneration, but these have yielded inconsistent results. One study found that gray matter atrophy was greater in men than women with MS.[11] T1 holes on MRI may also be more frequent in male patients,[9] although that has not been replicated.[11]

7.1.3 Possible Mechanisms for Sex Differences

Onset of Autoimmune Disorders and Inflammatory Activity

Overall, a growing body of evidence in animals and humans supports the notion that there are marked differences in innate as well as adaptive immune responses between males and females, which may underlie the observed sex differences in susceptibility to infections (males more than females), autoimmune diseases (females more than males), as well as vaccination responses (better responses in females) (see Ref. 20 for a current summary of this evidence). Both sex hormones[21] as well as sex chromosomes[22] have been implicated in the stronger immune responses seen in females. Thus both of these factors could potentially contribute to the higher risk of developing autoimmune neuroinflammation in females. In humans, sex hormones and sex chromosomes are closely linked. Thus only animal models can provide mechanistic insight into how hormonal and genetic factors might contribute to autoimmune disease.

Gonadectomy to remove endogenous testosterone only increases disease susceptibility in mouse strains characterized by a sex difference.[23] In line with this observation, several immunomodulatory and possibly neuroprotective mechanisms of testosterone have been described (reviewed in Ref. 24).

A series of studies have used a specific mouse model, called the "four core genotypes," to help dissect the effects of hormones and sex chromosomes on autoimmunity. In this model, the testis-determining gene *Sry* is "moved" from the Y chromosome to an autosome, thereby separating genetic and hormonal sex. Using this approach, it has been shown that the XX sex chromosome complement confers greater disease severity in experimental autoimmune encephalomyelitis (EAE) as well as lupus independent of sex hormones.[25,26] Sex chromosomes were also shown to affect autoantigen-specific immune responses.[25,27] In addition, several studies have suggested that the Y chromosome contains genes that may confer protection from EAE.[28,29]

Taken together, the higher susceptibility to autoimmune disorders and higher inflammatory activity in established disease observed in females may be a consequence of the combined effect of hormones (male hormones such as testosterone are protective) and genetic factors (disease promoting genes on XX sex chromosome complement and/or protective genes on the Y chromosome). However, more work is needed to unravel the precise immune mechanisms driven by these hormonal and genetic factors.

Faster Progression in Men

As reviewed previously, despite the higher incidence of MS in females and stronger (auto) immune responses, some evidence suggests that males have a faster disability progression. This observation raises the possibility that there might be biological resilience factors in females (or a vulnerability in males) regarding neurodegenerative processes. In an elegant animal study using a bone marrow chimera approach in the "four core genotype" model, Du and colleagues provided first evidence for a central nervous system (CNS)-specific sex chromosome effect on neurodegeneration in EAE.[30] When inducing EAE in XX versus XY bone marrow chimeras reconstituted with a common immune system of one sex chromosomal type, they could demonstrate that mice with an XY sex chromosome complement in the CNS had a more severe course of EAE with more pronounced neurodegeneration in the spinal cord, cerebellum, and cerebral cortex. This study is the first to highlight a potential sex-specific vulnerability mechanism to progression that could inform future strategies to intervene in both sexes.

Increase in Sex Ratio

The reason for the sex-specific rise in incidence is not clear but the time period over which it occurred would be considered too short for a genetic cause.[2] Thus this more likely reflects an interaction of environmental and lifestyle factors with biological sex (reviewed in Ref. 31). This is illustrated by an epidemiological study in Crete, a location where a sharp increase of MS incidence in women has been observed.[32] Here, the increased risk for developing MS in women over time was linked to dietary changes, smoking, urbanization, and later childbearing age. Since the female-to-male ratio within HLA-DRB1*15-positive patients is higher than in affected individuals with HLA-DRB1*15-negative genotypes, gene–environment interaction and epigenetic mechanisms related to major histocompatibility complex might also play a role.[33] However, more work is needed to identify modifiable as well nonmodifiable factors that might drive the increase in MS risk specifically in women.

7.2 THE EFFECTS OF PREGNANCY

7.2.1 The Effects of Pregnancy on Disease Incidence

Clinical Evidence

To date, no conclusive evidence is available to suggest that there is an inverse association between number of pregnancies and subsequent MS risk.[34–38] However, one study[39] demonstrated an approximate 50% reduced risk of a first clinical presentation for a demyelinating event for each birth. An association between parity and risk was seen in women with children, not in men with children, and the number of other children in the family had no effect. This indicates that prenatal factors in the mother, not postnatal factors in the family, were underlying the parity effect.[39] These observations warrant further investigation of mechanisms that could link parity with risk for neuroinflammatory attacks.

7.2.2 The Effects of Pregnancy on Disease Activity and Progression

Clinical Evidence

It is well known that disease activity of many cell-mediated autoimmune diseases including MS is ameliorated during pregnancy.[40] In MS, the relapse rate is reduced by up to 80% in the third trimester compared to prepregnancy levels.[41–45] However, in the postpartum period, a rebound effect with increased relapse risk can be observed. This effect has now been shown in many studies throughout the world and confirmed in a metaanalysis.[46]

Some evidence suggests that the "rebound" effect of relapses in the postpartum period can be ameliorated by exclusive breast-feeding.[47,48] However, since these studies are not randomized, potential confounding factors such as prepregnancy relapse rate and the choice to return to disease-modifying therapies quickly after delivery cannot be ruled out. Thus the role of breast-feeding for postpartum relapse risk remains controversial.[49,50]

One related field of research concerns the effect of assisted reproductive technologies (ART) in MS. All published studies in this area have shown an increased relapse risk during ART (reviewed in Ref. 51). While it is not clear if this reflects a biological or possibly a psychobiological effect (ie, due to stress), ART may provide another informative research paradigm to understand the triggers of inflammatory attacks in MS.[52]

In contrast to the clear effect on relapse risk, the effect of pregnancy on disability progression is less clear. In the short term, no effects of pregnancy can be observed with regard to irreversible disability.[53] One study with a longer term follow-up also showed no effect.[54] In contrast, some studies with defined disability endpoints (time to reach a given disability milestone as indicated by the Expanded Disability Status Scale) suggested a protective effect of pregnancy on disability.[55–58] However, not all of these studies have controlled for important confounds such as initial disability, disease duration, or age. Moreover, for obvious reasons, all of these studies are observational, and thereby prone to selection bias. For example, it is conceivable that women with more benign disease are more likely to opt to have children, thereby reversing the presumed cause–effect relationship. Thus taken together, currently available data do not provide convincing evidence that there is a significant beneficial effect of pregnancy on long-term disability.

7.2.3 Possible Biological Mechanisms of Pregnancy Protection

Pregnancy Hormones

Most studies in this area have focused on the role of pregnancy hormones and their immunomodulatory properties. Numerous hormones undergo pronounced shifts in their circulating levels during pregnancy including estrogens (estradiol and estriol), glucocorticoids, vitamin D, and progesterone,[59] and all of these have potent immunomodulatory properties. Evidence from animal models clearly shows the ameliorating effects of estrogens on EAE (see later). In contrast, treatment with progesterone had only small effects on EAE when used alone.[60,61] On the other hand, one study showed that progesterone treatment was deleterious in EAE in Lewis rats.[62] However, progesterone appeared to be additive with estrogen in offering protection when given in combination.[63] This complementary effect of estrogen and progesterone treatment may be due to potentially beneficial effects of progesterone such as axonal protection and decreased demyelination/increased remyelination.[64,65] While the best evidence to date is available for estrogens, other hormonal factors including vitamin D, glucocorticoids, and progesterone may also contribute to the protective effects of pregnancy in autoimmune neuroinflammation (reviewed in Ref. 66).

One aspect that has received relatively little attention is the question whether the "rebound" effect postpartum is simply due to the abrupt withdrawal of protective factors (such as pregnancy hormones) after delivery or if there are additional factors that drive the temporary "overshoot" in relapse activity above and beyond prepregnancy levels. This could be of particular relevance regarding preventive strategies for postpartum relapses.

Immunoregulation

A few studies in small samples have investigated immune changes during MS pregnancy. A genome-wide transcription analysis provided evidence for a decrease of several inflammation-related transcripts in peripheral blood mononuclear cells during pregnancy with a corresponding increase postpartum.[67] A shift from Th1 to Th2 was shown at both the mRNA and protein level in peripheral blood from MS subjects followed longitudinally during pregnancy.[42,68–70] One study found a decline in IFNγ producing $CD4^+$ T cells during pregnancy but no increase postpartum.[71] Serum IL-8 has been shown to decrease during MS pregnancy.[72] Two small studies in MS pregnancy showed an increase[73] or a decrease[74] in regulatory T cells and no change in Th17 T cells.[74] The larger Finish study showed no effect of MS pregnancy on regulatory T-cell percentage, but rather an increase in $CD56^{bright}$ natural killer (NK) cells during pregnancy with a decrease postpartum.[42]

Taken together, there is evidence for shifts in some immune markers during MS pregnancy; however, no marker has consistently been replicated. To better understand the effect of pregnancy on MS, it is necessary to move beyond cytokine shifts and enumerative changes in cell phenotype and include functional analyses. Very little evidence is available regarding Th17 cells, and potential contributions by the innate immune system have also not been explored.

In addition to the work on global shifts in immune function and T-cell phenotype during MS pregnancy described earlier, advances in our understanding of feto-maternal tolerance have provided evidence that fetal antigens directly interact with the maternal immune system, which results in specific immunomodulation such as fetal antigen–dependent induction of regulatory T cells.[75] Thus "shaping" of maternal immune responses by fetal antigens may represent an endogenous pathway by which antigen-specific immunomodulation might also contribute to reinstalling tolerance to autoantigens in MS (see Ref. 59).

Taken together, the available evidence would suggest that the protective effects of pregnancy on MS disease activity are mediated by a combination of immunomodulatory effects of pregnancy hormones and potentially antigen-specific regulation of immune responses as a result of establishing feto-maternal tolerance.

7.3 THERAPEUTIC APPLICATIONS

The clinical observations reviewed earlier provided a basis for the exploration of several hormone-based therapies for MS. In addition, a large number of animal studies have shown that sex hormones including testosterone (reviewed in Ref. 76), estrogens (reviewed in Ref. 77), progesterone, and allopregnanolon (reviewed in Refs 78,79) can ameliorate the course of MS animal models.

Studies using selective agonists for hormone receptors or cell-specific receptor knockouts (KOs) in peripheral and central nervous tissue have also highlighted the therapeutic mechanisms and cellular and molecular targets of sex hormones in autoimmune neuroinflammation. For example, estrogens such as estriol and estradiol have a potent effect on peripheral immune function mediated by estrogen-receptor-alpha (ERα) (reviewed in Ref. 80) but the expression of ERα in the periphery is dispensable for protection in EAE.[81]

Thus in recent years, the focus of this research has shifted toward central effects. Estrogens acting through ERα on astrocytes (but not on neurons) ameliorated EAE and protected against axonal loss in the CNS.[82] Studies using selective agonists as well as oligodendrocyte-specific KO of estrogen-receptor-beta (ERβ) suggest that signaling through this pathway can attenuate EAE by promoting remyelination.[83–85] However, ERβ KO in neurons or astrocytes had no impact on the therapeutic effects of estrogens.[86] Novel endogenous and exogenous ERβ agonists have also been shown to exert protective effects by modulating microglial activation during EAE.[85,87,88]

The therapeutic mechanisms of testosterone have been studied less extensively in EAE. Since dehydrotestosterone has also been shown to be effective in EAE, they appear to be independent from conversion to estrogen (reviewed in Ref. 76). A study using the cuprizone model of MS found that testosterone can promote remyelination.[89]

Based on the evidence from animal studies, several clinical trials of sex steroids have been conducted in patients with MS. To date, two small phase I/II trials of testosterone and estriol have been published by the Voskuhl group at the University of California Los Angeles (UCLA). Two larger phase II trials have been completed. Following, a brief summary of the available clinical studies and their results is provided.

7.3.1 Testosterone

Phase I/II Trial

A first pilot phase I/II trial of transdermal testosterone in male MS patients was conducted at UCLA by Voskuhl and colleagues (trial registration clinicaltrials.gov NCT00405353). In this open-label, baseline-to-treatment study,[90] patients were monitored for a 6-month period with monthly clinical visits and MRIs followed by a 12-month period of treatment with a testosterone gel (Androgel 100 mg). The treatment increased serum testosterone levels by approximately 50% from the low-normal to the high-normal range and significantly increased lean body mass (muscle mass). There were no subjective reports of adverse effects and no significant abnormalities in any blood test results. No effects were seen on contrast-enhancing lesions, likely due to the low MRI activity at baseline. However, the authors report a significant effect on a measure of cognition. In addition, a slowing of global brain atrophy rate was detected during the treatment phase, which was later confirmed in a secondary analysis using voxel-based morphometry and localized to gray matter regions in the right frontal cortex.[91] Moreover, testosterone treatment showed significant effects on immune function by decreasing delayed-type hypersensitivity; decreasing $CD4^+$ T cells and increasing NK cell frequencies; reducing IL-2 production; and increasing release of transforming growth factor-β, brain-derived neurotrophic factor, and platelet-derived growth factor by peripheral immune cells.[92]

Phase II Trials

A multicenter, prospective, placebo-controlled phase II clinical trial of testosterone in male MS patients is currently ongoing in the United States (trial registration clinicaltrials.gov NCT02317263). The primary endpoint of this trial is brain atrophy. The trial will enroll 114 patients and completion is expected in December 2018.

7.3.2 Estrogens

Phase I/II Trial

A small open-label pilot study of oral estriol was conducted in 12 female patients with RRMS and SPMS.[93] The study used a cross over design with a 6-month pretreatment period, followed by a 6-month treatment period with oral estriol (8 mg/day), and then a 6-month period without treatment, and finally a 4-month retreatment phase. Serum estriol reached levels comparable to concentrations typically seen in second trimester pregnancy. During the treatment phase, RRMS patients demonstrated significant decreases in delayed-type hypersensitivity responses to tetanus, and gadolinium-enhancing lesion numbers and volumes on monthly MRI. Enhancing lesions returned to pretreatment levels when estriol treatment was stopped and decreased again after resuming treatment. Secondary analyses from this trial revealed significant shifts in immune function with decreased production of TNFα and MMP-9 and increases in IL-5 and IL-10.[94,95]

Phase II Trials

Based on these initial results, a 2-year, randomized, double-blind, multicenter, placebo-controlled (1:1) trial of oral estriol 8 mg per day versus placebo (as an add-on to glatiramer acetate) was conducted in the United States (trial registration clinicaltrials.gov NCT00451204). The trial enrolled 164 female patients with RRMS from 16 sites and the primary endpoint was annualized relapse rate. The trial was completed in 2014 and recently published.[96] The annualized relapse rate for patients on the combined therapy was 0.25 relapses per year (95% CI 0.17–0.37) compared to 0.37 (95% CI 0.25–0.53) in the glatiramer acetate plus placebo group, which met the prespecified significance level of $p < 0.10$. Adverse effects were similar in both groups. There were, however, no effects on MRI measures of neuroinflammation. Secondary outcomes of cognitive function suggested some benefit.

In a related approach, a French consortium conducted a randomized controlled, double-blind multicenter study of Oral NOMA (10 mg/day) combined with transdermal Estradiol (75 µg, once a week) (trial registration clinicaltrials.gov NCT00127075) in women with MS in the postpartum phase. The goal of this trial was to reduce the relapses postpartum (primary endpoint: rate of relapses during the first 12 weeks after delivery). The study aimed to enroll $n = 300$ female MS patients during pregnancy and was completed in 2012. As of 2015, the results have not been published in a peer-reviewed journal. However, they were presented at the Annual Congress of the European Committee for Treatment and Research in Multiple Sclerosis 2012 in Lyon. According to the presentation, no significant effect was observed on postpartum annualized relapse rate compared to women receiving placebo.

In addition to these two completed phase II trials, a few other trials have been registered in publicly available databases: A Phase II Randomized Trial of Interferon-Beta-1a and Estroprogestins has been registered by an Italian group (trial registration clinicaltrials.gov NCT00151801), but results have not been published as of 2015 and the status of the trial is unknown. A double-blind, placebo-controlled monocenter trial of estriol treatment in women with MS is currently ongoing at the UCLA (trial registration clinicaltrials.gov NCT01466114). The primary endpoint of the trial is cognitive function as measured by the Paced Auditory Serial Addition Test. The trial will enroll 64 patients and completion is expected in April 2016.

7.4 CONCLUSIONS AND OUTLOOK

The clear sex differences in MS incidence and the well-established effect of pregnancy on disease activity in women with MS strongly suggest that sex-related factors have a major impact on the pathobiology of MS. Studies using innovative animal models have provided intriguing insights into the underlying modifiable and nonmodifiable mechanisms. This work is complemented by large epidemiological investigations. However, our understanding of the driving factors behind sex differences and pregnancy protection in MS is still limited and more work is needed.

The research so far clearly illustrates the potential of studying sex-related factors to discover novel treatment targets and develop new therapies. Given the strong preclinical evidence for protective effects of sex hormones, including testosterone and estrogens, several clinical trials have been undertaken in MS. Small phase I/II open label studies with estriol in women with MS and testosterone in men with MS showed promising results and provided the basis for phase II trials. The results of these trials appear to be less striking than the animal models or the phase I/II trials might have suggested. This could be due in part to methodological issues. For example, the US multicenter trial used estriol as an add-on to an already approved MS drug (glatiramer acetate), so that all patients in the trial received at least one active drug. It could also reflect the fact that the protective effects of pregnancy or mechanisms underlying sex differences are likely more complex and not attributable to a single hormonal factor.

Further basic and clinical research into the mechanisms of sex-related factors in MS could provide additional targets or mechanisms for the development of new therapies, hormone-based or otherwise.

KEY POINTS

- Sex-related factors have a major impact on MS susceptibility and activity.
- Women have a higher risk of developing MS (and other inflammatory autoimmune disorders).
- The sex ratio appears to be increasing. This is unlikely to be due to genetic causes and is possibly driven by environmental factors or gene–environment interactions.
- Some evidence indicates that women experience more inflammatory lesions and a higher relapse rate but men show faster disability progression. Evidence from animal models suggests that this might be driven by independent contributions from hormonal and genetic factors.
- Pregnancy has a pronounced effect on disease activity with a reduction in relapse rate of up to 80% in the third trimester but a rebound increase postpartum.
- While small open-label clinical trials have supported the notion that sex hormone-based therapies might be effective in MS, the more recent, larger, randomized, and placebo-controlled trials have yielded mixed results depending on the study design and the specific hormone(s) used.

References

1. Ngo ST, Steyn FJ, McCombe PA. Gender differences in autoimmune disease. *Front Neuroendocrinol*. 2014;35(3): 347–369.
2. Orton SM, Herrera BM, Yee IM, et al. Sex ratio of multiple sclerosis in Canada: a longitudinal study. *Lancet Neurol*. 2006;5(11):932–936.
3. Tremlett H, Zhao Y, Devonshire V, Neurologists UBC. Natural history comparisons of primary and secondary progressive multiple sclerosis reveals differences and similarities. *J Neurol*. 2009;256(3):374–381.
4. Dunn SE, Lee H, Pavri FR, Zhang MA. Sex-based differences in multiple sclerosis (Part I): biology of disease incidence. *Curr Top Behav Neurosci*. 2015;26:29–56.
5. Ramagopalan SV, Valdar W, Criscuoli M, et al. Age of puberty and the risk of multiple sclerosis: a population based study. *Eur J Neurol*. 2009;16(3):342–347.
6. Koch-Henriksen N, Sorensen PS. The changing demographic pattern of multiple sclerosis epidemiology. *Lancet Neurol*. 2010;9(5):520–532.
7. Confavreux C, Vukusic S, Adeleine P. Early clinical predictors and progression of irreversible disability in multiple sclerosis: an amnesic process. *Brain*. 2003;126(Pt 4):770–782.
8. Runmarker B, Andersson C, Odén A, Andersen O. Prediction of outcome in multiple sclerosis based on multivariate models. *J Neurol*. 1994;241(10):597–604.
9. Pozzilli C, Tomassini V, Marinelli F, Paolillo A, Gasperini C, Bastianello S. 'Gender gap' in multiple sclerosis: magnetic resonance imaging evidence. *Eur J Neurol*. 2003;10(1):95–97.
10. Weatherby SJ, Mann CL, Davies MB, et al. A pilot study of the relationship between gadolinium-enhancing lesions, gender effect and polymorphisms of antioxidant enzymes in multiple sclerosis. *J Neurol*. 2000;247(6):467–470.
11. Antulov R, Weinstock-Guttman B, Cox JL, et al. Gender-related differences in MS: a study of conventional and nonconventional MRI measures. *Mult Scler*. 2009;15(3):345–354.
12. Barkhof F, Held U, Simon JH, et al. Predicting gadolinium enhancement status in MS patients eligible for randomized clinical trials. *Neurology*. 2005;65(9):1447–1454.
13. Stone LA, Smith ME, Albert PS, et al. Blood–brain barrier disruption on contrast-enhanced MRI in patients with mild relapsing-remitting multiple sclerosis: relationship to course, gender, and age. *Neurology*. 1995;45(6): 1122–1126.
14. Libert C, Dejager L, Pinheiro I. The X chromosome in immune functions: when a chromosome makes the difference. *Nat Rev Immunol*. 2010;10(8):594–604.
15. Kantarci OH, Hebrink DD, Schaefer-Klein J, et al. Interferon gamma allelic variants: sex-biased multiple sclerosis susceptibility and gene expression. *Arch Neurol*. 2008;65(3):349–357.
16. Moldovan IR, Cotleur AC, Zamor N, Butler RS, Pelfrey CM. Multiple sclerosis patients show sexual dimorphism in cytokine responses to myelin antigens. *J Neuroimmunol*. 2008;193(1–2):161–169.
17. Pelfrey CM, Cotleur AC, Lee JC, Rudick RA. Sex differences in cytokine responses to myelin peptides in multiple sclerosis. *J Neuroimmunol*. 2002;130(1–2):211–223.
18. Wolinsky JS, Shochat T, Weiss S, Ladkani D. Glatiramer acetate treatment in PPMS: why males appear to respond favorably. *J Neurol Sci*. 2009;286(1–2):92–98.
19. Koch M, Kingwell E, Rieckmann P, Tremlett H. The natural history of secondary progressive multiple sclerosis. *J Neurol Neurosurg Psychiatry*. 2010;81(9):1039–1043.
20. Markle JG, Fish EN. SeXX matters in immunity. *Trends Immunol*. 2014;35(3):97–104.
21. Pennell LM, Galligan CL, Fish EN. Sex affects immunity. *J Autoimmun*. 2012;38(2–3):J282–J291.
22. Fish EN. The X-files in immunity: sex-based differences predispose immune responses. *Nat Rev Immunol*. 2008;8(9):737–744.
23. Palaszynski KM, Loo KK, Ashouri JF, Liu H, Voskuhl RR. Androgens are protective in experimental autoimmune encephalomyelitis: implications for multiple sclerosis. *J Neuroimmunol*. 2004;146(1–2):144–152.
24. Gold SM, Voskuhl RR. Estrogen and testosterone therapies in multiple sclerosis. *Prog Brain Res*. 2009;175:239–251.
25. Smith-Bouvier DL, Divekar AA, Sasidhar M, et al. A role for sex chromosome complement in the female bias in autoimmune disease. *J Exp Med*. 2008;205(5):1099–1108.
26. Sasidhar MV, Itoh N, Gold SM, Lawson GW, Voskuhl RR. The XX sex chromosome complement in mice is associated with increased spontaneous lupus compared with XY. *Ann Rheum Dis*. 2012;71(8):1418–1422.
27. Palaszynski KM, Smith DL, Kamrava S, Burgoyne PS, Arnold AP, Voskuhl RR. A Yin-Yang effect between sex chromosome complement and sex hormones on the immune response. *Endocrinology*. 2005;146(8): 3280–3285.

28. Spach KM, Blake M, Bunn JY, et al. Cutting edge: the Y chromosome controls the age-dependent experimental allergic encephalomyelitis sexual dimorphism in SJL/J mice. *J Immunol*. 2009;182(4):1789–1793.
29. Teuscher C, Noubade R, Spach K, et al. Evidence that the Y chromosome influences autoimmune disease in male and female mice. *Proc Natl Acad Sci USA*. 2006;103(21):8024–8029.
30. Du S, Itoh N, Askarinam S, Hill H, Arnold AP, Voskuhl RR. XY sex chromosome complement, compared with XX, in the CNS confers greater neurodegeneration during experimental autoimmune encephalomyelitis. *Proc Natl Acad Sci USA*. 2014;111(7):2806–2811.
31. Dunn SE, Gunde E, Lee H. Sex-based differences in multiple sclerosis (MS): Part II: rising incidence of multiple sclerosis in women and the vulnerability of men to progression of this disease. *Curr Top Behav Neurosci*. 2015;26:57–86.
32. Kotzamani D, Panou T, Mastorodemos V, et al. Rising incidence of multiple sclerosis in females associated with urbanization. *Neurology*. 2012;78(22):1728–1735.
33. Chao MJ, Ramagopalan SV, Herrera BM, et al. MHC transmission: insights into gender bias in MS susceptibility. *Neurology*. 2011;76(3):242–246.
34. Villard-Mackintosh L, Vessey MP. Oral contraceptives and reproductive factors in multiple sclerosis incidence. *Contraception*. 1993;47(2):161–168.
35. Thorogood M, Hannaford PC. The influence of oral contraceptives on the risk of multiple sclerosis. *Br J Obstet Gynaecol*. 1998;105(12):1296–1299.
36. Alonso A, Jick SS, Olek MJ, Ascherio A, Jick H, Hernan MA. Recent use of oral contraceptives and the risk of multiple sclerosis. *Arch Neurol*. 2005;62(9):1362–1365.
37. Hernan MA, Hohol MJ, Olek MJ, Spiegelman D, Ascherio A. Oral contraceptives and the incidence of multiple sclerosis. *Neurology*. 2000;55(6):848–854.
38. Weinshenker BG, Hader W, Carriere W, Baskerville J, Ebers GC. The influence of pregnancy on disability from multiple sclerosis: a population-based study in Middlesex County, Ontario. *Neurology*. 1989;39(11): 1438–1440.
39. Ponsonby AL, Lucas RM, van der Mei IA, et al. Offspring number, pregnancy and risk of first clinical demyelinating event: the Ausimmune study. *Neurology*. 2012;78(12):867–874.
40. Whitacre CC, Reingold SC, O'Looney PA. A gender gap in autoimmunity. *Science*. 1999;283(5406):1277–1278.
41. Confavreux C, Hutchinson M, Hours MM, Cortinovis-Tourniaire P, Moreau T. Rate of pregnancy-related relapse in multiple sclerosis. Pregnancy in multiple sclerosis group. [see comments] *N Engl J Med*. 1998;339(5):285–291.
42. Airas L, Saraste M, Rinta S, Elovaara I, Huang YH, Wiendl H. Immunoregulatory factors in multiple sclerosis patients during and after pregnancy: relevance of natural killer cells. *Clin Exp Immunol*. 2008;151(2):235–243.
43. De Las Heras V, De Andres C, Tellez N, Tintore M. Pregnancy in multiple sclerosis patients treated with immunomodulators prior to or during part of the pregnancy: a descriptive study in the Spanish population. *Mult Scler*. 2007;13(8):981–984.
44. Fernandez Liguori N, Klajn D, Acion L, et al. Epidemiological characteristics of pregnancy, delivery, and birth outcome in women with multiple sclerosis in Argentina (EMEMAR study). *Mult Scler*. 2009;15(5):555–562.
45. Finkelsztejn A, Fragoso YD, Ferreira ML, et al. The Brazilian database on pregnancy in multiple sclerosis. *Clin Neurol Neurosurg*. 2011;113(4):277–280.
46. Finkelsztejn A, Brooks JB, Paschoal Jr FM, Fragoso YD. What can we really tell women with multiple sclerosis regarding pregnancy? A systematic review and meta-analysis of the literature. *BJOG*. 2011;118(7):790–797.
47. Langer-Gould A, Huang SM, Gupta R, et al. Exclusive breastfeeding and the risk of postpartum relapses in women with multiple sclerosis. *Arch Neurol*. 2009;66(8):958–963.
48. Hellwig K, Haghikia A, Agne H, Beste C, Gold R. Protective effect of breastfeeding in postpartum relapse rate of mothers with multiple sclerosis. *Arch Neurol*. 2009;66(12):1580–1581. author reply 1581.
49. Langer-Gould A, Hellwig K. One can prevent post-partum MS relapses by exclusive breast feeding: yes. *Mult Scler*. 2013;19(12):1567–1568.
50. Vukusic S, Confavreux C. One can prevent post-partum MS relapses by exclusive breast feeding: no. *Mult Scler*. 2013;19(12):1565–1566.
51. Hellwig K, Correale J. Artificial reproductive techniques in multiple sclerosis. *Clin Immunol*. 2013;149(2):219–224.
52. Voskuhl RR. Assisted reproduction technology in multiple sclerosis: giving birth to a new avenue of research in hormones and autoimmunity. *Ann Neurol*. 2012;72(5):631–632.
53. Vukusic S, Hutchinson M, Hours M, et al. Pregnancy and multiple sclerosis (the PRIMS study): clinical predictors of post-partum relapse. *Brain*. 2004;127(Pt 6):1353–1360.

54. Roullet E, Verdier-Taillefer MH, Amarenco P, Gharbi G, Alperovitch A, Marteau R. Pregnancy and multiple sclerosis: a longitudinal study of 125 remittent patients. *J Neurol Neurosurg Psychiatry*. 1993;56(10):1062–1065.
55. Damek DM, Shuster EA. Pregnancy and multiple sclerosis. *Mayo Clin Proc*. 1997;72(10):977–989.
56. Runmarker B, Andersen O. Pregnancy is associated with a lower risk of onset and a better prognosis in multiple sclerosis. [see comments] *Brain*. 1995;118(Pt 1: 10):253–261.
57. Verdru P, Theys P, D'Hooghe MB, Carton H. Pregnancy and multiple sclerosis: the influence on long term disability. *Clin Neurol Neurosurg*. 1994;96(1):38–41.
58. D'Hooghe MB, Nagels G, Uitdehaag BM. Long-term effects of childbirth in MS. *J Neurol Neurosurg Psychiatry*. 2010;81(1):38–41.
59. Patas K, Engler JB, Friese MA, Gold SM. Pregnancy and multiple sclerosis: feto-maternal immune cross talk and its implications for disease activity. *J Reprod Immunol*. 2013;97(1):140–146.
60. Kim S, Liva SM, Dalal MA, Verity MA, Voskuhl RR. Estriol ameliorates autoimmune demyelinating disease: implications for multiple sclerosis. *Neurology*. 1999;52(6):1230–1238.
61. Yates MA, Li Y, Chlebeck P, Proctor T, Vandenbark AA, Offner H. Progesterone treatment reduces disease severity and increases IL-10 in experimental autoimmune encephalomyelitis. *J Neuroimmunol*. 2010;220(1–2):136–139.
62. Hoffman GE, Le WW, Murphy AZ, Koski CL. Divergent effects of ovarian steroids on neuronal survival during experimental allergic encephalitis in Lewis rats. *Exp Neurol*. 2001;171(2):272–284.
63. Garay L, Gonzalez Deniselle MC, Gierman L, et al. Steroid protection in the experimental autoimmune encephalomyelitis model of multiple sclerosis. *Neuroimmunomodulation*. 2008;15(1):76–83.
64. Garay L, Deniselle MC, Lima A, Roig P, De Nicola AF. Effects of progesterone in the spinal cord of a mouse model of multiple sclerosis. *J Steroid Biochem Mol Biol*. 2007;107(3–5):228–237.
65. Garay L, Deniselle MC, Meyer M, et al. Protective effects of progesterone administration on axonal pathology in mice with experimental autoimmune encephalomyelitis. *Brain Res*. 2009;1283:177–185.
66. Voskuhl RR, Gold SM. Sex-related factors in multiple sclerosis susceptibility and progression. *Nat Rev Neurol*. 2012;8(5):255–263.
67. Gilli F, Lindberg RL, Valentino P, et al. Learning from nature: pregnancy changes the expression of inflammation-related genes in patients with multiple sclerosis. *PLoS One*. 2010;5(1):e8962.
68. Al-Shammri S, Rawoot P, Azizieh F, et al. Th1/Th2 cytokine patterns and clinical profiles during and after pregnancy in women with multiple sclerosis. *J Neurol Sci*. 2004;222(1–2):21–27.
69. Gilmore W, Arias M, Stroud N, Stek A, McCarthy KA, Correale J. Preliminary studies of cytokine secretion patterns associated with pregnancy in MS patients. *J Neurol Sci*. 2004;224(1–2):69–76.
70. Lopez C, Comabella M, Tintore M, Sastre-Garriga J, Montalban X. Variations in chemokine receptor and cytokine expression during pregnancy in multiple sclerosis patients. *Mult Scler*. 2006;12(4):421–427.
71. Langer-Gould A, Gupta R, Huang S, et al. Interferon-γ-producing T cells, pregnancy, and postpartum relapses of multiple sclerosis. *Arch Neurol*. 2010;67(1):51–57.
72. Neuteboom RF, Verbraak E, Voerman JS, et al. First trimester interleukin 8 levels are associated with postpartum relapse in multiple sclerosis. *Mult Scler*. 2009;15(11):1356–1358.
73. Sanchez-Ramon S, Navarro AJ, Aristimuno C, et al. Pregnancy-induced expansion of regulatory T-lymphocytes may mediate protection to multiple sclerosis activity. *Immunol Lett*. 2005;96(2):195–201.
74. Neuteboom RF, Verbraak E, Wierenga-Wolf AF, et al. Pregnancy-induced fluctuations in functional T-cell subsets in multiple sclerosis patients. *Mult Scler*. 2010;16(9):1073–1078.
75. Erlebacher A. Mechanisms of T cell tolerance towards the allogeneic fetus. *Nat Rev Immunol*. 2013;13(1):23–33.
76. Gold SM, Voskuhl RR. Testosterone replacement therapy for the treatment of neurological and neuropsychiatric disorders. *Curr Opin Investig Drugs*. 2006;7(7):625–630.
77. Gold SM, Voskuhl RR. Estrogen treatment in multiple sclerosis. *J Neurol Sci*. 2009;286(1–2):99–103.
78. De Nicola AF, Coronel F, Garay LI, et al. Therapeutic effects of progesterone in animal models of neurological disorders. *CNS Neurol Disord Drug Targets*. 2013;12(8):1205–1218.
79. Noorbakhsh F, Baker GB, Power C. Allopregnanolone and neuroinflammation: a focus on multiple sclerosis. *Front Cell Neurosci*. 2014;8:134.
80. Spence RD, Voskuhl RR. Neuroprotective effects of estrogens and androgens in CNS inflammation and neurodegeneration. *Front Neuroendocrinol*. 2012;33(1):105–115.
81. Garidou L, Laffont S, Douin-Echinard V, et al. Estrogen receptor α signaling in inflammatory leukocytes is dispensable for 17β-estradiol-mediated inhibition of experimental autoimmune encephalomyelitis. *J Immunol*. 2004;173(4):2435–2442.

82. Spence RD, Hamby ME, Umeda E, et al. Neuroprotection mediated through estrogen receptor-α in astrocytes. *Proc Natl Acad Sci USA*. 2011;108(21):8867–8872.
83. Crawford DK, Mangiardi M, Song B, et al. Oestrogen receptor β ligand: a novel treatment to enhance endogenous functional remyelination. *Brain*. 2010;133(10):2999–3016.
84. Khalaj AJ, Yoon J, Nakai J, et al. Estrogen receptor (ER) β expression in oligodendrocytes is required for attenuation of clinical disease by an ERβ ligand. *Proc Natl Acad Sci USA*. 2013;110(47):19125–19130.
85. Moore SM, Khalaj AJ, Kumar S, et al. Multiple functional therapeutic effects of the estrogen receptor β agonist indazole-Cl in a mouse model of multiple sclerosis. *Proc Natl Acad Sci USA*. 2014;111(50):18061–18066.
86. Spence RD, Wisdom AJ, Cao Y, et al. Estrogen mediates neuroprotection and anti-inflammatory effects during EAE through ERα signaling on astrocytes but not through ERβ signaling on astrocytes or neurons. *J Neurosci*. 2013;33(26):10924–10933.
87. Saijo K, Collier JG, Li AC, Katzenellenbogen JA, Glass CK. An ADIOL-ERβ-CtBP transrepression pathway negatively regulates microglia-mediated inflammation. *Cell*. 2011;145(4):584–595.
88. Wu WF, Tan XJ, Dai YB, Krishnan V, Warner M, Gustafsson JA. Targeting estrogen receptor β in microglia and T cells to treat experimental autoimmune encephalomyelitis. *Proc Natl Acad Sci USA*. 2013;110(9):3543–3548.
89. Hussain R, Ghoumari AM, Bielecki B, et al. The neural androgen receptor: a therapeutic target for myelin repair in chronic demyelination. *Brain*. 2013;136(Pt 1):132–146.
90. Sicotte NL, Giesser BS, Tandon V, et al. Testosterone treatment in multiple sclerosis: a pilot study. *Arch Neurol*. 2007;64(5):683–688.
91. Kurth F, Luders E, Sicotte NL, et al. Neuroprotective effects of testosterone treatment in men with multiple sclerosis. *NeuroImage Clin*. 2014;4:454–460.
92. Gold SM, Chalifoux S, Giesser BS, Voskuhl RR. Immune modulation and increased neurotrophic factor production in multiple sclerosis patients treated with testosterone. *J Neuroinflammation*. 2008;5:32.
93. Sicotte NL, Liva SM, Klutch R, et al. Treatment of multiple sclerosis with the pregnancy hormone estriol. *Ann Neurol*. 2002;52(4):421–428.
94. Soldan SS, Retuerto AI, Sicotte NL, Voskuhl RR. Immune modulation in multiple sclerosis patients treated with the pregnancy hormone estriol. *J Immunol*. 2003;171(11):6267–6274.
95. Gold SM, Sasidhar MV, Morales LB, et al. Estrogen treatment decreases matrix metalloproteinase (MMP)-9 in autoimmune demyelinating disease through estrogen receptor alpha (ERα). *Lab Invest*. 2009;89(10):1076–1083.
96. Voskuhl RR, Wang H, Wu TC, et al. Estriol combined with glatiramer acetate for women with relapsing-remitting multiple sclerosis: a randomised, placebo-controlled, phase 2 trial. *Lancet Neurol*. 2016;15(1):35–46.

SECTION II

OTHER PATHO-MECHANISMS

CHAPTER

8

Environmental Factors and Their Regulation of Immunity in Multiple Sclerosis

M. Trojano
University of Bari, Bari, Italy

C. Avolio
University of Foggia, Foggia, Italy

OUTLINE

Translational Neuroimmunology in Multiple Sclerosis
http://dx.doi.org/10.1016/B978-0-12-801914-6.00008-8

8.1 INTRODUCTION

Multiple sclerosis (MS) is an inflammatory/neurodegenerative disease of the central nervous system (CNS) in which both genetic and environmental factors cooperate in the chronic activation of immune cells to produce oligodendrocyte and neuron damage. Epidemiological studies have identified several environmental risk factors in MS, such as exposure to certain viruses and smoking or even lack of exposure to sunlight with a subsequent reduced vitamin D production. These factors are associated with the susceptibility in developing MS but they could also influence the disease course. However, no single risk factor per se appears to be responsible for the development of the disease, but a multifactorial interplay is most likely. Because of this complex interplay, it is quite difficult to define the real impact of each single factor and in this respect the only way to proceed is to design large enough studies with high-quality data.[1,2] However, what is certainly even less known is the way in which these external factors are able to induce and sustain the internal pathology process of the disease. In this chapter we try to provide an overview of the most relevant environmental factors and how they may affect the immune response in MS.

8.2 MULTIPLE SCLEROSIS IMMUNOPATHOGENESIS

Though the etiology of MS remains as yet unknown,[3] its pathogenesis has been quite extensively investigated and mostly clarified since it is widely accepted that activated peripheral immune cells enter the CNS to produce the pathology.[4] The initial dysfunction can also occur within the CNS and it can include mitochondrial dysfunction in neurons or oligodendrocytes, axonal energy insufficiency, or even damage to other neural organelles such as peroxisomes.[5] In such a case, whichever the initial injury, the leakage of CNS antigens into draining lymph nodes activates T cells that address and enter the CNS, inducing inflammation, demyelination, and oligodendrocyte loss as well as axonal/neuronal injury and loss. Professional antigen-presenting cells (APCs) such as dendritic cells are needed to activate T cells. The APCs, either from the periphery or from the CNS, migrate to lymph nodes, carrying the antigen (a short segment of the pathogen) bound to major histocompatibility complex (MHC class I for $CD8^+$ and II for $CD4^+$ commitment) on their cell surface. In the lymph nodes, the antigen is presented to naïve T cells through a T-cell receptor (TCR) recognizing the antigen/MHC combination. This trimolecular complex (MHC/antigen/TCR) constitutes a first signal, but a second signal, mediated by costimulatory molecules (eg, B7 on APCs and CD28 on T cells) is needed for full activation of the T cells, their proliferation, and subsequent differentiation into effector cells. $CD4^+$ T cells are crucial in MS as they can differentiate into proinflammatory T helper (Th) 1 or 17 subsets, antiinflammatory Th2 cells, or into cells with regulatory/antiinflammatory properties (Tregs), depending on the microenvironment and cytokine milieu.[6] In MS patients there is a tendency to generate either Th1 or Th17 subsets, which in addition to being proinflammatory[7] may have neurotoxic effects,[8] whereas the regulatory/antiinflammatory Th2 and Tregs subsets are reported to be deficient in MS.[9]

$CD8^+$ T cells also have relevant roles in MS tissue damage.[10] B cells also importantly produce disease pathology in MS and this is supported by various evidence including the effectiveness of monoclonal antibody therapies that target the B-cell antigen such as CD20,[11,12] the oligoclonal bands in the cerebrospinal fluid commonly reported in MS patients, and B-cell follicular-like structures found in the meninges of secondary progressive MS patients.[13] In addition to the pathogenetic role in the production of antibodies targeting CNS structures,[14] B cells may play additional roles such as antigen presentation and help for T cells.[15] Once activated, immune cells upregulate different adhesion molecules and adhere to endothelial cells of postcapillary venules in the CNS. They then cross the endothelial cell barrier by means of the proteolytic activity of the matrix metalloproteinases (MMPs), first migrating across the endothelial basement membrane and then the parenchymal basement membrane or glia limitans, and finally they enter the CNS parenchyma. As a matter of fact MMPs have been reported to be upregulated in MS.[16] Upon entering the CNS parenchyma, T cells are reactivated through repeated antigen presentation by APCs such as microglia, macrophages, B cells, and dendritic cells. Activated immune cell subsets, as well as inflammation and demyelination, also induce neuronal injury and loss by producing free radicals, glutamate, and other excitotoxins, proteases, and cytokines.[8,17]

8.3 EPIGENETIC CHANGES IN MULTIPLE SCLEROSIS

It is therefore quite evident that MS has the characteristics of both an inflammatory/demyelinating and a neurodegenerative disease in terms of pathology but this is also clear in terms of clinical presentation, course, and accumulated disability in patients.

Even if not an inherited disorder, genetic factors are certainly implicated in the disease susceptibility and this is especially evident from studies demonstrating the increased risk of MS in relatives of patients with MS, with a higher risk the closer the individuals are related to the patients.[18,19] Several genetic loci, such as the *HLADRB1* on chromosome 6, have been reported to be associated with an increased risk for MS.[20] Nevertheless, effort has been focused on epigenetic mechanisms that may influence the pathophysiology of MS. Epigenetics is the study of mechanisms that alter the expression of genes without altering the DNA sequence. DNA methylation, histone modification, and microRNA (miRNA)-associated posttranscriptional gene silencing are the three most investigated epigenetic mechanisms. Even if epigenetic changes are passed from parent to offspring through the germ line, they are highly sensitive to environmental factors that therefore may really influence the susceptibility to the disease by acting through epigenetic modifications.[21,22]

8.3.1 DNA Methylation

DNA methylation[23] consists of the addition of a methyl group to the carbon-5 of a cytosine residue in DNA through the intervention of enzymes called DNA methyltransferase (DNMT). DNMT1 maintains DNA methylation patterns during DNA replication and localizes to the DNA replication fork, where it methylates nascent DNA strands at the same locations as in the template strand.[24] DNMT3a and DNMT3b intervene in the de novo methylation of unmethylated and hemimethylated sites in nuclear and mitochondrial DNA, respectively.[24,25] Especially in mammals, DNA methylation usually occurs at CpG sites (where a cytosine nucleotide is followed by a guanine nucleotide) that can be found with up to several hundred dinucleotide repeats, therefore called CpG islands and mostly found in gene promoter regions. The methylation or hypermethylation of CpG islands in promoter regions has been reported to block the expression of the associated gene.[26] DNA methylation is the best investigated physiological epigenetic mechanism so far.[27]

8.3.2 Histone Modification

Mainly in mammalian cells, histone proteins interact with DNA to form chromatin, the packaged form of DNA. Histones are octamers consisting of two copies of each of the four histone proteins: H2A, H2B, H3, and H4. Each histone octamer has 146 bp of the DNA strand wrapped around it to shape one nucleosome, the basic unit of the chromatin. Histone proteins can be modified[23] by posttranslational changes such as acetylation, methylation, phosphorylation, ubiquitination, and citrullination. Since these histone modifications produce changes to the structure of chromatin they may affect the accessibility of the DNA strand to transcriptional enzymes, therefore inducing either activation or repression of genes associated with the modified histone.[28] Acetylation, mediated by histone acetyltransferases and deacetylases, is currently the most investigated and hence the most clarified histone modification. Acetylation of histones generally results in the upregulation of transcriptional activity of the associated gene, whereas deacetylation of histones contributes to transcriptional silencing.[29]

8.3.3 MicroRNA-Associated Gene Silencing

Single-stranded, noncoding miRNAs are widely represented in cells either from plants or animals.[30] The transcripts undergo several posttranslation changes, either in the nucleus or in the cytoplasm, to generate mature and functional miRNAs. Moreover, in the cytoplasm itself, mature miRNAs associate with other proteins to form the RNA-induced silencing complex (RISC), in which the miRNA imperfectly pairs with cognate mRNA transcripts. The target mRNA is then degraded by the RISC, preventing its translation into protein.[31,32] Such miRNA-mediated repression of translation[23] is utilized in many cellular processes, namely differentiation, proliferation, and apoptosis, as well as other key cellular mechanisms.[33,34]

8.3.4 Role of Epigenetic Changes in Inflammatory Demyelination and Neuronal/Axonal Death

Current knowledge on the role of epigenetic mechanisms in MS mostly comes from pathological studies, either from biopsies or autopsies, focusing on active demyelinating or chronic lesions, but also from studies of patients with MS, either with a relapsing-remitting (RR), chronic primary progressive (PP), or secondary-progressive (SP) course.[35]

Patient brain biopsy samples show that active and inactive MS lesions have distinct miRNA profiles. As a matter of fact, the miRNAs miR-155, miR-34a, and miR-326 are highly upregulated in active MS lesions compared with inactive lesions and normal white matter from healthy controls.[36]

The differentiation of T cells, especially Th17 cells, is influenced by epigenetic mechanisms and miR-155 and miR-326 are also associated with T-cell differentiation.[37–40] The expression of miR-155 is upregulated in macrophages, T cells, and B cells in response to ligand binding to toll-like receptors (TLRs) and inflammatory cytokines, suggesting that it is involved in inflammatory processes.[41]

Mice that are deficient in miR-155 are highly resistant to the development of the experimental autoimmune encephalomyelitis (EAE), the animal model for MS,[41] and silencing of miR-155 by administering an antisense oligonucleotide before induction of EAE attenuates the severity of symptoms.[42] Moreover, expression of miR-326 is upregulated in mice with EAE; in vivo silencing of this miRNA results in attenuation of EAE symptoms and reduced numbers of Th17 cells.[43]

Others have shown that in untreated MS (PPMS, SPMS, or RRMS) and healthy controls, two other miRNAs, miR-17 and miR-20a, are downregulated in all three forms of MS.[44] These two miRNAs inhibit T-cell activation, and their downregulation in patients with MS, therefore, might contribute to a net increase in T-cell differentiation, including differentiation into Th17 cells.

Especially in progressive MS, the evidence for involvement of epigenetic changes comes from a study showing an association between DNA methylation and neuronal cell death and in fact the overexpression of DNMT3a, an enzyme involved in de novo DNA methylation, induced apoptosis.[45]

As far as histone modification is concerned, the citrullination of myelin basic protein (MBP) has an important role in the pathophysiology of MS.[46] MBP is a major component of myelin in the CNS, and can be modified in several ways after translation. In biopsy samples from MS patients, normal-appearing white matter shows increased levels of citrullinated MBP as compared with levels in healthy controls and patients with Alzheimer's disease.[47] Citrullinated MBP is less stable than unmodified MBP, which suggests that citrullination might contribute to myelin breakdown and eventually to the development of an autoimmune response to MBP.[48] Finally, brain biopsy material from progressive MS patients and controls without neurological disease show an increase in histone H3 acetylation in oligodendrocytes within chronic MS lesions, whereas oligodendrocytes within early-stage MS lesions show marked histone H3 deacetylation.[49] Increased histone H3 acetylation in oligodendrocytes is associated with impaired differentiation and, therefore, with impaired remyelination.

Since epigenetic changes are highly sensitive to environmental influences, it is likely that the effects of environmental risk factors in MS might be mediated by changes in patients' epigenetic profiles.

8.4 VIRAL INFECTIONS AND MULTIPLE SCLEROSIS

Migration studies have contributed to provide evidence that a viral infection may trigger the development of MS.[50] It has been shown that people migrating from a high-risk country for MS to a low-risk one are at lower risk of developing MS than they would be in their country of origin. Whereas those migrating from a low-risk country to a high-risk one keep the low risk of their country of origin, their children have a risk comparable to the country where they emigrate,[51] especially in those migrating before the age of 15,[52] suggesting that infection at a young age may predispose to the later development of MS. In addition to these migration studies, some classical studies on the incidence and prevalence of MS have suggested that there may have been MS epidemics in several locations, such as in the Faroe islands after the second world war,[53] and the increase of incidence in the Shetland Islands[54] and Sardinia[55] have been taken to suggest that an infectious agent may be involved in the pathogenesis of MS.

Different hypotheses have been proposed to explain how viral infections are associated with MS.[56] According to the bystander activation hypothesis, autoreactive T cells are activated by nonspecific inflammatory molecules occurring during infections, such as cytokines, superantigens, and TLR ligands.[4] The molecular mimicry hypothesis, instead, postulates that upon exposure to a pathogen, the pathogen/MHC conformation on an APC bears molecular similarity to that of an endogenous peptide, such as an MBP fragment presented within an MHC.[57] If appropriate costimulation occurs, it results in the expansion and differentiation not only of the pathogen-reactive T cells, a proper immune response, but also the expansion of MBP-reactive T cells, an improper response. If both pools differentiate into Th1 or Th17 proinflammatory subsets, these can become reactivated within the CNS to promote pathology. In fact, T-cell lines isolated from MS patients demonstrate cross-reactivity between MBP and coronavirus[58] or Epstein–Barr virus (EBV)[59] antigens. Furthermore, a significant degree of crystal structural similarity has been shown between the DRB5*0101-EBV peptide complex and the DRB1*1501-MBP peptide complex at the cell surface for TCR recognition.[60] Further immunological evidence in the association of EBV with MS has been provided. The follicular-like structures under the meninges include B cells that are infected with EBV in many patients.[61] MS patients have antibodies that cross-react between MBP and EBV, a possible additional mechanism by which anti-EBV antibodies may disrupt myelin.[62] Furthermore, EBV-reactive $CD8^+$ T cells that are restricted by HLA-B7, a common allele in MS, are dysregulated in MS[63] and the $CD8^+$ T-cell deficiency in MS impairs the capacity to control EBV infection with the result that EBV-infected B cells accumulate in the CNS where they produce pathogenic autoantibodies and provide survival signals to autoreactive T cells.[64]

EBV infection is certainly associated with changes in epigenetic profiles in infected cells but so far this has been evaluated especially in tumors and, as a result, several types of tumor are associated with prior EBV infection, probably due to promoter hypermethylation (and, therefore, repression) of tumor suppressor genes.[65] There is still a lack of evidence for these aspects in MS.

Despite molecular similarity between several other pathogens and a number of myelin peptides and other molecules within the CNS frequently occurring, there is a high probability that these pathogens can induce improper expansion of CNS-reactive T cells to promote pathology within the CNS and hence no single infectious agent may be uniquely associated with MS.

8.5 SMOKING AND MULTIPLE SCLEROSIS

Both epidemiological and clinical studies have recognized smoking as an environmental risk factor for MS.[50] Smoking increases the relative incidence rate of MS in current smokers compared to nonsmokers, with a dose–response dependent on the number of packs smoked per year.[66] Smoking also has an impact on inflammatory outcomes in MS. Patients with a clinically isolated syndrome have an increased risk of conversion to clinically definite MS in smokers compared to nonsmokers.[67] MS smokers have more gadolinium-enhancing lesions, a greater T2-lesion load, and more brain atrophy than nonsmokers,[68] as well as a quicker increase in T2-lesion volume and brain atrophy in an average follow-up period of time.[69] As far as the disease progression is concerned, the data are quite discordant since smoking is in some cases reported not to be associated with the risk of SP or with that of reaching Expanded Disability Status Scale (EDSS) 4.0 or 6.0[70]; in others it is reported to be associated with a greater risk of SP course[69,71] or even with an increase in EDSS scores during two years of follow-up.[72] In conclusion, smoking may have more influence in the early disease course than in the late disease stages of MS.

How smoking increases the risk of MS is still a matter of debate and even whether or not cigarette smoke contains mutagens that can affect long-lasting immunity, but smoking has been demonstrated to induce an immunosuppressant state.[73] Nevertheless, cigarette smoking induces immune functions and an interaction between smoking and genes regulating immune functions has been reported.[74] It would be relevant to figure out whether constituents of tobacco alter signaling through the aryl hydrocarbon receptor, a transcription factor affected by polycyclic aromatic hydrocarbons and polychlorinated dioxins, since the latter regulates T-cell polarization and alters the course of EAE.[75] It is almost certain that smoking affects MS by upregulating MMPs since immune cells and biological fluids of smokers tend to upregulate several MMPs[76] and these may facilitate immune-cell entry to the CNS parenchyma. When comparing MRI scans from smokers and nonsmokers with MS, more contrast-enhancing lesions are evident among the smokers, suggesting more severe blood–brain barrier damage.[68]

Smoking so far has been reported to be associated with changes in epigenetic profiles in patients with cancer, especially inducing silencing of tumor suppressor genes, mostly through DNA methylation.[77] Smoking is also associated with changes in miRNA expression profiles in spermatozoa,[78] and with altered histone modifications resulting from reduced levels of histone deacetylase 2 in macrophages.[79] In MS there is no evidence in this respect but no doubt these mechanisms are worth investigating in the disease.

8.6 SUNLIGHT EXPOSURE, VITAMIN D, AND MULTIPLE SCLEROSIS

MS is more prevalent in regions of higher latitude[80] where an increase of female/male rate incidence has been also demonstrated in the 2000s.[81] This phenomenon seems to be associated with a decreased sunlight (UV) exposure and the subsequent reduced vitamin D production.[82] It has been shown that the risk of developing MS decreases with increasing serum 25-hydroxy-vitamin D levels in a prospective case–control study.[83] Among various suspected environmental factors in MS, the lack of UV exposure has been found to be the most significant risk factor for MS.[50,84] Moreover, vitamin D may influence the disease course of MS since lower vitamin D levels have been demonstrated to be associated with higher levels of disability[85] and an association between higher levels of vitamin D and decreased risk of relapses has also been reported.[86] Finally, some authors provide data showing that vitamin D supplementation may be an effective treatment for MS since high-dose vitamin D treatment in MS tends to decrease relapses.[87]

The possible sequence of events linking sunlight exposure with MS is most likely based on the conversion, due to ultraviolet B radiation (290–320nm), of cutaneous 7-dehydrocholesterol to previtamin D_3, which then spontaneously gives origin to vitamin D_3.[88] The latter then undergoes two hydroxylations, by D-25-hydroxylase (CYP2R1) in the liver and

25-hydroxyvitamin D-1α-hydroxylase (CYP27B1) in the kidney, to produce the biologically active form of vitamin D, 1,25-dihydroxyvitamin D_3. Variants of the CYP27B1 gene have been reported to be associated with increased risk of MS[89] and others have confirmed the association of MS with two vitamin D-related genes, CYP27B1 and CYP24A1,[20] while a vitamin D response element lies close to the promoter region of HLA-DRB1, the main risk allele for MS.[90]

Different mechanisms of action of vitamin D that may impact different steps of the disease immunopathogenesis have been reported. Vitamin D either suppresses the maturation and activity of APCs, including dendritic cells, or increases their tolerogenic phenotype.[91] $CD4^+$ T helper cells are also affected by vitamin D, with a reduced production of proinflammatory Th1 and Th17 cells[92] while that of Th2 cells is increased.[93] Vitamin D treatment induces Treg activity[92] and reduces proinflammatory molecules produced by stimulated monocytes.[94] In EAE, vitamin D has proved to be effective either given as preventive[95] or therapeutic treatment.[96]

Vitamin D can enter the CNS to exert its immune-regulating properties while its possible neuroprotective role is more uncertain. Certainly, the enzymes necessary to synthesize the bioactive 1,25-dihydroxyvitamin D_3 are present in the brain[97] and abnormal brain development has been observed in rats deficient in vitamin D during gestation. Moreover, mice with gestational vitamin D deficiency have impaired learning in adulthood.[98] In vitro, vitamin D is able to reduce glutamate excitotoxicity to cortical, cerebellar, or hippocampal neurons.[99] Whether such vitamin D neuroprotective experimental evidence is valid in human MS still remains to be elucidated. Finally, it is quite evident that vitamin D may correct many of the immune abnormalities seen in MS, nevertheless which mechanisms are the most relevant to its therapeutic efficacy or whether such mechanisms include its actions within the CNS are as yet unclear.

Some evidence also exists to suggest vitamin D might influence epigenetic mechanisms. 1,25-hydroxyvitamin D_3 has been reported to affect histone modification in cancer: studies in human colon cancer cells have shown that vitamin D induces the expression of *JMJD3*, the gene encoding lysine-specific demethylase 6B, which specifically demethylates lysine 27 of histone H3.[100,101] As far as MS is concerned, the potential relevance of vitamin D-induced histone modification is suggested by a study showing that binding of 1,25-hydroxyvitamin D_3 to the vitamin D receptor leads to suppression of transcription of the proinflammatory cytokine IL-17, via recruitment of histone deacetylase 2 to the *IL17A* promoter region.[102]

8.7 MICROBIOTA AND MULTIPLE SCLEROSIS

Despite infection agents having long been investigated as possible triggers of autoimmunity in MS, their involvement still remains a matter of debate. Studies have focused on the involvement of resident commensal microbiota in CNS autoimmunity.[103]

Humans are colonized by a myriad of microbes, including bacteria, archaea, fungi, eukaryotes, and viruses both in mucosal surfaces and in the skin and are collectively termed microbiota.[104] Such microbial organisms mostly belong to two large phyla, the bacteroidetes and the firmicutes. The microbiota may generally have beneficial functions to the host, but may influence the physiology and/or pathology of the host.[105]

Studies in EAE have clarified that the microbial flora contributes to the CNS-specific autoimmune disease.[106,107] In fact, spontaneous EAE incidence has been found to be strongly reduced in TCR transgenic mice kept in germ-free (GF) conditions and therefore not having resident microbes.[108] But, EAE severity is also reduced in GF mice immunized with myelin peptide antigen in complete Freund's adjuvant.[109] Moreover, antibiotics have been found to affect disease severity by altering the gut flora.[110,111] Nevertheless, it remains unclear how and when these agents may become detrimental. Since the microbiota has an impact on the host's immune system,[105] it is likely to shift the balance between protective and pathogenic immune responses. Indeed, antibiotic-mediated protection from EAE has been associated with a decreased production of the proinflammatory cytokine IL-17 in the gut-associated lymphoid tissue, thus altering the function of invariant natural-killer T cells,[111] but also with an increase in the Tregs.[110]

CNS-reactive immune cells can be activated by commensal microbiota either through molecular mimicry or through a bystander activation mechanism, as proposed for other infectious pathogens. However, so far no CNS-mimicry epitope derived from gut bacteria has been identified, whereas the current data provide more evidence in favor of a bystander activation hypothesis. It is likely that the Th17 cells generated in the gut are a result of bystander activation of APCs and that their secreted cytokines can drive naïve T cells toward proinflammatory phenotypes. Nevertheless, it has been reported that specific commensal microbial species may induce either Th-17 or Tregs cells both in the intestine as well as at peripheral sites.[112,113]

So far, there is no clear evidence supporting the involvement of the gut microbiota either in the incidence or in the pathogenesis of MS; however, indirect data suggest a potential implication especially when considering dietary factors, which can rapidly alter gut microbial signatures.[114]

8.8 CONCLUSIONS

At the time of writing the pathophysiological mechanisms that mediate the effects of environmental risk factors on susceptibility to MS or the course of this disease are still unknown. It is quite intriguing though, that the most important environmental risk factors for MS seem to be clearly associated with changes in epigenetic profiles and more research is certainly required to establish whether epigenetic mechanisms can truly mediate the effects of these risk factors. Finally, the microbiota also deserves to be taken into consideration as an external factor favoring the disease, given the relevant implications it has in controlling the host's immune system.

References

1. Ascherio A, Munger KL. Environmental risk factors for multiple sclerosis. Part II: noninfectious factors. *Ann Neurol*. 2007;61:504–513.
2. Ascherio A, Munger KL. Environmental risk factors for multiple sclerosis. Part I: the role of infection. *Ann Neurol*. 2007;61:288–299.
3. Trapp BD, Nave K-A. Multiple sclerosis: an immune or neurodegenerative disorder? *Annu Rev Neurosci*. 2008;31:247–269.
4. Sospedra M, Martin R. Immunology of multiple sclerosis. *Annu Rev Immunol*. 2005;23:683–747.
5. Kassmann CM, Lappe-Siefke C, Baes M, et al. Axonal loss and neuroinflammation caused by peroxisome-deficient oligodendrocytes. *Nat Genet*. 2007;39(8):969–976.

6. Bettelli E, Carrier Y, Gao W, et al. Reciprocal developmental pathways for the generation of pathogenic effector TH17 and regulatory T cells. *Nature*. 2006;441(7090):235–238.
7. Becher B, Segal BM. T(H)17 cytokines in autoimmune neuro-inflammation. *Curr Opin Immunol*. 2011;23(6): 707–712.
8. Siffrin V, Radbruch H, Glumm R, et al. In vivo imaging of partially reversible th17 cell-induced neuronal dysfunction in the course of encephalomyelitis. *Immunity*. 2010;33(3):424–436.
9. Viglietta V, Baecher-Allan C, Weiner HL, Hafler DA. Loss of functional suppression by $CD4^{+}CD25^{+}$ regulatory T cells in patients with multiple sclerosis. *J Exp Med*. 2004;199(7):971–979.
10. Saxena A, Martin-Blondel G, Mars LT, Liblau RS. Role of CD8 T cell subsets in the pathogenesis of multiple sclerosis. *FEBS Lett*. 2011;585(23):3758–3763.
11. Hauser SL, Waubant E, Arnold DL, et al. B-cell depletion with rituximab in relapsing-remitting multiple sclerosis. *N Engl J Med*. 2008;358(7):676–688.
12. Kappos L, Li D, Calabresi PA, et al. Ocrelizumab in relapsing-remitting multiple sclerosis: a phase 2, randomised, placebo-controlled, multicentre trial. *Lancet*. 2011;378(9805):1779–1787.
13. Magliozzi R, Howell O, Vora A, et al. Meningeal B-cell follicles in secondary progressive multiple sclerosis associate with early onset of disease and severe cortical pathology. *Brain*. 2007;130(Pt 4):1089–1104.
14. Meinl E, Derfuss T, Krumbholz M, Pröbstel A-K, Hohlfeld R. Humoral autoimmunity in multiple sclerosis. *J Neurol Sci*. 2011;306(1–2):180–182.
15. von Büdingen H-C, Bar-Or A, Zamvil SS. B cells in multiple sclerosis: connecting the dots. *Curr Opin Immunol*. 2011;23(6):713–720.
16. Agrawal SM, Lau L, Yong VW. MMPs in the central nervous system: where the good guys go bad. *Semin Cell Dev Biol*. 2008;19(1):42–51.
17. Nikić I, Merkler D, Sorbara C, et al. A reversible form of axon damage in experimental autoimmune encephalomyelitis and multiple sclerosis. *Nat Med*. 2011;17(4):495–499.
18. Carton H, Vlietinck R, Debruyne J, et al. Risks of multiple sclerosis in relatives of patients in Flanders, Belgium. *J Neurol Neurosurg Psychiatr*. 1997;62:329–333.
19. Robertson NP, Fraser M, Deans J, Clayton D, Walker N, Compston DA. Age-adjusted recurrence risks for relatives of patients with multiple sclerosis. *Brain*. 1996;119:449–455.
20. Sawcer S, Hellenthal G, Pirinen M, et al. Genetic risk and a primary role for cell-mediated immune mechanisms in multiple sclerosis. *Nature*. 2011;476:214–219.
21. Jaenisch R, Bird A. Epigenetic regulation of gene expression: how the genome integrates intrinsic and environmental signals. *Nat Genet*. 2003;33(suppl):245–254.
22. Skinner MK, Manikkam M, Guerrero-Bosagna C. Epigenetic transgenerational actions of environmental factors in disease etiology. *Trends Endocrinol Metab*. 2010;21:214–222.
23. Koch MW, Metz LM, Kovalchuk O. Epigenetic changes in patients with multiple sclerosis. *Nat Rev Neurol*. 2013;9(1):35–43.
24. Goll MG, Bestor TH. Eukaryotic cytosine methyltransferases. *Annu Rev Biochem*. 2005;74:481–514.
25. Okano M, Bell DW, Haber DA, Li E. DNA methyltransferases Dnmt3a and Dnmt3b are essential for *de novo* methylation and mammalian development. *Cell*. 1999;99:247–257.
26. Klose RJ, Bird AP. Genomic DNA methylation: the mark and its mediators. *Trends Biochem Sci*. 2006;31:89–97.
27. Weber M, Schübeler D. Genomic patterns of DNA methylation: targets and function of an epigenetic mark. *Curr Opin Cell Biol*. 2007;19:273–280.
28. Dieker J, Muller S. Epigenetic histone code and autoimmunity. *Clin Rev Allergy Immunol*. 2010;39:78–84.
29. Brooks WH, Le Dantec C, Pers J-O, Youinou P, Renaudineau Y. Epigenetics and autoimmunity. *J Autoimmun*. 2010;34:J207–J219.
30. Bernstein E, Allis CD. RNA meets chromatin. *Genes Dev*. 2005;19:1635–1655.
31. Hwang H-W, Mendell JT. MicroRNAs in cell proliferation, cell death, and tumorigenesis. *Br J Cancer*. 2006;94:776–780.
32. Sevignani C, Calin GA, Siracusa LD, Croce CM. Mammalian microRNAs: a small world for fine-tuning gene expression. *Mamm Genome*. 2006;17:189–202.
33. Chang T-C, Mendell JT. MicroRNAs in vertebrate physiology and human disease. *Annu Rev Genomics Hum Genet*. 2007;8:215–239.
34. Fabbri M, Ivan M, Cimmino A, Negrini M, Calin GA. Regulatory mechanisms of microRNAs involvement in cancer. *Expert Opin Biol Ther*. 2007;7:1009–1019.

35. Koch M, Kingwell E, Rieckmann P, Tremlett H. The natural history of primary progressive multiple sclerosis. *Neurology*. 2009;73:1996–2002.
36. Junker A, Krumbholz M, Eisele S, et al. MicroRNA profiling of multiple sclerosis lesions identifies modulators of the regulatory protein CD47. *Brain*. 2009;132:3342–3352.
37. Haasch D, Chen YW, Reilly RM, et al. T cell activation induces a noncoding RNA transcript sensitive to inhibition by immunosuppressant drugs and encoded by the proto-oncogene, BIC. *Cell Immunol*. 2002;217:78–86.
38. Thai TH, Calado DP, Casola S, et al. Regulation of the germinal center response by microRNA-155. *Science*. 2007;316:604–608.
39. Teng G, Hakimpour P, Landgraf P, et al. MicroRNA-155 is a negative regulator of activation-induced cytidine deaminase. *Immunity*. 2008;28:621–629.
40. Teng G. Papavasiliou FN Shhh! Silencing by microRNA-155. *Philos Trans R Soc Lond B Biol Sci*. 2009;364:631–637.
41. O'Connell RM, Kahn D, Gibson WS, et al. MicroRNA-155 promotes autoimmune inflammation by enhancing inflammatory T cell development. *Immunity*. 2010;33:607–619.
42. Murugaiyan G, Beynon V, Mittal A, Joller N, Weiner HL. Silencing microRNA-155 ameliorates experimental autoimmune encephalomyelitis. *J Immunol*. 2011;187:2213–2221.
43. Du C, Liu C, Kang J, et al. MicroRNA miR-326 regulates TH-17 differentiation and is associated with the pathogenesis of multiple sclerosis. *Nat Immunol*. 2009;10:1252–1259.
44. Cox MB, Cairns MJ, Gandhi KS, et al. MicroRNAs miR-17 and miR-20a inhibit T cell activation genes and are under-expressed in MS whole blood. *PLoS One*. 2010;5:e12132.
45. Chestnut BA, Chang Q, Price A, Lesuisse C, Wong M, Martin LJ. Epigenetic regulation of motor neuron cell death through DNA methylation. *J Neurosci*. 2011;31:16619–16636.
46. Moscarello MA, Mastronardi FG, Wood DD. The role of citrullinated proteins suggests a novel mechanism in the pathogenesis of multiple sclerosis. *Neurochem Res*. 2007;32:251–256.
47. Moscarello MA, Wood DD, Ackerley C, Boulias C. Myelin in multiple sclerosis is developmentally immature. *J Clin Invest*. 1994;94:146–154.
48. Mastronardi FG, Noor A, Wood DD, Paton T, Moscarello MA. Peptidyl argininedeiminase 2 CpG island in multiple sclerosis white matter is hypomethylated. *J Neurosci Res*. 2007;85:2006–2016.
49. Pedre X, Mastronardi F, Bruck W, López-Rodas G, Kuhlmann T, Casaccia P. Changed histone acetylation patterns in normal-appearing white matter and early multiple sclerosis lesions. *J Neurosci*. 2011;31:3435–3445.
50. Koch MW, Metz LM, Agrawal SM, Yong VW. Environmental factors and their regulation of immunity in multiple sclerosis. *J Neurol Sci*. 2013;324:10–16.
51. Gale CR, Martyn CN. Migrant studies in multiple sclerosis. *Prog Neurobiol*. 1995;47(4–5):425–448.
52. Alter M, Leibowitz U, Speer J. Risk of multiple sclerosis related to age at immigration to Israel. *Arch Neurol*. 1966;15(3):234–237.
53. Joensen P. Multiple sclerosis: variation of incidence of onset over time in the Faroe Islands. *Mult Scler*. 2011;17(2):241–244.
54. Poskanzer DC, Sheridan JL, Prenney LB, Walker AM. Multiple sclerosis in the Orkney and Shetland Islands. II: the search for an exogenous aetiology. *J Epidemiol Community Health*. 1980;34(4):240–252.
55. Rosati G, Aiello I, Granieri E, et al. Incidence of multiple sclerosis in Macomer, Sardinia, 1912–1981: onset of the disease after 1950. *Neurology*. 1986;36(1):14–19.
56. Kakalacheva K, Münz C, Lünemann JD. Viral triggers of multiple sclerosis. *Biochim Biophys Acta*. 2011;1812(2):132–140.
57. Chastain EML, Miller SD. Molecular mimicry as an inducing trigger for CNS autoimmune demyelinating disease. *Immunol Rev*. 2012;245(1):227–238.
58. Talbot PJ, Paquette JS, Ciurli C, Antel JP, Ouellet F. Myelin basic protein and human coronavirus 229E cross-reactive T cells in multiple sclerosis. *Ann Neurol*. 1996;39(2):233–240.
59. Cheng W, Ma Y, Gong F, et al. Cross-reactivity of autoreactive T cells with MBP and viral antigens in patients with MS. *Front Biosci*. 2012;17:1648–1658.
60. Lang HL, Jacobsen H, Ikemizu S, et al. A functional and structural basis for TCR cross-reactivity in multiple sclerosis. *Nat Immunol*. 2002;3(10):940–943.
61. Serafini B, Severa M, Columba-Cabezas S, et al. Epstein-Barr virus latent infection and BAFF expression in B cells in the multiple sclerosis brain: implications for viral persistence and intrathecal B-cell activation. *J Neuropathol Exp Neurol*. 2010;69(7):677–693.

62. Gabibov AG, Belogurov Jr AA, Lomakin YA, et al. Combinatorial antibody library from multiple sclerosis patients reveals antibodies that cross-react with myelin basic protein and EBV antigen. *FASEB J*. 2011;25(12):4211–4221.
63. Jilek S, Schluep M, Harari A, et al. HLA-B7-restricted EBV-specific CD8+ T cells are dysregulated in multiple sclerosis. *J Immunol*. 2012;188(9):4671–4680.
64. Pender MP. CD8+ T-cell deficiency, Epstein–Barr virus infection, vitamin D deficiency, and steps to autoimmunity: a unifying hypothesis. *Autoimmune Dis*. 2012;2012:189096.
65. Niller HH, Wolf H, Minarovits J. Epigenetic dysregulation of the host cell genome in Epstein–Barr virus-associated neoplasia. *Semin Cancer Biol*. 2009;19:158–164.
66. Hernán MA, Olek MJ, Ascherio A. Cigarette smoking and incidence of multiple sclerosis. *Am J Epidemiol*. 2001;154(1):69–74.
67. Di Pauli F, Reindl M, Ehling R, et al. Smoking is a risk factor for early conversion to clinically definite multiple sclerosis. *Mult Scler*. 2008;14(8):1026–1030.
68. Zivadinov R, Weinstock-Guttman B, Hashmi K, et al. Smoking is associated with increased lesion volumes and brain atrophy in multiple sclerosis. *Neurology*. 2009;73(7):504–510.
69. Healy BC, Ali EN, Guttmann CRG, et al. Smoking and disease progression in multiple sclerosis. *Arch Neurol*. 2009;66(7):858–864.
70. Koch M, van Harten A, Uyttenboogaart M, De Keyser J. Cigarette smoking and progression in multiple sclerosis. *Neurology*. October 9, 2007;69(15):1515–1520.
71. Hernán MA, Jick SS, Logroscino G, Olek MJ, Ascherio A, Jick H. Cigarette smoking and the progression of multiple sclerosis. *Brain*. 2005;128(Pt 6):1461–1465.
72. Pittas F, Ponsonby A-L, van der Mei IAF, et al. Smoking is associated with progressive disease course and increased progression in clinical disability in a prospective cohort of people with multiple sclerosis. *J Neurol*. 2009;256(4):577–585.
73. Gonçalves RB, Coletta RD, Silvério KG, et al. Impact of smoking on inflammation: overview of molecular mechanisms. *Inflamm Res*. 2011;60(5):409–424.
74. Hedström AK, Sundqvist E, Bäärnhielm M, et al. Smoking and two human leukocyte antigen genes interact to increase the risk for multiple sclerosis. *Brain*. 2011;134(Pt 3):653–664.
75. Quintana FJ, Basso AS, Iglesias AH, et al. Control of T(reg) and T(H)17 cell differentiation by the aryl hydrocarbon receptor. *Nature*. 2008;453(7191):65–71.
76. Ozçaka O, Biçakci N, Pussinen P, Sorsa T, Köse T, Buduneli N. Smoking and matrixmetalloproteinases, neutrophil elastase and myeloperoxidase in chronic periodontitis. *Oral Dis*. 2011;17(1):68–76.
77. Wan ES, Qiu W, Baccarelli A, et al. Cigarette smoking behaviors and time since quitting are associated with differential DNA methylation across the human genome. *Hum Mol Genet*. 2012;21:3073–3082.
78. Marczylo EL, Amoako AA, Konje JC, Gant TW, Marczylo TH. Smoking induces differential miRNA expression in human spermatozoa: a potential transgenerational epigenetic concern? *Epigenetics*. 2012;7:432–439.
79. Ito K, Lim S, Caramori G, Chung KF, Barnes PJ, Adcock IM. Cigarette smoking reduces histone deacetylase 2 expression, enhances cytokine expression, and inhibits glucocorticoid actions in alveolar macrophages. *FASEB J*. 2001;15:1110–1112.
80. Simpson Jr S, Blizzard L, Otahal P, Van der Mei I, Taylor B. Latitude is significantly associated with the prevalence of multiple sclerosis: a meta-analysis. *J Neurol Neurosurg Psychiatr*. 2011;82(10):1132–1141.
81. Trojano M, Lucchese G, Graziano G, et al. Geographical variations in sex ratio trends over time in multiple sclerosis. *PLoS One*. 2012;7:e48078.
82. Ascherio A, Munger KL, Simon KC. Vitamin D and multiple sclerosis. *Lancet Neurol*. 2010;9(6):599–612.
83. Munger KL, Levin LI, Hollis BW, Howard NS, Ascherio A. Serum 25-hydroxyvitamin D levels and risk of multiple sclerosis. *JAMA*. December 20, 2006;296(23):2832–2838.
84. Sloka S, Silva C, Pryse-Phillips W, Patten S, Metz L, Yong VW. A quantitative analysis of suspected environmental causes of MS. *Can J Neurol Sci*. 2011;38(1):98–105.
85. Smolders J, Menheere P, Kessels A, Damoiseaux J, Hupperts R. Association of vitamin D metabolite levels with relapse rate and disability in multiple sclerosis. *Mult Scler*. 2008;14(9):1220–1224.
86. Simpson S, Taylor B, Blizzard L, et al. Higher 25-hydroxyvitamin D is associated with lower relapse risk in multiple sclerosis. *Ann Neurol*. 2010;68(2):193–203.
87. Burton JM, Kimball S, Vieth R, et al. A phase I/II dose-escalation trial of vitamin D3 and calcium in multiple sclerosis. *Neurology*. 2010;74(23):1852–1859.
88. Hart PH, Gorman S, Finlay-Jones JJ. Modulation of the immune system by UVradiation: more than just the effects of vitamin D? *Nat Rev Immunol*. 2011;11(9):584–596.

89. Ramagopalan SV, Dyment DA, Cader MZ, et al. Rare variants in the CYP27B1 gene are associated with multiple sclerosis. *Ann Neurol*. 2011;70(6):881–886.
90. Ramagopalan SV, Maugeri NJ, Handunnetthi L, et al. Expression of the multiple sclerosis-associated MHC class II Allele HLA-DRB1*1501 is regulated by vitamin D. *PLoS Genet*. 2009;5(2):e1000369.
91. Széles L, Keresztes G, Töröcsik D, et al. 1,25- dihydroxyvitamin D3 is an autonomous regulator of the transcriptional changes leading to a tolerogenic dendritic cell phenotype. *J Immunol*. 2009;182(4):2074–2083.
92. Correale J, Ysrraelit MC, Gaitán MI. Immunomodulatory effects of vitamin D in multiple sclerosis. *Brain*. May 2009;132(Pt 5):1146–1160.
93. Sloka S, Silva C, Wang J, Yong VW. Predominance of Th2 polarization by vitamin D through a STAT6-dependent mechanism. *J Neuroinflammation*. 2011;8:56.
94. Almerighi C, Sinistro A, Cavazza A, Ciaprini C, Rocchi G, Bergamini A. 1Alpha,25-dihydroxyvitamin D3 inhibits CD40L-induced pro-inflammatory and immunomodulatory activity in human monocytes. *Cytokine*. 2009;45(3):190–197.
95. Lemire JM, Archer DC. 1,25-dihydroxyvitamin D3 prevents the in vivo induction of murine experimental autoimmune encephalomyelitis. *J Clin Invest*. 1991;87(3):1103–1107.
96. Cantorna MT, Hayes CE, DeLuca HF. 1,25-Dihydroxyvitamin D3 reversibly blocks the progression of relapsing encephalomyelitis, a model of multiple sclerosis. *Proc Natl Acad Sci USA*. 1996;93(15):7861–7864.
97. Smolders J, Moen SM, Damoiseaux J, Huitinga I, Holmøy T. Vitamin D in the healthy and inflamed central nervous system: access and function. *J Neurol Sci*. December 15, 2011;311(1–2):37–43.
98. Fernandes de Abreu DA, Nivet E, Baril N, Khrestchatisky M, Roman F, Féron F. Developmental vitamin D deficiency alters learning in C57Bl/6J mice. *Behav Brain Res*. 2010;208(2):603–608.
99. Brewer LD, Thibault V, Chen KC, Langub MC, Landfield PW, Porter NM. Vitamin D hormone confers neuroprotection in parallel with downregulation of L-type calcium channel expression in hippocampal neurons. *J Neurosci*. 2001;21(1):98–108.
100. Pereira F, Barbáchano A, Singh PK, Campbell MJ, Muñoz A, Larriba MJ. Vitamin D has wide regulatory effects on histone demethylase genes. *Cell Cycle*. 2012;11:1081–1089.
101. Pereira F, Barbáchano A, Silva J, et al. KDM6B/JMJD3 histone demethylase is induced by vitamin D and modulates its effects in colon cancer cells. *Hum Mol Genet*. 2011;20:4655–4665.
102. Joshi S, Pantalena LC, Liu XK, et al. 1,25-dihydroxyvitamin D3 ameliorates Th17 autoimmunity via transcriptional modulation of interleukin-17A. *Mol Cell Biol*. 2011;31:3653–3669.
103. Berer K, Krishnamoorthy G. Microbial view of central nervous system autoimmunity. *FEBS Lett*. 2014;588:4207–4213.
104. Turnbaugh PJ, Ley RE, Hamady M, Fraser-Liggett CM, Knight R, Gordon JI. The human microbiome project. *Nature*. 2007;449:804–810.
105. Cerf-Bensussan N, Gaboriau-Routhiau V. The immune system and the gut microbiota: friends or foes? *Nat Rev Immunol*. 2010;10:735–744.
106. Wekerle H, Berer K, Krishnamoorthy G. Remote control-triggering of brain autoimmune disease in the gut. *Curr Opin Immunol*. 2013;25:683–689.
107. Berer K, Krishnamoorthy G. Commensal gut flora and brain autoimmunity: a love or hate affair? *Acta Neuropathol*. 2012;123:639–651.
108. Berer K, Mues M, Koutrolos M, et al. Commensal microbiota and myelin autoantigen cooperate to trigger autoimmune demyelination. *Nature*. 2011;479:538–541.
109. Lee YK, Menezes JS, Umesaki Y, Mazmanian SK. Proinflammatory T-cell responses to gut microbiota promote experimental autoimmune encephalomyelitis. *Proc Natl Acad Sci USA*. 2011;108(suppl 1): 4615–4622.
110. Ochoa-Repáraz J, Mielcarz DW, Ditrio LE, et al. Role of gut commensal microflora in the development of experimental autoimmune encephalomyelitis. *J Immunol*. 2009;183:6041–6050.
111. Yokote H, Miyake S, Croxford JL, Oki S, Mizusawa H, Yamamura T. NKT cell-dependent amelioration of a mouse model of multiple sclerosis by altering gut flora. *Am J Pathol*. 2008;173:1714–1723.
112. Atarashi K, Tanoue T, Shima T, et al. Induction of colonic regulatory T cells by indigenous Clostridium species. *Science*. 2011;331:337–341.
113. Ivanov II, Atarashi K, Manel N, et al. Induction of intestinal Th17 cells by segmented filamentous bacteria. *Cell*. 2009;139:485–498.
114. David LA, Maurice CF, Carmody RN, et al. Diet rapidly and reproducibly alters the human gut microbiome. *Nature*. 2014;505:559–563.

CHAPTER

9

Gut Microbiota in Multiple Sclerosis: A Bioreactor Driving Brain Autoimmunity

H. Wekerle
Max-Planck-Institute of Neurobiology, Martinsried, Munich, Germany

R. Hohlfeld
Institute of Clinical Neuroimmunology, Ludwig-Maximilian University, Munich, Germany

OUTLINE

9.1 INTRODUCTION

Worldwide there are about 2.5 million people with multiple sclerosis (MS) (www.msif.org/wp-content/uploads/2014/09/Atlas-of-MS.pdf).[1] Although MS is much less common than, for example, the prevalent neurodegenerative diseases of old age, there are other features that place MS among the most pressing challenges of present-day medicine: it is a chronic,

Translational Neuroimmunology in Multiple Sclerosis
http://dx.doi.org/10.1016/B978-0-12-801914-6.00009-X

debilitating disease; it typically attacks at young adult age, and then persists throughout the rest of the individual's lifetime, often for several decades; over time MS may result in severe physical disability and cognitive decline; and the disease prevalence tends to increase worldwide.[2] This creates to an enormous burden of individual distress, with severe socioeconomic consequences. Economic minds translate this into a drug market of over $10 billion worldwide per year.

Until quite recently, treatment of MS was largely symptomatic; with the recent advent of biologicals (recombinant interferons, monoclonal antibodies against immune cells) MS has become tractable. Unfortunately, the new drugs are of limited efficiency. While they mitigate the course and severity of the disease, they do not cure it. Clearly, new drugs with new pathogenic targets are needed.

Recent studies of experimental models of MS point to an unexpected element in the pathogenesis of MS: the microbial gut flora. As will be described here, bacteria contained within the bowel have the potential of activating brain-reactive immune cells. These become autoaggressive, invade the brain and spinal cord, and trigger an inflammatory chain reaction that ultimately is responsible for the variegated neurological defects diagnosed in an MS patient.

9.2 PATHOGENESIS AND ETIOLOGY

MS is commonly labeled as an autoimmune disease. This is not formally proven, but there are good reasons to assume that indeed MS is caused by an immune attack against the body's own brain white matter. The evidence comes from diverse areas: Early pathologists described neurodegenerative along with inflammatory changes as hallmarks of the MS plaque lesion.[3] The particular arrangement of immune cell infiltrates in lesions of myelin axon destruction was suggestive of an ongoing immunopathological process. This was supported more recently by studies applying molecular analyses, particularly a series of genome-wide association studies. These identified the infiltrate cells as cytotoxic T cells apparently committed in an acute immune attack.[4] Then, brain-specific antibodies are found in some (but not all) patients with MS.[5] Furthermore, large-scale genetic trials scanning patients' genomes discovered a large number of gene variants that coincide with an increased susceptibility (MS risk genes), and noted that most of these genes are involved in immune functions.[6] Finally, therapies interfering with immune cell activation and migration successfully mitigate the disease course.[7] Taken together, these features point to an autoimmune pathogenesis of MS, but they do not formally prove it (Box 9.1).

But how is MS triggered? Epidemiology may provide hints. MS is not evenly distributed over the globe (www.msif.org/wp-content/uploads/2014/09/Atlas-of-MS.pdf). It preferentially affects people with European ancestors (Caucasians) living in moderate climate zones. This could be explained by genetic factors as well as by factors contributed by the environment, including life style and milieu. Indeed, although the genetic risk factors identified by the genomic studies increase the disease risk, MS is by no means a hereditary disorder. Less than 3% of children of mothers with MS will come down with the disease. About the same concordance rate is seen in nonidentical twins (5%). Importantly, however, in genetically identical, monozygotic twin pairs concordance raises to more than 30%.[8] This discrepancy of concordance rates is telling. To put it simply, approximately 30% contribution of genes to disease risk is strong, but

BOX 9.1

IMMUNE PATHOGENESIS OF MULTIPLE SCLEROSIS

- Histology: T-cell and macrophage infiltrates in parenchyma and around small blood vessels
- T-cell infiltrates: T-cell receptor expansions, with silent mutations
- Abnormal oligoclonal immunoglobulin bands in cerebrospinal fluid
- In some patients: Autoantibodies against brain structures
- Genetics: Risk genes mostly related to (auto-)immune reactions
- Immunomodulatory therapies: Efficient in early, relapsing-remitting MS
- Animal models: Experimental autoimmune encephalomyelitis recapitulate some essential features of human disease

yet much smaller than the contribution of nongenetic, environmental factors. The role of the milieu has been further emphasized by studies of migrant populations. People migrating from high-risk areas to low-risk areas import disease susceptibility if immigrated as adolescents or adults.[9] Then, for unknown reasons, the incidence of MS is increasing, worldwide, a development that has been related to changes of life style, again to milieu factors.

This leads to a key question: If we accept the autoimmune pathogenesis and if we accept the role of the environment in triggering MS, how and where in the body is the disease triggered? While traditionally infectious agents were considered as disease triggers, we here present a new and unexpected suspect, our "alter ego," the commensal bacterial gut flora. We discuss fresh experimental observations indicating that indeed, the pathogenic autoimmune mechanisms can be triggered outside the central nervous system (CNS), in the gut.

9.3 SELF-TOLERANCE AND AUTOIMMUNITY

The immune system has been developed to protect the body from any potentially dangerous structure (Box 9.2). Classically, it identifies and kills infectious agents that have intruded into the organism, without creating undue collateral damage. To successfully accomplish this mission, immune cells must be able to distinguish the body's own tissues from foreign structures. The mechanisms that underlie self–nonself discrimination by immune cells are complex; they take place in the two primary immune organs, thymus and bone marrow, the sources of T and B cells, respectively. Within the thymus (the thymic medulla), self-reactive T cells are sorted out from the immune repertoire by local epithelial and dendritic cells specialized to produce or expose antigens typical of peripheral organs, such as liver, kidney, pancreatic islets, or brain cell. The thymic stroma cells act like an absorption column: freshly formed T cells pass through a network formed by the stroma cells. The T cells with self-reactive receptors bind to the ectopic autoantigen on the stroma cells and are arrested and absorbed. In contrast, the rest of the T cells—the ones with receptors for nonself-antigens—pass on and leave the thymus to settle in the peripheral immune organs.[10]

BOX 9.2

IMMUNOLOGICAL SELF-TOLERANCE AND AUTOIMMUNITY

- The immune system is composed of millions of different T- and B-cell families (clones), each characterized by one surface receptor for one antigen.
- The diversity of T cells is generated within the thymus, in two separate stages: initially, T-cell clones with broad affinity for local self-antigen are positively selected; then in a second step T cells with high affinity for self-antigens are sorted out, to avoid antiself, autoimmune reactions in the peripheral organism.
- The second negative sorting mechanism is leaky. A substantial number of self-reactive cells sneak through negative selection and arrive in the periphery. These include regulatory T cells (Tregs), as well as potential effector T cells.
- The self-reactive effector T cells persist in the peripheral immune system throughout life without doing any harm. They are kept in a resting state in the absence of activating signals and through active suppression by Tregs.
- Self-reactive effector T cells can turn autoaggressive following pathological activation.
- Activation can occur either through loss of suppression by Tregs, or through strong proinflammatory signals.
- Upon activation, the autoaggressive effector T cells move to their target organ (eg, the brain), where they recognize their specific autoantigen and start attacking the local tissue.

This process vindicates the Clonal Selection Theory, which predicted that self-tolerance is achieved by developmental elimination of self-reactive immune cells from the healthy repertoire. But few rules are without exceptions. In fact, elimination of self-reactive T cells in the thymus is leaky, allowing considerable numbers of tissue-specific T cells the exit to the peripheral immune system. Thus, the normal immune system (human and rodents equally) contains T cells with receptors for basically all tissues, among them liver, joints, pancreatic islets, and, especially intriguing, the CNS. These cells sit in immune organs throughout a healthy lifetime, doing no harm to the body. Their dormancy is warranted by several factors, passive and active. Autoimmune T cells are kept silent in the absence of stimulation; that is, during regular homeostasis. In addition, however, they are actively silenced by Tregs, which are formed either in the thymus (natural Tregs), or in the periphery after induction via specific signals (induced Tregs). The weakening of Treg subsets, as is the case in experimental models, after artificial ablation of preformed Tregs, or in human diseases with mutations that incapacitate Tregs, will lead to increased spontaneous autoimmune activity, with inflammation of various tissues and formation of autoantibodies.[11] Furthermore, self-reactive T cells can be aroused by pathological signals. When activated, the self-reactive immune cells unfold their autoaggressive potential: Dr. Jekyll mutates to Mr. Hyde. The signals driving self-reactive T cells toward autoaggression were traditionally sought in the context of microbial infections, such as molecular mimicry between microbial structures and self-antigens, microbial superantigens, and proinflammatory microenvironments activated by mechanisms of innate immunity.[12] As will be seen later, we introduce another driver of activation, namely the commensal microbiota.

What is the status of self-reactive T cells in the immune repertoire? Are they simply immunological time bombs left behind by an evolutionary error, or do they exert a positive biological function? It should be noted that one population of self-reactive T cells that sneaks through negative thymic selection are the natural Tregs.[13] In some among these, the regulatory potential may vanish under strong activating pressure, converting the Tregs to effector T cells, including myelin-specific effector cells mediating experimental autoimmune encephalomyelitis (EAE).[14] Another, more general concept proposed by Cohen speculated that T cells with receptors for tissue specific antigens could be involved in physiological responses like tissue repair and regeneration. The hypothesis postulated that the diversity of organotypic antigens is reflected by a set of complementary self-reactive T cells, forming an immunologic homunculus in the immune repertoire.[15]

9.4 EXPERIMENTAL MODEL OF MULTIPLE SCLEROSIS: TRIGGERING OF AUTOIMMUNE DISEASE IN THE GUT

The conversion of harmless self-reactive T cells to autoaggressive effector T cells has been first described in actively induced EAE, a rodent model representing essential inflammatory aspects of MS. EAE is actively induced by injecting susceptible inbred mice or rats with myelin proteins emulsified in a particularly strong immune adjuvant, Freund's complete adjuvant. This combination selects myelin autoaggressive T cells from the immune repertoire, and activates and expands them in lymph nodes draining the injection site. The activated myelin-autoimmune T cells have been isolated and propagated as homogeneous cell lines over many generations in culture. Myelin-autoimmune T cell lines have been isolated not only from immunized animals but also from perfectly naïve, untreated rodents,[16] and from the blood of humans with or without MS.[17,18] This established the presence of myelin self-reactive T cell clones in the healthy immune repertoire.

Importantly, myelin-specific T cell lines transfer EAE to perfectly healthy animals of the same inbred strain, provided they were activated before transfer by exposure to the specific myelin autoantigen. Nonstimulated T cells of the same lines would not trigger disease, even if transferred at monstrous numbers.[19] The activated autoimmune T cells reach the CNS after a tortuous journey through lymphatic organs and after transgression of the blood–brain barrier.[20] Within the white matter tissue, the T cells recruit macrophages to jointly attack and destroy myelin, myelin-forming glia cells, and myelinated axons.

Obviously, traditional EAE models are induced by highly artificial procedures, which definitely differ from the mechanisms leading to MS in humans, and thus are not useful for studying the initiation of the disease. Instead, such studies require EAE models that develop spontaneously, without any experimental manipulation. Several spontaneous EAE models have been developed over the past years.[21] They all are based on the insertion of the paired genes encoding a functional myelin-specific T cell receptor (TCR) pair into the germline. Expression of the transgenes results in an exaggerated proportion of myelin-autoimmune T cells in the otherwise regular immune cell repertoire, from lower than 1% to up to 70% of all $CD4^+$ T cells, respectively.

A particularly informative model of spontaneous EAE is the RR mouse.[22] RR mice are derived from the SJL/J strain. They express a transgenic myelin (MOG antigen) reactive TCR in about 70% of CD4 T cells. This proportion is important, because transgenic mice with lower rates (>20%) of anti-MOG T cells do not develop EAE. Also important for spontaneous EAE development is the genetic susceptibility of the mice. Relapsing EAE appears in close to 100% of transgenic SJL/J mice, while there is no spontaneous EAE in transgenic mice expressing the RR TCR on a major histocompatibility complex-compatible, but otherwise distinct, genetic background (B10.S). Also, in a transgenic strain, 2D2,[23] with a comparable anti-MOG T-cell proportion, but on the unrelated C57BL background, less than 10% develop EAE, and the disease differs from the one of RR mice, affecting exclusively the optic nerve and the spinal cord, and the course of the disease is chronic, with no remissions.

Within a time span of 4–8 months, practically all the RR mice come down with EAE spontaneously without any further treatment. Clinically, the disease proceeds in circumscribed bouts often with different neurological deficits, and the CNS lesions are characterized by areas of myelin and axon destruction and inflammatory T-cell and macrophage infiltrates. Commonly subsequent disease bouts differ by their neurological signs. There are bouts with classic paralytic defects, while in other episodes atypic ataxia predominates. Histologically, atactic mice have lesions in the cerebellum and midbrain, while in paralytic animals, the spinal cord is predominantly affected. With all this, the RR mouse model recapitulates much of the features typically seen in early human MS.

Spontaneous EAE develops in RR mice housed under specific pathogen-free (SPF) conditions, which should be distinguished from sterile, germfree conditions, and which roughly correspond to the hygiene standards of our industrialized world. Amazingly, however, in germfree environments, RR mice remain completely protected from neurological disease. EAE appears promptly in germfree RR mice recolonized with samples of SPF mouse-derived fecal microbiota. In addition, before onset of clinical EAE, the bacterially colonized RR mice start producing antimyelin autoantibodies, which bind to the surface of myelin sheaths and destroy them in cooperation with complement.

A detailed analysis of these events revealed a two-step pathogenesis of spontaneous RR EAE. It indicated that, in an initial step, bacteria from the gut flora activate transgenic myelin autoreactive T cells within the gut-associated lymphatic tissue (GALT). Then, in a second step, it appears that activated T cells enter the CNS tissue and there mobilize myelin material, which is exported via the lymphatic system to cervical brain-draining lymph nodes. There, the T cells in the presence of myelin autoantigens recruit myelin-specific B lymphocytes from the natural repertoire, and activate them to produce myelin-binding autoantibodies.[24]

Recolonization of germfree RR mice was done using samples of fresh feces, which represent organisms from all intestinal segments, summing up to a total of more than 1000 different species.[25] But not all bacteria share the same triggering potential. Notably, an experimental set of different bacterial strains, called Schaedler's flora, which had been developed to reconstitute germfree mice, did not elicit EAE. Unexpectedly, in gnotobiotics, segmented filamentous bacteria (SFBs) were also unable to efficiently trigger RR mouse EAE.[24]

9.5 POTENTIAL MECHANISMS OF IMMUNE MODULATION IN THE GUT

Traditionally our commensal gut flora was discounted as a bacterial mass filling our gut lumina without much function other than contributing to the digestion of foodstuff. This simplistic concept had to be revised radically over the past few years. The use of now available ultraefficient molecular technologies has disclosed an unexpected complexity of the gut flora, and revealed a true universe of functions of the gut flora that are essential for our body's healthy development and maintenance. We now know from nucleotide sequencing that healthy human feces regularly contain thousands of different bacterial species. These form a microbial society of an estimated 100 trillion of organisms compressed to a volume of a few liters.[26] This complexity of the gut microbial composition is reflected by a corresponding complexity of its functions (Box 9.3). Beside their established metabolic function (vide infra), the gut microbiota play a critical role in the development of GALT, which warrants confinement of the flora to the gut lumen, but in addition critically influences the function and maintenance of the entire immune system. Immune system and microbiota should be viewed as symbionts that have interacted throughout evolution and continue their interaction throughout healthy life.[27] The human microbiota is immensely dynamic. Microbial profiles differ not only between species, but even between individuals. Their composition can change substantially over time depending on early microbial exposure, age, gender, and dietary habits.[28,29]

Which are the bacteria that trigger EAE in RR mice, and where in the gut are the triggering events taking place? It is known from several models of autoimmune diseases that individual bacterial components exert specific effects on the immune system. Exemplary are the SFBs, relatives of clostridia, which are found in mice, not in humans. SFBs differ from the bulk

BOX 9.3

THE INTESTINAL MICROBIOTA

- The human gut flora (microbiome) contains a total of about 100×10^{12} organisms, thus among the densest bacterial communities known.
- Intestinal segments differ by their bacterial profile and density, with a steep increase from the small to the large intestine. The bacterial profiles differ between the gut segments.
- Microbiota functions, local: Digestion of foodstuff (eg, fibers) to small molecules (eg, short-chain fatty acids, ATP); induction of gut-associated lymphatic tissues.
- Microbiota functions, systemic: Maturation of immune system, hormonal homeostasis, central nervous system function.
- Microbiota changes in disease, local: Infection (eg, *Clostridium difficile*), allergy (eg, celiac disease), autoimmunity (eg, Crohn's disease, ulcerative colitis), tumors (eg, colon carcinoma).
- Microbiota changes in disease, systemic: Metabolic syndrome and obesity, autoimmunity (eg, type 1 diabetes? MS?).
- Therapies: Antibiotics, probiotics, phages, fecal transplants.

of gut bacteria by their preferred location beneath the mucus layer, directly on the epithelial lining, preferentially covering distal ileal Peyer's patches.[30] SFBs drive the GALT milieus toward the production of IL-17 and related immune mediators.[31,32] Implicitly, SFBs activated Th17-like autoimmune T cells to trigger disease in an experimental arthritis model.[33] The mechanism by which SFBs drive these local changes are not fully explored; they may involve the presentation of antigenic components to immune cells to create an IL-17-like milieu in the ileum, but it is not clear whether these processes depend on the formation of organized lymphatic tissues.[34,35]

Other commensal gut bacteria act the opposite way, driving antiinflammatory responses. Most studies place these interactions mostly into the large bowel (colon), not into the small intestine (ileum). *Bacteroides fragilis* releases a capsular structure, polysaccharide A, which activates Tregs via the innate pattern receptor TLR-2 to produce the antiinflammatory cytokine IL-10.[36] The polysaccharide A-activated Tregs suppress inflammatory responses locally in the gut, as well as by remote action, in actively induced EAE,[37] presumably via the innate TLR-2 pathway.[38] Another class of antiinflammatory commensals is formed by Clostridia, which acts via degrading digestible dietary fibers to short-chain fatty acids. These act directly on Tregs to activate their antiinflammatory potential.[39,40]

9.6 DIET AND GUT MICROBIOTA

The microbial composition of the gut flora is highly dynamic and flexible. It responds to numerous external influences, such as physical activity, microbial infections, and their treatments. An especially effective and permanent modifier is our daily diet. Dietary components can influence (auto-)immune responses in manifold ways: directly via the GALT immune reactivity, by acting on immune cells and their surrounding stromal milieu; or indirectly, by modifying the microbiota; or via both routes.

Highly publicized were studies showing that people with high fat consumption have microbial profiles that substantially differ from the microbiota in consumers of fiber-rich diets.[41] As mentioned, short-chain fatty acids of bacterial production profoundly impinge on the immune system and its regulatory equilibrium,[42] but they are not the only dietary immune signals. As examples, carotenes and their metabolite, vitamin A, act on epithelial cells and dendritic cells to support activation and expansion of Tregs (transforming growth factor-β). Then, there are ligands to the aryl-hydrocarbon receptor, AhR, a transcription factor strongly expressed in Th17, but also in Treg cells. The AhR binds, apart from the environmental toxin dioxin, vegetable components (*glucobrassicin*) found in cabbage species (including broccoli and mustard). AhR signaling is complex; on the one hand, it can activate pathogenic Th17effector cells, while in a different context, it may support the production of antiinflammatory mediators (eg, IL-10). In EAE models, deletion of AhR was beneficial, while treatment with AhR agonists exacerbated clinical disease.[43] Unexpectedly, dietary salt emerged as dietary factor with profound effect on immune reactivity. Several reports found that high-salt diets led to an impressive exacerbation of actively induced EAE,[44,45] involving an expansion of a Th17-like subset of $CD4^+$ T cells. One mechanism is via a cytoplasmic kinase, the Serum Glucocorticoid Kinase-1, SGK1, which activates proinflammatory cytokine cascades, although an additional effect through host microbiota has not been excluded (Box 9.4).

BOX 9.4

DIET EFFECTS ON INTESTINAL IMMUNE REACTIVITY

- Direct effects:
  Agonists of aryl-hydrocarbon (dioxin) receptor: Th1/Th17/Treg balance
  Vitamin D
  Fatty acids
  Salt level
- Effects via microbiota

9.7 MICROBES, MICROBIOTA, AND MULTIPLE SCLEROSIS

The idea that microbes could actually cause MS is not new. Indeed, over the past decades, numerous and diverse bacteria and viruses have been suspected of triggering the disease. The list of culprits included mostly banal viruses (including measles, mumps, varicella) and bacteria (chlamydia), but none of these stood the test of time. In contrast, there is epidemiological evidence that Epstein-Barr virus infection may be a risk factor, but not the actual trigger of MS.[46] The recent surfacing of the intestinal flora as a trigger of brain autoimmunity has appeared as counterintuitive for several reasons: first, there is the distance between gut and CNS. Why should a brain disease be ignited so far away? Second, usually the intestinal flora is not only harmless to the organism but has an essential function as a digestive partner. How could it turn so pernicious?

We should stress that presently, the evidence for intestinal triggering of brain autoimmunity rests on experimental models of MS. Is there any evidence favoring this pathogenesis in human disease? There is some evidence, but it is indirect.

Treatment of MS patients with antibiotics should provide direct proof of principle. Continued therapy with antibiotics definitely impacts the gut microbiota, but would it alter the course and/or severity of MS? Unfortunately, this approach is confounded by numerous factors inherent in drug actions and in the populations tested. Thus, unsurprisingly, epidemiological studies of antibiotic treatment in MS patients led to conflicting results. One study of MS patients treated with penicillin or other antibiotics found a decreased susceptibility in the penicillin-treated collectives, but a subsequent, larger study could not repeat this, suggesting underlying diseases as confounders.[47] Then, treatment attempts with macrolide antibiotics targeting chlamydia were negative.[48] A more actual case of importance is therapy with minocycline, a tetracycline, which, besides its antimicrobial effects, is known to strongly protect brain cells from degenerative assault. Minocycline indeed showed promising effects, but the mechanism of action–antimicrobial versus neuroprotective remained to be defined.[49] In none of these trials, antibiotic effects on the gut flora were considered as potential mechanisms of the therapeutic effect.

There is epidemiological evidence, with a worldwide trend of an increased incidence of MS, both in areas of high as well as low incidence. This cannot be ascribed to an increased awareness of disease, but seems to be real. The increase is particularly impressive in Japan. Traditionally, Japan has been considered an area of low occurrence, and the few cases diagnosed showed a particular opticospinal syndrome, rather than the variegated pattern typical in Western MS. However, over the past 20 years, there has been an astounding increase of MS, especially due to the appearance of Western-type disease.[50] This change has been attributed to a change of life style, most importantly a change of dietary habits. If confirmed, this will be an important test case to verify the relations between diet, microbiota, and the autoimmune pathogenesis of MS.

There is further support from other noninfectious, organ-specific diseases that share features with MS and are suspected to be of autoimmune nature. These include, most prominently, type 1 diabetes and rheumatoid arthritis. Their inflammatory pathogenesis involves T and B lymphocytes, a strong part of risk genes control immune functions (some of them shared with MS genes), and the disease responds well to immune modulatory therapies. In children with type 1 diabetes, disease was associated with particular microbial profiles.[51] In rheumatoid arthritis, gut bacteria have long been suspected to facilitate or trigger the disease.[52] A recent trial using sequencing technologies to compare the fecal microbiota of recent onset patients with healthy controls identified a disease-associated bacterium. The pathogenic potential of this organism was supported by transfer studies, where introduction of the bacterium into antibiotic pretreated mice enhanced experimental bowel inflammation.[53]

9.8 THERAPEUTIC PERSPECTIVES

It sounds so easy: A disease caused by a known microbial agent is best treated by neutralizing this microbe. Unfortunately, in the case of MS the identity of the suspected disease-triggering organism is unknown. Even if we accept a microbial disease initiation, we do not even know whether the pathogenic autoimmune response is set off by one individual microbe, or, more probably, by a disrupted homeostasis in the microbiota. Furthermore, we deal with bona fide beneficial commensal bacteria, instead of clearly classified exogenous infectious agents.

A number of metagenomic studies are underway to determine differences between the microbial profiles of healthy people and people affected by MS. Depending on the results, therapies will try either to eliminate specific pathogenic microbes, or, more realistic, restore lost microbial equilibria. In the case of a narrow set of triggering agents, specific antibiotic combinations could be designed, a strategy, which, applied over extended periods of time, hardly would spare innocent bacteria and thus could create considerable iatrogenic dysbiosis. More specific treatments could resort to bacterial phages, which have been ignored largely as therapeutic agents and wait to be resuscitated.[54]

Restoration of a lost microbial balance will require broader measures. A radical approach goes by the euphemistic term of fecal transplantation. It is based on the infusion of stool samples from a healthy person into a recipient with ailing bowel. This measure often leads to a long-term replacement of the microbiota, with the new microbiota outcompeting the pathogenic consortia. It has been applied to other intestinal disorders, such as ulcerative colitis.[56] The most spectacular success was noted in therapy-resistant infections with *C. difficile*, which profoundly affect the patient's gut flora (Box 9.5).[55]

Much milder and more palatable are dietary approaches, either relying on special food compositions or on probiotic organisms. Treatment of MS with particular dietary regimens

BOX 9.5

THERAPEUTIC MANIPULATION OF MICROBIOTA

- Diets: Fiber, (unsaturated) fatty acids, salt, vitamins, etc.
- Antimicrobial treatments: Antibiotics, xenobiotics, phages
- Probiotics
- Fecal transplants

has been practiced over decades, in particular at times when other medications were unavailable. The number of recommended dietary formulas is legion. Viewed with our present understanding, benefit of at least some of these MS diets should not be ruled out, but, unfortunately, these successes were rarely validated by rigorous trials.[58] The importance of diet and microbiota for MS and other chronic autoimmune diseases has not passed unrecognized by industry.[57] There is definitely a need for fresh and targeted explorations.

Acknowledgments

Hartmut Wekerle holds a senior professorship of the HERTIE Foundation. His research is additionally supported by funds from the Deutsche Forschungsgemeinschaft (Koselleck Award; SFB/Transregio 218, and SyNergy), Max-Planck-Society, and BMBF (KKNMS). Reinhard Hohlfeld is funded by the Deutsche Forschungsgemeinschaft (SyNergy, Transregio 128) and BMBF (KKNMS).

References

1. Beer S, Kesselring J. High prevalence of multiple sclerosis in Switzerland. *Neuroepidemiol*. 1994;13:14–18.
2. Confavreux C, Compston A. The natural history of multiple sclerosis. In: Compston A, Confavreux C, Lassmann H, Noseworthy J, Smith K, Wekerle H, eds. *McAlpine's Multiple Sclerosis*. 4th ed. Churchill Livingstone Elsevier; 2006:183–272.
3. Lassmann H. Multiple sclerosis pathology: evolution of pathogenetic concepts. *Brain Pathol*. 2005;15(3):217–222.
4. Junker A, Ivanidze J, Malotka J, et al. Multiple sclerosis: T-cell receptor expression in distinct brain regions. *Brain*. 2007;130(11):2789–2799.
5. Irani SR, Gelfand JM, Al-Diwani A, Vincent A. Cell-surface CNS autoantibodies: clinical relevance and emerging paradigms. *Ann Neurol*. 2014;76(2):168–184.
6. Beecham AH, Patsopoulos NA, Xifara DK, et al. Analysis of immune-related loci identifies 48 new susceptibility variants for multiple sclerosis. *Nat Genet*. 2013;45(11):1353–1360.
7. Steinman L. Immunology of relapse and remission in multiple sclerosis. *Annu Rev Immunol*. 2014;32:257–281.
8. Compston A, Coles A. Multiple sclerosis. *Lancet*. 2002;359(9313):1221–1231.
9. Dean G, Kurtzke JF. On the risk of multiple sclerosis related to age at immigration to South Africa. *Br Med J*. 1971;3:725–729.
10. Klein L, Kyewski B, Allen PM, Hogquist KA. Positive and negative selection of the T cell repertoire: what thymocytes see (and don't see). *Nat Rev Immunol*. 2014;14(6):377–391.
11. Sakaguchi S, Yamaguchi T, Nomura T, Ono M. Regulatory T cells and immune tolerance. *Cell*. 2008;133:775–787.
12. Wucherpfennig KW. Mechanisms for the induction of autoimmunity by infectious agents. *J Clin Invest*. 2001;108(8):1097–1104.
13. Wirnsberger G, Hinterberger M, Klein L. Regulatory T-cell differentiation versus clonal deletion of autoreactive thymocytes. *Immunol Cell Biol*. 2011;89(1):45–53.

14. Bailey-Bucktrout SL, Martinez-Llordella M, Zhou XY, et al. Self-antigen-driven activation induces instability of regulatory T cells during an inflammatory autoimmune response. *Immunity*. 2013;39(5):949–962.
15. Cohen IR, Young DB. Autoimmunity, microbial immunity and the immunological homunculus. *Immunol Today*. 1991;12(4):105–110.
16. Schluesener HJ, Wekerle H. Autoaggressive T lymphocyte lines recognizing the encephalitogenic region of myelin basic protein: in vitro selection from unprimed rat T lymphocyte populations. *J Immunol*. 1985;135:3128–3133.
17. Martin R, Jaraquemada D, Flerlage M, et al. Fine specificity and HLA restriction of myelin basic protein- specific cytotoxic T cell lines from multiple sclerosis patients and healthy individuals. *J Immunol*. 1990;145:540–548.
18. Pette M, Fujita K, Kitze B, et al. Myelin basic protein-specific T lymphocyte lines from MS patients and healthy individuals. *Neurol*. 1990;40:1770–1776.
19. Wekerle H. Myelin specific, autoaggressive T cell clones in the normal immune repertoire: their nature and their regulation. *Int Rev Immunol*. 1992;9:231–241.
20. Bartholomäus I, Kawakami N, Odoardi F, et al. Effector T cell interactions with meningeal vascular structures in nascent autoimmune CNS lesions. *Nature*. 2009;462:94–98.
21. Krishnamoorthy G, Wekerle H. EAE: an immunologist's magic eye. *Eur J Immunol*. 2009;39(8):2031–2035.
22. Pöllinger B, Krishnamoorthy G, Berer K, et al. Spontaneous relapsing-remitting EAE in the SJL/J mouse: MOG-reactive transgenic T cells recruit endogenous MOG-specific B cells. *J Exp Med*. 2009;206(6):1303–1316.
23. Bettelli E, Pagany M, Weiner HL, Linington C, Sobel RA, Kuchroo VK. Myelin oligodendrocyte glycoprotein-specific T cell receptor transgenic mice develop spontaneous autoimmune optic neuritis. *J Exp Med*. 2003;197(9):1073–1081.
24. Berer K, Mues M, Koutroulos M, et al. Commensal microbiota and myelin autoantigen cooperate to trigger autoimmune demyelination. *Nature*. 2011;479:538–541.
25. Carroll IM, Threadgill DW, Threadgill DS. The gastrointestinal microbiome: a malleable, third genome of mammals. *Mamm Genome*. 2009;20(7):395–403.
26. The Human Microbiome Project Consortium. Structure, function and diversity of the healthy human microbiome. *Nature*. 2012;486:207–214.
27. Maynard CL, Elson CO, Hatton RD, Weaver CT. Reciprocal interactions of the intestinal microbiota and immune system. *Nature*. 2012;489(7415):231–241.
28. The Human Microbiome Project Consortium. A framework for human microbiome research. *Nature*. 2012;486:215–221.
29. Yatsunenko T, Rey FE, Manary MJ, et al. Human gut microbiome viewed across age and geography. *Nature*. 2012;486:222–227.
30. Klaasen HL, Koopman JP, Poelma FG, Beynen AC. Intestinal, segmented, filamentous bacteria. *Fems Microbiol Rev*. 1992;88(3–4):165–179.
31. Gaboriau-Routhiau V, Rakotobe S, Lévuyer E, et al. The key role of segmented filamentous bacteria in the coordinated maturation of gut helper T cell responses. *Immunity*. 2009;31(4):677–689.
32. Ivanov II, Atarashi K, Manel N, et al. Induction of intestinal Th17 cells by segmented filamentous bacteria. *Cell*. 2009;139(3):485–498.
33. Wu H-J, Ivanov II, Darce D, et al. Gut-residing filamentous bacteria drive autoimmune arthritis via T helper 17 cells. *Immunity*. 2010;32(6):815–823.
34. Goto Y, Panea C, Nakato G, et al. Segmented filamentous bacteria antigens presented by intestinal dendritic cells drive mucosal Th17 cell differentiation. *Immunity*. 2014;40(4):594–607.
35. Lécuyer E, Rakotobe S, Lengliné-Garnier H, et al. Segmented filamentous bacterium uses secondary and tertiary lymphoid tissues to induce gut IgA and specific T helper 17 cell responses. *Immunity*. 2014;40(4):608–620.
36. Round JL, Lee SM, Li J, et al. The Toll-like receptor 2 pathway establishes colonization by a commensal of the human microbiota. *Science*. 2011;332:974–977.
37. Ochoa-Repáraz J, Mielcarz DW, Ditrio LE, et al. Central nervous system demyelinating disease protection by the human commensal *Bacteroides fragilis* depends on polysaccharide A expression. *J Immunol*. 2010;185(7):4101–4108.
38. Wang Y, Telesford KM, Ochoa-Repáraz J, et al. An intestinal commensal symbiosis factor controls neuroinflammation via TLR2-mediated CD39 signalling. *Nat Commun*. 2014;5:4432.
39. Atarashi K, Tanoue T, Oshima K, et al. T_{reg} induction by a rationally selected mixture of Clostridia strains from the human microbiota. *Nature*. 2013;500:232–236.
40. Furusawa Y, Obata Y, Fukuda S, et al. Commensal microbe-derived butyrate induces the differentiation of colonic regulatory T cells. *Nature*. 2013;504:446–470.

41. Wu GD, Chen J, Hoffmann C, et al. Linking long-term dietary patterns with gut microbial enterotypes. *Science*. 2011;334:105–108.
42. Trompette A, Gollwitzer ES, Yadava K, et al. Gut microbiota metabolism of dietary fiber influences allergic airway disease and hematopoiesis. *Nat Med*. 2014;20(2):159–166.
43. Stockinger B, Di Meglio P, Gialitakias M, Duarte JH. The aryl hydrocarbon receptor: multitasking in the immune system. *Annu Rev Immunol*. 2014;32:403–432.
44. Kleinewietfeld M, Manzel A, Titze J, et al. Sodium chloride drives autoimmune disease by the induction of pathogenic T_H17 cells. *Nature*. 2013;496:518–522.
45. Wu C, Yosef N, Thalhamer T, et al. Induction of pathogenic T_H17 cells by inducible salt-sensing kinase SGK1. *Nature*. 2013;496:513–517.
46. Ascherio A, Munger KL. Environmental risk factors for multiple sclerosis. Part I: the role of infection. *Ann Neurol*. 2007;61(4):288–299.
47. Norgaard M, Nielsen RB, Jacobsen JB, et al. Use of penicillin and other antibiotics and risk of multiple sclerosis: a population-based case-control study. *Am J Epidemiol*. 2011;174(8):945–948.
48. Woessner R, Grauer MT, Frese A, et al. Long-term antibiotic treatment with roxithromycin in patients with multiple sclerosis. *Infection*. 2006;34(6):342–344.
49. Zabad RK, Metz LM, Todoruk TR, et al. The clinical response to minocycline in multiple sclerosis is accompanied by beneficial immune changes: a pilot study. *Mult Scler*. 2007;13(4):517–526.
50. Yamamura T, Miyake S. Diet, gut flora, and multiple sclerosis: current research and future perspectives. In: Yamamura T, Gran B, eds. *Multiple Sclerosis Immunology*. NewYork, Heidelberg, Dordrecht, London: Springer; 2014:115–125.
51. de Goffau MC, Luopajarvi K, Knip M, et al. Fecal microbiota composition differs between children with β-cell autoimmunity and those without. *Diabetes*. 2013;62(4):1238–1244.
52. Toivanen P. Normal intestinal microbiota in the aetiopathogenesis of rheumatoid arthritis. *Ann Rheum Dis*. 2003;62(9):807–811.
53. Scher JU, Sczesnak A, Longman RS, et al. Expansion of intestinal *Prevotella copri* correlates with enhanced susceptibility to arthritis. *eLife*. 2013;2:e1202.
54. Reyes A, Semenkovich NP, Whiteson K, Rohwer F, Gordon JI. Going viral: next-generation sequencing applied to phage populations in the human gut. *Nat Rev Microbiol*. 2012;10(9):607–617.
55. Van Nood E, Vrieze A, Niewdorp M, et al. Duodenal infusion of donor feces for recurrent *Clostridium difficile*. *N Engl J Med*. 2013;368:407–415.
56. Borody TJ, Khoruts A. Fecal microbiota transplantation and emerging applications. *Nat Rev Gastroenterol Hepatol*. 2012;9(2):88–96.
57. Olle B. Medicines from microbiota. *Nat Biotech*. 2013;31(4):309–315.
58. Schwarz S, Leweling H, Meinck HM. Alternative and complementary therapies in multiple sclerosis. *Fortschritte Neurol Psychiatr*. 2005;73(8):451–462.

CHAPTER

10

Neuroendocrine Checkpoints of Innate Immune Responses in Multiple Sclerosis: Reciprocal Interactions Between Body and Brain

N. Deckx, Z.N. Berneman, N. Cools
University of Antwerp, Antwerp, Belgium

OUTLINE

10.1 THE INNATE IMMUNE SYSTEM

In order to confer our body with a balanced state of health, the chief function of a functional immune system is to discriminate between self and nonself. In doing so, infectious agents (bacteria, viruses, fungi, parasites) are effectively eliminated and malignancies controlled, while at the same time tolerance is maintained to harmless antigens.

http://dx.doi.org/10.1016/B978-0-12-801914-6.00010-6

The innate immune system is our body's first line of defense against invaders. The wide distribution of a diverse network of innate immune cells emphasizes their critical role to continuously patrol and scan the body for invading pathogens. For this, innate immune cells express a variety of pattern-recognition receptors recognizing so-called danger signals, including pathogen-associated molecular patterns (PAMPs) and damage-associated molecular patterns (DAMPs).[1–3] Upon antigen encounter, professional antigen-presenting cells (APCs) take up protein antigens in the peripheral tissues and process them into small peptides. Following migration toward the draining lymph nodes, APCs present these antigens to naïve or memory T cells. In steady-state conditions, this antigen presentation initiates T cell anergy, T cell apoptosis, and regulatory T cell (Treg) induction,[4,5] resulting in peripheral tolerance. However, when the antigen is recognized in combination with PAMP or DAMP, an effective adaptive immune response will be generated. Subsequently, APC-primed lymphocytes are directed to the site of inflammation where they will help carrying out the appropriate host attack on the pathogen in conjunction with resident or infiltrating innate immune cells, such as monocytes, macrophages, dendritic cells (DCs), and natural-killer cells. Innate immune cells, in particular DCs, were repeatedly reported to potentially play a key role in the immunopathogenesis of multiple sclerosis (MS).[6] This was demonstrated, among other methods, by the abundant presence of DCs in the inflamed central nervous system (CNS) lesions and cerebrospinal fluid of patients with MS.[6,7] In addition, circulating myeloid or conventional DCs of patients with relapsing-remitting MS (RRMS) as well as with secondary-progressive MS (SPMS) display a proinflammatory phenotype, as evidenced by increased expression of activation and costimulatory markers, in comparison with healthy controls.[8,9] Furthermore, a more pronounced secretion of proinflammatory cytokines, such as IL-12p70, tumor necrosis factor-α (TNF)-α, and IL-23p19, was found in both RRMS and SPMS patients[8,9] as compared to healthy controls.

While monocytes, macrophages, and DCs are the resident APCs in the periphery, microglia are the resident APCs in the CNS. While CNS-resident innate immune cells are vigilant guards directly dealing with invading pathogens and tissue damage,[10] microglia can also recruit immune cells from the periphery, such as T cells, in many circumstances. This is, for example, critical for protective host defense against infections and for repair after stroke or physical trauma.[11] Although microglial activity may promote CNS repair, as demonstrated by clearance of myelin debris and the production of growth factors in animal models,[12,13] if uncontrolled it leads to an exacerbated release of proinflammatory and toxic factors,[14] causing profound damage to the CNS. Hence, microglia can also be involved in the immunopathogenesis of neurological disorders, such as MS.[15,16] Indeed, in the chronic phase of experimental autoimmune encephalomyelitis (EAE), persistent activation of microglia has been found. Moreover, presence of activated microglia was correlated with loss of neuronal synapses.[17] Likewise, activated microglia have been reported in MS patients and are associated with white matter inflammation.[18] Accordingly, activated microglia contain myelin and axonal remnants, display high expression of major histocompatibility complex (MHC) and costimulatory molecules, and secrete large amounts of inflammatory and neurotoxic mediators in MS and EAE lesions.[19–21] Moreover, inhibition of microglial activation suppresses the development and maintenance of inflammatory lesions in the CNS[22–25] and delays EAE onset.[26] In contrast, this also resulted in increased disease severity and delayed recovery from neurological dysfunction,[26] in support of a neuroprotective phenotype of microglia in EAE. Altogether, microglia may have a dual role in the pathogenesis of MS and EAE, contributing to both neurodestruction and neuroprotection.[3]

10.2 NEUROENDOCRINE REGULATION OF THE INNATE IMMUNE SYSTEM

The neuroendocrine system is formed by cells that produce and release hormones into the bloodstream in response to chemical signals from other cells or messages from the nervous system. It can communicate bidirectionally with the immune system via shared receptors and shared messenger molecules, variously called hormones, neurotransmitters, or cytokines.[27] Several studies have addressed a possible role of the neuroendocrine system in susceptibility and severity of autoimmune diseases, such as MS. The hypothalamic–pituitary–adrenal (HPA) axis[28] constitutes a major part of the neuroendocrine system, controlling reactions to stress and regulating many body processes. Following various physical and psychological stimuli, the HPA axis is activated, which results in the release of glucocorticoids by the adrenal cortex. Glucocorticoids are powerful endogenous inhibitors of innate immune responses throughout the organism, including the CNS.[29] Because of their lipophylic nature, glucocorticoids can readily cross the blood–brain barrier and can avoid an exaggerated response by microglia during infection and injuries, thereby controlling inflammatory responses in the CNS. Also, the hypothalamic–pituitary–gonadal (HPG) axis plays an important role in the development and regulation of a number of the body's systems, including the immune system. Stimulation of the HPG axis results in the release of estrogens and progesterone by the reproductive organs. Like corticosteroids, estrogens are steroid hormones able to cross the blood–brain barrier and subsequently modulate innate immune responses in the CNS. The HPG axis is also shown to play a role in the susceptibility and severity of MS,[30] for instance, as demonstrated by gender differences in the susceptibility to autoimmunity. Indeed, whereas almost 8% of the world population develops an autoimmune disease, approximately 78% of them are women.[31]

Furthermore, neurotransmitters, such as catecholamines and acetylcholine (ACh), have also been shown to be involved in the pathogenesis of MS.[32–35] The catecholamines, epinephrine and norepinephrine, are released from sympathetic nerve terminals upon stimulation. Stress situations, such as a physical threat, excitement, a loud noise, or a bright light, are the major physiological triggers of the release of catecholamines. Through the release of catecholamines in lymphoid organs, the sympathetic nervous system has been demonstrated to exert a direct role in immunomodulation. On the other hand, ACh is the primary neurotransmitter of the parasympathetic nervous system. The parasympathetic nervous system modulates immune responses through the efferent and afferent fibers of the vagus nerve. Neurotransmitters can control inflammation and neurotoxicity in the CNS directly via neurotransmitter-expressing microglia.[14,36]

10.2.1 Regulation of the Peripheral Innate Immune System

Several anti-inflammatory effects have been attributed to the synthetic glucocorticoid dexamethasone, which binds the glucocorticoid receptor with a higher affinity than cortisol, the primary human glucocorticoid.[37] Indeed, human monocyte-derived DCs (moDCs) stimulated with CD40 ligand or lipopolysaccharide (LPS) in the presence of dexamethasone show reduced secretion of proinflammatory cytokines, such as IL-6, IL-12p70, and TNF-α, while secretion of the anti-inflammatory cytokine IL-10 is increased, as compared to untreated moDCs.[38–40] Moreover, dexamethasone-treated moDCs stimulated with CD40 ligand or LPS display lower expression levels of costimulatory (CD80, CD86), adhesion (CD54, CD58), and MHC class I and II molecules and failed to express the maturation marker, CD83.[38–41] In doing so, glucocorticoids can, besides direct effects on the adaptive immune system,[42] also indirectly affect T cell-mediated immunity via their effects on the innate immune system.

Also, 17β-estradiol, the most potent estrogen in humans, was reported to have anti-inflammatory effects on innate immune cells, including monocytes and macrophages.[43] In brief, 17β-estradiol controls the production of proinflammatory and/or immunosuppressive cytokines and growth factors. In particular, 17β-estradiol inhibits the expression of TNF-α, IL-1, IL-6, macrophage colony-stimulating factor, and granulocyte-macrophage colony-stimulating factor.[43] In contrast, immature moDCs cultured in the presence of 17β-estradiol show increased secretion of IL-6, IL-10, and of the chemokines IL-8 and monocyte-chemotactic protein, as compared to untreated moDCs.[44,45] Moreover, in the presence of 17β-estradiol, high IL-10 secretion is maintained in mature moDCs.[45] However, no changes with regard to the expression of CD40, CD83, CD86, and HLA-DR, an MHC class II molecule, were reported in immature and mature moDCs cultured in the presence of 17β-estradiol.[44–46] Nevertheless, murine bone marrow–derived DCs (BMDCs) treated with 17β-estradiol are able to inhibit activation of myelin-specific T cells and enhance Treg-induced suppression in EAE.[47] Overall, these findings indicate an anti-inflammatory role for estrogen on the function of innate immune cells, resulting in an immunosuppressive effect.

Treatment of murine BMDCs with epinephrine resulted in reduced secretion of proinflammatory cytokines, such as IL-6, IL-12p70, and IL-23, and increased IL-10 secretion following LPS stimulation, whereas the expression of CD80 and CD86 was upregulated.[48–51] Similar results were found following norepinephrine treatment of murine BMDCs and human cord blood-derived DCs.[48–52] The reported shift in the balance between the secretion of pro- and anti-inflammatory cytokines supports an immune-suppressive effect of catecholamines.

Binding of ACh to nicotinic receptors on macrophages and lymphocytes inhibits the production of proinflammatory cytokines.[53] Also in murine BMDCs, treatment with ACh resulted in reduced secretion of IL-12p70 following LPS stimulation, while secretion of IL-10 increased.[50] Similarly, treatment of human moDCs with nicotine, a psychoactive compound of tobacco products that binds to the ACh receptor,[54] resulted in decreased secretion of IL-10, IL-12p70, and TNF-α, while the expression of HLA-DR, CD40, CD80, CD83, and CD86 was not significantly altered.[57,58] Interestingly, following LPS stimulation, nicotine-treated human moDCs showed elevated levels of the coinhibitory molecules, programmed death-ligand 1 (PD-L1), and immunoglobulin-like transcript 4 (ILT-4).[57] In contrast, nicotine induced increased secretion of IL-12p70 by human moDC[55] and murine BMDC[56] in steady-state conditions. Moreover, in vitro generated DCs demonstrated upregulated expression of costimulatory (CD40, CD80, and CD86), adhesion (lymphocyte function-associated antigen 1, CD54, and CD11b), and MHC class II molecules, of CD83 and of the chemokine receptor CCR7 following nicotine exposure.[55,56] In conclusion, these observations may explain the opposing effects of cigarette smoke on the immune response, namely suppression of immunity against infectious agents and loss of tolerance to self.

10.2.2 Regulation of the Central Innate Immune System

Given their blood–brain barrier-crossing capacity most neuroendocrine messengers also affect the innate immune system in the CNS. As in the periphery, glucocorticoids exert an immune-suppressive effect on the innate immune system, which can be evidenced by reduced proinflammatory cytokine secretion, reduced expression of costimulatory and MHC class II molecules, and increased expression of coinhibitory molecules by in vitro generated microglia following treatment with glucocorticoids. Indeed, cortisol was shown to suppress secretion of TNF-α by in vitro generated murine microglia stimulated with LPS.[59] Similarly, dexamethasone and corticosterone, the main glucocorticoid in rodents, inhibit the secretion of proinflammatory cytokines, including IL-6, IL-12p70, and TNF-α, as well as the expression of MHC class II, CD40, and CD80 molecules by in vitro generated murine microglia stimulated with IFN-γ in a dose-dependent manner.[60,61] Furthermore, corticosterone treatment resulted in upregulation of the expression of the negative regulators, PD-L1 and PD-L2, by in vitro generated murine microglia.[61]

Also in the CNS, estrogens most likely have an immune-suppressive effect on the innate immune system. Indeed, both estriol and 17β-estradiol were shown to stimulate production of the anti-inflammatory cytokine IL-10 by murine microglia, while inhibition of the production of proinflammatory cytokines, such as IL-1α, IL-1β, IFN-γ, and TNF-α, and inhibition of the expression of costimulatory (CD40 and CD86) and MHC class I and II molecules were demonstrated, in vitro as well as in vivo.[62–65]

While in general neurotransmitters, including norepinephrine, ACh, and nicotine, exert an immune-suppressive effect in the CNS, as evidenced by reduced secretion of proinflammatory cytokines, such as IL-1β and TNF-α, by murine microglia following treatment with these neurotransmitters,[60,66–71] their effect is dependent on allosteric modulation of the respective receptors. Indeed, while agonists of the nicotinic ACh receptor α7-nAChR do not affect the release of proinflammatory cytokines by in vitro generated murine microglia, α7-nAChR antagonists reduce proinflammatory cytokine release by in vitro generated murine microglia. This indicates that α7-nAChR antagonism conveys anti-inflammatory properties on microglia thereby reducing neuroinflammation,[72] whereas classical activation (i.e., ion-flux through the α7-nAChR) does not reduce proinflammatory cytokine release from activated microglia.

10.3 THERAPEUTIC TARGETING OF THE INNATE IMMUNE SYSTEM OF MULTIPLE SCLEROSIS PATIENTS VIA NEUROENDOCRINE MODULATION

Since the 1950s, glucocorticoids are widely used for the suppression of inflammation in chronic inflammatory diseases like MS. Despite the introduction of disease-modifying therapies, glucocorticoid therapy remains the first-line treatment upon relapse in order to induce remission in MS patients sooner.[73] Methylprednisolone is among the most commonly used glucocorticoids in MS and reduces the number of gadolinium-enhancing lesions during MS exacerbations.[74] This effect is mediated by a general inhibition of the immune system.[75] Interestingly, also specific effects on the innate immune system have been reported. Indeed, following 5 days of intravenous administration of methylprednisolone for the treatment of a relapse, the number of circulating DCs was decreased in MS patients.[76,77] In accordance, patients with severe Graves' ophthalmopathy, rapidly progressive interstitial pneumonia, or with microscopic polyangiitis showed decreased numbers of circulating DCs following 3 days of intravenous administration of methylprednisolone.[78] Altogether, this reflects the suppressive capacity of glucocorticoids on innate immunity.

Several studies in EAE have shown the inhibitory effects of estrogens on disease pathogenesis.[79,80] Indeed, 17β-estradiol or estriol treatment before induction of EAE delays onset of disease and reduces disease activity. Moreover, the reduction in disease severity was accompanied by a coincident reduction in the number and size of inflammatory foci in the CNS of estrogen-treated mice.[79] Protective mechanisms of estrogen treatment in EAE involve anti-inflammatory processes, including enhanced suppressive capacity of DCs on pathogenic T cells.[47] In accordance, it was shown that splenic DCs isolated from 17β-estradiol-treated EAE mice produce lower levels of TNF-α and IFN-γ upon LPS exposure.[81] Furthermore, splenic DCs isolated from estriol-treated EAE mice display higher expression levels of costimulatory molecules (CD80 and CD86) and coinhibitory molecules (PD-L1 and PD-L2) as compared with untreated EAE mice. These DCs displayed decreased IL-6, IL-12p70, and IL-23 mRNA expression, while the mRNA expression of anti-inflammatory cytokines such as IL-10 and transforming growth factor β (TGF-β) was increased.[82] Based on these findings, several clinical trials investigating estrogen administration in MS are underway.

In a first pilot cross-over trial, six female RRMS patients and four female SPMS patients were treated with 8 mg estriol per day during 6 months, followed by a 6-month posttreatment period and subsequent retreatment during 4 months. The investigators reported reduced number and volume of gadolinium-enhancing lesions upon estriol treatment in all six RRMS patients but not in the SPMS patients.[30,83] An add-on study to extend these previous findings was recently completed. The investigators compared the combination of copaxone injection plus 8 mg estriol per day to copaxone injection plus placebo in a double-blind clinical trial over a treatment period of 2 years (www.clinicaltrials.gov/ct2/show/NCT00451204). A significant reduction in annualized relapse rate was reported following treatment with estriol in combination with copaxone as compared to copaxone treatment alone.[94] Similarly, the effect of estrogen treatment (8 mg per day for 1 year) in combination with standard MS anti-inflammatory drugs on cognitive testing in women with MS is evaluated in another double-blind, placebo-controlled clinical trial (www.clinicaltrials.gov/ct2/show/NCT01466114). Although the first results of therapeutic use of estrogen in MS are encouraging, more research is warranted in order to understand the estrogen-mediated underlying mechanisms. The outcomes of the currently ongoing MS trials may help to clarify the therapeutic use of estrogen in combination with first-line immunomodulatory drugs.

While selective depletion of norepinephrine levels in the CNS exacerbated clinical scores, increasing norepinephrine levels through the use of a selective norepinephrine reuptake inhibitor demonstrated beneficial effects on clinical signs in EAE.[84] Accordingly, administration of tri- and tetracyclic antidepressants and L-dopa, also resulting in an increase in norepinephrine levels, ameliorates the clinical course of MS.[85] Indeed, approximately 75% of patients experienced substantial improvements in sensory, motor, and autonomic symptoms after 1–2 months of treatment. Similarly, a small improvement in the Expanded Disability Status Scale (EDSS) score was observed in MS patients receiving a combination of lofepramine, a third-generation tricyclic antidepressant, with phenylalanine and vitamin B_{12}, whereas EDSS score deteriorated in a placebo group receiving only vitamin B_{12}. Moreover, MRI showed a significant reduction in lesion number that correlated with an improvement of symptoms in the active treatment group but not in the placebo group.[86] It was concluded that this combination treatment of lofepramine, phenylalanine, and vitamin B_{12} in MS patients is effective, works unexpectedly rapidly, may benefit patients with all types of MS and all degrees of severity, and affects all aspects of the syndrome.[87]

In EAE, administration of rivastigmine, a cholinesterase inhibitor that prevents the hydrolysis of endogenous ACh, alleviates neuroinflammatory responses, thereby reducing the clinical and pathological severity of EAE.[91] Accordingly, beneficial effects on cognitive deficits in MS were observed in a number of phase I/II clinical studies using cholinesterase inhibitor therapy.[92,93] Indeed, following treatment with rivastigmine, Shaygannejad et al. reported a modest but significant improvement of memory in MS patients with Wechsler Memory Scales confirmed mild verbal memory impairment.[93] Nevertheless, similar improvements were observed in placebo-treated MS patients. Additionally, treatment of MS patients with donepezil, an alternative ACh esterase inhibitor, showed significant improvement in memory performance on the selective reminding test, a test of verbal learning and memory, as compared to placebo-treated MS patients. Moreover, cognitive improvement was reported by clinicians in twice as many donepezil- versus placebo-treated MS patients. In addition, the donepezil-treated MS patients themselves reported more often memory improvement than placebo-treated MS patients.[92] Several studies have shown that exposure to nicotine significantly delays the onset and markedly attenuates the severity of disease symptoms in EAE.[88–90] Interestingly, it was shown that nicotine treatment of mice with EAE considerably diminished the number of innate immune cells, including microglia and DCs, in the CNS. Moreover, these cells expressed reduced levels of CD80, CD86, and MHC class II molecules.[88] While this points to an in vivo immunosuppressive effect of nicotine in EAE, exposure to cigarette smoke has been demonstrated to contribute to both disease susceptibility and more rapid disease progression of MS. Differences in disease pathogenesis between EAE and human MS may explain these discrepant observations.

10.4 CONCLUSION

Here we have provided an overview of neuroendocrine regulation of the principal cellular mediators of innate immunity that may contribute to the pathogenesis of MS. It was demonstrated that the neuroendocrine system plays an important immunoregulatory role in the periphery as well as in the CNS. In this perspective, several therapies targeting the neuroendocrine system have shown to be effective in modulating the disease course of MS. Moreover, some of these therapies may act at least in part through modulation of the innate immune system, including glucocorticoid and estrogen treatment. In summary, the innate immune system plays an important role in the immunopathogenesis of MS and can be targeted by the neuroendocrine system in order to modulate the disease course.

Acknowledgments

This work was supported by grant no. G.0168.09 of the Fund for Scientific Research—Flanders, Belgium (FWO-Vlaanderen). Further support was provided through the Special Research Fund (BOF), Medical Legacy Fund (UZA), the Methusalem Funding Program, the Belgian Hercules Foundation, the Belgian Charcot Foundation, the *Belgische Stichting Roeping*.

References

1. Adib-Conquy M, Scott-Algara D, Cavaillon JM, Souza-Fonseca-Guimaraes F. TLR-mediated activation of NK cells and their role in bacterial/viral immune responses in mammals. *Immunol Cell Biol*. 2014;92:256–262.
2. Janeway Jr CA, Medzhitov R. Innate immune recognition. *Annu Rev Immunol*. 2002;20:197–216.
3. Aravalli RN, Peterson PK, Lokensgard JR. Toll-like receptors in defense and damage of the central nervous system. *J Neuroimmune Pharmacol*. 2007;2:297–312.
4. Cools N, Petrizzo A, Smits E, et al. Dendritic cells in the pathogenesis and treatment of human diseases: a Janus Bifrons? *Immunotherapy*. 2011;3:1203–1222.
5. Bakdash G, Sittig SP, van DT, Figdor CG, de Vries IJ. The nature of activatory and tolerogenic dendritic cell-derived signal II. *Front Immunol*. 2013;4:53.
6. Nuyts AH, Lee WP, Bashir-Dar R, Berneman ZN, Cools N. Dendritic cells in multiple sclerosis: key players in the immunopathogenesis, key players for new cellular immunotherapies? *Mult Scler*. 2013;19:995–1002.
7. Serafini B, Rosicarelli B, Magliozzi R, et al. Dendritic cells in multiple sclerosis lesions: maturation stage, myelin uptake, and interaction with proliferating T cells. *J Neuropathol Exp Neurol*. 2006;65:124–141.
8. Thewissen K, Nuyts AH, Deckx N, et al. Circulating dendritic cells of multiple sclerosis patients are proinflammatory and their frequency is correlated with MS-associated genetic risk factors. *Mult Scler*. 2014;20:548–557.
9. Karni A, Abraham M, Monsonego A, et al. Innate immunity in multiple sclerosis: myeloid dendritic cells in secondary progressive multiple sclerosis are activated and drive a proinflammatory immune response. *J Immunol*. 2006;177:4196–4202.
10. Prinz M, Tay TL, Wolf Y, Jung S. Microglia: unique and common features with other tissue macrophages. *Acta Neuropathol*. 2014;128:319–331.
11. Ransohoff RM, Brown MA. Innate immunity in the central nervous system. *J Clin Invest*. 2012;122:1164–1171.
12. Kotter MR, Li WW, Zhao C, Franklin RJ. Myelin impairs CNS remyelination by inhibiting oligodendrocyte precursor cell differentiation. *J Neurosci*. 2006;26:328–332.
13. Miron VE, Boyd A, Zhao JW, et al. M2 microglia and macrophages drive oligodendrocyte differentiation during CNS remyelination. *Nat Neurosci*. 2013;16:1211–1218.
14. Fernandes A, Miller-Fleming L, Pais TF. Microglia and inflammation: conspiracy, controversy or control? *Cell Mol Life Sci*. 2014;71:3969–3985.
15. Gandhi R, Laroni A, Weiner HL. Role of the innate immune system in the pathogenesis of multiple sclerosis. *J Neuroimmunol*. 2010;221:7–14.
16. Bogie JF, Stinissen P, Hendriks JJ. Macrophage subsets and microglia in multiple sclerosis. *Acta Neuropathol*. 2014;128:191–213.
17. Rasmussen S, Wang Y, Kivisakk P, et al. Persistent activation of microglia is associated with neuronal dysfunction of callosal projecting pathways and multiple sclerosis-like lesions in relapsing–remitting experimental autoimmune encephalomyelitis. *Brain*. 2007;130:2816–2829.
18. Kutzelnigg A, Lucchinetti CF, Stadelmann C, et al. Cortical demyelination and diffuse white matter injury in multiple sclerosis. *Brain*. 2005;128:2705–2712.
19. Butovsky O, Ziv Y, Schwartz A, et al. Microglia activated by IL-4 or IFN-gamma differentially induce neurogenesis and oligodendrogenesis from adult stem/progenitor cells. *Mol Cell Neurosci*. 2006;31:149–160.
20. Huizinga R, van der Star BJ, Kipp M, et al. Phagocytosis of neuronal debris by microglia is associated with neuronal damage in multiple sclerosis. *Glia*. 2012;60:422–431.
21. Ponomarev ED, Veremeyko T, Barteneva N, Krichevsky AM, Weiner HL. MicroRNA-124 promotes microglia quiescence and suppresses EAE by deactivating macrophages via the C/EBP-α-PU.1 pathway. *Nat Med*. 2011;17:64–70.
22. Bhasin M, Wu M, Tsirka SE. Modulation of microglial/macrophage activation by macrophage inhibitory factor (TKP) or tuftsin (TKPR) attenuates the disease course of experimental autoimmune encephalomyelitis. *BMC Immunol*. 2007;8:10.
23. Goldmann T, Wieghofer P, Muller PF, et al. A new type of microglia gene targeting shows TAK1 to be pivotal in CNS autoimmune inflammation. *Nat Neurosci*. 2013;16:1618–1626.
24. Heppner FL, Greter M, Marino D, et al. Experimental autoimmune encephalomyelitis repressed by microglial paralysis. *Nat Med*. 2005;11:146–152.
25. Ponomarev ED, Shriver LP, Maresz K, Pedras-Vasconcelos J, Verthelyi D, Dittel BN. GM-CSF production by autoreactive T cells is required for the activation of microglial cells and the onset of experimental autoimmune encephalomyelitis. *J Immunol*. 2007;178:39–48.

26. Lu W, Bhasin M, Tsirka SE. Involvement of tissue plasminogen activator in onset and effector phases of experimental allergic encephalomyelitis. *J Neurosci*. 2002;22:10781–10789.
27. Deckx N, Lee WP, Berneman ZN, Cools N. Neuroendocrine immunoregulation in multiple sclerosis. *Clin Dev Immunol*. 2013;2013:705232.
28. Gold SM, Kruger S, Ziegler KJ, et al. Endocrine and immune substrates of depressive symptoms and fatigue in multiple sclerosis patients with comorbid major depression. *J Neurol Neurosurg Psychiatry*. 2011;82:814–818.
29. Buford TW, Willoughby DS. Impact of DHEA(S) and cortisol on immune function in aging: a brief review. *Appl Physiol Nutr Metab*. 2008;33:429–433.
30. Sicotte NL, Liva SM, Klutch R, et al. Treatment of multiple sclerosis with the pregnancy hormone estriol. *Ann Neurol*. 2002;52:421–428.
31. Fairweather D, Frisancho-Kiss S, Rose NR. Sex differences in autoimmune disease from a pathological perspective. *Am J Pathol*. 2008;173:600–609.
32. Zoukos Y, Leonard JP, Thomaides T, Thompson AJ, Cuzner ML. Beta-adrenergic receptor density and function of peripheral blood mononuclear cells are increased in multiple sclerosis: a regulatory role for cortisol and interleukin-1. *Ann Neurol*. 1992;31:657–662.
33. Rajda C, Bencsik K, Vecsei LL, Bergquist J. Catecholamine levels in peripheral blood lymphocytes from multiple sclerosis patients. *J Neuroimmunol*. 2002;124:93–100.
34. Kooi EJ, Prins M, Bajic N, et al. Cholinergic imbalance in the multiple sclerosis hippocampus. *Acta Neuropathol*. 2011;122:313–322.
35. Chiaravalloti ND, DeLuca J. Cognitive impairment in multiple sclerosis. *Lancet Neurol*. 2008;7:1139–1151.
36. Wake H, Moorhouse AJ, Jinno S, Kohsaka S, Nabekura J. Resting microglia directly monitor the functional state of synapses in vivo and determine the fate of ischemic terminals. *J Neurosci*. 2009;29:3974–3980.
37. Mulatero P, Panarelli M, Schiavone D, et al. Impaired cortisol binding to glucocorticoid receptors in hypertensive patients. *Hypertension*. 1997;30:1274–1278.
38. Woltman AM, de Fijter JW, Kamerling SW, Paul LC, Daha MR, van KC. The effect of calcineurin inhibitors and corticosteroids on the differentiation of human dendritic cells. *Eur J Immunol*. 2000;30:1807–1812.
39. Rea D, van KC, van Meijgaarden KE, Ottenhoff TH, Melief CJ, Offringa R. Glucocorticoids transform CD40-triggering of dendritic cells into an alternative activation pathway resulting in antigen-presenting cells that secrete IL-10. *Blood*. 2000;95:3162–3167.
40. Rozkova D, Horvath R, Bartunkova J, Spisek R. Glucocorticoids severely impair differentiation and antigen presenting function of dendritic cells despite upregulation of Toll-like receptors. *Clin Immunol*. 2006;120:260–271.
41. Piemonti L, Monti P, Allavena P, et al. Glucocorticoids affect human dendritic cell differentiation and maturation. *J Immunol*. 1999;162:6473–6481.
42. Barnes PJ. Anti-inflammatory actions of glucocorticoids: molecular mechanisms. *Clin Sci (London)*. 1998;94:557–572.
43. Harkonen PL, Vaananen HK. Monocyte-macrophage system as a target for estrogen and selective estrogen receptor modulators. *Ann N Y Acad Sci*. 2006;1089:218–227.
44. Bengtsson AK, Ryan EJ, Giordano D, Magaletti DM, Clark EA. 17beta-estradiol (E2) modulates cytokine and chemokine expression in human monocyte-derived dendritic cells. *Blood*. 2004;104:1404–1410.
45. Huck B, Steck T, Habersack M, Dietl J, Kammerer U. Pregnancy associated hormones modulate the cytokine production but not the phenotype of PBMC-derived human dendritic cells. *Eur J Obstet Gynecol Reprod Biol*. 2005;122:85–94.
46. Segerer SE, Muller N, van den Brandt J, et al. Impact of female sex hormones on the maturation and function of human dendritic cells. *Am J Reprod Immunol*. 2009;62:165–173.
47. Polanczyk MJ, Hopke C, Vandenbark AA, Offner H. Estrogen-mediated immunomodulation involves reduced activation of effector T cells, potentiation of Treg cells, and enhanced expression of the PD-1 costimulatory pathway. *J Neurosci Res*. 2006;84:370–378.
48. Maestroni GJ. Short exposure of maturing, bone marrow-derived dendritic cells to norepinephrine: impact on kinetics of cytokine production and Th development. *J Neuroimmunol*. 2002;129:106–114.
49. Maestroni GJ, Mazzola P. Langerhans cells beta 2-adrenoceptors: role in migration, cytokine production, Th priming and contact hypersensitivity. *J Neuroimmunol*. 2003;144:91–99.
50. Nijhuis LE, Olivier BJ, Dhawan S, et al. Adrenergic β2 receptor activation stimulates anti-inflammatory properties of dendritic cells in vitro. *PLoS One*. 2014;9:e85086.
51. Kim BJ, Jones HP. Epinephrine-primed murine bone marrow-derived dendritic cells facilitate production of IL-17A and IL-4 but not IFN-gamma by $CD4^+$ T cells. *Brain Behav Immun*. 2010;24:1126–1136.

52. Goyarts E, Matsui M, Mammone T, et al. Norepinephrine modulates human dendritic cell activation by altering cytokine release. *Exp Dermatol*. 2008;17:188–196.
53. Tracey KJ. Physiology and immunology of the cholinergic antiinflammatory pathway. *J Clin Invest*. 2007;117:289–296.
54. McAllister-Sistilli CG, Caggiula AR, Knopf S, Rose CA, Miller AL, Donny EC. The effects of nicotine on the immune system. *Psychoneuroendocrinology*. 1998;23:175–187.
55. Aicher A, Heeschen C, Mohaupt M, Cooke JP, Zeiher AM, Dimmeler S. Nicotine strongly activates dendritic cell-mediated adaptive immunity: potential role for progression of atherosclerotic lesions. *Circulation*. 2003;107:604–611.
56. Gao FG, Wan dF, Gu JR. Ex vivo nicotine stimulation augments the efficacy of therapeutic bone marrow-derived dendritic cell vaccination. *Clin Cancer Res*. 2007;13:3706–3712.
57. Yanagita M, Kobayashi R, Kojima Y, Mori K, Murakami S. Nicotine modulates the immunological function of dendritic cells through peroxisome proliferator-activated receptor-gamma upregulation. *Cell Immunol*. 2012;274:26–33.
58. Nouri-Shirazi M, Guinet E. Evidence for the immunosuppressive role of nicotine on human dendritic cell functions. *Immunology*. 2003;109:365–373.
59. Drew PD, Chavis JA. Female sex steroids: effects upon microglial cell activation. *J Neuroimmunol*. 2000;111:77–85.
60. Loughlin AJ, Woodroofe MN, Cuzner ML. Modulation of interferon-gamma-induced major histocompatibility complex class II and Fc receptor expression on isolated microglia by transforming growth factor-beta 1, interleukin-4, noradrenaline and glucocorticoids. *Immunology*. 1993;79:125–130.
61. Li M, Wang Y, Guo R, Bai Y, Yu Z. Glucocorticoids impair microglia ability to induce T cell proliferation and Th1 polarization. *Immunol Lett*. 2007;109:129–137.
62. Smith JA, Das A, Butler JT, Ray SK, Banik NL. Estrogen or estrogen receptor agonist inhibits lipopolysaccharide induced microglial activation and death. *Neurochem Res*. 2011;36:1587–1593.
63. Vegeto E, Belcredito S, Ghisletti S, Meda C, Etteri S, Maggi A. The endogenous estrogen status regulates microglia reactivity in animal models of neuroinflammation. *Endocrinology*. 2006;147:2263–2272.
64. Tapia-Gonzalez S, Carrero P, Pernia O, Garcia-Segura LM, Diz-Chaves Y. Selective oestrogen receptor (ER) modulators reduce microglia reactivity in vivo after peripheral inflammation: potential role of microglial ERs. *J Endocrinol*. 2008;198:219–230.
65. Dimayuga FO, Reed JL, Carnero GA, et al. Estrogen and brain inflammation: effects on microglial expression of MHC, costimulatory molecules and cytokines. *J Neuroimmunol*. 2005;161:123–136.
66. McNamee EN, Ryan KM, Kilroy D, Connor TJ. Noradrenaline induces IL-1ra and IL-1 type II receptor expression in primary glial cells and protects against IL-1beta-induced neurotoxicity. *Eur J Pharmacol*. 2010;626:219–228.
67. Dello RC, Boullerne AI, Gavrilyuk V, Feinstein DL. Inhibition of microglial inflammatory responses by norepinephrine: effects on nitric oxide and interleukin-1beta production. *J Neuroinflammation*. 2004;1:9.
68. Mori K, Ozaki E, Zhang B, et al. Effects of norepinephrine on rat cultured microglial cells that express alpha1, alpha2, beta1 and beta2 adrenergic receptors. *Neuropharmacology*. 2002;43:1026–1034.
69. Shytle RD, Mori T, Townsend K, et al. Cholinergic modulation of microglial activation by alpha 7 nicotinic receptors. *J Neurochem*. 2004;89:337–343.
70. De SR, Ajmone-Cat MA, Carnevale D, Minghetti L. Activation of alpha7 nicotinic acetylcholine receptor by nicotine selectively up-regulates cyclooxygenase-2 and prostaglandin E2 in rat microglial cultures. *J Neuroinflammation*. 2005;2:4.
71. Tyagi E, Agrawal R, Nath C, Shukla R. Inhibitory role of cholinergic system mediated via alpha7 nicotinic acetylcholine receptor in LPS-induced neuro-inflammation. *Innate Immun*. 2010;16:3–13.
72. Thomsen MS, Mikkelsen JD. The alpha7 nicotinic acetylcholine receptor ligands methyllycaconitine, NS6740 and GTS-21 reduce lipopolysaccharide-induced TNF-alpha release from microglia. *J Neuroimmunol*. 2012;251:65–72.
73. Sellebjerg F, Barnes D, Filippini G, et al. EFNS guideline on treatment of multiple sclerosis relapses: report of an EFNS task force on treatment of multiple sclerosis relapses. *Eur J Neurol*. 2005;12:939–946.
74. Frequin ST, Barkhof F, Lamers KJ, Hommes OR. The effects of high-dose methylprednisolone on gadolinium-enhanced magnetic resonance imaging and cerebrospinal fluid measurements in multiple sclerosis. *J Neuroimmunol*. 1992;40:265–272.
75. Sloka JS, Stefanelli M. The mechanism of action of methylprednisolone in the treatment of multiple sclerosis. *Mult Scler*. 2005;11:425–432.
76. Krystyna MS, Jacek T, Sebastian R, et al. Changes in circulating dendritic cells and B-cells in patients with multiple sclerosis relapse during corticosteroid therapy. *J Neuroimmunol*. 2009;207:107–110.

77. Navarro J, Aristimuno C, Sanchez-Ramon S, et al. Circulating dendritic cells subsets and regulatory T-cells at multiple sclerosis relapse: differential short-term changes on corticosteroids therapy. *J Neuroimmunol.* 2006;176:153–161.
78. Suda T, Chida K, Matsuda H, et al. High-dose intravenous glucocorticoid therapy abrogates circulating dendritic cells. *J Allergy Clin Immunol.* 2003;112:1237–1239.
79. Bebo Jr BF, Fyfe-Johnson A, Adlard K, Beam AG, Vandenbark AA, Offner H. Low-dose estrogen therapy ameliorates experimental autoimmune encephalomyelitis in two different inbred mouse strains. *J Immunol.* 2001;166:2080–2089.
80. Subramanian S, Yates M, Vandenbark AA, Offner H. Oestrogen-mediated protection of experimental autoimmune encephalomyelitis in the absence of $Foxp3^+$ regulatory T cells implicates compensatory pathways including regulatory B cells. *Immunology.* 2011;132:340–347.
81. Liu HY, Buenafe AC, Matejuk A, et al. Estrogen inhibition of EAE involves effects on dendritic cell function. *J Neurosci Res.* 2002;70:238–248.
82. Papenfuss TL, Powell ND, McClain MA, et al. Estriol generates tolerogenic dendritic cells in vivo that protect against autoimmunity. *J Immunol.* 2011;186:3346–3355.
83. Soldan SS, Alvarez Retuerto AI, Sicotte NL, Voskuhl RR. Immune modulation in multiple sclerosis patients treated with the pregnancy hormone estriol. *J Immunol.* 2003;171:6267–6274.
84. Simonini MV, Polak PE, Sharp A, McGuire S, Galea E, Feinstein DL. Increasing CNS noradrenaline reduces EAE severity. *J Neuroimmune Pharmacol.* 2010;5:252–259.
85. Berne-Fromell K, Fromell H, Lundkvist S, Lundkvist P. Is multiple sclerosis the equivalent of Parkinson's disease for noradrenaline? *Med Hypotheses.* 1987;23:409–415.
86. Puri BK, Bydder GM, Chaudhuri KR, et al. MRI changes in multiple sclerosis following treatment with lofepramine and L-phenylalanine. *Neuroreport.* 2001;12:1821–1824.
87. Loder C, Allawi J, Horrobin DF. Treatment of multiple sclerosis with lofepramine, L-phenylalanine and vitamin B(12): mechanism of action and clinical importance: roles of the locus coeruleus and central noradrenergic systems. *Med Hypotheses.* 2002;59:594–602.
88. Shi FD, Piao WH, Kuo YP, Campagnolo DI, Vollmer TL, Lukas RJ. Nicotinic attenuation of central nervous system inflammation and autoimmunity. *J Immunol.* 2009;182:1730–1739.
89. Nizri E, Irony-Tur-Sinai M, Lory O, Orr-Urtreger A, Lavi E, Brenner T. Activation of the cholinergic anti-inflammatory system by nicotine attenuates neuroinflammation via suppression of Th1 and Th17 responses. *J Immunol.* 2009;183:6681–6688.
90. Hao J, Simard AR, Turner GH, et al. Attenuation of CNS inflammatory responses by nicotine involves α7 and non-α7 nicotinic receptors. *Exp Neurol.* 2011;227:110–119.
91. Nizri E, Irony-Tur-Sinai M, Faranesh N, et al. Suppression of neuroinflammation and immunomodulation by the acetylcholinesterase inhibitor rivastigmine. *J Neuroimmunol.* 2008;203:12–22.
92. Christodoulou C, Melville P, Scherl WF, Macallister WS, Elkins LE, Krupp LB. Effects of donepezil on memory and cognition in multiple sclerosis. *J Neurol Sci.* 2006;245:127–136.
93. Shaygannejad V, Janghorbani M, Ashtari F, Dehghan H. Effects of adjunct low-dose vitamin d on relapsing-remitting multiple sclerosis progression: preliminary findings of a randomized placebo-controlled trial. *Mult Scler Int.* 2012;2012:452541.
94. Voskuhl RR, Wang H, Wu TC, et al. Estriol combined with glatiramer acetate for women with relapsing-remitting multiple sclerosis: a randomised, placebo-controlled, phase 2 trial. *Lancet Neurol.* Jan 2016;15(1):35–46. http://dx.doi.org/10.1016/S1474-4422(15)00322-1. Epub 2015 Nov 29. PubMed PMID: 26621682.

CHAPTER

11

Fighting Chronic Neuroinflammation by Boosting Autoimmunity: The Distinction Between Neurodegenerative Diseases and Multiple Sclerosis

M. Schwartz, K. Baruch
Weizmann Institute of Science, Rehovot, Israel

OUTLINE

Autoimmunity has long been viewed as synonymous with autoimmune disease, as if the mere existence of immunity that recognizes self-compounds/cells is pathological, and indicates failure of proper negative selection during immune ontogeny. It is now clear, however, that autoimmune activity, both humoral and cellular, exists in healthy individuals, as

http://dx.doi.org/10.1016/B978-0-12-801914-6.00011-8

well[45,47,52]; the function and outcome of such autoimmune responses are subject to debate, but this phenomenon has begun to receive increasing attention. Such autoimmune activity has attracted, for example, an interest in the context of tumor immunotherapy.[16,33,75] In addition to the understanding that autoimmune cells are part of the immune repertoire of healthy individuals, it became clear that regulatory T cells (Tregs) are needed to maintain such activity under tight control, a phenomenon collectively known as peripheral tolerance.[38,59,78] Moreover, while in the past it was commonly believed that autoimmunity should be deleted or avoided, anti-self-immune activity is now recognized as a purposeful defense response that requires a tight regulation to be on alert without a risk of causing autoimmune disease.[65] In the context of the central nervous system (CNS), as will be discussed next, autoimmune cells, as well as immunoregulatory cells, are needed for brain maintenance and repair, as emerged from our studies over more than a decade.[63]

11.1 AUTOIMMUNITY AND CENTRAL NERVOUS SYSTEM REPAIR

The experimental evidence demonstrating that autoimmunity has a beneficial role in the CNS first emerged from our studies utilizing experimental animal models of axonal injury, including optic nerve and spinal cord injury. Initially, we demonstrated that passive transfer of myelin-specific encephalitogenic T cells to axonally injured mice promotes neuronal repair.[29,49] Similarly, we found that systemic T-cell deficiency is associated with greatly increased neuronal loss following injury,[37,60,84] suggesting that T lymphocytes participate in mitigating the consequences of sterile injury to the CNS. These earlier studies also suggested that while the spontaneous activity of CNS-specific T cells is perhaps insufficient to contain detrimental consequences of neuronal injury, boosting their activity or their numbers in the circulation should be considered as a therapeutic approach.[28–30,50,53,81] These observations, which were initially entirely unexpected, led us to formulate the concept of "protective autoimmunity," describing a protective role for autoimmune T cells in CNS repair, a form of benign autoimmunity.[49,64] Yet, the nature of such T cells, and how they mediate their protective effect, remained elusive for many years. Moreover, among the effector T cells that accumulated at the CNS, it remained unknown which effector subtype of T cells participate in neuroprotection, and to what degree such a physiological antiself response is spontaneously evoked following injury.

A striking observation supporting the notion that an anti-self-immune response is induced following CNS injury emerged from a study demonstrating that a CNS injury (spinal cord injury) evokes a systemic immune response that enhances recovery from a second injury, remote from the primary injury site (optic nerve).[84] Additionally, significantly improved locomotor activity was observed in rats that received splenocytes derived from CNS-injured rats, compared to injured rats that received splenocytes from uninjured animals.[84] Notably, however, transfer of the same self-specific T cells to immunologically deficient transgenic mice did not result in any beneficial effect,[36] suggesting that autoimmune T cells could be neuroprotective only when they act in conjunction with additional immune cell types. Subsequent studies that spanned over a decade have helped explain that CNS repair following damage requires autoimmune effector memory T cells, which act as part of a cellular network, alongside FoxP3+ Tregs and circulating monocytes, the involvement of which is spatially and temporally regulated.[43,55,68,79]

11.2 AUTOIMMUNE T CELLS GUARD THE HEALTHY SELF

The role of autoimmune T cells in supporting recovery following CNS injuries highlighted the possibility that the primary role of circulating immune cells in the context of the CNS might be in maintenance of the healthy brain, in supporting brain's functional plasticity, and in restoration of homeostasis whenever it is perturbed. Specifically, our earlier studies revealed that T cells that recognize brain antigens are required for supporting hippocampal-dependent cognitive abilities and neurogenesis.[35,85] T cell–deficient mice and mice deficient in CNS-specific T cells were found to exhibit reduced spatial learning and memory capabilities, and reconstitution of the normal T-cell pool in T cell–deficient mice could overcome some of these spatial learning and memory deficits. Moreover, transgenic mice overexpressing a T-cell receptor specific to myelin basic protein (Tmbp mice) performed better than their wild-type counterparts in spatial learning and memory tasks, and a sudden imposition of immune deficiency in young mice was associated with spatial-memory impairments.[58,85] Similarly, systemic depletion of CD4+ T lymphocytes, but not of CD8 or B lymphocytes, was found to impair learning and memory functions in mice, and to result in decreased expression in the brain of brain-derived neurotrophic factor (BDNF), a key factor in maintaining brain plasticity.[82] Interestingly, we also found that CNS-specific T cells contribute to adult neurogenesis in nonconventional neurogenic niches, such as the intact adult spinal cord.[70] The fact that inflammatory activity within the CNS is typically a negative regulator of neurogenesis[19,51] suggested that the type of immune activation within the CNS critically determines the outcome.[39,46]

11.3 AUTOIMMUNITY THAT SHAPES BRAIN FUNCTION FROM AFAR

Searching for a mechanistic insight as to how T cells could affect healthy brain plasticity, while they are normally excluded from direct interaction with the brain parenchyma, led us to envision that such an effect might take place at the brain's territory, yet outside the neuronal parenchyma. We suggested that this brain-immune crosstalk might take place at immunologically "permissive" sites along the brains' borders, at the choroid plexus (CP), the meningeal spaces, and the cerebrospinal fluid (CSF).[8,66]

The CP consists of a monolayer of cuboidal epithelial cells which form the blood–CSF barrier, and surround a dense vascular stroma.[32] Strategically positioned at the brain ventricles, the CP is constantly exposed to brain-derived signals from its apical side, via the CSF, and to peripheral signals from its basal side, via the circulation. The common role attributed to the CP is the production of the CSF, providing the brain with a nutritive metabolic milieu, and forming a protective mechanical cushion. Over the last decade, however, this site was suggested to participate in various additional aspects of brain homeostasis, indicating that it plays a much greater role than previously thought.[20] The unique spatial position of the CP, and the fact that the CP has a fenestrated vasculature, which can support immune surveillance of the healthy CNS,[54,56] led us to consider this compartment as an active neuroimmunological interface between the brain and the circulation.

The CP expresses receptors for various parenchymal-derived "danger" signals, which endow it with the potential to sense and to respond to the needs of the CNS. Thus, for example, intracerebroventricular injection of Toll-like receptor (TLR) agonists directed at TLR-7 or TLR-9 were shown to affect CP function.[13] Similarly, in the hours following tumor necrosis factor-α (TNF-α) injection into the CSF, the CP was found to orchestrate synchronized cellular trafficking of different immune cell subsets from the blood to the CSF.[5]

Examining routes of leukocyte entry to the CNS following spinal cord injury revealed that as a result of damage to the CNS parenchyma, proinflammatory mediators are released to the CSF; these molecules activate the CP to function as a gateway for leukocyte recruitment to the injury site. In the days following the injury, expression of a wide array of adhesion molecules and chemokines essential for leukocyte trafficking is induced in the CP epithelium,[43,69] a process that was associated with trafficking of myeloid[69] and $CD4^+$ T cells[43,55] to the brain territory. The gateway activity of the CP to support recruitment of "healing" ("alternatively activated"/antiinflammatory (M2)) macrophages to the injury site was shown to be pivotal for CNS tissue recovery and repair.[43,69] Importantly, the fact that following localized injury to the spinal cord, immune cell recruitment to the CNS takes place through the CP, a site remote from the injury itself, suggested the functional significance of this route in the reparative process. Though infiltration of myeloid cells to the injury site takes place through additional routes,[69] entry through the CP–CSF pathway has a well orchestrated kinetics and seems to enable skewing the phenotype of the entering cells toward an antiinflammatory activity.[67]

In searching for a mechanism whereby the CP epithelium could get signaling from the periphery, we identified that the CP of healthy mice is constitutively populated by effector memory T cells, and is enriched with a subpopulation recognizing CNS antigens.[6] Further characterization of these T cells showed the presence of interleukin (IL-4)-expressing, interferon (IFN)-γ-expressing, and $FoxP3^+$ Tregs, at higher levels in comparison to those found in the circulation or in the lymphoid organs.[43] Such T cells, in synergy with CNS-derived "danger" signals, can modulate CP expression of trafficking molecules needed for CNS immune surveillance and for recruitment of leukocytes to the CNS parenchyma, upon need.[5,43] Specifically, IFN-γ is a key T cell–derived cytokine that is essential for CP gateway function,[4,43] and proinflammatory signaling that emerges from the CNS parenchyma can modulate its effect on CP activity, either synergistically or antagonistically.[5]

We therefore suggested that CP activity, as a neuroimmunological interface, might have a role in shaping the fate of the brain,[8,61] and thus dysfunction of this compartment could contribute to brain aging and neurodegenerative diseases.[62,63]

11.4 IMMUNOLOGICAL DYSFUNCTION IN NEURODEGENERATIVE DISEASES AND BRAIN AGING

The possibility that peripheral immune deviation affects the local effector cytokine milieu in the CP stroma, which in turn modulates CP activity, and possibly also brain function, led us to examine the immunological fate of the CP throughout life and along aging. We found that in the CP of aged mice, IFN-γ availability is reduced, leading to a shift in the cytokine composition of this compartment in favor of IL-4.[6] Further in vitro and in vivo characterizations of IL-4 effects on CP epithelial cells revealed that in higher concentration it can have a detrimental role in this compartment, yet in lower concentrations it can support the production of neurotrophic factors, such as BDNF, by the CP.[6] In the meninges, the membranes that enclose the CNS, T cell–derived IL-4 was found to play a key role in supporting cognitive functions.[18,24] These findings suggested that beyond controlling CP function as a gateway for immune cells, the cytokine balance at the CP critically affects this compartment activity in shaping brain function.

Multiple-organ high-throughput RNA-sequencing analysis of young and aged mouse tissues, including the CP, hippocampal tissues, and lymphoid organs, revealed that the CP of aged mice displays a unique expression profile among all the tested organs.[4] This expression profile was characterized by the induction of a type-I interferon (IFN-I) response program, most prominently represented by the cytokine, IFN-β. Importantly, this response profile was also characterized by downregulation of IFN-γ-dependent genes. Such local inverse relationships between type-I and type-II IFN response programs have been associated with unresolved inflammatory reactions outside the CNS.[73,74,80] Moreover, IFN-β expression by the aged CP was found to be induced by factors emerging from the aged brain, which are present in the CSF of aged mice. Neutralizing antibodies specific for the IFN-β receptor, when administered intracerebroventricularly could partially reverse age-associated cognitive loss in mice.[4] These findings, highlighting a role of the CP in affecting brain function in aging, prompted us to investigate how modulating the immune system could affect CP gateway activity in the context of neurodegenerative diseases, such as amyotrophic lateral sclerosis (ALS), and Alzheimer's disease (AD).

Multiple lines of evidence suggest that under neurodegenerative conditions, CNS-recruited immune cells, and the resident immune cells of the CNS, the microglia, play nonredundant roles with respect to their modulation of the chronic neuroinflammatory response.[12,15,44,68] In AD, for example, while microglia fail to ultimately clear Aβ deposits, CNS-infiltrating monocyte-derived macrophages play a beneficial role in facilitating Aβ plaque removal and fighting off AD-like pathology.[14,40,41,48,71,76] In mutant SOD1G93A (mSOD1) mice, an animal model of ALS, circulating immune cells were found to play a neuroprotective role, as life expectancy is shorter and disease emerges earlier in mSOD1 mice lacking T cells.[9,10,21] Likewise, increased survival of mSOD1 mice was associated with the enhanced infiltration of $CD4^+$ T cells to the spinal cord.[3,11,42] In our study, we showed that in mSOD1 mice, the CP is inadequately activated over the course of disease progression to enable leukocyte trafficking to the CNS; rather, the CP was characterized by reduced levels of IFN-γ, and reduced leukocyte numbers were observed in the CSF compartment.[42] Boosting systemic autoimmunity in these mice by immunization with a myelin-derived peptide resulted in recruitment of immunoregulatory cells to the spinal cord via the CP. These cells, which included both IL-10 producing monocyte-derived macrophages and Tregs, homed to sites of motor neuron degeneration in the spinal cord of the mice; the overall effect was correlated with attenuated disease progression and increased survival, suggesting a functional neuroprotective role for the recruited cells.[42]

Since an unresolved neuroinflammatory response characterizes all chronic neurodegenerative conditions,[23] these findings also suggested a more general approach, in which systemic autoimmunity should be boosted to allow recruitment of peripheral immune cells to fight neurodegeneration.[62] Indeed, when we examined trafficking of immune cells to the diseased brain in the context of AD, CP dysfunction was manifested by poor local availability of IFN-γ, in two transgenic mouse strains that develop AD-like pathology (5XFAD and APP/PS1).[7] Similar findings were recently reported in an additional AD mouse model, the PDGFB-APPSwInd (J20) transgenic line.[86] Using a genetic approach for transiently depleting Foxp3$^+$ Tregs in 5XFAD mice, we showed that similar to the situation in the mSOD1 mice, boosting systemic immunity (this time by targeting peripheral immune suppression) sets in motion a cascade of events that include augmenting availability of IFN-γ at the CP, expression of leukocyte trafficking molecules and chemokines in this compartment, and recruitment of immunoregulatory cells to the diseased CNS parenchyma.[7] In 5XFAD mice, this approach resulted in a dramatic effect on disease pathology—clearance of cerebral Aβ plaques, attenuation of the neuroinflammatory response, and reversal of cognitive decline.[7]

11.5 RELAPSING-REMITTING MULTIPLE SCLEROSIS AND NEURODEGENERATIVE DISEASES: COMMON NEUROINFLAMMATION, OPPOSITE THERAPEUTIC STRATEGIES

Decades of research and clinical practice in the field of autoimmune neuroinflammatory diseases, and primarily relapsing-remitting multiple sclerosis (RRMS), have suggested the use of systemic antiinflammatory/immunosuppressive therapy for attenuating brain inflammation. In contrast, attempts to adopt similar strategies for treating chronic neurodegenerative disease have largely failed.[1,2,17,22,25–27,57,83]

The findings described earlier, which attribute a neuroprotective role for a systemic autoimmune response under neurodegenerative conditions, may help reconcile the mystery of neuroinflammation in neurodegenerative diseases. Thus, while in all neuroinflammatory conditions there is a need to resolve brain inflammation, under chronic neurodegenerative conditions, reducing systemic immune suppression, rather than augmenting it, is needed in order to achieve this goal. Accordingly, CNS pathologies should not be simplistically characterized as diseases that would uniformly benefit from systemic antiinflammatory therapy. Moreover, a major issue that should be taken into consideration is the distinct effect of the state of inflammation inside and outside the brain, and thus the different consequences of an immunosuppressive approach outside and inside the CNS.[62]

As discussed earlier, peripheral immune tolerance is pivotal for maintaining immune system homeostasis. Foxp3$^+$ Tregs are important regulators of peripheral tolerance and are associated with protection against autoimmune diseases.[34] As a corollary, diseases like RRMS have been linked to systemic reduction or dysfunction in Tregs,[31,77] and therefore, augmenting immunosuppression could provide therapeutic benefit.[72] However, in neurodegenerative diseases, as described earlier, Treg-mediated immune suppression seems to interfere with the ability to mount "protective autoimmunity." We suggest that under neurodegenerative conditions, these peripheral immunoregulatory cells hinder the ability to mount a peripheral autoimmune response that is needed for setting in motion an immunological cascade of events, including augmenting CP gateway activity and recruitment of immunoregulatory cells to sites of brain pathology. This apparent paradox emphasizes how two classes of diseases with shared inflammatory pathological features, but of different etiology, necessitate opposite immune intervention and call for revisiting the simplistic approach of "one immunomodulation fits all."

References

1. Aisen PS, Schafer KA, Grundman M, et al. Effects of rofecoxib or naproxen vs placebo on Alzheimer disease progression: a randomized controlled trial. *JAMA*. 2003;289:2819–2826.
2. Arvanitakis Z, Grodstein F, Bienias JL, et al. Relation of NSAIDs to incident AD, change in cognitive function, and AD pathology. *Neurology*. 2008;70:2219–2225.
3. Banerjee R, Mosley RL, Reynolds AD, et al. Adaptive immune neuroprotection in G93A-SOD1 amyotrophic lateral sclerosis mice. *PLoS One*. 2008;3:e2740.
4. Baruch K, Deczkowska A, David E, et al. Aging. Aging-induced type I interferon response at the choroid plexus negatively affects brain function. *Science*. 2014;346:89–93.
5. Baruch K, Kertser A, Porat Z, Schwartz M. Cerebral nitric oxide represses choroid plexus NFκB-dependent gateway activity for leukocyte trafficking. *EMBO J*. 2015a;34:1816–1828.
6. Baruch K, Ron-Harel N, Gal H, et al. CNS-specific immunity at the choroid plexus shifts toward destructive Th2 inflammation in brain aging. *Proc Natl Acad Sci USA*. 2013;110:2264–2269.
7. Baruch K, Rosenzweig N, Kertser A, et al. Breaking immune tolerance by targeting Foxp3 regulatory T cells mitigates Alzheimer's disease pathology. *Nat Commun*. 2015b;6:7967.
8. Baruch K, Schwartz M. CNS-specific T cells shape brain function via the choroid plexus. *Brain Behav Immun*. 2013;34:11–16.
9. Beers DR, Henkel JS, Xiao Q, et al. Wild-type microglia extend survival in PU.1 knockout mice with familial amyotrophic lateral sclerosis. *Proc Natl Acad Sci USA*. 2006;103:16021–16026.
10. Beers DR, Henkel JS, Zhao W, Wang J, Appel SH. CD4$^+$ T cells support glial neuroprotection, slow disease progression, and modify glial morphology in an animal model of inherited ALS. *Proc Natl Acad Sci USA*. 2008;105:15558–15563.
11. Beers DR, Henkel JS, Zhao W, et al. Endogenous regulatory T lymphocytes ameliorate amyotrophic lateral sclerosis in mice and correlate with disease progression in patients with amyotrophic lateral sclerosis. *Brain*. 2011;134:1293–1314.
12. Britschgi M, Wyss-Coray T. Immune cells may fend off Alzheimer disease. *Nat Med*. 2007;13:408–409.
13. Butchi NB, Woods T, Du M, Morgan TW, Peterson KE. TLR7 and TLR9 trigger distinct neuroinflammatory responses in the CNS. *Am J Pathol*. 2011;179:783–794.
14. Butovsky O, Kunis G, Koronyo-Hamaoui M, Schwartz M. Selective ablation of bone marrow-derived dendritic cells increases amyloid plaques in a mouse Alzheimer's disease model. *Eur J Neurosci*. 2007;26:413–416.
15. Cameron B, Landreth GE. Inflammation, microglia, and Alzheimer's disease. *Neurobiol Dis*. 2010;37:503–509.
16. Caspi RR. Immunotherapy of autoimmunity and cancer: the penalty for success. *Nat Rev Immunol*. 2008;8: 970–976.
17. Cudkowicz ME, Shefner JM, Schoenfeld DA, et al. Trial of celecoxib in amyotrophic lateral sclerosis. *Ann Neurol*. 2006;60:22–31.
18. Derecki NC, Cardani AN, Yang CH, et al. Regulation of learning and memory by meningeal immunity: a key role for IL-4. *J Exp Med*. 2010;207:1067–1080.
19. Ekdahl CT, Claasen JH, Bonde S, Kokaia Z, Lindvall O. Inflammation is detrimental for neurogenesis in adult brain. *Proc Natl Acad Sci USA*. 2003;100:13632–13637.
20. Emerich DF, Skinner SJ, Borlongan CV, Vasconcellos AV, Thanos CG. The choroid plexus in the rise, fall and repair of the brain. *BioEssays*. 2005;27:262–274.
21. Finkelstein A, Kunis G, Seksenyan A, et al. Abnormal changes in NKT cells, the IGF-1 axis, and liver pathology in an animal model of ALS. *PLoS One*. 2011;6:e22374.
22. Fondell E, O'Reilly EJ, Fitzgerald KC, et al. Non-steroidal anti-inflammatory drugs and amyotrophic lateral sclerosis: results from five prospective cohort studies. *Amyotroph Lateral Scler*. 2012;13:573–579.
23. Frank-Cannon TC, Alto LT, McAlpine FE, Tansey MG. Does neuroinflammation fan the flame in neurodegenerative diseases? *Mol Neurodegener*. 2009;4:47.
24. Gadani SP, Cronk JC, Norris GT, Kipnis J. IL-4 in the brain: a cytokine to remember. *J Immunol*. 2012;189: 4213–4219.

25. Gordon PH, Moore DH, Miller RG, et al. Efficacy of minocycline in patients with amyotrophic lateral sclerosis: a phase III randomised trial. *Lancet Neurol*. 2007;6:1045–1053.
26. Group A-FR. Follow-up evaluation of cognitive function in the randomized Alzheimer's Disease Anti-inflammatory Prevention Trial and its Follow-up Study. *Alzheimer's Dement*. 2015;11:216–225.e1.
27. Group AR, Lyketsos CG, Breitner JC, et al. Naproxen and celecoxib do not prevent AD in early results from a randomized controlled trial. *Neurology*. 2007;68:1800–1808.
28. Hammarberg H, Lidman O, Lundberg C, et al. Neuroprotection by encephalomyelitis: rescue of mechanically injured neurons and neurotrophin production by CNS-infiltrating T and natural killer cells. *J Neurosci*. 2000;20:5283–5291.
29. Hauben E, Agranov E, Gothilf A, et al. Posttraumatic therapeutic vaccination with modified myelin self-antigen prevents complete paralysis while avoiding autoimmune disease. *J Clin Invest*. 2001;108:591–599.
30. Hofstetter HH, Sewell DL, Liu F, et al. Autoreactive T cells promote post-traumatic healing in the central nervous system. *J Neuroimmunol*. 2003;134:25–34.
31. Huan J, Culbertson N, Spencer L, et al. Decreased FOXP3 levels in multiple sclerosis patients. *J Neurosci Res*. 2005;81:45–52.
32. Johanson CE, Stopa EG, McMillan PN. The blood-cerebrospinal fluid barrier: structure and functional significance. *Methods Mol Biol*. 2011;686:101–131.
33. Kaufman HL, Wolchok JD. Is tumor immunity the same thing as autoimmunity? Implications for cancer immunotherapy. *J Clin Oncol*. 2006;24:2230–2232.
34. Kim JM, Rasmussen JP, Rudensky AY. Regulatory T cells prevent catastrophic autoimmunity throughout the lifespan of mice. *Nat Immunol*. 2007;8:191–197.
35. Kipnis J, Cohen H, Cardon M, Ziv Y, Schwartz M. T cell deficiency leads to cognitive dysfunction: implications for therapeutic vaccination for schizophrenia and other psychiatric conditions. *Proc Natl Acad Sci USA*. 2004;101:8180–8185.
36. Kipnis J, Mizrahi T, Yoles E, Ben-Nun A, Schwartz M. Myelin specific Th1 cells are necessary for post-traumatic protective autoimmunity. *J Neuroimmunol*. 2002;130:78–85.
37. Kipnis J, Yoles E, Porat Z, et al. T cell immunity to copolymer 1 confers neuroprotection on the damaged optic nerve: possible therapy for optic neuropathies. *Proc Natl Acad Sci USA*. 2000;97:7446–7451.
38. Kleinewietfeld M, Hafler DA. Regulatory T cells in autoimmune neuroinflammation. *Immunol Rev*. 2014;259: 231–244.
39. Kokaia Z, Martino G, Schwartz M, Lindvall O. Cross-talk between neural stem cells and immune cells: the key to better brain repair? *Nat Neurosci*. 2012;15:1078–1087.
40. Koronyo-Hamaoui M, Ko MK, Koronyo Y, et al. Attenuation of AD-like neuropathology by harnessing peripheral immune cells: local elevation of IL-10 and MMP-9. *J Neurochem*. 2009;111:1409–1424.
41. Koronyo Y, Salumbides BC, Sheyn J, et al. Therapeutic effects of glatiramer acetate and grafted $CD115^+$ monocytes in a mouse model of Alzheimer's disease. *Brain*. 2015;138.
42. Kunis G, Baruch K, Miller O, Schwartz M. Immunization with a myelin-derived antigen activates the Brain's choroid plexus for recruitment of immunoregulatory cells to the CNS and attenuates disease progression in a mouse model of ALS. *J Neurosci*. 2015;35:6381–6393.
43. Kunis G, Baruch K, Rosenzweig N, et al. IFN-gamma-dependent activation of the brain's choroid plexus for CNS immune surveillance and repair. *Brain*. 2013;136:3427–3440.
44. Lai AY, McLaurin J. Clearance of amyloid-beta peptides by microglia and macrophages: the issue of what, when and where. *Future Neurol*. 2012;7:165–176.
45. Madi A, Shifrut E, Reich-Zeliger S, et al. T-cell receptor repertoires share a restricted set of public and abundant CDR3 sequences that are associated with self-related immunity. *Genome Res*. 2014;24:1603–1612.
46. Martino G, Pluchino S, Bonfanti L, Schwartz M. Brain regeneration in physiology and pathology: the immune signature driving therapeutic plasticity of neural stem cells. *Physiol Rev*. 2011;91:1281–1304.
47. Merbl Y, Zucker-Toledano M, Quintana FJ, Cohen IR. Newborn humans manifest autoantibodies to defined self molecules detected by antigen microarray informatics. *J Clin Invest*. 2007;117:712–718.
48. Mildner A, Schlevogt B, Kierdorf K, et al. Distinct and non-redundant roles of microglia and myeloid subsets in mouse models of Alzheimer's disease. *J Neurosci*. 2011;31:11159–11171.
49. Moalem G, Leibowitz-Amit R, Yoles E, Mor F, Cohen IR, Schwartz M. Autoimmune T cells protect neurons from secondary degeneration after central nervous system axotomy. *Nat Med*. 1999;5:49–55.

50. Moalem G, Yoles E, Leibowitz-Amit R, et al. Autoimmune T cells retard the loss of function in injured rat optic nerves. *J Neuroimmunol*. 2000;106:189–197.
51. Monje ML, Toda H, Palmer TD. Inflammatory blockade restores adult hippocampal neurogenesis. *Science*. 2003;302:1760–1765.
52. Nagele EP, Han M, Acharya NK, DeMarshall C, Kosciuk MC, Nagele RG. Natural IgG autoantibodies are abundant and ubiquitous in human sera, and their number is influenced by age, gender, and disease. *PLoS One*. 2013;8:e60726.
53. Olsson T, Lidman O, Piehl F. Harm or heal–divergent effects of autoimmune neuroinflammation?. *Trends Immunol*. 2003;24:5–6. [author reply 7–8].
54. Ransohoff RM, Engelhardt B. The anatomical and cellular basis of immune surveillance in the central nervous system. *Nat Rev Immunol*. 2012;12:623–635.
55. Raposo C, Graubardt N, Cohen M, et al. CNS repair requires both effector and regulatory T cells with distinct temporal and spatial profiles. *J Neurosci*. 2014;34:10141–10155.
56. Redzic Z. Molecular biology of the blood–brain and the blood-cerebrospinal fluid barriers: similarities and differences. *Fluids Barriers CNS*. 2011;8:3.
57. Reines SA, Block GA, Morris JC, et al. Rofecoxib: no effect on Alzheimer's disease in a 1-year, randomized, blinded, controlled study. *Neurology*. 2004;62:66–71.
58. Ron-Harel N, Segev Y, Lewitus GM, et al. Age-dependent spatial memory loss can be partially restored by immune activation. *Rejuvenation Res*. 2008;11:903–913.
59. Sakaguchi S, Yamaguchi T, Nomura T, Ono M. Regulatory T cells and immune tolerance. *Cell*. 2008;133:775–787.
60. Schori H, Lantner F, Shachar I, Schwartz M. Severe immunodeficiency has opposite effects on neuronal survival in glutamate-susceptible and -resistant mice: adverse effect of B cells. *J Immunol*. 2002;169:2861–2865.
61. Schwartz M, Baruch K. Vaccine for the mind: immunity against self at the choroid plexus for erasing biochemical consequences of stressful episodes. *Hum Vaccin Immunother*. 2012;8:1465–1468.
62. Schwartz M, Baruch K. Breaking peripheral immune tolerance to CNS antigens in neurodegenerative diseases: boosting autoimmunity to fight-off chronic neuroinflammation. *J Autoimmun*. 2014a;54:8–14.
63. Schwartz M, Baruch K. The resolution of neuroinflammation in neurodegeneration: leukocyte recruitment via the choroid plexus. *EMBO J*. 2014b;33:7–22.
64. Schwartz M, Cohen IR. Autoimmunity can benefit self-maintenance. *Immunol Today*. 2000;21:265–268.
65. Schwartz M, Kipnis J. Autoimmunity on alert: naturally occurring regulatory CD4(+)CD25(+) T cells as part of the evolutionary compromise between a 'need' and a 'risk'. *Trends Immunol*. 2002;23:530–534.
66. Schwartz M, Shechter R. Protective autoimmunity functions by intracranial immunosurveillance to support the mind: the missing link between health and disease. *Mol Psychiatry*. 2010;15:342–354.
67. Shechter R, London A, Schwartz M. Orchestrated leukocyte recruitment to immune-privileged sites: absolute barriers versus educational gates. *Nat Rev Immunol*. 2013a;13:206–218.
68. Shechter R, London A, Varol C, et al. Infiltrating blood-derived macrophages are vital cells playing an anti-inflammatory role in recovery from spinal cord injury in mice. *PLoS Med*. 2009;6:e1000113.
69. Shechter R, Miller O, Yovel G, et al. Recruitment of beneficial M2 macrophages to injured spinal cord is orchestrated by remote brain choroid plexus. *Immunity*. 2013b;38:555–569.
70. Shechter R, Ziv Y, Schwartz M. New GABAergic interneurons supported by myelin-specific T cells are formed in intact adult spinal cord. *Stem Cells*. 2007;25:2277–2282.
71. Simard AR, Soulet D, Gowing G, Julien JP, Rivest S. Bone marrow-derived microglia play a critical role in restricting senile plaque formation in Alzheimer's disease. *Neuron*. 2006;49:489–502.
72. Stankiewicz JM, Kolb H, Karni A, Weiner HL. Role of immunosuppressive therapy for the treatment of multiple sclerosis. *Neurotherapeutics*. 2013;10:77–88.
73. Teijaro JR, Ng C, Lee AM, et al. Persistent LCMV infection is controlled by blockade of type I interferon signaling. *Science*. 2013;340:207–211.
74. Teles RM, Graeber TG, Krutzik SR, et al. Type I interferon suppresses type II interferon-triggered human anti-mycobacterial responses. *Science*. 2013;339:1448–1453.
75. Toomer KH, Chen Z. Autoimmunity as a double agent in tumor killing and cancer promotion. *Front Immunol*. 2014;5:116.
76. Town T, Laouar Y, Pittenger C, et al. Blocking TGF-β-Smad2/3 innate immune signaling mitigates Alzheimer-like pathology. *Nat Med*. 2008;14:681–687.

77. Venken K, Hellings N, Thewissen M, et al. Compromised CD4+ CD25(high) regulatory T-cell function in patients with relapsing-remitting multiple sclerosis is correlated with a reduced frequency of FOXP3-positive cells and reduced FOXP3 expression at the single-cell level. *Immunology*. 2008;123:79–89.
78. Walker LS, Abbas AK. The enemy within: keeping self-reactive T cells at bay in the periphery. *Nat Rev Immunol*. 2002;2:11–19.
79. Walsh JT, Zheng J, Smirnov I, Lorenz U, Tung K, Kipnis J. Regulatory T cells in central nervous system injury: a double-edged sword. *J Immunol*. 2014;193:5013–5022.
80. Wilson EB, Yamada DH, Elsaesser H, et al. Blockade of chronic type I interferon signaling to control persistent LCMV infection. *Science*. 2013;340:202–207.
81. Wolf SA, Fisher J, Bechmann I, Steiner B, Kwidzinski E, Nitsch R. Neuroprotection by T-cells depends on their subtype and activation state. *J Neuroimmunol*. 2002;133:72–80.
82. Wolf SA, Steiner B, Akpinarli A, et al. CD4-positive T lymphocytes provide a neuroimmunological link in the control of adult hippocampal neurogenesis. *J Immunol*. 2009;182:3979–3984.
83. Wolinsky JS, Narayana PA, O'Connor P, et al. Glatiramer acetate in primary progressive multiple sclerosis: results of a multinational, multicenter, double-blind, placebo-controlled trial. *Ann Neurol*. 2007;61:14–24.
84. Yoles E, Hauben E, Palgi O, et al. Protective autoimmunity is a physiological response to CNS trauma. *J Neurosci*. 2001;21:3740–3748.
85. Ziv Y, Ron N, Butovsky O, et al. Immune cells contribute to the maintenance of neurogenesis and spatial learning abilities in adulthood. *Nat Neurosci*. 2006;9:268–275.
86. Mesquita SD, Ferreira AC, Gao F, et al. The choroid plexus transcriptome reveals changes in type I and II interferon responses in a mouse model of Alzheimer's disease. *Brain Behav Immun*. 2015;49:280–292. PMID: 26092102.

CHAPTER

12

Neuroactive Steroids and Neuroinflammation

S. Giatti, R.C. Melcangi

Department of Pharmacological and Biomolecular Sciences, University of Milan, Milan, Italy

OUTLINE

12.1 INTRODUCTION

Steroid molecules play a key role in the physiological control of the central nervous system (CNS) and peripheral nervous system (PNS). These molecules, produced by peripheral steroidogenic glands, such as gonads or adrenal glands (ie, steroid hormones) or by the nervous system (ie, neurosteroids),[1] are now included in the neuroactive steroid family.[2] In the developing CNS, neuroactive steroids are responsible for the sex differences observed in several brain areas.[3,4] Indeed, synaptic connectivity, microtubules, and growth of axons are also affected by neuroactive steroids in a sex-specific manner. In the adult brain, they are involved in the control of sex behavior and in neuroendocrine regulation. Moreover, neuroactive steroids may control the release of neurotransmitters and the expression of their receptors, as

http://dx.doi.org/10.1016/B978-0-12-801914-6.00012-X

well as the remodeling of synapse and glial processes.[5,6] Neuroactive steroids exert important functions also in learning, memory formation, and consolidation, as well as in regulation of mood.[7–10] Furthermore, in the PNS, these molecules cooperate to regulate myelin synthesis, proliferation of Schwann cells (ie, the glial cell in the PNS), neuronal genes transcription, and GABA-glutamate pathways (reviewed in Ref. 11). Results obtained in the last decade have clearly indicated that these neuroactive steroids are also protective agents in neurodegenerative diseases. Neuroinflammation is a common feature shared by different clinical conditions such as Alzheimer's disease (AD), Parkinson's disease (PD), multiple sclerosis (MS), and traumatic brain injury (TBI). The state of the art of the protective effects exerted by neuroactive steroids on this important target will be reported here.

12.2 SYNTHESIS AND MECHANISM OF ACTION FOR NEUROACTIVE STEROIDS

As we just mentioned, the nervous system represents a steroidogenic organ. Both glial cells and neurons are involved in the synthesis and metabolism of neuroactive steroids, even if with different roles, depending on the enzymatic repertoire they express. However, due to the lipophilic nature of such compounds, a molecule could be produced in a cell and be metabolized in another one. This mechanism enables control of neuroactive steroid production and the ability to synthesize molecules that are specifically requested. Indeed, the steroidogenic process is a highly compartmentalized, chained sequence of reactions. The involved enzymes, even if poorly specific for substrates, are able to produce precise molecules thanks to the exact cellular localization. The first, limiting, and hormonal-controlled step of steroidogenesis is the accumulation of cholesterol from cytosol to the mitochondrial membrane and its subsequent internalization into this organelle. The process, schematically summarized in Fig. 12.1, is mediated by the steroidogenic acute regulatory protein (StAR) and also by the translocator protein of 18 kDa (TSPO), even if observations suggest that this channel is not essential for this process, as previously assumed.[12,13] In the inner mitochondrial membrane, cholesterol is converted by the cytochrome P450 side chain cleavage (P450scc) into pregnenolone (PREG), and further metabolized into progesterone (PROG) or dehydroepiandrosterone (DHEA) in the endoplasmic reticulum. The enzymes responsible for these conversions are the 3beta-hydroxysteroid dehydrogenase (3β-HSD) and the cytochrome P450c17, respectively. These enzymes are only expressed by astrocytes and neurons. DHEA may then be substrate for other androgens (ARs), such as androstenedione or testosterone (T). PROG and T can be metabolized by the enzyme 5alpha-reductase (5α-R) into dihydroprogesterone (DHP) and dihydrotestosterone (DHT), respectively. DHP is then converted into tetrahydroprogesterone (THP) and DHT into 3alpha, 5alpha-androstane, 17beta-diol (3α-diol). Interestingly, microglial cells are also able to exert these enzymatic conversions even if they seem not able to synthesize PROG and PREG. T through the action of the enzyme aromatase (ARO) may also be converted estradiol (17β-E_2; Fig. 12.1). This metabolic pathway is present in astrocytes and neurons but not in microglial cells.[14]

The biological effects exerted by neuroactive steroids are the consequence of the interaction with specific receptors. However, depending on the receptor utilized, the biological effect could be extremely different. For instance, PROG, T, and 17β-E_2 bind their classical steroid

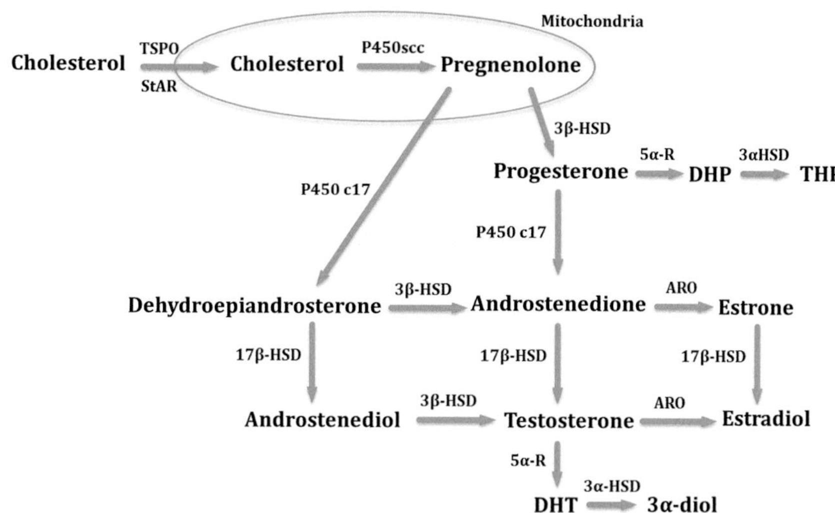

FIGURE 12.1 Synthesis of neuroactive steroids in the nervous system. Steroidogenic Acute Regulatory protein (StAR); Translocator protein of 18kDa (TSPO); Cytochrome P450 side chain cleavage (P450scc); 5alpha-reductase (5α-R); Cytochrome P450 17 hydroxylase (P450c17); 3beta-hydroxysteroid dehydrogenase (3β-HSD); 17beta-hydroxysteroid dehydrogenase (17β-HSD); Aromatase (ARO); 3alpha-hydroxysteroid dehydrogenase (3α-HSD). Dihydroprogesterone (DHP); Tetrahydroprogesterone (THP); Dihydrotestosterone (DHT); 3alpha, 5alpha-androstane, 17beta-diol (3α-diol).

receptors, such as PROG, AR, and estrogen (ER) receptors, promoting the transcription of target genes. Additionally, PROG could also bind to membrane PRs and to the membrane-associated protein PGRMC1, also termed 25-Dx.[15–17] Moreover, the metabolic conversion of PROG and T into their metabolites has a deep impact in their mechanisms of action. Indeed, the first metabolite of T, DHT, has a higher affinity for AR than its precursor, but the further metabolite, 3α-diol, is a ligand for a nonclassical steroid receptor, such as the gamma amino-butyric acid receptor type A ($GABA_A$-R). A similar situation also occurs in case of PROG. Indeed, PROG and DHP can bind to PR, whereas THP is a potent ligand for $GABA_A$-R.[18,19] Also other neuroactive steroids, such as PREG, DHEA, and their derivatives sulfates, exert their effects by nonclassical steroid receptors, like ion channels or other membrane receptors, such as sigma-1 receptor, alpha-1 receptor, nicotine receptor, dopamine 1 (D1) receptor, N-methyl-D-aspartate (NMDA) receptor, $GABA_A$-R, and L-type Ca^{2+} channels (also known as dihydropyridine channel).[20–22]

Similar to what was mentioned for steroidogenic enzymes, the expression of steroid receptors is dependent on cell type, regional localization, and stage of development. For example, PR, AR, and an isoform of ER (ie, ERβ) are not expressed in physiological conditions in microglial cells that, however, express another isoform of ER, such as the ERα. Astrocytes in male rat display a different pattern of expression for the AR. Thus, AR is expressed in cerebral cortex but only at postnatal day 10, whereas in hippocampus and arcuate nucleus its expression is reported only in adult animals. Different insults and injury may also affect the expression of receptors and enzymes. For example, ERα expression was increased in reactive astrocytes after brain injury, whereas the same insult was able to increase AR expression in reactive microglia. Enzymatic machinery may be controlled by inflammation. For instance, a decreased expression of 17β-HSD (ie, the enzyme that converts DHEA into androstenediol) was detected in glial cultures after the exposure to lipopolysaccharide (LPS).[23]

12.3 NEUROACTIVE STEROIDS AS NEUROPROTECTIVE AGENTS IN PATHOLOGICAL CONDITIONS: FOCUS ON NEUROINFLAMMATION

Neuroinflammation is an inflammatory reaction that occurs locally into the nervous system, with the aim to facilitate recovery and promote tissue repair. To reach this goal, neuroinflammatory response should be highly regulated and orchestrated; otherwise, the uncontrolled response leads to pathological conditions. The major players in neuroinflammation are microglial cells and astrocytes for the CNS, as well as Schwann cells for the PNS.[24–27]

As mentioned earlier, neuroinflammation is a common feature shared by different neurodegenerative events. Clinical and preclinical investigations have suggested a possible influence of the hormonal status on the etiopathogenesis of these pathological conditions. In particular, the incidence, susceptibility, and outcome of treatment are known to be different between men and women in many neurodegenerative diseases and after trauma.[28] Moreover, changes in neuroactive steroid levels during the estrous cycle, pregnancy, or after castration may have an impact in the outcome for such pathologies.[28] Furthermore, neuroactive steroid levels are affected in postmortem tissues from AD, PD, and MS brain patients.[29–32] A similar situation also occurs in experimental models of these disorders and after trauma.[33–39] It is interesting to note that, in the situation of acute neuroinflammation (ie, LPS injection), the brain steroidogenic machinery (eg, expression of StAR, 3β-HSD, and ARO) was stimulated,[40] possibly representing a strategy to cope with inflammatory events. In fact, the antiinflammatory and protective effects of neuroactive steroids in neuroinflammatory conditions have been extensively proved, in both in vitro and in vivo experimental models. In particular, the effects of ERs have been reported in several reviews.[41–45] Therefore, here we focus our attention on the effects of PROG, T, and particularly their metabolites in neuroinflammation.

In vitro observations, mainly obtained in glial cultures exposed to LPS, represent a useful tool to dissect the molecular mechanism of action for neuroactive steroids. In particular, it has been recently reported that PROG attenuated neuroinflammation in an in vitro model for acute brain injury, when administered to microglial cell line before LPS stimulation. Thus, PROG reduced tumor necrosis factor alpha (TNF-α), inducible nitric oxide synthase (iNOS), and cyclooxygenase 2 (COX-2) expression. Moreover, this neuroactive steroid suppressed nuclear factor KAPPA B (NF-κB) activation by decreasing inhibitory κBα, NF-κB p65 phosphorylation and p65 nuclear translocation, and finally, decreased the phosphorylation of p38, c-Jun N-terminal kinase and extracellular regulated kinase MAPKs.[46] Interestingly, these effects were counteracted by mifepristone, a PR antagonist, suggesting that the reported effects could be ascribed to the interaction of PROG to its classical steroid receptor.[46]

Also in animal models for brain trauma, PROG exerts protective effects. In a report obtained in the TBI model, endogenous brain levels of PROG are positively correlated to neurological recovery,[47] suggesting, as earlier, that increasing PROG levels may represent a strategy to cope with the trauma. In agreement, in the same experimental model, the treatment with

PROG decreases edema, acting on astrocytes in the cerebral cortex, and it is able to decrease the expression of NF-κB, of active C3 fragments, of the proinflammatory interleukin 1 beta (IL1β) and TNF-α.[48–52] Similarly, in a different animal model of trauma, such as the spinal cord injury, PROG exerts its action not only in astrocytes but also in macrophages/microglia, by decreasing their activation.[53,54]

As mentioned earlier, it is also important to note that the metabolism of PROG could be extremely relevant, since treatment with PROG metabolites, at least in some case, could be more effective than PROG administration per se. For example, the reduction of reactive gliosis, exerted by PROG in the hilus of hippocampus in ovariectomized rats after the injection of kainic acid,[55] was completely abolished by inhibitors of the enzymes that are able to metabolize PROG into DHP (ie, 5α-R) and subsequently this neuroactive steroid into THP (ie, 3α-HSD).[56] Moreover, in the TBI model, THP has been reported to reduce the expression of IL1β, TNF-α[50] and to promote the expression of CD55, a potent inhibitor of the complement convertases (ie, activators of the inflammatory cascade).[57] As mentioned earlier, the mechanism of action of THP involves the $GABA_A$-R[18,19]; however, since THP could be also retroconverted into DHP,[2] a possible action mediated by this last pathway could not be excluded.

Also T and its metabolites have been proven to be effective for the treatment of neuroinflammation. In particular, in the experimental model of brain injury, T administration was able to reduce the expression of vimentin, glial reactivity (eg, major histocompatibility complex of class II (MHC-II)), and glial fibrillary acidic protein staining.[48,58] Similar to that observed for PROG, the effects of T may also be mediated by the conversion into metabolites, such as 17β-E_2 or DHT. Indeed, the treatment with these two steroids seems to produce the same effects of T in glial cells.[58] However, as recently reported by Vincent and collaborators, the control of gliosis seems to be under AR (ie, by DHT). Thus, a novel AR-interacting protein, named p44/WDR77, has been identified. In particular, its deletion caused premature death and dramatic astrogliosis in mouse brain, while its loss in astrocyte in vitro produce the halt of cell cycle and astrogliosis.[59]

Another strategy for neuroprotection could be the induction, by pharmacological tools, of neuroactive steroids directly in the nervous system. A similar strategy could be very useful in order to bypass possible side effects related to the systemic treatment with neuroactive steroids. Examples of them may be provided by TSPO and liver X receptor (LXR) ligands. As mentioned earlier, activation of TSPO is reported to be involved in the translocation of cholesterol into mitochondria and consequently in the first step of steroidogenesis (ie, in the enzymatic conversion of this molecule into PREG), even if observations seem to deny this feature.[12,13,60] Despite that, at least in some experimental models, like for instance in diabetic animals, TSPO ligands are able to increase neuroactive steroid levels both in CNS[61] and PNS,[62] and to exert neuroprotective effects. Moreover, in an in vitro model, treatment with TSPO ligands, such as Ro5-4864 or PK11195, seemed to reduce inflammation.[63] Furthermore, results obtained in an animal model of MS showed that another TSPO ligand (ie, etifoxine) improves the clinical manifestation of the disease, decreases inflammation, and promotes remyelination.[64] Finally, in the 3xTg-AD mice model for AD, a different ligand, Ro5-4864, reduced the development of the pathology and β amyloid accumulation both in young and aged mice, also improving neuroactive steroid levels.[65]

In the PNS, treatment with GW3965, an LXR ligand, has been reported to improve diabetic neuropathy, thus, improving behavioral tests, expression of myelin protein and Na^+,

K^+-ATPase pump activity that are impaired by diabetic neuropathy. All these effects were associated with restoration of neuroactive steroid levels.[66] Nevertheless, LXR ligands have also been successfully applied in pathological conditions affecting the CNS, like in animal models of AD, PD, MS, and Huntington's disease. It is interesting to note that these ligands, together with the involvement of neuroactive steroids, could also mediate other processes, such as the activation of antiinflammatory pathways mediated by glial cells and the regulation of lipid metabolism. Thus, these molecules represent valuables tools, thanks to their broad spectrum of action. A review discusses these possible mechanisms within the CNS.[67]

12.4 NEUROACTIVE STEROIDS AND MULTIPLE SCLEROSIS

Several observations indicate that neuroactive steroids may exert an important role in MS. Indeed, the sex difference observed in the incidence and outcome of this pathology may support this concept. Thus, the ratio of incidence (female to male) is 4:1 during adolescence, and declines with age, until reaching a male prevalence, over 50 years of age.[68] It has been proposed that the decline of T levels in aging man could be permissive for disease onset,[69] and related to this, testicular hypofunction has been reported to be strongly correlated to the pathological onset.[70] Moreover, men present a worse prognosis and clinical outcome than women, who present a more benign clinical course.

Other differences between male and female patients are related to cognitive impairment, where women seem to perform better than men. MRI studies on functional connectivity and brain volumetric measures underlined gender differences.[71] Other clinical observations also suggest molecular sex differences. For instance, analyses performed on circulating cells revealed that, at least in male MS patients, regulatory T-cell (Treg) function was reduced. In particular, $CD4^+CD25^{+hi}$ Treg showed a significantly higher expression of ERβ, ERα, and PR with respect to responder T cells ($CD4^+CD25^-$), both in male MS patients and controls. It is interesting to note that the pathology is able to decrease plasma levels of 17β-E_2 and to reduce the expression of ERβ in $CD4^+CD25^-$ cells so that these cells are less prone to express a Treg phenotype.[72] Sex-specific alterations were also observed in autoptic tissues. Indeed, in male MS lesions, an increased gene expression of ARO, ERβ, and TNF was detected, whereas 3β-HSD and PR expression was upregulated in female tissues. Thus, these data described differences that possibly represent contributing factors for the gender bias observed in the pathology.[73]

A further support to a role of neuroactive steroids in the MS may be provided by the modifications of the disease course observed in female patients in relation to the hormonal milieu, for instance like those occurring during estrous cycle, menopause, and pregnancy. Thus, MS exacerbation immediately before the menstruation cycle, when ER and PROG levels are very low, has been reported.[74] Moreover, menopause worsens, while the hormone replacement therapy may possibly improve, the symptomatology.[68] Pregnant MS patients experience an improvement of the symptomatology, in particular during late pregnancy, when 17β-E_2 and PROG levels are high. On the contrary, the rate for relapses doubled during the first three months after delivery,[75] when the levels of neuroactive steroids dramatically drop. To date, the mechanism(s) favoring the protective effects of pregnancy remain elusive. A study obtained in an experimental model reported that pregnancy and pregnancy levels of 17β-E_2 improved the disease course and reduced demyelination in the CNS. In particular, these conditions seem to favor Treg and Th2 differentiation, increasing the expression of Foxp3 and GATA3, while reducing the expression of Th1 and Th17 markers including T-bet and ROR-γt (ie, two transcription factors exclusively expressed by Th1 and Th17 cells) and the cytokines IFN-γ, TNF-α, IL-17, and IL-23.[76]

Beside the observations that neuroactive steroids may alter disease course, the opposite is also observed. Thus, as observed in plasma and cerebrospinal fluid (CSF) of relapsing-remitting (RR) MS male patients, the levels of these molecules are modified.[77] It is interesting to note that a decrease in the levels of the active metabolite of T (ie, DHT) was observed in CSF of these patients,[77] providing a possible explanation for the protective effects observed in male MS patients treated with T.[78,79] In particular, after 12 months of transdermal T treatment, an improvement in the cognitive performance and a slowing of brain atrophy was reported.[78] Moreover, in a different clinical trial, with a similar schedule of treatment, an increase in the volume of gray matter in the frontal cortex of patients was observed.[79]

The best characterized and generally accepted model for MS is the experimental autoimmune encephalomyelitis (EAE). This experimental model, sharing with MS several features, is characterized by neuroinflammation, demyelination, and peripheral infiltration in the spinal cord. In the EAE model, with clinical manifestation similar to the human RR progression, alteration of neuroactive steroid levels was reported. In particular, the levels of neuroactive steroids were differently affected depending on the sex, phase of disease (ie, during the acute peak of inflammatory infiltration into the spinal cord or during the chronic phase of the disease), and nervous region considered.[36,37] These observations suggest that neuroactive steroids might represent a therapeutic and/or diagnostic strategy for this disease. Indeed, in the EAE model, ER administration was able to reduce the expression of several proinflammatory cytokine and chemokine, as well as the expression of their receptors, and to increase the expression of antiinflammatory cytokine.[80] Other observations also indicate a suppression of Th1 and Th17 and the promotions of Treg functions after estradiol administration.[76,81–83] ERs exert their neuroprotective effects through the two different isoform receptors. In particular, the ERβ-mediated neuroprotection does not involve the reduction of inflammation. On the contrary, activation of ERα in astrocytes was able to reduce the expression of different proinflammatory chemokine.[84]

Also PROG exerts protective effects in the EAE model. For instance, it has been reported that PROG exerts both antiinflammatory and promyelinating actions. In particular, the treatment with this neuroactive steroid was able to decrease the activation of microglial cell and infiltration of peripheral lymphocytes, to decrease the expression of proinflammatory cytokine and chemokine, and to ameliorate biochemical parameters affected by the disease.[85,86] Moreover, PROG promotes the expression of myelin proteins, increasing the expression of their transcription factors and the number of oligodendrocytes responsible for myelin production.[85,87]

Interestingly, in a different animal model of MS such as the demyelinating model induced by a toxic agent (eg, cuprizone), PROG is still effective in promoting myelin repair and decreasing neuroinflammation, both in the corpus callosum (ie, the main target in this model) and in the cerebral cortex.[88]

Similar to that described earlier for other neurodegenerative disorders, also in the EAE model the metabolism of PROG and T seems to be crucial in term of neuroprotection. In particular, the PROG metabolite THP exerts multiple actions in the contest of the MS (reviewed in Ref. 89). Indeed, in the spinal cord of EAE mice, the treatment with THP decreased Iba1 and CD33, markers for monocytoid cells and lymphocytes, respectively, improving the clinical manifestation of the disease, possibly reducing the recruitment of macrophages and circulating lymphocytes into the CNS.[30]

Another metabolite that seems to be effective in the context of EAE is the 5α-reduced product of T, DHT. It has been reported that in a mice EAE model, DHT treatment exerts a beneficial effect on the disease, decreasing expression for the proinflammatory cytokine IFN-γ and increasing antiinflammatory cytokine IL10.[90] In an RR experimental model for the pathology, it has been demonstrated that DHT administration to intact males improved the clinical manifestation of the disease, decreasing the expression of the proinflammatory cytokine IL1β, the number of astrocytes and microglial cells, and the oxidative stress in spinal cord.[91] Moreover, this neuroactive steroid exerts beneficial effects in the impairment of mitochondria, increasing the reduced mitochondrial content and the protein levels of mitochondrial respiratory chain complexes.[91]

12.5 CONCLUSIONS

Neuroactive steroids are important physiological regulators in the nervous system. In addition, they show interesting neuroprotective effects in several experimental models of neurodegenerative diseases affecting, among the different targets, an important pathological event like the neuroinflammation. In particular, possible therapeutic strategies based on PROG or T metabolites as well as by molecules able to increase the levels of neuroactive steroids (ie, TSPO or LXR ligands) seem to be very promising and need to be further explored in the future.

References

1. Baulieu EE. Neurosteroids: a novel function of the brain. *Psychoneuroendocrinology*. 1998;23(8):963–987. Available at: http://www.ncbi.nlm.nih.gov/pubmed/9924747. Accessed 10.03.15.
2. Melcangi RC, Garcia-Segura LM, Mensah-Nyagan AG. Neuroactive steroids: state of the art and new perspectives. *Cell Mol Life Sci*. 2008;65(5):777–797. http://dx.doi.org/10.1007/s00018-007-7403-5.
3. Rhodes ME, Rubin RT. Functional sex differences ('sexual diergism') of central nervous system cholinergic systems, vasopressin, and hypothalamic-pituitary-adrenal axis activity in mammals: a selective review. *Brain Res Brain Res Rev*. 1999;30(2):135–152. Available at: http://www.ncbi.nlm.nih.gov/pubmed/10525171. Accessed 23.03.15.
4. McCarthy MM, De Vries GFN. Sexual differentiation of the brain: mode, mechanisms and meaning. In: Pfaff D, Arnold AP, Etgen AM, Fahrbach SE, Rubin RT, eds. *Hormones, Brain and Behavior*. San Diego, CA: Academic Press; 2009:1707–1744.
5. García-Segura LM, Chowen JA, Párducz A, Naftolin F. Gonadal hormones as promoters of structural synaptic plasticity: cellular mechanisms. *Prog Neurobiol*. 1994;44(3):279–307. Available at: http://www.ncbi.nlm.nih.gov/pubmed/7886228. Accessed 13.03.15.
6. Garcia-Segura LM, Luquín S, Párducz A, Naftolin F. Gonadal hormone regulation of glial fibrillary acidic protein immunoreactivity and glial ultrastructure in the rat neuroendocrine hypothalamus. *Glia*. 1994;10(1):59–69. http://dx.doi.org/10.1002/glia.440100108.
7. Parducz A, Hajszan T, Maclusky NJ, et al. Synaptic remodeling induced by gonadal hormones: neuronal plasticity as a mediator of neuroendocrine and behavioral responses to steroids. *Neuroscience*. 2006;138(3):977–985. http://dx.doi.org/10.1016/j.neuroscience.2005.07.008.
8. Walf AA, Frye CA. A review and update of mechanisms of estrogen in the hippocampus and amygdala for anxiety and depression behavior. *Neuropsychopharmacology*. 2006;31(6):1097–1111. http://dx.doi.org/10.1038/sj.npp.1301067.

9. Frye CA. Progestins influence motivation, reward, conditioning, stress, and/or response to drugs of abuse. *Pharmacol Biochem Behav*. 2007;86(2):209–219. http://dx.doi.org/10.1016/j.pbb.2006.07.033.
10. Rupprecht R, Holsboer F. Neuroactive steroids in neuropsychopharmacology. *Int Rev Neurobiol*. 2001;46:461–477. Available at: http://www.ncbi.nlm.nih.gov/pubmed/11599310. Accessed 23.03.15.
11. Melcangi RC, Giatti S, Pesaresi M, et al. Role of neuroactive steroids in the peripheral nervous system. *Front Endocrinol (Lausanne)*. December 2011;2(104).
12. Morohaku K, Pelton SH, Daugherty DJ, Butler WR, Deng W, Selvaraj V. Translocator protein/peripheral benzodiazepine receptor is not required for steroid hormone biosynthesis. *Endocrinology*. 2014;155(1):89–97. http://dx.doi.org/10.1210/en.2013-1556.
13. Tu LN, Morohaku K, Manna PR, et al. Peripheral benzodiazepine receptor/translocator protein global knock-out mice are viable with no effects on steroid hormone biosynthesis. *J Biol Chem*. 2014;289(40):27444–27454. http://dx.doi.org/10.1074/jbc.M114.578286.
14. Garcia-Segura LM, Veiga S, Sierra A, Melcangi RC, Azcoitia I. Aromatase: a neuroprotective enzyme. *Prog Neurobiol*. 2003;71(1):31–41. Available at: http://www.ncbi.nlm.nih.gov/pubmed/14611865. Accessed 13.03.15.
15. Brinton RD, Thompson RF, Foy MR, et al. Progesterone receptors: form and function in brain. *Front Neuroendocrinol*. 2008;29(2):313–339. http://dx.doi.org/10.1016/j.yfrne.2008.02.001.
16. Guennoun R, Meffre D, Labombarda F, et al. The membrane-associated progesterone-binding protein 25-Dx: expression, cellular localization and up-regulation after brain and spinal cord injuries. *Brain Res Rev*. 2008;57(2): 493–505. http://dx.doi.org/10.1016/j.brainresrev.2007.05.009.
17. Schumacher M, Guennoun R, Stein DG, De Nicola AF. Progesterone: therapeutic opportunities for neuroprotection and myelin repair. *Pharmacol Ther*. 2007;116(1):77–106. http://dx.doi.org/10.1016/j.pharmthera.2007.06.001.
18. Belelli D, Lambert JJ. Neurosteroids: endogenous regulators of the GABA(A) receptor. *Nat Rev Neurosci*. 2005;6(7):565–575. http://dx.doi.org/10.1038/nrn1703.
19. Lambert JJ, Belelli D, Peden DR, Vardy AW, Peters JA. Neurosteroid modulation of GABAA receptors. *Prog Neurobiol*. 2003;71(1):67–80. Available at: http://www.ncbi.nlm.nih.gov/pubmed/14611869. Accessed 16.03.15.
20. Zheng P. Neuroactive steroid regulation of neurotransmitter release in the CNS: action, mechanism and possible significance. *Prog Neurobiol*. 2009;89(2):134–152. http://dx.doi.org/10.1016/j.pneurobio.2009.07.001.
21. Mellon SH, Griffin LD. Neurosteroids: biochemistry and clinical significance. *Trends Endocrinol Metab*. 2002;13(1):35–43. Available at: http://www.ncbi.nlm.nih.gov/pubmed/11750861. Accessed 10.03.15.
22. Melcangi RC, Panzica GC. Neuroactive steroids: old players in a new game. *Neuroscience*. 2006;138(3):733–739. http://dx.doi.org/10.1016/j.neuroscience.2005.10.066.
23. Saijo K, Collier JG, Li AC, Katzenellenbogen JA, Glass CK. An ADIOL-ERβ-CtBP transrepression pathway negatively regulates microglia-mediated inflammation. *Cell*. 2011;145(4):584–595. http://dx.doi.org/10.1016/j.cell.2011.03.050.
24. Meyer zu Hörste G, Hu W, Hartung H-P, Lehmann HC, Kieseier BC. The immunocompetence of Schwann cells. *Muscle Nerve*. 2008;37(1):3–13. http://dx.doi.org/10.1002/mus.20893.
25. Farina C, Aloisi F, Meinl E. Astrocytes are active players in cerebral innate immunity. *Trends Immunol*. 2007;28(3):138–145. http://dx.doi.org/10.1016/j.it.2007.01.005.
26. Traish AM, Guay AT, Zitzmann M. 5α-Reductase inhibitors alter steroid metabolism and may contribute to insulin resistance, diabetes, metabolic syndrome and vascular disease: a medical hypothesis. *Horm Mol Biol Clin Invest*. 2014;20(3):73–80. http://dx.doi.org/10.1515/hmbci-2014-0025.
27. Hanisch U-K, Kettenmann H. Microglia: active sensor and versatile effector cells in the normal and pathologic brain. *Nat Neurosci*. 2007;10(11):1387–1394. http://dx.doi.org/10.1038/nn1997.
28. Melcangi RC, Garcia-Segura LM. Sex-specific therapeutic strategies based on neuroactive steroids: in search for innovative tools for neuroprotection. *Horm Behav*. 2010;57(1):2–11. http://dx.doi.org/10.1016/j.yhbeh.2009.06.001.
29. Naylor JC, Kilts JD, Hulette CM, et al. Allopregnanolone levels are reduced in temporal cortex in patients with Alzheimer's disease compared to cognitively intact control subjects. *Biochim Biophys Acta*. 2010;1801(8):951–959. http://dx.doi.org/10.1016/j.bbalip.2010.05.006.
30. Noorbakhsh F, Ellestad KK, Maingat F, et al. Impaired neurosteroid synthesis in multiple sclerosis. *Brain*. 2011;134(Pt 9):2703–2721. http://dx.doi.org/10.1093/brain/awr200.
31. Luchetti S, Huitinga I, Swaab DF. Neurosteroid and GABA-A receptor alterations in Alzheimer's disease, Parkinson's disease and multiple sclerosis. *Neuroscience*. 2011;191:6–21. http://dx.doi.org/10.1016/j.neuroscience.2011.04.010.
32. Luchetti S, Bossers K, Van de Bilt S, et al. Neurosteroid biosynthetic pathways changes in prefrontal cortex in Alzheimer's disease. *Neurobiol Aging*. 2011;32(11):1964–1976. http://dx.doi.org/10.1016/j.neurobiolaging.2009.12.014.

33. Labombarda F, Pianos A, Liere P, et al. Injury elicited increase in spinal cord neurosteroid content analyzed by gas chromatography mass spectrometry. *Endocrinology*. 2006;147(4):1847–1859. http://dx.doi.org/10.1210/en.2005-0955.
34. Meffre D, Pianos A, Liere P, et al. Steroid profiling in brain and plasma of male and pseudopregnant female rats after traumatic brain injury: analysis by gas chromatography/mass spectrometry. *Endocrinology*. 2007;148(5): 2505–2517. http://dx.doi.org/10.1210/en.2006-1678.
35. Melcangi RC, Caruso D, Levandis G, Abbiati F, Armentero M-T, Blandini F. Modifications of neuroactive steroid levels in an experimental model of nigrostriatal degeneration: potential relevance to the pathophysiology of Parkinson's disease. *J Mol Neurosci*. 2012;46(1):177–183. http://dx.doi.org/10.1007/s12031-011-9570-y.
36. Caruso D, D'Intino G, Giatti S, et al. Sex-dimorphic changes in neuroactive steroid levels after chronic experimental autoimmune encephalomyelitis. *J Neurochem*. 2010;114(3):921–932.
37. Giatti S, D'Intino G, Maschi O, et al. Acute experimental autoimmune encephalomyelitis induces sex dimorphic changes in neuroactive steroid levels. *Neurochem Int*. 2010;56(1):118–127.
38. Caruso D, Barron AM, Brown MA, et al. Age-related changes in neuroactive steroid levels in 3xTg-AD mice. *Neurobiol Aging*. 2013;34(4):1080–1089. http://dx.doi.org/10.1016/j.neurobiolaging.2012.10.007.
39. Giatti S, Boraso M, Abbiati F, et al. Multimodal analysis in acute and chronic experimental autoimmune encephalomyelitis. *J Neuroimmune Pharmacol*. 2012:238–250. http://dx.doi.org/10.1007/s11481-012-9385-9.
40. Sadasivam M, Ramatchandirin B, Balakrishnan S, Selvaraj K, Prahalathan C. The role of phosphoenolpyruvate carboxykinase in neuronal steroidogenesis under acute inflammation. *Gene*. 2014;552(2):249–254. http://dx.doi.org/10.1016/j.gene.2014.09.043.
41. Arevalo M-A, Azcoitia I, Garcia-Segura LM. The neuroprotective actions of oestradiol and oestrogen receptors. *Nat Rev Neurosci*. 2014;16(1):17–29. http://dx.doi.org/10.1038/nrn3856.
42. Elkabes S, Nicot AB. Sex steroids and neuroprotection in spinal cord injury: a review of preclinical investigations. *Exp Neurol*. 2014;259:28–37. http://dx.doi.org/10.1016/j.expneurol.2014.01.008.
43. Petrone AB, Simpkins JW, Barr TL. 17β-estradiol and inflammation: implications for ischemic stroke. *Aging Dis*. 2014;5(5):340–345. http://dx.doi.org/10.14336/AD.2014.0500340.
44. Barton M, Prossnitz ER. Emerging roles of GPER in diabetes and atherosclerosis. *Trends Endocrinol Metab*. 2015. http://dx.doi.org/10.1016/j.tem.2015.02.003.
45. Acaz-Fonseca E, Sanchez-Gonzalez R, Azcoitia I, Arevalo MA, Garcia-Segura LM. Role of astrocytes in the neuroprotective actions of 17β-estradiol and selective estrogen receptor modulators. *Mol Cell Endocrinol*. 2014;389(1–2): 48–57. http://dx.doi.org/10.1016/j.mce.2014.01.009.
46. Lei B, Mace B, Dawson HN, Warner DS, Laskowitz DT, James ML. Anti-inflammatory effects of progesterone in lipopolysaccharide-stimulated BV-2 microglia. *PLoS One*. 2014;9(7):e103969. http://dx.doi.org/10.1371/journal.pone.0103969.
47. Lopez-Rodriguez AB, Acaz-Fonseca E, Giatti S, et al. Correlation of brain levels of progesterone and dehydroepiandrosterone with neurological recovery after traumatic brain injury in female mice. *Psychoneuroendocrinology*. 2015. http://dx.doi.org/10.1016/j.psyneuen.2015.02.018.
48. Garcia-Estrada J, Del Rio JA, Luquin S, Soriano E, Garcia-Segura LM. Gonadal hormones down-regulate reactive gliosis and astrocyte proliferation after a penetrating brain injury. *Brain Res*. 1993;628(1–2):271–278. Available at: http://www.ncbi.nlm.nih.gov/pubmed/8313156. Accessed 13.03.15.
49. Grossman KJ, Goss CW, Stein DG. Effects of progesterone on the inflammatory response to brain injury in the rat. *Brain Res*. 2004;1008(1):29–39. http://dx.doi.org/10.1016/j.brainres.2004.02.022.
50. He J, Evans C-O, Hoffman SW, Oyesiku NM, Stein DG. Progesterone and allopregnanolone reduce inflammatory cytokines after traumatic brain injury. *Exp Neurol*. 2004;189(2):404–412. http://dx.doi.org/10.1016/j.expneurol.2004.06.008.
51. Pettus EH, Wright DW, Stein DG, Hoffman SW. Progesterone treatment inhibits the inflammatory agents that accompany traumatic brain injury. *Brain Res*. 2005;1049(1):112–119. http://dx.doi.org/10.1016/j.brainres.2005.05.004.
52. Feeser VR, Loria RM. Modulation of traumatic brain injury using progesterone and the role of glial cells on its neuroprotective actions. *J Neuroimmunol*. 2011;237(1–2):4–12. http://dx.doi.org/10.1016/j.jneuroim.2011.06.007.
53. Garay L, Tüngler V, Deniselle MCG, Lima A, Roig P, De Nicola AF. Progesterone attenuates demyelination and microglial reaction in the lysolecithin-injured spinal cord. *Neuroscience*. 2011;192:588–597. http://dx.doi.org/10.1016/j.neuroscience.2011.06.065.
54. Labombarda F, González S, Lima A, et al. Progesterone attenuates astro- and microgliosis and enhances oligodendrocyte differentiation following spinal cord injury. *Exp Neurol*. 2011;231(1):135–146. http://dx.doi.org/10.1016/j.expneurol.2011.06.001.

55. Ciriza I, Azcoitia I, Garcia-Segura LM. Reduced progesterone metabolites protect rat hippocampal neurones from kainic acid excitotoxicity in vivo. *J Neuroendocrinol.* 2004;16(1):58–63. Available at: http://www.ncbi.nlm.nih.gov/pubmed/14962077. Accessed 16.03.15.
56. Ciriza I, Carrero P, Frye CA, Garcia-Segura LM. Reduced metabolites mediate neuroprotective effects of progesterone in the adult rat hippocampus. The synthetic progestin medroxyprogesterone acetate (Provera) is not neuroprotective. *J Neurobiol.* 2006;66(9):916–928. http://dx.doi.org/10.1002/neu.20293.
57. VanLandingham JW, Cekic M, Cutler S, Hoffman SW, Stein DG. Neurosteroids reduce inflammation after TBI through CD55 induction. *Neurosci Lett.* 2007;425(2):94–98. http://dx.doi.org/10.1016/j.neulet.2007.08.045.
58. Barreto G, Veiga S, Azcoitia I, Garcia-Segura LM, Garcia-Ovejero D. Testosterone decreases reactive astroglia and reactive microglia after brain injury in male rats: role of its metabolites, oestradiol and dihydrotestosterone. *Eur J Neurosci.* 2007;25(10):3039–3046. http://dx.doi.org/10.1111/j.1460-9568.2007.05563.x.
59. Vincent B, Wu H, Gao S, Wang Z. Loss of the androgen receptor cofactor p44/WDR77 induces astrogliosis. *Mol Cell Biol.* 2012;32(17):3500–3512. http://dx.doi.org/10.1128/MCB.00298-12.
60. Wolf L, Bauer A, Melchner D, et al. Enhancing neurosteroid synthesis – relationship to the pharmacology of translocator protein (18kDa) (TSPO) ligands and benzodiazepines. *Pharmacopsychiatry.* 2015;48(2):72–77. http://dx.doi.org/10.1055/s-0034-1398507.
61. Mitro N, Cermenati G, Giatti S, et al. LXR and TSPO as new therapeutic targets to increase the levels of neuroactive steroids in the central nervous system of diabetic animals. *Neurochem Int.* 2012;60(6):616–621.
62. Giatti S, Pesaresi M, Cavaletti G, et al. Neuroprotective effects of a ligand of translocator protein-18kDa (Ro5-4864) in experimental diabetic neuropathy. *Neuroscience.* 2009;164:520–529. http://dx.doi.org/10.1016/j.neuroscience.2009.08.005.
63. Bae K-R, Shim H-J, Balu D, Kim SR, Yu S-W. Translocator protein 18 kDa negatively regulates inflammation in microglia. *J Neuroimmune Pharmacol.* 2014;9(3):424–437. http://dx.doi.org/10.1007/s11481-014-9540-6.
64. Daugherty DJ, Selvaraj V, Chechneva OV, Liu X-B, Pleasure DE, Deng W. A TSPO ligand is protective in a mouse model of multiple sclerosis. *EMBO Mol Med.* 2013;5(6):891–903. http://dx.doi.org/10.1002/emmm.201202124.
65. Barron AM, Garcia-Segura LM, Caruso D, et al. Ligand for translocator protein reverses pathology in a mouse model of Alzheimer's disease. *J Neurosci.* 2013;33(20):8891–8897. http://dx.doi.org/10.1523/JNEUROSCI.1350-13.2013.
66. Cermenati G, Giatti S, Cavaletti G, et al. Activation of the liver X receptor increases neuroactive steroid levels and protects from diabetes-induced peripheral neuropathy. *J Neurosci.* 2010;30(36):11896–11901.
67. Xu P, Li D, Tang X, et al. LXR agonists: new potential therapeutic drug for neurodegenerative diseases. *Mol Neurobiol.* 2013;48:715–728. http://dx.doi.org/10.1007/s12035-013-8461-3.
68. Kipp M, Amor S, Krauth R, Beyer C. Multiple sclerosis: neuroprotective alliance of estrogen-progesterone and gender. *Front Neuroendocrinol.* 2012;33(1):1–16. http://dx.doi.org/10.1016/j.yfrne.2012.01.001.
69. Weinshenker BG. Natural history of multiple sclerosis. *Ann Neurol.* 1994;36(suppl):S6–S11. Available at: http://www.ncbi.nlm.nih.gov/pubmed/8017890. Accessed 13.03.15.
70. Pakpoor J, Goldacre R, Schmierer K, Giovannoni G, Goldacre MJ. Testicular hypofunction and multiple sclerosis risk: a record-linkage study. *Ann Neurol.* 2014;76(4):625–628. http://dx.doi.org/10.1002/ana.24250.
71. Schoonheim MM, Hulst HE, Landi D, et al. Gender-related differences in functional connectivity in multiple sclerosis. *Mult Scler.* 2012;18(2):164–173. http://dx.doi.org/10.1177/1352458511422245.
72. Aristimuño C, Teijeiro R, Valor L, et al. Sex-hormone receptors pattern on regulatory T-cells: clinical implications for multiple sclerosis. *Clin Exp Med.* 2012;12(4):247–255. http://dx.doi.org/10.1007/s10238-011-0172-3.
73. Luchetti S, van Eden CG, Schuurman K, van Strien ME, Swaab DF, Huitinga I. Gender differences in multiple sclerosis: induction of estrogen signaling in male and progesterone signaling in female lesions. *J Neuropathol Exp Neurol.* 2014;73(2):123–135. http://dx.doi.org/10.1097/NEN.0000000000000037.
74. Smith R, Studd JW. A pilot study of the effect upon multiple sclerosis of the menopause, hormone replacement therapy and the menstrual cycle. *J R Soc Med.* 1992;85(10):612–613. Available at: http://www.pubmedcentral.nih.gov/articlerender.fcgi?artid=1293688&tool=pmcentrez&rendertype=abstract. Accessed 04.02.15.
75. Confavreux C, Hutchinson M, Hours MM, Cortinovis-Tourniaire P, Moreau T. Rate of pregnancy-related relapse in multiple sclerosis. Pregnancy in Multiple Sclerosis Group. *N Engl J Med.* 1998;339(5):285–291. http://dx.doi.org/10.1056/NEJM199807303390501.
76. Haghmorad D, Amini AA, Mahmoudi MB, Rastin M, Hosseini M, Mahmoudi M. Pregnancy level of estrogen attenuates experimental autoimmune encephalomyelitis in both ovariectomized and pregnant C57BL/6 mice through expansion of Treg and Th2 cells. *J Neuroimmunol.* 2014;277(1–2):85–95. http://dx.doi.org/10.1016/j.jneuroim.2014.10.004.

77. Caruso D, Melis M, Fenu G, et al. Neuroactive steroid levels in plasma and cerebrospinal fluid of male multiple sclerosis patients. *J Neurochem*. 2014;130(4):591–597. http://dx.doi.org/10.1111/jnc.12745.
78. Sicotte NL, Giesser BS, Tandon V, et al. Testosterone treatment in multiple sclerosis: a pilot study. *Arch Neurol*. 2007;64(5):683–688. http://dx.doi.org/10.1001/archneur.64.5.683.
79. Kurth F, Luders E, Sicotte NL, et al. Neuroprotective effects of testosterone treatment in men with multiple sclerosis. *NeuroImage Clin*. 2014;4:454–460. http://dx.doi.org/10.1016/j.nicl.2014.03.001.
80. Spence RD, Voskuhl RR. Neuroprotective effects of estrogens and androgens in CNS inflammation and neurodegeneration. *Front Neuroendocrinol*. 2012;33(1):105–115. http://dx.doi.org/10.1016/j.yfrne.2011.12.001.
81. Bebo BF, Schuster JC, Arthur A. Androgens alter the cytokine profile and reduce encephalitogenicity of myelin-reactive T cells. *J Immunol*. 2012;162.
82. Offner H. Neuroimmunoprotective effects of estrogen and derivatives in experimental autoimmune encephalomyelitis: therapeutic implications for multiple sclerosis. *J Neurosci Res*. 2004;78(5):603–624. http://dx.doi.org/10.1002/jnr.20330.
83. Wang C, Dehghani B, Li Y, Kaler LJ, Vandenbark AA, Offner H. Oestrogen modulates experimental autoimmune encephalomyelitis and interleukin-17 production via programmed death 1. *Immunology*. 2009;126(3):329–335. http://dx.doi.org/10.1111/j.1365-2567.2008.03051.x.
84. Spence RD, Wisdom AJ, Cao Y, et al. Estrogen mediates neuroprotection and anti-inflammatory effects during EAE through ERα signaling on astrocytes but not through ERβ signaling on astrocytes or neurons. *J Neurosci*. 2013;33(26):10924–10933. http://dx.doi.org/10.1523/JNEUROSCI.0886-13.2013.
85. Giatti S, Caruso D, Boraso M, et al. Neuroprotective effects of progesterone in chronic experimental autoimmune encephalomyelitis. *J Neuroendocrinol*. 2012;24:851–861. http://dx.doi.org/10.1111/j.1365-2826.2012.02284.x.
86. Yates MA, Li Y, Chlebeck P, Proctor T, Vandenbark AA, Offner H. Progesterone treatment reduces disease severity and increases IL-10 in experimental autoimmune encephalomyelitis. *J Neuroimmunol*. 2010;220(1–2):136–139. http://dx.doi.org/10.1016/j.jneuroim.2010.01.013.
87. Garay LI, Gonza MC, Brocca ME, Lima A, Roig P, De Nicola AF. Progesterone down-regulates spinal cord inflammatory mediators and increases myelination in experimental autoimmune encephalomyelitis. *Neuroscience*. 2012;226:40–50.
88. El-Etr M, Rame M, Boucher C, et al. Progesterone and nestorone promote myelin regeneration in chronic demyelinating lesions of corpus callosum and cerebral cortex. *Glia*. 2014. http://dx.doi.org/10.1002/glia.22736.
89. Noorbakhsh F, Baker GB, Power C. Allopregnanolone and neuroinflammation: a focus on multiple sclerosis. *Front Cell Neurosci*. June 2014;8(134). http://dx.doi.org/10.3389/fncel.2014.00134.
90. Dalal M, Kim S, Voskuhl RR. Testosterone therapy ameliorates experimental autoimmune encephalomyelitis and induces a T helper 2 bias in the autoantigen-specific T lymphocyte response. *J Immunol*. 1997;159(1):3–6. Available at: http://www.ncbi.nlm.nih.gov/pubmed/9200430. Accessed 11.03.15.
91. Giatti S, Rigolio R, Romano S, et al. Dihydrotestosterone as protective agent in chronic experimental autoimmune encephalomyelitis. *Neuroendocrinology*. 2015. http://dx.doi.org/10.1159/000381064.

SECTION III

SURROGATE MARKERS IN MULTIPLE SCLEROSIS

CHAPTER

13

Surrogate Markers in Multiple Sclerosis: The Role of Magnetic Resonance Imaging

M.A. Rocca, G. Comi, M. Filippi
Vita-Salute San Raffaele University, Milan, Italy

OUTLINE

Translational Neuroimmunology in Multiple Sclerosis
http://dx.doi.org/10.1016/B978-0-12-801914-6.00013-1

13.1 INTRODUCTION

Magnetic resonance imaging (MRI) has a high sensitivity in revealing macroscopic tissue abnormalities in patients with multiple sclerosis (MS). Conventional magnetic resonance (MR) sequences, such as dual-echo, fluid-attenuated inversion recovery (FLAIR), and T1-weighted, both with and without gadolinium (Gd) contrast agent administration (Fig. 13.1), provide important pieces of information for diagnosing MS, understanding its natural history, and assessing treatment efficacy. Dual-echo and FLAIR imaging have a high sensitivity in detecting MS lesions, which appear as hyperintense focal areas on these scans. However, they lack specificity to the heterogeneous pathological substrates of individual lesions. Gd-enhanced T1-weighted images allow active lesions to be distinguished from inactive lesions, since enhancement occurs as a result of increased blood–brain barrier permeability and corresponds to areas with ongoing inflammation. Finally, lesions that persistently appear dark on postcontrast T1-weighted images are associated with more severe tissue damage (both demyelination and axonal loss) compared to lesions that do not appear dark on such images (Fig. 13.1).

Disappointingly, the strength of the association between conventional MRI findings and the clinical manifestations of the disease remains modest in patients with MS. This is likely due to the relative lack of specificity of conventional MRI in the evaluation of the heterogeneous pathological substrates of the disease, its inability to provide accurate estimates of damage outside focal lesions, and the fact that it cannot provide information on central nervous system (CNS) functional reorganization after tissue injury has occurred. Structural, metabolic, and functional MRI techniques have allowed the identification of novel markers of the disease, more closely linked to its pathological features, which may in part overcome the limitations of conventional MRI.

This chapter discusses the main insights derived from the application of MRI-based techniques to diagnose MS, define its pathophysiology, and monitor the efficacy of disease-modifying treatments.

13.2 MAGNETIC RESONANCE IMAGING AND THE DIAGNOSIS OF MULTIPLE SCLEROSIS

13.2.1 Features of Multiple Sclerosis Lesions

The characterization of lesion features suggestive of MS on conventional MR scans is central in the diagnostic work-up of patients suspected of having this condition. Brain MS lesions are frequently located, asymmetrically between the two hemispheres, in the periventricular

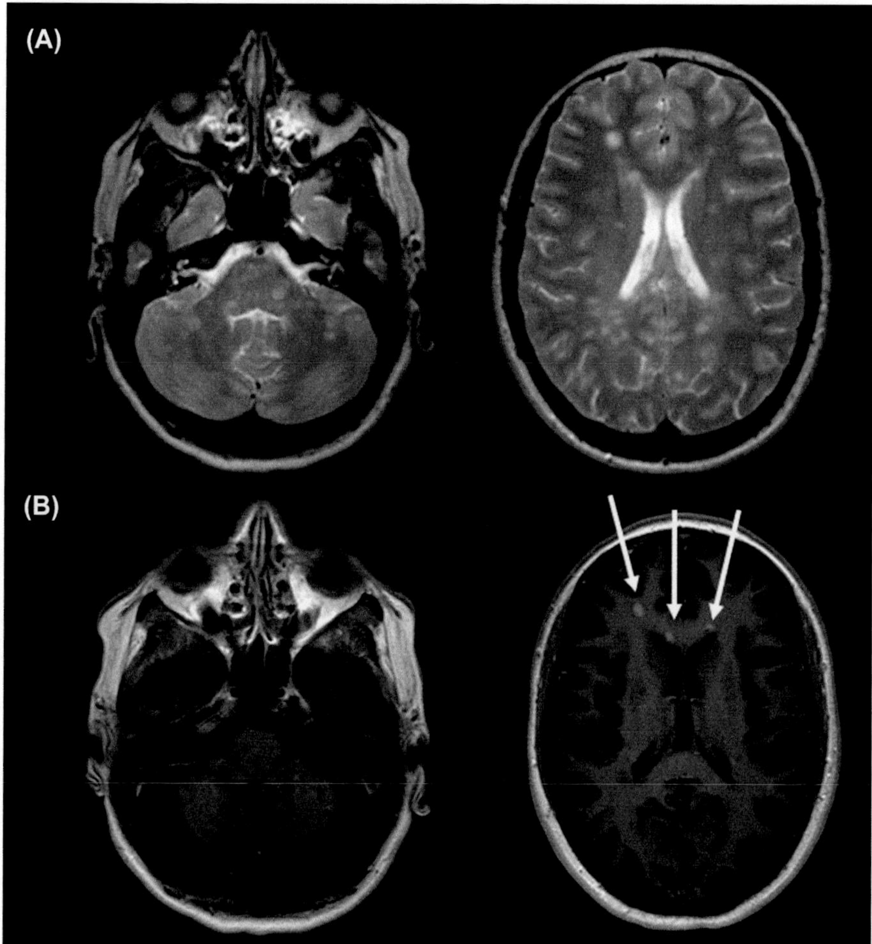

FIGURE 13.1 Brain axial T2-weighted (A) and postcontrast T1-weighted (B) spin-echo magnetic resonance images from a 40-year-old patient with relapsing-remitting multiple sclerosis. In A, multiple hyperintense lesions are visible, which suggest multifocal white matter pathology. In B, several of these lesions (*white arrows*) are clearly contrast-enhanced, which indicates the presence of a local disruption of the blood–brain barrier.

and juxtacortical white matter (WM), the corpus callosum (CC) (where the so-called Dawson's fingers can be seen) and infratentorial areas (with the pons and cerebellum more frequently affected than the medulla and midbrain). Such lesions can have oval or elliptical shapes.[1] Consensus has also been reached on criteria useful to identify T2-hyperintense[2] and T1-enhancing lesions.[3]

13.2.2 Magnetic Resonance Imaging Diagnostic Criteria

In 2001, MRI has been formally included in the diagnostic work-up of patients suspected of having MS by an International Panel (IP) of MS experts.[4] The definition of MRI criteria for a diagnosis of MS is based, on the one hand, on the demonstration of lesion dissemination in space (DIS) and dissemination in time (DIT), and, on the other, on the exclusion of alternative neurological conditions. The original criteria have been revised several times to simplify them and to implement their use in the clinical setting. According to the Swanton criteria,[5] at least one subclinical T2 lesion in at least two of the four locations defined as characteristic for MS (ie, juxtacortical, periventricular, infratentorial, and spinal cord) is required for DIS. Rovira et al.[6] suggested that a single brain MRI study performed early (ie, <3 months) after the onset of a clinically isolated syndrome (CIS) is highly specific for predicting the development of MS in the presence of both Gd-enhancing and nonenhancing lesions, which when present suggest DIT. Both the previous criteria have been included in the most recently published revision of the IP criteria. In addition, according to this revision, a new or enhancing lesion at any time with respect to the baseline scan suffices to demonstrate DIT.[7]

One aspect that has not been considered yet in the available sets of MRI diagnostic criteria is the role of cortical lesions (CL), some of which can be visualized using double-inversion recovery (DIR) sequences[8] (Fig. 13.2). The sensitivity of such lesion detection in the context of MS diagnosis has been considered by Filippi and coworkers,[9] who proposed a model for DIS that includes the presence of at least one intracortical lesion in addition to the presence of at least one infratentorial and one spinal cord or Gd-enhancing lesion.

Another aspect that has been investigated is the contribution of spinal cord involvement in CIS patients in diagnosing MS according to the revised 2010 IP criteria and in predicting conversion to definite MS.[10] The presence of spinal cord lesions facilitated the diagnosis of MS and was predictive of conversion to definite MS, especially in those patients without spinal CIS and who did not fulfill brain MRI criteria.

13.2.3 Magnetic Resonance Imaging and Prognosis in Clinically Isolated Syndrome

In several studies, MRI findings at disease onset that showed the strongest predictive value for the subsequent development of definite MS were the number and extent of brain T2 lesions,[11–13] the presence of infratentorial lesions,[13] and the presence of Gd-enhancing lesions.[14] For patients with CIS and brain MRI lesions, the chance of developing definite MS was >80% over the subsequent 20 years, in the longest follow-up study to date.[11,12] Baseline T2 lesion load and the accumulation of such lesions in the first 5 years after clinical onset are strong predictors of disability accumulation over time in these patients.[12] In a longitudinal study,[15] baseline T1 hypointense lesion number and volume predicted the severity of executive deficits, and new T2 lesions at a 3-month follow-up predicted slowed information processing 7 years later.

13.3 MAGNETIC RESONANCE IMAGING AND PATHOPHYSIOLOGY OF MULTIPLE SCLEROSIS

13.3.1 Conventional Magnetic Resonance Imaging

T2 Lesions

Lesion burden on T2-weighted MRI scans of MS patients increases by about 5–10% per year. The magnitude of the correlation between T2 lesion burden and disability in cross-sectional

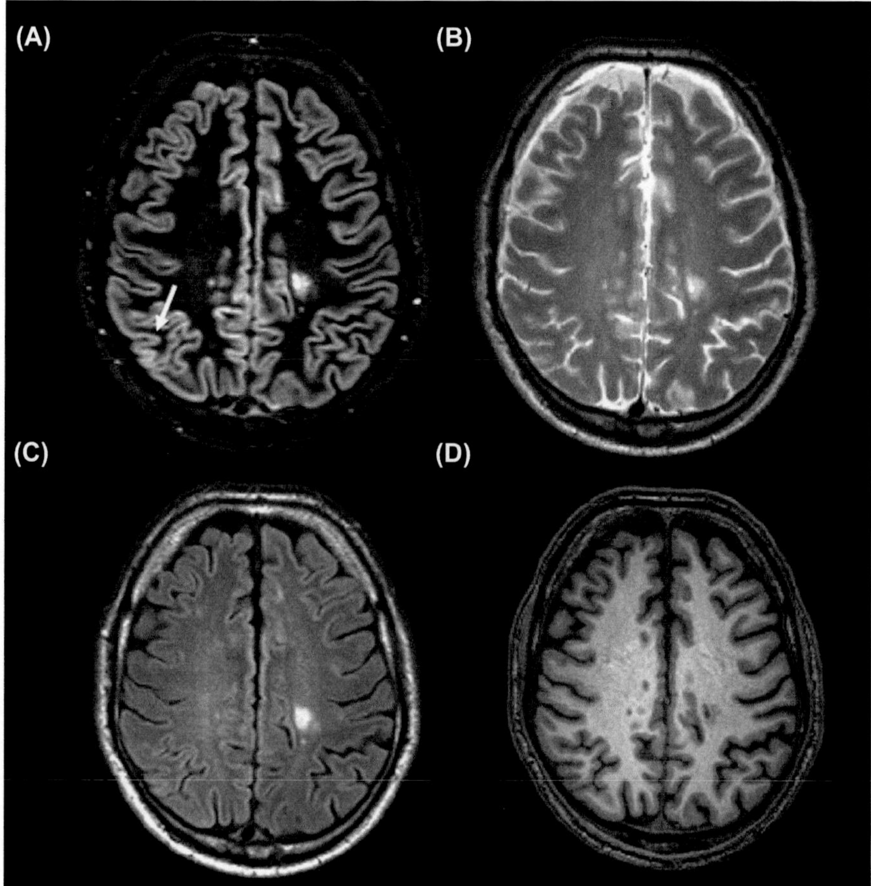

FIGURE 13.2 Brain axial double-inversion recovery (A), T2-weighted (B), fluid-attenuated inversion recovery (C), and T1-weighted (D) sequences of a patient with a clinically isolated syndrome. A cortical lesion (*white arrow*) is clearly visible in A and can be retrospectively identified in B, C, and D.

studies is disappointing. The presence of this so-called clinicoradiological paradox might be due to the poor pathological specificity of T2 hyperintense lesions that do not distinguish edema and inflammation from irreversible demyelination and axonal loss. This may also simply be a reflection of the fact that many T2 lesions are clinically silent.[16]

A plateauing relationship between T2 lesion load and disability has been suggested for Expanded Disability Status Scale (EDSS) scores higher than 4.5[17]; however, this finding has not been confirmed by subsequent studies.[18,19]

Gd-Enhancing Lesions

Serial MRI studies have shown that enhancement occurs in virtually all new lesions in patients with relapsing-remitting MS (RRMS) or secondary-progressive MS (SPMS) and can sometimes be detected even before the onset of clinical symptoms.[20] The frequency of MRI activity varies according to the clinical phenotype of the disease, being higher in RRMS and SPMS than in primary progressive MS (PPMS) and benign MS (BMS).[21]

The number of enhancing lesions increases shortly before and during clinical relapses and predicts subsequent MRI activity. As noted for T2 lesions, contrast-enhancing lesions also show a relatively modest correlation with disability accumulation.[22]

Recently, using dynamic contrast-enhanced MRI and ultrahigh field magnets, different patterns of centrifugal and centripetal Gd enhancement have been described,[23–25] suggesting a possible replacement of the previously accepted definition of nodular and ring-like lesions based on single post-Gd T1-weighted scans in favor of a paradigm based on spatiotemporal enhancement dynamics.

T1-Hypointense Lesions

A subset of T2 lesions (around 30–40%) appears persistently dark on postcontrast T1-weighted images on serial scans and represent regions where irreversible axonal loss, demyelination, and gliosis has occurred.[26] T1-hypointense lesions are only a few in the early stage of MS and increase over the course of the disease. Studies assessing the correlations between T1-hypointense lesion burden and disability provided conflicting results, since some of them found such a correlation to be stronger than for T2 lesions, while others did not.

Cortical Lesions

DIR sequences (which use two inversion times to suppress the signal from both WM and cerebrospinal fluid) have improved the sensitivity of MRI to detect CL lesions in vivo (a gain of 538% has been reported versus the use of T2-weighted sequences).[8] Using DIR sequences, CLs have been detected in all the major MS clinical phenotypes, including CIS patients.[27] CLs are more frequently seen in patients with SPMS than in those with CIS or RRMS.[27] Such lesions have also been visualized in the hippocampus.[28] Longitudinal studies have shown that new CLs continue to form in patients with early RRMS and in those with the progressive disease phenotypes over 1- to 2-year periods of follow-up.[29–32] An association has been found between CL burden and progression of disability over the subsequent 2–5 years[29,30] in patients with different MS phenotypes, as well as between CL burden and the severity of cognitive impairment.[29,32–34]

Brain Atrophy

In MS patients, brain volume decreases on average by about 0.7–1% yearly[35] (Fig. 13.3). Although brain atrophy measurements are more pathologically specific than T2 lesion load measurements, they are at best only moderately correlated with disability in RRMS and SPMS.[35,36] The strength of such a correlation increases when neuropsychological impairment is considered[37] and in longitudinal studies.[38,39] A large-scale, 14-month follow-up study[40] of untreated MS patients showed that the progression of brain atrophy is independent of clinical phenotype, when such an assessment is corrected for brain volume at baseline.

Improvements in methods of analysis have allowed quantification of the extent of tissue loss in gray matter (GM) and WM, separately, and also to define the distribution of atrophy at a regional level. Cross-sectional and longitudinal studies showed that GM atrophy occurs from the early stages of the disease[41,42] is associated with MS clinical disability[43–45] and cognitive deterioration,[46,47] and tends to worsen over time.[48] Chen et al.[49] measured

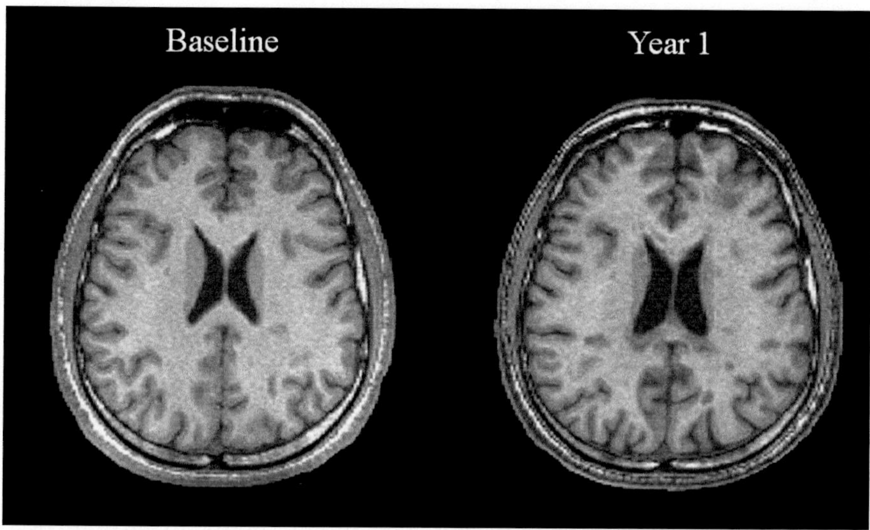

FIGURE 13.3 Two T1-weighted magnetic resonance imaging scans from a relapsing-remitting multiple sclerosis patient acquired at baseline and after 1 year of follow-up. Significant ventricular and cortical atrophy is visible at follow-up.

cortical thickness from patients with stable disease or progressing disability, and showed an increased rate of cortical tissue loss in the latter group. Fisher et al.[50] compared atrophy rates over 4 years across the main MS clinical phenotypes and found that GM atrophy rate increases with disease stage, from 3.4-fold normal in CIS patients converting to RRMS to 14-fold normal in SPMS.

The topography of atrophy varies in the different brain structures and in different phases of the disease, as suggested by several voxel-based morphometry studies. In CIS patients, GM atrophy involves mainly the deep GM nuclei.[51] In RRMS, atrophy of the frontotemporal lobes is typically detected.[52] In SPMS patients, atrophy of deep GM structures, brain stem, cerebellum, and several cortical regions (virtually in all lobes) is observed.[53]

Assessment of atrophy of strategic GM structures could contribute to explain deficits in selective cognitive domains, as well as the occurrence of specific symptoms and disability progression. Hippocampal atrophy has been associated with deficits in memory encoding and retrieval,[54,55] atrophy of the frontal and parietal lobes has been correlated with fatigue,[56] and thalamic atrophy has been correlated with accumulation of disability after an 8-year follow-up period in patients with relapse-onset MS[57] and after a 5-year follow-up period in PPMS patients.[58]

Despite these promising findings, brain atrophy has several important limitations that prevent this measure from being an optimal marker of MS progression. First, measurements of brain atrophy do not distinguish axonal injury from myelin loss. Second, reactive gliosis has the potential to mask considerable tissue loss. Third, brain atrophy fluctuates considerably with the amount of tissue water, which, in turn, can be affected by the presence of vasogenic edema from active inflammation, the administration of anti-inflammatory treatments such as steroids and dehydration. Fourth, atrophy is an end-stage phenomenon. Although detection of atrophy may be considered as a robust end point, the ability to monitor MS at a stage prior to irreversible tissue loss would seem preferable to simply measuring its end result.

13.3.2 Quantitative Structural and Metabolic Magnetic Resonance Imaging Techniques

In quantitative MR-based techniques, including magnetization transfer (MT)[59] and diffusion tensor (DT),[60] MRI can quantify the extent and improve the characterization of the nature of structural changes occurring within and outside focal MS lesions. Proton MR spectroscopy (^{1}H-MRS)[61] can add information on the biochemical nature of such abnormalities. T2 hypointense areas and reduced T2 relaxation time (RT) are thought to reflect iron deposition, which is believed to be a sign of neurodegeneration.

Magnetization Transfer Magnetic Resonance Imaging

MT MRI, which is based on the interactions between protons in free fluid and protons bound to macromolecules, allows the calculation of an index, the MT ratio (MTR), which, when reduced, indicates a diminished capacity of the protons bound to the brain tissue matrix to exchange magnetization with the surrounding free water. As a consequence, this index provides an estimate of the extent of MS tissue disruption. Variable degrees of MTR reduction have been reported in acute and chronic MS lesions, with the most prominent changes found in T1-hypointense lesions. Several studies with serial scanning showed that at least in some lesions dramatic changes in normal-appearing WM (NAWM) areas can be seen days to weeks before the development of enhancing lesions.[62] Reduced MTR values have been found in NAWM and GM of MS patients, including those with CIS. Such MT MRI abnormalities are more severe in patients with the progressive clinical phenotypes and tend to worsen over time.[62] MT MRI changes in the NAWM and GM correlate with the severity of clinical disability and cognitive impairment.[63] In addition, in patients with relapse-onset MS, GM MTR was found to be an independent predictor of cognitive deterioration over the subsequent 13 years.[64] In PPMS patients, GM MTR decline was shown to reflect the rate of clinical deterioration over 3 years[65] and it was the best predictor of poor cognitive performance after 5.5 years.[66]

Voxel-wise procedures have been applied to track longitudinal changes of MTR values within individual, newly formed MS lesions, and to map the regional distribution of microscopic damage to the NAWM and GM. Chen et al.[67] developed a method to monitor the evolution of MTR changes of individual lesion voxels and found significant changes of lesional MTR consistent with demyelination and remyelination that followed different temporal evolutions and that were still present in some lesions 3 years after their formation. Recently, a postmortem study reported the correlation between MTR values and focal cortical demyelination, supporting the notion that this technique is sensitive to demyelination/remyelination processes also in the cortex.[68]

Diffusion Tensor Magnetic Resonance Imaging

Diffusion-weighted MRI is a quantitative technique that exploits the diffusion of water within biological tissues. The diffusion coefficient measures the ease of this translational motion of water. In biological tissues, this coefficient is lower than that in free water because the various structures of the tissues (membranes, macromolecules, etc.) impede the free movement of water molecules. For this reason, the measured diffusion coefficient in biological systems is referred to as the apparent diffusion coefficient (ADC). Pathological processes that alter tissue integrity typically reduce the impediments to free water motion and, as a result, these processes tend to increase the measured ADC values. A full characterization of diffusion can be provided in terms of a tensor,[69] which has a principal axis and two smaller axes that describe its width and depth. The diffusivity along the principal axis is also called parallel or axial diffusivity (AD), while the diffusivities in the two minor axes are often averaged to produce a measure of radial diffusivity (RD). It is also possible to calculate the magnitude of diffusion, reflected by the mean diffusivity; and the degree of anisotropy, which is a measure of tissue organization that can be expressed by several indexes, including a dimensionless one named fractional anisotropy (FA). Clearly, the different and heterogeneous pathological elements of MS have the potential to alter the microarchitecture of brain tissue influencing water diffusion in the CNS. In particular, axonal damage has been shown to alter predominantly FA and AD,[69] while myelin breakdown has been associated with an increased RD.[69,70]

In line with MT MRI findings, DT MRI studies confirmed the heterogeneity of MS-related damage to T2 lesions, NAWM and GM, and showed that DT MRI abnormalities may precede lesion formation. Using DT MRI, an increased FA has been found in CLs of MS patients, suggesting local loss of dendrites and microglial activation.[71–73] DT MRI abnormalities of the NAWM, cortex, and deep GM nuclei are present from the earliest stages of MS and become more pronounced with increasing disease duration and neurological impairment.[60] Longitudinal DT MRI studies demonstrated a worsening of GM damage over time in patients with RRMS,[74] SPMS, and PPMS.[75–77] The severity of GM damage has been correlated with the degree of cognitive impairment in mildly disabled RRMS patients,[78] and has been found to predict accumulation of disability over a 5-year period in patients with PPMS.[76]

Several approaches have been developed to investigate damage to selected WM tracts, with the ultimate goal of improving the correlation with clinical measures. These approaches include the use of DT tractography, and the quantification of abnormalities at a voxel level, by means of voxel-based or tract-based spatial statistics (TBSS) analyses. In patients with CIS and definite MS, diffusivity measures of the corticospinal tract (CST) correlate with clinical measures of motor impairment.[79,80] Using TBSS, FA[81] and RD[82] abnormalities of the CC and CST have been related to clinical disability in RRMS patients. Diffusivity abnormalities in optic radiations have been related to trans-synaptic degeneration secondary to optic nerve damage and Wallerian degeneration due to local lesions in a study in which patients were classified according to the presence of previous optic neuritis and lesions along these tracts.[83] TBSS studies[84,85] have found a correlation of impaired attention, working memory, and speed of information processing with decreased FA in the CC and other tracts mainly connecting prefrontal cortical regions.

Advances in DT MRI and tractography have spurred the development of brain neuroconnectivity techniques, which define and quantify anatomical links between remote brain regions by axonal fiber pathways. The use of these approaches has revealed reduced network efficiency in the WM structural networks of MS patients,[64,86] including those at the earliest stages of the disease.[87]

^{1}H-MR Spectroscopy

Water-suppressed, proton MR spectra of the healthy human brain at long echo times reveal four major resonances: choline-containing phospholipids (Cho); creatine and phosphocreatine (Cr); *N*-acetyl groups, mainly *N*-acetylaspartate (NAA); and lactate (Lac). Because NAA is a metabolite that is found almost exclusively in neurons and neuronal processes in the normal adult brain, a decrease in its concentration is thought to be secondary to axonal dysfunction, injury, or loss. By contrast, increases in Cho and Lac are thought to reflect acute inflammatory and demyelinating changes. ^{1}H-MRS with shorter echo times can detect additional metabolites, such as lipids and myo-inositol (mI), which are also thought to indicate ongoing myelin damage and glial proliferation, respectively.

Using ^{1}H-MRS, dynamic changes in metabolite profiles have been shown in NAWM areas that will become lesions[88] and in acute MS lesions, extending from the first days of lesion formation over the subsequent months.[61] Several studies have found metabolite abnormalities, including reduced levels of NAA and Cho (a marker of membrane turnover), and increased levels of mI in the NAWM, cortex[89–91] and subcortical GM tissue[92–94] from MS patients, including those with CIS.[95,96] NAWM mI increase within 3 months of a CIS has been found to predict a poor performance on executive functions 7 years later.[97]

A longitudinal study showed that whole-brain NAA declines significantly (5%/year) in RRMS patients over a 2-year time period.[98] Another study[99] assessed longitudinally for 3 years the metabolic abnormalities in the GM and WM from early RRMS, and found that WM Cr, Cho, and mI concentrations were higher while WM NAA were lower at all the time points in RRMS compared to controls. No difference was found in GM metabolites. Inglese et al.[100] showed the feasibility of sodium MRI at 3.0 T in RRMS patients.

Imaging Iron Deposition

Abnormal iron deposition is thought to be the substrate of T2 hypointense areas and reduced T2 RT seen in the basal ganglia, thalamus, dentate nucleus, and cortical regions of MS patients.[101] In MS patients, GM T2 hypointensities have been correlated with the severity of clinical disability and cognitive impairment,[101,102] as well as with clinical progression.[103]

The ability to detect abnormal iron deposition increased with the availability of 3.0 T scanners. Higher basal ganglia transverse relaxation rate (R2*) values were found in RRMS than in CIS patients.[104] Increased magnetic field correlation was demonstrated in the deep GM of RRMS patients, which was associated with the T2 lesion burden and severity of neuropsychological deficits.[105]

Susceptibility-weighted imaging (SWI) has also been used to assess iron concentration in MS and confirmed the presence of an abnormal iron concentration in the deep GM nuclei of MS patients compared to healthy controls.[106,107]

13.3.3 Functional Imaging Techniques

Functional imaging techniques allow assessment of hemodynamic abnormalities in MS patients and are improving the understanding of the role of cortical reorganization following tissue injury.

Functional Magnetic Resonance Imaging

Functional MRI (fMRI) is a noninvasive technique that allows us to study CNS function, and to define abnormal patterns of activation and/or functional connectivity (FC) caused by injury or disease. The signal changes seen during fMRI studies depend on the blood oxygenation level-dependent mechanism,[108] which, in turn, involves changes of the transverse magnetization RT—either T2* in a gradient-echo sequence, or T2 in a spin-echo sequence. Local increases in neuronal activity result in a rise of blood flow and oxygen consumption. The increase of blood flow is greater than oxygen consumption, thus determining an increased ratio between oxygenated and deoxygenated hemoglobin, which enhances the MRI signal.[109] By analyzing these data with appropriate statistical methods, it is possible to obtain information about the location and extent of activation as well as connectivity of specific areas involved in the performance of a given task in healthy subjects and in patients with different neurological conditions. Recently, a completely task-free approach, based on the assessment of functional correlations of neural networks at rest (resting-state (RS) fMRI) has been developed (for a review see Refs. 110,111).

Functional cortical abnormalities have been demonstrated consistently in all MS phenotypes using different active paradigms. The correlation found by the majority of these studies between measures of abnormal activations and quantitative MR metrics of disease burden suggests that, at least at some stages of the disease, functional reorganization might play an adaptive role, which limits the clinical consequences of disease-related structural damage. The results of a cross-sectional study of the motor network in patients with different clinical MS phenotypes[112] support the notion of a natural history of brain adaptive mechanisms in MS. Such a study showed, at the beginning of the disease, an increased recruitment of those areas normally devoted to the performance of a motor task, such as the primary sensorimotor cortex and the supplementary motor area. At a later stage, a bilateral activation of these regions is seen, followed by a widespread recruitment of additional areas, which are usually recruited in normal people to perform novel/complex tasks. The preservation of a focused and strictly lateralized movement-associated pattern of cortical activations has been suggested as a possible mechanism to explain the favorable clinical outcome of patients with pediatric MS[113] and BMS.[114]

There is also evidence supporting a maladaptive role of cortical functional abnormalities in MS. In patients with progressive MS,[114–116] reduced activations of classical regions of the sensorimotor network and an increased recruitment of high-order regions, such as the superior temporal sulcus and the insula, have been found with motor tasks. In patients with cognitive decline, a reallocation of neuronal resources and an inefficiency of neuronal processes have been associated with the extent of structural damage.

The combination of measures of FC with measures of structural damage to specific WM tracts is likely to improve our understanding of the relationship between structural and functional abnormalities, as suggested by studies in patients with RRMS[117] and BMS.[118]

The analysis of brain activity at rest has shown an increased synchronization of the majority of the resting-state networks (RSNs) in patients with CIS[119] and a reduced activity of the anterior regions of the default-mode network in patients with progressive MS[120] and cognitive impairment.[121] Distributed abnormalities of RS FC within and between large-scale neuronal networks have been shown in RRMS patients and have been related to the extent of T2 lesions and severity of disability.[122]

Positron Emission Tomography

Positron emission tomography (PET) is an imaging technique that uses radiolabeled molecules to provide information about function and metabolism of different tissues. This technique is based on the detection and quantification of pairs of gamma rays emitted indirectly by a positron-emitting radionuclide (tracer), which is introduced into the body on a biologically active molecule. Three-dimensional images of tracer concentration within the body are then constructed by computer analysis. Many radiotracers are used in PET to image tissue concentration of many types of molecules of interest in different neurological conditions.

Functional abnormalities in MS patients have also been investigated using PET.[123] A seminal investigation[124] showed an association between CC atrophy and decreased metabolic activity, especially in the association cortices of the left hemisphere. Sorensen et al.[125] found a reduced cerebral metabolic rate of glucose in the cortex, putamen, thalamus, and hippocampus, which was correlated with cognitive impairment and T2 lesion load. A 2-year longitudinal study[126] detected a decrease of global cortical metabolism, especially in frontal and parietal areas, which was not correlated with changes of T2 lesion load and clinical disability.

Other radioligands have been used to study patients with MS. Among them, [C11]-1(-2-chlorophenyl)-N-methyl-N-(-1methylpropyl)-3-isoquinolinecarboxamide (11C-PK11195), which binds specifically to translocator protein 18 KDa (TSPO), a protein that is upregulated upon exposure to various insults, might be useful to evaluate neuroinflammation and microglia activation. A recent PK11195 PET study[127] found that MS patients have increased cortical GM PK11195 binding relative to controls. This increase was correlated with disability in patients with SPMS, but not in those with RRMS. No binding in the WM was detected. Interestingly, the binding patterns suggested the presence of regional pathology, with involvement detected in the postcentral, middle frontal, anterior orbital, fusiform, and parahippocampal gyri. Patients with SPMS showed additional binding in the precentral, superior parietal, lingual and anterior superior, medial and inferior temporal gyri. A preliminary study in nine RRMS patients has suggested the potential of this technique to monitor the effects of disease-modifying treatments, by showing a significant decrease of PK11195 uptake in cortical GM and WM after a year of treatment with glatiramer acetate (GA).[128]

In order to better define the microstructural tissue abnormalities in SPMS, a selective radioligand to adenosine A2A receptor ([C11]-TMSX), which is a potent regulator of inflammation, has been used.[129] Compared to controls, patients showed increased NAWM [C11]-TMSX values, which were correlated with the EDSS score.

13.4 IMAGING THE SPINAL CORD

MRI features of MS cord lesions have been identified.[130] MS cord lesions are more frequently observed in the cervical than in other regions, are usually peripheral, limited to two vertebral segments in length or less, occupy less than half the cross-sectional area of the cord, and typically are not T1-hypointense (Fig. 13.4).[130] Asymptomatic spinal cord lesions have been described in 30–40% of CIS patients and in up to 90% of patients with definite MS.[130] A set of new strategies has been proposed to improve the detection of spinal cord lesions, including the use of an axial 3D gradient echo sequence with or without MT-prepulse[131] and of an optimized T1 magnetization-prepared rapid acquisition gradient-echo sequence.[132]

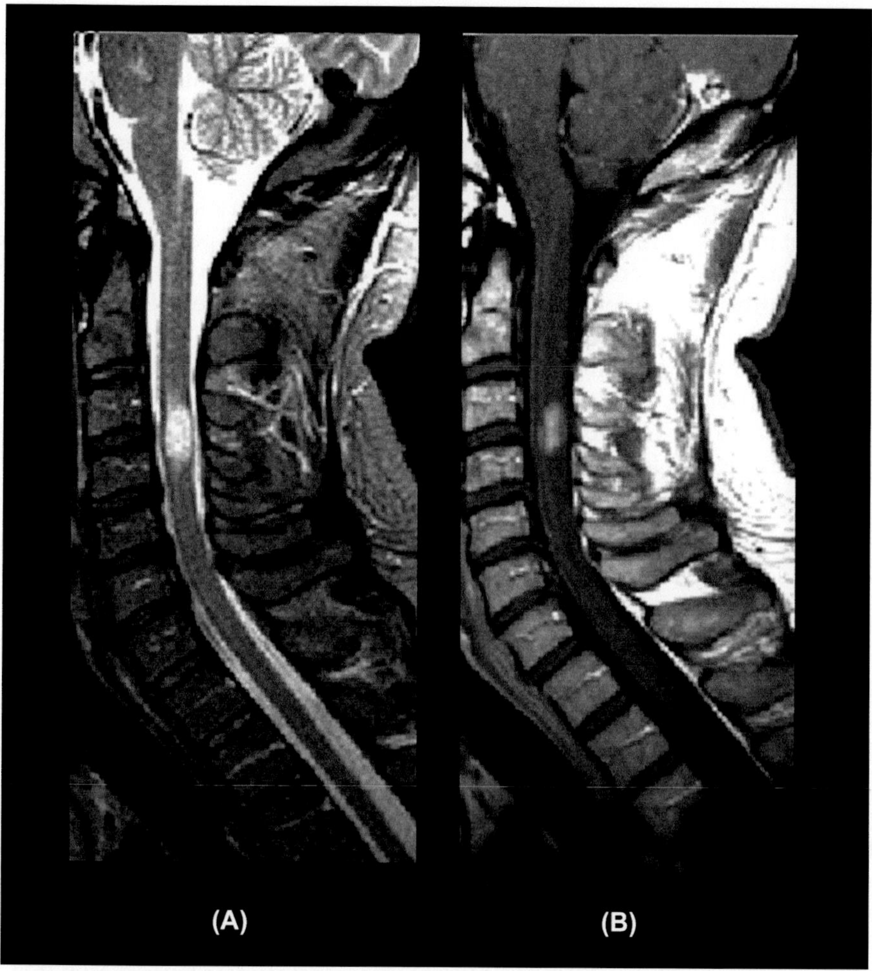

FIGURE 13.4 Spinal cord sagittal T2-weighted (A) and postcontrast T1-weighted (B) magnetic resonance images from a patient with relapsing-remitting multiple sclerosis. An oval-shape cervical cord lesion, hyperintense on T2-weighted scan and enhancing after contrast administration, is evident at C3–C4 level.

Although a significant reduction of cervical cord size can be observed since the early phase of MS,[133] cord atrophy is more severe in the progressive forms of the disease.[130] Abnormalities at a given time point and changed time of cord cross-sectional area correlate better with clinical disability than changes of T2 lesion burden.[134] A semiautomatic method,[135] which allows segmentation of long portions of the cord, has been developed. The use of this method in a multicenter study of a large sample of MS patients has demonstrated that cord area differs significantly among the main MS clinical phenotypes and is correlated with the EDSS, with a differential effect among phenotypes: no association in either CIS or BMS patients, but association in RRMS, SPMS, and PPMS.[82] A voxel-vise approach has also been applied to define the regional distribution of cervical cord damage of MS patients.[136,137]

MTR can also be measured in the cervical cord.[138,139] A reduction of MTR of the cervical cord has been described in all phenotypes of the disease, except for CIS. These changes have been suggested to correlate with disability. MTR abnormalities located in the dorsal and lateral columns of the spinal cord correlate with deficits of vibration sensation and strength, respectively.[140] In patients with RRMS, reduced cervical cord GM average MTR was correlated with the degree of disability.[141]

Abnormal DT MRI quantities from the cervical cord have been shown in patients with definite MS, but not in those with CIS.[142] A 2-year follow-up study of patients with relapse-onset MS found that baseline cord area and FA correlated with increase in disability at follow-up.[143] Compared to controls, MS patients with a cervical cord relapse have reduced NAA and lower structural connectivity in the lateral CST and posterior tracts, and such abnormalities were found to be correlated with disability.[144] A cross-sectional study that combined different MR modalities (DT MRI, MTR, and atrophy) to quantify cervical cord damage in a large sample of MS patients showed that a multiparametric MR approach contributes to discriminate patients with high from those with low levels of disability.[145]

13.5 ULTRAHIGH FIELD IMAGING

Magnets operating at 3.0–4.0T detect a greater number and volume of MS T2 and enhancing brain lesions than those operating at 1.5T. However, one study compared the performance of the MRI diagnostic criteria for MS at 1.5 and 3.0T in CIS patients and found that, despite increased lesion detection, 3.0T imaging led only to a little gain in term of showing DIS.[146]

Ultrahigh field MRI (7.0T or more) provides a better definition of lesion location in the WM and GM, their morphology, and their association with vasculature.[147–149] Using 7.0T T2*-weighted MRI, the presence of a central vein in more than 40% of lesions contributed to discriminate patients from those without definite MS.[150,151] SWI (examining both T2*-weighted magnitude and phase) has been used to detect features that may be more closely linked to important aspects of MS pathology. Myelin, iron, deoxyhemoglobin, and free radicals influence susceptibility due to their paramagnetic properties and can thereby determine image contrast. Several combined susceptibility MRI-pathological studies have shown that the presence of a hypointense rim on T2*/phase contrast in chronic MS lesions appears to be influenced by the topography of iron-laden macrophages and ferritin at the lesion's edge.[152,153]

The cortical cyto- and myeloarchitecture[154,155] as well as location and characteristics of CLs can also be visualized accurately at these field strengths in postmortem brain samples[152,156,157] and in vivo.[158–163] Such a characterization of CL subtypes might contribute to identify novel biomarkers of MS clinical status, as suggested by the demonstration that leukocortical (type I) and subpial (III-IV) CLs are correlated with cognitive impairment and disability.[164]

13.6 MAGNETIC RESONANCE IMAGING IN THE MONITORING OF MULTIPLE SCLEROSIS TREATMENT

At present, in MS clinical trials, MRI is used as a primary outcome measure in phase-II studies, in which serial scans are acquired to detect disease activity (new/enlarged T2 lesions and enhancing lesions).[14] In phase-III trials, given the uncertainty of conventional

MRI in predicting clinical evolution, MRI measures (absolute or percentage increase in total T2 lesion load) are used as secondary outcomes.[2] During the mid-2010s, a large effort has been spent to validate these metrics as surrogate measures for treatment effect on relapses and disability progression in MS.[16,165,166] Two meta-analysis studies of randomized, placebo-controlled clinical trials of disease-modifying drugs in RRMS[167,168] have shown that a large proportion (around 80%) of the variance in treatment effect on relapses is explained by the variance of the effect on active MRI lesions. Analysis of trials that tested the same drugs in phase II and III studies showed that the effect on MRI lesions over short follow-up (6–9 months) can also predict that on relapses over longer follow-up (12–24 months), with the exception of fingolimod (the effect of this drug on relapses was 67% higher than that predicted on MRI[168]). Sample size calculations for trials of biosimilar or generic versions of interferon (IFN) beta or GA demonstrated a significant reduction of the number of patients needed to be recruited as well as of the duration of the trial when an MRI outcome replaces relapse rate as the primary end point measure. Recent sample size estimations also showed that 6-month phase-II trials using new T2 lesion counts as the end point are feasible in pediatric MS patients.[169]

MRI surrogacy in predicting treatment effect on disability progression has also been validated both at trial[170] and individual patient[165,166,171] levels in RRMS[166,170,171] and SPMS patients.[165] An analysis of RRMS patients treated with IFN beta-1a subcutaneously[171] found that a combined measure of 1-year change of MRI lesions and relapses during therapy fully estimates the effect on 2-year EDSS worsening. Clearly, the validation of surrogates is treatment specific. Therefore, similar studies are now needed for other approved treatments.

Halting inflammation in MS does not necessarily modify the disease course favorably, since neurodegeneration, which is not only a mere reflection of inflammatory-mediated abnormalities, is also known to occur from the earliest stages of the disease.

MRI markers sensitive to irreversible tissue damage have been introduced, including monitoring the evolution of active lesions into permanent black holes (PBH) and quantification of brain atrophy. Over 6–12 months, about 30–40% of active lesions evolve into PBH lesions (Fig. 13.1). A few trials have assessed the evolution of active lesions into PBH. Significant effect in modifying the fate of active lesions evolving into PBH in RRMS patients have been described for GA,[172] natalizumab,[173] and laquinimod.[174] Two comparative studies, BECOME and BEYOND, suggested that RRMS patients treated with IFN beta-1b might have a lower proportion of new lesions evolving into PBH than those treated with GA.[175,176]

Only a subset of available MS clinical trials has assessed treatment effect on brain atrophy progression. IFNs,[174,177,178] GA,[179] natalizumab,[180] fingolimod,[181] laquinimod,[182] and hematopoietic stem cell transplantation[183–185] were shown to slow significantly the progression of brain atrophy. A recent meta-analysis of randomized clinical trials in RRMS lasting at least 2 years demonstrated that treatment effect on brain atrophy is correlated with the effect on disability progression ($R^2=0.48$) and is independent of the effect on active MRI lesions. However, these two MRI measures predicted treatment effect on disability more closely when used in combination ($R^2=0.75$).[186]

A paradoxical loss of brain volume has been observed in the early phase of treatment following administration of steroids,[187] intramuscular (i.m.) IFN beta-1a,[177] and natalizumab,[180] while such an effect has not been detected with fingolimod.[181] Such an acute reduction of brain volume is likely to reflect the resolution of inflammation-associated edema, a phenomenon described as pseudoatrophy. This suggests that, at least for some treatments, quantification of brain atrophy might not be the ideal outcome measure of treatment response in studies of short duration (1 year or less).

Recent trials have included GM atrophy quantification as an outcome measure. In RRMS patients, laquinimod was shown to slow the rate of GM and thalamic atrophy progression over a 2-year period, and such an effect was more pronounced during the first year of treatment. In SPMS patients, GM atrophy progression was reduced following treatment with pioglitazione used as add-on therapy to i.m. IFN beta-1a.[188] Such an effect was not found for lamotrigine.[189]

A few single-center trials have used cord atrophy as an exploratory end point.[190–193] Reduction of cord CSA over a 4-year period did not differ between RRMS patients treated with various formulations of INF beta-1a versus those initially on placebo.[191] In PPMS patients, riluzole slowed the rate of progression of cord atrophy,[190] whereas no effect was detected after administration of i.m. IFN beta-1a[192] or -1b.[193]

Clearly, a rigorous and valid strategy for MR-based longitudinal monitoring of MS response to treatment requires the use of standardized imaging protocols (including consistency in slice thickness and imaging planes, field strength, and patient repositioning) and evaluation procedures. As a consequence, the definition of individual patient response to MS treatment based upon routine clinical MRI scanning remains a challenging task. Patients treated with IFN beta who developed new MRI lesions after 2 years have a higher risk of poor treatment response than those who did not.[194] In patients with RRMS treated with IFN beta, Rio et al. showed that the number of active MRI lesions detected on an MRI scan performed 12 months after treatment initiation was the most important factor related to the progression of disability after 2 years.[195] In a subsequent study, the same group found that a combination of clinical and MRI measures of disease activity during the first year of treatment allows identification of responders to IFN beta treatment after 2 and 3 years.[196] Tomassini et al.[197] demonstrated that the formation of T1-hypointense and enhancing lesions during the first year of therapy with IFN beta is associated with relapse rate and disability progression after 6 years. Based on the degree of clinical and MRI activity, guidelines have been proposed to identify responders to disease-modifying treatments.[196,198,199]

The utility of advanced MR techniques in the context of clinical trial monitoring still needs to be validated. For instance, the acquisition of DIR sequences has not been standardized yet across different centers, and their performance has not been tested in the setting of MS clinical trials. Consensus recommendations for scoring CLs using DIR sequences nonetheless have been formulated and tested.[200] Sample size calculations to use CL formation as an outcome of clinical trials of RRMS have also been performed.[201] However, only two single-center trials[202,203] have included CL quantification. In RRMS patients, over a 2-year period, natalizumab slowed significantly new CL formation compared to other disease-modifying drugs,[202] and sc IFN beta-1a was shown to have a faster and more pronounced effect on this measure than i.m. IFN beta-1a or GA.[203]

Only a few clinical trials have included quantitative MRI metrics as outcome measures.[204–207] Recovery of MTR in focal T2 lesions has been proposed as an outcome measure to assess the effect of remyelinating agents,[208,209] and sample size calculations have been performed for multicenter clinical trials using this metric.[208,209] A technique to quantify longitudinal changes in MTR has been developed and applied to patients treated with autologous stem cell transplant.[209] A single-center MT MRI study has suggested that alemtuzumab protects against GM damage.[210] A combined MT MRI and ^{1}H-MRS study[174] showed that compared to placebo, patients treated with laquinimod tend to accumulate less microscopic WM and GM damage. Another study showed that lamotrigine has no effect on the evolution of WM and GM damage measured using MTR in SPMS patients.[211]

At present, only a limited number of single-center studies have applied fMRI to monitor drug or rehabilitation effects in MS. Despite having no effect on cognitive performance, chronic administration of rivastigmine was shown to be able to modify the patterns of activations and FC of frontal lobe areas in MS patients.[212] Modifications of the pattern of brain functional recruitment and resting activity have been associated with a better clinical outcome following motor[213] and cognitive[214] rehabilitation.

13.7 CONCLUSIONS

Conventional and advanced MR-based techniques have been applied extensively to the study of MS and such an effort has contributed to improve our ability to diagnose and monitor the disease, as well as our understanding of its pathophysiology and treatment efficacy. Nevertheless, many challenges remain. Quantitative, metabolic, and functional imaging techniques need to be optimized and standardized across multiple centers to monitor adequately disease evolution, either natural or modified by treatment. With the increased availability of high-field and ultrahigh field MR scanners, such an issue is now becoming extremely critical. Furthermore, some of the MR approaches discussed here are in their infancy and their practical utility, from a research setting to daily-life clinical practice, still needs to be investigated.

References

1. Ormerod IE, Miller DH, McDonald WI, et al. The role of NMR imaging in the assessment of multiple sclerosis and isolated neurological lesions. A quantitative study. *Brain*. 1987;110(Pt 6):1579–1616.
2. Filippi M, Gawne-Cain ML, Gasperini C, et al. Effect of training and different measurement strategies on the reproducibility of brain MRI lesion load measurements in multiple sclerosis. *Neurology*. 1998;50:238–244.
3. Barkhof F, Filippi M, van Waesberghe JH, et al. Improving interobserver variation in reporting gadolinium-enhanced MRI lesions in multiple sclerosis. *Neurology*. 1997;49:1682–1688.
4. McDonald WI, Compston A, Edan G, et al. Recommended diagnostic criteria for multiple sclerosis: guidelines from the international panel on the diagnosis of multiple sclerosis. *Ann Neurol*. 2001;50:121–127.
5. Swanton JK, Fernando K, Dalton CM, et al. Modification of MRI criteria for multiple sclerosis in patients with clinically isolated syndromes. *J Neurol Neurosurg Psychiatry*. 2006;77:830–833.
6. Rovira A, Swanton J, Tintore M, et al. A single, early magnetic resonance imaging study in the diagnosis of multiple sclerosis. *Arch Neurol*. 2009;66:587–592.
7. Polman CH, Reingold SC, Banwell B, et al. Diagnostic criteria for multiple sclerosis: 2010 revisions to the McDonald criteria. *Ann Neurol*. 2011;69:292–302.
8. Geurts JJ, Pouwels PJ, Uitdehaag BM, Polman CH, Barkhof F, Castelijns JA. Intracortical lesions in multiple sclerosis: improved detection with 3D double inversion-recovery MR imaging. *Radiology*. 2005;236:254–260.
9. Filippi M, Rocca MA, Calabrese M, et al. Intracortical lesions: relevance for new MRI diagnostic criteria for multiple sclerosis. *Neurology*. 2010;75:1988–1994.
10. Sombekke MH, Wattjes MP, Balk LJ, et al. Spinal cord lesions in patients with clinically isolated syndrome: a powerful tool in diagnosis and prognosis. *Neurology*. 2013;80:69–75.
11. Brex PA, Ciccarelli O, O'Riordan JI, Sailer M, Thompson AJ, Miller DH. A longitudinal study of abnormalities on MRI and disability from multiple sclerosis. *N Engl J Med*. 2002;346:158–164.
12. Fisniku LK, Brex PA, Altmann DR, et al. Disability and T2 MRI lesions: a 20-year follow-up of patients with relapse onset of multiple sclerosis. *Brain*. 2008;131:808–817.

13. Minneboo A, Barkhof F, Polman CH, Uitdehaag BM, Knol DL, Castelijns JA. Infratentorial lesions predict long-term disability in patients with initial findings suggestive of multiple sclerosis. *Arch Neurol*. 2004;61:217–221.
14. Barkhof F, Filippi M, Miller DH, et al. Comparison of MRI criteria at first presentation to predict conversion to clinically definite multiple sclerosis. *Brain*. 1997;120(Pt 11):2059–2069.
15. Summers M, Fisniku L, Anderson V, Miller D, Cipolotti L, Ron M. Cognitive impairment in relapsing-remitting multiple sclerosis can be predicted by imaging performed several years earlier. *Mult Scler*. 2008;14:197–204.
16. Goodin DS. Magnetic resonance imaging as a surrogate outcome measure of disability in multiple sclerosis: have we been overly harsh in our assessment? *Ann Neurol*. 2006;59:597–605.
17. Li DK, Held U, Petkau J, et al. MRI T2 lesion burden in multiple sclerosis: a plateauing relationship with clinical disability. *Neurology*. 2006;66:1384–1389.
18. Sormani MP, Rovaris M, Comi G, Filippi M. A reassessment of the plateauing relationship between T2 lesion load and disability in MS. *Neurology*. 2009;73:1538–1542.
19. Caramanos Z, Francis SJ, Narayanan S, Lapierre Y, Arnold DL. Large, nonplateauing relationship between clinical disability and cerebral white matter lesion load in patients with multiple sclerosis. *Arch Neurol*. 2012;69:89–95.
20. Kermode AG, Thompson AJ, Tofts P, et al. Breakdown of the blood–brain barrier precedes symptoms and other MRI signs of new lesions in multiple sclerosis. Pathogenetic and clinical implications. *Brain*. 1990;113 (Pt 5):1477–1489.
21. Thompson AJ, Miller D, Youl B, et al. Serial gadolinium-enhanced MRI in relapsing/remitting multiple sclerosis of varying disease duration. *Neurology*. 1992;42:60–63.
22. Kappos L, Moeri D, Radue EW, et al. Predictive value of gadolinium-enhanced magnetic resonance imaging for relapse rate and changes in disability or impairment in multiple sclerosis: a meta-analysis. Gadolinium MRI meta-analysis Group. *Lancet*. 1999;353:964–969.
23. Gaitan MI, Shea CD, Evangelou IE, et al. Evolution of the blood–brain barrier in newly forming multiple sclerosis lesions. *Ann Neurol*. 2011;70:22–29.
24. Gaitan MI, Sati P, Inati S, Reich DS. Initial investigation of the blood–brain barrier in MS lesions at 7 tesla. *Mult Scler*. 2013;19:1068–1073.
25. Absinta M, Sati P, Gaitan MI, et al. Seven-tesla phase imaging of acute multiple sclerosis lesions: a new window into the inflammatory process. *Ann Neurol*. 2013;74:669–678.
26. Filippi M, Rocca MA, Barkhof F, et al. Association between pathological and MRI findings in multiple sclerosis. *Lancet Neurol*. 2012;11:349–360.
27. Calabrese M, Filippi M, Gallo P. Cortical lesions in multiple sclerosis. *Nat Rev Neurol*. 2010;6:438–444.
28. Roosendaal SD, Moraal B, Vrenken H, et al. In vivo MR imaging of hippocampal lesions in multiple sclerosis. *J Magn Reson Imaging*. 2008;27:726–731.
29. Calabrese M, Rocca M, Atzori M, et al. A three-year MRI study of cortical lesions in relapse-onset multiple sclerosis. *Ann Neurol*. 2010;67:376–383.
30. Calabrese M, Rocca MA, Atzori M, et al. Cortical lesions in primary progressive multiple sclerosis: a 2-year longitudinal MR study. *Neurology*. 2009;72:1330–1336.
31. Calabrese M, Filippi M, Rovaris M, et al. Morphology and evolution of cortical lesions in multiple sclerosis. A longitudinal MRI study. *NeuroImage*. 2008;42:1324–1328.
32. Roosendaal SD, Moraal B, Pouwels PJ, et al. Accumulation of cortical lesions in MS: relation with cognitive impairment. *Mult Scler*. 2009;15:708–714.
33. Calabrese M, Agosta F, Rinaldi F, et al. Cortical lesions and atrophy associated with cognitive impairment in relapsing-remitting multiple sclerosis. *Arch Neurol*. 2009;66:1144–1150.
34. Calabrese M, Poretto V, Favaretto A, et al. Cortical lesion load associates with progression of disability in multiple sclerosis. *Brain*. 2012;135:2952–2961.
35. Miller DH, Barkhof F, Frank JA, Parker GJ, Thompson AJ. Measurement of atrophy in multiple sclerosis: pathological basis, methodological aspects and clinical relevance. *Brain*. 2002;125:1676–1695.
36. Giorgio A, Battaglini M, Smith SM, De Stefano N. Brain atrophy assessment in multiple sclerosis: importance and limitations. *Neuroimaging Clin N Am*. 2008;18:675–686, [xi].
37. Lanz M, Hahn HK, Hildebrandt H. Brain atrophy and cognitive impairment in multiple sclerosis: a review. *J Neurol*. 2007;254(suppl 2):II43–48.
38. Fisher E, Rudick RA, Simon JH, et al. Eight-year follow-up study of brain atrophy in patients with MS. *Neurology*. 2002;59:1412–1420.
39. Khaleeli Z, Ciccarelli O, Manfredonia F, et al. Predicting progression in primary progressive multiple sclerosis: a 10-year multicenter study. *Ann Neurol*. 2008;63:790–793.

40. De Stefano N, Giorgio A, Battaglini M, et al. Assessing brain atrophy rates in a large population of untreated multiple sclerosis subtypes. *Neurology*. 2010;74:1868–1876.
41. Chard DT, Griffin CM, Parker GJ, Kapoor R, Thompson AJ, Miller DH. Brain atrophy in clinically early relapsing-remitting multiple sclerosis. *Brain*. 2002;125:327–337.
42. De Stefano N, Matthews PM, Filippi M, et al. Evidence of early cortical atrophy in MS: relevance to white matter changes and disability. *Neurology*. 2003;60:1157–1162.
43. Tedeschi G, Lavorgna L, Russo P, et al. Brain atrophy and lesion load in a large population of patients with multiple sclerosis. *Neurology*. 2005;65:280–285.
44. Sanfilipo MP, Benedict RH, Sharma J, Weinstock-Guttman B, Bakshi R. The relationship between whole brain volume and disability in multiple sclerosis: a comparison of normalized gray vs. white matter with misclassification correction. *NeuroImage*. 2005;26:1068–1077.
45. Prinster A, Quarantelli M, Lanzillo R, et al. A voxel-based morphometry study of disease severity correlates in relapsing– remitting multiple sclerosis. *Mult Scler*. 2010;16:45–54.
46. Amato MP, Portaccio E, Goretti B, et al. Association of neocortical volume changes with cognitive deterioration in relapsing-remitting multiple sclerosis. *Arch Neurol*. 2007;64:1157–1161.
47. Benedict RH, Bruce JM, Dwyer MG, et al. Neocortical atrophy, third ventricular width, and cognitive dysfunction in multiple sclerosis. *Arch Neurol*. 2006;63:1301–1306.
48. Valsasina P, Benedetti B, Rovaris M, Sormani MP, Comi G, Filippi M. Evidence for progressive gray matter loss in patients with relapsing-remitting MS. *Neurology*. 2005;65:1126–1128.
49. Chen JT, Narayanan S, Collins DL, Smith SM, Matthews PM, Arnold DL. Relating neocortical pathology to disability progression in multiple sclerosis using MRI. *NeuroImage*. 2004;23:1168–1175.
50. Fisher E, Lee JC, Nakamura K, Rudick RA. Gray matter atrophy in multiple sclerosis: a longitudinal study. *Ann Neurol*. 2008;64:255–265.
51. Henry RG, Shieh M, Okuda DT, Evangelista A, Gorno-Tempini ML, Pelletier D. Regional grey matter atrophy in clinically isolated syndromes at presentation. *J Neurol Neurosurg Psychiatry*. 2008;79:1236–1244.
52. Bendfeldt K, Radue EW, Borgwardt SJ, Kappos L. Progression of gray matter atrophy and its association with white matter lesions in relapsing-remitting multiple sclerosis. *J Neurol Sci*. 2009;285:268–269. author reply 269.
53. Ceccarelli A, Rocca MA, Pagani E, et al. A voxel-based morphometry study of grey matter loss in MS patients with different clinical phenotypes. *NeuroImage*. 2008;42:315–322.
54. Sicotte NL, Kern KC, Giesser BS, et al. Regional hippocampal atrophy in multiple sclerosis. *Brain*. 2008;131: 1134–1141.
55. Longoni G, Rocca MA, Pagani E, et al. Deficits in memory and visuospatial learning correlate with regional hippocampal atrophy in MS. *Brain Struct Funct*. 2015;220:435–444.
56. Sepulcre J, Masdeu JC, Goni J, et al. Fatigue in multiple sclerosis is associated with the disruption of frontal and parietal pathways. *Mult Scler*. 2009;15:337–344.
57. Rocca MA, Mesaros S, Pagani E, Sormani MP, Comi G, Filippi M. Thalamic damage and long-term progression of disability in multiple sclerosis. *Radiology*. 2010;257:463–469.
58. Mesaros S, Rocca MA, Pagani E, et al. Thalamic damage predicts the evolution of primary-progressive multiple sclerosis at 5 years. *AJNR Am J Neuroradiol*. 2011;32:1016–1020.
59. Ropele S, Fazekas F. Magnetization transfer MR imaging in multiple sclerosis. *Neuroimaging Clin N Am*. 2009;19:27–36.
60. Rovaris M, Agosta F, Pagani E, Filippi M. Diffusion tensor MR imaging. *Neuroimaging Clin N Am*. 2009;19:37–43.
61. Sajja BR, Wolinsky JS, Narayana PA. Proton magnetic resonance spectroscopy in multiple sclerosis. *Neuroimaging Clin N Am*. 2009;19:45–58.
62. Filippi M, Rocca MA, De Stefano N, et al. Magnetic resonance techniques in multiple sclerosis: the present and the future. *Arch Neurol*. 2011;68:1514–1520.
63. Amato MP, Portaccio E, Stromillo ML, et al. Cognitive assessment and quantitative magnetic resonance metrics can help to identify benign multiple sclerosis. *Neurology*. 2008;71:632–638.
64. Filippi M, Preziosa P, Copetti M, et al. Gray matter damage predicts the accumulation of disability 13 years later. *Neurology*. 2013;81:1759–1767.
65. Khaleeli Z, Altmann DR, Cercignani M, Ciccarelli O, Miller DH, Thompson AJ. Magnetization transfer ratio in gray matter: a potential surrogate marker for progression in early primary progressive multiple sclerosis. *Arch Neurol*. 2008;65:1454–1459.
66. Penny S, Khaleeli Z, Cipolotti L, Thompson A, Ron M. Early imaging predicts later cognitive impairment in primary progressive multiple sclerosis. *Neurology*. 2010;74:545–552.

67. Chen JT, Collins DL, Atkins HL, Freedman MS, Arnold DL. Magnetization transfer ratio evolution with demyelination and remyelination in multiple sclerosis lesions. *Ann Neurol*. 2008;63:254–262.
68. Chen JT, Easley K, Schneider C, et al. Clinically feasible MTR is sensitive to cortical demyelination in MS. *Neurology*. 2013;80:246–252.
69. Pierpaoli C, Barnett A, Pajevic S, et al. Water diffusion changes in Wallerian degeneration and their dependence on white matter architecture. *NeuroImage*. 2001;13:1174–1185.
70. Wheeler-Kingshott CA, Cercignani M. About "axial" and "radial" diffusivities. *Magn Reson Med*. 2009;61:1255–1260.
71. Poonawalla AH, Hasan KM, Gupta RK, et al. Diffusion-tensor MR imaging of cortical lesions in multiple sclerosis: initial findings. *Radiology*. 2008;246:880–886.
72. Calabrese M, Rinaldi F, Seppi D, et al. Cortical diffusion-tensor imaging abnormalities in multiple sclerosis: a 3-year longitudinal study. *Radiology*. 2011;261:891–898.
73. Filippi M, Preziosa P, Pagani E, et al. Microstructural MR imaging of cortical lesion in multiple sclerosis. *Mult Scler*. 2012;19:418–426.
74. Oreja-Guevara C, Rovaris M, Iannucci G, et al. Progressive gray matter damage in patients with relapsing-remitting multiple sclerosis: a longitudinal diffusion tensor magnetic resonance imaging study. *Arch Neurol*. 2005;62:578–584.
75. Rovaris M, Bozzali M, Iannucci G, et al. Assessment of normal-appearing white and gray matter in patients with primary progressive multiple sclerosis: a diffusion-tensor magnetic resonance imaging study. *Arch Neurol*. 2002;59:1406–1412.
76. Rovaris M, Judica E, Gallo A, et al. Grey matter damage predicts the evolution of primary progressive multiple sclerosis at 5 years. *Brain*. 2006;129:2628–2634.
77. Rovaris M, Gallo A, Valsasina P, et al. Short-term accrual of gray matter pathology in patients with progressive multiple sclerosis: an in vivo study using diffusion tensor MRI. *NeuroImage*. 2005;24:1139–1146.
78. Rovaris M, Iannucci G, Falautano M, et al. Cognitive dysfunction in patients with mildly disabling relapsing-remitting multiple sclerosis: an exploratory study with diffusion tensor MR imaging. *J Neurol Sci*. 2002;195:103–109.
79. Lin X, Tench CR, Morgan PS, Niepel G, Constantinescu CS. 'Importance sampling' in MS: use of diffusion tensor tractography to quantify pathology related to specific impairment. *J Neurol Sci*. 2005;237:13–19.
80. Pagani E, Filippi M, Rocca MA, Horsfield MA. A method for obtaining tract-specific diffusion tensor MRI measurements in the presence of disease: application to patients with clinically isolated syndromes suggestive of multiple sclerosis. *NeuroImage*. 2005;26:258–265.
81. Giorgio A, Palace J, Johansen-Berg H, et al. Relationships of brain white matter microstructure with clinical and MR measures in relapsing-remitting multiple sclerosis. *J Magn Reson Imaging*. 2010;31:309–316.
82. Rocca MA, Horsfield MA, Sala S, et al. A multicenter assessment of cervical cord atrophy among MS clinical phenotypes. *Neurology*. 2011;76:2096–2102.
83. Rocca MA, Mesaros S, Preziosa P, et al. Wallerian and trans-synaptic degeneration contribute to optic radiation damage in multiple sclerosis: a diffusion tensor MRI study. *Mult Scler*. 2013;19:1610–1617.
84. Roosendaal SD, Geurts JJ, Vrenken H, et al. Regional DTI differences in multiple sclerosis patients. *NeuroImage*. 2009;44:1397–1403.
85. Dineen RA, Vilisaar J, Hlinka J, et al. Disconnection as a mechanism for cognitive dysfunction in multiple sclerosis. *Brain*. 2009;132:239–249.
86. Shu N, Liu Y, Li K, et al. Diffusion tensor tractography reveals disrupted topological efficiency in white matter structural networks in multiple sclerosis. *Cereb Cortex*. 2011;21:2565–2577.
87. Li Y, Jewells V, Kim M, et al. Diffusion tensor imaging based network analysis detects alterations of neuroconnectivity in patients with clinically early relapsing-remitting multiple sclerosis. *Hum Brain Mapp*. 2012;34:3376–3391.
88. Narayana PA, Doyle TJ, Lai D, Wolinsky JS. Serial proton magnetic resonance spectroscopic imaging, contrast-enhanced magnetic resonance imaging, and quantitative lesion volumetry in multiple sclerosis. *Ann Neurol*. 1998;43:56–71.
89. Sharma R, Narayana PA, Wolinsky JS. Grey matter abnormalities in multiple sclerosis: proton magnetic resonance spectroscopic imaging. *Mult Scler*. 2001;7:221–226.
90. Chard DT, Griffin CM, McLean MA, et al. Brain metabolite changes in cortical grey and normal-appearing white matter in clinically early relapsing-remitting multiple sclerosis. *Brain*. 2002;125:2342–2352.

91. Sarchielli P, Presciutti O, Tarducci R, et al. Localized (1)H magnetic resonance spectroscopy in mainly cortical gray matter of patients with multiple sclerosis. *J Neurol*. 2002;249:902–910.
92. Cifelli A, Arridge M, Jezzard P, Esiri MM, Palace J, Matthews PM. Thalamic neurodegeneration in multiple sclerosis. *Ann Neurol*. 2002;52:650–653.
93. Inglese M, Liu S, Babb JS, Mannon LJ, Grossman RI, Gonen O. Three-dimensional proton spectroscopy of deep gray matter nuclei in relapsing-remitting MS. *Neurology*. 2004;63:170–172.
94. Geurts JJ, Reuling IE, Vrenken H, et al. MR spectroscopic evidence for thalamic and hippocampal, but not cortical, damage in multiple sclerosis. *Magn Reson Med*. 2006;55:478–483.
95. Filippi M, Bozzali M, Rovaris M, et al. Evidence for widespread axonal damage at the earliest clinical stage of multiple sclerosis. *Brain*. 2003;126:433–437.
96. Fernando KT, McLean MA, Chard DT, et al. Elevated white matter myo-inositol in clinically isolated syndromes suggestive of multiple sclerosis. *Brain*. 2004;127:1361–1369.
97. Summers M, Swanton J, Fernando K, et al. Cognitive impairment in multiple sclerosis can be predicted by imaging early in the disease. *J Neurol Neurosurg Psychiatry*. 2008;79:955–958.
98. Rigotti DJ, Inglese M, Kirov II, et al. Two-year serial whole-brain N-acetyl-L-aspartate in patients with relapsing-remitting multiple sclerosis. *Neurology*. 2012;78:1383–1389.
99. Kirov II, Tal A, Babb JS, Herbert J, Gonen O. Serial proton MR spectroscopy of gray and white matter in relapsing-remitting MS. *Neurology*. 2012;80:39–46.
100. Inglese M, Madelin G, Oesingmann N, et al. Brain tissue sodium concentration in multiple sclerosis: a sodium imaging study at 3 tesla. *Brain*. 2010;133:847–857.
101. Neema M, Stankiewicz J, Arora A, et al. T1- and T2-based MRI measures of diffuse gray matter and white matter damage in patients with multiple sclerosis. *J Neuroimaging*. 2007;17(suppl 1):16S–21S.
102. Bermel RA, Puli SR, Rudick RA, et al. Prediction of longitudinal brain atrophy in multiple sclerosis by gray matter magnetic resonance imaging T2 hypointensity. *Arch Neurol*. 2005;62:1371–1376.
103. Neema M, Arora A, Healy BC, et al. Deep gray matter involvement on brain MRI scans is associated with clinical progression in multiple sclerosis. *J Neuroimaging*. 2009;19:3–8.
104. Khalil M, Enzinger C, Langkammer C, et al. Quantitative assessment of brain iron by R(2)* relaxometry in patients with clinically isolated syndrome and relapsing-remitting multiple sclerosis. *Mult Scler*. 2009;15:1048–1054.
105. Ge Y, Jensen JH, Lu H, et al. Quantitative assessment of iron accumulation in the deep gray matter of multiple sclerosis by magnetic field correlation imaging. *AJNR Am J Neuroradiol*. 2007;28:1639–1644.
106. Haacke EM, Garbern J, Miao Y, Habib C, Liu M. Iron stores and cerebral veins in MS studied by susceptibility weighted imaging. *Int Angiol*. 2010;29:149–157.
107. Zivadinov R, Schirda C, Dwyer MG, et al. Chronic cerebrospinal venous insufficiency and iron deposition on susceptibility-weighted imaging in patients with multiple sclerosis: a pilot case-control study. *Int Angiol*. 2010;29:158–175.
108. Vanzetta I, Grinvald A. Increased cortical oxidative metabolism due to sensory stimulation: implications for functional brain imaging. *Science*. 1999;286:1555–1558.
109. Ogawa S, Menon RS, Tank DW, et al. Functional brain mapping by blood oxygenation level-dependent contrast magnetic resonance imaging. A comparison of signal characteristics with a biophysical model. *Biophys J*. 1993;64:803–812.
110. Biswal BB. Resting state fMRI: a personal history. *NeuroImage*. 2012;62:938–944.
111. Filippi M, Agosta F, Spinelli EG, Rocca MA. Imaging resting state brain function in multiple sclerosis. *J Neurol*. 2012;260:1709–1713.
112. Rocca MA, Colombo B, Falini A, et al. Cortical adaptation in patients with MS: a cross-sectional functional MRI study of disease phenotypes. *Lancet Neurol*. 2005;4:618–626.
113. Rocca MA, Absinta M, Ghezzi A, Moiola L, Comi G, Filippi M. Is a preserved functional reserve a mechanism limiting clinical impairment in pediatric MS patients? *Hum Brain Mapp*. 2009;30:2844–2851.
114. Rocca MA, Ceccarelli A, Rodegher M, et al. Preserved brain adaptive properties in patients with benign multiple sclerosis. *Neurology*. 2010;74:142–149.
115. Filippi M, Rocca MA, Falini A, et al. Correlations between structural CNS damage and functional MRI changes in primary progressive MS. *NeuroImage*. 2002;15:537–546.
116. Rocca MA, Matthews PM, Caputo D, et al. Evidence for widespread movement-associated functional MRI changes in patients with PPMS. *Neurology*. 2002;58:866–872.
117. Rocca MA, Pagani E, Absinta M, et al. Altered functional and structural connectivities in patients with MS: a 3-T study. *Neurology*. 2007;69:2136–2145.

118. Rocca MA, Valsasina P, Ceccarelli A, et al. Structural and functional MRI correlates of Stroop control in benign MS. *Hum Brain Mapp*. 2009;30:276–290.
119. Roosendaal SD, Schoonheim MM, Hulst HE, et al. Resting state networks change in clinically isolated syndrome. *Brain*. 2010;133:1612–1621.
120. Bonavita S, Gallo A, Sacco R, et al. Distributed changes in default-mode resting-state connectivity in multiple sclerosis. *Mult Scler*. 2011;17:411–422.
121. Rocca MA, Valsasina P, Absinta M, et al. Default-mode network dysfunction and cognitive impairment in progressive MS. *Neurology*. 2010;74:1252–1259.
122. Rocca M, Valsasina P, Martinelli V, et al. Large-scale neuronal network dysfunction in relapsing-remitting multiple sclerosis. *Neurology*. 2012;79:1449–1457.
123. Kiferle L, Politis M, Muraro PA, Piccini P. Positron emission tomography imaging in multiple sclerosis-current status and future applications. *Eur J Neurol*. 2011;18:226–231.
124. Pozzilli C, Fieschi C, Perani D, et al. Relationship between corpus callosum atrophy and cerebral metabolic asymmetries in multiple sclerosis. *J Neurol Sci*. 1992;112:51–57.
125. Sorensen PS, Jonsson A, Mathiesen HK, et al. The relationship between MRI and PET changes and cognitive disturbances in MS. *J Neurol Sci*. 2006;245:99–102.
126. Blinkenberg M, Jensen CV, Holm S, Paulson OB, Sorensen PS. A longitudinal study of cerebral glucose metabolism, MRI, and disability in patients with MS. *Neurology*. 1999;53:149–153.
127. Politis M, Giannetti P, Su P, et al. Increased PK11195 PET binding in the cortex of patients with MS correlates with disability. *Neurology*. 2012;79:523–530.
128. Ratchford JN, Endres CJ, Hammoud DA, et al. Decreased microglial activation in MS patients treated with glatiramer acetate. *J Neurol*. 2012;259:1199–1205.
129. Rissanen E, Virta JR, Paavilainen T, et al. Adenosine A2A receptors in secondary progressive multiple sclerosis: a [(11)C]TMSX brain PET study. *J Cereb Blood Flow Metab*. 2013;33:1394–1401.
130. Lycklama G, Thompson A, Filippi M, et al. Spinal-cord MRI in multiple sclerosis. *Lancet Neurol*. 2003;2:555–562.
131. Ozturk A, Aygun N, Smith SA, Caffo B, Calabresi PA, Reich DS. Axial 3D gradient-echo imaging for improved multiple sclerosis lesion detection in the cervical spinal cord at 3T. *Neuroradiology*. 2013;55:431–439.
132. Nair G, Absinta M, Reich DS. Optimized T1-MPRAGE sequence for better Visualization of spinal cord multiple sclerosis lesions at 3T. *AJNR Am J Neuroradiol*. 2013;34:2215–2222.
133. Brex PA, Leary SM, O'Riordan JI, et al. Measurement of spinal cord area in clinically isolated syndromes suggestive of multiple sclerosis. *J Neurol Neurosurg Psychiatry*. 2001;70:544–547.
134. Losseff NA, Kingsley DP, McDonald WI, Miller DH, Thompson AJ. Clinical and magnetic resonance imaging predictors of disability in primary and secondary progressive multiple sclerosis. *Mult Scler*. 1996;1:218–222.
135. Horsfield MA, Sala S, Neema M, et al. Rapid semi-automatic segmentation of the spinal cord from magnetic resonance images: application in multiple sclerosis. *NeuroImage*. 2010;50:446–455.
136. Rocca MA, Valsasina P, Damjanovic D, et al. Voxel-wise mapping of cervical cord damage in multiple sclerosis patients with different clinical phenotypes. *J Neurol Neurosurg Psychiatry*. 2012;84:35–41.
137. Valsasina P, Rocca MA, Horsfield MA, et al. Regional cervical cord atrophy and disability in multiple sclerosis: a voxel-based analysis. *Radiology*. 2013;266:853–861.
138. Filippi M, Bozzali M, Horsfield MA, et al. A conventional and magnetization transfer MRI study of the cervical cord in patients with MS. *Neurology*. 2000;54:207–213.
139. Rovaris M, Gallo A, Riva R, et al. An MT MRI study of the cervical cord in clinically isolated syndromes suggestive of MS. *Neurology*. 2004;63:584–585.
140. Zackowski KM, Smith SA, Reich DS, et al. Sensorimotor dysfunction in multiple sclerosis and column-specific magnetization transfer-imaging abnormalities in the spinal cord. *Brain*. 2009;132:1200–1209.
141. Agosta F, Pagani E, Caputo D, Filippi M. Associations between cervical cord gray matter damage and disability in patients with multiple sclerosis. *Arch Neurol*. 2007;64:1302–1305.
142. Agosta F, Filippi M. MRI of spinal cord in multiple sclerosis. *J Neuroimaging*. 2007;17(suppl 1):46S–49S.
143. Agosta F, Absinta M, Sormani MP, et al. In vivo assessment of cervical cord damage in MS patients: a longitudinal diffusion tensor MRI study. *Brain*. 2007;130:2211–2219.
144. Ciccarelli O, Wheeler-Kingshott CA, McLean MA, et al. Spinal cord spectroscopy and diffusion-based tractography to assess acute disability in multiple sclerosis. *Brain*. 2007;130:2220–2231.
145. Oh J, Saidha S, Chen M, et al. Spinal cord quantitative MRI discriminates between disability levels in multiple sclerosis. *Neurology*. 2013;80:540–547.

146. Wattjes MP, Harzheim M, Lutterbey GG, et al. Does high field MRI allow an earlier diagnosis of multiple sclerosis? *J Neurol*. 2008;255:1159–1163.
147. Ge Y, Zohrabian VM, Grossman RI. Seven-Tesla magnetic resonance imaging: new vision of microvascular abnormalities in multiple sclerosis. *Arch Neurol*. 2008;65:812–816.
148. Hammond KE, Metcalf M, Carvajal L, et al. Quantitative in vivo magnetic resonance imaging of multiple sclerosis at 7 tesla with sensitivity to iron. *Ann Neurol*. 2008;64:707–713.
149. Tallantyre EC, Brookes MJ, Dixon JE, Morgan PS, Evangelou N, Morris PG. Demonstrating the perivascular distribution of MS lesions in vivo with 7-Tesla MRI. *Neurology*. 2008;70:2076–2078.
150. Tallantyre EC, Dixon JE, Donaldson I, et al. Ultra-high-field imaging distinguishes MS lesions from asymptomatic white matter lesions. *Neurology*. 2011;76:534–539.
151. Mistry N, Dixon J, Tallantyre E, et al. Central veins in brain lesions visualized with high-field magnetic resonance imaging: a pathologically specific diagnostic biomarker for inflammatory demyelination in the brain. *JAMA Neurol*. 2013;70:1–6.
152. Pitt D, Boster A, Pei W, et al. Imaging cortical lesions in multiple sclerosis with ultra-high-field magnetic resonance imaging. *Arch Neurol*. 2010;67:812–818.
153. Mehta V, Pei W, Yang G, et al. Iron is a sensitive biomarker for inflammation in multiple sclerosis lesions. *PLoS One*. 2013;8:e57573.
154. Duyn JH, van Gelderen P, Li TQ, de Zwart JA, Koretsky AP, Fukunaga M. High-field MRI of brain cortical substructure based on signal phase. *Proc Natl Acad Sci USA*. 2007;104:11796–11801.
155. Cohen-Adad J, Polimeni JR, Helmer KG, et al. T(2)* mapping and B(0) orientation-dependence at 7 T reveal cyto- and myeloarchitecture organization of the human cortex. *NeuroImage*. 2012;60:1006–1014.
156. Kangarlu A, Bourekas EC, Ray-Chaudhury A, Rammohan KW. Cerebral cortical lesions in multiple sclerosis detected by MR imaging at 8 Tesla. *AJNR Am J Neuroradiol*. 2007;28:262–266.
157. Schmierer K, Parkes HG, So PW, et al. High field (9.4 tesla) magnetic resonance imaging of cortical grey matter lesions in multiple sclerosis. *Brain*. 2010;133:858–867.
158. Mainero C, Benner T, Radding A, et al. In vivo imaging of cortical pathology in multiple sclerosis using ultra-high field MRI. *Neurology*. 2009;73:941–948.
159. Tallantyre EC, Morgan PS, Dixon JE, et al. 3 Tesla and 7 Tesla MRI of multiple sclerosis cortical lesions. *J Magn Reson Imaging*. 2010;32:971–977.
160. Bluestein KT, Pitt D, Sammet S, et al. Detecting cortical lesions in multiple sclerosis at 7 T using white matter signal attenuation. *Magn Reson Imaging*. 2012;30:907–915.
161. Nielsen AS, Kinkel RP, Tinelli E, Benner T, Cohen-Adad J, Mainero C. Focal cortical lesion detection in multiple sclerosis: 3 Tesla DIR versus 7 Tesla FLASH-T2. *J Magn Reson Imaging*. 2012;35:537–542.
162. de Graaf WL, Zwanenburg JJ, Visser F, et al. Lesion detection at seven Tesla in multiple sclerosis using magnetisation prepared 3D-FLAIR and 3D-DIR. *Eur Radiol*. 2012;22:221–231.
163. Kilsdonk ID, de Graaf WL, Soriano AL, et al. Multicontrast MR imaging at 7T in multiple sclerosis: Highest lesion detection in cortical Gray matter with 3D-FLAIR. *AJNR Am J Neuroradiol*. 2013;34:791–796.
164. Nielsen AS, Kinkel RP, Madigan N, Tinelli E, Benner T, Mainero C. Contribution of cortical lesion subtypes at 7T MRI to physical and cognitive performance in MS. *Neurology*. 2013;81:641–649.
165. Sormani MP, Bruzzi P, Beckmann K, et al. MRI metrics as surrogate endpoints for EDSS progression in SPMS patients treated with IFN beta-1b. *Neurology*. 2003;60:1462–1466.
166. Sormani MP, Bruzzi P, Comi G, Filippi M. MRI metrics as surrogate markers for clinical relapse rate in relapsing-remitting MS patients. *Neurology*. 2002;58:417–421.
167. Sormani MP, Bonzano L, Roccatagliata L, Cutter GR, Mancardi GL, Bruzzi P. Magnetic resonance imaging as a potential surrogate for relapses in multiple sclerosis: a meta-analytic approach. *Ann Neurol*. 2009;65:268–275.
168. Sormani MP, Bruzzi P. MRI lesions as a surrogate for relapses in multiple sclerosis: a meta-analysis of randomised trials. *Lancet Neurol*. 2013;12:669–676.
169. Verhey LH, Signori A, Arnold DL, et al. Clinical and MRI activity as determinants of sample size for pediatric multiple sclerosis trials. *Neurology*. 2013;81:1215–1221.
170. Sormani MP, Bonzano L, Roccatagliata L, Mancardi GL, Uccelli A, Bruzzi P. Surrogate endpoints for EDSS worsening in multiple sclerosis. A meta-analytic approach. *Neurology*. 2010;75:302–309.
171. Sormani MP, Li DK, Bruzzi P, et al. Combined MRI lesions and relapses as a surrogate for disability in multiple sclerosis. *Neurology*. 2011;77:1684–1690.
172. Filippi M, Rovaris M, Rocca MA, Sormani MP, Wolinsky JS, Comi G. Glatiramer acetate reduces the proportion of new MS lesions evolving into "black holes". *Neurology*. 2001;57:731–733.

173. Dalton CM, Miszkiel KA, Barker GJ, et al. Effect of natalizumab on conversion of gadolinium enhancing lesions to T1 hypointense lesions in relapsing multiple sclerosis. *J Neurol*. 2004;251:407–413.
174. Filippi M, Rocca MA, Pagani E, et al. Placebo-controlled trial of oral laquinimod in multiple sclerosis: MRI evidence of an effect on brain tissue damage. *J Neurol Neurosurg Psychiatry*. 2013;85:851–858.
175. Cadavid D, Cheriyan J, Skurnick J, Lincoln JA, Wolansky LJ, Cook SD. New acute and chronic black holes in patients with multiple sclerosis randomised to interferon beta-1b or glatiramer acetate. *J Neurol Neurosurg Psychiatry*. 2009;80:1337–1343.
176. Filippi M, Rocca MA, Camesasca F, et al. Interferon beta-1b and glatiramer acetate effects on permanent black hole evolution. *Neurology*. 2011;76:1222–1228.
177. Hardmeier M, Wagenpfeil S, Freitag P, et al. Rate of brain atrophy in relapsing MS decreases during treatment with IFNbeta-1a. *Neurology*. 2005;64:236–240.
178. Rudick RA, Fisher E, Lee JC, Duda JT, Simon J. Brain atrophy in relapsing multiple sclerosis: relationship to relapses, EDSS, and treatment with interferon beta-1a. *Mult Scler*. 2000;6:365–372.
179. Ge Y, Grossman RI, Udupa JK, et al. Glatiramer acetate (Copaxone) treatment in relapsing-remitting MS: quantitative MR assessment. *Neurology*. 2000;54:813–817.
180. Miller DH, Soon D, Fernando KT, et al. MRI outcomes in a placebo-controlled trial of natalizumab in relapsing MS. *Neurology*. 2007;68:1390–1401.
181. Radue EW, O'Connor P, Polman CH, et al. Impact of fingolimod therapy on magnetic resonance imaging outcomes in patients with multiple sclerosis. *Arch Neurol*. 2012;69:1259–1269.
182. Comi G, Jeffery D, Kappos L, et al. Placebo-controlled trial of oral laquinimod for multiple sclerosis. *N Engl J Med*. 2012;366:1000–1009.
183. Chen JT, Collins DL, Atkins HL, Freedman MS, Galal A, Arnold DL. Brain atrophy after immunoablation and stem cell transplantation in multiple sclerosis. *Neurology*. 2006;66:1935–1937.
184. Inglese M, Mancardi GL, Pagani E, et al. Brain tissue loss occurs after suppression of enhancement in patients with multiple sclerosis treated with autologous haematopoietic stem cell transplantation. *J Neurol Neurosurg Psychiatry*. 2004;75:643–644.
185. Roccatagliata L, Rocca M, Valsasina P, et al. The long-term effect of AHSCT on MRI measures of MS evolution: a five-year follow-up study. *Mult Scler*. 2007;13:1068–1070.
186. Sormani MP, Arnold DL, De Stefano N. Treatment effect on brain atrophy correlates with treatment effect on disability in multiple sclerosis. *Ann Neurol*. 2013;75:43–49.
187. Zivadinov R, Reder AT, Filippi M, et al. Mechanisms of action of disease-modifying agents and brain volume changes in multiple sclerosis. *Neurology*. 2008;71:136–144.
188. Kaiser CC, Shukla DK, Stebbins GT, et al. A pilot test of pioglitazone as an add-on in patients with relapsing remitting multiple sclerosis. *J Neuroimmunol*. 2009;211:124–130.
189. Kapoor R, Furby J, Hayton T, et al. Lamotrigine for neuroprotection in secondary progressive multiple sclerosis: a randomised, double-blind, placebo-controlled, parallel-group trial. *Lancet Neurol*. 2010;9:681–688.
190. Kalkers NF, Barkhof F, Bergers E, van Schijndel R, Polman CH. The effect of the neuroprotective agent riluzole on MRI parameters in primary progressive multiple sclerosis: a pilot study. *Mult Scler*. 2002;8:532–533.
191. Lin X, Tench CR, Turner B, Blumhardt LD, Constantinescu CS. Spinal cord atrophy and disability in multiple sclerosis over four years: application of a reproducible automated technique in monitoring disease progression in a cohort of the interferon beta-1a (Rebif) treatment trial. *J Neurol Neurosurg Psychiatry*. 2003;74:1090–1094.
192. Leary SM, Miller DH, Stevenson VL, Brex PA, Chard DT, Thompson AJ. Interferon beta-1a in primary progressive MS: an exploratory, randomized, controlled trial. *Neurology*. 2003;60:44–51.
193. Montalban X, Sastre-Garriga J, Tintore M, et al. A single-center, randomized, double-blind, placebo-controlled study of interferon beta-1b on primary progressive and transitional multiple sclerosis. *Mult Scler*. 2009;15:1195–1205.
194. Rudick RA, Lee JC, Simon J, Ransohoff RM, Fisher E. Defining interferon beta response status in multiple sclerosis patients. *Ann Neurol*. 2004;56:548–555.
195. Rio J, Rovira A, Tintore M, et al. Relationship between MRI lesion activity and response to IFN-beta in relapsing-remitting multiple sclerosis patients. *Mult Scler*. 2008;14:479–484.
196. Rio J, Castillo J, Rovira A, et al. Measures in the first year of therapy predict the response to interferon beta in MS. *Mult Scler*. 2009;15:848–853.
197. Tomassini V, Paolillo A, Russo P, et al. Predictors of long-term clinical response to interferon beta therapy in relapsing multiple sclerosis. *J Neurol*. 2006;253:287–293.

198. Freedman MS, Patry DG, Grand'Maison F, Myles ML, Paty DW, Selchen DH. Treatment optimization in multiple sclerosis. *Can J Neurol Sci*. 2004;31:157–168.
199. Sormani MP, De Stefano N. Defining and scoring response to IFN-beta in multiple sclerosis. *Nat Rev Neurol*. 2013;9:504–512.
200. Geurts JJ, Roosendaal SD, Calabrese M, et al. Consensus recommendations for MS cortical lesion scoring using double inversion recovery MRI. *Neurology*. 2011;76:418–424.
201. Sormani M, Stromillo ML, Battaglini M, Signori A, De Stefano N. Modelling the distribution of cortical lesions in multiple sclerosis. *Mult Scler*. 2011;18:229–231.
202. Rinaldi F, Calabrese M, Seppi D, Puthenparampil M, Perini P, Gallo P. Natalizumab strongly suppresses cortical pathology in relapsing-remitting multiple sclerosis. *Mult Scler*. 2012;18:1760–1767.
203. Calabrese M, Bernardi V, Atzori M, et al. Effect of disease-modifying drugs on cortical lesions and atrophy in relapsing-remitting multiple sclerosis. *Mult Scler*. 2012;18:418–424.
204. Inglese M, van Waesberghe JH, Rovaris M, et al. The effect of interferon beta-1b on quantities derived from MT MRI in secondary progressive MS. *Neurology*. 2003;60:853–860.
205. Filippi M, Rocca MA, Pagani E, et al. European study on intravenous immunoglobulin in multiple sclerosis: results of magnetization transfer magnetic resonance imaging analysis. *Arch Neurol*. 2004;61:1409–1412.
206. Narayanan S, De Stefano N, Francis GS, et al. Axonal metabolic recovery in multiple sclerosis patients treated with interferon beta-1b. *J Neurol*. 2001;248:979–986.
207. Sajja BR, Narayana PA, Wolinsky JS, Ahn CW. Longitudinal magnetic resonance spectroscopic imaging of primary progressive multiple sclerosis patients treated with glatiramer acetate: multicenter study. *Mult Scler*. 2008;14:73–80.
208. van den Elskamp IJ, Knol DL, Vrenken H, et al. Lesional magnetization transfer ratio: a feasible outcome for remyelinating treatment trials in multiple sclerosis. *Mult Scler*. 2010;16:660–669.
209. Brown RA, Narayanan S, Arnold DL. Segmentation of magnetization transfer ratio lesions for longitudinal analysis of demyelination and remyelination in multiple sclerosis. *NeuroImage*. 2013;66C:103–109.
210. Button T, Altmann D, Tozer D, et al. Magnetization transfer imaging in multiple sclerosis treated with alemtuzumab. *Mult Scler*. 2012;19:241–244.
211. Hayton T, Furby J, Smith KJ, et al. Longitudinal changes in magnetisation transfer ratio in secondary progressive multiple sclerosis: data from a randomised placebo controlled trial of lamotrigine. *J Neurol*. 2012;259:505–514.
212. Cader S, Palace J, Matthews PM. Cholinergic agonism alters cognitive processing and enhances brain functional connectivity in patients with multiple sclerosis. *J Psychopharmacol*. 2009;23:686–696.
213. Tomassini V, Johansen-Berg H, Jbabdi S, et al. Relating brain damage to brain plasticity in patients with multiple sclerosis. *Neurorehabil Neural Repair*. 2012;26:581–593.
214. Filippi M, Riccitelli G, Mattioli F, et al. Effects of cognitive rehabilitation on structural and functional MRI measures in multiple sclerosis: an explorative study. *Radiology*. 2012;262:932–940.

SECTION IV

CURRENTLY APPROVED THERAPIES—INJECTABLE

CHAPTER

14

Interferon β in Multiple Sclerosis: A Review

R.H. Gross, F. Lublin

Icahn School of Medicine at Mount Sinai, New York, NY, United States

OUTLINE

14.1 INTRODUCTION

Since they were first described in 1957,[1] there has long been interest in harnessing the biological properties of interferons (IFNs) as therapy for human disease. They have been used in a variety of disorders, including hepatitis B and C, certain types of leukemia and lymphoma, chronic granulomatous disease, melanoma, and multiple sclerosis (MS). Recombinant interferon beta (IFN β) was the first disease-modifying therapy (DMT) to become available for MS: its approval in 1993 heralded a new era in MS management, in which the aim of therapy would not simply be the alleviation of symptoms or the transient suppression of relapse-related inflammation, but rather the prevention of new disease activity. There are currently three IFN subtypes in clinical use: IFN β-1a, IFN β-1b, and pegylated IFN β-1a. IFN β-1a is available in 30 μg IM weekly (Avonex®, Biogen Idec, Cambridge, MA) and in 22 or 44 μg SC three times weekly (Rebif®, EMD Serono, Rockland, MA) formulations. IFN β-1b is available as 250 μg SC every other day (Betaseron®/Betaferon®, Bayer HealthCare, Whippany, NJ; Extavia®, Novartis, East Hanover, NJ). Pegylated IFN β-1a is sold under the trade name Plegridy® (Biogen Idec, Cambridge, MA) and administered subcutaneously every other week.

Translational Neuroimmunology in Multiple Sclerosis
http://dx.doi.org/10.1016/B978-0-12-801914-6.00014-3

14.2 MECHANISM OF ACTION

IFN β is a type-1 IFN naturally produced by fibroblasts.[2] It is encoded by a single IFN β gene on the short arm of chromosome 9.[3] Type-1 IFNs, which include IFN β and IFN α, comprise a family of structurally related proteins and compete for binding at the same complex of receptors.[4] IFN γ, by contrast, is a type-2 IFN produced by immune cells that binds to a completely different cytokine receptor. The number of IFN binding sites on most cell types is relatively low, with the highest being around 1000 sites/cell.[5] Binding of IFNs to these cell surface receptors induces cytoplasmic transcription factors to translocate to the nucleus, ultimately resulting in altered gene transcription. IFN β-1a is produced in mammalian cells (Chinese hamster ovary) using a natural human gene sequence and is glycosylated, while IFN β-1b is produced in *Escherichia coli* cells using a modified human gene sequence containing a genetically engineered cysteine-to-serine substitution at position 17.[6,7]

Myriad in vitro and in vivo studies have provided insights into the antiviral, antineoplastic, and immunoregulatory activities of IFNs. Their early identification as antiviral agents is reflected in the use of international units of antiviral activity to describe their potency. It was subsequently appreciated that IFNs have the ability to regulate growth and differentiation of a variety of cell types and are effective in inhibiting tumor growth and proliferation.[4] Despite the wealth of information accumulated about its biomolecular mechanisms, it remains incompletely understood precisely why IFN β has a beneficial effect in MS (Table 14.1). As immune dysregulation is believed to underlie MS pathogenesis, it seems logical that IFN β's immunomodulatory effect is what is responsible for its effectiveness in MS; IFN β-1a and IFN β-1b do not appear to differ substantively in this regard. Undeniably, our understanding of IFN β's immunomodulatory behavior has progressed in parallel with the recognition of its efficacy, demonstrated in numerous randomized clinical trials (RCTs) and in years of clinical practice in treating MS. IFN β has been shown to decrease antigen presentation, enhance suppressor T cells, reduce the production of proinflammatory cytokines and matrix metalloproteinases (MMPs), and inhibit lymphocyte trafficking into the central nervous system (CNS).

IFN β, like other IFNs, augments the expression of major histocompatibility complex (MHC) class I, which is present on all cells. IFN γ has a stimulatory effect on the surface expression of MHC class II, the antagonism of which by IFN β might prevent epitope spreading and contribute to the latter's efficacy in MS.[8] In addition to reducing the expression of MHC class II on

TABLE 14.1 Various Immunological Properties of Interferon Beta that are Thought to be Responsible for its Effectiveness in Multiple Sclerosis

Immune Mechanisms of Interferon β
Regulation of antigen presentation
Inhibition of T-cell activation and proliferation
Apoptosis of autoreactive T cells
Reduction of proinflammatory cytokines and matrix metalloproteinases
Enhancement of regulatory T cells
Decrease in lymphocyte trafficking into the central nervous system/stabilization of blood–brain barrier

antigen-presenting cells (APCs), IFN β downregulates costimulatory molecules like CD80 and CD28 on APCs and lymphocytes, thereby resulting in decreased overall T-cell activation.[9] Meanwhile, IFN β upregulates intracellular CTLA4 and surface Fas[10] and downregulates an antiapoptotic protein, FLIP,[11] which has the overall effect of promoting apoptosis of autoreactive T cells.

Until relatively recently, much of our understanding about immune dysregulation in MS focused on the creation of a proinflammatory milieu by Th1 cells, while Th2 cells have been thought of as playing a protective role. Along these lines, much has been written about the ability of IFN β to shift the adaptive immune response from a proinflammatory Th1 to an antiinflammatory Th2 profile.[12] The IFNs as a group participate in an immense network of cytokines, chemokines, metalloproteinases, and other mediators of inflammation, whose exceptionally complex pathways have been the subject of intense investigation over the years. It is now known that a distinct class of CD4+ lymphocytes identified as Th17, which produces IL-17, is involved in autoimmunity and MS disease activity.[13] In vitro studies have shown that IFN β has the capacity—mediated by dendritic cells, B cells, and T cells—to prevent Th17 differentiation through the inhibition of cytokines IL-1β, IL-23, and TGF β and the promotion of IL-27, IL-12p35, and IL-10.[14,15] High IL-17 serum concentrations have been shown to be associated with lack of response to IFN β therapy in MS patients.[16] Enhanced function of regulatory T cells has also been suggested to contribute to IFN β's efficacy in MS.[17] Following 12 months of treatment with IFN β, CD45RA+ naïve (Th0) cells were increased from baseline in a group of relapsing-remitting MS (RRMS) patients, and CD4+CD45RO+CCR7+ central memory cells were decreased compared to untreated RRMS controls but not compared to baseline.[18]

The disruption of the blood–brain barrier (BBB) has been recognized as a critical factor in MS pathophysiology. After crossing the BBB, activated T lymphocytes recruit more immune cells, such as other T cells, B cells, and macrophages, which results in further inflammation, demyelination, and neuronal damage. IFN β has an impact on the ability of lymphocytes to migrate into the CNS, both by preventing their adhesion to the endothelium and by preventing subsequent extravasation into the CNS.[19,20] It accomplishes this through downregulation of the surface adhesion molecules VLA-4 and LFA-1[21–23] and of MMP-2, -3, -7, -9[24–28] on activated immune cells, and perhaps also via a direct stabilizing effect on BBB permeability.[29]

14.3 EFFICACY

As first-generation DMTs, the various IFN β formulations are, in general, modestly effective at reducing MS disease activity. Early phase studies in MS patients conducted between 1977 and 1980 involving the intrathecal injection of natural human IFN β through serial lumbar punctures suggested a beneficial effect in terms of the number of clinical relapses.[30,31] The initial multicenter phase III clinical trial that led to the approval of IFN β-1b showed that IFN β-1b at doses of either 50 or 250 μg SQ every other day was superior to placebo at reducing relapse rates and the proportion of relapse-free patients, as well as reducing new and active lesions on MRI and proportion of scans with activity.[32–34] At 2 years, the annualized relapse rates (ARRs) in the 250 μg, 50 μg, and placebo groups were 0.84, 1.17, and 1.27, respectively, ($p = 0.0001$ for 250 μg vs. placebo, $p = 0.01$ for 50 μg vs. placebo); and the proportion of relapse-free patients was 31%, 21%, 16% ($p = 0.007$ for 250 μg vs. placebo, not significant for other comparisons). In its phase III clinical trial, patients treated with IFN β-1a 30 μg IM weekly had an

ARR of 0.61, compared to 0.90 in the placebo arm during the 2 years of the study.[35] In PRISMS, IFN β-1a 22 and 44 μg SQ three times weekly both outperformed placebo over 2 years in clinical and radiographic efficacy measures, with 3- and 4-year extension studies showing similar results.[36,37] Over the 2 years of the original study there were fewer relapses per patient in the 22 μg (mean 1.82) and 44 μg (mean 1.73) groups compared to placebo (mean 2.56, $p<0.005$ for both) and a higher proportion of relapse-free patients (27% for 22 μg ($p<0.05$), 32% for 44 μg ($p<0.005$), and 16% for placebo).

Pegylated IFN β-1a is the newest IFN β product to enter the MS arena. Pegylation involves the attachment of a polyethylene glycol polymer to the IFN β-1a, which has the effect of increasing the apparent mass of the molecule, which in turn reduces its renal clearance and prolongs its half-life. PEG-IFN β-1a 125 μg is self-administered via subcutaneous injection every 2 weeks. Clinical experience with this agent outside of clinical trials is limited given its recent approval. In a phase III clinical trial, PEG-IFN β-1a reduced relapse rates in the first 48 weeks compared to placebo—the primary outcome of the trial—with respective ARRs of 0.26 and 0.40, a relative reduction of 36% ($p=0.0007$).[38] This effect size is what we have come to expect from IFN β.

Clinically isolated syndrome (CIS) is defined as an initial presentation of a disease with characteristics of inflammatory demyelination, which has not demonstrated dissemination in time required for the diagnosis of MS.[39] CIS patients with brain lesions have been noted to be at a much greater risk for developing MS.[40,41] IFN β was shown in a series of pivotal CIS trials—CHAMPS,[42] BENEFIT,[43,44] ETOMS[45]—to delay the conversion to clinically definite multiple sclerosis (CDMS) and reduce MRI activity. These CIS trials, along with natural history studies of CIS, provided evidence for DMT initiation earlier in the course of the disease. In CHAMPS, 383 patients between ages 18 and 50 who had an initial demyelinating episode (optic neuritis, partial transverse myelitis, brain stem or cerebellar syndrome) and two or more clinically silent lesions measuring 3 mm or greater, were randomized to IFN β-1a 30 mcg IM weekly or placebo. The trial was terminated early following a preplanned interim efficacy analysis; during 3 years of follow-up, the hazard ratio of conversion to CDMS was 0.56 (95% CI 0.38–0.81). BENEFIT, which had similar enrollment criteria and compared IFN β-1b 250 μg to placebo, reported a hazard ratio of 0.50 (95% CI 0.36–0.70) for CDMS after 2 years and 0.59 (95% CI 0.44–0.80) after 3 years favoring IFN. In ETOMS, IFN β-1a 22 μg reduced the rate of conversion to CDMS with a hazard ratio of 0.65 (95% CI 0.45–0.94). However, there was no benefit to early treatment in terms of long-term disability found in follow-up studies.[46,47]

In deciding on appropriate treatment for MS, we must weigh the risks and benefits of proposed therapies. In order to do this, it is important to have some idea of the comparative efficacies among the different options. Most of the controlled trials of the MS DMTs did not, however, directly compare two active treatments—the most scientifically sound way of determining relative efficacy. Comparing ARRs across trials, for instance, is fraught with peril due to differences in the enrolled populations, declining relapse rates in the placebo arm over the past 20 years and other methodological issues. There are a handful of head-to-head RCTs that did compare two different IFN β formulations or, more recently, IFN β to a newer DMT. INCOMIN (Independent Comparison of Interferons), a nonpharmaceutical-sponsored multicenter RCT, randomized 188 RRMS patients to IFN β-1b 250 μg SC every other day or IFN β-1a 30 μg IM weekly.[48] In the primary outcome measure—the proportion of patients free from relapse during the 24 months of study—51% of patients in the high-dose/high-frequency group versus 36%

of patients in the low-dose/low-frequency group remained relapse-free, a relative risk of 0.76 ($p=0.03$). These clinical data were supported by MRI metrics demonstrating the superiority of high-dose over low-dose IFN in the proportion of patients free from new proton density/T2 lesions at 24 months: 55% versus 26%, a relative risk of 0.60 ($p<0.001$). EVIDENCE compared IFN β-1a 44 μg SC to IFN β-1a 30 μm IM.[49] Though of relatively brief duration (the primary endpoint was the proportion of patients free of relapses by 24 weeks), the trial showed superiority of IFN β-1a 44 μg SC thrice weekly over IFN β-1a 30 μm IM weekly in the proportion of patients who remained relapse-free by 24 weeks (75% vs. 63%, OR 1.9 ($p=0.0005$)), an effect that was maintained over 48 weeks of treatment (62% vs. 52%, OR 1.5 ($p=0.009$)).

The second generation of MS DMTs includes some that have been shown in clinical trials to be superior to IFNs. Fingolimod (Gilenya®, Novartis, East Hanover, NJ), a sphingosine-1-phosphate receptor modulator, was the first oral medication to be approved for the treatment of MS. TRANSFORMS evaluated fingolimod against IFN β-1a 30 μg IM weekly.[50] In this 12-month, double-blind, double-dummy head-to-head trial, fingolimod reduced the relapse rate compared to IFN β-1a 30 μg: the ARR in the fingolimod 0.5 mg daily arm was 0.16, whereas the ARR in the IFN β-1a 30 μg group was 0.33, a 52% relative reduction. Alemtuzumab (Lemtrada®, Genzyme, Cambridge, MA), a humanized monoclonal antibody against CD52 given as an intravenous infusion, was compared against IFN β-1a 44 μg thrice weekly in two head-to-head RCTs: CARE-MS I and CARE-MS II.[51,52] In CARE-MS I, the ARR was reduced by 55% in the alemtuzumab group versus the IFN group (0.18 vs. 0.39, $p<0.0001$), while in CARE-MS II, the ARR was decreased by 49% (0.26 vs. 0.52, $p<0.0001$). Both CARE-MS I and CARE-MS II also demonstrated that alemtuzumab was more effective than IFN at reducing MRI markers of disease activity, though IFN was associated with fewer serious adverse events.

Substantial evidence indicates that IFNs are not effective in halting progression in patients with established primary or secondary progressive multiple sclerosis (PPMS and SPMS). A Cochrane Review of the RCTs comparing IFN to placebo in SPMS patients showed that in the five trials meeting inclusion criteria, analyzing in total 3122 patients (1829 IFN vs. 1293 placebo), IFN did not decrease the risk of progression sustained at 6 months following 3 years of treatment.[53] Though having no effect on the progression of disability, IFN did decrease the risk of developing new relapses and new MRI activity in this time period. In PPMS, a Cochrane Review identified only two placebo-controlled RCTs involving IFN β in which a total of 123 patients were included.[54] Neither showed a beneficial effect of IFN β on disease progression, though the one that included MRI metrics as secondary endpoints (Montalban 2004) demonstrated that at 2 years, there were fewer active lesions in the IFN β arm than in the placebo arm (weighted mean difference −1.3, $p=0.003$) and fewer patients with active lesions on MRI scans (22% vs. 51%, $p=0.02$).[55] The authors suggested that the trials were underpowered to detect any difference in disease progression.

CombiRx employed a double-blind, double-dummy design to test whether combined IFN β-1a 30 μg IM weekly and glatiramer acetate (GA) 20 mg SC daily improved outcomes compared to either treatment alone.[56] In this multicenter trial, which ran from 2005 to 2012, it was found that combined therapy did not confer any significant clinical benefit compared to the better of the two single agents alone (GA), but that GA alone or in combination with IFN β was superior to IFN β alone at reducing the ARR (31% and 25% relative reduction, respectively, $p=0.027$ and $p=0.022$). Disease activity-free status (DAFS, sometimes referred to as "no evidence of disease activity") is a composite outcome measure combining relapse,

TABLE 14.2 A Review of the Major Clinical Trials in Multiple Sclerosis Involving Interferon Beta

Trial	Arms	Relapse Rates[a]	Percent Relapse-free
IFN β-1b study group	IFN β-1b vs. placebo	0.84 vs. 1.27 (250 μg vs. placebo)	31% vs. 16% (250 μg vs. placebo)
MSCRG	IFN β-1a IM vs. placebo	0.61 vs. 0.9 (30 μg vs. placebo)	Not reported
PRISMS	IFN β-1a SC vs. placebo	1.82 vs. 1.73 vs. 2.56 (22 μg vs. 44 μg vs. placebo)	32% vs. 27% vs. 16%
INCOMIN	IFN β-1b vs. IFN β-1a IM	0.5 vs. 0.7 (250 μg vs. 30 μg)	51% vs. 36%
EVIDENCE	IFN β-1a SC vs. IFN β-1a IM	0.29 vs. 0.40 (24 w) 0.54 vs. 0.64 (48 w) (44 μg vs. 30 μg)	75% vs. 63% (24 w) 62% vs. 52% (48 w)
TRANSFORMS	Fingolimod vs. IFN β-1a IM	0.16 vs. 0.33 (0.5 mg vs. placebo)	83% vs. 69%
CARE-MS I	Alemtuzumab vs. IFN β-1a SC	0.18 vs. 0.39 (12 mg vs. 44 mcg)	78% vs. 59%
CARE-MS II	Alemtuzumab vs. IFN β-1a SC	0.26 vs. 0.52 (12 mg vs. 44 mcg)	65% vs. 47%
CombiRx	IFN β-1a IM vs. glatiramer acetate vs. IFN β-1a IM + glatiramer acetate	0.16 vs. 0.11 vs. 0.12 (30 μg for IFN β, 20 mg for GA)	74% vs. 80% vs. 77%

[a]*Given as annualized relapses except for PRISMS, where it is for 2 years.*

MRI, and disability (EDSS) metrics that has been increasingly employed because of its greater sensitivity to detect treatment effects in an era of increasingly effective agents.[57] When DAFS was analyzed post hoc in CombiRx, it was found that combined treatment with IFN β and GA produced more patients with DAFS (33.3%) than did either treatment alone (21.2% for IFN β and 19.4% for GA)—an effect driven by the MRI results. In REGARD, a multicenter open-label randomized study comparing IFN β-1a 44 μg SC to GA, the investigators found no significant difference between the two groups in terms of the primary outcome measure, time to first relapse, nor was there any significant difference in the number or change in volume of T2 active lesions or volume of gadolinium-enhancing lesions, though IFN β-treated patients had significantly fewer gadolinium-enhancing lesions per scan (0.24 vs. 0.41, $p = 0.0002$).[58]

While the principal aim of treatment is to limit the chance of patients developing the irreversible disability, RCTs, because of their usual duration of two to three years, have only been able to show a positive effect on the risk of disability through the prevention of inflammatory activity and relapses. These trials do not study patients for the decades it takes to determine the behavior of DMTs over the long term. It has been hypothesized that the use of DMTs during the relapsing phase of the disease will limit the amount of disability that patients ultimately accumulate, though this has been difficult to prove (Table 14.2).[59,60]

14.4 IMMUNOGENICITY

IFN β products have immunogenic properties: neutralizing antibodies (nABs) may develop that prevent IFN β from binding its receptor and having its intended biological effect.[61–63,65–67] Their presence in the serum has varied from 7% to as high as 80%, depending on the sensitivity of the assay used, the cut-off used to demark positivity, and also on the variability in antibody levels over time within a given patient.[64–66] A study involving prospectively gathered serum from RRMS patients in Denmark who began IFN β from 1996 to 1999 found that a significantly lower proportion of nAB-positive patients treated with IFN β-1b at 36 months compared to 12 months ($p = 0.023$).[65] Relapse rates were significantly higher during periods of antibody positivity (0.64–0.70) than during periods of antibody negativity (0.43–0.46; $p < 0.03$) in this study. Different clinical trials, however, have found different relationships between nAB presence and disease activity, though nABs were more consistently noted to correspond to breakthrough MRI activity than to clinical relapses, perhaps as a result of lack of power in many of these trials.

In the pivotal IFN β-1b trial, nAB presence was associated with an ARR of 1.16, while nAB negativity was associated with an ARR of 0.50 ($p < 0.05$).[32,67] Similarly in PRISMS, ARR was higher among nAB-positive than among nAB-negative patients (0.85–0.52, $p < 0.05$).[36] Hartung et al. analyzed subjects from the BENEFIT study by nAB status and titer (low: ≥20–100 NU/mL, medium: >100–400 NU/mL, high: >400 NU/mL) and determined that nAB status was not associated with time to CDMS, time to confirmed disability progression, or ARR.[67] MRI activity in the form of newly active lesion number, T2 lesion volume, and conversion to McDonald MS, however, was associated with nAB positivity and was seen in greater frequency in patients with higher nAB titers.

nABs are more likely to be present with more frequent, higher dosed IFN β preparations than in low-dose weekly IFN β, as was demonstrated in the INCOMIN and EVIDENCE trials comparing low-dose to high-dose IFN β; nevertheless, the higher dose IFN β, as indicated earlier, was demonstrated to be superior in both of these head-to-head trials. In INCOMIN, 30% of IFN β-1b-treated patients developed nABs, compared to 7% of IFN β-1a treated patients ($p = 0.0004$). After 48 weeks in the EVIDENCE trial, 25% of patients treated with the higher dose IFN β-1a had developed nABs, while only 2% of patients treated with lower dose IFN β-1a had ($p < 0.001$). Whether the increase in immunogenicity is related to the route or frequency of administration, or to the dose itself, has not been fully elucidated.

14.5 SAFETY/SIDE-EFFECT PROFILE

The side-effect profile of the different classes of IFN β are comparable and include flu-like symptoms, injection site reactions, liver function test abnormalities, decreased peripheral blood counts (eg, leukopenia), and worsening of depression. Oral analgesics and antipyretics are generally recommended on injection days to prevent or mitigate flu-like symptoms, which can consist of fever, chills, headaches, fatigue, malaise, nausea, and anorexia. That they remain a mainstay treatment for MS owes much to the fact that, along with GA, the other main first-generation injectable DMT, IFN β has the longest and most benign safety record of all the MS DMTs.

Periodic complete blood count with differential, liver function, and thyroid testing is recommended for patients taking IFNs.

IFNs are pregnancy category C: they have not been adequately studied in pregnant women, but animal studies suggest that they may be harmful. Specifically, doses considered supratherapeutic for humans that were given to animals had abortifacient effects. A review of safety data from clinical trials and postmarketing reports on pregnancy outcomes following in utero exposure to IFN β-1a SC identified 425 pregnancies, of which 324 (76.2%) resulted in normal live births.[68] The rates of spontaneous abortions and major congenital anomalies in live births were similar to those seen in the general population.

14.6 PHARMACOGENOMICS

Treatment response in MS is variable and unpredictable; indeed, experience has shown that a proportion of MS patients do not respond to IFN β. The search for reliable biomarkers to measure disease activity and treatment response is ongoing. Pharmacogenetics and pharmacogenomics attempt to identify genotypes that are associated with better response, with the ultimate goal of creating rational treatment algorithms. In the process, they may also provide more information about IFN β's mechanism of action. A genome-wide study investigating response to IFN identified single nucleotide polymorphisms (SNPs) in several genes that distinguished responders from nonresponders.[69] These genes included *glypican 5*, collagen type XXV α1, hyaluronan proteoglycan link protein, calpastatin, TAFA1, neuronal PAS domain protein 3, and *LOC442331*. Glypicans are heparin sulfate proteoglycans that are involved in synapse formation and axon regeneration; potential modulation of *glypican 5* expression by IFN may affect neuronal growth and repair. Polymorphisms found in extracellular matrix proteins suggest that differential response may be related to altered binding of MMPs and impaired migration of leukocytes through basement membranes. The investigators also performed a gene ontologic analysis, which demonstrated an overrepresentation of genes that encode for glutamate and GABA receptors. Though these specific genetic associations were not validated, a subsequent genome-wide scan of 500,000 SNPs identified a polymorphism in *GRIA3*, encoding an AMPA-type glutamate receptor.[70] These results provide an intriguing, but not yet validated, link to the previous observation that oligodendroglial excitotoxicity may be involved in MS pathogenesis. Genetic variability may also contribute to worse treatment response through the differential production of nABs: HLA-DRB1*0401 and HLA-DRB1*0408 alleles were found to be associated with the development of nABs.[71] Though helping to inform our understanding of heterogeneity in treatment response and providing insights into the molecular mechanisms of IFN, MS pharmacogenomics have yet to find their way into widespread clinical practice.

14.7 CONCLUSION

Given the ever-growing number of DMT options now available for MS patients, clinicians who treat them have an increasingly difficult decision to make. There are several newer agents that are probably more efficacious than any IFN β, the various formulations of which have been shown in their pivotal trials and in clinical practice to be moderately effective at reducing disease activity in MS. Of course, it is necessary to bear in mind that MS is a heterogeneous disease, and it is difficult at this point to predict who will respond to a given DMT, or even who would have a benign course off therapy. Statistics in clinical trials are meant to draw inferences about a population, not about the fate of the individual patient sitting in one's office. Moreover, it remains uncertain what the primary goal of treatment should be. Should we strive for complete freedom from disease? Or is the rare blip on the radar screen in an otherwise stable patient to be tolerated? Besides questions of efficacy, there are several other factors involved in DMT selection, including patient convenience and comfort, age and comorbidities, family planning, and drug availability and cost. One patient might be attracted to IFN β's long safety record while another patient might prefer the convenience of a monthly infusion. Within this context, clinicians owe their patients an honest discussion of risks and benefits of available DMTs, including IFN β.

References

1. Isaac A, Lindenmann J. Virus interference. I. The interferon. *Proc R Soc Lonc Biol*. 1957;147:258–267.
2. Markowitz CE. IFN-beta: mechanism of action and dosing issues. *Neurology*. 2007;68:S8–S11.
3. Weinstock-Guttman B, Ransohoff RM, et al. The interferons: biological effects, mechanisms of action, and use in multiple sclerosis. *Ann Neurol*. 1995;37:7–15.
4. De Maeyer E, De Maeyer-Guignard J. Type I interferons. *Int Rev Immunol*. 1998;17:53–73.
5. Shearer M, Taylor-Papadimitriou J. Regulation of cell growth by interferon. *Cancer Metas Rev*. 1987;6:199–221.
6. Dhib-Jalbut S, Marks S. Interferon-β mechanisms of action in multiple sclerosis. *Neurol*. 2010;74:S17–S24.
7. Betaseron® [package insert]. Whippany, NJ: Bayer HealthCare Pharmaceuticals Inc.; 2014.
8. Barna BP, Chou SM, et al. Interferon-β impairs induction of HLA-DR antigen expression in cultured adult human astrocytes. *J Neuroimmunol*. 1989;23:45–53.
9. Chofflon M. Mechanisms of action for treatments in multiple sclerosis: does a heterogeneous disease demand a multi-targeted therapeutic approach? *Bio Drugs*. 2005;19:299–308.
10. Hallal-Longo DE, Mirandola SR, et al. Diminished myelin-specific T cell activation associated with increase in CTLA4 and Fas molecules in multiple sclerosis patients treated with IFN-beta. *J Interferon Cytokine Res*. 2007;27:865–873.
11. Sharief MK, Semra YK, et al. Intereferon-β therapy downregulates the anti-apoptosis protein FLIP in T cells in patients with multiple sclerosis. *J Neuroimmunol*. 2001;120:199–207.
12. Yong VW, Chabot S, et al. Interferon beta in the treatment of multiple sclerosis: mechanisms of action. *Neurol*. 1998;51:682–689.
13. McKenzie BS, Kastelein RA, Cua DJ. Understanding the IL-23-IL-17 immune pathway. *Trends Immunol*. 2006;27:17–23.
14. Zhang X, Markovic-Plese S. Interferon beta inhibits the Th17 cell-mediated autoimmune response in patients with relapsing-remitting multiple sclerosis. *Clin Neurol Neurosurg*. 2010;112:641–645.
15. Ramgolam VS, Markovic-Plese S. Interferon-beta inhibits Th17 cell differentiation in patients with multiple sclerosis. *Endocr Metab Immune Disord Drug Targets*. 2010;10:161–167.
16. Axtell RC, de Jong BA, et al. Helper type 1 and 17 cells determine efficacy of interferon-beta in multiple sclerosis and experimental encephalomyelitis. *Nat Med*. 2010;16:406–412.
17. de Andrés C, Aristimuño C, et al. Interferon beta-1a therapy enhances CD4+ regulatory T-cell function: an ex vivo and in vitro longitudinal study in relapsing-remitting multiple sclerosis. *J Neuroimmunol*. 2007;182:204–211.
18. Praksova P, Stourac P, et al. Immunoregulatory T cells in multiple sclerosis and the effect of interferon beta and glatiramer acetate treatment on T cell subpopulations. *Jour Neurol Sci*. 2012;319:18–23.
19. Graber J, Zhan M, et al. Interferon-beta-1a induces increases in vascular cell adhesion molecule: implications for its mode of action in multiple sclerosis. *J Neuroimmunol*. 2005;161:169–176.
20. Dhib-Jalbut S. Mechanisms of action of interferons and glatiramer acetate in multiple sclerosis. *Neurol*. 2002;58(S4):S3–S9.
21. Murano PA, Leist T, et al. VLA-4/CD49d downregulated on primed T lymphocytes during interferon beta therapy in multiple sclerosis. *J Immunol*. 2000;111:186–194.
22. Murano PA, Liberati L, et al. Decreased integrin gene expression in patients with MS responding to interferon-β treatment. *J Neuroimmunol*. 2004;150:123–131.
23. Jensen J, Krakaner M, et al. Cytokines and adhesion molecules in multiple sclerosis patients treated with interferon-beta 1b. *Cytokine*. 2005;29:24–30.
24. Trojano M, Avolio C, et al. Changes of serum sICAM-1 and MMP-9 induced by rIFNbeta-1b treatment in relapsing-remitting MS. *Neurol*. 1999;53:1402–1408.

25. Stuve O, Dooley NP, et al. Interferon beta-1b decreases the migration of T lymphocytes in vitro: effects on matrix metalloproteinase-9. *Ann Neurol*. 1996;40:853–863.
26. Leppert D, Waubant E, et al. Interferon beta-1b inhibits gelatinase secretion and in vitro migration of human T cells: a possible mechanism for treatment efficacy in multiple sclerosis. *Ann Neurol*. 1996;40:846–852.
27. Ozenci V, Kouwenhoven M, et al. Multiple sclerosis: pro- and anti-inflammatory cytokines and metalloproteinases are affected differentially by treatment with IFN-beta. *J Neuroimmunol*. 2000;108:236–243.
28. Galboiz Y, Shapiro S, et al. Matrix metalloproteinases and their tissue inhibitors as markers of disease subtype and response to interferon-β therapy in relapsing and secondary-progressive multiple sclerosis patients. *Ann Neurol*. 2001;50:443–451.
29. Kraus J, Ling AK, et al. Interferon-beta stabilizes barrier characteristics of brain endothelial cells in vitro. *Ann Neurol*. 2004;56:192–205.
30. Jacobs L, O'Malley J, et al. Intrathecal interferon reduces exacerbations of multiple sclerosis. *Science*. 1981;214: 1026–1028.
31. Jacobs L, O'Malley J, et al. Intrathecal interferon in multiple sclerosis. *Arch Neurol*. 1982;39:609–615.
32. The IFNB Multiple Sclerosis Study Group. IFN β-1b is effective in relapsing-remitting multiple sclerosis. *Neurol*. 1993;43:655–661.
33. Paty DW, Li DKB, UBC MS/MRI Study Group, IFNB Multiple Sclerosis Study Group. Interferon beta-1b is effective in relapsing-remitting multiple sclerosis, 2: MRI analysis results of a multicenter, randomized, double-blind, placebo-controlled trial. *Neurol*. 1993;43:662–667.
34. The IFNB Multiple Sclerosis Study Group of British Columbia MS/MRI Analysis Group. IFN β-1b in the treatment of multiple sclerosis: final outcome of the randomized control trial. *Neurol*. 1995;45:1277–1285.
35. Jacobs LD, Cookfair DL, et al. Intramuscular interferon beta-1a for disease progression in relapsing multiple sclerosis. The Multiple Sclerosis Collaborative Research Group (MSCRG). *Ann Neurol*. 1996;39:285–294.
36. PRISMS Study Group. Randomised double-blind placebo-controlled study of interferon beta-1a in relapsing/remitting multiple sclerosis. *Lancet*. 1998;352:1498–1504.
37. PRISMS Study Group, University of British Columbia MS/MRI Analysis Group. PRISMS-4: long-term efficacy of interferon beta-1a in relapsing MS. *Neurol*. 2001;56:1623–1636.
38. Calabresi PA, Kieseier BC, et al. Pegylated interferon β-1a for relapsing-remitting multiple sclerosis (ADVANCE): a randomised, phase 3, double-blind study. *Lancet Neurol*. 2014;13:657–665.
39. Lublin FD, Reingold SC, et al. Defining the clinical course of multiple sclerosis: the 2013 revisions. *Neurol*. 2014;83:1–9.
40. Brex PA, Ciccarelli O, et al. A longitudinal study of abnormalities on MRI and disability from multiple sclerosis. *NEJM*. 2002;346:158–164.
41. Fisniku LK, Brex PA, et al. Disability and T2 MRI lesions: a 20-year follow-up of patients with relapse onset of multiple sclerosis. *Brain*. 2008;131:808–817.
42. Jacobs LD, Beck RW, et al. Intramuscular IFN β-1a therapy initiated during a first demyelinating event in multiple sclerosis: CHAMPS study group. *N Engl J Med*. 2000;343:898–904.
43. Kappos L, Polman CH, et al. Treatment with interferon beta-1b delays conversion to clinically definite Mcdonald MS in patients with clinically isolated syndromes. *Neurol*. 2006;67:1242–1249.
44. Kappos L, Freedman MS, et al. Effect of early versus delayed interferon beta-1a treatment on disability after a first clinical event suggestive of multiple sclerosis: a 3-year follow-up analysis of the benefit study. *Lancet*. 2007;370:389–397.
45. Comi G, Filippi M, et al. Effect of early IFN treatment on conversion to definite multiple sclerosis: a randomised study. *Lancet*. 2001;35:1576–1582.
46. Kappos L, Freedman MS, et al. Long-term effect of early treatment with IFN β-1b after a first clinical event suggestive of multiple sclerosis: 5-year active treatment extension of the phase 3 benefit trial. *Lancet Neurol*. 2009;8:987–997.
47. Champions Study Group. IM interferon β-1a delays definite multiple sclerosis 5 years after a first demyelinating event. *Neurol*. 2006;66:678–684.
48. Durelli L, Verdun E, et al. Every-other-day interferon beta-1b versus once-weekly interferon beta-1a for multiple sclerosis: results of a 2-year prospective randomised multicentre study (INCOMIN). *Lancet*. 2002;359:1453–1460.
49. Panitch H, Goodin DS, et al. Randomized, comparative study of interferon β-1a treatment regimens in MS: the evidence trial. *Neurol*. 2002;59:1496–1506.

50. Cohen JA, Barkhof F, et al. Oral fingolimod or intramuscular IFN for relapsing multiple sclerosis. *N Engl J Med.* 2010;362:402–415.
51. Cohen JA, Coles AJ, et al. Alemtuzumab versus interferon beta 1a as first-line treatment for patients with relapsing-remitting multiple sclerosis: a randomized controlled phase 3 trial. *Lancet.* 2012;380:1819–1828.
52. Coles AJ, Twyman CL, et al. Alemtuzumab for patients with relapsing multiple sclerosis after disease-modifying therapy: a randomized controlled phase 3 trial. *Lancet.* 2012;380:1829–1839.
53. La Mantia L, Vacchi L, et al. Interferon beta for secondary progressive multiple sclerosis. *Cochrane Database Syst Rev.* 2012. http://dx.doi.org/10.1002/14651858.CD005181.pub3. Issue 1. Art. No.: CD005181.
54. Rojas JI, Romano M, et al. Interferon beta for primary progressive multiple sclerosis (review). *Cochrane Collab.* 2010;1:1–22.
55. Montalban X. Overview of European pilot study of interferon β-1b in primary progressive multiple sclerosis. *Mult Scler.* 2004;10:S62–S64.
56. Lublin FD, Cofield SS, et al. Randomized study combining interferon and glatiramer acetate in multiple sclerosis. *Ann Neurol.* 2013;73:327–340.
57. Lublin FD. Disease activity free status in MS. *Mult Scler Relat Dis.* 2012;1:6–7.
58. Mikol DD, Barkhof F, et al. Comparison of subcutaneous interferon beta-1a with glatiramer acetate in patients with relapsing multiple sclerosis (the REbif versus glatiramer acetate in relapsing MS disease [REGARD] study): a multicentre, randomised, parallel, open-label trial. *Lancet Neurol.* 2008;10:903–914.
59. Ebers GC. Disease evolution in multiple sclerosis. *J Neurol.* 2006;253:VI/3–VI/8.
60. Vukusic S, Confavreux C. Natural history of multiple sclerosis: risk factors and prognostic indicators. *Curr Opin Neurol.* 2007;20:269–274.
61. The IFNB Multiple Sclerosis Study Group of British Columbia MS/MRI Analysis Group. Neutralizing antibodies during treatment of multiple sclerosis with IFN β-1b: experience during the first three years. *Neurol.* 1996;47:889–894.
62. Abdul Ahad AK, Galazka AR, et al. Incidence of antibodies to IFN-beta in patients treated with recombinant human IFN-beta 1a from mammalian cells. *Cytokines Cell Mol.* 1997;3:27–32.
63. Rudick RA, Simonian NA, et al. Incidence and significance of neutralizing antibodies to IFN β-1a in multiple sclerosis. *Neurol.* 1998;50:1266–1272.
64. Sorensen PS, Ross C, et al. Clinical importance of neutralising antibodies against IFN β in patients with relapsing-remitting multiple sclerosis. *Lancet.* 2003;362:1184–1191.
65. Goodin DS, Frohman EM, et al. Neutralizing antibodies to interferon beta: assessment of their clinical and radiographic impact: an evidence report: report of the therapeutics and technology assessment subcommittee of the American academy of neurology. *Neurol.* 2007;68:977–984.
66. Polman CH, Bertolotto A, et al. Recommendations for clinical use of data on neutralising antibodies to interferon-beta therapy in multiple sclerosis. *Lancet Neurol.* 2010;9:740–750.
67. Hartung HP, Freedman MS, et al. Interferon β-1b-neutralizing antibodies 5years after clinically isolated syndrome. *Neurol.* 2011;77:835–843.
68. Sandberg-Wollheim M, Alteri E, et al. Pregnancy outcomes in multiple sclerosis following subcutaneous interferon beta-1a therapy. *MSJ.* 2011;17:423–430.
69. Byun E, Caillier SJ, et al. Genome-wide pharmacogenomic analysis of the response to interferon beta therapy in multiple sclerosis. *JAMA Neurol.* 2008;65:337–344.
70. Comabella M, Craig DW, et al. Genome-wide scan of 500,000 single-nucleotide polymorphisms among responders and nonresponders to interferon beta therapy in multiple sclerosis. *JAMA Neurol.* 2009;66:972–978.
71. Hoffmann S, Cepok S, et al. HLA-DRB1*0401 and HLA-DRB1*0408 are strongly associated with the development of antibodies against interferon-beta therapy in multiple sclerosis. *Am J Hum Genet.* 2008;83:219–227.

CHAPTER

15

Glatiramer Acetate: From Bench to Bed and Back

R. Arnon, R. Aharoni
Weizmann Institute of Science, Rehovot, Israel

OUTLINE

15.1 HISTORY OF DEVELOPMENT

15.1.1 Discovery of Therapeutic Potential

The discovery and development of Copolymer 1 began in the late 1960s as a basic research project into the mechanisms involved in the induction and suppression of experimental autoimmune encephalomyelitis (EAE), which is the primary animal model for multiple sclerosis (MS). This is an induced neurological autoimmune disease mediated by autoreactive T cells

Translational Neuroimmunology in Multiple Sclerosis
http://dx.doi.org/10.1016/B978-0-12-801914-6.00015-5

that recognize the encephalitogenic antigen(s) in association with major histocompatibility complex (MHC) class II molecules, which migrate into the central nervous system (CNS) and mediate the pathogenic process. Three main myelin proteins were demonstrated to be encephalitogenic: myelin basic protein (MBP), proteolipid protein (PLP), and myelin oligodendrocyte glycoprotein (MOG). These three proteins or their corresponding peptides were also implicated as putative autoantigens in MS.[1] However, when our research began, with Michael Sela, in 1967 the only encephalitogenic material identified in the CNS was MBP.

Our approach was to use synthetic copolymers of amino acids whose composition resembled that of MBP, assuming that they would simulate the ability of MBP to induce EAE, thus serving as a research tool for studying the mechanism of induction of the disease. However, none of the synthesized copolymers proved to be encephalitogenic, but some, particularly Copolymer 1 (Cop 1), showed high efficacy in suppressing both incidence and severity of EAE.[2] These results led to a shift of direction in our research, toward the study of disease suppression, and eventually to the development of the drug, Copaxone.

15.1.2 Preclinical Research

In the first suppression experiment, in guinea pigs, the incidence of EAE was reduced from 75% in the control group to 20% in those treated by Cop 1, and in a repeated experiment, using a different but identical batch of Cop 1, the reduction was from 80% to 22%.[2] Further experiments demonstrated that Cop 1 is effective in suppressing EAE also in several other animal species, including rabbits, mice, and primates,[3] with a remarkable degree of suppression ranging between 60% and 100%, even though different encephalitogenic determinants of MBP are involved in disease induction in the different species.[4] The studies in primates are of particular relevance to the treatment of MS in humans. It was known that rhesus monkeys and baboons were very sensitive to MBP-induced EAE and typically died within 2 weeks of the onset of symptoms. Treatment with Cop 1 was found to reverse EAE in both species when administered after the appearance of symptoms.[5,6] Furthermore, Cop 1 was effective not only in the acute model of EAE, but also in the chronic relapsing model (CR-EAE), which is characterized by two or more discrete periods with clinical or neurological signs, and resembles the appearance of clinical relapses in relapsing-remitting MS (RRMS). Cop 1 was effective in both preventing and suppressing CR-EAE induced in both juvenile guinea pigs and mice.[7,8]

Preclinical research also included studies on the immunological mechanisms involved in the suppression of EAE, by the demonstration of immunological cross-reactivity, at both the T-cell and B-cell levels, as will be described later in this chapter. The resultant findings indicate immunomodulation by Cop 1, which correlates well with its ability to suppress EAE. These promising cumulative preclinical results prompted the initiation of clinical trials.

15.1.3 Initial Clinical Trials in Multiple Sclerosis Patients

In view of the putative resemblance between EAE and MS and the assumption that MBP may be involved in the pathogenesis of MS, preliminary clinical trials using Cop 1 were conducted in MS patients. These were begun after toxicity studies in experimental animals showed that Cop 1 was nontoxic after both acute and subchronic administration to mice, rats, rabbits, and beagle dogs, and that there was no significant uptake by any of the animal organs. Our clinical trials have included three preliminary open trials and two double-blind phase II trials, one involving exacerbating-remitting patients and another one including chronic-progressive patients. These were followed by a phase III trial.

The first preliminary trial, at Hadassah Medical Center in Israel, included four patients in terminal stages of MS with severe disabilities. They received 2 mg Cop 1 for 5 months. All four were stable during treatment; Cop 1 was well tolerated and no side effects or toxicity were observed.[9] The second preliminary trial was performed in Gottingen, Germany. This was also an open trial involving 21 patients, 10 of them receiving 2 mg Cop 1 daily and 11 receiving 20 mg daily, for 1 month. Its importance is the further demonstration of safety—only a few minor local reactions were observed.[10] The third preliminary trial, performed at the Einstein College of Medicine in New York, included 16 MS patients: 12 with chronic-progressive disease and 4 with RRMS; each received 20 mg Cop 1 for 6 months. At the end of the study, two of the four RRMS and three of the 12 chronic-progressive disease showed improvement, showing reduced numbers of relapses or slowed progression.[11]

The first phase II trial was a double-blind one, performed at the Albert Einstein College of Medicine, and included 48 RRMS patients, pair-matched, 24 in each arm, and lasted for 24 months. The average number of relapses per patient in 2 years dropped from 2.7 in the control group to 0.6 in those treated with Cop 1. This was accompanied by a difference in the Expanded Disability Status Scale (EDSS) score (0.75 units).[12] A second phase II double-blind trial, with chronic-progressive patients, was performed in two centers—the Albert Einstein College of Medicine in New York and the Baylor College in Houston—and included 169 patients and lasted for 24 months. The results showed a trend for less progression in the groups receiving Cop 1 compared to the placebo (17.6% vs. 25.5%).[13]

Based on these cumulative results, the TEVA Pharmaceutical Company has undertaken the development of Cop 1 as a drug, and performed a phase III trial, which was a double-blind placebo-controlled multicenter trial involving 251 exacerbating-remitting MS patients in 11 medical centers in the United States. The enrolled patients had clinically definite MS (EDSS 0–5) with two or more well-defined relapses in the 2 years prior to randomization. They were randomly assigned to receive daily injections of 20 mg Cop 1 or placebo control. At the end of 24 months an overall reduction of 29% in relapse rate was observed in the Cop 1 group compared to placebo ($p = 0.007$). Some beneficial effect was also observed in the disability status in favor of Cop 1 ($p = 0.024$). Furthermore, the positive effects of Cop 1 on neurological disability persisted during a 9-month extension of this trial. Most importantly, the side effects observed were minimal, consisting mainly of mild injection site reactions.[14] A subset of the patients participated in a pilot MRI study, and the percent of patients who had static or improved scan versus worse scans also favored Cop 1 treatment.

15.1.4 FDA Approval and Follow-Up Trials

On the basis of the successful results of the phase III trial and the cumulative results of the two double-blind phase II trials, Cop 1 was submitted to the FDA under the commercial name Copaxone®. The approval for the treatment of RRMS was obtained in December 1996 and the drug has been marketed worldwide since 1997. It is also known by the generic name glatiramer acetate (GA).

A large group of patients, who have participated in the phase II and phase III trials and continued to receive GA, were followed up for a period of 15–22 years, during which they were evaluated every 6 months. The results show that relapse rates declined from 1.18/year prestudy to approximately 0.2 after 10 years and remained steady, or improved, reaching 0.12 after 15 years. Mean EDSS change was 0.50 points and 57% of the patients had stable/improved EDSS scores. Furthermore, patients who withdrew from the study had greater disability than the ongoing patients. These data provide clear evidence for the long-term efficacy and adequate safety of Copaxone.[15]

In parallel to the pharmacological development, extensive efforts were devoted to elucidate the mechanism of activity of Copaxone. The major modes by which this drug exerts its therapeutic effect are detailed in the following.

15.2 IMMUNOMODULATORY MECHANISMS

15.2.1 Peripheral Immunomodulation

GA induces a broad immunomodulatory effect on various subsets of the immune system (Fig. 15.1). The initial prerequisite step is the binding of GA to MHC class II molecules. In vitro studies on murine and human antigen-presenting cells (APCs) showed that GA undergoes a rapid and efficient binding to various MHC class II molecules, and even displaces other peptides from the MHC binding groove.[16] This competitive binding for the histocompatibility molecules can prevent the presentation of other antigens and hinder their T-cell activation. Several groups have demonstrated that GA induces generalized alterations of various types of APCs, such as dendritic cells and monocytes, so that they preferentially stimulate protective antiinflammatory responses. Indeed, dendritic cells from GA-treated MS patients produce less tumor necrosis factor (TNF)-α, less IL-12, and more IL-10, compared to those of untreated patients.[17] GA induces a broad inhibitory effect on monocyte reactivity,[18] and promotes the development of antiinflammatory type II monocytes characterized by increased secretion of IL-10 and transforming growth factor (TGF)-β, as well as by decreased production of IL-12 and TNF.[19] This modulation on the level of the innate immune system is the least specific step in the immunological processes affected by GA, and can be beneficial for inhibiting of the response to several myelin antigens. In addition, GA acts in a strictly antigen-specific manner in the case of the MBP immunodominant encephalitogenic epitope (comprising amino acids 82–100). Using MBP 82–100 specific T-cell clones from MS patients and from EAE-induced mice, it was shown that GA inhibits their activation by T-cell receptor antagonism, acting as an altered peptide ligand.[20]

Most studies attribute the primary mechanism of GA activity to its ability to shift the T-cell response from the proinflammatory to the antiinflammatory pathway. It has been long known that GA-treated animals develop specific T cells in their peripheral immune system.[21,22] These cells act as modulatory suppressor cells, as they inhibit the response of MBP-specific effector cells in vitro, and adoptively transfer protection against EAE in vivo. T-cell lines induced by GA progressively polarize toward the T-helper (Th) 2/3 subtype, secreting high amounts of antiinflammatory cytokines such as interleukin (IL)-4, -5, -10, and TGF-β, until they completely lose the ability to secrete Th1 proinflammatory cytokines such as INF-γ.[22] In several cases, the secretion of Th2/3 cytokines by the GA-induced T-cell lines was obtained in response to either GA or MBP. Other myelin antigens such as PLP and MOG do

Peripheral immunomodulation	Competition for MHC	Promiscuous binding to various MHC class II molecules, displacement of myelin antigens from the MHC binding groove.
	Alteration of the innate immune response	Inhibitory effect on monocytes reactivity, deviation of dendritic cells and monocytes to produce less TNF-α and IL-12, more IL-10 and TGF-β, and to stimulate Th2 anti-inflammatory responses.
	T-cell receptor antagonism	Inhibition of the activation of T-cells specific to the 82-100 epitope of MBP.
	T-cell deviation	Induction of specific Th2/3-cells that secrete high amounts of IL-4, IL-5, IL-10, and TGF-β. Elevation of the prevalence and function of T-regulatory cells, activation of the transcription factor Foxp3. Reduction of Th-17 cells and their transcription factors RORγt. Improvement of the regulatory function of $CD8^+$ T-cells.
	Modification of B-cells	Induction of antibodies with beneficial rather than neutralizing activity. Bias towards production of anti-inflammatory cytokines such as IL-10. Down-regulating of chemokine receptors.
Immunomodulation in the CNS	Secretion of anti-inflammatory cytokine	GA-specific Th2/3 cells cross the BBB and secrete *in situ* anti-inflammatory cytokines. Bystander expression of IL-10 and TGF-β by resident astrocyte and microglia. Reduction in the overall expression of IFN-γ. (A) 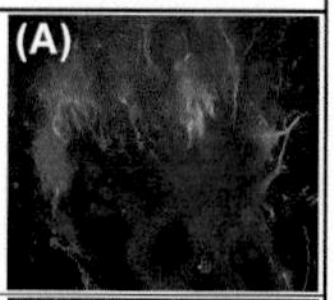
	Th-17 and T-regulatory cells	Decrease in the amount of Th-17 cells. Increase in T-regulatory cells. (B) 
Neuroprotection	Elevation of neurotrophic factors	GA-specific T-cells express BDNF in the brain. Restoration of the impaired expressions of BDNF, NT-3, NT4, IGF-1, and IGF-2. (C) 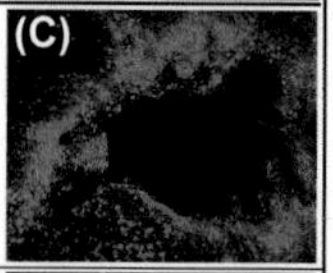
	Reduced CNS injury	Prevention of demyelination. Preservation of retinal ganglion cells. Inhibition of motor neuron loss. Preservation of brain tissue integrity by the MRI parameters MTR and DTI. Reduced formation of "black holes." Increase in NAA:Cr ratio. (D) 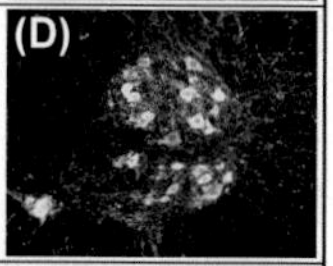
	Remyelination	Augmented remyelination. Increased proliferation, maturation and survival of oligodendrocyte progenitor cells and their accumulation in the lesions. (E) 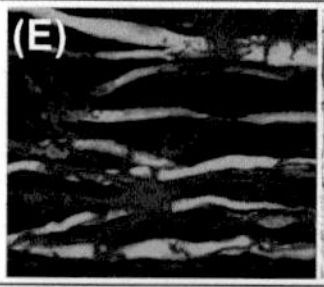 (F) 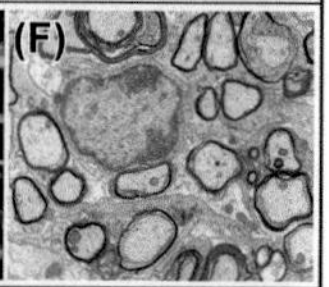
	Neurogenesis	Elevated proliferation, migration and differentiation of neuronal progenitor cells and their recruitment into injury sites. (G) 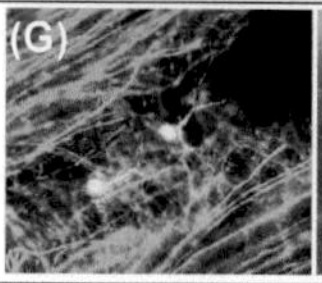 (H) 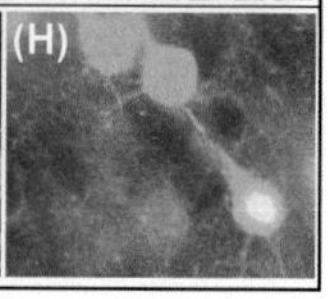

FIGURE 15.1 Immunomodulatory and neuroprotective effects of glatiramer acetate (GA) in multiple sclerosis and experimental autoimmune encephalomyelitis. Inserts demonstrate the in situ consequences of GA in the central nervous system: (A) GA-specific T cells (blue) expressing IL-10 (red). (B) Infiltration of Foxp3 expressing T cells (yellow). (C) GA-specific T cells (blue) expressing brain-derived neurotrophic factor (red). (D) Preservation of motor neurons. (E) Oligodendrocyte progenitor cells (red) extending processes between transected neuronal fibers (green). (F) Remyelination zone with newly myelinated axons surrounding an oligodendrocyte. (G) Neuronal progenitor cells (yellow) in a lesion site following GA treatment. (H) A BrdU-expressing neuron (yellow) born during BrdU/GA injection, also expressing a mature neuronal marker in the cortex (green).

not activate the GA-specific T cells, yet EAE induced by PLP and MOG can be suppressed by GA as well as by GA-induced T cells.[23] These results are indicative of bystander suppression mechanisms induced by GA, which are especially important in view of the epitope spreading occurring in MS/EAE.[24] A shift from a proinflammatory Th1-biased cytokine profile toward an antiinflammatory Th2-biased profile was also observed in GA-treated MS patients,[25,26] indicating that such cells are involved in the therapeutic effect of GA in MS.

Several studies demonstrated the effect of GA on Th17 and on regulatory T cells (Tregs), which are pivotal effectors of disease exacerbation and suppression, respectively.[27,28] Thus, it was shown that in vitro exposure of peripheral $CD4^+$ T cells, from healthy humans or from GA-immunized mice, to GA, results in elevated level of Tregs, through activation of the transcription factor forkhead box P3 (Foxp3).[29] Furthermore, GA treatment leads to increased Foxp3 expression in $CD4^+$ T cells of MS patients, whose Foxp3 level is low at baseline. Pretreatment of mice with GA, before EAE induction, results in increased Foxp3 expression on Tregs during the mild disease that subsequently develops. These Tregs are more effective in EAE prevention than Tregs isolated from untreated mice.[30] GA treatment in EAE-induced mice results in elevated levels of Tregs and a reduction of Th17 cells, as demonstrated by the detection of their specific transcription factors, Foxp3 and RORγt, respectively, on both the mRNA and the protein levels.[31,32] In addition to its effect on the $CD4^+$ T-cell subset, GA also affects $CD8^+$ T cells. The regulatory function of these cells, which is impaired in MS untreated patients, is drastically improved after several months of GA treatment, to levels observed in healthy individuals.[33,34]

B cells are involved in both the pathogenesis and modulation of MS and EAE by secreting antibodies and cytokines as well as by their efficient antigen presentation.[35] GA-treated patients develop GA-specific antibodies that do not interfere with GA activity in terms of MHC binding or T-cell stimulation, and eventually decline 6 months after treatment initiation.[36] The high proportion of IgG1 versus IgG2 antibody isotypes as well as the switch to IgG4 observed in the treated patients reflects the Th1 to Th2 shift induced by GA. Moreover, relapse-free patients develop higher GA-antibody titers, suggesting a beneficial rather than neutralizing activity of anti-GA antibodies. Recently, it has been demonstrated that the effect of GA on B cells contributes to its therapeutic activity, by biasing toward antiinflammatory cytokines such as IL-10.[37] These B cells ameliorate EAE by downregulating chemokine receptors associated with trafficking inflammatory cells into the CNS.[38]

15.2.2 Immunomodulation in the Central Nervous System

A therapy efficacy is obviously measured by its effect in the diseased organ, in the case of an MS therapy, by its ability to modulate the pathological processes in the CNS. The initial immunological activity of GA apparently occurs in the periphery (at the injection sites and in the corresponding draining lymph nodes). An indication for dendritic uptake of GA and its delivery to the CNS has been demonstrated.[39] However, since GA is rapidly degraded in the periphery, it is unlikely that sufficient amounts reach the CNS and effectively compete with myelin antigens or initiate specific immune response. The majority of researchers agree that the therapeutic effect of GA is mediated by GA-induced immune cells that penetrate the CNS. The presence of GA-specific T cells in the CNS has been demonstrated by their actual isolation from the brains of actively sensitized mice, as well as by their localization in the brain following passive transfer to the periphery.[40] Specific ex vivo reactivity to GA, manifested by cell proliferation and by Th2 cytokine secretion, was found in whole lymphocyte populations isolated from brains of EAE-induced mice treated by GA. Moreover, highly reactive GA-specific T-cell lines that secrete IL-4, IL-5, IL-10, and TGF-β in response to GA, and cross-react with MBP at the level of Th2 cytokine secretion, were obtained from brains and spinal cords of GA-treated mice. The ability of the GA-induced cells to cross the blood–brain barrier (BBB) and accumulate in the CNS was confirmed by the injection of labeled GA-specific T cells into the periphery and their subsequent detection in the brain.[40] There is currently a consensus that the brain is not an immune privileged site and that activated T cells, regardless of their specificity, penetrate the CNS, especially in the course of MS/EAE when the BBB is disrupted. While the cross-reactivity of the GA-specific T cells with MBP[21,22] is not essential for the entering the CNS, it may enable their in situ reactivation.

In the CNS of EAE-induced mice, GA-specific T cells highly express two potent regulatory antiinflammatory cytokines, IL-10 and TGF-β.[41] Of special interest is the finding that IL-10 and TGF-β are expressed not only by the GA-specific T cells but also by CNS-resident cells in their vicinity, such as astrocytes and microglia. In contrast, the overall expression of IFN-γ in the brain tissue is drastically reduced. In addition, in mice with either chronic or relapsing-remitting EAE, GA treatment results in a drastic reduction in the proinflammatory Th17 cells, and a parallel elevation of Tregs in the CNS.[32] Importantly, analysis of GA-reactive T cells from the cerebrospinal fluid of GA-treated MS patients revealed a pronounced antiinflammatory profile.[42] These cumulative results indicate that GA induces a bystander immunomodulatory effect in the CNS and generates an in situ antiinflammatory cytokine shift, thus restraining the proinflammatory pathological disease progression.

15.3 NEUROPROTECTION

An essential challenge for MS therapy is to target not only the inflammatory characteristic of the disease but also its neuroaxonal pathology, inducing neuroprotective outcomes. Neuroprotection is broadly defined as an effect that results in salvage, recovery, or regeneration of the nervous system, its cells, structure, and function. Accumulated findings indicate that GA treatment indeed generates neuroprotective consequences in the CNS[43,44] (Fig. 15.1).

15.3.1 Increase in Neurotrophic Factors

The initial indication for neuroprotective activity was the ability of GA-induced cells to secrete brain-derived neurotrophic factor (BDNF). This was demonstrated for murine GA-specific T cells originating from the periphery or the CNS, as well as for human T-cell lines.[41,45–47] Furthermore, GA-specific T cells demonstrated extensive BDNF expression in the brain of EAE-induced mice.[41] BDNF was shown to be elevated in brains of mice that were injected daily with GA.[48] A similar elevation was found in brains of GA-treated mice for additional neurotrophic factors such as neurotrophin (NT)-3 and NT-4,[48] as well as for insulin-like growth factor (IGF)-1[49] and IGF-2.[50]

Members of the NT family such as BDNF, NT-3, and NT-4 are important regulators of neuronal function and survival.[51] Besides their role in neuronal development, they were shown to promote axonal outgrowth, remyelination, and regeneration. Of special significance is the elevation by GA of neurotrophic factor levels even when treatment starts late, in the chronic disease phase, when their levels are diminished. Furthermore, in Mecp2 knockout mice in which BDNF levels are low, GA treatment induced elevation of BDNF expression to the level found in brains of healthy mice.[52] The relevance of this effect to human therapy has been shown in reversing the reduced BDNF levels in the serum and in the cerebrospinal fluid of MS patients following GA treatment.[53]

15.3.2 Reduced Central Nervous System Injury

The neuroprotective effect of GA is manifested by the actual preservation of the CNS and reduced typical tissue damage. Several studies, utilizing immunohistochemistry and electron microscopy in different EAE models, demonstrated the protective outcome of GA on the primary disease target, the myelin.[54–56] Furthermore, in mice inflicted by MOG-induced EAE, in which chronic disease with extensive neurodegeneration is typically manifested, GA treatment results in reduced neuroaxonal damage. This was evident by the preservation of retinal ganglion cells,[57] less axonal deterioration, and fewer deformed neurons.[58] Motor neuron loss that occurs in this model was also prevented by GA treatment.[56] In a study that employed MRI parameters such as magnetization transfer ratio and diffusion tensor imaging for the assessment of the whole brain, as well as for the detection of specifically affected regions, GA restored all the MRI parameters in both chronic and relapsing-remitting EAE models.[59] These effects were obtained by various regimens of GA administration. When treatment was applied before the appearance of clinical manifestations, thus blocking the development of the pathological processes (prevention regimen), mice displayed nearly no damage. Moreover, when treatment was applied by a therapeutic schedule, after disease exacerbation (suppression regimen) or even late in the chronic phase, when substantial injury was already manifested (delayed suppression regimen), significant reduction in myelin and neuroaxonal damages was obtained. This suggested the induction of genuine repair mechanisms.

15.4 REPAIR PROCESSES

The central elements of the CNS, the myelinating oligodendrocytes as well as the neurons, are terminally differentiated cells with a limited capacity to respond to injury. They depend for renewal on the availability of their precursors, the oligodendrocyte progenitor cells (OPCs) and the neuronal progenitor cells, which need to undergo proliferation, migration, and differentiation into the defined progeny. It should be noted that CNS injury by itself triggers repair processes.[60] Thus, in MS and EAE, subsequent to the demyelination and degeneration, repair processes, remyelination and neurogenesis, are stimulated and progenitor cells migrate into damage sites.[58,61] However, these repair processes are characteristic mostly to the early disease phase.[62] As the disease progresses oligodendrocytes succumb to the hostile conditions within the inflamed lesions and self-repair mechanisms drastically decline. Promoting repair beyond the body's limited spontaneous extent is, therefore, a major goal for MS therapy. In this context the results demonstrated with GA, as delineated herewith, are promising.

15.4.1 Remyelination

Reduced myelin damages detected by scanning electron microscopy and immunohistochemistry in EAE-inflicted mice treated by a delayed suppression therapeutic GA regimen suggested the induction of actual repair processes.[55] In one study we established the ability of GA to augment remyelination by applying transmission electron microscopy, which facilitates the visualization of newly myelinated axons.[56] Ultrastructural quantitative analysis in the spinal cord of mice induced with relapsing-remitting EAE provided evidence for significant increase in remyelination after GA treatment. This was manifested by seven- and threefold increases in the relative remyelination compared to demyelination over untreated mice, when GA was applied during the first or the second disease exacerbation, respectively. The mode of action of GA in this system is attributed to an increase in proliferation and survival of OPCs and their recruitment into injury sites.[50,55]

It was subsequently questioned if the remyelination demonstrated in EAE-induced mice is solely due to the antiinflammatory activity of GA in the inflamed CNS. In one study we therefore investigated whether GA can affect postnatal myelinogenesis in the developing nervous system under nonpathological conditions, when injected at postnatal days 7–21.[63] Immunohistological and ultrastructural analyses revealed significant elevation in the number of myelinated axons as well as in the thickness of the myelin encircling them in spinal cords of GA-injected mice compared to their Phosphate buffered saline (PBS)-injected littermates at postnatal day 14. A prominent elevation in the amount of progenitor oligodendrocytes and their proliferation, as well as in mature oligodendrocytes, indicated that similarly to the findings in EAE, the effect of GA in postnatal myelination is linked to increased proliferation and differentiation along the oligodendroglial maturation cascade. In addition, as previously found in EAE, GA injection resulted in increased expression of IGF-1 and BDNF, which were shown to promote myelination during development as well as under pathological conditions.[51,64] Furthermore, GA-injected mice exhibited better performance in a rotating rod test than their PBS-injected littermates, suggesting that the accelerated myelin development results in functional advantage in sensorimotor functions.[63] During remyelination, features of developmental myelination are recapitulated. The effect of GA on postnatal myelinogenesis in healthy mice is, therefore, relevant to its effect in EAE and MS, supporting the notion that the myelination process in the CNS can be upregulated by therapy.

15.4.2 Neurogenesis

Concerning the functional cells in the CNS, the neurons, current opinion claims that stem cells with the potential to give rise to new neurons reside in different regions of the mammalian brain and contribute to repair after injury.[60] EAE induction as such triggers increased neuroprogenitor proliferation in the neuroproliferative zones, but this effect is of short duration and subsequently declines to levels below that of naïve mice. In contrast, GA treatment, applied at various disease stages, augments neuronal proliferation to a higher level than that observed in EAE mice, and this effect persists for a prolonged duration.[58] Furthermore, neuronal progenitors were seen diverging from the classic migratory streams and spreading to damage sites in adjacent brain regions that do not normally undergo neurogenesis. After 1 month, neurons born during GA treatment were found in the cortex of the treated mice, expressing mature neuronal markers and displaying mature morphology. These newly formed neurons could constitute a pool for the replacement of dead or dysfunctional cells and/or induce a growth-promoting environment that supports neuroprotection and axonal growth. The latter activity is evidenced by the extensive BDNF expression manifested by these new neurons.

Findings from human studies support the notion that GA confers neuroprotection in MS patients. GA treatment reduced the formation of permanent T1 hypointense lesions that evolve into black holes, which have been associated with irreversible neurological disability.[65] Since T1 hypointensity is also associated with reduced myelin content, the finding of fewer black holes in GA-treated patients can be related to less axonal damage and/or improved myelin repair. Using quantitative MRI analysis, it was also shown that GA treatment for 1 year leads to a significant increase in the NAA:Cr ratio compared to pretreatment values. These results imply an axonal metabolic recovery and protection from sublethal axonal injury.[66] Altogether, these findings may explain the long-term beneficial effect of GA in patients followed up for 15 years and more.[15]

15.4.3 Potential for Neurodegenerative Diseases

Several laboratories have reported on beneficial effects of GA in animal models of neuronal trauma such as loss of retinal ganglion cells[67] and crush injury of the optic nerve,[45] as well as in models of the neurodegenerative diseases Parkinson's[68] and amyotrophic lateral sclerosis.[69] Recently, we demonstrated that continuous GA treatment to Mecp2 knockout mice, a model for Rett syndrome, results in elevation of BDNF expression in their brain, to the level found in healthy mice.[52] GA is currently tested in a small clinical trial for its effect in Rett syndrome patients.

These cumulative results indicate that GA may have neuroprotective and neurogenerative effects on a broader spectrum of neurodegenerative disorders.

15.5 CONCLUDING REMARK

GA, Copaxone®, a synthetic random copolymer of amino acids, is an approved drug for the treatment of MS since 1997, administered by daily injection of 20 mg. Recently, a new version was introduced, for injection of 40 mg three times weekly. The efficacy of GA in reducing the frequency of exacerbations, its high safety profile, as well as its unique mechanism of action establish GA as a first-line treatment for MS.

References

1. Grau-Lòpez L, Raich D, Ramo-Tello C, et al. Myelin peptides in multiple sclerosis. *Autoimmun Rev*. 2009;8(8):650–653.
2. Teitelbaum D, Meshorer A, Hirshfeld T, Arnon R, Sela M. Suppression of experimental allergic encephalomyelitis by a synthetic polypeptide. *Eur J Immunol*. 1971;1:242–248.
3. Teitelbaum D, Arnon R, Sela M. Copolymer 1: from basic research to clinical application. *Cell Mol Life Sci*. 1997;53:24–28.
4. Teitelbaum D, Webb C, Meshorer A, Arnon R, Sela M. Suppression of several synthetic polypeptides of experimental allergic encephalomyelitis induced in guinea pigs and rabbits with bovine and human basic encephalitogen. *Eur J Immunol*. 1973;3:273.
5. Teitelbaum D, Webb C, Bree M, Meshorer A, Arnon R. Suppression of experimental allergic encephalomyelitis in rhesus monkey. *J Clin Immunopathol*. 1974;3:256–262.
6. Teitelbaum D, Meshorer A, Arnon R. Abstract. *Isr J Med Sci*. 1977;13:1038.

7. Keith AB, Arnon R, Teitelbaum D, Caspary EA, Wisniewski HM. The effect of Cop1, a synthetic polypeptide, on chronic relapsing experimental allergic encephalomyelitis in guinea pigs. *J Neurol Sci*. 1979;42:267–274.
8. Teitelbaum D, Fridkis-Hareli M, Arnon R, Sela M. Copolymer 1 inhibits, chronic relapsing experimental allergic encephalomyelitis induced by proteolipid, protein (PLP) peptides in mice and interferes with PLP-specific T cell responses. *J Neuroimmunol*. 1996;64:209–217.
9. Abramsky O, Teitelbaum D, Arnon R. Effect of a synthetic polypeptide (Cop 1) on patients with multiple sclerosis and with acute disseminated encephalitis. *J Neurol Sci*. 1977;31:433.
10. Arnon R. The development of Cop 1 (Copaxone®), an innovative drug for the treatment of multiple sclerosis – personal reflections. *Immun Lett*. 1996;50:1–15.
11. Bornstein MB, Miller AL, Teitelbaum D, Arnon R, Sela M. Multiple sclerosis: trial of a synthetic polypeptide. *Ann Neurol*. 1982;11:317–319.
12. Bornstein M, Miller A, Slagle S, et al. A pilot trial of Cop 1 in exacerbating-remitting sclerosis. *N Engl J Med*. 1987;317:4080414.
13. Miller M, Borenstein M, Slagle S, et al. Clinical trial of Cop 1 in chronic-progressive multiple sclerosis. *Neurology*. 1988;(suppl 1):356.
14. Johnson KP, Brooks BR, Cohen JA, et al. Copolymer 1 reduces relapse rate and improves disability in relapsing-remitting multiple sclerosis: results of a phase III multicenter, double-blind placebo-controlled trial. The Copolymer 1 Multiple Sclerosis Study Group. *Neurology*. 1995;45(7):1268–1276.
15. Ford C, Goodman AD, Johnson K, et al. Continuous long-term immunomodulatory therapy in relapsing multiple sclerosis: results from the 15-year analysis of the US prospective open-label study of glatiramer acetate. *Mult Scler J*. 2010;16:342–350.
16. Fridkis-Hareli M, Teitelbaum D, Gurevich E, et al. Direct binding of myelin basic protein and synthetic copolymer 1 to class II major histocompatibility complex molecules on living antigen-presenting cells–specificity and promiscuity. *Proc Natl Acad Sci USA*. 1994;91:4872–4876.
17. Vieira PL, Heystek HC, Wormmeester J, Wierenga EA, Kapsenberg ML. Glatiramer acetate (copolymer-1, copaxone) promotes Th2 cell development and increased IL-10 production through modulation of dendritic cells. *J Immunol*. 2003;170:4483–4488.
18. Weber MS, Starck M, Wagenpfeil S, Meinl E, Hohlfeld R, Farina C. Multiple sclerosis: glatiramer acetate inhibits monocyte reactivity in vitro and in vivo. *Brain J Neurol*. 2004;127:1370–1378.
19. Weber MS, Prod'homme T, Youssef S, et al. Type II monocytes modulate T cell-mediated central nervous system autoimmune disease. *Nat Med*. 2007;13:935–943. Proceedings of the National Academy of Sciences of the United States of America, 1999;96:634–9.
20. Aharoni R, Teitelbaum D, Arnon R, Sela M. Copolymer 1 acts against the immunodominant epitope 82-100 of myelin basic protein by T cell receptor antagonism in addition to major histocompatibility complex blocking. *Proc Natl Acad Sci USA*. 1999;96:634–639.
21. Aharoni R, Teitelbaum D, Arnon R. T suppressor hybridomas and interleukin-2-dependent lines induced by copolymer 1 or by spinal cord homogenate down-regulate experimental allergic encephalomyelitis. *Eur J Immunol*. 1993;23:17–25.
22. Aharoni R, Teitelbaum D, Sela M, Arnon R. Copolymer 1 induces T cells of the T helper type 2 that crossreact with myelin basic protein and suppress experimental autoimmune encephalomyelitis. *Proc Natl Acad Sci USA*. 1997;94:10821–10826.
23. Aharoni R, Teitelbaum D, Sela M, Arnon R. Bystander suppression of experimental autoimmune encephalomyelitis by T cell lines and clones of the Th2 type induced by copolymer 1. *J Neuroimmunol*. 1998;91:135–146.
24. McMahon EJ, Bailey SL, Castenada CV, Waldner H, Miller SD. Epitope spreading initiates in the CNS in two mouse models of multiple sclerosis. *Nat Med*. 2005;11:335–339.
25. Neuhaus O, Farina C, Yassouridis A, et al. Multiple sclerosis: comparison of copolymer-1- reactive T cell lines from treated and untreated subjects reveals cytokine shift from T helper 1 to T helper 2 cells. *Proc Natl Acad Sci USA*. 2000;97:7452–7457.
26. Duda PW, Schmied MC, Cook SL, Krieger JI, Hafler DA. Glatiramer acetate (Copaxone) induces degenerate, Th2-polarized immune responses in patients with multiple sclerosis. *J Clin Invest*. 2000;105:967–976.
27. Becher B, Segal BM. T(H)17 cytokines in autoimmune neuro-inflammation. *Curr Opin Immunol*. 2011;23:707–712.
28. Miyara M, Gorochov G, Ehrenstein M, Musset L, Sakaguchi S, Amoura Z. Human FoxP3^{+} regulatory T cells in systemic autoimmune diseases. *Autoimmun Rev*. 2011;10:744–755.

29. Hong J, Li N, Zhang X, Zheng B, Zhang JZ. Induction of $CD4^{+}CD25^{+}$ regulatory T cells by copolymer-I through activation of transcription factor Foxp3. *Proc Natl Acad Sci USA*. 2005;102:6449–6454.
30. Jee Y, Piao WH, Liu R, et al. CD4(+)CD25(+) regulatory T cells contribute to the therapeutic effects of glatiramer acetate in experimental autoimmune encephalomyelitis. *Clin Immunol*. 2007;125:34–42.
31. Begum-Haque S, Sharma A, Kasper IR, et al. Downregulation of IL-17 and IL-6 in the central nervous system by glatiramer acetate in experimental autoimmune encephalomyelitis. *J Neuroimmunol*. 2008;204:58–65.
32. Aharoni R, Eilam R, Stock A, et al. Glatiramer acetate reduces Th-17 inflammation and induces regulatory T-cells in the CNS of mice with relapsing-remitting or chronic EAE. *J Neuroimmunol*. 2010;225:100–111.
33. Karandikar NJ, Crawford MP, Yan X, et al. Glatiramer acetate (Copaxone) therapy induces CD8(+) T cell responses in patients with multiple sclerosis. *J Clin Invest*. 2002;109:641–649.
34. Tennakoon DK, Mehta RS, Ortega SB, Bhoj V, Racke MK, Karandikar NJ. Therapeutic induction of regulatory, cytotoxic $CD8^{+}$ T cells in multiple sclerosis. *J Immunol*. 2006;176:7119–7129.
35. Fraussen J, Vrolix K, Martinez-Martinez P, et al. B cell characterization and reactivity analysis in multiple sclerosis. *Autoimmun Rev*. 2009;8:654–658.
36. Teitelbaum D, Brenner T, Abramsky O, Aharoni R, Sela M, Arnon R. Antibodies to glatiramer acetate do not interfere with its biological functions and therapeutic efficacy. *Mult Scler*. 2003;9:592–599.
37. Kala M, Miravalle A, Vollmer T. Recent insights into the mechanism of action of glatiramer acetate. *J Neuroimmunol*. 2011;235:9–17.
38. Begum-Haque S, Christy M, Ochoa-Reparaz J, et al. Augmentation of regulatory B cell activity in experimental allergic encephalomyelitis by glatiramer acetate. *J Neuroimmunol*. 2011;232:136–144.
39. Liu J, Johnson TV, Lin J, et al. T cell independent mechanism for copolymer-1-induced neuroprotection. *Eur J Immunol*. 2007;37:3143–3154.
40. Aharoni R, Teitelbaum D, Leitner O, Meshorer A, Sela M, Arnon R. Specific Th2 cells accumulate in the central nervous system of mice protected against experimental autoimmune encephalomyelitis by copolymer 1. *Proc Natl Acad Sci USA*. 2000;97:11472–11477.
41. Aharoni R, Kayhan B, Eilam R, Sela M, Arnon R. Glatiramer acetate-specific T cells in the brain express T helper 2/3 cytokines and brain-derived neurotrophic factor in situ. *Proc Natl Acad Sci USA*. 2003;100:14157–14162.
42. Hestvik AL, Skorstad G, Price DA, Vartdal F, Holmoy T. Multiple sclerosis: glatiramer acetate induces anti-inflammatory T cells in the cerebrospinal fluid. *Mult Scler*. 2008;14:749–758.
43. Aharoni R, Arnon R. Linkage between immunomodulation, neuroprotection and neurogenesis. *Drug News Perspect*. 2009;22:301–312.
44. Arnon R, Aharoni R. Neuroprotection and neurogeneration in MS and its animal model EAE effected by glatiramer acetate. *J Neural Transm*. 2009;116:1443–1449.
45. Kipnis J, Yoles E, Porat Z, et al. T cell immunity to copolymer 1 confers neuroprotection on the damaged optic nerve: possible therapy for optic neuropathies. *Proc Natl Acad Sci USA*. 2000;97:7446–7451.
46. Ziemssen T, Kumpfel T, Klinkert WE, Neuhaus O, Hohlfeld R. Glatiramer acetate-specific T-helper 1- and 2-type cell lines produce BDNF: implications for multiple sclerosis therapy. Brain-derived neurotrophic factor. *Brain J Neurol*. 2002;125:2381–2391.
47. Chen M, Valenzuela RM, Dhib-Jalbut S. Glatiramer acetate-reactive T cells produce brain-derived neurotrophic factor. *J Neurol Sci*. 2003;215:37–44.
48. Aharoni R, Eilam R, Domev H, Labunskay G, Sela M, Arnon R. The immunomodulator glatiramer acetate augments the expression of neurotrophic factors in brains of experimental autoimmune encephalomyelitis mice. *Proc Natl Acad Sci USA*. 2005;102:19045–19050.
49. Skihar V, Silva C, Chojnacki A, et al. Promoting oligodendrogenesis and myelin repair using the multiple sclerosis medication glatiramer acetate. *Proc Natl Acad Sci USA*. 2009;106:17992–17997.
50. Zhang Y, Jalili F, Ouamara N, et al. Glatiramer acetate-reactive T lymphocytes regulate oligodendrocyte progenitor cell number in vitro: role of IGF-2. *J Neuroimmunol*. 2010;227:71–79.
51. Lessmann V, Gottmann K, Malcangio M. Neurotrophin secretion: current facts and future prospects. *Prog Neurobiol*. 2003;69:341–374.
52. Ben-Zeev B, Aharoni R, Nissenkorn A, Arnon R. Glatiramer acetate (GA, Copolymer-1) an hypothetical treatment option for Rett syndrome. *Med Hypotheses*. 2011;76:190–193.
53. Azoulay D, Vachapova V, Shihman B, Miler A, Karni A. Lower brain-derived neurotrophic factor in serum of relapsing remitting MS: reversal by glatiramer acetate. *J Neuroimmunol*. 2005;167:215–218.

54. Gilgun-Sherki Y, Panet H, Holdengreber V, Mosberg-Galili R, Offen D. Axonal damage is reduced following glatiramer acetate treatment in C57/bl mice with chronic-induced experimental autoimmune encephalomyelitis. *Neurosci Res*. 2003;47:201–207.
55. Aharoni R, Herschkovitz A, Eilam R, et al. Demyelination arrest and remyelination induced by glatiramer acetate treatment of experimental autoimmune encephalomyelitis. *Proc Natl Acad Sci USA*. 2008;105:11358–11363.
56. Aharoni R, Vainshtein A, Stock A, et al. Distinct pathological patterns in relapsing-remitting and chronic models of experimental autoimmune enchephalomyelitis and the neuroprotective effect of glatiramer acetate. *J Autoimmun*. 2011;37:228–241.
57. Maier K, Kuhnert AV, Taheri N, et al. Effects of glatiramer acetate and interferon-beta on neurodegeneration in a model of multiple sclerosis: a comparative study. *Am J Pathol*. 2006;169:1353–1364.
58. Aharoni R, Arnon R, Eilam R. Neurogenesis and neuroprotection induced by peripheral immunomodulatory treatment of experimental autoimmune encephalomyelitis. *J Neurosci*. 2005;25:8217–8228.
59. Aharoni R, Sasson E, Blumenfeld-Katzir T, et al. Magnetic resonance imaging characterization of different experimental autoimmune encephalomyelitis models and the therapeutic effect of glatiramer acetate. *Exp Neurol*. 2013;240:130–144.
60. Magavi SS, Leavitt BR, Macklis JD. Induction of neurogenesis in the neocortex of adult mice. *Nature*. 2000;405:951–955.
61. Picard-Riera N, Decker L, Delarasse C, et al. Experimental autoimmune encephalomyelitis mobilizes neural progenitors from the subventricular zone to undergo oligodendrogenesis in adult mice. *Proc Natl Acad Sci USA*. 2002;99:13211–13216.
62 Hagemeier K, Bruck W, Kuhlmann T. Multiple sclerosis – remyelination failure as a cause of disease progression. *Histol Histopathol*. 2012;27:277–287.
63. From R, Eilam R, Bar-Lev DD, et al. Oligodendrogenesis and myelinogenesis during postnatal development effect of glatiramer acetate. *Glia*. 2014;62:649–665.
64. O'Kusky J, Ye P. Neurodevelopmental effects of insulin-like growth factor signaling. *Front Neuroendocrinol*. 2012;33:230–251.
65. Filippi M, Rovaris M, Rocca MA, Sormani MP, Wolinsky JS, Comi G. Glatiramer acetate reduces the proportion of new MS lesions evolving into "black holes". *Neurology*. 2001;57:731–733.
66. Khan O, Shen Y, Caon C, et al. Axonal metabolic recovery and potential neuroprotective effect of glatiramer acetate in relapsing-remitting multiple sclerosis. *Mult Scler*. 2005;11:646–651.
67. Schori H, Kipnis J, Yoles E, et al. Vaccination for protection of retinal ganglion cells against death from glutamate cytotoxicity and ocular hypertension: implications for glaucoma. *Proc Natl Acad Sci USA*. 2001;98:3398–3403.
68. Laurie C, Reynolds A, Coskun O, Bowman E, Gendelman HE, Mosley RL. $CD4^+$ T cells from Copolymer-1 immunized mice protect dopaminergic neurons in the 1-methyl-4-phenyl-1,2,3,6-tetrahydropyridine model of Parkinson's disease. *J Neuroimmunol*. 2007;183:60–68.
69. Angelov DN, Waibel S, Guntinas-Lichius O, et al. Therapeutic vaccine for acute and chronic motor neuron diseases: implications for amyotrophic lateral sclerosis. *Proc Natl Acad Sci USA*. 2003;100:4790–4795.

CHAPTER

16

Currently Approved Disease-Modifying Drugs: Monoclonal Antibody Natalizumab

B.C. Kieseier, V.I. Leussink, C. Warnke
Heinrich-Heine University, Düsseldorf, Germany

OUTLINE

Multiple sclerosis (MS) is an inflammatory demyelinating disorder of the central nervous system (CNS),[1] in which the presence of leukocytes in cerebral perivascular spaces in areas of disease activity is one of the pathological hallmarks.[2–4] An absolute requirement for the influx of leukocytes from the peripheral blood into the CNS is their expression of adhesion

Translational Neuroimmunology in Multiple Sclerosis
http://dx.doi.org/10.1016/B978-0-12-801914-6.00016-7

molecules, which are composed of integrins. The alpha4-beta1 (α4-β1) integrin (very late activation antigen-4 (VLA-4)) is one of the four main integrins required for the firm arrest of leukocytes following their rolling adhesion.[5]

16.1 TARGETING THE α4 INTEGRIN

Natalizumab (Tysabri®) is a recombinant humanized monoclonal IgG4-antibody that binds, among others, to the α4-subunit of the α4-β1 integrin, and interferes with the α4-mediated binding to its natural ligands of the extracellular matrix and endothelial lining, vascular cell adhesion molecule-1 (VCAM1), and fibronectin.[6–7] Therefore, inhibition of leukocyte migration and extravasation is believed to be the leading mode of action although additional mechanisms might modulate the therapeutic and adverse effects of natalizumab (see Fig. 16.1). Apart from this obvious effect related to cell migration, various other effects on the immune system have been reported. For instance, natalizumab seems to exhibit effects on B-cell function as well.[8] On the other hand, natalizumab seems to spare the regulatory T-cell population,[9] decreases the overall frequency of plasmacytoid dendritic cells, key regulators in the development of both innate and adaptive immune responses, but increases their phenotype toward more activated and mature cells.[10]

In vivo, antibodies against VLA-4 interfere with the binding of leukocytes to cerebral blood vessels, and effectively prevent symptoms of experimental autoimmune encephalomyelitis (EAE), an animal model mimicking certain features of MS.[11] In this model system blocking the α4 integrin reduced the influx of T cells and monocytes into the CNS and substantially ameliorated the clinical course of EAE.[12] The efficacy seen in preclinical model systems based on a pan immunological concept relevant for lymphocyte migration into the CNS prompted

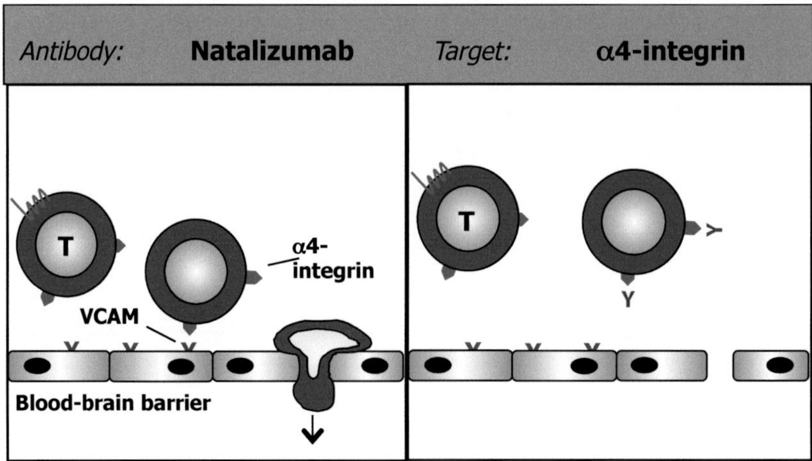

FIGURE 16.1 Natalizumab interferes with the α4 integrin, which is critically involved in the process of cell migration across the blood–brain barrier. By blocking the interaction between VLA-4 and VCAM, natalizumab inhibits the transmigration of immunocompetent cells, especially of T cells, out of the blood vessel into the central nervous system.

that translation of this therapeutic concept from bench to bedside by initiating a clinical trial program in patients with MS.

16.2 EFFICACY OF NATALIZUMAB

16.2.1 Clinical Trials

After the first successful phase-II and phase-IIb trials,[13,14] natalizumab was tested in two large multicenter randomized controlled phase-III studies.[15,16] In the AFFIRM trial, a total of 942 patients received either natalizumab 300mg every 4weeks intravenously or placebo.[15] Treatment with natalizumab resulted in a relapse rate reduction (the primary end point of the study) of 68%. Seventy-six percent of natalizumab-treated but only 53% of placebo-treated patients remained relapse free. Confirmed disability progression, as measured by the Expanded Disability Status Scale (EDSS), was reduced over the 2years of treatment by 42% as compared to placebo (17% vs. 29%), when disability progression was confirmed after 12weeks. These clinical data were mirrored by paraclinical measures assessed by magnetic resonance imaging (MRI): gadolinium (Gd)-enhancing lesions on T1-weighted images were reduced by over 90%, in line with the previous results from the phase-IIb study (see Fig. 16.2).[14] Positive effects of natalizumab treatment were also reported for quality of life assessments, requirement for concomitant glucocorticosteroid pulse treatments, or MS-related hospitalizations.[17] Furthermore, a post hoc analysis of the disability assessment revealed that one out of four patients with an EDSS ≥2.0 improved in their level of disability as measured by the EDSS.[18] Further analyses demonstrated that natalizumab treatment decreased clinical severity of relapses and improved recovery from disability induced by relapses, suggesting that these beneficial effects might limit the stepwise accumulation of disability.[19]

At the same time the AFFIRM trial was conducted, a second phase-III study, the SENTINEL trial, included 1171 patients pretreated with interferon beta1a (IFNβ1a) that had at least one relapse under this treatment. Natalizumab was administered in combination with IFNβ1a and compared to IFNβ1a treatment alone.[16] The results of SENTINEL corroborated the data from the AFFIRM study. Natalizumab treatment resulted in a relapse rate reduction of 53% in comparison to treatment with IFNβ1a alone. Similarly, MRI data revealed a reduction of 89% of Gd-enhancing lesions and 83% reduction of new or new enlarging T2 lesions over 2years.

Since there were two cases of progressive multifocal leukoencephalopathy (PML) as severe side effects in the SENTINEL trial, natalizumab was licensed as monotherapy only for patients with highly active RRMS based mainly on the data from the AFFIRM trial. Highly active RRMS was defined by the licensing bodies as at least two severe relapses per year of patients that do not respond to first-line therapy (IFNβ or glatiramer acetate (GA)). This restricted approval was introduced as a consequence to the PML cases for risk–benefit considerations (see later).

16.2.2 No Evidence of Disease Activity

The results of the AFFIRM trial changed substantially the expectations toward pharmacotherapies today. By combining clinical as well as paraclinical measures we can assess the

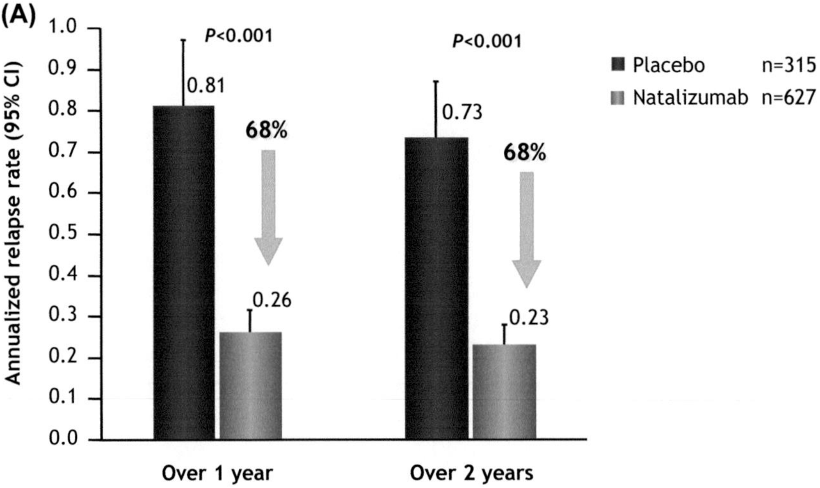

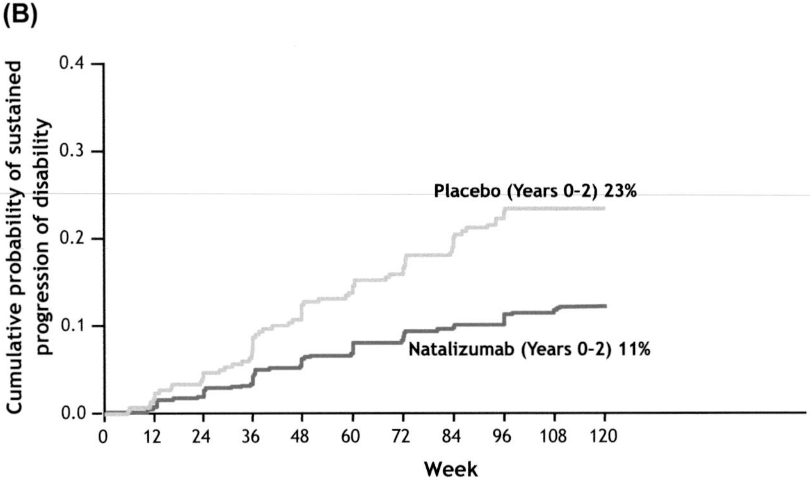

FIGURE 16.2 Natalizumab: Clinical efficacy on annualized relapse rate (A) and disability progression confirmed after 24 weeks (B) against placebo in the AFFIRM trial.[15]

percentage of patients that did not reveal activity on various outcome measures, such as relapses and disability progression measured by the EDSS as well as various MRI measures.[20] A post hoc analysis of this pivotal natalizumab trial revealed that more than 37% of patients in the natalizumab group remained free of disease activity as defined by the absence of relapses, no sustained disease progression, no Gd-enhancing lesions, and no new or enlarging lesions, while only 7% in the placebo group remained free of these combined disease activity measures.[21] Consequently, the terms "freedom of measurable disease activity" and "no evidence of disease activity (NEDA)" were introduced as possible outcome measures for MS clinical trials (Fig. 16.3). Some authors suggest that cognition should be added as an additional assessment to this composite measure in daily practice.[22]

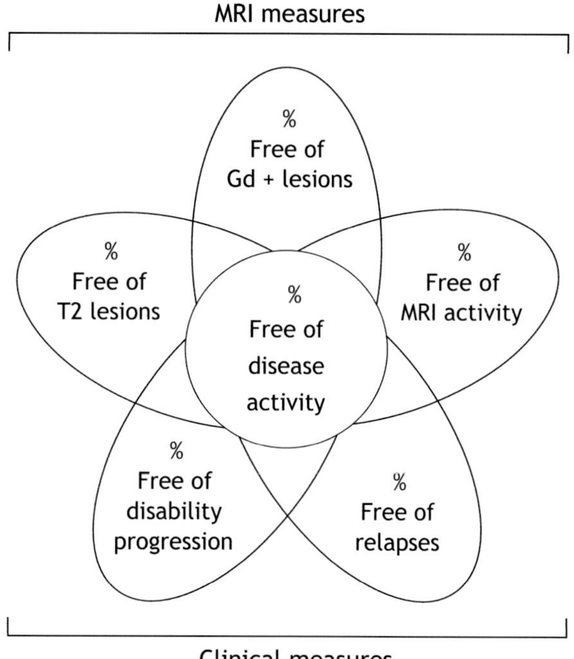

FIGURE 16.3 A novel approach to study the overall efficacy of a treatment is the assessment of the absence of clinical and radiological disease activity. Absence of activity is defined as no activity on clinical measures (no relapses and no sustained disability progression) and radiological measures (eg, no Gd-enhancing lesions, no new or enlarging T2-hyperintense lesions, no overall activity on cranial magnetic resonance imaging, MRI). The composite of the two is classified as the absence of combined activity, also coined "freedom of disease activity" or "no evidence of disease activity."

NEDA seems intuitively appealing and has been compared with disease remission in rheumatoid arthritis, which clearly highlights the remarkable progress in MS therapy over the past two decades. However, before NEDA becomes a new standard efficacy measure in MS trials, its prognostic utility in assessing MS disability will need to be established.[23]

16.2.3 Clinical Efficacy in the Real World

In numerous observational studies the profound clinical and paraclinical efficacy of natalizumab seen in the phase-III trial program could be replicated in a real-world setting. The level of efficacy appeared even more pronounced since the label for the drug given by the authorities preselects for a more clinically active patient population compared to the inclusion criteria in AFFIRM. One large observational study corroborating such findings is the so-called Tysabri Observational Program (TOP), recruiting patients outside the United States.[24] Interestingly, the clinical benefit seen in patients treated with natalizumab seems to be associated with an increase in the percentage of patients showing stable or even ameliorated electrophysiological parameters assessed by evoked potentials.[25]

Besides positive effects on relapse rate and disability progression an increase in walking speed in MS patients treated with natalizumab could be demonstrated in a prospective open-label study (TIMER).[26] Furthermore, it was shown that natalizumab, as used in a real-life setting, might improve MS-related fatigue based on the results from a one-armed uncontrolled study. In addition, other parameters related to patients' quality of life seemed to improve with natalizumab treatment.[27]

16.3 SAFETY

Overall, natalizumab was well tolerated in the large phase-III trials.[15,16] Side effects included headaches and fatigue on infusion days. Allergic reactions occurred in about 4% of natalizumab-treated patients, most of them after the second or third infusion with urticaria, headache, flush, and hypotonia.[28] However, besides this type-I hypersensitivity reaction delayed type-III reactions also have been reported.[29,30] The occurrence of allergic reactions is linked to the presence of anti-natalizumab antibodies. In the AFFIRM trial 68% of the patients with infusion-related reactions were positive for anti-natalizumab antibodies.[31] Approximately 10% of natalizumab-treated patients develop transiently anti-natalizumab antibodies that persist in 6%. Antibodies usually develop within the first 12 weeks of treatment. In patients tested positive for anti-natalizumab antibodies further infusions should be halted and retesting should be performed after 4–6 weeks. Patients persistently positive for anti-natalizumab antibodies should not be exposed again to the drug due to an increased risk for allergic reactions and the neutralizing effect on natalizumab efficacy.

Elevations of liver enzymes and bilirubin were observed in 0.1% of natalizumab-treated patients. However in nearly all cases this side effect was clinically not relevant.

As already mentioned, there were two cases of PML in the SENTINEL trial and another case in a study of natalizumab in Crohn's disease.[32–34] Since these patients had concomitant treatment with IFNβ or previous immunosuppressive treatment, it was originally suspected that PML occurs as a result of a combination therapy, which is why the approval was and still is restricted to monotherapy.

16.3.1 Progressive Multifocal Leukoencephalopathy

PML is an opportunistic demyelinating disease of the brain. It is caused by reactivation of the John Cunningham virus (JCV), a double-stranded DNA virus belonging, together with the SV-40 and BK viruses, to the family of polyomaviridae.[35] PML is not a natalizumab-specific side effect, albeit integrin blocking therapies seem to specifically heighten its risk; it has been diagnosed in the context of many other immunoactive drugs as well, specifically in immunocompromised patients, in particular in patients with reduced cellular immunity, including in patients with HIV, patients with hematological diseases, or patients receiving immunosuppressive medication (see Fig. 16.4).[36] Primary infection by JCV takes place in childhood and is asymptomatic. This virus persists in the tonsillar tissue, bone marrow, kidney, and spleen.[37,38] The presence of JCV-DNA has been detected in urine, different blood compartments, and in cerebrospinal fluid (CSF).[39,40] JCV viruria was found as frequently in HIV-positive individuals as in control subjects, suggesting that its detection has no clinical value.[41] JCV-DNA has

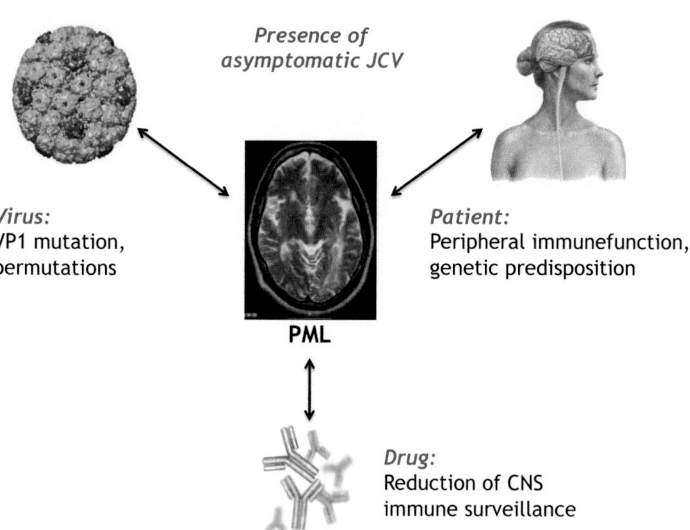

FIGURE 16.4 Progressive multifocal leukoencephalopathy (PML) is caused by the John Cunningham virus that needs to mutate before it can infect glial cells in the central nervous system. Predisposing factors for the development of PML are an immunodeficiency (either in the context of other diseases or iatrogenic by the application of immunosuppressive drugs) and potentially host genetics. More details in the text.

been detected in peripheral blood mononuclear cells (PBMCs) in HIV-positive patients with and without PML but not in HIV-negative control subjects. There was no specificity for any leukocyte subtype, but JCV-DNA detection correlated with low $CD4^+$ lymphocyte counts.[41] Thus, JCV-DNA load in PBMCs has no predictive value as a screening tool for PML.

Currently, our knowledge about transmission of JCV infection and its circle of life in the healthy human population is limited. The transmittable form of JCV is commonly referred to as the JCV archetype, as it is thought that all other genotypes originate from it. The JCV-archetype is detectable in the urine and sewage. In contrast, the JCV-PML type appears more neurotropic, and can be isolated from brains of PML patients. This pathological variant is characterized by deletions, duplications, and point mutations in a specific JCV regulatory region. Most likely, episodically and on low-level replicated JCV archetype within the kidney explains urine excretion and JCV-DNA in plasma of immunocompromised patients. There is growing evidence on persistency of JCV in the CNS. Usually JCV DNA can be found in oligodendrocytes and astrocytes, and the current pathogenic concept implies a lytic infection of glial cells causing PML.[42,43] A newly described phenotype of this disease comprises a direct infection of neurons. This distinct clinical and radiological syndrome is named JCV granule cell neuronopathy, characterized by exclusive or predominant cerebellar atrophy.[44] In the context of natalizumab only a single case of this variant has been reported so far.[45]

No pathognomonic initial symptoms of PML have been defined, which often makes an early clinical diagnosis of this disorder very challenging. Some of the classic clinical signs and symptoms of PML include rapidly progressive dementia, motor dysfunction, and vision loss, which can be difficult to differentiate from MS relapses.

The clinical outcome of natalizumab-treated PML patients seems much better than in patients with HIV-associated PML, but early diagnosis (see later) and consequent treatment appear relevant.[46] Critical for the diagnosis of PML are the presence of JCV-DNA in the CSF assessed by polymerase chain reaction and MRI. Present therapeutic strategies in the context of natalizumab-associated PML include discontinuation of natalizumab and plasmapheresis/immunoadsorption (PLEX/IA). This may accelerate the occurrence of immune reconstitution inflammatory syndrome (IRIS), which is relevant to eliminate the virus from the CNS. During IRIS corticosteroids are applied.[47] In addition, other treatment strategies have been reported, all of which require further studies before treatment recommendations can be given.[48]

16.3.2 The Anti-JCV-Antibody

So far, three different risk factors could be identified that are associated with a risk of developing PML while receiving treatment with natalizumab: (1) treatment duration beyond 24 months and (2) prior treatment with an immunosuppressive drug (independent of the immunosuppressant used, its exposition time, and the time frame between last application of the immunosuppressive therapy and treatment initiation with natalizumab).[49]

Of specific interest is the risk stratification based on the presence of the so-called anti-JCV antibody. Since JCV infection is a prerequisite for the development of PML, an enzyme-linked immunosorbent assay (ELISA) that detects JCV antibodies in human serum or plasma was developed, optimized, and validated. The seroprevalence reported for MS patients (and controls) is in a range of 50–60% for most of the countries.[50] Based on this assay PML risk stratification became possible (see Fig. 16.5).[51]

In case the anti-JCV antibody test turns out negative, the risk of developing PML remains very low (≤0.09/1000), given a false negative rate of 2–3% of the ELISA. The test should be repeated every 6 months to detect possible seroconversion (rate: 2–3% per year).

In case of a positive anti-JCV antibody test the PML risk during the first 2 years of treatment is calculated with 1/1000 independent of a prior immunosuppressive treatment course (0.56/1000 without prior immunosuppression; 1.6/1000 in case of prior immunosuppression). Beyond 24 months of treatment prior use of immunosuppressive therapy becomes relevant in risk determination: in case of such a therapy the calculated risk is 11.1/1000; if not the risk estimate is 4.6/1000. These numbers are based on PML incidences calculated in 2012. The PML risk may indeed cumulate over time with extended therapy, and the denominator may decrease due to risk stratification in clinical practice. Hence, these incidence numbers may not exactly mirror the individual risk, but could help clinicians in advising their patient.

Furthermore, an analysis was conducted to analyze whether anti-JCV antibody levels, measured as index, may further define PML risk in seropositive patients.[52] This analysis was based on the association between serum or plasma anti-JCV antibody levels and PML risk in anti-JCV antibody-positive MS patients from natalizumab clinical studies and postmarketing sources. For PML and non-PML patients, the probabilities of having an index below and above a range of anti-JCV antibody index thresholds were calculated using all available data and applied to the PML risk stratification algorithm (Table 16.1).

Given the longitudinal stability of the anti-JCV antibody index in the majority of patients, anti-JCV antibody levels in serum/plasma, measured as index, may differentiate PML risk in anti-JCV antibody-positive MS patients, however only in those with no prior

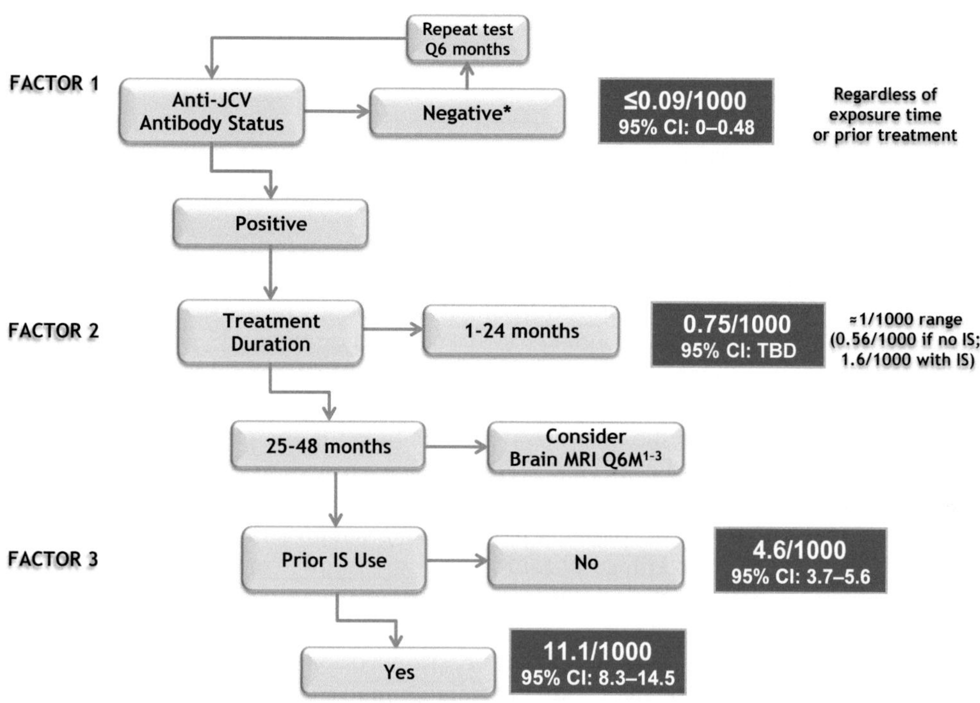

FIGURE 16.5 Risk estimates for developing progressive multifocal leukoencephalopathy (PML).[49] The estimate of PML incidence in anti-John Cunningham virus (JCV) antibody-negative patients assumes that all patients received at least one dose of natalizumab and that one hypothetical PML case tested negative for anti-JCV antibodies prior to the onset and diagnosis of PML. *The estimate of PML incidence in anti-JCV antibody-negative patients assumes that all patients received at least 1 dose of natalizumab and that 1 hypothetical PML case tested negative for anti-JCV antibodies prior to the onset and diagnosis of PML. IS, immunosuppressant.

TABLE 16.1 PML Risk Estimates by Index Threshold in Anti-JCV Antibody Positive Patients With No Prior Immunosuppressants Use. PML Risk Estimates for Anti-JCV Antibody Index Thresholds Were Calculated Based on the Current PML Risk Stratification Algorithm (see Fig. 16.5)[52]

	PML Risk Estimates (95% CI) Per 1000 Patients (No Prior IS Use)		
Index Threshold	**1–24 Months**	**25–48 Months**	**49–72 Months**
≤0.9	0.1 (0–0.41)	0.3(0.04–1.13)	0.4(0.01–2.15)
≤1.1	0.1 (0–0.34)	0.7(0.21–1.53)	0.7 (0.08–2.34)
≤1.3	0.1 (0.01–0.39)	1.0(0.48–1.98)	1.2 (0.31–2.94)
≤1.5	0.1(0.03–0.42)	1.2(0.64–2.15)	1.3 (0.46–2.96)
>1.5	1.0 (0.64–1.41)	8.1(6.64–9.8)	8.5 (6.22–11.28)

immunosuppressant use. This additional information might help to stratify the PML risk in patients at risk. However, further data should be collected to better understand the real value of this measurement.

16.3.3 Other Potential Biomarkers Predicting Progressive Multifocal Leukoencephalopathy

Additional biomarkers have been proposed that might be helpful in the process of predicting the PML risk in patients treated with natalizumab.[53] One proposed biomarker is the presence of lipid-specific immunoglobulin M oligoclonal bands in the CSF (IgM bands). In a study based on 24 MS patients who developed PML and another 343 who did not while treated with natalizumab the presence of IgM bands was associated with a reduced risk of developing PML.

A biomarker accessible from the peripheral venous blood is the percentage of L-selectin (CD62L) expressing $CD4^+$ T lymphocytes.[54] The surface expression of this adhesion molecule was analyzed and found to be significantly lower in patients treated long-term with natalizumab when compared with patients not receiving natalizumab treatment or healthy controls. An unusually low percentage (9-fold lower) was highly correlated with the risk of developing PML in the patient group with available pre-PML samples when compared with non-PML natalizumab-treated patients.

All these findings underline that there is an urgent need for reliable biomarkers helping in stratifying the risk for PML in the context of natalizumab treatment in daily practice. All attempts appear very interesting; however, there is a need for larger numbers and assay validation prior to an implementation in the clinical routine.

16.3.4 Pharmacovigilance

As there is currently no proven treatment for patients suffering from PML under natalizumab treatment other than accelerated clearance of therapy,[55] the early establishment of a diagnosis remains crucial. Upon the (re-) approval of natalizumab, each country initiated a risk management program to closely monitor patients at risk, which usually includes a three-step diagnostic algorithm for natalizumab-treated patients with new or worsening neurological signs and symptoms. Early suspension of natalizumab treatment and strategies for clinical, MRI, and laboratory assessments have been proposed.[56]

Based on our current understanding, younger age at diagnosis, less functional disability prior to PML diagnosis, lower JC viral load at diagnosis in the CSF, and more localized brain involvement by MRI at the time of diagnosis seems to predict improved survival in natalizumab-associated PML.[57] In addition, PML patients asymptomatic at diagnosis have a better survival and less functional disability than those who were symptomatic at diagnosis.[58] Thus, early detection of changes on MRI suspicious for PML is critical. Asymptomatic natalizumab-associated PML manifestations on MRI show a rather localized disease, frequently located in the frontal lobes, affecting the cortical gray matter and adjacent juxtacortical white matter.[59] Diffusion-weighted imaging and fluid-attenuated inversion recovery appear very sensitive and helpful in early detection and discrimination from MS lesions on MRI.[60]

16.4 NATALIZUMAB IN THE CURRENT TREATMENT ALGORITHM

In the phase-III study AFFIRM, natalizumab was investigated in a large population of treatment-naïve patients. In clinical practice, however, natalizumab is commonly used as a second-line therapy predominantly driven by the label defined by the authorities. Since head-to-head studies have not been performed mimicking treatment decision scenarios in clinical practice, collected evidence from the real world can be helpful in obtaining insights into the feasibility of treatment strategies. One powerful data source for addressing such specific questions is MSBase, an ongoing, international, observational registry acquiring real-world data from patients with MS. Using quasirandomization with propensity score-based matching specific patient subpopulations with comparable baseline characteristics can be selected and investigated.[61]

16.4.1 Switching to Natalizumab

In patients with ongoing disease activity despite treatment with a platform therapy, specifically IFNβ or GA, clinicians either switch among different platform therapies or to either fingolimod or natalizumab.

Using large, real-world, propensity-matched datasets from MSBase it could be demonstrated that compared to changing the treatment regimen between IFNβ and GA, switching to natalizumab reduced the annualized relapse rate in year 1 by 65–75%, treatment discontinuation events by 48–65%, and the risk of confirmed disability progression by 26%. The results were consistent regardless of the prior treatment identity.[62]

An alternative scenario is the switch from an injectable therapy either to fingolimod or natalizumab. Addressing this question MSBase demonstrated that in active MS during treatment with injectable disease-modifying therapies, switching to natalizumab is more effective than switching to fingolimod in reducing relapse rate and short-term disability burden.[63]

These results demonstrate that, although formally never tested in the clinical trial program, natalizumab is a powerful therapy not only in treatment-naïve patients, but specifically also in patients failing on a platform therapy.

16.4.2 Stopping Natalizumab

Given the mode of action of natalizumab and its half-life as an IgG4 monoclonal antibody it is plausible that after treatment cessation disease activity will return. Thus, a discussion of how to continue MS therapy after stopping treatment with natalizumab is ongoing.

The RESTORE study investigated to which extent other treatment modalities could compensate for the efficacy of natalizumab:[64] Eligible patients, relapse-free through the prior year on natalizumab, were randomized to continue natalizumab, to switch to placebo, or to receive alternative immunomodulatory therapy (IFNβ, GA, or methylprednisolone). MRI and clinical disease activity recurred despite the use of other therapies. A specific MRI analysis revealed that in most patients recurring radiological disease activity during natalizumab interruption did not exceed prenatalizumab levels or levels seen in historical control patients,[65] emphasizing that there is no collective evidence for an MS rebound after treatment discontinuation.[66]

In MSBase the risk of relapse after stopping natalizumab and switching to fingolimod was investigated and compared with experience switching from IFNβ to GA and those previously treatment-naive. Furthermore, predictors of time to first relapse on fingolimod were determined. Relapse rates were generally low across all patient groups in the first 9 months on fingolimod; however, 30% of patients with disease activity on natalizumab relapsed within the first 6 months on fingolimod. Independent predictors of time to first relapse on fingolimod were the number of relapses in the prior 6 months.[67] Similar data were obtained in a French study, which described an increased risk of MS reactivation during the washout period or shortly after fingolimod initiation. Based on these observations a washout period shorter than 3 months is recommended when switching from natalizumab to fingolimod.[68]

16.4.3 Natalizumab and Pregnancy

Natalizumab is usually withdrawn 3 months before pregnancy; however, data on exposure to natalizumab during pregnancy are accumulating. In one study 101 women with RRMS exposed to natalizumab during the first trimester of pregnancy were identified. Exposure to natalizumab in early pregnancy did not seem to increase the risk of adverse pregnancy outcomes in comparison to a group of patients not exposed to natalizumab.[69] In another case series of 12 women with 13 pregnancies and highly active MS who were treated with natalizumab during their third trimester of pregnancy mild-to-moderate hematologic alterations in 10 of 13 infants including thrombocytopenia and anemia were observed.[70] These findings were transient and resolved during the 4 months after birth; none of the infants needed any specific treatment.

16.5 OUTLOOK

Given its mode of action natalizumab could also be used as a treatment in other immune-driven disorders or disease entities in which the immune system plays a critical role for tissue damage, such as stroke. Therefore, it is not surprising that natalizumab is currently explored in different indications.

A current unmet need in MS therapy is the lack of efficient treatment options for progressive forms of this disease. In an open-label, phase-IIA, proof-of-concept study the efficacy of natalizumab in progressive MS was investigated.[71] Seventeen patients completed this 60-week study; primary end point was changed in CSF osteopontin, a biomarker of intrathecal inflammation, which decreased significantly when compared to baseline under natalizumab therapy. Magnetization transfer ratio increased in both cortical gray and normal-appearing white matter and correlated with decreases in CSF neurofilament light chain, a marker for tissue damage. Additional experimental as well as clinical evidence prompted a large phase-III study investigating safety and clinical efficacy of natalizumab in secondary-progressive MS (ASCEND). This study, however, did not achieve statistical significance on the primary or secondary endpoints.[72]

Chronic inflammatory demyelinating polyradiculoneuropathy is an autoimmune disease, most likely T-cell-driven, affecting the peripheral nervous system. Conceptionally, natalizumab might be clinically effective in this indication as well. Whereas a first-case report could not reveal any benefit,[73] another series of three cases documented disease stabilization and even clinical improvement.[74] Further studies are warranted investigating the efficacy of natalizumab in this disabling disease.

Experimental evidence points to a potential role of VLA-4 in T-cell trafficking during brain ischemia. In various model systems the blockade of the α4 integrin prompted beneficial effects,[75–77] although also controversial data exist.[78] A phase-II study was just completed, in which natalizumab was tested in patients with acute ischemic stroke when given at ≤6h or at >6 to ≤9h from when they were last known normal (ACTION).[79] Primary end point of the study is the change in infarct volume from baseline to day 5 assessed by magnetic resonance imaging (MRI) in participants. The results of this study have not been published yet.

16.6 CONCLUSION

Natalizumab is a highly effective treatment for patients with relapsing forms of MS with a risk of developing PML, prompting a thorough risk–benefit assessment. Today, neurologists have to make complex treatment decisions together with their patients, sometimes on the basis of very limited clinical data and evidence. It becomes even more complex, since risk perceptions between patients and treating physician can differ dramatically, as shown for the PML risk perception in the context of natalizumab.[80]

Thus, optimal assessment of risks and benefits remain challenging.[81] However, withholding a potent therapy with proven efficacy on disease activity (see Fig. 16.6), prevention of disability, and improvement in quality of life can also cause harm to patients with a disabling disease, such as MS.

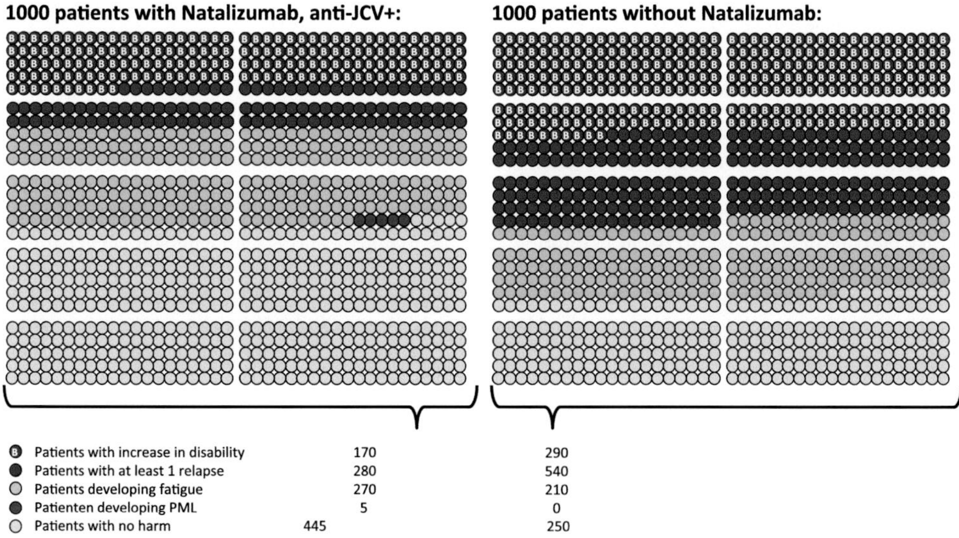

FIGURE 16.6 Balancing risk and benefit. This cereal-box plot was compiled based on the results of the AFFIRM study.[15] The left part demonstrates risk and benefit in an estimated 1000 patients positive for anti-John Cunningham virus antibody. In contrast, on the right, an estimated 1000 patients who did not receive natalizumab treatment. It becomes evident that no treatment increases the number of patients who do not experience no harm. *Adapted from Wolfgang Gaissmaier, with thanks.*

References

1. Compston A, Coles A. Multiple sclerosis. *Lancet*. 2002;359:1221–1231.
2. Martin R, McFarland HF, McFarlin DE. Immunological aspects of demyelinating diseases. *Annu Rev Immunol.* 1992;10:153–187.
3. Lucchinetti C, Brück W, Parisi J, et al. Heterogeneity of multiple sclerosis lesions: implications for the pathogenesis of demyelination. *Ann Neurol*. 2000;47:707–717.
4. Lassmann H, Brück W, Lucchinetti CF. The immunopathology of multiple sclerosis: an overview. *Brain Pathol.* 2007;17:210–218.
5. Luster AD, Alon R, von Andrian UH. Immune cell migration in inflammation: present and future therapeutic targets. *Nat Immunol*. 2005;6:1182–1190.
6. Ulbrich H, Eriksson EE, Lindbom L. Leukocyte and endothelial cell adhesion molecules as targets for therapeutic interventions in inflammatory disease. *Trends Pharmacol Sci*. 2003;24:640–647.
7. von Andrian UH, Engelhardt B. Alpha4 integrins as therapeutic targets in autoimmune disease. *N Engl J Med.* 2003;348:68–72.
8. Warnke C, Stettner M, Lehmensiek V, et al. Natalizumab exerts a suppressive effect on surrogates of B cell function in blood and CSF. *Mult Scler*. 2014. [in press].
9. Stenner MP, Waschbisch A, Buck D, et al. Effects of natalizumab treatment on Foxp3+ T regulatory cells. *PLoS One*. 2008;3:e3319.
10. Kivisäkk P, Francois K, Mbianda J, Gandhi R, Weiner HL, Khoury SJ. Effect of natalizumab treatment on circulating plasmacytoid dendritic cells: a cross-sectional observational study in patients with multiple sclerosis. *PLoS One*. 2014;9:e103716.
11. Yednock TA, Cannon C, Fritz LC, Sanchez-Madrid F, SteinmanL KN. Prevention of experimental autoimmune encephalomyelitis by antibodies against alpha 4 beta 1 integrin. *Nature*. 1992;356:63–66.
12. Kanwar JR, Harrison JE, Wang D, et al. Beta7 integrins contribute to demyelinating disease of the central nervous system. *J Neuroimmunol*. 2000;103:146–152.
13. Tubridy N, Behan PO, Capildeo R, et al. The effect of anti-alpha4 integrin antibody on brain lesion activity in MS. The UK Antegren Study Group. *Neurology*. 1999;53:466–472.
14. Miller DH, Khan OA, Sheremata WA, et al. A controlled trial of natalizumab for relapsing multiple sclerosis. *N Engl J Med*. 2003;348:15–23.
15. Polman CH, O'Conner PW, Havrdova E, et al. A randomized, placebo-controlled trial of natalizumab for relapsing multiple sclerosis. *N Engl J Med*. 2006;354:899–910.
16. Rudick RA, Stuart WH, Calabresi PA, et al. Natalizumab plus interferon beta-1a for relapsing multiple sclerosis. *N Engl J Med*. 2006;354:911–923.
17. Rudick RA, Miller D, Hass S, et al. Health-related quality of life in multiple sclerosis: effects of natalizumab. *Ann Neurol*. 2007;62:335–346.
18. Phillips JT, Giovannoni G, Lublin FD, et al. Sustained improvement in expanded disability status scale as a new efficacy measure of neurological change in multiple sclerosis: treatment effects with natalizumab in patients with relapsing multiple sclerosis. *Mult Scler*. 2011;17:970–979.
19. Lublin FD, Cutter G, Giovannoni G, Pace A, Campbell NR, Belachew S. Natalizumab reduces relapse clinical severity and improves relapse recovery in MS. *Mult Scler Relat Disord*. 2014;3:705–711.
20. Kieseier BC, Wiendl H, Hartung HP, Leussink VI, Stüve O. Risks and benefits of multiple sclerosis therapies: need for continual assessment? *Curr Opin Neurol*. 2011;24:238–243.
21. Havrdova E, Galetta S, Hutchinson M, et al. Effect of natalizumab on clinical and radiological disease activity in multiple sclerosis: a retrospective analysis of the Natalizumab Safety and Efficacy in Relapsing-Remitting Multiple Sclerosis (AFFIRM) study. *Lancet Neurol*. 2009;8:254–260.
22. Stangel M, Penner IK, Kallmann BA, Lukas C, Kieseier BC. Towards the implementation of "no evidence of disease activity" in multiple sclerosis treatment: the multiple sclerosis decision model. *Ther Adv Neurol Disord*. 2015;8:1–3.
23. Bevan CJ, Cree BA. Disease activity free status. A new end point for a new era in multiple sclerosis clinical research ? *JAMA Neurol*. 2014;71:803.
24. Butzkueven H, Kappos L, Pellegrini F, et al. TYSABRI Observational Program (TOP) Investigators. Efficacy and safety of natalizumab in multiple sclerosis: interim observational programme results. *J Neurol Neurosurg Psychiatry*. 2014;85:1190–1197.

25. Meuth SG, Bittner S, Seiler C, Göbel C, Wiendl H. Natalizumab restores evoked potential abnormalities in patients with relapsing–remitting multiple sclerosis. *Mult Scler J*. 2011;17:198–203.
26. Voloshyna N, Havrdová E, Hutchinson M, et al. Natalizumab improves ambulation in relapsing-remitting multiple sclerosis: results from the prospective TIMER study and a retrospective analysis of AFFIRM. *Eur J Neurol*. 2015;22:570–577.
27. Svenningsson A, Falk E, Celius EG, et al. Natalizumab treatment reduces fatigue in multiple sclerosis. Results from the TYNERGY trial; a study in the real life setting. *PLoS ONE*. 2013;8:e58643.
28. Phillips JT, O'Connor PW, Havrdova E, et al. Infusion-related hypersensitivity reactions during natalizumab treatment. *Neurology*. 2006;67:1717–1718.
29. Krumbholz M, Pellkofer H, Gold R, Hoffmann LA, Hohlfeld R, Kumpfel T. Delayed allergic reaction to natalizumab associated with early formation of neutralizing antibodies. *Arch Neurol*. 2007;64:1331–1333.
30. Leussink VI, Lehmann HC, Hartung HP, Gold R, Kieseier BC. Type III systemic allergic reaction to natalizumab. *Arch Neurol*. 2008;65:851–852.
31. Calabresi PA, Giovannoni G, Confavreux C, et al. The incidence and significance of anti-natalizumab antibodies: results from AFFIRM and SENTINEL. *Neurology*. 2007;69:1391–1403.
32. Kleinschmidt-DeMasters BK, Tyler KL. Progressive multifocal leukoencephalopathy complicating treatment with natalizumab and interferon beta-1a for multiple sclerosis. *N Engl J Med*. 2005;353:369–374.
33. Langer-Gould A, Atlas SW, Green AJ, Bollen AW, Pelletier D. Progressive multifocal leukoencephalopathy in a patient treated with natalizumab. *N Engl J Med*. 2005;353:375–381.
34. Van Assche G, Van Ranst M, Sciot R, et al. Progressive multifocal leukoencephalopathy after natalizumab therapy for Crohn's disease. *N Engl J Med*. 2005;353:362–368.
35. Frisque RJ, Bream GL, Cannella MT. Human polyomavirus JC virus genome. *J Virol*. 1984;51:458–469.
36. Koralnik IJ. Progressive multifocal leukoencephalopathy revisited: has the disease outgrown its name? *Ann Neurol*. 2006;60:162–173.
37. Ferrante P, Caldarelli-Stefano R, Omodeo-Zorini E, et al. Comprehensive investigation of the presence of JC virus in AIDS patients with and without progressive multifocal leukoencephalopathy. *J Med Virol*. 1997;52:235–242.
38. Major EO, Amemiya K, Tornatore CS, Houff SA, Berger JR. Pathogenesis and molecular biology of progressive multifocal leukoencephalopathy, the JC virus-induced demyelinating disease of the human brain. *Clin Microbiol Rev*. 1992;5:49–73.
39. Azzi A, De Santis R, Ciappi S, et al. Human polyomaviruses DNA detection in peripheral blood leukocytes from immunocompetent and immunocompromised individuals. *J Neurovirol*. 1996;2:411–416.
40. Kitamura T, Sugimoto C, Kato A, et al. Persistent JC virus (JCV) infection is demonstrated by continuous shedding of the same JCV strains. *J Clin Microbiol*. 1997;35:1255–1257.
41. Koralnik IJ, Schmitz JE, Lifton MA, Forman MA, Letvin NL. Detection of JC virus DNA in peripheral blood cell subpopulations of HIV-1-infected individuals. *J Neurovirol*. 1999;5:430–435.
42. Perez-Liz G, Del Valle L, Gentilella A, Croul S, Khalili K. Detection of JC virus DNA fragments but not proteins in normal brain tissue. *Ann Neurol*. 2008;64:379–387.
43. Rollison DEM, Utaipat U, Ryschkewitsch C, et al. Investigation of human brain tumors for the presence of polyomavirus genome sequences by two independent laboratories. *Int J Cancer*. 2005;113:769–774.
44. Koralnik IJ, Wüthrich C, Dang X, et al. JC virus granule cell neuronopathy: a novel clinical syndrome distinct from progressive multifocal leukoencephalopathy. *Ann Neurol*. 2005;57:576–580.
45 Schippling S, Kempf C, Büchele F, et al. JC virus granule cell neuronopathy and GCN-IRIS under natalizumab treatment. *Ann Neurol*. 2013;74:622–626.
46. Dahlhaus S, Hoepner R, Chan A, et al. Disease course and outcome of 15 monocentrically treated natalizumab-associated progressive multifocal leukoencephalopathy patients. *J Neurol Neurosurg Psychiatry*. 2013;84:1068–1074.
47. Tan IL, McArthur JC, Clifford DB, Major EO, Nath A. Immune reconstitution inflammatory syndrome in natalizumab-associated PML. *Neurology*. 2011;77:1061–1067.
48. Giacomini PS, Rozenberg A, Metz I, Araujo D, Arbour N, Bar-Or A. Maraviroc in Multiple Sclerosis–Associated PML–IRIS (MIMSAPI) Group. Maraviroc and JC virus-associated immune reconstitution inflammatory syndrome. *N Engl J Med*. 2014;370:486–488.
49. Sorensen PS. New management algorithms in multiple sclerosis. *Curr Opin Neurol*. 2014;27:246–259.
50. Lee P, Plavina T, Castro A, et al. A second-generation ELISA (STRATIFY JCV™ DxSelect™) for detection of JC virus antibodies in human serum and plasma to support progressive multifocal leukoencephalopathy risk stratification. *J Clin Virol*. 2013;57:141–146.

51. Bloomgren G, Richman S, Hotermans C, et al. Risk of natalizumab-associated progressive multifocal leukoencephalopathy. *N Engl J Med*. 2012;366:1870–1880.
52. Plavina T, Subramanyam M, Bloomgren G, et al. Anti-JC virus antibody levels in serum or plasma further define risk of natalizumab-associated progressive multifocal leukoencephalopathy. *Ann Neurol*. 2014;76:802–812.
53. Villar LM, Costa-Frossard L, Masterman T, et al. Lipid-specific immunoglobulin M bands in cerebrospinal fluid are associated with a reduced risk of developing progressive multifocal leukoencephalopathy during treatment with natalizumab. *Ann Neurol*. 2015;77:447–457.
54. Schwab N, Schneider-Hohendorf T, Posevitz V, et al. L-selectin is a possible biomarker for individual PML risk in natalizumab-treated MS patients. *Neurology*. 2013;81:865–871.
55. Khatri BO, Man S, Giovannoni G, et al. Effect of plasma exchange in accelerating natalizumab clearance and restoring leukocyte function. *Neurology*. 2009;72:402–409.
56. Kappos L, Bates D, Edan G, et al. Natalizumab treatment for multiple sclerosis: updated recommendations for patient selection and monitoring. *Lancet Neurol*. 2011;10:745–758.
57. Dong-Si T, Gheuens S, Gangadharan A, et al. Predictors of survival and functional outcomes in natalizumab-associated progressive multifocal leukoencephalopathy. *J Neurovirol*. 2015. [in press].
58. Dong-Si T, Richman S, Wattjes MP, et al. Outcome and survival of asymptomatic PML in natalizumab-treated MS patients. *Ann Clin Transl Neurol*. 2014;1:755–764.
59. Wattjes MP, Vennegoor A, Steenwijk MD, et al. MRI pattern in asymptomatic natalizumab-associated PML. *J Neurol Neurosurg Psychiatry*. 2014;86(7):793–798. [in press].
60. Cosottini M, Tavarelli C, Del Bono L, et al. Diffusion-weighted imaging in patients with progressive multifocal leukoencephalopathy. *Eur Radiol*. 2008;18:1024–1030.
61. Butzkueven H, Chapman J, Cristiano E, et al. MSBase: an international, online registry and platform for collaborative outcomes research in multiple sclerosis. *Mult Scler*. 2006;12:769–774.
62. Spelman T, Kalincik T, Zhang A, et al. Comparative efficacy of switching to natalizumab in active multiple sclerosis. *Ann Clin Transl Neurol*. 2015;2:373–387.
63. Kalincik T, Horakova D, Spelman T, et al. Switch to natalizumab versus fingolimod in active relapsing-remitting multiple sclerosis. *Ann Neurol*. 2015;77:425–435.
64. Fox RJ, Cree BA, De Sèze J, et al. MS disease activity in RESTORE: a randomized 24-week natalizumab treatment interruption study. *Neurology*. 2014;82:1491–1498.
65. Kaufman M, Cree BA, De Sèze J, et al. Radiologic MS disease activity during natalizumab treatment interruption: findings from RESTORE. *J Neurol*. 2015;262:326–336.
66. O'Connor PW, Goodman A, Kappos L, et al. Disease activity return during natalizumab treatment interruption in patients with multiple sclerosis. *Neurology*. 2011;76:1858–1865.
67. Jokubaitis VG, Li V, Kalincik T, et al. Fingolimod after natalizumab and the risk of short-term relapse. *Neurology*. 2014;82:1204–1211.
68. Cohen M, Maillart E, Tourbah A, et al. Switching from natalizumab to fingolimod in multiple sclerosis: a French prospective study. *JAMA Neurol*. 2014;71:436–441.
69. Ebrahimi N, Herbstritt S, Gold R, Amezcua L, Koren G, Hellwig K. Pregnancy and fetal outcomes following natalizumab exposure in pregnancy. A prospective, controlled observational study. *Mult Scler*. 2015;21:198–205.
70. Haghikia A, Langer-Gould A, Rellensmann G, et al. Natalizumab use during the third trimester of pregnancy. *JAMA Neurol*. 2014;71:891–895.
71. Romme Christensen J, Ratzer R, Börnsen L, et al. Natalizumab in progressive MS: results of an open-label, phase 2A, proof-of-concept trial. *Neurology*. 2014;82:1499–1507.
72. https://clinicaltrials.gov/ct2/show/NCT01416181; Accessed 12.01.16.
73. Wolf C, Menge T, Stenner MP, et al. Natalizumab treatment in a patient with chronic inflammatory demyelinating polyneuropathy. *Arch Neurol*. 2010;67:881–883.
74. Vallat JM, Mathis S, Ghorab K, Milor MA, Richard L, Magy L. Natalizumab as a Disease-Modifying Therapy in Chronic Inflammatory Demyelinating Polyneuropathy - A Report of Three Cases. *Eur Neurol*. 2015;73:294–302.
75. Liesz A, Zhou W, Mracskó É, et al. Inhibition of lymphocyte trafficking shields the brain against deleterious neuroinflammation after stroke. *Brain*. 2011;134:704–720.
76. Becker K, Kindrick D, Relton J, Harlan J, Winn R. Antibody to the alpha4 integrin decreases infarct size in transient focal cerebral ischemia in rats. *Stroke*. 2001;32:206–211.
77. Relton JK, Sloan KE, Frew EM, Whalley ET, Adams SP, Lobb RR. Inhibition of alpha4 integrin protects against transient focal cerebral ischemia in normotensive and hypertensive rats. *Stroke*. 2001;32:199–205.

78. Langhauser F, Kraft P, Göb E, et al. Blocking of α4 integrin does not protect from acute ischemic stroke in mice. *Stroke*. 2014;45:1799–1806.
79. https://clinicaltrials.gov/ct2/show/NCT01955707?term=natalizumab +stroke&rank=1; Accessed 29.05.15.
80. Heesen C, Kleiter I, Nguyen F, et al. Risk perception in natalizumab-treated multiple sclerosis patients and their neurologists. *Mult Scler*. 2010;16:1507–1512.
81. Kieseier BC, Stüve O. A critical appraisal of treatment decisions in multiple sclerosis–old versus new. *Nat Rev Neurol*. 2011;7:255–262.

CHAPTER

17

Alemtuzumab (Campath-1H)

A. Coles, J. Jones, A. Compston
University of Cambridge, Cambridge, United Kingdom

OUTLINE

17.1 INTRODUCTION

Alemtuzumab was first used in people with multiple sclerosis (MS) in 1991, at a time when many innovative therapies were being evaluated as possible treatments for the disease: prominent at that time were trials of antibodies against T cells,[1,2] cyclophosphamide,[3] cyclosporin,[4] and cladribine.[5] Patients chosen for these trials already had a developed significant disability and were in the chronic progressive phase, an ambiguous classification that included both secondary and primary progressive disease. Results from these studies were mixed; for instance, the antibody trials were all regarded as negative whereas cyclophosphamide was thought to be effective at reducing disability, but with adverse effects too severe for regular use. Experience from using alemtuzumab first helped to clarify the distinction between mechanisms underlying relapsing and progressive disease; and the

Translational Neuroimmunology in Multiple Sclerosis
http://dx.doi.org/10.1016/B978-0-12-801914-6.00017-9

important clinical application that immunotherapies are effective only when given early in the relapsing-remitting phase. Reverse translation—understanding basic mechanisms by evaluating the experience of treating patients—is prominent in the narrative of alemtuzumab treatment of MS. Understanding symptom production in MS and the biology of lymphopenia-associated autoimmunity have all been advanced by studying the adverse effects of alemtuzumab.

Alemtuzumab (originally designated Campath-1H in the Department of Pathology in Cambridge, UK)[6] is the first therapeutic monoclonal antibody that was ever humanized and therefore usable, in principle, as a medicine. Targeting CD52, an antigen of unknown function, Campath-1H was found to effectively deplete human T and B lymphocytes. Originally used to prevent graft versus host disease in transplantation,[7] it was also explored in lymphoid malignancy[8] and then in autoimmunity[9,10] before being used in MS.

17.2 ALEMTUZUMAB INFORMS THE MECHANISM OF SYMPTOM PRODUCTION IN MULTIPLE SCLEROSIS

Alemtuzumab was first given as a single cycle of 20 mg daily infusions for 5 days. Within 1–2 h of starting the first infusion, recipients developed systemic symptoms including rash, bradycardia, hypotension, headache, nausea, and pyrexia. Unexpectedly, patients also experienced neurological symptoms. Lasting for hours only, and with full recovery, these constituted a rehearsal of symptoms experienced during earlier relapses, or a worsening of current fixed disabilities with, in one instance, reversible conduction block demonstrated on visual evoked potentials from an eye previously affected by demyelination.[11] More than just an Uhthoff phenomenon, due to a rise in temperature, the transient deterioration in symptoms was almost completely abolished by pretreatment with high-dose intravenous methylprednisolone, but not by concomitant infusion of a monoclonal antibody that inhibited the availability of soluble tumor necrosis factor (TNF)α. We concluded that some component (but not TNFα) of the cytokines released during alemtuzumab infusion leads to physiological conduction block at sites of previous demyelination, and speculated (at the suggestion of Ken Smith, UCL) that this might be nitric oxide but were unable to confirm this experimentally. Later studies showed that cytokine release results not from cell lysis by alemtuzumab, but from cross-linking-induced ligation of CD16 on natural-killer cells.[12]

After this experience, all patients were routinely pretreated with corticosteroids and antihistamines, and after some experimentation with dosing regimens, alemtuzumab is now licensed to be given as an infusion of 12 mg over 4 h, daily for 5 days at month 0 and then repeated for 3 days at month 12.

The phenomenon of symptom production with alemtuzumab, caused by cytokine-induced conduction block at sites of previous demyelination, suggests more generally that the dynamic of symptoms occurring during an acute MS relapse may be due more to the transient effect of soluble inflammatory mediators than demyelination; that is, a functional rather than a structural change. This also provides a possible explanation for how systemic inflammatory responses may cause deterioration in MS symptoms in the absence of focal inflammation in the brain.

17.3 THE FIRST USE OF ALEMTUZUMAB: TREATING INFLAMMATION IN PROGRESSIVE MULTIPLE SCLEROSIS

The first group of patients exposed to alemtuzumab had established secondary progressive disease, of 11 years' duration on average, and they had accumulated significant disability (mean Expanded Disability Status Scale (EDSS) 5.8 at time of treatment). A single cycle of alemtuzumab dramatically reduced relapses and MRI lesion formation for several years, but their disability nonetheless progressed.[13] Seven years after first treatment MRI scans showed no new lesion formation, attesting to alemtuzumab's efficacy at suppressing focal inflammation, but there had been progressive brain atrophy on MRI, correlating with worsening disability.[14] This dissociation challenged the concept that inflammation and plaque formation drive the accumulation of disability in MS. At the same time evidence was emerging from pathological studies and brain imaging that neuronal loss is the basis for disability and progression in chronic MS.[15,16]

Although these findings converged to suggest that postinflammatory neurodegeneration underlies disease progression in MS, some commentators retained the view, supported by epidemiological studies,[17,18] that inflammation and neurodegeneration are independent mechanisms contributing to the evolving and complex course of MS. We always preferred the explanation that neurodegeneration is secondary to persistent demyelination induced by inflammation, months or years earlier (it has been further suggested that progressive MS is caused by inflammation confined to the central nervous system (CNS), trapped behind a closed blood–brain barrier and therefore not accessible to immunotherapies that are delivered systemically[19]).

17.4 THE WINDOW OF THERAPEUTIC OPPORTUNITY IN MULTIPLE SCLEROSIS REVEALED BY ALEMTUZUMAB

The corollary of our original hypothesis was that immunotherapies would only be effective if given early in the relapsing-remitting phase of the disease. By this stage it was clear that alemtuzumab had significant adverse effects; despite inducing prolonged lymphopenia due to depletion rather than suppression of lymphocyte turnover and production, the most significant of these was not infection as initially predicted but, rather, thyroid autoimmunity.[20] Nonetheless, we felt it justifiable to study a small number of people with early relapsing-remitting disease. The first experience was encouraging. Twenty-two individuals with mean disease duration of 2.7 years and EDSS of 4.8 had a substantial reduction in relapse rate after receiving alemtuzumab, as expected from our previous experience with more advanced cases. However, in contrast to what had been seen with the progressive group, the new cohort also experienced a mean improvement in disability.[14] Over the next few years we continued to treat people with early relapsing-remitting disease using alemtuzumab, so that in 2015 we were able to publish a long-term follow-up of 87 patients who had a mean disease duration of 3.0 years and mean EDSS of 3.8 at last observation. At a median of 7 years from first treatment, 59/87 (68%) of patients still had an improved or unchanged disability compared with baseline.[21]

As these encouraging open-label data were emerging we proposed that future trials should be designed under the hypothesis that, "patients receiving effective immunological treatment before the cascade of events leading to uncontrolled destruction of the axon-glial unit is irretrievably established will not subsequently accumulate disability… develop cerebral atrophy or enter the secondary progressive phase of the illness." This was the basis for the design of one phase-II[22] and two phase-III trials of alemtuzumab,[23,24] sponsored at first by Ilex Oncology, then Genzyme (now a Sanofi company). This comprehensive program has a number of unique features. First, in order to ensure that only people with early MS were included, eligibility was limited to those with limited disease duration (3 years in CAMMS223, 5 years in CARE MS1, and 10 years in CARE MS2). Second, recognizing the importance that alemtuzumab should unequivocally demonstrate efficacy, we set a high hurdle by comparing it against an active comparator (interferon beta-1a SC) and by having a clinical disability and relapse rate as coprimary outcomes in all three trials and their extension phases. The disease characteristics of recruited cohorts, and results of the primary and extension outcome measures, are given in Table 17.1. Taken as a whole these trials demonstrate clear superiority of alemtuzumab over interferon beta-1a SC in the medium-term (2–3 years at first analysis) at reducing relapse rate, reducing the proportion of people developing worsening disability, and increasing the number who experience improvement in disability.

Extension studies of the sponsored trials have shown that these effects are maintained for 5 years (CAMMS223)[25] and 4 years (CARE MS1[26] and MS2[30]). Perhaps most impressively, in the 2 years of the phase-III trials, alemtuzumab significantly slowed the rate of cerebral atrophy compared to that seen in interferon beta-1a-treated patients (by 42% in CARE MS1, $p<0.0001$; and 24% in CARE MS2, $p=0.0121$) and this reduced rate of atrophy was maintained for a further 2 years despite the lack of need for retreatment with alemtuzumab in more than 70% patients.[27]

None of these data speak to the key question: Does early aggressive immunotherapy slow or prevent transition from the relapsing-remitting phase of the disease to secondary progression? This is not easy to answer: A robust test would require follow-up of a relapsing-remitting cohort for 20 years or more, and with placebo controls. The sponsored trials of alemtuzumab are too recent to be useful. Some answer comes from the open-label studies. Of the 87 patients in the alemtuzumab-treated group, just four (5%) fulfilled the criteria for secondary progression defined by worsening of EDSS on two consecutive sets of observation, each confirmed at 6 months, with the second event occurring from the new baseline EDSS established by the first.[21] In an attempt to put this in context, 87 people with MS were matched from a natural history cohort in Cardiff, UK (unpublished); of these, 19 (22%) fulfilled the criteria for secondary progression (odds ratio of 0.17, $p=0.013$ by Fisher's exact test). Over the next few years it will become clearer whether early alemtuzumab treatment of MS impacts on the transition to clinically progressive disease.

17.5 THE MECHANISM OF DISABILITY IMPROVEMENT AFTER ALEMTUZUMAB

The improvement in disability seen after alemtuzumab has not been described consistently after exposure to any other immunotherapy in MS. Several mechanisms may be responsible. Any physiological conduction block induced by soluble inflammatory mediators will be relieved by alemtuzumab. Remyelination may also be involved. In 20 patients

TABLE 17.1 Baseline Characteristics of Recruited Patients, and a Summary of the Key Results from the Industry-Sponsored Studies

	Phase 2 CAMMS 223[22] N=333	Phase 3 CARE MS1[23] N=581	Phase 3 CARE MS2[24] N=840
Previous DMT	No	No	Yes
Age, year mean (SD)	32.3 (8.5)	33.0 (8.2)	35.1 (8.4)
EDSS score mean (SD)	2.0 (0.7)	2.0 (0.8)	2.7 (1.2)
Disease duration, year median (range)	1.3 (0.1–6.3)	1.7 (0.1–6.0)	3.7 (0.2–16.9)
Relapses in past 2 years mean (range)	2.3 (1–7)	2.4 (1–7)	2.7 (1–9)
Pivotal results	**3-year study**	**2-year study**	
Relapse rate reduction	69%**	55%***	50%***
Annualized relapse rate (alemtuzumab vs interferon)	0.10 vs 0.36	0.18 vs 0.39	0.26 vs 0.52
Proportion relapse free	77% vs 52%**	78% vs 59%***	65% vs 47%***
% with sustained disability confirmed at 6 months	9% vs 26%**	8% vs 11% nonsignificant	13% vs 21%**
Change in mean EDSS from baseline	Improvement of 0.39 compared to deterioration of 0.38 on IFNb1a**	No significant change	Improvement of 0.17 compared to deterioration of 0.24 on IFNb1a***
Reduction in brain atrophy on alemtuzumab vs interferon		42%***	24%*
Extension	**5-year extension[25]**	**4-year extension[26–29]**	
N on alemtuzumab entering extension	107	325	361
% of patients not requiring retreatment	Not applicable[a]	74%	68%
% free from relapse	80%	76%	
% of patients with improved or stable EDSS	71%	73.5%	71.3%
% of patients with 6-month confirmed reduction in disability	Not available	30%	41%
MRI activity free at year 4	Not available	70%	70%

*, $p<0.05$; **, $p<0.001$; ***, $p<0.001$.

[a] *Retreatment rates were very low in this extension, largely because of a dosing suspension following the death of the first person affected by ITP in June 2005.*

after alemtuzumab, an MRI biomarker of remyelination (gray matter magnetization ratio) improved, unlike 18 controls on interferon beta.[31] This effect may not be specific for alemtuzumab, but rather may reflect the shorter disease duration of recruited patients which would lead to greater endogenous capacity for repair.

However, in a post hoc subgroup analysis of the CAMMS223 trial, we showed that among those participants with no clinical disease activity immediately before treatment, or any clinical or radiological disease activity on trial, disability improved after alemtuzumab but not following interferon beta-1a. This suggests that disability improvement after alemtuzumab may not solely be attributable to its antiinflammatory effect. We hypothesized that lymphocytes reconstituting after alemtuzumab may permit or promote brain repair. In that context, we showed that after alemtuzumab, and only when specifically stimulated with myelin basic protein (MBP), peripheral blood mononuclear cell cultures produced increased concentrations of brain-derived neurotrophic factor, platelet-derived growth factor, and ciliary neurotrophic factor.[32] Media from such cultures promoted survival of rat neurons and increased axonal length in vitro, effects that were partially reversed by neutralizing antibodies against brain-derived nerve growth factor and ciliary neurotrophic factor. These conditioned media also enhanced oligodendrocyte precursor cell survival, maturation, and myelination.

Presumably neurotrophins are released by lymphocytes after alemtuzumab as part of the homeostatic response to lymphopenia, although that does not easily explain the apparent MBP-specific response. Speculatively, MBP-specific lymphocytes could enter the CNS and promote repair, under the putative mechanism of neuroprotective autoimmunity.[33] Some support for this proposal comes from a small study ($n = 19$) comparing the MRI myelin water fraction of lesions acquired while on alemtuzumab therapy. Lesions occurring 6 months or more after treatment showed a recovery of myelin water fraction, not seen in lesions acquired earlier, suggesting there may be a delayed proremyelinating effect of alemtuzumab.[34]

17.6 LYMPHOPENIA, INFECTION, AND AUTOIMMUNITY

Alemtuzumab causes prolonged T-cell lymphopenia. In a study of the first people to receive alemtuzumab for MS, who had only one cycle a median of 12 years earlier, B-cell counts had returned to the lower limit of normal ($\geq 0.1 \times 10^9$/L) by 7 months. In contrast, median recovery times for $CD8^+$ (LLN $\geq 0.2 \times 10^9$ cells/L) and $CD4^+$ lymphocytes (LLN $\geq 0.4 \times 10^9$ cells/L) were 20 months and 35 months, respectively. However, T-cell counts rarely returned to their pretreatment levels; only 30% and 21% patients had recovered to baseline $CD8^+$ and $CD4^+$ counts, respectively.[35] Despite this, infections are not a prominent adverse effect of alemtuzumab; in the pivotal trials grade 1 or 2 infections, especially of the upper respiratory and urinary tracts, were only slightly increased compared to interferon controls. In the immediate month after alemtuzumab there was an increased risk of simple herpetic infections, especially cold sores, which was reduced by prophylactic acyclovir treatment. Infection rates tended to diminish with increasing time from the last infusion (Fig. 17.1). A small study of responses to vaccination confirmed that patients are immunocompetent after alemtuzumab, except in the immediate month after infusion.[36]

Paradoxically, in the late 1990s it became clear that one-third of people with MS treated using alemtuzumab developed autoimmune thyroid disease, predominantly Graves' hyperthyroidism,

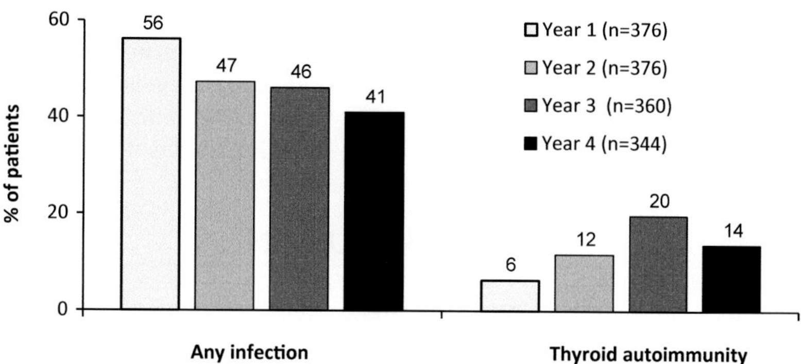

FIGURE 17.1 The yearly rate of any infective adverse event or new thyroid autoimmunity in the pivotal study and extension of CARE MS1.[26]

due to the production of antithyroid stimulating hormone receptor antibodies.[37] In 2005 a patient in the phase-II trial died of an intracranial hemorrhage due to thrombocytopenia; despite symptoms of easy bleeding for the previous 2 weeks he had not sought medical attention. Soon more cases of ITP following alemtuzumab were identified, at a current rate of 2%.[38] Seven cases of immune renal disease, including four examples of antiglomerular basement membrane disease (Goodpasture's syndrome, but without lung hemorrhage), have also been seen after alemtuzumab; two treated outside of the sponsored trials ended up requiring renal transplants despite prompt treatment, whereas the two treated in the phase-III trials, where there was monitoring for renal dysfunction, recovered full renal function after vigorous treatment.[39,40] There have been several isolated cases of autoimmune cytopenias, vitiligo, and alopecia. As of 2015 there have been no cases of diabetes, myasthenia, lupus, rheumatoid arthritis, or other systemic autoimmune disease.

This collection of autoimmune diseases is also seen after autologous bone marrow transplantation and antiretroviral treatment of HIV-AIDS. This and their timing, appearing typically 6–48 months after alemtuzumab (Fig. 17.1), suggest that this secondary autoimmunity arises not from a direct effect of alemtuzumab, but rather in response to reconstitution of the immune repertoire. Inadvertently we have generated a human equivalent of the lymphopenia-associated autoimmunity that is well studied in animal models. Initially we proposed there was a selective depletion of regulatory T cells after alemtuzumab, but the opposite seems to be true: for 6 months after treatment T cells with a regulatory phenotype dominate the depleted lymphocyte pool.[41] We are currently investigating whether they function as regulatory cells, or are defective in some way. Because the autoimmune manifestations are largely autoantibody-mediated, we examined B cells and B-cell activating factors after alemtuzumab[42] but found no association with autoimmunity.

Instead, there are T-cell changes associated with autoimmunity. We have shown that T-cell recovery after alemtuzumab is driven by homeostatic proliferation, leading to the generation of chronically activated ($CD28^-CD57^+$), highly proliferative ($Ki67^+$), oligoclonal, memory-like CD4 and CD8 T cells ($CCR7^-CD45RA^-$ or $CCR7^-CD45RA^+$) capable of producing proinflammatory cytokines. Individuals who develop autoimmunity after treatment are no more lymphopenic than their nonautoimmune counterparts, but they show reduced thymopoiesis and generate a more restricted T-cell repertoire. Taken together, these findings demonstrate that homeostatic proliferation drives lymphopenia-associated autoimmunity in humans.[43] IL-21, a cytokine involved in homeostatic proliferation, is found in high levels in pretreatment serum of people who subsequently develop autoimmunity after alemtuzumab[44] although the commercial assays currently available are insufficiently sensitive for this to be clinically useful.[45]

17.7 CONCLUSION

We take three main lessons from our 25-year experience of using alemtuzumab in MS:

- In the modern age the emergence of increasingly effective treatments for MS has been at the expense of more safety profiles—efficacy has been traded for safety.
- The treatment of MS is a highly specialist activity in which choice and decision depends on a meeting of two experts—the patient with MS and the neurologist with a special interest in the disease—who must debate the merits and demerits of the licensed therapies, weighing them in terms of efficacy, safety, convenience, and personal attitude to risk.
- Timing is not negotiable—this disease has to be treated early and aggressively—the comfort of safe but inadequate escalation therapy should now give way gracefully to induction and maintenance with close follow-up of cases to anticipate and manage risk.

More specifically, now that alemtuzumab is licensed in 40 countries worldwide including the European Union, United States and Canada, South America, the Middle East, Asia, China, and Australia and New Zealand, there is both a greater need to understand the mechanisms of its efficacy and safety profile and an increased opportunity to learn about the pathogenesis of MS and autoimmunity from its effects. The boundaries of the window of therapeutic opportunity for effective treatment of MS should be explored further, with the goal of optimizing disability reversal. The mechanisms underlying sustained improvement in disability require further investigation, and autoimmunity after alemtuzumab is an excellent opportunity to study lymphopenia-associated autoimmunity.

Acknowledgments

The authors have received research support, honoraria, and travel support, and the University of Cambridge has received gifts, from Genzyme (a Sanofi company). Much of this work is supported by the Cambridge Biomedical Research Centre.

References

1. Hafler DA, Fallis RJ, Dawson DM, Schlossman SF, Reinherz EL, Weiner HL. Immunologic responses of progressive multiple sclerosis patients treated with an anti-T-cell monoclonal antibody, anti-T12. *Neurology*. 1986;36(6):777–784.
2. Weinshenker BG, Bass B, Karlik S, Ebers GC, Rice GP. An open trial of OKT3 in patients with multiple sclerosis. *Neurology*. 1991;41(7):1047–1052.
3. Weiner HL, Mackin GA, Orav EJ, et al. Intermittent cyclophosphamide pulse therapy in progressive multiple sclerosis: final report of the Northeast Cooperative Multiple Sclerosis Treatment Group. *Neurology*. 1993;43(5):910–918.

4. Rudge P, Koetsier JC, Mertin J, et al. Randomised double blind controlled trial of cyclosporin in multiple sclerosis. *J Neurol Neurosurg Psychiatry*. 1989;52(5):559–565.
5. Sipe JC, Romine JS, Koziol JA, McMillan R, Zyroff J, Beutler E. Cladribine in treatment of chronic progressive multiple sclerosis. *Lancet*. 1994;344(8914):9–13.
6. Waldmann H, Hale G. CAMPATH: from concept to clinic. *Philos Trans R Soc Lond B Biol Sci*. 2005;360(1461): 1707–1711.
7. Waldmann H, Polliak A, Hale G, et al. Elimination of graft-versus-host disease by in-vitro depletion of alloreactive lymphocytes with a monoclonal rat anti-human lymphocyte antibody (CAMPATH-1). *Lancet*. 1984;2(8401): 483–486.
8. Hale G, Dyer MJ, Clark MR, et al. Remission induction in non-Hodgkin lymphoma with reshaped human monoclonal antibody CAMPATH-1H. *Lancet*. 1988;2(8625):1394–1399.
9. Lockwood CM. Monoclonal antibody therapy for vasculitis. *Adv Exp Med Biol*. 1993;336:235–238.
10. Lockwood CM, Thiru S, Isaacs JD, Hale G, Waldmann H. Long-term remission of intractable systemic vasculitis with monoclonal antibody therapy. *Lancet*. 1993;341(8861):1620–1622.
11. Moreau T, Coles A, Wing M, et al. Transient increase in symptoms associated with cytokine release in patients with multiple sclerosis. *Brain*. 1996;119(Pt 1):225–237.
12. Wing MG, Moreau T, Greenwood J, et al. Mechanism of first-dose cytokine-release syndrome by CAMPATH 1-H: involvement of CD16 (FcgammaRIII) and CD11a/CD18 (LFA-1) on NK cells. *J Clin Invest*. 1996;98(12):2819–2826.
13. Coles AJ, Wing MG, Molyneux P, et al. Monoclonal antibody treatment exposes three mechanisms underlying the clinical course of multiple sclerosis. *Ann Neurol*. 1999;46(3):296–304.
14. Coles AJ, Cox A, Le Page E, et al. The window of therapeutic opportunity in multiple sclerosis: evidence from monoclonal antibody therapy. *J Neurol*. 2006;253(1):98–108.
15. Ferguson B, Matyszak MK, Esiri MM, Perry VH. Axonal damage in acute multiple sclerosis lesions. *Brain*. 1997;120(Pt 3):393–399.
16. Trapp BD, Peterson J, Ransohoff RM, Rudick R, Mork S, Bo L. Axonal transection in the lesions of multiple sclerosis. *N Engl J Med*. 1998;338(5):278–285.
17. Vukusic S, Confavreux C. Prognostic factors for progression of disability in the secondary progressive phase of multiple sclerosis. *J Neurol Sci*. 2003;206(2):135–137.
18. Confavreux C, Vukusic S. Age at disability milestones in multiple sclerosis. *Brain*. 2006;129(Pt 3):595–605.
19. Frischer JM, Bramow S, Dal-Bianco A, et al. The relation between inflammation and neurodegeneration in multiple sclerosis brains. *Brain*. 2009;132(Pt 5):1175–1189.
20. Coles AJ, Wing M, Smith S, et al. Pulsed monoclonal antibody treatment and autoimmune thyroid disease in multiple sclerosis. *Lancet*. 1999;354(9191):1691–1695.
21. Tuohy O, Costelloe L, Hill-Cawthorne G, et al. Alemtuzumab treatment of multiple sclerosis: long-term safety and efficacy. *J Neurol Neurosurg Psychiatry*. 2015;86(2):208–215.
22. Coles AJ, Compston DA, Selmaj KW, et al. Alemtuzumab vs. interferon beta-1a in early multiple sclerosis. *N Engl J Med*. 2008;359(17):1786–1801.
23. Cohen JA, Coles AJ, Arnold DL, et al. Alemtuzumab versus interferon beta 1a as first-line treatment for patients with relapsing-remitting multiple sclerosis: a randomised controlled phase 3 trial. *Lancet*. 2012; 380(9856):1819–1828.
24. Coles AJ, Twyman CL, Arnold DL, et al. Alemtuzumab for patients with relapsing multiple sclerosis after disease-modifying therapy: a randomised controlled phase 3 trial. *Lancet*. 2012;380(9856):1829–1839.
25. Coles AJ, Fox E, Vladic A, et al. Alemtuzumab more effective than interferon beta-1a at 5-year follow-up of CAMMS223 clinical trial. *Neurology*. 2012;78(14):1069–1078.
26. Compston A, Giovannoni G, Arnold D, et al. Durable effect of alemtuzumab on clinical outcomes in treatment-naive relapsing-remitting multiple sclerosis patients: four-year follow-up of CARE-MS I (S4.007). *Neurology*. 2015;84(14 suppl).
27. Coles A, Arnold D, Cohen J, et al. Alemtuzumab slows brain volume loss over 4 years despite most relapsing-remitting multiple sclerosis patients not receiving treatment for 3 years (P7.263). *Neurology*. 2015;84(14 suppl).
28. Arnold D, Traboulsee A, Coles A, et al. Durable effect of alemtuzumab on MRI activity in treatment-naive active relapsing-remitting multiple sclerosis patients: 4-year follow-up of CARE-MS I (P7.246). *Neurology*. 2015;84(14 suppl).
29. Havrdova E, Giovannoni G, Arnold D, et al. Durable effect of alemtuzumab on clinical outcomes in patients with relapsing-remitting multiple sclerosis who relapsed on prior therapy: 4-year follow-up of CARE-MS II (P7.276). *Neurology*. 2015;84(14 suppl).

30. LaGanke C, Hughes B, Berkovich R, et al. Durable effect of alemtuzumab on disability improvement in patients with relapsing-remitting multiple sclerosis who relapsed on a prior therapy (P3.261). *Neurology*. 2015;84(14 suppl).
31. Button T, Altmann D, Tozer D, et al. Magnetization transfer imaging in multiple sclerosis treated with alemtuzumab. *Mult Scler*. 2013;19(2):241–244.
32. Jones JL, Anderson JM, Phuah CL, et al. Improvement in disability after alemtuzumab treatment of multiple sclerosis is associated with neuroprotective autoimmunity. *Brain*. 2010;133(Pt 8):2232–2247.
33. Moalem G, Gdalyahu A, Shani Y, et al. Production of neurotrophins by activated T cells: implications for neuroprotective autoimmunity. *J Autoimmun*. 2000;15(3):331–345.
34. Vavasour I, MacKay A, Li D, Laule C, Traboulsee A. Advanced imaging in lesion and normal-appearing white matter over 2 years in MS patients treated with alemtuzumab (S29.009). *Neurology*. 2015;84(14 suppl).
35. Hill-Cawthorne GA, Button T, Tuohy O, et al. Long term lymphocyte reconstitution after alemtuzumab treatment of multiple sclerosis. *J Neurol Neurosurg Psychiatry*. 2012;83(3):298–304.
36. McCarthy CL, Tuohy O, Compston DA, Kumararatne DS, Coles AJ, Jones JL. Immune competence after alemtuzumab treatment of multiple sclerosis. *Neurology*. 2013;81(10):872–876.
37. Costelloe L, Jones J, Coles A. Secondary autoimmune diseases following alemtuzumab therapy for multiple sclerosis. *Expert Rev Neurother*. 2012;12(3):335–341.
38. Cuker A, Coles AJ, Sullivan H, et al. A distinctive form of immune thrombocytopenia in a phase 2 study of alemtuzumab for the treatment of relapsing-remitting multiple sclerosis. *Blood*. 2011;118(24):6299–6305.
39. Clatworthy MR, Friend PJ, Calne RY, et al. Alemtuzumab (CAMPATH-1H) for the treatment of acute rejection in kidney transplant recipients: long-term follow-up. *Transplantation*. 2009;87(7):1092–1095.
40. Meyer D, Coles A, Oyuela P, Purvis A, Margolin DH. Case report of anti-glomerular basement membrane disease following alemtuzumab treatment of relapsing-remitting multiple sclerosis. *Mult Scler Relat Disord*. 2013;2(1):60–63.
41. Cox AL, Thompson SA, Jones JL, et al. Lymphocyte homeostasis following therapeutic lymphocyte depletion in multiple sclerosis. *Eur J Immunol*. 2005;35(11):3332–3342.
42. Thompson SA, Jones JL, Cox AL, Compston DA, Coles AJ. B-cell reconstitution and BAFF after alemtuzumab (Campath-1H) treatment of multiple sclerosis. *J Clin Immunol*. 2009;30(1):99–105.
43. Jones JL, Thompson SA, Loh P, et al. Human autoimmunity after lymphocyte depletion is caused by homeostatic T-cell proliferation. *Proc Natl Acad Sci USA*. 2013;110(50):20200–20205.
44. Jones JL, Phuah CL, Cox AL, et al. IL-21 drives secondary autoimmunity in patients with multiple sclerosis, following therapeutic lymphocyte depletion with alemtuzumab (Campath-1H). *J Clin Invest*. 2009;119(7):2052–2061.
45. Azzopardi L, Thompson SA, Harding KE, et al. Predicting autoimmunity after alemtuzumab treatment of multiple sclerosis. *J Neurol Neurosurg Psychiatry*. 2014;85(7):795–798.

CHAPTER

18

Currently Approved Injectable Disease-Modifying Drugs: Cytotoxic Immunosuppressive Drugs

G. Edan

University Hospital, Rennes, France; West Neurosciences Network of Excellence (WENNE), Rennes, France

OUTLINE

Translational Neuroimmunology in Multiple Sclerosis
http://dx.doi.org/10.1016/B978-0-12-801914-6.00018-0

18.1 INTRODUCTION

Mitoxantrone is the only cytotoxic immunosuppressive drug among the currently approved disease-modifying drugs (DMDs) for the treatment of multiple sclerosis (MS).

In the late 1980s, two cytotoxic immunosuppressive drugs, azathioprine and cyclophosphamide,[1,2] were evaluated in controlled phase-2 and -3 trials but failed to respond to the primary end point: demonstrating clinical efficacy on disability progression. Such results led to disappointing conclusions by some MS experts: "It might be that the immunologic alterations in MS are not pathologically related to the progression of the disease and that suppressing immune function is therefore of little help to patients."[3] Nevertheless, it is important to underlie that these phase-3 trials included patients at a late stage of the disease, most of them having a progressive phenotype and presenting moderate to severe disability (Table 18.1). No well-designed randomized phase-3 trials devoted specifically to relapsing-remitting MS (RRMS) or CIS have been conducted with these old immunosuppressive drugs. The only immunosuppressant that demonstrated clinical as well as MRI efficacies in the RRMS population was cladribine,[4] an oral immunosuppressant not FDA- or EMEA-approved for MS due to potential safety concerns after its long-term use.

Mitoxantrone, mitox (1,4-dihydroxy-5,8-bis([2-([2-hydroxyethyl]amino)ethyl]amino)-9,10-anthracenedione), molecular weight 517 Da, is a synthetic antineoplastic agent first discovered in 1978. Just after its introduction as a cytotoxic agent in cancer chemotherapy, it was found to be immunosuppressive. Mitox was first tested as a potential disease-modifying therapy in MS in 1990.[5] Subsequently, on the basis of two controlled efficacy studies,[6,7] the US Food and Drug Administration first approved mitox for worsening RRMS, secondary-progressive MS (SPMS), and progressive-relapsing MS in October 2000.

TABLE 18.1 Summary of the Clinical Impacts of Azathioprine and Cyclophosphamide in Two Randomized Placebo-Controlled Trials[1,2]

Nb Patients	Azathioprine 174 Patients	Placebo 180 Patients	Cyclophosphamide 54 Patients	Placebo 54 Patients
Mean age	39	38	41	42.5
Mean EDSS at entry	3.7	3.7	5.8	5.8
Mean EDSS deterioration	+0.62	+0.80	+0.81	+0.68
	p=NS		p=NS	

18.2 TRIALS SUPPORTING REGULATORY APPROVAL OF MITOXANTRONE TO TREAT MULTIPLE SCLEROSIS

18.2.1 The Phase-2 French and British Multicenter Controlled Trial of Mitox in Multiple Sclerosis with Active Disease (Table 18.2)[6]

Forty-two patients with RRMS and SPMS, two relapses with sequelae, or progression of ≥2 Expanded Disability Status Scale (EDSS) points in the preceding 12 months were enrolled in this trial. All patients initially received three monthly infusions of MP 1 g and had three monthly Gd-enhanced MRI scans. Patients who had at least one new active lesion on the baseline scans were randomly assigned to therapy, mitox 20 mg plus MP 1 g or MP 1 g alone, monthly for 6 months. Patients who initiated therapy completed monthly Gd-enhanced and T2-weighted scans.

Baseline clinical characteristics were similar in both groups. Six patients in the control group and four patients in the mitox group had SPMS. No mitox + MP recipient discontinued treatment. A significant treatment effect was observed on the primary end point, the proportion of patients by treatment group without new Gd-enhancing MRI lesions (Table 18.1). Treatment benefits were also observed on secondary end points, the mean

TABLE 18.2 The French and British Mitox Trial: Study Efficacy End Points

	Mitox ($n=21$)	Control ($n=21$)	p	Reduction vs Placebo
Clinical Impact				
Relapse rate	0.33	1.47	<0.001	−81%
% free of relapse	66%	33%	<0.001	
Nb of relapses treated with steroid	5	19	<0.01	−80%
1 pt worsening EDSS (nb of patients)	1	6	<0.01	−84%
1 pt improvement EDSS (nb of patients)	12	3	<0.01	
EDSS change	−1.1	+0.1	<0.05	
MRI Impact (Primary End Points)				
% of patients free of MRI activity at month 6	90%	30%	<0.001	66%
Mean number of accumulated T2 lesions between month 6 and baseline	1.1	5.5	<0.05	80%
Mean number of accumulated Gad$^+$ lesions on monthly MRI for 6 months	1.2	5.6	<0.001	84%
Mean number of Gad$^+$ lesions at month 6	0.1	2.9	<0.001	97%

number of new Gd-enhancing lesions at month 6 ($p<0.01$) and the mean number of new T2 lesions from baseline to the end of treatment ($p<0.01$). Globally there was an 85% reduction of new lesions in the mitox + MP group. Unblinded clinical assessments of the patients showed a benefit for mitox recipients. Improvements in mean EDSS scores from month 0 to months 2–6 were significant for mitox recipients (all $p<0.005$). In contrast, the MP recipients generally deteriorated. During the 2-month baseline period, the mitox + MP and MP recipients had annualized relapse rates of 3.1 and 2.9, respectively. These rates were similar for the 12 months preceding therapy (3.1 vs 2.4). During the treatment period, there were fewer relapses in the mitox + MP group as compared with the MP group (relapse rate 0.33 vs 1.47). This effect was even more pronounced during the last 4 months of treatment (1 relapse in mitox group vs 19 relapses in control group). During the treatment period, the proportion of exacerbation-free patients was 66% in the mitox + MP group and 33% in the MP group.

18.2.2 The Phase-3 Randomized, Double-Blind, Placebo-Controlled, Multicenter Trial of Mitox in Progressive Multiple Sclerosis (Table 18.3)[7]

In the Mitoxantrone in Multiple Sclerosis (MIMS) study,[7] 194 patients were enrolled between 1993 and 1997 at 17 centers in Belgium, Germany, Hungary, and Poland, and randomly assigned to treatment with mitox 12 mg/m^2 ($n=63$), 5 mg/m^2 ($n=66$), or placebo ($n=65$), administered intravenously every 3 months for 24 months. One hundred and ninety-one patients received at least one dose, and 188 patients completed at least one clinical evaluation and were available for efficacy analyses. All patients met the following entry criteria: age 18–55 years, documentation of stepwise progression (worsening RRMS) or gradual progression of disability with or without superimposed relapses (SPMS), EDSS 3.0–6.0, worsening of ≥1.0 EDSS point over 18 months prior to enrollment, no clinical relapse or treatment with glucocorticoids within 8 weeks of enrollment. Severe relapses were prospectively defined as the occurrence of new symptoms lasting for more than 48 h, with a change in Functional System score of more than two points or a deterioration of existing symptoms with a change of more than one point in at least one of the four following systems: pyramidal, brain stem, cerebellar, or visual. EDSS, Ambulation Index (AI), and Standardized Neurologic Status (SNS) scores were determined at each scheduled and unscheduled visit by a neurologist who was blinded to treatment assignment (assessing physician). A separate treating physician, not blinded to treatment assignment, performed all medical evaluations, reviewed laboratory data, adjusted the dose of study drug according to protocol, provided symptomatic therapies, and diagnosed and graded the severity of clinical relapses. Baseline clinical and MRI characteristics were similar for evaluable patients across treatment groups.

A significant treatment effect ($p<0.0001$) was detected with the primary efficacy outcome, a multivariate comparison of 12 mg/m^2 versus placebo on five clinical measures tested in one combined hypothesis of stochastic ordered alternatives (Table 18.3). The preplanned ordered analyses of each of the five components of the composite outcome showed significant treatment effects for change in EDSS, change in AI, number of relapses treated with corticosteroids, time to first severe relapse (treated with corticosteroids), and change in SNS. Time to first severe relapse differed significantly between the placebo and 12-mg/m^2 mitox groups ($p=0.0004$, log rank test). The median time to the first severe relapse was 14.2 months for the placebo group, but

TABLE 18.3 MIMS Study Primary Efficacy End Points

	Number of Patients			
Variable	**Placebo ($n=64$)**	**Mitoxantrone 5 mg/m² ($n=64$)**	**Mitoxantrone 12 mg/m² ($n=60$)**	p^{a}
EDSS deterioration (>1 from baseline)	16 (25%)	10 (16%)	5 (8%)	0.013[b]
3-month confirmed EDSS deterioration during study	14 (22%)	9 (14%)	5 (8%)	0.036[b]
Without relapses	23 (36%)	25 (39%)	34 (57%)	0.021[b]
Adjusted total relapses irrespective of severity[c]	129.4	77.4	48.2	0.0002
Admitted to hospital	43 (67%)	36 (56%)	24 (40%)	0.002[b]
Total admissions[c]	89	85	50	0.0082[b]
Total MS-related admissions[c]	81	64	34	0.0018[b]

Total for group, not number of patients, Wilcoxon–Mann–Whitney test.
[a]*MS = multiple sclerosis.*
[b]*Placebo versus mitoxantrone 12 mg/m².*
[c]*χ^2 test.*

was not reached in 24 months by either mitox group. A highly significant difference ($p=0.005$) was also demonstrated for the 5-mg/m² group as compared with placebo with the multivariate efficacy analysis. One hundred and thirty-eight of 188 patients (73%) who were included in the intent-to-treat analysis of efficacy at 24 months (24-month cohort) completed an additional clinical evaluation at 36 months (36-month cohort) for safety assessment. Comparing disability levels at 36 months relative to baseline, the mean EDSS change was 0.10 (SD = 1.22) in the 12-mg/m² group and 0.46 in placebo recipients. Six of 42 (16.2%) 12-mg/m² recipients and 16 of 40 (42.1%) placebo recipients deteriorated by at least one point on the EDSS. Similarly, the mean change in AI was 0.61 (±1.78) in the 12 mg/m² and 1.13 (±1.64) in the placebo group. The mean change in SNS was 0.19 (±10.00) and 3.28 (±9.08), respectively. The number of severe relapses decreased from 66 in placebo to 26 in 12-mg/m² recipients.

Significant treatment effects were observed for most of the preplanned secondary outcomes of efficacy. Treatment effects for the 5-mg/m² recipients were generally intermediate between those observed in 12 mg/m² and placebo recipients. The difference between groups in EDSS change at 24 months reflected fewer patients demonstrating deterioration of at least one point (25% for placebo and 8% for 12 mg/m², $p=0.03$). Over 24 months, confirmed neurological progression was observed in significantly fewer patients receiving 12 mg/m² relative to placebo (5 (8.3%) vs 14 (22%), $p=0.04$). Patients in the 12-mg/m² mitox group showed a significant advantage in the analysis of time to confirmed EDSS deterioration at 3 months ($p=0.027$) and 6 months ($p=0.034$). Annualized relapse rates were significantly lower in the 12-mg/m² group relative to placebo at year 1 (0.42 vs 1.15, $p<0.0001$) and year 2 (0.27 vs 0.85, $p=0.0001$), a reduction by 63% and 68%, respectively. Moreover, significantly more patients in the 12-mg/m^2 group did not experience any relapse over 24 months relative to the placebo group (34 (57%) vs 23 (36%), $p=0.021$).

Significantly more patients in the placebo group were hospitalized for reasons other than administration of study medication. Only 15 patients showed progression that required the use of a wheelchair (corresponding to an EDSS of 7.0). No significant difference between groups was apparent, but fewer 12-mg/m^2 recipients than placebo recipients progressed to EDSS 7.0 (3 (5%) vs 7 (11%), $p=0.23$). Quality of life assessment was conducted with the validated Stanford Health Assessment Questionnaire (HAQ). The placebo group mean score increased (0.26), with significantly less change observed in the 12-mg/m^2 mitox group (0.09; $p=0.024$). Moreover, significantly more patients in the placebo group ($n=41$) showed deterioration in HAQ index relative to the 12-mg/m^2 mitox group ($n=25$, $p=0.012$).

18.3 CONTROLLED TRIALS FOLLOWING APPROVAL OF MITOXANTRONE TO TREAT MULTIPLE SCLEROSIS

18.3.1 Glatiramer Acetate After Induction Therapy with Mitox in Relapsing Multiple Sclerosis[8]

In this controlled study,[8] 40 relapsing MS patients with 1–15 Gd-enhancing lesions on screening brain MRI and EDSS score 0–6.5 were randomized to receive short-term induction therapy with mitox (three monthly 12 mg/m^2 infusions) followed by 12 months of daily glatiramer acetate (GA) therapy 20 mg/day subcutaneously for a total of 15 months (M-GA, $n=21$) or daily GA 20 mg/day for 15 months (GA, $n=19$). MRI scans were performed at months 6, 9, 12, and 15. The primary measure of outcome was the incidence of adverse events; secondary measures included a number of Gd-enhancing lesions, confirmed relapses, and EDSS change. Except age, baseline demographic characteristics were well matched in the two treatment arms. Both treatments were safe and well tolerated. M-GA induction produced an 89% greater reduction (relative risk 0.11, 95% confidence interval (CI) 0.04–0.36, $p=0.0001$) in the number of Gd-enhancing lesions at months 6 and 9 and a 70% reduction (relative risk 0.30, 95% CI 0.11–0.86, $p=0.0147$) at months 12 and 15 versus GA alone. Mean relapse rates were 0.16 and 0.32 in the M-GA and GA groups, respectively. Short-term immunosuppression with mitox followed by daily GA for up to 15 months was found to be safe and effective with an early and sustained decrease in MRI disease activity.

18.3.2 Interferon Beta After Induction Therapy with Mitox in Relapsing Multiple Sclerosis[9]

A consortium of Italian and French academic neurologists conducted a trial[9] aiming to determine whether a treatment strategy combining induction treatment with mitox prior to IFNβ-1b can delay disability progression over 3 years as compared with IFNβ-1b alone in patients with aggressive RRMS. This randomized controlled trial included patients with RRMS who had experienced at least two relapses with incomplete recovery the previous year, and who displayed Gd^+ lesions on MRI. Fifty-five patients were randomly allocated to the 6-month mitoxantrone induction regimen used in the previous studies, followed by a 3-month wash-out period, then interferon beta-1b (Table 18.2). The other study arm (54 patients) received interferon beta-1b for 3 years, combined with methylprednisolone at 1 g for the first 6 months. The two patient groups had comparable clinical and demographic features at inclusion. Patients underwent a complete neurological examination every 3 months, and spin-echo MRI at inclusion and at months 9, 24, and 36.

Compared with the interferon beta group, the time to sustained worsening of EDSS by one point was delayed in the mitoxantrone group ($p<0.012$). The 3-year risk of sustained worsening of disability was reduced by 65% (12% vs 34%). Mitoxantrone delayed the time of confirmed worsening of disability by 18 months, compared with the interferon beta group. The raw percentage of patients with sustained worsening of disability was also reduced (9.1% vs 25.9%). The mean EDSS score had improved by 0.45 point at the last observation point (M36) in the mitoxantrone group (from 4.16 to 3.66, $p=0.007$), but remained unchanged in the interferon beta group (from 3.86 to 3.76, $p=0.771$). Post hoc analyses suggested that this induction strategy using mitoxantrone is more effective in patients with a lower disability: the risk of sustained worsening of disability was reduced by 86.9% in patients with a low baseline EDSS (<4). Throughout the study, from baseline to month 36, the mitoxantrone group enjoyed a lower annualized relapse rate both per patient (0.4 vs 1.1, $p<0.03$) and per group ($p<0.001$). The time to first relapse after treatment institution was delayed in the mitoxantrone group by 21 months, compared with the interferon beta group ($p<0.001$). The proportion of patients who remained free of relapses throughout the follow-up period was 53% in the induction group and 26% in the monotherapy group ($p<0.01$).

The mean number of new lesions on T2 MRI was significantly lower in the mitoxantrone group at each of the measurement points. The mean cumulative number of new T2 lesions over 36 months was lower in patients in the mitoxantrone group than in patients in the interferon beta group (3.6 vs 9.8, $p=0.04$). The mean number of Gd^+ lesions was also lower in the mitoxantrone group at month 9 (0.36 vs 2.1, $p=0.012$), and the percentage of patients without any Gd^+ lesions was higher (88% vs 57% at month 9, $p=0.010$). This study demonstrates the usefulness of mitox prior to IFNβFNi as an induction therapy in aggressive RRMS patients.

18.4 TOLERABILITY OF MITOXANTRONE (TABLE 18.4)

Substantial tolerability data are available from oncology studies in which mitox was generally used in combination and from MS studies in which mitox was used as monotherapy (Table 18.4).

18.4.1 Cardiotoxicity

A well-recognized side effect related to mitox is the reduction of the left ventricular ejection fraction (LVEF) and the occurrence of congestive heart failure. Importantly, this is a cumulative dose risk. In oncology, this risk is associated with a prior treatment with anthracyclines, prior mediastinal radiotherapy, preexisting cardiovascular disease, and the

TABLE 18.4 Summary of Safety Profile in 802 Multiple Sclerosis Patients Over 5 Years After Use of Mitox (Mean Cumulative Dose 70 mg/m^2)[11]

Side Effect		Frequency *N* (%)
Cardiac	Symptomatic cardiac failure	1/802 (0.1%)
	Decrease of the LVEF<50%	39/794
	Transitory decrease of the LVEF	27/39 (69%)
	Persistent decrease of the LVEF	11/39 (28%)
Hematological	Acute leukemia	2/802 (0.25%)
Reproductive functions	Amenorrhea <35 years	9/167 (5.4%)
	Amenorrhea >35 years	46/150 (30.7%)
	Teratogenicity/52 births	0
Death attributed to mitox	I Case (leukemia)	1/802 (0.1%)

Mitox, mitoxantrone; *LVEF*, left ventricular ejection fraction.

cumulative dose administered. In MS, Ghalie et al.[10] reported two fatal cases of congestive heart failure in the pooled data on 1378 patients with a follow-up of 4084 patient-years (0.15%; 95% CI 0.02–0.52%). In the French prospective safety study of 802 MS patients treated with mitox[11] all 802 patients had clinically healthy cardiac conditions before the first course of mitox. Within the 5 years following mitox treatment, only one, a 54-year-old woman, presented with acute heart failure (0.1%). In the US prospective RENEW study,[12] 509 MS patients received a mean of six infusions of 12 mg/m^2 and were prospectively assessed for at least 5 years after the initiation of mitox. Among them, 10 cases of congestive heart failure occurred (2%) but the American cohort might have more additional cardiovascular risk factors than the French cohort (0.1%). The prevalence of cardiotoxicity of mitox occurring in these cohorts should also be compared with the risk of heart failure in the general European population, which is estimated from 0.4% to 2%.[26] In the Ghalie study[10] asymptomatic decrease in the LVEF <50% tended to be higher when the cumulative dose of mitox was ≥100 mg/m^2 (5%) versus <100 mg/m^2 (1.8%).

In the prospective RENEW study,[12] about 5% of the MS population treated with mitox presented an asymptomatic decrease in the LVEF < 50% during or after mitox treatment, similar to the French cohort (4.9%). More importantly, in the French cohort, in a majority of the cases this decrease of LVEF was transient (69%) and rarely permanent (28%), both during (46%) or after (54%) mitox treatment.[11] The definition of cardiotoxicity by a decrease of the LVEF was inconsistent between studies reporting mitox cardiac toxicity. An absolute reduction >10% between two evaluations has been considered by some authors (Kingwell et al.,[13] Marriott et al.[14]) leading to estimate the risk of cardiotoxicity at about 12% of the treated population. This estimation might be uncertain, however, because of the possible variations of the measure, especially when the LVEF is superior to 50%. Indeed in the RENEW study,[12] among the 509 patients regularly monitored during treatment, if a decrease of the LVEF by 10% was observed in 10% of patients, an increase by 10% of the

LVEF was observed in 8.4% of patients. A decrease of the LVEF under 50% might be the best red flag to indicate the risk of cardiac side effect. *We recommend monitoring carefully cardiac tolerance of MS patients receiving mitoxantrone with an echocardiogram before each perfusion and yearly up to 5 years after MITOX withdrawal.*

18.4.2 Risk of Therapy-Related Acute Myeloid Leukemia

The risk of therapy-related acute myeloid leukemia (AML) when using mitox is known in oncology where mitox was administered as combination therapy with other chemotherapeutic agents with or without radiotherapy, making it difficult to estimate its precise contribution to the risk. In this context, Ghalie et al.[15] pooled the data on 2973 patients from seven independent oncological studies: 31 patients (1.04%) developed therapy-related AML within 4 years after starting on mitox. Marriott et al.[14] pooled data from sporadic and retrospective studies and calculated an incidence of 33/4076 (about 0.8%). Martinelli et al.[16] reported a retrospective study of 3220 MS patients from 40 Italian MS centers and found an incidence of AML of 0.9% and a mortality rate of 36% of patients with AML. In the prospective RENEW US study, two therapy-related AMLs (0.4%) and one chronic myeloid leukemia were reported. The association between the hematologic risk and the cumulative dose of mitox administered is unclear, but a higher risk at a cumulative dose of more than 60 mg/m^2 was reported by Ellis and Boggild.[17] *We recommend performing white blood cell count tests before each perfusion and every 3 months for 5 years after mitoxantrone withdrawal.*

18.4.3 Gonadal Dysfunction and Fertility

In men treated for Hodgkin's disease, mitox in combination with other chemotherapeutic agents (vincristine, vinblastine, prednisone) caused significant decreases in sperm counts and motility, but these parameters usually recovered within 3–4 months after completion of chemotherapy,[18] and there is generally complete recovery of sperm production without morphological changes in vitro or genotoxic effects on germinal cells in vivo.

In women treated with mitoxantrone, secondary amenorrhea can be a delayed side-effect of chemotherapy. This risk was evaluated in two prospective MS cohorts. In the French prospective study[11] of 802 MS patients, transient chemotherapy-induced amenorrhea (CIA) (duration several months) was reported in 27% and persistent CIA in 17.3%. The frequency of persistent CIA increased with increasing age. Persistent CIA was not observed in women treated before the age of 25 years (0/54 MS women) and in few women <30 years (2/99). So, in women <35 years old, the global risk of persistent CIA (9/167, 5.4%) was much lower than in women >35 years old (46/150, 30.7%). Furthermore, 27 women gave birth to 32 normal babies (66% were females), between 1 and 7 years after their last course of mitox. Fifteen men fathered 20 normal babies (50% were females). The FEMIMS[19] study reported a cohort of 189 MS women (mean age 37 years old), followed for a median period of 26 months after mitox administration, a global CIA risk of 26% that varies with dose, age, and estroprogestinic (EP) treatment. The probability of CIA increased by 2% for each increase of 1 mg/m^2 cumulative dose and by 18% for each year of age. Interestingly, EP treatment exerted a protective effect against CIA. At the age of 35 years, EP decreased the risk of permanent amenorrhea from 20% to 7% at the cumulative dose of 60 mg/m^2. Compared with the French cohort of 317 women aged <45 years at mitox start, the global relative risk of persistent amenorrhea was lower (17.3% in the French study vs 26% in the FEMIMS study). For that reason, *we recommend performing cryopreservation of ovarian tissue in nulliparous women before starting mitoxantrone.*

18.5 DISCUSSION ABOUT THE USE OF MITOXANTRONE IN MULTIPLE SCLEROSIS

Data from clinical trials indicate consistently that mitox is an effective and generally well-tolerated disease-modifying therapy for patients with MS. Benefit has been shown for relapse rate, progression of disability, and MRI activity in controlled clinical trials[6–9,20] and in observational studies.[21–23] Nonetheless, several important questions remain concerning the safety and use of mitox in MS.

18.5.1 What Is the Long-Term Safety Profile of Mitox in Multiple Sclerosis?

Although mitox is generally well tolerated, and the risk of clinically significant cardiac dysfunction and therapy-related AML have been low in clinical trials, the long-term risk of these potential drug-related side effects needs to be considered before its use. Such long-term data were collected prospectively in two studies that started in 2001: one study from 802 MS patients followed at least 5 years after mitox treatment onset in an open-label study conducted by the consortium of French MS centers,[11] and a US registry study[12] from 504 MS patients followed for 5 years. It is important to recall that in these two studies, the mean cumulative dose of mitoxantrone was moderate (about 70 mg/m^2) and that the magnitude of side effects was dose-dependent. They gave the magnitude of the risks (Table 18.3) and helped to determine the risk/benefit ratio of the drug that should be compared with other long-term DMDs (Table 18.4).

18.5.2 The Role of Mitoxantrone for Patients With Primary- or Secondary-Progressive Multiple Sclerosis

The potential relationship between inflammatory aspects of MS pathology (focal vs diffuse inflammation; innate vs adaptive immunology) and disease-modifying treatment effects was not fully appreciated in any of the published controlled trials. We believe that the efficacy of mitox in progressive MS (with or without relapses) can only be answered definitively by a study designated to answer that question.

18.5.3 The Best Use of Mitoxantrone in the Therapeutic Strategy: Escalation or Induction?[24]

From a theoretical point of view, there are two contrasting treatment regimens in MS induction versus escalating. The rationale behind escalating therapy is that treatment starts with safe drugs, and moves on to more aggressive ones only if the ongoing treatment fails. In the escalating approach, glatiramer acetate, interferon betas, and tecfidera and teriflunomide are regarded as first-line drugs; immunosuppressants (mitoxantrone, natalizumab, fingolimod, alemtuzumab) as second-line ones; and very intensive immunosuppression (autologous bone marrow transplantation, high-dose cyclophosphamide) as third-line ones. The key to the success of escalation therapy is to define up front with the patient the exact suboptimal response threshold at which the next-level therapeutic option should be introduced. The decision to adopt a second-line therapy in patients who respond poorly to first-line therapy should not be delayed until severe and irreversible disability has set in. Given that all the immunosuppressants that are currently available present potentially serious side effects, the induction strategy has generally been reserved for patients with very active and aggressive disease. In these patients, there is an acknowledged risk of early disability, and once neurological function is lost, it cannot be regained. In such patients, this disease-inherent risk can be assumed to outweigh that associated with the use of powerful immunosuppressants. This treatment strategy involves the use of immunosuppressants for the minimum amount of time needed to gain adequate control over disease activity. Once disease control has been achieved, treatment can be switched to maintenance therapy with a better tolerated drug. This approach may be a useful and conservative means of using these highly effective therapies while minimizing exposure and the attendant safety risk.

An observational study on the long-term impact of this induction strategy with mitoxantrone (6 monthly courses) was performed in our center.[22] Prospectively 100 consecutive very active relapsing MS patients were assessed over a 5-year period. The majority of patients (73%) were assigned to a maintenance treatment following induction with mitoxantrone. In the year preceding treatment, the annualized relapse rate in the patient group was 3.3, and the EDSS score progressed by a mean of 2.2 points. In addition, Gd^+ lesions were visible on MRI for 84% of the sample. These patients thus presented very active disease.

In the first year following the initial administration of mitoxantrone, the relapse rate declined to 0.29 relapses per year, a reduction of 91%. This reduction was maintained across the 5-year observation period, oscillating between 0.3 and 0.4 relapses/year. Around one-third of patients remained free of relapses throughout the 5-year period, and the median time to first relapse was 2.72 years. Likewise, the proportion of patients whose disability deteriorated by at least one point on the EDSS, confirmed at 3 months, decreased from 88% in the year preceding treatment to 5% for the first year. Furthermore, the clinical benefit at 1 year in terms of disability was maintained in 59% of patients throughout the 5-year observation period.

Potential determinants of a good treatment response were assessed. The five patients whose disability worsened during the first year of treatment were significantly ($p<0.02$) older when treatment was started (41 years) than those whose disability stabilized or improved (32 years). The 20 patients who converted to SPMS during the follow-up period were significantly ($p<0.02$) more disabled (mean EDSS score: 4.8) at the start of treatment than those who did not (mean EDSS score: 3.9).

Apart from clinically aggressive RRMS[24,25] for which induction treatment should certainly be regarded as the first line of treatment, there is a lack of biomarkers to guide early choices between induction or escalating strategies at an individual level. In the coming years, new MRI techniques (brain and spinal cord imaging) should help us to identify those RRMS patients, especially individuals without any real disability, who are most at risk of developing

TABLE 18.5 Clinical and MRI Definition of Aggressive Relapsing-Remitting Multiple Sclerosis Suitable for an Induction Strategy[24]

1. Pure RRMS
2. Age <40 years
3. Highly active disease with at least two or more relapses within the previous 12 months
4. Severe relapse resulting in EDSS score ≥4
5. Worsening EDSS score due to relapses (increase of 2 or more points within the previous 12 months)
6. Two or more additional lesions on MRI (new T2 or Gad^+ T1 lesions)

RRMS, relapsing remitting multiple sclerosis; *EDSS*, Expanded Disability Status Scale.

destructive CNS lesions with or without first-line therapy, and who are therefore more eligible for an early and more aggressive treatment strategy (Table 18.5).

18.6 CONCLUSION

The available data from studies in patients with MS suggest that mitox may have a role to play in the management of aggressive disease, particularly as an induction therapy. The key remaining question is whether cytotoxic agents offer real advantages over more recent therapies that have more specific, targeted effects on the immune system.

References

1. Double-masked trial of azathioprine in multiple sclerosis. British and Dutch Multiple Sclerosis Azathioprine Trial Group. *Lancet*. 1988;23(2):179–183.
2. The Canadian cooperative trial of cyclophosphamide and plasma exchange in progressive multiple sclerosis. The Canadian Cooperative Multiple Sclerosis Study Group. *Lancet*. 1991;23(337):441–446.
3. Goodin DS. The use of immunosuppressive agents in the treatment of multiple sclerosis: a critical review. *Neurology*. 1991;41:980–985.
4. Giovannoni G, Comi G, Cook S, et al. A placebo-controlled trial of oral cladribine for relapsing multiple sclerosis. *N Engl J Med*. 2010;362:416–426.
5. Gonsette RE, Demonty L. Immunosuppression with mitoxantrone in multiple sclerosis: a pilot study for 2 years in 22 patients (P573). *Neurology*. 1990;40(suppl 1):261.
6. Edan G, Miller D, Clanet M, et al. Therapeutic effect of mitoxantrone combined with methylprednisolone in multiple sclerosis: a randomised multicentre study of active disease using MRI and clinical criteria. *J Neurol Neurosurg Psychiatry*. 1997;62:112–118.
7. Hartung HP, Gonsette R, König N, et al. Mitoxantrone in Multiple Sclerosis Study Group (MIMS). Mitoxantrone in progressive multiple sclerosis: a placebo-controlled, double-blind, randomised, multicentre trial. *Lancet*. 2002;360:2018–2025.
8. Vollmer T, Panitch H, Bar-Or A, et al. Glatiramer acetate after induction therapy with mitoxantrone in relapsing multiple sclerosis. *Mult Scler*. 2008;14:663–670.
9. Edan G, Comi G, Le Page E, Leray E, Rocca MA, Filippi M. French–Italian Mitoxantrone Interferon-beta-1b Trial Group. Mitoxantrone prior to interferon beta-1b in aggressive relapsing multiple sclerosis: a 3-year randomised trial. *J Neurol Neurosurg Psychiatry*. 2011;82:1344–1350.
10. Ghalie RG, Edan G, Laurent M, et al. Cardiac adverse effects associated with mitoxantrone (Novantrone) therapy in patients with MS. *Neurology*. 2002;59:909–913.
11. Le Page E, Leray E, Edan G. French Mitoxantrone Safety Group. Long-term safety profile of mitoxantrone in a French cohort of 802 multiple sclerosis patients: a 5-year prospective study. *Mult Scler*. 2011;17:867–875.

12. Rivera VM, Jeffery DR, Weinstock-Guttman B, Bock D, Dangond F. Results from the 5-year, phase IV RENEW (Registry to Evaluate Novantrone Effects in Worsening Multiple Sclerosis) study. *BMC Neurol*. 2013;13:80.
13. Kingwell E, Koch M, Leung B, et al. Cardiotoxicity and other adverse events associated with mitoxantrone treatment for MS. *Neurology*. 2010;74:1822–1826.
14. Marriott JJ, Miyasaki JM, Gronseth G, O'Connor PW, Therapeutics and Technology Assessment Subcommittee of the American Academy of Neurology. Evidence Report: the efficacy and safety of mitoxantrone (Novantrone) in the treatment of multiple sclerosis: Report of the Therapeutics and Technology Assessment Subcommittee of the American Academy of Neurology. *Neurology*. 2010;74:1463–1470.
15. Ghalie RG, Mauch E, Edan G, et al. A study of therapy-related acute leukemia after mitoxantrone therapy for multiple sclerosis. *Mult Scler*. 2002;8:441–445.
16. Martinelli V, Cocco E, Capra R, et al. Incidence rate of acute myeloid leukemia and related mortality in Italian MS patients treated with mitoxantrone. *Mult Scler*. 2010;16:S160.
17. Ellis R, Boggild M. Therapy-related acute leukemia with mitoxantrone: what is the risk and can we minimize it? *Mult Scler*. 2009;15:505–508.
18. Meistrich ML, Wilson G, Mathur K, et al. Rapid recovery of spermatogenesis after mitoxantrone, vincristine, vinblastine, and prednisone chemotherapy for Hodgkin's disease. *J Clin Oncol*. 1997;15:3488–3495.
19. Cocco E, Sardu C, Gallo P, et al. Frequency and risk factors of mitoxantrone-induced amenorrhea in multiple sclerosis: the FEMIMS study. *Mult Scler*. 2008;14:1225–1233.
20. Millefiorini E, Gasperini C, Pozzilli C, et al. Randomized-placebo-controlled trial of mitoxantrone in relapsing-remitting multiple sclerosis. 24 month clinical and MRI outcome. *J Neurol*. 1997;244:153–159.
21. Debouverie M, Taillander L, Pittion-Vouyovitch S, et al. Clinical follow-up of 304 patients with multiple sclerosis three years after mitoxantrone treatment. *Mult Scler*. 2007;13:626–631.
22. Le Page E, Leray E, Taurin G, et al. Mitoxantrone as induction treatment in aggressive relapsing remitting multiple sclerosis: treatment response factors in a 5 year follow-up observational study of 100 consecutive patients. *J Neurol Neurosurg Psychiatry*. 2008;79:52–56.
23. Esposito F, Radaelli M, Martinelli V, et al. Comparative study of mitoxantrone efficacy profile in patients with relapsing-remitting and secondary progressive multiple sclerosis. *Mult Scler*. 2010;16:1490–1499.
24. Edan G, Le Page E. Induction therapy for patients with multiple sclerosis: why? When? How? *CNS Drugs*. 2013;27:403–409.
25. Boster A, Edan G, Frohman E, et al. Intense immunosuppression in patients with rapidly worsening multiple sclerosis: treatment guidelines for the clinician. *Lancet Neurol*. 2008;7:173–183.
26. The Euro Heart Failure Survey Programme. A survey on the quality of care among patients with heart failure in Europe. *Eur. Heart J*. 2003;24:422–474.

SECTION V

CURRENTLY APPROVED DMDs-ORAL

CHAPTER

19

Fingolimod (Gilenya®)

S. Schippling

Neuroimmunology and Multiple Sclerosis Research, Department of Neurology, University Hospital Zurich and University of Zurich, Zurich, Switzerland

OUTLINE

19.1 INTRODUCTION

Fingolimod (2-amino-2-[2-(4-octylphenyl)ethyl]-1,3-propanediol hydrochloride) is a chemical derivative of myriocin, which itself is a metabolite of the fungus *Isaria sinclairii*.[1,2] Japanese scientists first synthesized fingolimod in 1992, as an attempt to reduce the toxicities that myriocin, a highly immunosuppressive substance, showed in vivo.[3,4] Fingolimod was first tested clinically as a new immunosuppressant to prevent graft rejection in patients following renal transplantation. As a result of these trials, however, Fingolimod turned out not to be superior to

Translational Neuroimmunology in Multiple Sclerosis
http://dx.doi.org/10.1016/B978-0-12-801914-6.00019-2

the existing classical immunosuppressants used in this indication.[5,6] Using lower doses, fingolimod's effects seemed to be rather immunomodulating than immunosuppressive, which, in the end, led to it being further evaluated in the field of multiple sclerosis (MS).

Here, it turned out to be a first-in-class, orally available sphingosine 1-phosphate (S1P) receptor modulator and functional antagonist that was highly efficacious in phase-2 as well as in three major phase-3 trials, both against placebo and an active comparator (INF-β-1a i.m.)

Consequently, in September 2010, fingolimod was approved by the US Food and Drug Administration (FDA) and became the first oral drug available for the treatment of relapsing-remitting forms of MS. Approvals in the European Union and several other countries followed.

19.2 MODE OF ACTION OF FINGOLIMOD

Sphingolipids represent major components of cell membranes as well as of myelin in the central nervous system (CNS). During degradation of sphingomyelin, a major sphingolipid, sphingosine is formed. In vivo, sphingosine is phosphorylated by sphingosine kinases (SphK) to S1P.[7] Physiologically, S1P binds to five different G-protein-coupled S1P receptor isoforms, the S1P receptor subtypes 1–5.[7] S1P signaling is involved in a variety of processes within different cells of the human system, given its wide expression. In the immune system, the cardiovascular system, and the CNS the receptor subtypes S1P1, S1P2, and S1P3 are abundantly expressed. S1P1 can also be found on lymphocytes/leukocytes.[8] Adaptive immune responses depend on the circulation of lymphocytes between lymphoid tissues and the periphery. In the immune system, retention and egress of lymphocytes from secondary lymphoid tissues such as lymph nodes into the circulation follows an S1P gradient. Expression of S1P receptors on the cell surface of lymphocytes, allowing the binding of its physiological ligand, appears to be a prerequisite in this process and regulates recirculation of lymphocytes as well as their homing to lymph nodes upon antigen presence.[9–13]

In vivo fingolimod is phosphorylated by SphK-2 to yield the biologically active fingolimod-phosphate, which shows close structural similarity to S1P.[9,14] Phosphorylated fingolimod is able to bind to four out of five S1P receptor subtypes with high affinity, namely S1P1, S1P3, S1P4, and S1P5.[14] Its binding to the S1P1 receptor on lymphocytes leads to the internalization and—as opposed to the effect of the physiological counterpart S1P—functional antagonism by degradation of the S1P receptor, thus rendering lymphocyte subsets unresponsive to the S1P gradient. As a consequence, lymphocytes—including autoreactive subsets—are retained from sites of inflammation in the CNS.[9,11–18] This process does not seem to affect all lymphocyte subsets but spares, for example, the effector memory T-cell pool.[19] This is of relevance as effector memory T cells are thought to be involved in maintaining local immunological responses in peripheral tissues.[17,19,20]

In addition and opposed to other therapeutics used in MS, fingolimod is capable of passing the blood–brain barrier due to its size and lipophilic properties.[21] Within the CNS S1P receptors are abundantly expressed on a variety of CNS resident cells, especially on neurons and both macro- as well as microglial cells.[10,22,23] By modulation of these receptors fingolimod in principle has the potential to modulate functions of these cells, theoretically impacting on mechanisms relevant to the disease such as neurodegeneration, astrogliosis, microglia activation, as well as endogenous repair mechanisms.

19.3 CLINICAL TRIAL PROGRAM IN MULTIPLE SCLEROSIS

The effects of fingolimod have been tested extensively in the animal model of MS in a number of preclinical studies.[14,24] Here, fingolimod has been reported efficacious both in prophylactic and in therapeutic application at different stages in experimental autoimmune encephalomyelitis (EAE). After fingolimod had demonstrated its capability to prevent disease progression and to attenuate disease-related symptoms after onset in different rodent models, a clinical trial program in patients with relapsing-remitting MS (RRMS) was initiated.

With more than 3000 patients enrolled, the fingolimod phase-3 clinical trial program finally turned out to be one of the largest in the field of MS so far. In the following section, results of the placebo-controlled FREEDOMS and FREEDOMS II studies as well as the actively controlled TRANSFORMS trial will be summarized.[25–27]

19.3.1 Placebo-Controlled Trials

FREEDOMS

The 24-month, double-blind, placebo-controlled study FREEDOMS (FTY720 Research Evaluating Effects of Daily Oral Therapy in Multiple Sclerosis) enrolled 1272 patients with RRMS.[25] To be included patients needed to have had at least one or more relapses in the year prior to study enrollment or two or more in the previous 2years. Patients were randomly assigned 1:1:1 to receive either oral fingolimod (0.5 or 1.25mg daily) or placebo. The primary end point of the study was the annualized relapse rate over 2years. The key secondary end point was the time to confirmed disability progression. Other secondary end points included a range of different MRI parameters.

A total of 1033 patients completed the study. Over 24months, fingolimod 0.5mg significantly reduced the annualized relapse rate by 54% compared to placebo ($p<0.001$). The higher dosage of 1.25mg fingolimod daily reduced the annualized relapse rate by 60% compared to placebo. The risk of disability progression was significantly reduced in both fingolimod arms (hazard ratio, 0.70 and 0.68, respectively; $p=0.02$ vs. placebo, for both comparisons). Furthermore, fingolimod 0.5 and 1.25mg daily proved to be superior to placebo with regard to the number of new or enlarged T2-lesions as well as gadolinium-enhancing T1-lesions and brain-volume loss ($p<0.001$ for all comparisons at 24months), a measure supposedly reflecting the degenerative component of the disease.

Overall, similar proportions of patients receiving fingolimod or placebo reported adverse events (93–94%). Adverse events reported at higher frequency among patients treated with fingolimod as compared to placebo treatment included bradycardia (fingolimod 1.25mg: 3.3%; fingolimod 0.5mg: 2.1%; placebo: 0.7%) and atrioventricular conduction block (first degree 1.2%/0.5%/0.5%; second degree 0.2%/0%/0.5%) at the time of fingolimod treatment initiation, macular edema (1.6%/0%/0%), elevated liver-enzyme levels (18.6%/15.8%/5.0%), and mild hypertension (6.3%/6.1%/3.8%).

FREEDOMS II

FREEDOMS II was a placebo-controlled double-blind phase-3 study predominantly conducted in the United States.[26] Its aim was to further assess the efficacy and safety of fingolimod in patients with RRMS. Between June 2006 and March 2009, 1083 patients were randomly allocated (1:1:1) to receive either fingolimod (1.25 or 0.5 mg) or placebo. In November 2009, after data review from other phase-3 trials and upon recommendation of the data and safety monitoring board, all patients assigned to fingolimod 1.25 mg were switched to 0.5 mg in a blinded fashion. For the primary outcome analysis, these patients were analyzed as being in the 1.25 mg group (ie, according to the original assignment = intention to treat). FREEDOMS II showed a mean annualized relapse rate of 0.40 (95% confidence interval [CI] 0.34–0.48) in patients on placebo and of 0.21 (0.17–0.25) in patients treated with fingolimod 0.5 mg. This corresponded to a relative risk reduction of 48% with fingolimod 0.5 mg compared to placebo ($p < 0.0001$). The mean percentage brain volume change was significantly lower in patients treated with fingolimod 0.5 mg compared to the placebo group (−0.86 vs. −1.28, treatment difference −0.41, 95% CI −0.62 to −0.20; $p = 0.0002$). There was no statistically significant between-group difference with regard to confirmed disability progression rates.

Fingolimod 1.25 and 0.5 mg compared to placebo led to more of the following adverse events: lymphopenia (10%/8% vs. 0%), increased alanine aminotransferase (10%/8% vs. 2%), herpes zoster infection (3%/3% vs. 1%), hypertension (13%/9% vs. 3%), first-dose bradycardia (6%/1% vs. <0.5%), and first-degree atrioventricular block (10%/5% vs. 2%).

19.3.2 Actively Controlled Phase-3 Trial

TRANSFORMS

TRANSFORMS (Trial Assessing Injectable Interferon vs. FTY720 Oral in RRMS) was a randomized, 12-month, double-blind, actively controlled, double-dummy study.[27] A total of 1280 patients with a confirmed diagnosis of RRMS according to the revised McDonald criteria 2005 and at least one documented relapse during the final year or at least two documented relapses during the 2 years prior to enrollment were randomly assigned to receive either daily doses of 0.5 mg fingolimod ($n = 429$) or 1.25 mg fingolimod ($n = 420$), or a weekly dose of 30 μg intramuscular IFN β-1a (i.m., $n = 431$), while the respective other groups received the comparator as placebo (double dummy). The primary end point of TRANSFORMS was the annualized relapse rate. The number of new or enlarged lesions on T2-weighted MRI scans at 12 months and progression of disability that was sustained for at least 3 months were defined as key secondary end points.

A total of 1153 patients (89%) completed the study (1123 on study drug, 30 off study drug). The annualized relapse rate was significantly lower in both groups treated with fingolimod—0.20 (95% CI, 0.16–0.26) in the 1.25-mg group and 0.16 (95% CI, 0.12–0.21) in the 0.5-mg group—compared to the interferon group with 0.33 (95% CI, 0.26–0.42; $p < 0.001$ for both comparisons). Comparing fingolimod 0.5 mg to IFN β-1a i.m., this corresponds to a reduction of the annualized relapse rate by 52% over 12 months. After 1 year, there were significantly fewer new or enlarged T2-lesions and gadolinium-enhancing T1-lesions on MRI among patients in the two fingolimod groups compared to the scans of the patients in the interferon group. In particular, fingolimod 0.5 mg reduced the number of new or enlarged T2-lesions by 35% compared to IFN β-1a i.m. and the number of gadolinium-enhancing T1-lesions was reduced by 55%. With respect to progression of disability there were no significant differences between the study groups.

In TRANSFORMS, the mean percent reduction in brain volume from baseline to 12 months was significantly lower in the two fingolimod groups compared to the interferon group.[27]

The rate of adverse events reported during TRANSFORMS was comparable between study groups, ranging from 86% to 92%. Of note, two deaths occurred in the fingolimod arms: one fatal case of disseminated primary varicella zoster infection and one fatal case of herpes simplex encephalitis, both in the group receiving 1.25 mg fingolimod. Further adverse events reported in patients treated with fingolimod were, for example, nonfatal herpesvirus infections, cardiovascular adverse events like bradycardia and atrioventricular block, hypertension, macular edema, skin cancer, and elevated liver-enzyme levels.

TRANSFORMS Extension Trial

TRANSFORMS was followed by an extension trial.[28] Patients who had received fingolimod during the core study stayed on the same dosage for another 12 months. Patients who had originally received 30 μg IFN βa-1a i.m. once weekly were rerandomized (1:1) to receive either 0.5 or 1.25 mg fingolimod. The extension study aimed to assess the effects of switching from IFN β-1a i.m. in year 1 to fingolimod in year 2 by comparing the second year of treatment with the first year in patients who had switched their therapy as compared to those who stayed on either dose of the drugs for both the core and the extension phase of the study. Furthermore, the effect of delaying the start of therapy with fingolimod was assessed by comparing the treatment groups over 24 months based on the original assignment at baseline.

Of the 1123 patients that had completed the core study on study drug, 1027 (92%) entered the extension, and 882 completed 24 months of treatment. The study showed a decrease in the annualized relapse rate in patients switching from IFN β-1a to fingolimod in year 2 as well as a significant reduction of new or newly enlarging T2 and gadolinium-enhancing T1 lesions compared to year 1. Continuous treatment with fingolimod over 2 years compared to patients who had switched their therapy provided a sustained treatment effect with improvement in clinical parameters as well as in MRI outcomes. Regarding the safety of fingolimod, the extension study revealed no new signals beyond what had been described in year 1 of the study.

19.4 RECOMMENDED PRECAUTIONS BEFORE FINGOLIMOD THERAPY

During the clinical trial program distinct adverse events were recorded in patients treated with fingolimod. Some of these can be explained by its effects on S1P receptors. For example, S1P receptors are also expressed on cardiac myocytes, where they are involved in the regulation of heart rate and conduction.[29] Binding of fingolimod to S1P1 receptors in atrial myocytes causes a transient, dose-dependent, and usually mild-to-moderate negative chronotropic effect.[30] This effect reaches its maximum 4–5 h after the first dose. Due to desensitization of the receptor the effect of fingolimod on cardiac S1P1 receptors is self-limiting. Upon continued dosing the observed cardiac effects are attenuating over time.[30] As a consequence, marketing authorities like the FDA issued some precautions to be taken when administering fingolimod for the first time.[31] Patients should have an electrocardiogram prior to and 6 h after first-dose treatment and all patients should be monitored for signs and symptoms of bradycardia for 6 h following the first tablet of fingolimod (first dose observation).

The exact pathogenesis underlying the cases of macular edema (frequency range 0.4–0.5%) seen during treatment with fingolimod is currently unknown. An ophthalmological exam is recommended (and mandatory in patients with an increased risk of developing macular edema, ie, patient with diabetes or a history of uveitis) before starting fingolimod and 3–4 months after treatment initiation, as fingolimod-induced macular edema are rarely seen thereafter.[31] In addition, visual symptoms and acuity should be monitored at routine evaluations.

Before initiating fingolimod therapy, patients without a history of chicken pox or without vaccination against varicella zoster virus (VZV) should be tested for antibodies to VZV.[31] VZV vaccination of antibody-negative patients has to be performed prior to commencing treatment, following which initiation of treatment with fingolimod should be postponed for 1 month to allow the full effect of vaccination to occur. Live attenuated vaccines should be avoided during, and for 2 months after stopping, fingolimod treatment due to risk of infection.

Due to reversible sequestration of lymphocytes in lymphoid tissues fingolimod leads to a dose-dependent reduction in peripheral lymphocyte counts to 20–30% of baseline levels. Therefore, it may increase the overall risk of infections. Hence a recent complete blood count (ie, within 6 months) should be available before starting treatment. In case a patient develops a serious infection during treatment, suspension and reassessment of the benefits and risks of the therapy should be considered prior to its reinitiation. As the elimination of fingolimod after discontinuation may take up to 2 months, monitoring for infections should be continued throughout this period.

19.5 S1P RECEPTOR MODULATORS UNDER INVESTIGATION

Inspired by the successful development of fingolimod S1P-mediated signaling in MS, it has become a focus of intensive clinical research. Partly due to the role of the S1P1 receptors in the observed cardiac side effects, some of the research concentrated on the development of fingolimod-analogs with increased S1P3 selectivity.[32] Extensive work finally led to the discovery of (E)-1-(4-(1-(((4-cyclohexyl-3-(trifluoromethyl)benzyl)oxy)imino)ethyl)-2-ethylbenzyl) azetidine-3-carboxylic acid (32, BAF312, Siponimod). In BOLD, an adaptive, dose-ranging, randomized, phase-2 study, siponimod was sequentially tested in two cohorts of patients with RRMS.[33] Patients in cohort 1 ($n=188$) were randomly allocated (1:1:1:1) to receive once-daily siponimod 10, 2, or 0.5 mg, or placebo for 6 months. Patients in cohort 2 ($n=109$) were randomized (4:4:1) to receive siponimod 1.25 or 0.25 mg, or placebo once daily for 3 months. The primary end point was dose-responsiveness, assessed by percentage reductions in the monthly number of combined unique active lesions at 3 months for the different doses of siponimod compared to placebo. Safety was assessed in all patients who received at least one dose of the study drug. The observed therapeutic effects of siponimod on MRI lesion activity as well as its tolerability warrant further investigations. So far, there is a single phase-3 trial ongoing in patients with secondary progressive MS.[34]

Another orally active selective S1P1 agonist is ponesimod (Z,Z)-5-[3-chloro-4-((2R)-2,3-dihydroxy-propoxy)-benzylidene]-2-propylimino-3-o-tolyl-thiazolidin-4-one.[35] In models of lymphocyte-mediated tissue inflammation ponesimod (ACT128800) led to blood lymphocyte count reduction.[35] A double-blind, placebo-controlled, dose-finding phase-2b study evaluated the efficacy and safety of ponesimod in patients with RRMS.[36] The study showed that once-daily treatment with ponesimod (10, 20, or 40 mg) significantly reduced the number of new gadolinium-enhancing T1-lesions and a beneficial effect on clinical end points. It was generally well tolerated. As the study duration was only 24 weeks and provision of long-term safety and clinical efficacy data was therefore not possible, an extension study of up to 5 years' duration is currently ongoing.[37] In addition, ponesimod has also shown significant clinical benefits in patients with moderate to severe chronic plaque psoriasis.[38] And in a mouse model of autoimmune diabetes ponesimod in combination with CD3 antibodies was able to restore self-tolerance.[39]

19.6 CONCLUSION

With the development of fingolimod, a first-in-class oral S1P modulator became available for the treatment of MS that, for the first time, had proven superiority over an active comparator first-line drug in a head-to-head trial (TRANSFORMS). As of the conception of this book more than 135,000 patients have been treated with fingolimod worldwide. While the overall safety experience of the postmarketing period has largely confirmed the data generated within the phase-3 core and extension programs of fingolimod, confirmed cases of progressive multifocal leucencephalopathy (PML) without pretreatment with the monoclonal antibody natalizumab (known to increase the risk of PML) have been reported.[40] Although this does not compromise the solid efficacy experience with fingolimod, it emphasizes the need for continuous safety monitoring programs including real-world registries like PANGAEA in case of fingolimod to generate long-term efficacy but also safety data to reassure the use of potent new compounds in the treatment and assist in the consulting of MS patients.[41]

Acknowledgment

This chapter has been drafted with the independent financial support of Novartis, which has no involvement whatsoever with respect to its content.

References

1. Adachi K, Kohara T, Nakao N, et al. Design, synthesis, and structure-activity relationships of 2-substituted-2-amino-1,3-propanediols: discovery of a novel immunosuppressant, FTY720. *Bioorg Med Chem Lett*. 1995;5: 853–856.
2. Fujita T, Hirose R, Yoneta M, et al. Potent immunosuppressants, 2-alkyl-2-aminopropane-1,3-diols. *J Med Chem*. 1996;39:4451–4459.
3. Hiestand PC, Rausch M, Meier DP, Foster CA. Ascomycete derivative to MS therapeutic: S1P receptor modulator FTY720. *Prog Drug Res*. 2008;66(361):363–381.
4. Adachi K, Chiba K. FTY720 story. Its discovery and the following accelerated development of sphingosine 1-phosphate receptor agonists as immunomodulators based on reverse pharmacology. *Perspect Med Chem*. 2008;1:11–23.
5. Tedesco-Silva H, Pescovitz MD, Cibrik D, et al. Randomized controlled trial of FTY720 versus MMF in de novo renal transplantation. *Transplantation*. 2006;82:1689–1697.
6. Salvadori M, Budde K, Charpentier B, et al. FTY720 versus MMF with cyclosporine in de novo renal transplantation: a 1-year, randomized controlled trial in Europe and Australasia. *Am J Transpl*. 2006;6:2912–2921.
7. Hla T. Physiological and pathological actions of sphingosine 1-phosphate. *Semin Cell Dev Biol*. 2004;15:513–520.
8. Chae SS, Proia RL, Hla T. Constitutive expression of the S1P1 receptor in adult tissues. *Prostagl Other Lipid Mediat*. 2004;73:141–150.

9. Mandala S, Hajdu R, Bergstrom J, et al. Alteration of lymphocyte trafficking by sphingosine-1-phosphate receptor agonists. *Science*. 2002;296:346–349.
10. Brinkmann V. Sphingosine 1-phosphate receptors in health and disease: mechanistic insights from gene deletion studies and reverse pharmacology. *Pharmacol Ther*. 2007;115:84–105.
11. Brinkmann V, Cyster JG, Hla T. FTY720: sphingosine 1-phosphate receptor-1 in the control of lymphocyte egress and endothelial barrier function. *Am J Transpl*. 2004;4:1019–1025.
12. Matloubian M, Lo CG, Cinamon G, et al. Lymphocyte egress from thymus and peripheral lymphoid organs is dependent on S1P receptor 1. *Nature*. 2004;427:355–360.
13. Schwab SR, Cyster JG. Finding a way out: lymphocyte egress from lymphoid organs. *Nat Immunol*. 2007;8: 1295–1301.
14. Brinkmann V, Davis MD, Heise CE, et al. The immune modulator FTY720 targets sphingosine 1-phosphate receptors. *J Biol Chem*. 2002;277:21453–21457.
15. Oo ML, Thangada S, Wu MT, et al. Immunosuppressive and anti-angiogenic sphingosine 1-phosphate receptor-1 agonists induce ubiquitinylation and proteasomal degradation of the receptor. *J Biol Chem*. 2007; 282:9082–9089.
16. Graler MH, Goetzl EJ. The immunosuppressant FTY720 down-regulates sphingosine 1-phosphate G-protein-coupled receptors. *Faseb J*. 2004;18:551–553.
17. Pinschewer DD, Ochsenbein AF, Odermatt B, Brinkmann V, Hengartner H, Zinkernagel RM. FTY720 immunosuppression impairs effector T cell peripheral homing without affecting induction, expansion, and memory. *J Immunol*. 2000;164:5761–5770.
18. Fujino M, Funeshima N, Kitazawa Y, et al. Amelioration of experimental autoimmune encephalomyelitis in Lewis rats by FTY720 treatment. *J Pharmacol Exp Ther*. 2003;305:70–77.
19. Mehling M, Brinkmann V, Antel J, et al. FTY720 therapy exerts differential effects on T cell subsets in multiple sclerosis. *Neurology*. 2008;71:1261–1267.
20. Masopust D, Vezys V, Marzo AL, Lefrançois L. Preferential localization of effector memory cells in nonlymphoid tissue. *Science*. 2001;291:2413–2417.
21. Foster CA, Howard LM, Schweitzer A, et al. Brain penetration of the oral immunomodulatory drug FTY720 and its phosphorylation in the central nervous system during experimental autoimmune encephalomyelitis: consequences for mode of action in multiple sclerosis. *J Pharmacol Exp Ther*. 2007;323:469–475.
22. Herr DR, Chun J. Effects of LPA and S1P on the nervous system and implications for their involvement in disease. *Curr Drug Targets*. 2007;8:155–167.
23. Dev KK, Mullershausen F, Mattes H, et al. Brain sphingosine-1-phosphate receptors: implication for FTY720 in the treatment of multiple sclerosis. *Pharmacol Ther*. 2008;117:77–93.
24. Aktas O, Küry P, Kieseier B, Hartung HP. Fingolimod is a potential novel therapy for multiple sclerosis. *Nat Rev Neurol*. 2010;6:373–382.
25. Kappos L, Radue EW, O'Connor P, et al. A placebo-controlled trial of oral fingolimod in relapsing multiple sclerosis. *N Engl J Med*. 2010;362:387–401.
26. Calabresi PA, Radue EW, Goodin D, et al. Safety and efficacy of fingolimod in patients with relapsing-remitting multiple sclerosis (FREEDOMS II): a double-blind, randomised, placebo-controlled, phase 3 trial. *Lancet Neurol*. 2014;13:545–556.
27. Cohen JA, Barkhof F, Comi G, et al. Oral fingolimod or intramuscular interferon for relapsing multiple sclerosis. *N Engl J Med*. 2010;362:402–415.
28. Khatri B, Barkhof F, Comi G, et al. Comparison of fingolimod with interferon beta-1a in relapsing-remitting multiple sclerosis: a randomised extension of the TRANSFORMS study. *Lancet Neurol*. 2011;10:520–529.
29. Peters SLM, Alewijnse AE. Sphingosine-1-phosphate signaling in the cardiovascular system. *Curr Opin Pharmacol*. 2007;7:186–192.
30. Schmouder R, Serra D, Wang Y, et al. FTY720: placebo-controlled study of the effect on cardiac rate and rhythm in healthy subjects. *J Clin Pharmacol*. 2006;46:895–904.
31. NDA 02257—FDA Approved Labeling Text for Gilenya (Fingolimod) Capsules. Available at: http://www.accessdata.fda.gov/drugsatfda_docs/label/2010/022527s000lbl.pdf; September 21, 2010 Accessed 18.10.15.
32. Pan S, Gray NS, Gao W, et al. Discovery of BAF312 (Siponimod), a potent and selective S1P receptor modulator. *ACS Med Chem Lett*. 2013;4:333–337.
33. Selmaj K, Li DK, Hartung HP, et al. Siponimod for patients with relapsing-remitting multiple sclerosis (BOLD): an adaptive, dose-ranging, randomised, phase 2 study. *Lancet Neurol*. 2013;12:756–767.

34. Efficacy and Safety of Siponimod in Patients with Secondary Progressive Multiple Sclerosis (EXPAND). NCT01665144. Available at: www.clinicaltrial.gov; Accessed 18.10.15.
35. Piali L, Froidevaux S, Hess P, et al. The selective sphingosine 1-phosphate receptor 1 agonist ponesimod protects against lymphocyte-mediated tissue inflammation. *J Pharmacol Exp Ther*. 2011;337:547–556.
36. Olsson T, Boster A, Fernández Ó, et al. Oral ponesimod in relapsing-remitting multiple sclerosis: a randomised phase II trial. *J Neurol Neurosurg Psychiatry*. 2014;85:1198–1208.
37. Ponesimod in patients with relapsing-remitting multiple sclerosis -Extension study. NCT01093326. Available at: www.clinicaltrial.gov; Accessed 18.10.15.
38. Vaclavkova A, Chimenti S, Arenberger P, et al. Oral ponesimod in patients with chronic plaque psoriasis: a randomised, double-blind, placebo-controlled phase 2 trial. *Lancet*. 2014;384:2036–2045.
39. You S, Piali L, Kuhn C, et al. Therapeutic use of a selective S1P1 receptor modulator ponesimod in autoimmune diabetes. *PLoS One*. 2013;8:e77296.
40. Gilenya Information Center. Notification from August 17, 2015. Available at: www.novartis.com/news/statements/gilenya-information-center; Accessed 18.10.15.
41. Ziemssen T, Schwarz HJ, Fuchs A, Cornelissen C. 36 month PANGAEA: a 5-year non-interventional study of safety, efficacy and pharmacoeconomic data for fingolimod patients in daily clinical practice. In: *Annual Meeting of the American Academy of Neurology*; April 18–25, 2015. Washington, DC. Abstract P3.251.

CHAPTER

20

Oral Dimethyl Fumarate (BG-12; Tecfidera®) for Multiple Sclerosis

A. Salmen

Ruhr-University Bochum, Bochum, Germany

R.A. Linker

Friedrich Alexander University Erlangen-Nürnberg, Erlangen, Germany

R. Gold

Ruhr-University Bochum, Bochum, Germany

OUTLINE

Translational Neuroimmunology in Multiple Sclerosis
http://dx.doi.org/10.1016/B978-0-12-801914-6.00020-9

20.1 INTRODUCTION

Within the past few years, therapeutic options for the treatment of multiple sclerosis (MS) have expanded including orally available disease-modifying drugs (DMDs). Among the latter, dimethyl fumarate (DMF) has been approved as a first-line DMD in relapsing-remitting MS (RRMS) in 2013 by the US Food and Drug Administration and in 2014 by the European Medicines Agency after the results of two pivotal phase-3 trials.[1,2]

MS pathology comprises both autoimmune inflammatory and neurodegenerative processes as at least partially independent mechanisms.[3–5] Targeting both inflammation and degeneration is thus a valuable goal for modern DMD treatment.

After a description of the discovery of DMF as a potential treatment option in MS, hypotheses on antiinflammatory and putatively cytoprotective characteristics of DMF based on experimental studies will be depicted.

Efficacy and safety data of the pivotal clinical trials in early and established RRMS will be described including patient-related outcomes. First available data on DMF treatment in progressive forms of MS will be reviewed.

Available postmarketing safety data and safety data of other formulations of fumaric acid esters are critically discussed.

20.2 THE DISCOVERY OF DIMETHYL FUMARATE AS A POTENTIAL TREATMENT OPTION FOR MULTIPLE SCLEROSIS

Fumarate is part of the citric acid cycle in the aerobic metabolism of eu- and prokaryotic cells. Under the hypothesis of a metabolic impact on autoimmune cells, a German chemist used fumaric acid esters as a topic and oral treatment for his severe psoriasis,[6] later resulting in two double-blind trials.[7,8] A compound of DMF and ethylhydrogen fumarate (Fumaderm®) was thus approved for the treatment of severe therapy-refractory psoriasis in Germany in 1994.[9]

After the serendipitous observation of the stabilization of MS in two patients with psoriasis treated with Fumaderm®, a small proof-of-concept study was performed on 10 RRMS patients with proven MRI activity. The evaluable study population comprised only seven patients, yet significant results were shown for number and volume of gadolinium-enhancing lesions on cerebral MRI. Drop-outs may have been driven by gastrointestinal side effects and flushing.[10]

DMF in delayed-release, enteric-coated capsules (BG-12) was thus designed to improve galenics and tolerability of the drug and investigated in a multicenter, randomized, double-blind phase-2b trial.[11] This trial included 257 patients with RRMS, aged 18–55 years who were treated with (1) once daily 120 mg DMF, (2) thrice daily 120 mg DMF, (3) thrice daily 240 mg DMF, or (4) placebo for 24 weeks. The placebo group was switched to regimen (3) in the 24-week extension of the study. The primary end point, the number of new gadolinium-enhancing lesions, was met showing a significant reduction of 69% in the high-dose regimen versus placebo. Moreover, comparing these groups, a significant reduction of new/enlarging T2-hyperintense and new T1-hypointense lesions was shown comparing the high-dose group versus placebo.[11,12] The significance of the primary end point was corroborated in a subgroup analysis stratifying for Expanded Disability Status Scale (EDSS) score, age, gender, disease duration, and gadolinium-enhancing lesions at baseline.[13]

With generally favorable safety data, adverse events attributed to DMF or regarded as dose-related were again gastrointestinal side effects, flushing including hot flush, headache, fatigue, and feeling hot.[11] In this study, dose titration was performed for the high-dose group receiving 120 mg DMF thrice daily for 1 week with consecutive full-dose treatment of thrice 240 mg DMF from the second week both exceeding current treatment recommendations.

On the basis of these data, the conduction of larger phase-3 trials was encouraged.

20.3 HYPOTHESES ON MECHANISMS OF ACTION AND EXPERIMENTAL DATA OF DIMETHYL FUMARATE

20.3.1 Direct Interactions of Dimethyl Fumarate with the Immune System

Both DMF (though possibly short-lived) and its main metabolite, monomethyl fumarate (MMF), are attributed antiinflammatory and antioxidative, thus cytoprotective effects.[14–16]

Antiinflammatory mechanisms of action include the induction of T-cell apoptosis, shown in vitro,[17] and a reduction of peripheral T cells, predominantly CD8+ lymphocytes, initially shown in a small population of psoriasis patients.[18] Recently, a serial analysis of peripheral blood cells of MS patients under DMF treatment at different time points confirmed the pronounced reduction of CD8+ and to smaller extent CD4+ T cells, but also showed a reduction of CD19+ B cells and eosinophils with unaltered counts of other immune-cell subsets.[19] The reduction in total lymphocyte count reached the suggested clinically relevant threshold of <500 cells per μL in about 50% of patients,[19] which may imply consequences for clinical practice.

Immunohistochemical studies have demonstrated a reduction of CD4+ cells in epidermal inflammatory infiltrates,[20] which may hint at lesion-specific immunomodulatory effects.

Besides effects on T-cell counts, fumaric acid esters induce a cytokine shift of T cells to a T helper-type-2 (Th2) response with increased interleukin (IL)-4 and IL-5 and reduced interferon-gamma production, shown in vitro both in psoriatic and in MS patients.[21–24] Often a transient eosinophilia is observed.

20.3.2 Effects of Dimethyl Fumarate on Other Cell Types

Antigen-presenting cells and especially dendritic cells (DCs) are potential modulators of the immune response. The differentiation of precursor cells to DCs was inhibited by the application of DMF in vitro,[25] which may result in an altered interaction with the immune system. Fumarate treatment has also been shown to shift the cytokine response of DCs from IL-12/IL-23 to IL-10. In a murine experimental model of MS, this DC pattern after fumarate application induced a Th2 response and ameliorated the model disease.[24]

Likewise, keratinocytes that are involved in the cutaneous inflammatory response in psoriasis and release cytokines increase IL-10 expression and inhibit expression of numerous chemokines in vitro when exposed to DMF.[21,26]

Considering central nervous system (CNS) cells, DMF attenuated a proinflammatory response in rat glial cocultures of microglia and astrocytes activated by lipopolysaccharide.[27] In astrocytes, DMF treatment results in a reduced expression of nitric oxide synthase 2 (NOS2) and thus exerts antiinflammatory effects.[28] Fumarate application on astrocytes, but not on microglia, also resulted in an increased release of vascular endothelial growth factor that has been suggested to be a pseudo-hypoxic response and to be modulated by several transcription factors including nuclear (erythroid-derived 2)-related factor (Nrf2).[29]

An inhibitory effect of DMF on the expression of adhesion molecules in a human umbilical vein endothelial cell-culture model resulted in reduced monocyte adhesion.[30] Initially evaluated as relevant for treatment of psoriasis, this may represent another mechanism contributing to therapeutic effects of DMF in MS when being reproduced for the blood–brain barrier (BBB). Data on this are thus far contradictory: Although DMF and MMF induced Nrf2 expression also in human endothelial cells, an influence on tight junctions and BBB integrity was observed neither in the human cell-culture model nor in the murine model of experimental autoimmune encephalomyelitis (EAE).[31] However, in a recent study using a murine stroke model, thus modeling tissue hypoxia and oxidative stress with secondary inflammation, systemic DMF treatment prevented disruption of the BBB. In vitro, endothelial tight junctions were stabilized, inflammatory cytokine responses were diminished, and leukocyte transmigration was attenuated by DMF.[32]

20.3.3 Potential Pathways for the Effects of Dimethyl Fumarate

Mechanisms of action of DMF are not fully understood, but various pathways DMF possibly interacts with have been described. An involvement of cellular redox systems, modulation of glutathione levels, and therefore an interaction with (signaling ways of) the transcription factor nuclear factor-kappa B (NF-κB) has been shown by several groups in different contexts and is thus of high interest.

In the context of the induction of type II DCs, DMF depleted glutathione levels, which resulted in a higher hemoxygenase-1 (HO-1) expression and lower phosphorylation of Signal Transducer and Activator of Transcription 1 (STAT1). After cleavage, a fragment of HO-1 is translocated to the nucleus and interacts with NF-κB sites leading to lower IL-23 transcription. The STAT1 inactivation causes lower IL-12 expression.[24] Yet, a different publication demonstrated a more complex kinetic of glutathione in astrocytes.[28]

DMF also directly inhibits the nuclear entry of activated NF-κB and leads to decreased NF-κB-dependent gene expression.[28,33] This results in reduced expression of proinflammatory cytokines and adhesion molecules.[26,30] The DMF-induced lower NF-κB activation has also been demonstrated to decrease activity of NOS2 and thus reduced nitrite accumulation.[28] Besides these direct antiinflammatory effects, DMF also enhances the activity of detoxification enzymes such as the NAD(P)H quinone reductase.[27] Putative cytoprotective properties of DMF may thus be mediated via an improved detoxification, but also reduced oxidative stress. Investigating mechanisms for the latter, Nrf2 is of interest as a transcription factor inducing transcription of several genes involved in antioxidative pathways. DMF leads to increased intranuclear translocation of Nrf2, which enhances Nrf2-dependent transcription; this may be mediated via succination of Kelch-like ECH-associated protein 1 (KEAP1).[15,16,32,34,35]

In vitro, a functional relevance of the Nrf2 pathway has been demonstrated: DMF-induced prolonged survival of neurons and glial cells was absent in Nrf2-deficient cells.[35] Further proof was found using an EAE model in which DMF treatment augmented Nrf2 levels in CNS combined with an ameliorated disease course, especially in late stages of the model disease.[15]

Other pathways and receptors that might be involved in DMF-associated antioxidative effects include the hydroxycarboxylic acid receptor 2 as well as hypoxia-inducible transcription factor 1 alpha, at least in subtypes of CNS cells.[29,36]

20.4 DATA OF CLINICAL TRIALS IN DIFFERENT MULTIPLE SCLEROSIS DISEASE COURSES

20.4.1 Dimethyl Fumarate in Relapsing-Remitting Multiple Sclerosis: The Pivotal Phase-3 Trials DEFINE and CONFIRM

DEFINE was a randomized, double-blind, placebo-controlled trial of DMF twice and thrice 240 mg daily.[2] Primary end point of the study was the proportion of patients without relapses after the study duration of 2 years, secondary end points comprised clinical and paraclinical outcome parameters (relapse rate, disability progression, MRI parameters) (Table 20.1). After randomization of 1237 patients, 77% completed the trial, demographics and withdrawal rates were similar between the two treatment groups and the placebo group.

Whereas overall there were no clear advantages for the thrice- versus the twice-daily dosage, a significant reduction of the proportion of patients with relapse(s) favoring treatment (46% (placebo); 26% (2 × 240 mg DMF); 27% (3 × 240 mg)) and of the annualized relapse rate (ARR; 53% (2 × 240 mg DMF); 48% (3 × 240 mg DMF) vs placebo)) was shown. Progression of EDSS as a measure of disability was reduced (27% (placebo); 16% (2 × 240 mg DMF); 18% (3 × 240 mg)). MRI end points (reduction of number of new/enlarging T2 lesions, reduction of number of gadolinium-enhancing T1 lesions in treatment groups vs placebo) were met. The evolution of MRI changes over time was evaluated in a distinct analysis[37]: After 6 months of DMF treatment and sustained after 1 and 2 years, a reduction in the number of lesions (T2 hyperintense, gadolinium-enhancing T1 and T1 hypointense) versus placebo was significant. A positive effect on the evolution of brain atrophy was demonstrated for 2 × 240 mg DMF.

Magnetization transfer ratio (MTR) as a putative measure of myelin density may reflect CNS damage in more detail than numbers and volumes of different lesion formations. In a subset of DEFINE patients ($n = 392$), MTR was measured in whole brain (WB) and normal-appearing brain tissue (NABT) at different time points[38]: At the end of the study period after

TABLE 20.1 Summary of Efficacy Parameters of the DEFINE[2] and CONFIRM[1] Trials After 2 Years

	DEFINE			CONFIRM			
	2 × 240 mg DMF	**3 × 240 mg DMF**	**Placebo**	**2 × 240 mg DMF**	**3 × 240 mg DMF**	**Glatiramer Acetate**	**Placebo**
(Estimated) patients with relapse(s) (%) (primary end point in DEFINE)	27	26	46	29	24	32	41
(Adjusted) annualized relapse rate (rate (95% CI)) (primary end point in CONFIRM)	0.17 (0.14–0.21)	0.19 (0.15–0.23)	0.36 (0.30–0.44)	0.22 (0.18–0.28)	0.20 (0.16–0.25)	0.29 (0.23–0.35)	0.40 (0.33–0.49)
(Estimated) patients with sustained disability progression (%)	16	18	27	13	13	16	17
Adjusted number of new/enlarging T2 lesions (mean (95% CI))	2.6 (2.0–3.5)	4.4 (3.2–5.9)	17.0 (12.9–22.4)	5.1 (3.9–6.6)	4.7 (3.6–6.2)	8.0 (6.3–10.2)	17.4 (13.5–22.4)
Number of gadolinium-enhancing T1 lesions (mean ± SD)	0.1 ± 0.6	0.5 ± 1.7	1.8 ± 4.2	0.5 ± 1.7	0.4 ± 1.2	0.7 ± 1.8	2.0 ± 5.6
Adjusted number of new T1 hypointense lesions (mean (95% CI))	n.a.	n.a.	n.a.	3.0 (2.3–4.0)	2.4 (1.8–3.2)	4.1 (3.2–5.3)	7.0 (5.3–9.2)

CI, confidence interval; *DMF*, dimethyl fumarate; *n.a.*, not available/applicable; *SD*, standard deviation.

2 years, a small, but significant change in median WB MTR (reduction of 0.386% (placebo) vs increase of 0.129% (2 × 240 mg DMF) and 0.096% (3 × 240 mg DMF)) and in median NABT MTR (reduction of 0.392% (placebo) vs increase of 0.190% (2 × 240 mg DMF) and 0.115% (3 × 240 mg DMF)) favors DMF.

Subgroup analyses of DEFINE stratified the included patients by different characteristics (gender, age, relapse history, McDonald criteria, treatment history, EDSS, T2 lesion volume, gadolinium-enhancing lesions).[39] Clinical efficacy was consistently shown across these subgroups confirming both the primary end point of the original study and positive effects on ARR and disability progression.

In addition to the groups of the DEFINE trial, CONFIRM included a comparative treatment group (glatiramer acetate; GA).[1] Evaluation of the study results was slightly different from DEFINE: ARR after 2 years was defined as the primary end point, therefore secondary end points included the proportion of patients with relapse(s) and again clinical and paraclinical measures (Table 20.1). CONFIRM randomized 1430 patients with an 80% completion rate. In the placebo group, therapy discontinuation occurred slightly more often than in treatment groups. The primary end point of ARR reduction was met in all treatment groups versus placebo (0.22 (2 × 240 mg DMF); 0.20 (3 × 240 mg DMF); 0.29 (GA); 0.40 (placebo)). Yet, the study design was not powered to detect differences between the treatment groups. However, the relative reduction of relapses was 29% (GA), 44% (2 × 240 mg DMF), and 51% (3 × 240 mg DMF). The proportion of patients with relapse(s) was significantly reduced in treatment groups (32% (GA); 29% (2 × 240 mg DMF); 24% (3 × 240 mg DMF); 41% (placebo)). The subgroup analyses of CONFIRM, likewise stratified for baseline characteristics (see earlier), supported the positive influence of DMF on these clinical relapse-associated outcome parameters across the groups.[40]

Disability progression was not significantly different between the groups (16% (GA); 13% (2 × 240 mg DMF); 13% (3 × 240 mg DMF); 17% (placebo)). MRI end points were met by the treatment groups versus placebo (reduction of number of new/enlarging T2 lesions, reduction of number of new hypointense T1 lesions, reduction of number of gadolinium-enhancing T1 lesions).

Adverse events (AEs) and serious adverse events (SAEs) in both the DEFINE and CONFIRM trials were comparable in frequencies across the groups (Table 20.2).[1,2] In DMF groups, gastrointestinal events and flushing as well as abnormal laboratory values (lymphocyte counts, liver enzymes) were more frequently observed. Flushing and gastrointestinal symptoms occurred predominantly within the treatment initiation phases and decreased over time. These symptoms thus seem to be dose-dependent and habituating. Still, flushing may be a more severe reaction (hot flush) with redness and itching. Serious gastrointestinal events were rare, but slightly more frequent in the DMF groups. Safety signals in terms of malignancies and opportunistic infections were not observed during the phase-3 studies.

Monitoring of blood cell counts showed a decrease in total leukocytes (10–12% from baseline) and lymphocytes (28–32% from baseline) plateauing after 1 year of treatment. Up to 5% of study participants experienced a grade 3 or higher lymphopenia (according to National Cancer Institute Common Toxicity Criteria). Asymptomatic increases of liver enzymes occurred early within the first 6 months of treatment. These data derived from the phase-3 trials were thus favorable in terms of efficacy and safety.

TABLE 20.2 Overview of Adverse and Serious Adverse Events of the DEFINE[2] and CONFIRM[1] Trials (%)

	DEFINE			CONFIRM			
	2×240 mg DMF	3×240 mg DMF	Placebo	2×240 mg DMF	3×240 mg DMF	Glatiramer Acetate	Placebo
Back pain	n.a.	n.a.	n.a.	10	10	9	9
Depression	n.a.	n.a.	n.a.	7	4	9	10
Fatigue	n.a.	n.a.	n.a.	10	10	9	9
Flushing	38	32	5	31	24	2	4
Gastro-intestinal symptoms	10–15	7–19	5–13	10–13	10–15	1–4	5–8
Headache	n.a.	n.a.	n.a.	14	13	13	13
Infections	n.a.	n.a.	n.a.	10–17	12–18	8–15	9–16
MS relapse	27	27	46	31	25	34	43
Proteinuria	9	12	8	8	10	9	7
Pruritus	10	8	5	n.a.	n.a.	n.a.	n.a.
Therapy discontinuation	16	16	13	12	12	10	10
Any AE	**96**	**95**	**95**	**94**	**92**	**87**	**92**
Anaphylactic reaction	n.a.	n.a.	n.a.	0	0	<1	0
Cellulitis	n.a.	n.a.	n.a.	<1	<1	0	0
Convulsion	n.a.	n.a.	n.a.	0	0	0	<1
Death	<1	<1	0	0	<1	<1	<1
Depression	n.a.	n.a.	n.a.	0	<1	<1	0
Gastritis	0	<1	0	n.a.	n.a.	n.a.	n.a.
Gastroenteritis	<1	<1	0	<1	<1	0	0
Headache	0	<1	0	n.a.	n.a.	n.a.	n.a.
Malignant neoplasm	<1	<1	<1	n.a.	n.a.	n.a.	n.a.
MS relapse	10	8	15	11	9	10	14
Other serious infection	2	2	2	n.a.	n.a.	n.a.	n.a.
Ovarian cyst	<1	<1	<1	n.a.	n.a.	n.a.	n.a.
Pain: abdominal, back, muscle strain	n.a.	n.a.	n.a.	0 to <1	0 to <1	0	0
Pneumonia	<1	0	<1	0	0	<1	<1
Spontaneous abortion	n.a.	n.a.	n.a.	0	0	0	<1
Any SAE	**18**	**16**	**21**	**17**	**16**	**17**	**22**

Total rates of AEs and SAEs were described as similar between groups. *AE*, adverse event; *DMF*, dimethyl fumarate; *MS*, multiple sclerosis; *n.a.*, not available/applicable; *SAE*, serious adverse event.

Patient-related outcomes have been investigated in both DEFINE and CONFIRM and analyzed in an integrated analysis (n=2301 patients in total; n=769 (2×240mg DMF); n=761 (3×240mg DMF); n=771 (placebo))[41]: Physical and Mental Component Summaries (PCS; MCS) of the 36-item Short Form Health Survey in patients were lower as compared to the general US population, especially with increasing impairment (ie, higher EDSS) at baseline. After 2 years of treatment, PCS and MCS increased in the DMF groups whereas they decreased as compared to baseline in the placebo group with a significant difference for both DMF groups and placebo. Predictive factors for improved PCS and MCS besides DMF treatment were lower baseline EDSS and corresponding lower baseline PCS and MCS and younger age (below 40 years) for PCS. These results indicate positive effects of DMF treatment on health-related quality of life.

20.4.2 Dimethyl Fumarate in Newly Diagnosed, Treatment-Naïve Relapsing-Remitting Multiple Sclerosis: A Post Hoc Analysis of DEFINE and CONFIRM

An integrated analysis of DEFINE and CONFIRM focused on newly diagnosed (within 1 year before enrolment) DMD-naïve patients (n=678 patients in total; n=221 (2×240mg DMF); n=234 (3×240mg DMF); n=223 (placebo))[42]: After 2 years, ARR was significantly reduced in both treatment groups and placebo (56% (2×240mg DMF); 60% (3×240mg DMF)). Accordingly, the risk of a relapse was significantly reduced (54% (2×240mg DMF); 57% (3×240mg DMF)). The risk of sustained disability progression by EDSS was significantly reduced by 71% (2×240mg DMF) and 47% (3×240mg DMF).

In the subset of patients with MRI data (n=308 patients in total; n=99 (2×240mg DMF); n=109 (3×240mg DMF); n=100 (placebo)), a significant effect of DMF versus placebo was shown for the reduction of the number of new/enlarging T2 lesions (80% (2×240mg DMF); 81% (3×240mg DMF)), the reduction of the odds of having gadolinium-enhancing lesion after 2 years (92% (2×240mg DMF); 92% (3×240mg DMF)), and the reduction of the number of new T1 hypointense lesions (68% (2×240mg DMF); 70% (3×240mg DMF)).

The overall occurrence of AEs was comparable between the groups (97% (2×240mg DMF); 95% (3×240mg DMF); 92% (placebo)), again with a higher incidence of flushing, gastrointestinal symptoms, nasopharyngitis, and headaches. Treatment discontinuation as a consequence of an AE was observed in 12% (2×240mg DMF), 11% (3×240mg DMF), and 5% (placebo).

This post hoc analysis thus reveals similar efficacy and safety results for the subgroup of newly diagnosed, treatment-naïve patients.

20.4.3 Fumarate Treatment in Progressive Stages of Multiple Sclerosis: Results of a Small Observational Study

Fumarate treatment (mixed compound of DMF and ethylhydrogen fumarate as approved for treatment of psoriasis in Germany or DMF by pharmaceutical preparation) was evaluated in 26 patients (n=12 (primary progressive MS; PPMS); n=14 (secondary progressive MS; SPMS)).[43] These patients were older than respective RRMS populations in DEFINE and CONFIRM (mean age 56.6±10 (standard deviation; SD) years). Fumarate treatment was initiated in cases of failure to standard therapies and progression of disease according to the treating physician. In this study, patients were observed for safety, efficacy, and adherence to treatment including evaluation of EDSS, AEs/SAEs, full blood cell counts, renal and hepatic laboratory parameters. Over a mean follow-up period of 13.2 months (range 6–30 months), EDSS remained stable in 15 patients and decreased in 5 patients whereas 6 patients progressed despite treatment with an EDSS deterioration of more than 0.5 points.

The occurrence of AEs was generally low (5 of 26 patients) and of mild character, again mainly including gastrointestinal symptoms, flushing, and rhinorrhea. One patient stopped treatment for formal reasons (health insurance). Transient leukopenia (≤3000/μL) and lymphopenia (≤500/μL) were observed in one patient each. However, it has to be noted that dose titration was adapted to the occurrence of relative lymphopenia with a daily dose range of 215–860 mg (mixed compound) and 120–600 mg (compounded DMF). Other remarkable laboratory changes were not observed.

Major limitations of this observational study are obviously the small size and study design impeding a further stratification of PPMS versus SPMS or the inclusion of a proper control group. Moreover, the follow-up period is rather short and no data on the time course of disease progression before treatment initiation is shown. Still, being aware of the experimental data on cytoprotective effects of DMF, this study might be regarded as a first proof of concept warranting further controlled studies in progressive stages of MS.

20.5 SAFETY DATA IN THE POSTMARKETING SETTING AND OTHER APPLICATIONS OF FUMARIC ACID ESTERS

Due to the recent approval of DMF for RRMS, frequencies of rare potential severe side effects are as yet hard to determine. Single case descriptions of such events should therefore raise appropriate attention and require further observation if similar events come up in the future. Although different fumarate formulations and different conditions for treatment may hamper a direct comparison, interesting information can still be drawn from these experiences.

For instance, in the treatment of psoriasis with mixed compounds of DMF and ethylhydrogen fumarate, cases of progressive multifocal leukoencephalopathy (PML) have been observed. Two of these cases experienced a prolonged lymphopenia that was not followed by dose adaption or treatment interruption.[44,45] Another two PML cases were described in patients with a malignancy or previous treatment with immunosuppressive drugs or monoclonal antibodies.[46] The first case of PML has been described in an MS patient treated with DMF over 4.5 years, again with prolonged lymphopenia preceding PML.[47] As of November 2015, three additional cases of PML after treatment with DMF were reported. All four cases were aged 50–70 years, three of them had documented lymphopenia below 500/μl preceding PML diagnosis. This emphasizes the importance of the assessment of disease and medication history as a part of the risk–benefit assessment. Moreover, a safety monitoring plan should include regular full blood cell counts, especially during the first year of treatment with six to eight weekly intervals, thus more frequently than currently suggested in the summary of medical product characteristics and recommendations for treatment interruption or dose adaption.

Acute renal failure has been attributed to treatment with the mixed compound as approved for psoriasis in one patient with MS and three patients with psoriasis.[48] As monoethylesters of fumaric acid seem to bury a higher nephrotoxic potential,[49] it is thus far not clear whether this risk has to be addressed in DMF treatment as well; to our experience we do not assume that this is relevant for DMF.

20.6 CONCLUSIONS

DMF as an oral treatment option for RRMS can be used both as first-line treatment in DMD-naïve patients and patients with established RRMS diagnosis after a previous DMD regimen. Although not investigated in a larger trial setting and uncovered by respective treatment recommendations and thus off-label, it may be considered to use a fumarate treatment in selected cases of progressive forms of MS.

Experimental and MRI data support the concept of not only antiinflammatory, but also cytoprotective properties. Two pivotal phase-3 trials proved efficacy with favorable safety outcomes. Especially side effects during treatment initiation (flushing, gastrointestinal symptoms) can be partially mitigated by slow titration toward the full dosage of 2 × 240 mg DMF.

In the postmarketing setting and in other indications of fumarate treatment, effects on leuko- and lymphocyte counts appear to be of special clinical relevance and may be a factor contributing to the occurrence of rare severe adverse reactions such as PML. Safety monitoring plans are needed to address this risk, propose sensible laboratory check-up intervals, and give explicit recommendations for dose adaption or interruption. The incidences of other rare events will have to be determined in the postmarketing setting with growing patient numbers and patient years on DMF.

Although the overall positive profile of DMF in RRMS treatment and the comfortable application of the drug make DMF an attractive option for both patients and physicians, thorough counseling of patients including alternative DMD options and potential risks of each therapy is necessary to take individualized therapeutic decision in RRMS treatment.

References

1. Fox RJ, Miller DH, Phillips JT, et al. Placebo-controlled phase 3 study of oral BG-12 or glatiramer in multiple sclerosis. *N Engl J Med*. 2012;367(12):1087–1097.
2. Gold R, Kappos L, Arnold DL, et al. Placebo-controlled phase 3 study of oral BG-12 for relapsing multiple sclerosis. *N Engl J Med*. 2012;367(12):1098–1107.
3. Hafler DA, Slavik JM, Anderson DE, O'Connor KC, De Jager P, Baecher-Allan C. Multiple sclerosis. *Immunol Rev*. 2005;204:208–231.
4. Hohlfeld R, Wekerle H. Autoimmune concepts of multiple sclerosis as a basis for selective immunotherapy: from pipe dreams to (therapeutic) pipelines. *Proc Natl Acad Sci USA*. 2004;101(suppl 2):14599–14606.
5. van Horssen J, Witte ME, Schreibelt G, de Vries HE. Radical changes in multiple sclerosis pathogenesis. *Biochim Biophys Acta*. 2011;1812(2):141–150.
6. Schweckendiek W. Treatment of psoriasis vulgaris. *Med Monatsschr*. 1959;13(2):103–104.
7. Altmeyer PJ, Matthes U, Pawlak F, et al. Antipsoriatic effect of fumaric acid derivatives. Results of a multicenter double-blind study in 100 patients. *J Am Acad Dermatol*. 1994;30(6):977–981.
8. Nieboer C, de Hoop D, Langendijk PN, van Loenen AC, Gubbels J. Fumaric acid therapy in psoriasis: a double-blind comparison between fumaric acid compound therapy and monotherapy with dimethylfumaric acid ester. *Dermatologica*. 1990;181(1):33–37.
9. Mrowietz U, Christophers E, Altmeyer P. Treatment of severe psoriasis with fumaric acid esters: scientific background and guidelines for therapeutic use. The German fumaric acid ester consensus conference. *Br J Dermatol*. 1999;141(3):424–429.

10. Schimrigk S, Brune N, Hellwig K, et al. Oral fumaric acid esters for the treatment of active multiple sclerosis: an open-label, baseline-controlled pilot study. *Eur J Neurol*. 2006;13(6):604–610.
11. Kappos L, Gold R, Miller DH, et al. Efficacy and safety of oral fumarate in patients with relapsing-remitting multiple sclerosis: a multicentre, randomised, double-blind, placebo-controlled phase IIb study. *Lancet*. 2008;372(9648):1463–1472.
12. MacManus DG, Miller DH, Kappos L, et al. BG-12 reduces evolution of new enhancing lesions to T1-hypointense lesions in patients with multiple sclerosis. *J Neurol*. 2011;258(3):449–456.
13. Kappos L, Gold R, Miller DH, et al. Effect of BG-12 on contrast-enhanced lesions in patients with relapsing–remitting multiple sclerosis: subgroup analyses from the phase 2b study. *Mult Scler*. 2012;18(3):314–321.
14. Albrecht P, Bouchachia I, Goebels N, et al. Effects of dimethyl fumarate on neuroprotection and immunomodulation. *J Neuroinflamm*. 2012;9:163.
15. Linker RA, Lee DH, Ryan S, et al. Fumaric acid esters exert neuroprotective effects in neuroinflammation via activation of the Nrf2 antioxidant pathway. *Brain*. 2011;134(Pt 3):678–692.
16. Papadopoulou A, D'Souza M, Kappos L, Yaldizli O. Dimethyl fumarate for multiple sclerosis. *Expert Opin Investig Drugs*. 2010;19(12):1603–1612.
17. Treumer F, Zhu K, Glaser R, Mrowietz U. Dimethylfumarate is a potent inducer of apoptosis in human T cells. *J Invest Dermatol*. 2003;121(6):1383–1388.
18. Hoxtermann S, Nuchel C, Altmeyer P. Fumaric acid esters suppress peripheral CD4- and CD8-positive lymphocytes in psoriasis. *Dermatology*. 1998;196(2):223–230.
19. Spencer CM, Crabtree-Hartman EC, Lehmann-Horn K, Cree BA, Zamvil SS. Reduction of CD8(+) T lymphocytes in multiple sclerosis patients treated with dimethyl fumarate. *Neurol Neuroimmunol Neuroinflamm*. 2015;2(3):e76.
20. Bacharach-Buhles M, Pawlak FM, Matthes U, Joshi RK, Altmeyer P. Fumaric acid esters (FAEs) suppress CD 15- and ODP 4-positive cells in psoriasis. *Acta Derm Venereol Suppl Stockh*. 1994;186:79–82.
21. Ockenfels HM, Schultewolter T, Ockenfels G, Funk R, Goos M. The antipsoriatic agent dimethylfumarate immunomodulates T-cell cytokine secretion and inhibits cytokines of the psoriatic cytokine network. *Br J Dermatol*. 1998;139(3):390–395.
22. Zoghi S, Amirghofran Z, Nikseresht A, Ashjazadeh N, Kamali-Sarvestani E, Rezaei N. Cytokine secretion pattern in treatment of lymphocytes of multiple sclerosis patients with fumaric acid esters. *Immunol Invest*. 2011;40(6):581–596.
23. de Jong R, Bezemer AC, Zomerdijk TP, van de Pouw-Kraan T, Ottenhoff TH, Nibbering PH. Selective stimulation of T helper 2 cytokine responses by the anti-psoriasis agent monomethylfumarate. *Eur J Immunol*. 1996;26(9):2067–2074.
24. Ghoreschi K, Bruck J, Kellerer C, et al. Fumarates improve psoriasis and multiple sclerosis by inducing type II dendritic cells. *J Exp Med*. 2011;208(11):2291–2303.
25. Zhu K, Mrowietz U. Inhibition of dendritic cell differentiation by fumaric acid esters. *J Invest Dermatol*. 2001;116(2):203–208.
26. Stoof TJ, Flier J, Sampat S, Nieboer C, Tensen CP, Boorsma DM. The antipsoriatic drug dimethylfumarate strongly suppresses chemokine production in human keratinocytes and peripheral blood mononuclear cells. *Br J Dermatol*. 2001;144(6):1114–1120.
27. Wierinckx A, Breve J, Mercier D, Schultzberg M, Drukarch B, Van Dam AM. Detoxication enzyme inducers modify cytokine production in rat mixed glial cells. *J Neuroimmunol*. 2005;166(1–2):132–143.
28. Lin SX, Lisi L, Dello Russo C, et al. The anti-inflammatory effects of dimethyl fumarate in astrocytes involve glutathione and haem oxygenase-1. *ASN Neuro*. 2011;3(2).
29. Wiesner D, Merdian I, Lewerenz J, Ludolph AC, Dupuis L, Witting A. Fumaric acid esters stimulate astrocytic VEGF expression through HIF-1alpha and Nrf2. *PLoS One*. 2013;8(10):e76670.
30. Vandermeeren M, Janssens S, Borgers M, Geysen J. Dimethylfumarate is an inhibitor of cytokine-induced E-selectin, VCAM-1, and ICAM-1 expression in human endothelial cells. *Biochem Biophys Res Commun*. 1997;234(1):19–23.
31. Benardais K, Pul R, Singh V, et al. Effects of fumaric acid esters on blood–brain barrier tight junction proteins. *Neurosci Lett*. 2013;555:165–170.
32. Kunze R, Urrutia A, Hoffmann A, et al. Dimethyl fumarate attenuates cerebral edema formation by protecting the blood–brain barrier integrity. *Exp Neurol*. 2015;266:99–111.
33. Loewe R, Holnthoner W, Groger M, et al. Dimethylfumarate inhibits TNF-induced nuclear entry of NF-kappa B/p65 in human endothelial cells. *J Immunol*. 2002;168(9):4781–4787.

34. Liu Y, Kern JT, Walker JR, Johnson JA, Schultz PG, Luesch H. A genomic screen for activators of the antioxidant response element. *Proc Natl Acad Sci USA*. 2007;104(12):5205–5210.
35. Scannevin RH, Chollate S, Jung MY, et al. Fumarates promote cytoprotection of central nervous system cells against oxidative stress via the nuclear factor (erythroid-derived 2)-like 2 pathway. *J Pharmacol Exp Ther*. 2012;341(1):274–284.
36. Chen H, Assmann JC, Krenz A, et al. Hydroxycarboxylic acid receptor 2 mediates dimethyl fumarate's protective effect in EAE. *J Clin Invest*. 2014;124(5):2188–2192.
37. Arnold DL, Gold R, Kappos L, et al. Effects of delayed-release dimethyl fumarate on MRI measures in the Phase 3 DEFINE study. *J Neurol*. 2014;261(9):1794–1802.
38. Arnold DL, Gold R, Kappos L, et al. Magnetization transfer ratio in the delayed-release dimethyl fumarate DEFINE study. *J Neurol*. 2014;261(12):2429–2437.
39. Bar-Or A, Gold R, Kappos L, et al. Clinical efficacy of BG-12 (dimethyl fumarate) in patients with relapsing-remitting multiple sclerosis: subgroup analyses of the DEFINE study. *J Neurol*. 2013;260(9):2297–2305.
40. Hutchinson M, Fox RJ, Miller DH, et al. Clinical efficacy of BG-12 (dimethyl fumarate) in patients with relapsing-remitting multiple sclerosis: subgroup analyses of the CONFIRM study. *J Neurol*. 2013;260(9):2286–2296.
41. Kita M, Fox RJ, Gold R, et al. Effects of delayed-release dimethyl fumarate (DMF) on health-related quality of life in patients with relapsing-remitting multiple sclerosis: an integrated analysis of the phase 3 DEFINE and CONFIRM studies. *Clin Ther*. 2014;36(12):1958–1971.
42. Gold R, Giovannoni G, Phillips JT, et al. Efficacy and safety of delayed-release dimethyl fumarate in patients newly diagnosed with relapsing-remitting multiple sclerosis (RRMS). *Mult Scler*. 2014;21(1):57–66.
43. Strassburger-Krogias K, Ellrichmann G, Krogias C, Altmeyer P, Chan A, Gold R. Fumarate treatment in progressive forms of multiple sclerosis: first results of a single-center observational study. *Ther Adv Neurol Disord*. 2014;7(5):232–238.
44. Ermis U, Weis J, Schulz JB. PML in a patient treated with fumaric acid. *N Engl J Med*. 2013;368(17):1657–1658.
45. van Oosten BW, Killestein J, Barkhof F, Polman CH, Wattjes MP. PML in a patient treated with dimethyl fumarate from a compounding pharmacy. *N Engl J Med*. 2013;368(17):1658–1659.
46. Sweetser MT, Dawson KT, Bozic C. Manufacturer's response to case reports of PML. *N Engl J Med*. 2013;368(17):1659–1661.
47. Anonymous. *Rote-Hand-Brief: Tecfidera (Dimethylfumarat): Bei einer Patientin mit schwerer und lang anhaltender Lymphopenie trat eine progressive multifokale Leukenzephalopathie (PML) auf.* http://akdae.de/Arzneimittelsicherheit/RHB/20141204.pdf; 2014. Last accessed 15.12.14.
48. Arzneimittelkommission der deutschen Ärzteschaft. Akutes Nierenversagen unter der Behandlung mit Fumarsäure bei Multipler Sklerose. *Dtsch Ärzteblatt*. 2014;111(25):1177–1178.
49. Hohenegger M, Vermes M, Sadjak A, Egger G, Supanz S, Erhart U. Nephrotoxicity of fumaric acid monoethylester (FA ME). *Adv Exp Med Biol*. 1989;252:265–272.

CHAPTER

21

Emerging Therapies for Multiple Sclerosis

M.S. Freedman

University of Ottawa, Ottawa, ON, Canada

OUTLINE

21.1 OVERVIEW

The treatment landscape for multiple sclerosis (MS) is rapidly evolving and a number of new therapies are poised to enter the marketplace. There are several of these that are still in early phase II (proof of concept) studies with others already entering phase III. This review

Translational Neuroimmunology in Multiple Sclerosis
http://dx.doi.org/10.1016/B978-0-12-801914-6.00021-0

will walk through some of these therapies and offer an opinion on their potential role in treating MS. Importantly, where to introduce the therapies within the treatment cascade will be rather subjective and typically is based on weighing the risks and benefits for any particular treatment in a given patient.

21.2 INTRODUCTION

The notion of disease-modifying medications for MS arose as late as the 1990s, when interferon-β (IFNβ) and glatiramer acetate (GA) were both shown to alter the natural history of MS by reducing relapses and stalling the build-up of disability. Now, after nearly three decades of treatment with the disease-modifying therapies, we are seeing many new agents entering or poised to enter the world of current therapy for relapsing or even progressive forms of MS. ClinicalTrials.gov lists >800 ongoing trials for MS with many new agents in development or in phase I and II testing. An improved understanding of the pathoimmunology of the disease has revealed many new potential targets for therapeutic intervention. Monoclonal antibodies directed at specific cell types via selective cell expression molecules, surface adhesion molecules, or activation markers, or even against the inflammatory mediators such as cytokines, are actively being developed, tested, and proven. Discovery of molecules that may inhibit remyelination or repair are being identified and antagonists tested. Cell-based therapies and even vaccines are all in the early exploratory stages.

There is no way of knowing just what immune mechanism is dominating in any given individual's central nervous system (CNS) at any time, but hopefully with the development of some biomarkers, we might yet be able to better select a particular therapy as being more beneficial or discount others as being futile or less likely to produce a treatment response. Treatment choices now depend on a full understanding of a patient's clinical condition, weighing prognostic factors, comorbidities, and personal acceptance of risk. Therapy has really evolved into a truly personalized approach.

A concept of MS immunopathogenesis (see Fig. 21.1) forms the basis for understanding why many of the therapies have been developed and where and how in the disease process they might make a difference. There are many options for interfering with the perceived disease process at almost every step along the pathway:

1. Interference with antigen (Ag) presentation (eg, GA)
2. Immunomodulation (eg, IFNβ)
3. Interference with T-cell trafficking (eg, sphingosine-1-phosphate receptor antagonists)
4. T-cell depletion (eg, alemtuzumab)
5. B-cell depletion (eg, ocrelizumab)
6. Binding and traversing the blood–cerebrospinal fluid (CSF) barrier (eg, natalizumab)
7. Anticytokine or chemokine therapy (eg, daclizumab)
8. Antioxidants (eg, dimethyl fumarate)
9. Neuroprotection (eg, laquinimod)
10. Remyelination (eg, anti-LINGO-1)

As each new treatment becomes available, we will have to try and position it in our current management scheme, which isn't always easy given that most of the trials that lead to

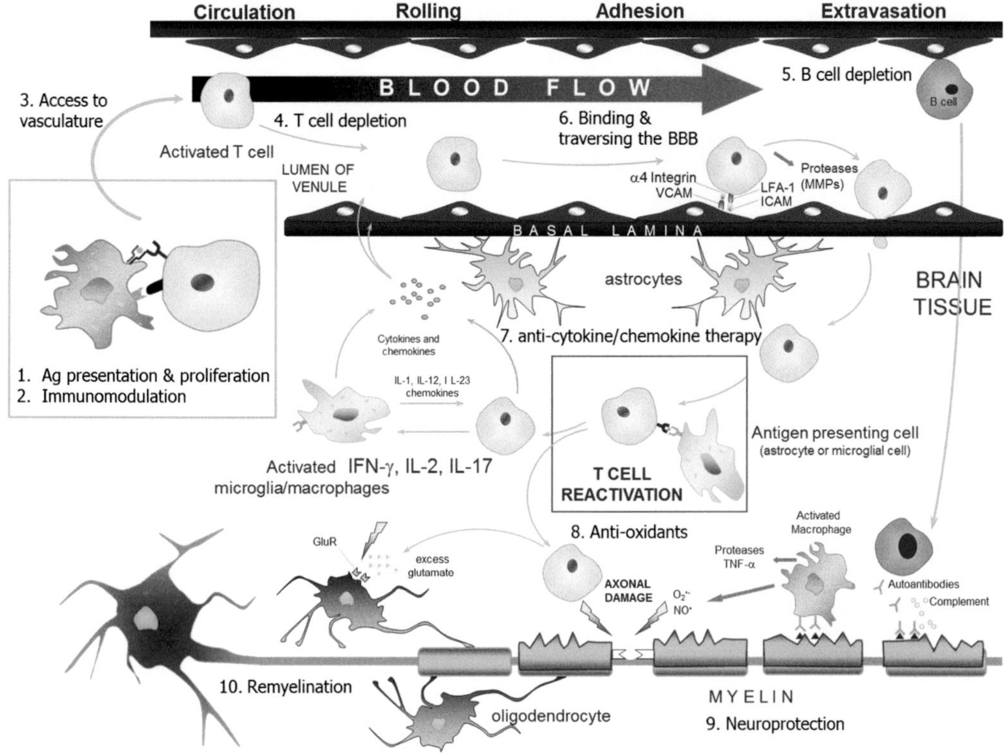

FIGURE 21.1 Theories of immunopathogenesis.

their approval are either against a placebo or a different active comparator such as IFNβ or teriflunomide (Fig. 21.2).

There has been a flurry of activity concerning MS therapies and a number of the important trials have now been published or presented. Most of these address relapsing MS, with one focused on primary-progressive MS (PPMS) and ongoing trials in secondary-progressive MS (SPMS).

21.3 TREATMENTS FOR RELAPSING FORMS OF MULTIPLE SCLEROSIS

21.3.1 S1p Receptor Agonists

The agents siponimod, ponesimod, ceralifimod, and ozanimod are similar in principle to fingolimod, currently in use for relapsing MS, but are all prephosphorylated and differ in their specificity for the diverse sphingosine-1-phosphate receptors. Fingolimod suffers from having some specificity for S1p1 and S1p4 found on lymphocytes, but also has considerable effect on S1p3 receptors on the heart where it induces first-dose bradycardia. So far, all of these have completed phase II studies and some of have moved into phase III registry studies.

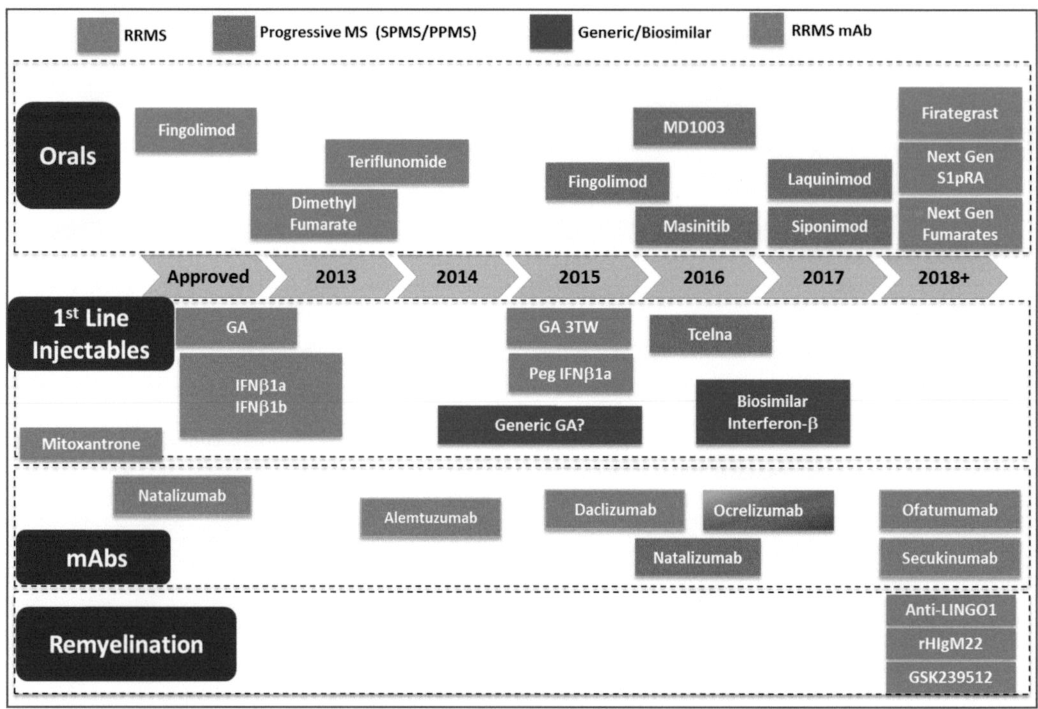

FIGURE 21.2 The evolving treatment landscape.

Siponimod

S1p signaling is important for immune, cardiovascular, and CNS function. Like its FDA approved predecessor fingolimod, siponimod is an oral S1p signaling modulator. While fingolimod binds four out of five known S1p receptors, siponimod is selective for receptor subtypes 1 and 5. This selectivity may allow for an improved adverse event profile.

Even though it is currently being assessed in a phase III placebo-controlled study in SPMS, it demonstrated efficacy in the way of reducing new lesion formation in the BOLD study, a phase II randomized double-blind placebo-controlled (RDBPC) trial that utilized a novel dose-finding design[1] to examine three doses (0.5, 2, and 10 mg). Then following an interim analysis at 3 months (0.25 and 1.25 mg), it was ultimately determined that 2 mg was optimal. Similar to fingolimod, siponimod induced a dose-dependent reduction in lymphocyte counts reaching a nadir at 7 days before stabilizing. The primary outcome (a significant reduction in combined unique active lesions at 3 months) was observed at doses 0.5 mg or higher, with a peak at 2 mg. Annualized relapse rate (ARR) reductions calculated at 6 months showed a statistically significant (0.58 vs 0.23 placebo vs 2 mg siponimod; $p<0.05$) 61% reduction with the 2 mg dose or a 90% increase in relapse-free patients (vs 73% placebo). Of interest were the adverse events, generally dose dependent, occurring most frequently at the 10 mg dose. Bradycardia (4–28% in siponimod vs 2–6% in placebo) was minimized using a dose-titration approach. There was one case of myocardial infarction (10 mg dose), second-degree atrioventricular block (one case at 10 mg and three cases at 2 mg of siponimod), and one drug-related coronary artery disease-associated death (1.25 mg siponimod). Headache also figured in as a prominent complaint (31% vs 8% 2 mg vs placebo) that was dose dependent.

Ponesimod

This S1p receptor agonist is similar to the other S1p receptor agonists, but unlike fingolimod, which is a structural analog of sphingosine, ponesimod is at least 10-fold more potent on the S1p1 receptor than on any other S1p receptor subtype. Ponesimod has a rapid onset of action, with maximal plasma concentrations observed 2.5–4 h after dosing and maximum lymphocyte count reduction achieved.

Six hours after dosing, ponesimod has a half-life of approximately 32 h, and following treatment discontinuation, mean lymphocyte count returns to normal range within 7 days. It too was tested in a phase II dose-finding and proof-of-concept study examining 10, 20, and 40 mg taken orally once a day.[2]

In a RDBPC study, 464 patients were randomized 1:1:1:1 to receive once-daily oral ponesimod 10, 20, or 40 mg, or placebo for 24 weeks with primary outcome being cumulative number of gadolinium-enhancing lesions (GELs) per patient on scans performed every 4 weeks from weeks 12–24. Secondary end points were the ARR along with safety. All patients randomized to ponesimod initially received ponesimod 10 mg (days 1–7), then on day 8, patients randomized to receive ponesimod 20 or 40 mg were titrated up to the 20 mg dose and patients randomized to the 10 mg dose were mock titrated. On day 15, patients randomized to receive ponesimod 40 mg were titrated up to the 40 mg dose, but patients randomized to ponesimod 10 or 20 mg were mock titrated. Placebo-randomized patients were mock titrated on days 8 and 15.

The mean cumulative number of GELs at weeks 12–24 was significantly lower in all the ponesimod groups compared with placebo: 10 mg (43% reduction; $p = 0.0318$), 20 mg (83% reduction; $p < 0.0001$), and 40 mg (77% reduction; $p < 0.0001$). Compared with placebo, the mean cumulative number of combined unique active lesions at weeks 12–24 was reduced by 42% ($p = 0.0318$), 80% ($p < 0.0001$), and 73% ($p < 0.0001$), with ponesimod 10, 20, and 40 mg, respectively. There was even a small mean increase in brain volume at week 24 observed with ponesimod (0.02%, 0.05%, and 0.23% in the 10, 20, and 40 mg groups, respectively) compared with a mean decrease (−0.26%) seen with the placebo. The mean annualized rate was lower with all doses of ponesimod; however, only the 40 mg dose reached statistical significance with a 52% reduction (0.25, 40 mg vs 0.53, placebo; $p = 0.0363$). Though the other doses were associated with lower ARRs (0.33, 10 mg and 0.42, 20 mg) corresponding to a 67% and 58% reduction relative to the placebo group, these reductions did not reach statistical significance.

The proportion of patients with ≥1 treatment-emergent adverse events was similar across all the ponesimod and the placebo groups, with more frequently reported anxiety, dizziness, dyspnea, increased alanine aminotransferase (ALT), influenza, insomnia, and peripheral edema in the ponesimod groups with both dyspnea and peripheral edema somewhat seen in a more dose-dependent fashion. The onset of dyspnea leading to treatment discontinuation in the ponesimod group usually occurred within the first month of treatment and resolved following discontinuation. Correlating with dose-dependent dyspnea was parallel decrease in forced expiratory volume in 1 s (FEV1) in ponesimod-treated patients (mean decrease in FEV1 from baseline to week 24 was 0.6%, 5.2%, 6.0%, and 10.3% in the placebo and ponesimod

10, 20, and 40mg groups, respectively). FEV1 returned to within normal range within 1week of treatment discontinuation. Four cases of macular edema were reported, all starting within 3months of treatment initiation; two in the ponesimod 20mg group (only one confirmed by optical coherence tomography) resolving after treatment discontinuation, one in the 20mg group, also resolving during study treatment, and a fourth case actually in the placebo group also resolving during the study treatment. Two malignancies were reported: one case of breast cancer in the 10-mg ponesimod group and one case of carcinoma of the cervix in the placebo group.

Ponesimod 20mg is currently being studied in a 2-year phase III RDBPC trial with an active comparator, teriflunomide 14mg.

Ceralifimod

Another new selective $S_{1,5}$ sphingosine-1-phosphate receptor agonist known as ceralifimod completed a 26-week phase II RDBPC trial examining three potential doses (0.05, 0.1, and 0.15mg) with a primary outcome of MRI.[3] All doses reduced GEL formation by 72–92% and new or enlarging T2 lesions by 71–82%, with the maximal seen with 0.1mg (92% and 82%; $p<0.0001$). ARRs were lower with all three doses compared with placebo, reaching significance only with the 0.10-mg dose only (0.56 placebo vs 0.17 0.1-mg ceralifimod; $p=0.0069$), which showed a 70% reduction relative to placebo.

Safety issues included first-dose bradycardia, which was dose dependent. Others included a small (1–2mm Hg) increase in mean systolic and diastolic blood pressure after week 6 and mild bronchoconstriction was seen on pulmonary function tests at the 0.05 and 0.15mg doses. Infections were overall similar with ceralifimod and placebo, but nondisseminated *Herpes* infections were more frequent with the 0.05 and 0.10mg doses. Gastrointestinal disorders were more common with all doses of ceralifimod than with placebo (23% vs 11% 0.1mg ceralifimod vs placebo). Grade 4 lymphopenia occurred in five patients only at the higher doses: 0.10mg, one patient (1%); 0.15mg, four patients (4%).

Ozanimod

Ozanimod is another S1p receptor agonist with little affinity for the S1p3 and S1p4 receptors and a short half-life of some 19h compared with the other molecules, perhaps offering it the advantage of a more rapid lymphocyte recovery once it is stopped. It also crosses the blood–CSF barrier. A higher volume of distribution and delayed absorption may make it more tolerable in terms of fewer first-dose cardiac effects such as bradycardia.

The phase II portion of a combined phase II/III trial called RADIANCE[3] concerned the safety, efficacy, and tolerability of two doses of ozanimod (0.5 or 1mg) versus placebo once daily for 24weeks. To minimize first-dose cardiac effects, the drug was titrated from 0·25mg up to 1mg over 7days. MRI was performed monthly after week 8 and the primary end point was the cumulative number of total GELs in weeks 12–24. In total, 258 participants were randomly assigned: 88 to placebo, 87 to ozanimod 0.5mg, and 83 to ozanimod 1mg; 252 (97.7%) completed their assigned treatment. The mean cumulative number of GELs in weeks 12–24 was reduced 86% from 11.1 with placebo to 1.5 with both ozanimod doses ($p<0.0001$). At week 24, participants treated with ozanimod had fewer total number of GELs compared with placebo 91% reduction, 0.5mg; 94% reduction, 1mg dose (both $p<0.0001$). The cumulative number of new or enlarging T2 lesions in weeks 12–24 was decreased by 84% and

91% with ozanimod 0.5 and 1 mg, respectively, compared with placebo (both $p<0.0001$). In terms of clinical secondary outcomes, the proportions of patients who were relapse-free were 77%, 83%, and 89% (placebo, 0.5, 1 mg ozanimod, respectively); however, ARR reduction only trended in favor of the higher 1 mg dosage (0.5 placebo vs 0.35, 0.5 mg, $p=0.2714$; 0.24, $p=0.0531$).

There were no serious infectious or cardiac adverse effects, and no cases of macular edema with ozanimod. Aside from the nonspecific side effects of headache and nasopharyngitis (actually less than in placebo), there was some orthostatic hypotension reported on the first day of dosing more commonly with ozanimod.

21.3.2 Firategrast

Firategrast is an oral small-molecule α4β-integrin antagonist thought to be an alternative to natalizumab, with a shorter half-life (2.5–4.5 h vs 11 days). However, in a multicenter phase II RDBPC study of 343 relapsing MS patients receiving either 900 mg (women) or 1200 mg (men) twice daily there was only a 49% reduction in GELs relative to placebo.[4] This effect was less pronounced with the 600 mg dose and paradoxically reversed with the 150 mg dose (79% increase in GELs relative to placebo), but this worsening was not seen with any other MRI metric. Notably, there was individual variability in blood firategrast levels not entirely explicable by dose administered and higher blood levels were associated with better MRI activity response. Further investigation is required to understand these results. Though relapse rates were lower in all firategrast doses (0.35, 0.32, 0.3 for the 150, 600, or 900/1200 bid doses), they were not statistically reduced compared with the placebo (0.41).

Most adverse events were similar across study groups; however, a dose-dependent increase in urinary tract infections and vomiting was noted in the firategrast groups. Another study of 46 firategrast-treated patients showed that lymphocytic access to CSF was only modestly reduced.[5] No cases of progressive multifocal leukoencephalopathy were observed during the study period; however, exposure was quite limited to a short 24 weeks.

With only modest effects on MRI and insignificant effects on relapses, there was no indication that firategrast would achieve similar results to natalizumab.

21.3.3 Ofatumumab

There is particular interest in anti-B-cell therapies of late, given their potential for interacting differently compared with current therapies that target mainly the T cell. The mechanism of these antibodies is not fully understood, but it is unlikely that benefit belies the ability to reduce antibody (Ab) formation. Instead, the postulate involves particular reduction in B-cell Ag presentation to T cells.[6]

Ofatumumab is a fully humanized Ab, thus theoretically reducing the immunogenicity compared with other anti-CD20 Abs. In an early phase II RDBPC trial of ofatumumab in 38 relapsing MS patients, there was a >99% reduction in total GELs ($p<0.001$) and a significant reduction in the number of new GELs and new or enlarging T2 lesions within the first 24 weeks,[7] which is similar to the MRI effect seen with ocrelizumab (see later). Ofatumumab was administered as two IV infusions 2 weeks apart at 100, 300, or 700 mg per infusion and was relatively well tolerated, with two subjects discontinuing the study, one with rash with bronchospasm and cough, and the other for pharyngeal edema with erythema and pruritus. B-cell suppression and MRI findings were similar across treatment groups, suggesting that lower doses of ofatumumab may have sufficient therapeutic effect. A larger study is needed to more definitively determine MRI activity suppression rates, relapse modulation, duration of effect, and optimal dose selection as well as to elucidate the spectrum of adverse events.

A Phase IIb study to further characterize ofatumumab safety and efficacy involved 232 subjects with RRMS randomized to 3, 30, or 60 mg of subcutaneous ofatumumab every 12 weeks or 60 mg of subcutaneous ofatumumab every 4 weeks, or placebo followed by 3 mg of subcutaneous ofatumumab at week 12. Found was a clear separation from placebo on the cumulative number of GELs over 12 weeks in all dose groups of ofatumumab compared to placebo ($p < 0.001$). For the primary end point, a 65% reduction in the cumulative number of new GELs over weeks 0–12 was observed for all doses ($p < 0.001$). In weeks 4–12, there was a ≥90% reduction in the cumulative number of new GELs for all the cumulative doses of ofatumumab ≥30 mg ($p < 0.001$).

There were no unexpected safety findings in the study. From weeks 0–12, injection-related reactions were the most common adverse reaction and were observed in 52% of subjects receiving ofatumumab compared to 15% of subjects receiving placebo. There were five serious adverse events reported; all subjects received the 60 mg ofatumumab dose yet none withdrew from the study.

Given the strong results with ocrelizumab (see next), ofatumumab will be tested in a phase III trial, likely against an active comparator, with the advantage being it is self-administered compared with ocrelizumab.

21.3.4 Ocrelizumab

Ocrelizumab is a recombinant monoclonal anti-CD20 Ab that is structurally similar to rituximab (nearly 89% homologous), but more humanized. In the phase II study, 218 RRMS patients were randomized to placebo, low dose (600 mg), or high dose (2000 mg) i.v. ocrelizumab given on days 1 and 15, or IFNβ-1a i.m. (30 μg weekly) and followed for 24 weeks.[8] The primary end point was the total number of GELs counted at four weekly intervals starting from weeks 12–24. Ocrelizumab (both doses) significantly reduced the total number of GELs at weeks 12–24 ($p < 0.0001$) compared to placebo and IFNβ-1a (relative risk reductions 89% 600 mg, 96% 2000 mg groups). ARRs were also significantly reduced: 80% in 600 mg (0.13, $p = 0.0005$); 73% in 2000 mg (0.17, $p = 0.0014$) compared to placebo patients (0.64).

Safety

Though overall rates of adverse events were similar between the treatment arms, there was one unusual death in the higher dose 2000 mg group described as a severe inflammatory response syndrome. Both groups of ocrelizumab suffered from infusion-related reactions.

The strong results obtained in the phase II study strongly recommended that a phase III study be conducted. Two studies (OPERA I and II) have recently been completed. Both produced nearly similar results in two very similar populations of patients. In both studies, ocrelizumab was given at first as two doses of 300 mg 2 weeks apart followed by a single dose of 600 mg every 6 months. The active comparator was IFNβ-1a 44 μg sc tiw. The studies were both double-blind and double-dummy controlled, meaning all patients received both infusions and sc injections. The primary outcome was ARR at 96 weeks with numerous secondary outcomes including Expanded Disability Status Scale (EDSS) progression, MRI, and quality of life.

Approximately 400 patients were randomized to all arms in both studies and approximately 40% had GEL at baseline. Both studies demonstrated highly statistically significant reductions in all outcome measures relative to the interferon treatment. ARRs were reduced by 46% and 47% in both studies relative to IFNβ-1a (0.16 vs 0.29, $p<0.0001$). Though the intention was to pool the results from the two studies to look at EDSS progression due to perceived underpowering, the results were independently significant in both studies such that pooling was unnecessary. In the pooled results, both 12- and 24-week confirmed disability progression was reduced significantly by 40%. The proportion of 12- or 24-week progressions was reduced: 15.2 IFNβ-1a versus 9.8% ocrelizumab, $p=0.0006$; 12% versus 7.6% ocrelizumab, $p=0.0025$; respectively. A highly significant 94% and 95% reduction ($p<0.0001$) in GELs relative to IFNβ-1a was seen in both studies. Other outcomes include a 77% and 83% reduction in new or enlarging T2 lesions in the two studies ($p<0.0001$) and a 24% reduction in loss of brain volume ($p=0.0001$). No evidence of disease activity (NEDA) for ocrelizumab-treated patients was improved by 64% (OPERA I) or 89% (OPERA II) ($p<0.0001$), going from 25% to 29% for IFNβ-1a to 48% for ocrelizumab.

There were few safety issues outside of infusion reactions being more common with ocrelizumab and injection reactions with the interferon. There were no opportunistic infections. Of concern, however, were two cases of breast cancer in the ocrelizumab arms, along with renal cancer and melanoma and no such organ cancers in the IFNβ-1a arm. Added to the four breast cancers noted with ocrelizumab in the primary-progressive study ORATORIO (see later), this raises the possibility that treatment with this agent may interfere with tumor surveillance.

21.3.5 Anti-LINGO-1

The *l*eucine-rich repeat and *i*mmunoglobulin domain-containing *n*eurite out*g*rowth inhibitor N*o*go receptor-interacting protein-1 (LINGO-1) is a glycoprotein expressed on neurons and oligodendrocyte progenitor cells that negatively regulates myelination through distinct mechanisms involving activation of RhoA-GTPase as well as nerve growth factor and the tyrosine kinase A receptor. Blockade of LINGO-1 has been shown to result in remyelination in various animal models of CNS demyelination. A human aglycosyl IgG_1 mAb that binds LINGO-1 with high affinity may be able to enhance remyelination via blocking LINGO-1.[9]

Two phase II trials examining dose, efficacy, and safety of anti-LINGO-1 are ongoing. The first, called RENEW, assessed the effects of anti-LINGO-1 in recovery from optic neuritis, while the second, called SYNERGY, is examining the benefits in relapsing forms of MS when added to IFNβ-1a.

RENEW was an RDBPC study looking at optic nerve recovery neurophysiologically following acute optic neuritis. Following treatment of both groups of 41 patients with high-dose methylprednisolone, half the patients received 100 mg/kg anti-LINGO-1 monthly for 6 months. The primary end point was optic nerve conduction latency at week 24 in the affected eye compared with the unaffected fellow eye at baseline, as measured using full-field visual-evoked potential (FF-VEP). The trial showed that in the per-protocol group analysis at week 24 there was a 34% improvement in the recovery of optic nerve latency as measured by FF-VEP relative to placebo (−3.48 ms, $p=0.05$), whereas the difference did not reach significance in the intention to treat (ITT) population (−3.48 ms, $p=0.33$). There were also no significant effects on visual acuity. Severity and incidence of adverse events were comparable across the treatment arms.

This agent and the study design is the first to demonstrate a therapy that changes an outcome consistent with remyelination. Though optic neuritis is not always MS, the ongoing study in MS (SYNERGY) is hopeful to show similar evidence.

21.3.6 Daclizumab

Daclizumab is a humanized monoclonal Ab against the α-chain of the high affinity IL-2 receptor (CD25). CD25 is expressed at low levels on resting T cells but is rapidly upregulated after T-cell activation, which enhances high-affinity interleukin-2 signal transduction. Because CD25 antagonism selectively inhibits activated T cells, daclizumab treatment was postulated to be useful in patients with autoimmune conditions characterized by abnormal T-cell responses, such as MS. CD25 antagonism causes expansion of a regulatory subset of natural killer (NK) cells known as $CD56^{bright}$ NK cells. This expansion was associated with a reduction in MS disease activity, presumably through the $CD56^{bright}$ NK cell-mediated lysis of autologous activated T cells that would be responsible for attacking the CNS in MS.[10]

The earliest large study with daclizumab was known as CHOICE[11] and was an add-on to IFNβ. That study did show that the combination was effective at reducing Gd+ lesions. Since then, the molecule underwent a refinement in production and changed company hands. The SELECT study looked at daclizumab high-yield process (HYP) in a 52-week study of once-monthly sc treatment of 150 or 300 mg daclizumab in an RDBPC trial of monotherapy in 600 patients with RRMS.[12] Patients had at least one relapse in the 12 months before randomization or one new Gd+ lesion in the prior 6 weeks. The agent was well tolerated with >90% of entered patients completing the study.

The primary end point was ARR (0.46 placebo, 0.21 150-mg daclizumab, 0.23 300-mg daclizumab) corresponding to a 50–54% reduction in ARR relative to placebo ($p<0.001$ and $p<0.0002$, respectively). Additionally, more than 80% of the daclizumab-treated patients were relapse-free at 52 weeks compared with 64% of placebo patients. There was a trend toward an effect on confirmed 3-month EDSS progression; a 57% reduction in the 150-mg group ($p=0.021$) and a 43% reduction with 300 mg dose that did not reach statistical significance ($p=0.09$).

Results of an MRI substudy in 309 patients showed that compared with placebo, there was a 69–78% reduction in new or enlarging Gd+ lesions between weeks 8 and 24 ($p<0.0001$, 150 and 300 mg daclizumab). Looking at this outcome in the overall population of 600 patients, there was a 79–86% reduction in new Gd+ lesions at week 52 versus placebo. Similarly, there was an impact on new or enlarging T2 lesions, with a reduction of 70–79% in the whole cohort at week 52.

Adverse events and serious adverse events were similar in all the groups. There was one death in the trial in a patient with a psoas abscess that was not diagnosed before death. The most common adverse events were nasopharyngitis, upper respiratory tract infection, and headache. Serious infections occurred in 1% and 3% of the daclizumab groups. There were four malignancies: one in each of the placebo and 150 mg groups, and two in the 300 mg group. Raised liver function tests LFTs were also noted, with levels >5× upper limit of normal (ULN) in 4% of daclizumab-treated patients. Interestingly, these increases generally occurred later in treatment, with a median of 308 days; all the increases resolved spontaneously.

The encouraging phase II study showing no real advantage of the higher dose of daclizumab HYP prompted a subsequent phase III (DECIDE)[13] conducted with only 150 mg every 4 weeks sc compared to IFNβ-1a 30 μg i.m. weekly. Slightly more than 900 patients were randomized to each arm of the study that ran for 2–3 years with median treatment duration of 109 and 111 weeks in the daclizumab and IFNβ-1a arms, respectively. The primary outcome of ARR demonstrated a significant 45% reduction with daclizumab compared with IFNβ-1a (0.22 vs 0.39, $p<0.001$). At week 144, the proportion of patients experiencing sustained 12 week EDSS progression was not significantly different between the treatment arms (16% daclizumab vs 20% IFNβ-1a, $p=0.16$). The number of new or newly enlarged hyperintense lesions on T2-weighted images at week 96 was 54% lower in the daclizumab HYP group than IFNβ-1a ($p<0.001$).

Adverse events in this study unique to daclizumab were serious skin reactions and possible autoimmune-type hepatitis. Cutaneous events (eg, rash, eczema, seborrheic dermatitis, acne, erythema, and pruritus) were reported in 37% of the daclizumab versus 19% IFNβ-1a. Twice as many daclizumab patients experienced ALT elevations >5× ULN. There were no significant signals of an increased risk to opportunistic infections.

The convenience of daclizumab as an injectable (once monthly) is superior to any of the current injectables including the higher dosed GA or the pegylated version of IFNβ-1a; however, the efficacy is only slightly better than IFNβ-1a but is greatly offset by tolerability issues and potential side effects, particularly the skin. It is therefore difficult for most researchers to position this medication. Perhaps it will be reserved for patients with a suboptimal response to front-line treatments, but too risk averse to consider fingolimod, natalizumab, or even alemtuzumab. Maybe it will be used temporarily to gain control over disease before reverting back to front-line (safer) medications.

21.4 TREATMENTS FOR PROGRESSIVE FORMS OF MULTIPLE SCLEROSIS

21.4.1 Laquinimod

Laquinimod was derived from roquinimex (linomide), an immunomodulatory drug that was first developed in the early 1980s. Phase III studies of linomide were halted in the 1990s because of the occurrence of two cardiovascular-related deaths and at least eight nonfatal myocardial infarctions. Modifications were made that were thought to reduce the side effects without compromising the potential beneficial immunomodulatory properties. The exact mechanism of action is unknown, but many properties have been demonstrated in at least in vitro and animal studies. Of great interest is that it appears laquinimod can cross into the CNS, achieving concentrations 7–8% that of peripheral blood,[14] an amount that might double in the presence of some blood–brain barrier permeability as might be seen in experimental autoimmune encephalomyelitis (EAE) or MS. Laquinomod was effective in EAE experiments, possibly due to a reduction of leukocyte trafficking into the CNS or the modulation of inflammatory cytokine production. Other properties include the possible protection of axonal integrity, possibly via the production of brain-derived neurotrophic factor.

The first phase III study, ALLEGRO,[15] compared laquinimod 0.6 mg once daily to placebo in 1106 patients with RRMS. Mean EDSS at baseline was 2.6 in both groups. Disease duration prior to the study was 8.7 and 8.6 years, respectively. One or more Gd+ lesions were present in 45.7% of placebo and 40.4% of the laquinimod arm. The primary end point was the number of confirmed relapses (and the ARR); secondary end points included the cumulative number of Gd+ and new or newly enlarging T2 lesions, as well as disability (EDSS) and the Multiple Sclerosis Functional Composite (MSFC) at 24 months. There was some drop-out of patients in this study, with 79% and 77% of laquinimod and placebo-treated patients completing the follow-up. The main reasons for discontinuation were adverse events and then the perception of patients that their disease was not under control.

Laquinimod produced a modest, yet statistically significant 23% reduction in ARR rate compared with placebo (see Table 21.1). The absolute reduction was quite small (0.091), yielding a number needed to treat (NNT) of 11.

Treatment was also associated with a 36% reduction in EDSS disability progression versus placebo for 3-month confirmed progression ($p=0.0122$) (but amounting to only a roughly 4% absolute change in the percent progressing compared to placebo). This reduction in progression increased to a 48% reduction if a 6-month confirmed progression ($p=0.0023$) was used.

Both the mean cumulative number of Gd+ lesions and new T2 lesions were significantly lower in the laquinimod-treated patients. Cumulative Gd+ lesions were reduced by 37% ($p=0.0003$) and new T2 lesions by 30% ($p=0.0002$). Similarly, new T1 hypointense lesions, which are thought to reflect more severe tissue damage, were reduced by 27% during the 2 years of the study. Additional exploratory analyses showed that laquinimod reduced severe relapses (requiring hospitalization) by 38% and the requirement for steroids by 27%.

Serious adverse events occurred in 22.2% of laquinimod versus 16.2% of placebo patients. In particular, there were no cardiovascular or pulmonary problems. The most common adverse event occurring more often with treatment was elevations in LFTs (ALT) in 6.9% of laquinimod and 2.7% of the placebo group. Most elevations were more than three times the upper limit of normal, 4.9% in the laquinimod group and 2.0% of those taking placebo. There was some excess of abdominal pain (5.3% vs 2.9%) and back pain (16% vs 9%).

The second phase III study, called BRAVO,[16] was also a 2-year, multicenter, randomized, double-blind, parallel-group, placebo-controlled study comparing the safety, efficacy, and tolerability of a once daily, oral, 0.6-mg dose of laquinimod with placebo along with a single-blinded group of patients receiving IFNβ-1a as an active comparator. Patients were required to have at least one relapse in the prior year or two relapses in the prior

TABLE 21.1 ALLEGRO Primary End Point

End Point	Placebo	Laquinimod	Relative Risk (95% CI)	p
Annualized relapse rate	0.395	0.304	0.770 (0.650–0.911)	0.0024
EDSS disability progression			0.641 (0.452–0.908)	0.0122

2 years, or one relapse within 2 years and a Gd+ lesion in the prior 2 years. An eligible 1331 patients were randomly assigned in a 1:1:1 manner to receive laquinimod ($n=434$), placebo ($n=450$), or 30 mcg once weekly of IFNβ-1a ($n=447$). The primary end point was the efficacy of 0.6 mg laquinimod daily, measured by ARR versus placebo. Secondary outcome measures included effect on the accumulation of disability (EDSS) and brain atrophy. Gd+ lesions were found at baseline in 39.6% of laquinimod, 33.4% of placebo, and 38.1% of IFNβ-1a treated patients. Mean EDSS was similar among the three groups at ~2.6 and disease duration from first onset of symptoms was 6.6, 6.9, and 7.0 for laquinimod, placebo, and IFNβ-1a i.m. groups, respectively. The three groups overall appeared similar to the patients recruited to the ALLEGRO study.

At 24 months in the unadjusted primary analysis, the ARR for laquinimod versus placebo did not reach statistical significance, but showed an 18% trend in reduction compared with the 25% statistically significant reduction achieved using IFNβ-1a versus placebo (see Table 21.2). However, after a predefined adjustment was made using covariates accounting for differences in baseline MRI activity between laquinimod and placebo, the laquinimod comparison with placebo did reach statistical significance, but still a modest 21% reduction in ARR compared with the 29% corrected value for IFNβ-1a.

MRI outcomes were also more modest with laquinimod compared with IFNβ-1a with only a 22% relative reduction ($p=0.062$) in Gd+ lesions (vs 60% with IFNβ-1a) and a 19% reduction in new T2 lesions ($p=0.037$) (vs 52% with IFNβ-1a) against placebo. A reduction in brain atrophy by 32.8% was noted at 24 months ($p<0.0001$), though the significance of this finding in RRMS is still controversial (ie, should brain volume decrease as inflammation is brought into check, or should it increase to reflect regeneration or repair?).

Overall, 9.7% of laquinimod, 13.3% of placebo, and 10.5% of IFNβ-1a patients met the definition of sustained 3-month confirmed EDSS progression, yielding a 33.5% reduction relative to placebo for laquinimod ($p=0.04$) and a 28.7% reduction ($p=0.09$) with IFNβ-1a, in post hoc corrected analyses.

Safety Data

Adverse events were infrequent and were balanced through the groups. There were two malignant neoplasms in each of the laquinimod and IFNβ-1a groups, with one case of thyroid cancer in each group, one case of skin squamous cell carcinoma in the laquinimod group, and one case of colon cancer in the interferon group. Back pain was seen again in this study, though it generally resolved on treatment. An increase in LFTs on laquinimod was also noted, with elevations that were for the most part mild—<5× ULN. Of patients receiving laquinimod, 28.9% had increased LFTs between 1 and 3 × ULN, and 4.2% had increases >3× ULN. A similar proportion of patients receiving IFNβ-1a showed liver enzyme elevations in this same range.

TABLE 21.2 BRAVO: Primary End Point in Unadjusted and Adjusted Analyses

End Point	Placebo	Laquinimod	p	IFNβ-1a	p
Annualized relapse rate (unadjusted analysis)	0.34	0.28	0.075	0.26	0.007
Annualized relapse rate (adjusted analysis)	0.37	0.29	0.026	0.27	0.002

Discussion

Based on the results of these two large studies with laquinimod and its modest effects on relapses and MRI and seemingly more robust effect on brain volume changes and perhaps on EDSS progression, it is unlikely that this agent will offer a competitive benefit over either teriflunomide or BG-12. However, studies are examining whether higher doses of laquinimod might offer greater benefit, especially in PPMS, while maintaining a satisfactory safety and tolerability profile. A phase III trial in PPMS is underway.

21.4.2 Masitinib

Masitinib mesilate is an orally bioavailable, small-molecule, tyrosine kinase inhibitor that interferes with mast cell survival, migration, cytokine production, and degranulation through its actions on growth and activation pathways. In MS, CNS mast cells are located perivascularly and influence both blood–CSF barrier permeability and possibly T-cell activation.[17]

Recently, an RDBPC phase IIa trial was carried out to determine whether masitinib might hold promise as a treatment for progressive forms of MS. PPMS and relapse-free (for 2 years) SPMS patients were administered an initial oral dose of 3–6 mg/kg/day in two daily doses. Dosing was either increased or decreased by 1.5 mg/kg/day due to perceived lack of efficacy or toxicity. The drug was known to cause a rash, so mandatory cotreatment with cetirizine 10 mg/day was implemented for the first 30 days of treatment. The primary outcome was the average change in the MSFC relative to baseline.

Thirty-five patients were randomized (27 to masitinib vs 8 to placebo; 9 PPMS and 15 SP in the ITT population). At 12 months, the average change in MSFC score from baseline was 103% ± 189 in the masitinib group versus 60 ± 190 in the placebo group, which was not statistically significant, though the difference was present throughout from months 3 through 18. There was a greater improvement seen in the timed 25-foot walk (T25FW) part of the MSFC for the PP versus SP (+13% vs −1%) patients, but the reverse was seen with respect to the PASAT (+19% for PP vs +55% for SP).

The most frequent side effects were asthenia (41%), rash (26%), nausea (22%), edema (19%), and diarrhea (11%). Leukopenia (22%), lymphopenia (15%), and neutropenia (4%) were also observed, which was in keeping with the known effects of masitinib. Rash was the leading cause of discontinuation in this and other nononcology masitinib trials, prompting researchers to suggest having a dermatologist on board should future trials be considered. With only modest short-term efficacy which did not reach statistical significance, together with the difficulty encountered with skin reactions, enthusiasm has dulled somewhat for organizing a larger phase III study in this population.

21.4.3 Ibudilast

Ibudilast is a small-molecule inhibitor of macrophage migration inhibitor factor and a nonselective phosphodiesterase (4 and 10) inhibitor, which, as deduced from animal studies, may reduce the clinical and pathological severity of MS by mediating an increase in nitric oxide production, suppression of both IFNγ and TNFα production in macrophages, and a reduction in Ag-specific proliferation of T cells. Its mechanism might be neuroprotective.

A small 12-month phase II RDBPC study was performed in RRMS ($n=297$) randomized to receive either 30 or 60 mg versus placebo, looking at the appearance of newly active T2 or GELs as well as other clinical and exploratory MRI measures.[18]

There was no difference between doses and placebo in terms of the number of newly active T2 or GELs, however, with the higher dose (60 mg) of ibudilast there was a statistically significant reduction in the brain volume change relative to placebo (placebo −1.20% vs ibudilast 60 mg −0.79%; $p=0.04$) over 12 months (effect sustained at 24 months) as well as a reduction in the proportion of lesions that evolved into T1 black holes for 60 mg (0.14, $p=0.004$) and 30 mg (0.17, $p=0.036$) compared with the placebo (0.24). Both the latter measures suggest that there may be a neuroprotective effect of ibudilast. Over 2 years there was also less confirmed EDSS progression in those patients on active treatment throughout the study (10.4%) versus those initially on placebo (21%, $p=0.026$).

The most frequently occurring side effects over the entire study were nasopharyngitis (20%), headache (14%), urinary tract infection (9%), pharyngitis (6%), and nausea (5%). There was more mild to moderate self-limiting nausea, vomiting, and diarrhea in the 60-mg/d dose compared with the other treatment groups as well as a possible increase in depression.

Though there was little evidence for an antiinflammatory effect of ibudilast, there was a possibility the evidence was pointing to a more neuroprotective benefit. There was therefore enthusiasm for testing ibudilast in progressive disease. An ongoing study of 255 PPMS and SPMS (52% and 48%, respectively) patients given 100 mg ibudilast daily versus placebo will examine efficacy as determined by the effect on 96-week progression as determined by brain atrophy measured and determined by MRI and the evaluation of the brain parenchymal fraction.

21.4.4 Simvastatin

Statins have many immunomodulatory and potentially neuroprotective properties that could make them appealing candidates for treating MS. Several small studies have been conducted in relapsing disease with at best some modest results compared with current approved therapies. In an attempt to test their neuroprotective qualities, it was decided to look at simvastatin in treating patients with SPMS.

An RDBPC trial was conducted in SPMS patients to receive either 80 mg of simvastatin (after 1 month of 40 mg) or placebo.[19] The primary outcome was the annualized rate of whole-brain atrophy measured from serial volumetric MRI studies. Clinical outcomes were assessed at baseline, 12 and 24 months, and were EDSS, MSFC, and other patient-related outcomes. Patients were evenly split between simvastatin ($n=70$) and placebo ($n=70$). On the primary outcome, mean annualized atrophy rate was 43% lower in the simvastatin group (0.29% simvastatin vs 0.58 placebo; $p=0.003$). The mean EDSS difference changed less for the simvastatin versus placebo patients ($p<0.05$), but there was no difference in the MSFC.

Overall, simvastatin was well tolerated and there were no significant safety issues. The results support the advancement of this treatment to at least a phase III study.

21.4.5 Ocrelizumab

Ocrelizumab is a humanized mAb directed against immature B cells, more specifically CD20, as was described earlier in RRMS. A previous study (OLYMPUS)[20] of a similar mAb rituximab in PPMS demonstrated a trend toward slowing disease progression in a subset of younger male patients with PPMS, especially those with GELs at baseline. ORATORIO was an RDBPC study examining 600 mg of ocrelizumab (given as two doses 300 mg 2 weeks apart) every 6 months in patients with PPMS showing evidence of disease progression and having oligoclonal bands in their CSF.

Patients were stratified by age (older or younger than 45 years old) and randomized 2:1 to ocrelizumab ($n=488$) or placebo ($n=244$). The study was event driven, and powered to show a difference of at least 25% between treatment and placebo. (The OLYMPUS study was a 96-week study, powered to show a difference of 50% between treatment and placebo.) The primary outcome of the study was confirmed disability progression (CDP) at 12 weeks.

Ocrelizumab demonstrated a 24% reduction in CDP compared with placebo ($p=0.0321$). A similar result was noted if a 24-week CDP was examined. There was also a 29% reduction in progression as noted by the T25FW ($p=0.04$). The most important of the MRI metrics in the context of PPMS was the rate of brain atrophy, which was reduced by 17.5% ($p=0.0206$) relative to placebo.

In terms of safety, there were no issues different from that seen with relapsing disease; however, a number of malignancies, including four cases of breast cancer, were observed only in the ocrelizumab arm.

Ocrelizumab is the first treatment ever to demonstrate a significant slowing of disease progression in PPMS. It is still not clear whether all patients with PPMS benefited from the treatment or whether a subgroup (such as males with GELs at baseline) carried most of the benefit. Further analysis will be required to understand the results of this study and to better assess whether the malignancies, especially breast cancer, are truly a concern.

21.4.6 MD 1003 (High-Dose Biotin)

Biotin (vitamin B7 or vitamin H) is a water-soluble vitamin found in many foods. Very high doses have been observed to possibly change the course of several rare neurodegenerative diseases, so researchers decided to test whether there may be some benefit to treating MS. The exact mechanism underlying neurodegeneration in MS is unknown, but one theory of virtual hypoxia[21] created by a mismatch of energy needs from a demyelinated axon and its injured mitochondria may contribute to this damage. High doses of biotin may help in countering the loss of mitochondrial energy metabolism or help by stimulating the basic pathways for myelin formation via its effect as a coenzyme for the numerous carboxylases involved in both these processes.[22]

Researchers started by giving 100–300 mg daily of biotin to 23 consecutive patients with PPMS or SPMS[23] and noted unexpected improvements in visual acuity or mobility in a few patients. This provided the impetus for a larger study of 154 PPMS or SPMS patients, 103 receiving 300 mg of biotin daily versus 51 taking placebo. There was a mismatch in terms of the breakdown of SPMS and PPMS patients: in the biotin arm 41% and 59% had PPMS versus SPMS, whereas in the placebo arm the breakdown was more 26% and 74%. The primary end point of the study was the proportion of patients showing improvement at month 9 compared with month 12 of the 1-year study, as determined by a one-point change in the EDSS or a 20% reduction in the T25FW. Both baseline values were determined as the best result obtained at baseline or baseline −1 month. On the primary end point, 13% of the biotin-treated group versus none of the placebo group reached the end point ($p=0.0051$); 77% doing so by EDSS change and 39% by changes in the T25FW. Changes were evident at month 3 and persisted for the remainder of the year and on to at least 18 months. Most of the changes were noted in the SPMS group.

There were few to no side effects of concern in the biotin-treated arm of the study.

The study was overall small and short, but the results somewhat dramatic in that none of the patients with PPMS or SPMS had evidence of MRI activity or relapse in the prior 1–2 years, unlike those of the ORATORIO study. Many view these results with cautious optimism and require a much greater study, possibly with some additional objective measures such as MRI or neurophysiology to back up the important clinical observations.

21.5 RECONCILING THE NEW TREATMENTS

It is very difficult to compare across clinical trials given the changing scenario of patients, newer diagnostic criteria for MS, and most importantly, the ever-changing behavior of the placebo groups in terms of their risk for events. Indeed, even between the now-required two-registry phase III trials for each of the new therapies, the groups, recruited around the same time, behaved differently. With more therapies being approved for relapsing disease, newer trials are focused on showing superiority with current first- or even second-line treatments. Presumably if a medication is superior to a current therapy, then it will no doubt be better than placebo. How do we compare then the efficacy of one treatment versus another if they are all tested against different agents? The dilemma is highlighted in Table 21.3, listing the outcomes of recent comparative trials. In the case of daclizumab and ocrelizumab, the comparator is active (another therapy) but for laquinimod it is a placebo. Most recent trials are almost all designed to compare the test medication against an active comparator due to the fact that it is unethical in most Western countries today to not offer medication to patients with relapsing MS. For progressive studies, given there is no approved or proven therapies, placebo-controlled studies will continue. Still the clinician seeking to decide upon a particular

TABLE 21.3 Summary of New Study Results for Daclizumab, Ocrelizumab, and Laquinimod

Study Agent	Daclizumab[a]	Ocrelizumab[b]	Laquinimod
Relapse rate reduction	45%	54%	23%
Annualized relapse rate	0.22	0.16	0.28
Absolute reduction in relapse rate	0.18	0.14	0.09
Number needed to treat (2-year relapse)	6	7	11
Relative reduction in new			
T2	54%	80%	30%
Gd+ MRI activity	65%	95%	37%
Relative reduction in EDSS progression	25% (ns)	40%	36%
Absolute reduction in proportion progressing	0.02	0.055	0.036
Number needed to treat (2-year progression)	50	18	28

[a] *Compared to IFNβ-1a 30 μg i.m. weekly.*
[b] *Compared to IFNβ-1a 44 μg sc tiw;* ns, *not statistically significant; laquinimod results compared to placebo.*

treatment must have some way of weighing in the efficacy data obtained from analyzing the clinical trials.

Calculating the NNT could offer some degree of comparison by looking at absolute differences among agents, assuming that even though the behavior of placebo groups change, the overall magnitude of an effect should be consistent. However, when event rates become very low, the NNT will reflect not only the magnitude of the treatment effect but will also factor in the rarity of the event in question. All studies had to choose a primary end point and were powered to allow a determination of a treatment effect, if there was one. All primary outcomes were relapse related, but powering for relapse usually overpowers for MRI, which is a much more frequent event, but underpowers for progression—a much rarer event. It is therefore not surprising that all the successful studies for current therapies in relapsing MS were able to show an effect on relapse, but not all on progression. Still the overall reduction in ARR is somewhat difficult to comprehend in terms of benefit, since reducing relapse rates by 0.2 translates into saving about one attack every 5 years. When the cost of preventing only one attack every 5 years is factored in, it is important to have both secondary outcomes and patient-related outcomes to appreciate the overall benefit patients will receive from treatment.

There are other problems that arise when you compare across clinical studies, not just because of different patient groups, but also different inclusion/exclusion criteria, definitions of relapse, blinding of evaluators for relapse or progression, or even in the way MRI studies are performed and analyzed, their frequency, and the metrics used.

Studies of progressive forms of MS are hampered by other problems, including the lack of sensitivity of the primary outcome measure, the EDSS. Some studies have reverted to a composite measure such as combining the EDSS with the T25FW, but even with this we have learned that fingolimod had no effect whatsoever in PPMS and natalizumab had no effect in SPMS. Effects on brain atrophy may not be clearly a surrogate for disease progression, but perhaps another biomarker will be found that more closely follows or predicts disease progression and is more sensitive than current clinical measures. Cognition has more closely aligned its changes with brain atrophy, but is too cumbersome and logistically difficult to administer as a primary outcome measure in a large multicenter study.

21.6 CONCLUSIONS

Already, there is a wide range of agents to choose from for the treatment of relapsing MS, and in the near future there are anticipated to be even more, with improved and perhaps enduring efficacy. Neurologists must be prepared to understand the merits of each along with the recognized risks. Without a cure, physicians must accept that some disease activity is inevitable, but deciding upon when to switch and what to switch to will be the issue. Various treatment algorithms have been proposed for choosing and monitoring the response to therapy[24] and even for identifying the patient with aggressive disease, warranting a more urgent approach.[25] Setting up a treatment algorithm for choosing, switching, and sequencing therapies is going to be a moving target as newer, more effective agents become available, or if new toxicities are realized. We might well be able to start talking about attaining a disease-activity-free state or reaching NEDA, at least in the short term. We have come to learn that there is no "one shoe fits all" empiric treatment for patients today and much needs to be considered before embarking on a given treatment—the duration, course, and burden (clinical and MRI) of disease, previous response to disease-modifying drugs, and most importantly, the perceived risk of progression all weigh in to profiling the patient.

The cost of current medications also precludes the ability to combine medications, though some studies have shown an additive effect of doing so without sacrificing safety.[26] It may also be possible to escalate therapy to a riskier, more effective medication for a period of time in order to gain control of the disease, then later back down to a safer long-term medication. With newer options and potential problems with sequencing therapies the treatment of MS is no doubt going to become more complex and likely will require some additional training for many general neurologists interested in optimizing the treatment of their MS patients. Still, with no treatment proven to alter the natural history of the disease until nearly 20 years ago, we have made incredible strides. We anticipate soon that treatments that are vaccine-based, cell-based, or even those using nanomolecules will enter the picture and we will be able to talk about not only controlling or limiting disease, but fixing the damage with remyelinating or regenerative strategies. The future is truly bright for the treatment of MS.

References

1. Selmaj K, Li DK, Hartung HP, et al. Siponimod for patients with relapsing-remitting multiple sclerosis (BOLD): an adaptive, dose-ranging, randomised, phase 2 study. *Lancet Neurol*. 2013;12:756–767.
2. Olsson T, Boster A, Fernandez O, et al. Oral ponesimod in relapsing-remitting multiple sclerosis: a randomised phase II trial. *J Neurol Neurosurg Psychiatry*. 2014;85:1198–1208.
3. Subei AM, Cohen JA. Sphingosine 1-phosphate receptor modulators in multiple sclerosis. *CNS Drugs*. 2015;29:565–575.
4. Miller DH, Weber T, Grove R, et al. Firategrast for relapsing remitting multiple sclerosis: a phase 2, randomised, double-blind, placebo-controlled trial. *Lancet Neurol*. 2012;11:131–139.
5. Grove RA, Shackelford S, Sopper S, et al. Leukocyte counts in cerebrospinal fluid and blood following firategrast treatment in subjects with relapsing forms of multiple sclerosis. *Eur J Neurol*. 2013;20:1032–1042.
6. Hauser SL. The Charcot Lecture I beating MS: a story of B cells, with twists and turns. *Mult Scler*. 2015;21:8–21.
7. Sorensen PS, Lisby S, Grove R, et al. Safety and efficacy of ofatumumab in relapsing-remitting multiple sclerosis: a phase 2 study. *Neurology*. 2014;82:573–581.
8. Kappos L, Li D, Calabresi PA, et al. Ocrelizumab in relapsing-remitting multiple sclerosis: a phase 2, randomised, placebo-controlled, multicentre trial. *Lancet*. 2011;378:1779–1787.
9. Tran JQ, Rana J, Barkhof F, et al. Randomized phase I trials of the safety/tolerability of anti-LINGO-1 monoclonal antibody BIIB033. *Neurol Neuroimmunol Neuroinflamm*. 2014;1:e18.
10. Sheridan JP, Zhang Y, Riester K, et al. Intermediate-affinity interleukin-2 receptor expression predicts CD56(bright) natural killer cell expansion after daclizumab treatment in the CHOICE study of patients with multiple sclerosis. *Mult Scler*. 2011;17:1441–1448.
11. Wynn D, Kaufman M, Montalban X, et al. Daclizumab in active relapsing multiple sclerosis (CHOICE study): a phase 2, randomised, double-blind, placebo-controlled, add-on trial with interferon beta. *Lancet Neurol*. 2010;9:381–390.
12. Gold R, Giovannoni G, Selmaj K, et al. Daclizumab high-yield process in relapsing-remitting multiple sclerosis (SELECT): a randomised, double-blind, placebo-controlled trial. *Lancet*. 2013;381:2167–2175.
13. Kappos L, Wiendl H, Selmaj K, et al. Daclizumab HYP versus interferon Beta-1a in relapsing multiple sclerosis. *N Engl J Med*. 2015;373:1418–1428.
14. Thone J, Ellrichmann G, Seubert S, et al. Modulation of autoimmune demyelination by laquinimod via induction of brain-derived neurotrophic factor. *Am J Pathol*. 2012;180:267–274.
15. Comi G, Jeffery D, Kappos L, et al. Placebo-controlled trial of oral laquinimod for multiple sclerosis. *N Engl J Med*. 2012;366:1000–1009.
16. Vollmer TL, Sorensen PS, Selmaj K, et al. A randomized placebo-controlled phase III trial of oral laquinimod for multiple sclerosis. *J Neurol*. 2014;261:773–783.

17. Theoharides TC, Kempuraj D, Kourelis T, Manola A. Human mast cells stimulate activated T cells: implications for multiple sclerosis. *Ann N Y Acad Sci*. 2008;1144:74–82.
18. Barkhof F, Hulst HE, Drulovic J, et al. Ibudilast in relapsing-remitting multiple sclerosis: a neuroprotectant? *Neurology*. 2010;74:1033–1040.
19. Chataway J, Schuerer N, Alsanousi A, et al. Effect of high-dose simvastatin on brain atrophy and disability in secondary progressive multiple sclerosis (MS-STAT): a randomised, placebo-controlled, phase 2 trial. *Lancet*. 2014;383:2213–2221.
20. Hawker K, O'Connor P, Freedman MS, et al. Rituximab in patients with primary progressive multiple sclerosis: results of a randomized double-blind placebo-controlled multicenter trial. *Ann Neurol*. 2009;66:460–471.
21. Koch MW, Cutter G, Stys PK, Yong VW, Metz LM. Treatment trials in progressive MS–current challenges and future directions. *Nat Rev Neurol*. 2013;9:496–503.
22. Sedel F, Bernard D, Mock DM, Tourbah A. Targeting demyelination and virtual hypoxia with high-dose biotin as a treatment for progressive multiple sclerosis. *Neuropharmacology*. 2015. http://dx.doi.org/10.1016/j.neuropharm.2015.08.028.
23. Sedel F, Papeix C, Bellanger A, et al. High doses of biotin in chronic progressive multiple sclerosis: a pilot study. *Mult Scler Relat Disord*. 2015;4:159–169.
24. Freedman MS, Selchen D, Arnold DL, et al. Treatment optimization in MS: Canadian MS Working Group updated recommendations. *Can J Neurol Sci*. 2013;40:307–323.
25. Rush CA, MacLean HJ, Freedman MS. Aggressive multiple sclerosis: proposed definition and treatment algorithm. *Nat Rev Neurol*. 2015;11:379–389.
26. Freedman MS, Wolinsky JS, Wamil B, et al. Teriflunomide added to interferon-β in relapsing multiple sclerosis: a randomized phase II trial. *Neurology*. 2012;78:1877–1885.

SECTION VI

SYMPTOMATIC & COMPLEMENTARY TREATMENTS

CHAPTER

22

Treatment of Acute Relapses in Multiple Sclerosis

R. Berkovich

University of Southern California, Los Angeles, CA, United States

OUTLINE

22.1 MULTIPLE SCLEROSIS RELAPSE DEFINITION AND CLINICAL PRESENTATION

Relapsing types[1] or, according to the new descriptive classification,[2a] active types of multiple sclerosis (MS), affect the majority of patients. About 85% of MS cases originate with relapsing-remitting disease, and relapse events average about 1.1 per year early in the disease course and appear to decrease with advancing disease, according to the 1980s publication.[2b] According to the recent clinical trials in MS, however, the annual relapse rate in placebo

Translational Neuroimmunology in Multiple Sclerosis
http://dx.doi.org/10.1016/B978-0-12-801914-6.00022-2

groups tends to fluctuate around 0.4, which still may not accurately represent the actual MS population at large.

MS relapses are typically defined as a new or worsening neurologic deficit lasting longer than 24h in the absence of fever or infection. Relapses are a hallmark of MS.[3a] Charcot's original definition of clinical MS relapse[3b] is focal disturbance of function, affecting a white matter tract, lasting for more than 24h, that does not have an alternative explanation and is preceded by >30 days of clinical stability. Clinical trial criteria for relapse require a minimum of 24–48h of symptom duration and changes in functional measures assessed by disability and functional scores; this definition may not always effectively distinguish between clinical relapses and pseudo relapses, and although change in the scores is intended to provide objective assessment to support the diagnosis, in actual practice this may not be feasible. From a clinical perspective, the wide interpatient variability, timing, and duration of relapses suggests a need to study relapses to address important questions: What they reflect in terms of disease activity and whether they may predict disease prognosis over a period of years.

From a patient's perspective an MS relapse is associated with emotional burden significant increase in the economic costs as well as a decline in health-related quality of life and functional ability.[4] For the vast majority of MS patients relapses are one of the biggest concerns associated with the disease. Notorious unpredictability of MS exacerbations complicates the potential impact on patients quality of life.[5]

MS exacerbations may clinically reflect formation of new demyelinating activity or reactivation of previously existing demyelinating lesion(s) located in any segment of the central nervous system (CNS).[6,7] Commonly seen symptom complexes may thus be related to acute inflammatory processes affecting optic nerve, spinal cord, cerebellum, and/or cerebrum. Therefore, presenting symptoms may include various combination of visual disturbances, motor and sensory deficits, coordination and balance dysfunctions, and cognitive impairments.[6,7]

Cognitive and psychiatric presentations of MS relapses have attracted deserved attention, as those may occur in the absence of routinely expected neurologic exam findings.[8–11] Relapses were found to be substantiated by decline in neuropsychological testing, observed or reported cognitive change with significant decline on the Symbol Digit Modalities Test, and in a subset of patients, gadolinium-enhancing MRI lesions.[10] On the other hand, so-called isolated cognitive relapses, while not associated with subjective cognitive deficits or depression, were found to be accompanied with significantly reduced objective cognitive performance.[11] Thus, isolated cognitive relapses should not be expected to be associated with changes in neurologic symptoms, mood, fatigue levels, or cognitive performance self-evaluations.

High rates of depression were observed during MS exacerbations. Although depression scores reduced significantly postrelapse, rates of possible depression at follow-ups remained high, suggesting that although improvements in disability may influence depression, once its symptoms are elevated at relapse then depression becomes persistent.[12]

Among rare clinical presentations of MS exacerbation it is worth mentioning cases of altered visual perception in the absence of abnormalities on ophthalmological examination,[13] restless legs syndrome presenting acutely,[14] paroxysmal dysarthria,[15] and Jacksonian seizure,[16] as the sole presentation of MS relapse. In addition, cardiovascular disorders in the setting of MS relapse can be observed; that is, acute onset of cerebellar symptomatology along with sinus bradycardia in the absence of underlying cardiac pathology.[17] Finally, stabbing headache may present as the only sign of MS relapse.[18]

Clinically it is important to exclude certain conditions that may trigger so-called pseudo-exacerbations; those include fever, infections (most commonly urinary tract and upper respiratory infections), as well as heat exposure and stress.[3,5–7]

Remarkably, while infections may mimic relapses and cause pseudo-exacerbations, the global role of infections and pathogens in MS activity may not be as simple. Pathogens that may predispose to the initial development or exacerbation of MS include bacteria (such as *Mycoplasma pneumonia*, *Chlamydia pneumonia*, and *Staphylococcus*), viruses (such as Epstein–Barr virus (EBV) and human herpesvirus 6), human endogenous retrovirus families, and the protozoa *Acanthamoeba castellanii*.[19] Specifically, *Staphylococcus aureus*-produced enterotoxins function as super antigens and can activate CD4 cells to target the myelin basic protein.[20] Herpes viruses' and human endogenous retroviruses' epitopes are expressed in higher quantities on the surface of B cells and monocytes in patients with active MS, and the expression of these proteins may be associated with exacerbation of the disease.[21] DNA from *varicella zoster* virus was found in the cerebrospinal fluid (CSF) from 100% of MS patients studied during relapse versus only 31% of samples collected during remission.[22] Marked differences in the prevalence of EBV-specific CD8+ T cell responses were observed in MS patients; expansion of CD8+ T cells specific for EBV lytic antigens was revealed during active disease in untreated MS patients but not in relapse-free, natalizumab-treated patients. It was proposed that inability to control EBV infection during inactive MS could set the stage for intracerebral viral reactivation and disease relapse.[23] On the other hand, infection with certain parasites such as helminthes (*Schistosoma mansoni, Fasciola hepatica, Hymenolepis nana, Trichuris trichiura, Ascaris lumbricoides, Strongyloides stercoralis, Enterobius vermicularis*) seems to protect against the development or exacerbation of MS.[19] Thus, complex interfaces between the CNS, various infectious pathogens, and the immune changes they trigger need to be better understood.

Various clinical scenarios related to different pharmacotherapy stages may potentially predispose to MS relapse. Among those, most attention is given to discontinuation of previously effective disease-modifying therapy (ie, natalizumab).[24] Tumor necrosis factor-blocking drugs, which have shown a well-known ability to promote onset or exacerbation of MS,[25] also deserve vigilant attention. Notably, reproductive hormones have an important role in regulating immune responses, therefore it is clinically important to remember that assisted reproduction technologies may significantly increase the risk of MS exacerbation in such patients.[26]

Some environmental factors were linked to increased clinical and radiological disease activity in patients with MS. Among them it is worth mentioning higher sodium intake[27] and low serum vitamin D levels.[28]

Research evaluating association of stress with MS exacerbation show a fairly consistent direct correlation; higher stress levels appear to increase the risk of development of gadolinium-enhancing lesions as well. Stress management therapy and cognitive-behavioral therapy may prove important.[29] Association was described between direct coping and planning and reduced MS relapse rate during wartime.[30] This corresponds well with World War II physicians observations, reporting low frequency of exacerbations during the Leningrad blockade.[31] It has been shown that in MS patients perceived stress and mood disturbances correlate with induced production of interleukin (IL)-6 and IL-10; in addition, compared to controls, MS subjects exhibited a significant fourfold increase in the production of IL-12.[32]

Seasonal variation of relapse rate in MS noted to be latitude dependent,[33] with peaks in early spring and lows in autumn in both hemispheres. Increased IFN-γ levels have been associated with increasing relapse risk, which were greatly attenuated by immunomodulatory therapies, by summer season and by higher serum vitamin D level.[34] Relapse phenotype tends to be associated with demographic and clinical characteristics, and with common symptomatic recurrence.[35] Whereas visual, brain stem relapses, and sensory relapses tend to occur more frequently in early disease, the pyramidal, sphincter, and cerebellar relapses were more common later in the disease and in more progressive types. Female patients were observed to have more frequently sensory or visual symptoms; men presented more frequently with pyramidal, brain stem, and cerebellar relapses. Previous relapse phenotype tends to predict more common future relapses. Sensory, visual, and brain stem relapses tend to recover better than other relapse phenotypes. Relapse severity increased and the ability to recover decreased with age or more advanced disease. It was observed that female patients overall are predisposed to higher relapse activity than males. Decline in relapse activity over time is more closely related to patient age than disease duration.[36]

In early pediatric MS, optic nerve involvement is associated with a severe initial demyelinating event, while non-White race, localization to the cerebral hemispheres, or encephalopathy showed a trend toward increased disease severity; similar association with race was found for severe second events. Severe initial demyelinating event was associated with incomplete recovery, with similar trends for second and third events. Incomplete recovery from the first event predicted incomplete second-event recovery, helping identify children at risk for a more aggressive disease course.[37]

22.2 APPROACH TO MULTIPLE SCLEROSIS RELAPSE IMMUNOBIOLOGY

Whereas chronic inflammation in the CNS plays a role in MS-associated progression and disability, acute inflammatory exacerbations are the mechanism by which demyelination and axonal loss are believed to occur.[38] The activation of the immune process is likely to be initiated peripherally, resulting in migration of activated immune cells into the CNS where they are reactivated and result in central inflammation; the acute inflammation in MS may follow focal, multifocal, or diffuse patterns; it is characterized by infiltration of activated lymphocytes, macrophages, and microglia, it may involve cortex, white matter, and deep gray matter and cause myelin destruction, axonal, neuronal, and synaptic loss; astroglial reaction; remyelination; and synaptic rearrangement.[6,38] As phrased by L. Steinman, the mechanisms of action of the approved drugs for relapsing-remitting MS (RRMS) provide a strong foundation for understanding the pathobiology of the relapse.[39] Deregulated immune response, including inflammatory cells (eg, T cells, B cells, macrophages) and mediators (eg, cytokines, chemokines, matrix metalloproteinases, complement), contributes to the expansion of autoreactive T cells; proinflammatory shifts promote blood–brain barrier lymphocyte and monocyte extravasation.[38] It was found that expression of microRNA-320a is decreased in B cells of MS patients, which may contribute to increased blood–brain barrier permeability.[40] Regulatory T cells (Tregs) normally control the intensity of an immune response; however, their regulatory function in MS exacerbations dramatically decreases.[41] Accumulating evidence indicates an immunosuppressive role for CD4+ CD25+ Tregs and that their suppressive capacity is more affected in the early phase of the disease. Consistent with this, there are differences in function and expression of FOXP3 (a master regulator in the development and function of Tregs). Disease exacerbation of MS is characterized by loss of terminally differentiated autoregulatory CD8+ T cells.[41] Recently, Correale and Villa investigated the role of CD8+ CD25+ FOXP3 in RRMS patients, and found that it is suppressed to a greater degree in the peripheral blood during MS exacerbations than during remission or in healthy controls.[42] Likewise, in the CSF of MS patients during exacerbations, lower levels of CD8+ CD25+ FOXP3 T cells are detected.[42] It was observed that Th17 expansion in MS patients is counterbalanced by an expanded CD39+ Treg population during remission but not during relapse.[43] Novel perforin-expressing regulatory B cell subsets were higher during relapse as compared to nonclinically active MS patients.[44] There is a growing body of evidence that antibodies play an important role in the pathobiology of MS and MS relapse; immunoglobulin G (IgG) antibodies purified from an MS patient and transferred to mice with experimental autoimmune encephalomyelitis (EAE) caused a dramatic clinical improvement during relapse after selective IgG removal with immunoadsorption, whereas passive transfer of patient's IgG exacerbated motor deficits in animals.[45] These new data provide evidence for a previously unknown mechanism involved in immune regulation in acute MS.

22.2.1 Reliable Surrogate Markers of Multiple Sclerosis Relapse

A quantitative relation between the treatment effects on MRI lesions and clinical relapses has been shown.[46a] Analysis of trials that tested the same drugs in phase-2 and phase-3 studies showed that the effects on MRI lesions over 6- to 9-month follow-up periods can predict the effects on relapses over 12- to 24-month follow-up periods with the 95% prediction intervals in eight of nine trials. These findings indicate that the effect of a treatment on relapses can be accurately predicted by the effect of that therapy on MRI lesions.[46a] According to the classic research of B. Trapp, the number of transected axons increases with level of activity in MS lesions, and in active MS lesions can be more than 11,000.[46b] These data call for important clinical questions, such as whether can we effectively translate this information to the level of every individual patient? Should new gadolinium-enhancing lesions be recognized as a surrogate marker for active MS process? And if so, can new MRI activity be seen as a radiologic surrogate of MS relapse? This clearly has been the case in enrollment strategies of clinical trials. And finally, if these are indicative of acute MS inflammation, should we treat patients with enhancing lesions as we would treat MS relapse? Some experts would say yes,[38] but there is no consensus in respect to these aspects at the present, not unlike in the case of radiologically-isolated syndrome matters.

Research emerged on potential biological markers of acute MS. It has been shown, for example, that the levels of antibodies against KIR4—a potassium channel that shares functional properties with AQP4—differed in MS patients during relapse and remission; as such, they may represent a marker of disease exacerbation.[47] Another example: RRMS patients in clinical relapse had lower endogenous secretory receptor for advanced glycation end-product serum levels than clinically stable patients, which may potentially serve as a new biomarker of clinical relapse.[48] Third, IL-8 might provide utility in determining the presence of active intrathecal inflammation, and could be important in diagnostically undefined cases of radiologically isolated syndrome.[49] In addition, microRNAs (a class of noncoding RNAs that participate in posttranscriptional regulation of gene expression) were reported to play a role in induction of Th17 differentiation and progress of autoimmune diseases; specifically, miR-326 and miR-26a are involved in progress of Th17 and MS disease. Upregulation of both miR-326 and miR-26a in relapsing phase of MS patients compared with remitting phase and healthy controls was observed, suggesting that there is a potential of miR-326 to discriminate between relapsing and remitting phases of MS.[50]

22.3 APPROACH TO MULTIPLE SCLEROSIS RELAPSE TREATMENT: RESEARCH AND PRACTICE

Expected natural course of most of MS exacerbations usually ends with a period of repair followed by clinical remission and most favorably—usually early in the disease course—to a complete recovery. Residual deficit after MS relapse may persist and contribute to the stepwise MS progression.[3] Treatment of MS relapses helps to shorten and lessen the disability associated with exacerbations. Successful treatment of MS relapse helps MS patients to gain most important sense of trust, of being capable to take control over this disease.[5] Remarkably patients with relapsing MS who receive treatment for exacerbations report better outcomes than those who are simply observed.[51]

In early 20th century bed rest was the only treatment of choice for acute relapse; it was commonly implemented as a useful measure to help recovery and shorten the duration of attack. Several publications from the 1930s and 1940s supported this approach.[52–54]

New period of pharmacologic treatment of MS began with adrenocorticotropic hormone (ACTH), which happened to be the first medication successfully studied and approved for MS relapse treatment, remaining for a short while the "gold standard." One of the first peer-reviewed publications in English and arguably the very first controlled clinical trial in MS reported results of a controlled study;[55] 40 MS patients with acute exacerbations were treated with either corticotrophin (ACTH) or saline. ACTH gel dosage was 60 units twice daily the first week; 40 units twice daily the second week; and 60, 40, and 20 units on the second, fourth, and sixth days of the third week. The study "has confirmed the clinical impression that the hormone exercises a favorable effect on the outcome of some of such episodes." These and other similar data provided evidence that ACTH is beneficial in treatment of acute MS.[56–58]

A much larger, well-designed, controlled, double-blind, multicenter study results were published by Rose et al.[59,60] Patients were randomly assigned to either ACTH gel or placebo. From 10 centers throughout the United States, 197 patients were enrolled. This study was the first one to introduce the Disability Status Scale (DSS), which was later modified to the Expanded DSS (EDSS). In this study 40 units of ACTH gel given as intramuscular injections twice daily for 7 days, then 20 units twice daily for 4 days, and 20 units twice daily for 3 days, demonstrated beneficial effects in comparison to the similarly administered placebo gel. These data lead to acceptance of ACTH as a treatment for MS relapse and eventually to the FDA approval of the ActharGel for this indication in 1978.[61] It is worth mentioning that back in that time mechanisms of ACTH action were attributed solely to steroidogenic potentials of the compound.[59]

It used to be generally accepted as a fact that the efficacy of ACTH in MS relapses is determined by its corticotrophic effects only. However, more recent data in other disease states like nephrotic syndrome,[62] opsoclonus-myoclonus,[63] and infantile spasms[64] suggest that steroidogenic mechanisms alone fail to explain the efficacy of ACTH in these conditions. For example, in infantile spasms, a disease confined to the CNS, ACTH has been shown remarkable outcomes, which cannot be explained by glucocorticoid actions since corticosteroid treatment has suboptimal efficacy in infantile spasms.[64] It has been shown that ACTH has direct antiinflammatory and immunomodulatory effects via activation of central and peripheral melanocortin receptors (MCRs), in addition to the effects achieved by systems originating in the adrenal gland.[65] As strong melanocortin agonist, ACTH binds to all five known classes of MCRs, of which only the adrenal MC2R is involved in steroidogenesis.[65] The other classes are distributed broadly in the organs and systems. MC1R is expressed in melanocytes, epithelial cells, monocytes, neutrophils, lymphocytes, podocytes, periaqueductal gray matter in the CNS, and in microvascular endothelial cells, astrocytes, and Schwann cells. MC2R, as mentioned, is the receptor in the adrenal glands underlying the steroidogenic actions of ACTH, and has also been localized to osteoblasts and skin. MC3R and MC4R have been identified in the CNS; MC3R is expressed primarily in the hypothalamus and limbic system, while MC4R is the prevalent receptor in the CNS, with wide expression in the cortex, thalamus, hypothalamus, brain stem, spinal cord, and astrocytes. MC5R is widely distributed and found in exocrine glands and lymphocytes.[65,66] It was found that ACTH stimulates proliferation of oligodendrocyte progenitor cells (OPCs), and provides benefits by increasing the number of OPCs, accelerating their development into mature oligodendrocytes, and reducing OPC death from toxic insults.[67]

Recent clinical reports support therapeutic merit of ACTH: For example, it has been shown that treatment with ACTH gel 80 U for five consecutive days resulted in patient functional improvement, including vision and gait.[68]

Presumption that the efficacy of ACTH depends exclusively on its corticotrophic effects back in the early 1980s led to increased interest to high-dose corticosteroids for MS exacerbations treatment. Since that time focus shifts to systemic steroids as the first line and preferred treatment option for MS relapse.[69]

Therefore, the second, and at this point the only medication other medication approved by the FDA for the MS relapse treatment is intravenous methylprednisolone (IVMP).[69] Today systemic steroids are considered the first line of MS relapse treatment and are the most commonly used treatment for this indication. Mechanisms of action of systemic corticosteroids in the treatment of acute relapse initially were thought to be attributed to immunologic alterations they cause and "likely that the main, if not the sole, mechanism is the resolution of edema."[70] Furthermore, it was found that acute corticosteroid administration "brings about a lymphocytopenia,"[71] including the reduction of B-lymphocyte counts and their availability at the inflammatory sites, which could result in decreased number of IgG synthesizing cells in CNS.[71] This may lead to reduction of blood–brain barrier abnormally increased permeability (and decrease of active lesions on MRI).[6,71,72] Thus, the main mechanism of glucocorticoid action is attributed to induction of T-cell apoptosis leading to reduced lymphocyte infiltration into the CNS. Research has shown that IVMP decreases the overall proportion of Tregs and increases proportions of CD39-expressing Tregs and monocytes.[73] Other experimental data revealed that the redirection of T-cell migration in response to chemokines seems to be the essential principle of corticoids mechanisms.[74]

In addition, it has been reported that rapid functional changes in the excitability of cortical circuits involved in motor control can be induced by steroids, and those effects seem to balance between intracortical GABAergic inhibition and glutamatergic facilitation, which could lead to improving the motor performance in MS patients.[75]

However, as remarked by S. Krieger et al., corticosteroids may act in unpredictable ways in the context of autoimmune conditions; and it is difficult to predict when patients will respond favorably to corticosteroids, both in terms of therapeutic response and tolerability profile.[76]

A study of corticosteroids on the expression of cellular and molecular markers of spontaneous endogenous remyelination in the toxic nonimmune cuprizone animal model at early and intermediate remyelination, as well as steroidal effects in primary astrocytes and oligodendrocyte progenitor cultures has shown that the steroids have, in addition to the well-known beneficial effects on inflammatory processes, a negative impact on remyelination.[77]

Several clinical trials were done comparing IVMP to the ACTH,[70,78–80] and to the placebo.[71,81,82] Interestingly, several comparison trials failed to demonstrate differences in efficacy of the systemic steroids and ACTH, although different protocols were used. For example, in the study by Abbruzzese et al.[78] 60 patients were randomly assigned to either synthetic ACTH (1 mg/day in two 250 mL saline infusions, for 15 days) or the bolus IVMP (20 mg/kg/day for 3 days, then 10 mg/kg/day on days 4–7, then 5 mg/kg/day on days 8–10, and 1 mg/kg/day on days 11–15). Authors concluded that "these two treatments seem to have the same effectiveness." It was noticed that MP "showed prompt effectiveness" and "high-dose intravenous steroids may, therefore be regarded as a useful alternative in the management of acute demyelinating disease." Another study[79] randomized 25 patients either IVMP given 1 g daily for 7 days, or intramuscular (IM) ACTH, 80, 60, 40, and then 20 units daily, each for 7 days. It has been observed that "the methylprednisolone group improved faster than the ACTH group over the first 3 days and this difference was maintained at Day 28. However, at 3 months there was no longer any significant difference between the two groups." Yet another study[80] had 30 patients, which were randomized to one of three groups: (1) ACTH, 50 units/day for 7 days, 25 units/day for 4 days, 12.5 units/day for 3 days; (2) dexamethasone, 8 mg/day for 7 days, 4 mg/day for 4 days, 2 mg/day for 3 days; (3) methylprednisolone, 40 mg/day for 7 days, 20 mg/day for 4 days, 10 mg/day for 3 days. It was observed that "dexamethasone and ACTH-treated patients showed clinical improvement" and "dexamethasone seems to be the most effective of the three drugs studied" while "methylprednisolone was not as effective as dexamethasone and ACTH." Thus, the stronger effects of high versus low dose of corticosteroids have been shown. Finally, in the double-blind, randomized, controlled study by Thompson et al.[70] relative efficacy of IVMP and ACTH in the treatment of acute relapse was evaluated. Sixty-one patients were randomly allocated to two groups: one group received IVMP 1 g daily for 3 days and IM placebo injections for 14 days and another group received IV placebo daily for 3 days and IM ACTH over 14 days (80 units for 7 days, 40 units for 4 days, and 20 units for 3 days). "A clear improvement in both groups" was observed, but "no significant difference between the two groups in either rate of recovery or final outcome at 3 months" was found. Still, it was remarked that "3-day course of IV treatments rather than 14 days of IM injections has obvious advantages." Thus, it is fair to conclude that these four different trials following diverse protocols of using IV steroids versus ACTH all suggested similar efficacy of these two compounds.

In a placebo-controlled trial of IVMP,[71] 23 patients were randomly allocated to either the IVMP group (15mg/kg/day on days 1–3, 10mg/kg/day on days 4–6, 5mg/kg/day on days 7–9, 2.5mg/kg/day on days 10–12, and 1mg/kg/day on days 13–15) or the placebo group (500mL of physiologic solution for 15days). At the end of the controlled phase of the study, patients of group B were administered IVMP according to the schedule of group A. After the IV treatment period, both groups received oral prednisone starting with 100mg/day, slowly tapered over 120days. During the whole double-blind part of the study, the ratio of improved/unimproved patients was significantly higher in the oral MP group than in the placebo group. In another double-blind, placebo-controlled trial of high-dose IVMP,[81] 50 patients were randomized to IVMP 500mg daily for 5days or to an equivalent volume of saline given IV for 5days. MP-treated patients showed decreased disability scores at 4weeks in 19 of 26 patients: "significant effect in favor of methylprednisolone treatment."[81]

The relatively large Optic Neuritis Treatment Trial, although not done on MS patients, certainly had "a bearing on the treatment of multiple sclerosis with corticosteroids."[82] The study compared oral prednisone (1mg/kg/day for 14days; $n=156$), IVMP (1g/d for 3days followed by oral prednisone 1mg/kg/day for 11 days; $n=151$), and oral placebo ($n=150$) for 14days in the treatment of 457 patients with optic neuritis. It was found that visual function recovered faster in the group receiving IVMP than in the placebo group. Although the differences between the groups decreased with time, at 6months the IVMP group had better visual fields, contrast sensitivity, and color vision, although not better visual acuity. The outcome in the oral prednisone group did not differ from that in the placebo group, and disturbingly, the rate of new episodes of optic neuritis was higher in the group receiving oral prednisone, but not the group receiving IVMP. It was concluded that the IVMP followed by oral prednisone speeds the recovery of visual loss due to optic neuritis, but the oral prednisone alone is an ineffective treatment and increases the risk of new episodes of optic neuritis.

As we see, the dosages of IVMP used in these studies differed considerably: from as low as 40mg/day[80] to 500mg[81] and to 15mg/kg/day IV[71] (which is close to 900–1200mg/day) and finally, 1g a day.[70,71,82] The low dosages were found to be ineffective[80] and the dosages from 500mg to 1g of IVMP per day became a widely accepted regiment.

The prolongation of treatment also varied in the studies, and the general consensus on how long the MS relapse should be treated has undergone a notable change over the years. While back in the 1960s to 1980s it was a common practice to treat MS exacerbation for 4weeks and even for up to 35days,[58,60,80] in more recent times significantly shorter courses of 3–7days were found to be quite adequate.[5–7,69,70,82]

Beginning 1990s the concept of oral high-dose MP received intense attention, which in most trials was found to be comparable to the effects of IVMP.[72,83,84] Financial, logistical, and other practical advantages associated with the oral route of administration are apparent.[6] A double-blind, randomized, placebo-controlled study[26] with 51 patients randomized to either the group of oral MP (500mg once a day for 5days followed by a tapering period) or the placebo group showed that there were significantly more responders in the oral MP group, concluding that oral high-dose MP is efficacious in managing attacks of MS, and although adverse effects were common, no serious side effects were observed. In the study by Barnes et al.[84] 42 patients with MS relapse were assigned to receive oral MP, and 38 patients to receive IVMP. The primary outcome was a difference between the two treatment groups of one or more EDSS grades at 4weeks; there were no significant differences found between the two groups. The randomized MRI study of high-dose oral MP versus IVMP enrolled 40 patients, who were randomized to receive either 1g/day of oral MP for 5days or 1g/day of IVMP for the same 5days. It was concluded that the oral MP was as effective as IVMP in reducing gadolinium-enhancing lesions with similar clinical, safety, and tolerability profiles.[72] More recently Ramo-Tello et al. published the results of a randomized clinical trial of oral MP versus IVMP for MS relapses, with the primary endpoint of noninferiority assessment of EDSS improvement at 4weeks, which was achieved, and no differences were found.[85] High rate of compliance was reported with high-dose (1250mg) oral prednisone.[86] The Cochrane Database of Systemic Review analyzed five different studies, including the aforementioned ones, and concluded that oral versus IV administration of MP does not demonstrate any significant differences in clinical, radiological, and pharmacological outcomes.[87]

22.3.1 Adverse Events During Steroid and ACTH Therapy

Steroid- or ACTH-induced adverse effects and their relative frequencies vary from patient to patient and depend on several factors, including the individual patient's age, comorbidities and the dose, duration, and possibly type and route of administration.[7] Short-term use of either steroids or ACTH has been associated with relatively minor side effects.[55,59,69–72] The adverse effects that may occur with ACTH are thought to be related primarily to its steroidogenic effects and are similar to corticosteroids. There may be increased susceptibility to new infection and increased risk of reactivation of latent infections; adrenal insufficiency may occur after abrupt withdrawal of the drug following prolonged therapy; Cushing's syndrome, elevated blood pressure, salt and water retention, and hypokalemia may be seen; masking of symptoms of other underlying disease/disorders may occur; there is a risk of gastrointestinal perforation and bleeding with increased risk of perforation in patients with certain GI disorders; acne; onset or worsening of euphoria, insomnia, irritability, mood swings, personality changes, depression, and psychosis may occur.[61] Caution should be used when prescribing ACTH to patients with diabetes or myasthenia gravis; prolonged use may produce cataracts, ocular infections, or glaucoma; use in patients with hypothyroidism or liver cirrhosis may result in an enhanced effect; there may be negative effects on growth and physical development and decreases in bone density.[61] Cass et al. cataloged adverse events in 47 patients with MS undergoing ACTH treatment[88] and reported hyperglycemia, steroid diabetes, and osteoporosis.

The most frequently observed corticosteroid side effects were gastrointestinal symptoms, weight gain, edema, mood changes, dysphoria, anxiety, insomnia, musculoskeletal pain, palpitations, edema, acne, weight gain, headache, and unpleasant (metallic) taste. Less frequently reported were hyperglycemia, hypertension, moon face, hirsutism, and unusual taste during or after intravenous infusion.[5–7,69–71] Among adverse events involving the musculoskeletal system, osteoporosis has been estimated to develop in at least 50% of individuals requiring long-term corticosteroid therapy.[89] However, short-term steroid treatment for relapse does not seem to reduce bone density in fully ambulatory patients with MS.[90,91] Still, osteopenia was observed more frequently in MS patients than healthy controls; found only in patients treated for relapses who also had a significantly higher EDSS score, it is suggesting that decreased mobility may contribute to bone loss more than corticosteroid use.[91]

Severe psychiatric disorders such as psychosis, depression, or manic episodes are reported in up to a third of steroid-treated patients. Insomnia has been reported by ~50% of patients on corticosteroid treatment. The risk of psychosis appears to be highest within the first days or weeks of starting therapy and may be higher in women.[92]

Other frequently reported side effects are infections; the most common infections include pneumonia, septic arthritis/bursitis, and complicated urinary tract infections.[3,5–7,70–72,92] The development of posterior and subcapsular cataracts has been reported to occur mostly with low steroid dosage.[92]

Effects of steroid treatment on carbohydrate metabolism and glucose tolerance appear to be proportional to the patients' preexisting status. A case control study of more than 20,000 patients reported that the relative risk for development of hyperglycemia requiring treatment was significantly increased in patients taking corticosteroids compared to nonusers.[93] Feldman-Billard et al. assessed short-term tolerance of 3-day IVMP therapy in 80 patients with type-2 diabetes treated for eye disorders, and reported that each IVMP treatment induced a mean twofold increase of peak blood glucose levels 10h later.[94]

Aseptic necrosis has been reported to be a not-infrequent complication of systemic steroid use, and its development is unpredictable and may occur within the first few weeks of therapy.[92,95]

Several reports on rare side effects associated with MS relapse treatment deserve our attention. Venous thrombosis in MS patients after high-dose IVMP has been documented, suggesting that prophylactic anticoagulant treatment in this setting is warranted.[96]

Steroid resistance is a known problem frequently reported in the settings of relapse treatment.[5] Steroid resistance phenomenon has been described in both autoantigen-specific and nonspecific responses of T cells obtained from mice with EAE. T cells obtained during EAE were resistant to glucocorticoid-induced apoptosis, and this was linked to downregulation of glucocorticoid receptor expression. Steroid resistance in T cells was also seen in MS patients with radiological evidence for ongoing inflammation; it was not found in the MS patients during pregnancy, when relapse risk is decreased, but recurred postpartum, a time of increased relapse risk.[97]

According to the patient-reported outcomes from the North American Research Committee on Multiple Sclerosis Registry, 32% of IVMP-treated and 34% of oral corticosteroid-treated patients indicated their symptoms were worse 1month after treatment than prerelapse, as did 39% of observation-only patients; 30% of IVMP-treated patients indicated their treatment made relapse symptoms worse or had no effect, as did 38% of oral corticosteroid-treated patients and 76% of observation-only patients. Thus, a sizeable percentage of patients feel that their symptoms following corticosteroid treatment are worse than prerelapse symptoms and that treatment had no effect or worsened symptoms.[51]

Clearly, options other than systemic steroids are needed for MS relapse treatment. This necessity is one of the explanations for why there is such an increased attention to the ACTH; another being the increased volume of knowledge on ACTH mechanisms and its direct anti-inflammatory and immunomodulatory effects via activation of central and peripheral MCRs, in addition to the well-known steroidogenic effects.[65] Research into melanocortin peptides and their receptors argue against the longstanding belief that the beneficial effect of ACTH depends solely on its ability to stimulate the release of endogenous corticosteroids in adrenal glands. As discussed earlier, the melanocortin system has many diverse regulatory functions in the human body, including glucocorticoid production, control of food intake and energy expenditure, control of behavioral effects, memory, and especially important for MS, neuroprotection, immune modulation, and antiinflammatory effects.[65,66,98] Without a doubt, in the 1970s ACTH was abandoned in the understudied position. Successful arrival of robustly effective and relatively inexpensive corticosteroids helped deviate professional attentions from ACTH for decades. New research is clearly needed, including placebo-controlled studies, as it may contain a perspective on the potential value of melanocortin system agonists for the treatment of MS.[98]

Although data from clinical trials have not demonstrated a clear difference in efficacy of ACTH and corticosteroids,[5,70] there have been anecdotal reports of patients who do not respond to steroids but do respond to ACTH.[38,99,100] Likewise, some patients who cannot tolerate steroids may tolerate ACTH.[100] Still, more data are needed before firm recommendations can be made to support the use of ACTH versus IVMP or oral steroids, considering the high price of ACTH gel.

An important question that at this point remains unanswered is that of potential differences in the safety of ACTH relative to the corticosteroids. For instance, risk for bone loss is particularly important in the context of high-dose or lengthened use of corticosteroids, which lead to reduced estrogen, testosterone, and renal androgens; reduced gastrointestinal calcium absorption, increased calcium excretion, and increased parathyroid hormone, which promotes excessive osteoclastic bone removal[101,102]; and corticosteroids also induce osteonecrosis supposedly via increased apoptosis of osteoblasts.[102] In contrast, in experimental studies ACTH shows potential osteoprotective properties of ACTH.[101] Well-designed clinical trials are needed to evaluate potential ACTH, feasibility in the setting of comorbid osteoporosis; today, however, the existing data favoring ACTH in this respect are purely experimental. Furthermore, ACTH may be a reasonable option for MS patients with diabetes and other comorbid autoimmune conditions,[103] but again, this hypothesis requires clinical evidence. Thus, despite the confounding amount of preclinical data currently available on ACTH and melanocortins, many clinical questions remain to be answered. Clinical studies evaluating immunological aspects of ACTH are emerging[67] and further trials are needed to determine its mechanisms of action.

22.4 SECOND LINE OF TREATMENT FOR MULTIPLE SCLEROSIS RELAPSE CASES UNRESPONSIVE TO CORTICOSTEROIDS OR ACTH

There are cases of MS relapse responding to neither corticosteroids nor ACTH. Several alternatives, including plasmapheresis,[104–111] cyclophosphamide,[104,112,113] intravenous immunoglobulin (IVIG),[114–119] and natalizumab[120] have been studied; at this point, plasmapheresis is the only option supported by strong clinical evidence.

As previously discussed, B cells and humoral antibody-driven mechanisms have their important role in MS exacerbation. The 2011 AAN guideline recommends using plasma exchange as a secondary treatment for severe flares in relapsing MS.[108] A large multicenter, randomized, double-blind controlled trial of an 8-week course of 11 plasma exchange treatments was examined for MS relapse treatment by Weiner et al.[104] One-hundred-sixteen subjects were randomized to either sham or true plasmapheresis, and both groups received identical treatment with intramuscular ACTH and oral cyclophosphamide. Plasmapheresis produced significant reductions in IgG, IgA, IgM, C3, and fibrinogen. The results suggested that plasma exchange given with ACTH plus cyclophosphamide enhances recovery from an exacerbation of disease in relapsing-remitting patients.[104] Results of a randomized, double-blind, sham-controlled study of either plasmapheresis or sham treatment in patients who did not respond to IVMP showed significant efficacy of plasma exchange in this category of MS patients. In the study 12 subjects with MS and 10 with other acute inflammatory demyelinating conditions were randomized to either plasmapheresis or sham treatment, and were given seven exchanges every other day for 14 days. Nineteen courses of plasmapheresis were performed, resulting in eight moderate or marked improvements, and only one moderate improvement was noted across 17 courses of sham treatment.[105] In a subsequent retrospective review of 59 steroid-unresponsive demyelinating events treated by plasmapheresis, Keegan et al.[106] found that male gender, preserved or brisk reflexes, and early initiation of treatment (within less than 60 days) were associated with improvement. More recently, Keegan et al.[107] correlated the pathologic features of demyelination with responsiveness to plasma exchange. Subjects who responded to plasma exchange in their series (10 of 19) had a particular pattern of demyelination, characterized by the presence of antibodies and complement, whereas nonresponders had pathologic characteristics of T-cell/macrophage-associated demyelination or distal oligodendrogliopathy.[107] In recent observational study of plasma exchange for steroid-refractory relapses, 93.3% patients showed a marked to moderate clinical improvement, and 46.7% recovered their baseline EDSS score 3 months postplasmapheresis. On the postplasmapheresis MRI, 60% showed radiologic resolution, 20% had partial resolution, and 20% had no resolution. It was observed that marked to moderate clinical improvement postplasmapheresis accompanied by a lack of radiologic resolution of the active lesion is not indicative of poor prognosis.[109]

Importantly, it was reported that plasmapheresis is well tolerated and associated with a favorable outcome in pediatric patients similar to reports in adults.[110,111]

Another purifying procedure—tryptophan immunoadsorption—was found to be safe, well-tolerated, and effective in the treatment of MS and neuromyelitis optica (NMO) relapses during pregnancy and breastfeeding, sometimes without preceding glucocorticoid pulse therapy.[121]

The role of IVIG in MS relapse treatment remains to be defined. While there are many anecdotal observations of beneficial effects of IVIG in the treatment of MS relapses, most published clinical studies provide no clear evidence to support this. Part of the controversy is a result of the fact that IVIG is being attempted usually after the IVMP administration; thus possible delayed effects of IVMP may overlap and obscure the pure IVIG effects.

Results of a large, double-blind, placebo-controlled trial by Noseworthy et al.[114] suggest that delayed IVIG administration had no effect on recovery from optic neuritis. However, an open-label study by Tselis et al.[115] showed that some cases of steroid-unresponsive optic neuritis may respond to IVIG administered 0.4 mg/kg/day for 5 days, followed by monthly 400 mg/kg infusion for 5 months. Visual acuity response as defined by a change to 20/30 or better 1 year later was significantly better in subjects with IVIG than in those who received only IVMP. Other studies did not show IVIG effects when administered concomitantly with, or immediately subsequent to IVMP.[116–118]

In a retrospective review of 10 patients treated with IVIG for acute relapses of NMO where IVIG was used in the majority of cases because of lack of response to steroids with or without plasmapheresis, improvement was noted in 5 of 11 (45.5%) events; the remaining had no further worsening. One patient, a 79-year-old woman, had a myocardial infarction 7 days after IVIG. It was concluded that IVIG may have a role in treating acute NMO relapses.[119]

A possible role of IVIG as a therapeutic option for treating MS relapse will require further study.

22.5 SUMMARY AND PRACTICAL CONSIDERATIONS

MS relapse treatment is important part of MS patient management. While it is generally accepted that while mild exacerbations may not require immediate treatment, the moderate-to-severe MS relapses with disabling symptoms should be treated using first-line treatments (systemic steroids or ACTH).[5] Starting the treatment as early as possible (within 5–7 days) of the MS relapse symptoms onset is considered appropriate, although there is no direct evidence to support this approach. It has been observed that relapse treatment can be successfully initiated as late as 1–2 months into a relapse.[6] Varied treatment regimens for MS relapses have been adopted based on the many clinical trials in MS relapses performed in the last 20 years.

We propose the following algorithm (which we developed on generally accepted principles) (Fig. 22.1):

- Evaluate patients with possible MS relapse within a week (or 5 working days) of the new or worsened symptoms onset rule out pseudo-exacerbation, most importantly the one imposed by infection.
- If MS relapse is confirmed (relapse is a clinical diagnosis. MRIs are not needed to either confirm or rule out MS relapse; MRI at the time of MS relapse can be done to assess the DMT adequacy, but not for the purposes of MS relapse diagnosis), start the treatment as soon as possible.
  - First line of treatment, IVMP 1 g/day for 3–5 days, is generally recommended as a first choice.
  - The need for oral prednisone taper following the IVMP should be considered on an individual basis (although there are data[122] suggesting no additional benefit associated with the use of the oral taper).
  - Although not FDA approved, oral administration of high-dose MP instead of IVMP may be suggested.[6]
- Patients who did not improve or could not tolerate the MP may be offered another FDA-approved option—ACTH. It should be noted that effects of IVMP or oral high-dose MP may be delayed, therefore in general as a rule in our practice we may wait 2–3 weeks after the last dose of high-dose corticosteroids before initiating the ACTH gel therapy administered either IM or SQ[100,123,124] at a dose of 80 units a day for at least 5 days and up to 15 days.[5,6,59] Our experience indicates that the majority of MS patients in acute exacerbation who previously failed (did not improve) or could not tolerate the MP treatment may have positive clinical outcomes and fewer adverse events with ACTH gel treatment.[100]

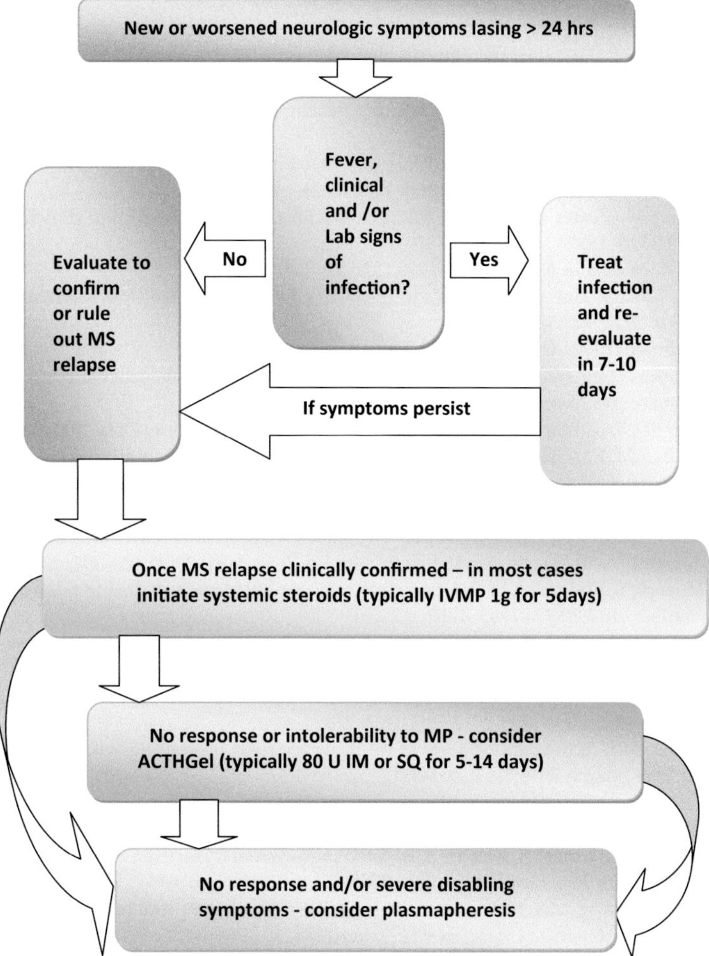

FIGURE 22.1 Proposed Algorithm for MS Relapse Management.

- For patients with disabling MS relapse symptoms not responding to the initial treatment, especially those patients who experience severe relapses or clinical worsening of symptoms following first line treatment, the plasmapheresis options should be considered on an individual basis. It should be administered as every-other-day procedures to as many as 5–10 exchanges on average.[105,108]
- Patients experiencing new or worsened neurologic symptoms after recent relapse treatment still have to be evaluated to rule out infection and other causes of pseudo-exacerbation.

As a conclusion, we would like to highlight once again that although there is no absolute consensus on long-term impact of relapses, their prognostic significance,[125] or even projecting importance of their treatment, from the patient's perspective MS exacerbations bring about a severe fall in quality of life, and therefore every attempt should be undertaken to shorten this episodes in the disease course.

References

1. Lublin FD, Reingold SC. Defining the clinical course of multiple sclerosis: results of an international survey. *Neurology*. 1996;46:907–911.
2. a. Lublin FD, Reingold SC, Cohen JA, et al. Defining the clinical course of multiple sclerosis: the 2013 revisions. *Neurology*. July 15, 2014;83(3):278–286.
   b. Patzold U, Pocklington PR. Course of multiple sclerosis. First results of a prospective study carried out of 102 MS patients from 1976–1980. *Acta Neurol Scand*. 1982;65:248–266.
3. a. Lublin FD, Baier M, Cutter G. Effect of relapses on development of residual deficit in multiple sclerosis. *Neurology*. 2003;61:1528–1532.
   b. Schumacher GA, et al. *Ann NY Acad Sci*. 1968;122:552–568.
4. Oleen-Burkey M, Castelli-Haley J, Lage MJ, Johnson KP. Burden of a multiple sclerosis relapse: the patient's perspective. *Patient*. 2012;5(1):57–69.
5. Berkovich R. Treatment of acute MS relapses. *Neurotherapeutics*. 2013;10.
6. Frohman EM, Shah A, Eggenberger E, et al. Corticosteroids for multiple sclerosis: I. Application for treating exacerbations. *Am Soc Exp Neurother*. 2007;4:618–626.
7. Repovic P, Lublin FD. Treatment of multiple sclerosis exacerbations. *Neurol Clin*. 2011;29:389–400.
8. Penner IK, Hubacher M, Rasenack M, Sprenger T, Weber P, Naegelin Y. Cognitive relapse in the absence of new neurologic symptoms: clinical utilty of neuropsychological assessment in a case of juvenile multiple sclerosis. *Mult Scler*. October 2012;18(4 suppl 1):303. 28th ECTRIMS in Lyon, France. Conference Publication.
9. Gabelik T, Adames I, Mrden A, Rados M, Brinar VV, Habek M. Psychotic reaction as a manifestation of multiple sclerosis relapse treated with plasma exchange. *Neurol Sci*. 2012 April;33(2):379–382.
10. Benedict RH, Morrow S, Rodgers J, et al. Characterizing cognitive function during relapse in multiple sclerosis. *Mult Scler*. November 2014;20(13):1745–1752.
11. Pardini M, Uccelli A, Grafman J, Yaldizli Ö, Mancardi G, Roccatagliata L. Isolated cognitive relapses in multiple sclerosis. *J Neurol Neurosurg Psychiatry*. September 2014;85(9):1035–1037.
12. Moore P, Hirst C, Harding KE, Clarkson H, Pickersgill TP, Robertson NP. Multiple sclerosis relapses and depression. *J Psychosom Res*. October 2012;73(4):272–276.
13. Anbarasan D, Howard J. Acute exacerbation of multiple sclerosis presenting with facial metamorphopsia and palinopsia. *Mult Scler*. March 2013;19(3):369–371.
14. Bernheimer JH. Restless legs syndrome presenting as an acute exacerbation of multiple sclerosis. *Mult Scler Int*. 2011;2011:872948.
15. Codeluppi L, Bigliardi G, Chiari A, Meletti S. Isolated paroxysmal dysarthria caused by a single demyelinating midbrain lesion. *BMJ Case Rep*. October 16, 2013;2013.
16. Najafi MR, Chitsaz A, Najafi MA. Jacksonian seizure as the relapse symptom of multiple sclerosis. *J Res Med Sci*. March 2013;18(suppl 1):S89–S92.
17. Jurić S, Mišmaš A, Mihić N, Barać AM, Habek M. Newly onset sinus bradycardia in the context of multiple sclerosis relapse. *Intern Med*. 2012;51(9):1121–1124.
18. Klein M, Woehrl B, Zeller G, Straube A. Stabbing headache as a sign of relapses in multiple sclerosis. *Headache*. July–August 2013;53(7):1159–1161.
19. Libbey JE, Cusick MF, Fujinami RS. Role of pathogens in multiple sclerosis. *Int Rev Immunol*. July–August 2014;33(4):266–283.
20. Mulvey MR, Doupe M, Prout M, et al. *Staphylococcus aureus* harbouring Enterotoxin A as a possible risk factor for multiple sclerosis exacerbations. *Mult Scler*. April 2011;17(4):397–403.
21. Brudek T, Christensen T, Aagaard L, Petersen T, Hansen HJ, Møller-Larsen A. B cells and monocytes from patients with active multiple sclerosis exhibit increased surface expression of both HERV-H Env and HERV-W Env, accompanied by increased seroreactivity. *Retrovirology*. November 16, 2009;6:104.
22. Sotelo J, Ordoñez G, Pineda B, Flores J. The participation of varicella zoster virus in relapses of multiple sclerosis. *Clin Neurol Neurosurg*. April 2014;119:44–48.
23. Angelini DF, Serafini B, Piras E, et al. Increased CD8+ T cell response to Epstein-Barr virus lytic antigens in the active phase of multiple sclerosis. *PLoS Pathog*. 2013;9(4):e1003220.
24. Beume LA, Dersch R, Fuhrer H, Stich O, Rauer S, Niesen WD. Massive exacerbation of multiple sclerosis after withdrawal and early restart of treatment with natalizumab. *J Clin Neurosci*. February 2015;22(2):400–401.

25. Gregory AP, Dendrou CA, Attfield KE, et al. TNF receptor 1 genetic risk mirrors outcome of anti-TNF therapy in multiple sclerosis. *Nature*. August 23, 2012;488(7412):508–511.
26. Correale J, Farez MF, Ysrraelit MC. Increase in multiple sclerosis activity after assisted reproduction technology. *Ann Neurol*. November 2012;72(5):682–694.
27. Farez MF, Fiol MP, Gaitán MI, Quintana FJ, Correale J. Sodium intake is associated with increased disease activity in multiple sclerosis. *J Neurol Neurosurg Psychiatry*. January 2015;86(1):26–31.
28. Runia TF, Hop WC, de Rijke YB, Buljevac D, Hintzen RQ. Lower serum vitamin D levels are associated with a higher relapse risk in multiple sclerosis. *Neurology*. July 17, 2012;79(3):261–266.
29. Lovera J, Reza T. Stress in multiple sclerosis: review of new developments and future directions. *Curr Neurol Neurosci Rep*. November 2013;13(11):398.
30. Somer E, Golan D, Dishon S, Cuzin-Disegni L, Lavi I, Miller A. Patients with multiple sclerosis in a war zone: coping strategies associated with reduced risk for relapse. *Mult Scler*. April 2010;16(4):463–471.
31. Berkovich RR. Personal communications.
32. Sorenson M, Janusek L, Mathews H. Psychological stress and cytokine production in multiple sclerosis: correlation with disease symptomatology. *Biol Res Nurs*. April 2013;15(2):226–233.
33. Spelman T, Gray O, Trojano M, et al. Seasonal variation of relapse rate in multiple sclerosis is latitude dependent. *Ann Neurol*. December 2014;76(6):880–890.
34. Simpson Jr S, Stewart N, van der Mei I, et al. Stimulated PBMC-produced IFN-γ and TNF-α are associated with altered relapse risk in multiple sclerosis: results from a prospective cohort study. *J Neurol Neurosurg Psychiatry*. February 2015;86(2):200–207.
35. Kalincik T, Buzzard K, Jokubaitis V, et al. Risk of relapse phenotype recurrence in multiple sclerosis. *Mult Scler*. October 2014;20(11):1511–1522.
36. Kalincik T, Vivek V, Jokubaitis V, et al. Sex as a determinant of relapse incidence and progressive course of multiple sclerosis. *Brain*. December 2013;136(Pt 12):3609–3617.
37. Fay AJ, Mowry EM, Strober J, Waubant E. Relapse severity and recovery in early pediatric multiple sclerosis. *Mult Scler*. July 2012;18(7):1008–1012.
38. Berkovich R, Agius MA. Mechanisms of action of ACTH in the management of relapsing forms of multiple sclerosis. *Ther Adv Neurol Disord*. March 2014;7(2):83–96.
39. Steinman L. Immunology of relapse and remission in multiple sclerosis. *Annu Rev Immunol*. 2014;32:257–281.
40. Aung LL, Mouradian MM, Dhib-Jalbut S, Balashov KE. MMP-9 expression is increased in B lymphocytes during multiple sclerosis exacerbation and is regulated by microRNA-320a. *J Neuroimmunol*. January 15, 2015;278:185–189.
41. Cunnusamy K, Baughman EJ, Franco J, et al. Disease exacerbation of multiple sclerosis is characterized by loss of terminally differentiated autoregulatory CD8+ T cells. *Clin Immunol*. May–June 2014;152(1–2):115–126.
42. Correale J, Villa A. Role of CD8+ CD25+ Foxp3+ regulatory T cells in multiple sclerosis. *Ann Neurol*. 2010;67:625–638.
43. Peelen E, Damoiseaux J, Smolders J, et al. Th17 expansion in MS patients is counterbalanced by an expanded CD39+ regulatory T cell population during remission but not during relapse. *J Neuroimmunol*. December 15, 2011;240–241:97–103.
44. de Andrés C, Tejera-Alhambra M, Alonso B, et al. New regulatory CD19(+)CD25(+) B-cell subset in clinically isolated syndrome and multiple sclerosis relapse. Changes after glucocorticoids. *J Neuroimmunol*. May 15, 2014;270(1–2):37–44.
45. Pedotti R, Musio S, Scabeni S, et al. Exacerbation of experimental autoimmune encephalomyelitis by passive transfer of IgG antibodies from a multiple sclerosis patient responsive to immunoadsorption. *J Neuroimmunol*. September 15, 2013;262(1–2):19–26.
46. a. Sormani MP, Bruzzi P. MRI lesions as a surrogate for relapses in multiple sclerosis: a meta-analysis of randomised trials. *Lancet Neurol*. July 2013;12(7):669–676.
    b. Trapp B, et al. Axonal transection in the lesions of multiple sclerosis. *N Engl J Med*. 1998;338:278.
47. Brill L, Goldberg L, Karni A, et al. Increased anti-KIR4.1 antibodies in multiple sclerosis: could it be a marker of disease relapse? *Mult Scler*. November 12, 2014;21. pii:1352458514551779. [Epub ahead of print].
48. Sternberg Z, Sternberg D, Drake A, Chichelli T, Yu J, Hojnacki D. Disease modifying drugs modulate endogenous secretory receptor for advanced glycation end-products, a new biomarker of clinical relapse in multiple sclerosis. *J Neuroimmunol*. September 15, 2014;274(1–2):197–201.

49. Rossi S, Motta C, Studer V, et al. Subclinical central inflammation is risk for RIS and CIS conversion to MS. *Mult Scler*. January 12, 2015;21. pii:1352458514564482. [Epub ahead of print].
50. Honardoost MA, Kiani-Esfahani A, Ghaedi K, Etemadifar M, Salehi M. miR-326 and miR-26a, two potential markers for diagnosis of relapse and remission phases in patient with relapsing-remitting multiple sclerosis. *Gene*. July 10, 2014;544(2):128–133.
51. Nickerson M, Marrie RA. The multiple sclerosis relapse experience: patient-reported outcomes from the North American Research Committee on Multiple Sclerosis (NARCOMS) Registry. *BMC Neurol*. September 10, 2013;13:119.
52. Hoch E. *Ergebnisse verschiedener Behandlungsversuche der multiplen Sklerose an der neurologischen Abteilung der Medizinische und Nervenklinik der Universitat Wiirzburg wahrend der Jahre 1935/1936* Würzburg: Inaugural-Dissertation. Leipzig: Schneider and Mischkewitz, Naunhof; 1937.
53. Pette H. *Die akut entziindlichen Erkrankungen des Nervensystems*. Leipzig: Thieme; 1942:516–517.
54. Schaltenbrand G. *Die multiple Sklerose des Menschen*. Leipzig: Thieme; 1943:238–246.
55. Miller H, Newell DJ, Ridley A. Multiple sclerosis. Treatment of acute exacerbations with corticotrophin (ACTH). *Lancet*. 1961;2:1120–1122.
56. Alexander L, Cass LJ. The present status of ACTH therapy in multiple sclerosis. *Ann Intern Med*. 1963;58:454–471.
57. Millar JHD, Belf MD, Vas CJ. Long-term treatment of multiple sclerosis with corticotrophin. *Lancet*. 1967;7513: 429–431.
58. Rinne UK, Sonninen V, Tuovinen T. Corticotrophin treatment in multiple sclerosis. *Acta Neurol Scand*. 1967;31: 185–186.
59. Rose AS, Kuzma JW, Kurtzke JF, Sibley WA, Tourtellotte WW. Cooperative study in the evaluation of therapy in multiple sclerosis: ACTH vs placebo in acute exacerbation. *Trans Am Neurol Assoc*. 1969;94:126–133.
60. Rose AS, Kuzma JW, Kurtzke JF, Namerow NS, Sibley WA, Tourtellotte WW. Cooperative study in the evaluation of therapy in multiple sclerosis: ACTH vs. placebo—final report. *Neurology*. 1970;20:1–59.
61. Questcor Pharmaceuticals, Inc. *H.P. Acthar® Gel (Repository Corticotrophin Injection) [Prescribing Information]*. Hayward, CA: Questcor Pharmaceuticals, Inc.; June 2011.
62. Bomback AS, Tumlin JA, Baranski J, et al. Treatment of nephrotic syndrome with adrenocorticotropic horone (ACTH) gel. *Drug Des Devel Ther*. 2011;5:147–153.
63. Pranzatelli MR, Chun KY, Moxness M, Tate ED, Allison TJ. Cerebrospinal fluid ACTH and cortisol in opsoclonus-myoclonus: effect of therapy. *Pediatr Neurol*. 2005;33:121–126.
64. Stafstrom CE, Arnason BG, Baram TZ, et al. Treatment of infantile spasms: emerging insights from clinical and basic science perspectives. *J Child Neurol*. 2011;26. [Epub ahead of print].
65. Catania A, Gatti S, Colombo G, Lipton JM. Targeting melanocortin receptors as a novel strategy to control inflammation. *Pharmacol Rev*. 2004;56:1–29.
66. Catania A. Neuroprotective actions of melanocortins: a therapeutic opportunity. *Trends Neurosci*. 2008;31:353–360.
67. Benjamins JA, Nedelkoska L, Lisak RP. Adrenocorticotropin hormone 1-39 promotes proliferation and differentiation of oligodendroglial progenitor cells and protects from excitotoxic and inflammation-related damage. *J Neurosci Res*. October 2014;92(10):1243–1251.
68. Napoli S. ACTH gel in the treatment of multiple sclerosis exacerbation: a case study. *Int Med Case Rep J*. January 7, 2015;8:23–27.
69. Miller DM, Weinstock-Guttman B, Bethoux F, et al. A meta-analysis of methylprednisolone in recovery from multiple sclerosis exacerbations. *Mult Scler*. 2000;6:267–273.
70. Thompson AJ, Kennard C, Swash M, et al. Relative efficacy of intravenous methylprednisolone and ACTH in the treatment of acute relapse in MS. *Neurology*. 1989;39:969–971.
71. Durelli L, Cocito D, Riccio A, et al. High-dose intravenous methylprednisolone in the treatment of multiple sclerosis: clinical-immunologic correlations. *Neurology*. 1986;36:238–243.
72. Martinelli V, Rocca MA, Annovazzi P, et al. A short-term randomized MRI study of high-dose oral vs intravenous methylprednisolone in MS. *Neurology*. 2009;73:1842–1848.
73. Muls NG, Dang HA, Sindic CJ, van Pesch V. Regulation of Treg-associated CD39 in multiple sclerosis and effects of corticotherapy during relapse. *Mult Scler*. February 6, 2015;21. pii: 1352458514567215. [Epub ahead of print].
74. Schweingruber N, Fischer HJ, Fischer L, et al. Chemokine-mediated redirection of T cells constitutes a critical mechanism of glucocorticoid therapy in autoimmune CNS responses. *Acta Neuropathol*. May 2014;127(5):713–729.
75. Ayache SS, Créange A, Farhat WH, et al. Relapses in multiple sclerosis: effects of high-dose steroids on cortical excitability. *Eur J Neurol*. April 2014;21(4):630–636.

76. Krieger S, Sorrells SF, Nickerson M, Pace TW. Mechanistic insights into corticosteroids in multiple sclerosis: war horse or chameleon? *Clin Neurol Neurosurg*. April 2014;119:6–16.
77. Clarner T, Parabucki A, Beyer C, Kipp M. Corticosteroids impair remyelination in the corpus callosum of cuprizone-treated mice. *J Neuroendocrinol*. July 2011;23(7):601–611.
78. Abbruzzese G, Gandolfo C, Loeb C. "Bolus" methylprednisolone versus ACTH in the treatment of multiple sclerosis. *Ital J Neurol Sci*. 1983;2:169–172.
79. Barnes M, Bateman D, Cleland P, et al. Intravenous methylprednisolone for multiple sclerosis in relapse. *J Neuro Neurosurg Psychiatry*. 1985;48:157–159.
80. Milanese C, La Mantia L, Salmaggi A, et al. Double-blind randomized trial of ACTH versus dexamethasone versus methylprednisolone in multiple sclerosis bouts. Clinical, cerebrospinal fluid and neurophysiological results. *Eur Neurol*. 1989;29:10–14.
81. Milligan NM, Newcombe R, Compston DA. A double-blind controlled trial of high dose methylprednisolone in patients with multiple sclerosis: 1. Clinical effects. *J Neurol Neurosurg Psychiatry*. 1987;50:5.
82. Beck RW, Cleary PA, Anderson MM, et al. A randomized, controlled trial of corticosteroids in the treatment of acute optic neuritis. *N Engl J Med*. 1992;9:581–588.
83. Sellebjerg F, Frederiksen JL, Nielsen PM, Olesen J. Double-blind, randomized, placebo-controlled study of oral, high-dose methylprednisolone in attacks of MS. *Neurology*. 1998;51:529–534.
84. Barnes D, Hughes RAC, Morris RW, et al. Randomised trial of oral and intravenous methylprednisolone in acute relapses of multiple sclerosis. *Lancet*. 1997;349:902–906.
85. Ramo-Tello C, Grau-López L, Tintoré M, et al. A randomized clinical trial of oral versus intravenous methylprednisolone for relapse of MS. *Mult Scler*. May 2014;20(6):717–725.
86. Morrow SA, McEwan L, Alikhani K, Hyson C, Kremenchutzky M. MS patients report excellent compliance with oral prednisone for acute relapses. *Can J Neurol Sci*. May 2012;39(3):352–354.
87. Burton JM, O'Connor PW, Hohol M, Beyene J. Oral versus intravenous steroids for treatment of relapses in multiple sclerosis. *Cochrane Database Syst Rev*. December 12, 2012;12:CD006921.
88. Cass LJ, Alexander L, Enders M. Complications of corticotropin therapy in multiple sclerosis. *JAMA*. 1966;197:173–178.
89. Lukert BP, Raisz LG. Glucocorticoid-induced osteoporosis: pathogenesis and management. *Ann Intern Med*. 1990;112:352–364.
90. Schwid SR, Goodman AD, Puzas JE, et al. Sporadic corticosteroid pulses and osteoporosis in multiple sclerosis. *Arch Neurol*. 1996;8:753–757.
91. Zorzon M, Zivadinov R, Locatelli L, et al. Long-term effects of intravenous high dose methylprednisolone pulses on bone mineral density in patients with multiple sclerosis. *Eur J Neurol*. 2005;7:550–556.
92. Fardet L, Kassar A, Cabane J, et al. Corticosteroid-indused adverse events in adults: frequency, screening and prevention. *Drug Saf*. 2007;30:861–881.
93. Gurwitz JH, Bohn RL, Glynn RJ, et al. Glucocorticoids and the risk for initiation of hypoglycemic therapy. *Arch Intern Med*. 1994;154:97–101.
94. Feldman-Billard S, Lissak B, Kassaei R, et al. Short-term tolerance of pulse methylprednisolone therapy in patients with diabetes mellitus. *Ophthalmology*. 2005;112:511–515.
95. Weiner ES, Abeles M. Aseptic necrosis and glucocorticois in systemic lupus erythematosus: a reevaluation. *J Rheumatol*. 1989;16:604–608.
96. Kalanie H, Harandi AA, Alidaei S, Heidari D, Shahbeigi S, Ghorbani M. Venous thrombosis in multiple sclerosis patients after high-dose intravenous methylprednisolone: the preventive effect of enoxaparin. *Thrombosis*. 2011;2011:785459.
97. Gold SM, Sasidhar MV, Lagishetty V, et al. Dynamic development of glucocorticoid resistance during autoimmune neuroinflammation. *J Clin Endocrinol Metab*. August 2012;97(8):E1402–E1410.
98. Arnason B, Berkovich R, Catania A, et al. Therapeutic mechanisms of action of adrenocorticotropic hormone (ACTH) and other melanocortin peptides for the clinical management of patients with MS. *Mult Scler*. 2012.
99. Poser CM. Corticotropin is superior to corticosteroids in the treatment of MS. *Arch Neurol*. 1989;46:946.
100. Berkovich R, Fernandez M, Subhani D. Adrenocorticotropic hormone treatment of multiple sclerosis exacerbations. *CMSC-ACTRIMS*. 2012. [Abstract DX66].
101. Zaidi M, Sun L, Robinson LJ, et al. ACTH protects against glucocorticoid-induced osteonecrosis of bone. *Proc Natl Acad Sci USA*. 2010;107:8782–8787.
102. Weinstein RS. Clinical practice. Glucocorticoid-induced bone disease. *N Engl J Med*. 2011;365:62–70.

103. Berkovich R, Subhani D, Steinman L. Autoimmune comorbid conditions in multiple sclerosis. *US Neurol.* 2011;7:132–138.
104. Weiner HL, Dau PC, Khatri BO, et al. Double-blind study of true vs sham plasma exchange in patients treated with immunosuppression for acute attacks of multiple sclerosis. *Neurology.* 1989;38:1143–1149.
105. Weinshenker BG, O'Brien PC, Petterson TM, et al. A randomized trial of plasma exchange in acute central nervous system inflammatory demyelinating disease. *Ann Neurol.* 1999;46:878–886.
106. Keegan M, Pineda AA, McClelland RL, et al. Plasma exchange for severe attacks of CNS demyelination: predictors of response. *Neurology.* 2002;58:143–146.
107. Keegan M, Konig F, Mcclelland R, et al. Relation between humoral pathological changes in multiple sclerosis and response to therapeutic plasma exchange. *Lancet.* 2005;366:579–582.
108. Cortese V, Chaudhry YT, So F, et al. Evidence-based guideline update: plasmapheresis in neurologis disorders: report of the therapeutics and technology assessment subcommittee of the American academy of neurology. *Neurology.* 2011;76:294–300.
109. Meca-Lallana JE, Hernández-Clares R, León-Hernández A, Genovés Aleixandre A, Cacho Pérez M, Martín-Fernández J. Plasma exchange for steroid-refractory relapses in multiple sclerosis: an observational, MRI pilot study. *J Clin Ther.* April 2013;35(4):474–485.
110. Koziolek M, Mühlhausen J, Friede T, et al. Therapeutic apheresis in pediatric patients with acute CNS inflammatory demyelinating disease. *Blood Purif.* 2013;36(2):92–97.
111. Mogami Y, Yamada K, Toribe Y, Yanagihara K, Mano T, Suzuki Y. Successful treatment with additional plasmapheresis for the exacerbation of acute neurological symptoms in a girl with multiple sclerosis. [Article in Japanese] *No To Hattatsu.* January 2011;43(1):36–40.
112. Barile-Fabris L, Ariza-Andraca R, Olguín-Ortega L, et al. Controlled clinical trial of IV cyclophosphamide versus IV methylprednisolone in severe neurological manifestations in systemic lupus erythematosus. *Ann Rheum Dis.* 2005;64:620–625.
113. Greenberg BM, Thomas KP, Krishnan C, et al. Idiopathic transverse myelitis: corticosteroids, plasma exchange, or cyclophosphamide. *Neurology.* 2007;68:1614–1617.
114. Noseworthy JH, O'Brien PC, Petterson TM, et al. A randomized trial of intravenous immunoglobulin in inflammatory demyelinating optic neuritis. *Neurology.* 2001;56:1514–1522.
115. Tselis A, Perumal J, Caon C, et al. Treatment of corticosteroid refractory optic neuritis in multiple sclerosis patients with intravenous immunoglobulin. *Eur J Neurol.* 2008;15:1163–1167.
116. Visser LH, Beekman R, Tijssen CC, et al. A randomized, double-blind, placebo-controlled pilot study of IV immune globulins in combination with IV Methylprednisolone in the treatment of relapses in patients with MS. *Mult Scler.* 2004;10:89–91.
117. Sorensen PS, Haas J, Sellebjerg F, et al. IV immunoglobulins as add-on treatment to methylprednisolone for acute relapses in MS. *Neurology.* 2004;63:2028–2033.
118. Roed HG, Langkilde A, Sellebjerg F, et al. A double-blind, randomized trial of IV immunoglobulin treatment in acute optic neuritis. *Neurology.* 2005;64:804–810.
119. Elsone L, Panicker J, Mutch K, Boggild M, Appleton R, Jacob A. Role of intravenous immunoglobulin in the treatment of acute relapses of neuromyelitis optica: experience in 10 patients. *Mult Scler.* April 2014;20(4):501–504.
120. O'Connor PW, Goodman A, Willmer-Hulme AJ, et al. Randomized multicenter trial of natalizumab in acute MS relapses: clinical and MRI effects. *Neurology.* 2004;62:2038–2043.
121. Hoffmann F, Kraft A, Heigl F, et al. Tryptophan immunoadsorption for multiple sclerosis and neuromyelitis optica: therapy option for acute relapses during pregnancy and breastfeeding. [Article in German] *Nervenarzt.* February 2015;86(2):179–186.
122. Perumal JS, Caon C, Hreha S, et al. Oral prednisone taper following intravenous steroids fails to improve disability or recovery from relapses in multiple sclerosis. *Eur J Neurol.* 2008;7:677–680.
123. Brod SA, Morales MM. Bio-equivalence of intramuscular and SQ H.P. Acthar gel. *Biomed Pharmacother.* 2009;63:251–253.
124. Simsarian JP, Saunders C, Smith DM. Five-day regimen of intramuscular or subcutaneous self-administered adrenocorticotropic hormone gel for acute exacerbations of multiple sclerosis: a prospective, randomized, open-label pilot trial. *Drug Des Devel Ther.* 2011;5:381–389.
125. Hutchinson M. There is no such thing as a mild MS relapse. The mild relapse is an Anglo-Saxon delusion – commentary. *Mult Scler.* July 2012;18(7):930–931.

CHAPTER

23

Shedding Light on Vitamin D and Multiple Sclerosis

J. Smolders

Canisius Wilhelmina Hospital, Nijmegen, The Netherlands; Zuyderland Medical Center, Sittard, The Netherlands

R. Hupperts

Zuyderland Medical Center, Sittard, The Netherlands; Maastricht University Medical Center, Maastricht, The Netherlands

J. Damoiseaux

Maastricht University Medical Center, Maastricht, The Netherlands

OUTLINE

Translational Neuroimmunology in Multiple Sclerosis
http://dx.doi.org/10.1016/B978-0-12-801914-6.00023-4

23.1 BACKGROUND

The observation that the incidence of multiple sclerosis (MS) is higher in regions more remote from the equator sparked the interest in vitamin D exposure as a risk factor for developing MS.[1] Since then, many epidemiological studies reproduced a negative correlation of sunlight exposure, distance to the poles, and vitamin D intake with MS incidence. These observations are the origin of the hypothesis that a poor vitamin D status increases the risk of developing MS. Most notably, data from the Nurses' Health Studies and US Department of Defense Serum Depository provided valuable support for this hypothesis. Among the subjects included in the Nurses' Health Studies, 173 subjects developed MS in later life. Systematic assessment of dietary intake prior to disease onset revealed that subjects in the highest quintiles of total vitamin D intake as well as subjects taking vitamin D supplements per se had a lower risk of developing MS in later life.[2] Among US army personnel, whites within the lowest quintile of vitamin D status measured in blood during adolescence showed a higher risk of developing MS in later life.[3] Intervention studies on vitamin D supplementation and populational risk of developing MS are challenging to undertake and have thus far not been performed. Further support for a mechanistic role for vitamin D in the risk of MS is provided by a genome-wide association study, in which two genetic polymorphisms of key enzymes in vitamin D metabolism (25(OH)D-1α-hydroxylase and 1,25$(OH)_2$D-24-hydroxylase, vide infra) were associated with MS.[4]

Besides disease incidence, also disease activity in established MS has been associated with vitamin D status. In cohort studies, lower blood vitamin D status has been measured in relapsing-remitting MS (RRMS) patients during relapses when compared to remission.[5–8] A poor vitamin D status has been associated with a higher subsequent relapse risk in adults with RRMS,[9–11] as well as in children with MS.[12] Also, a negative correlation of vitamin D status with Expanded Disability Status Scale (EDSS) score has been described in RRMS,[7,13] and a poor vitamin D status may predict subsequent EDSS progression in Clinically Isolated Syndrome (CIS) patients.[11] A negative correlation between vitamin D status and depressive symptoms has been described in subjects with MS.[14,15] Not only clinical outcomes in MS have been associated with vitamin D status. In cohorts of RRMS patients, a high vitamin D status predicts a subsequent lower risk of new T2 lesions and new gadolinium-enhancing T1 lesions,[11,16,17] as well as a reduced progression of brain atrophy in CIS.[11] Several clinical studies on vitamin D supplementation in MS patients have been performed,[18–21] but none of these were adequately powered to assess the efficacy of vitamin D supplementation on the disease outcomes as discussed earlier.

The causality of the associations between (correlates of) vitamin D status and MS outcomes in these studies remains to be revealed. There are several hypotheses. First, an increase in physical disability correlates with a decrease in sun exposure,[13] which could underlie a lower vitamin D status in subjects with more disability. Second, a consumopathy of vitamin D by an activated immune system has been postulated,[22] as may also be the case in sarcoidosis.[23] Third, not specifically vitamin D but rather correlates may affect the disease process of MS, such as UV-light exposure per se.[24] Fourth, modulation of the disease process of MS by vitamin D has been proposed.[25] The latter hypothesis is clinically the most favorable, since this implies that vitamin D supplementation may be a disease-modulating treatment in MS and may even prevent MS. For the association between disease incidence and vitamin D, solving this issue is challenging. For the association between disease activity and vitamin D, upcoming clinical trials on vitamin D supplementation in RRMS may provide answers.

Many studies focused on the underlying mechanism by which vitamin D may modulate the disease course of MS. Although vitamin D is classically known for its roles in calcium homeostasis, the discovery of the vitamin D receptor (VDR) within leukocytes[26] prompted many experimental studies by different groups, which revealed vitamin D metabolites to be potent immunomodulating molecules. In the present chapter, we will discuss proposed mechanisms by which vitamin D status may influence disease pathophysiology of MS. First, vitamin D metabolism will be discussed, followed by immunomodulating functions of vitamin D in the neurological periphery and within the borders of the central nervous system (CNS) in relation with MS pathobiology. In conclusion, the challenges to incorporate these outcomes in clinical trials will be discussed.

23.2 VITAMIN D METABOLISM

Vitamin D has been discovered as cure for rachitis, and has a well-consolidated role in bone health and calcium metabolism.[27] Most vitamin D is acquired by exposure of the skin to sunlight, and to a lesser extent by dietary intake (Fig. 23.1). Both plant-derived vitamin D_2 (ergocalciferol) and animal-derived vitamin D_3 (cholecalciferol) are acquired by dietary intake. Vitamin D_3 is presumably most relevant, since only a negligible proportion of human circulating vitamin D metabolites is derived from vitamin D_2.[28,29] Vitamin D metabolism has been reviewed extensively elsewhere.[30] In short, the vast majority of circulating vitamin D and its metabolites are bound to the carrier molecule group component (or vitamin D binding protein (DBP)). The most abundant circulating vitamin D metabolite is 25(OH)D, which is also measured to assess vitamin D status.[31] However, this is not the biologically most active metabolite. Activation of 25(OH)D is achieved by hydroxylation toward $1,25(OH)_2D$ by the enzyme 25(OH)D-1α-hydroxylase (1αOHase, gene CYP27B1). The latter molecule has a high affinity for the intracellular VDR. Ligation of $1,25(OH)_2D$ to the VDR induces transcription or transrepression of vitamin D responsive genes. One of the promoted genes is the $1,25(OH)_2D$-24-hydroxylase (24OHase, gene CYP24A1), by which $1,25(OH)_2D$ induces autocatabolism. The kidney is the most important site of endocrine vitamin D metabolism, with a vital role in maintaining calcium and bone homeostasis.[27] However, VDR and vitamin D metabolizing enzymes are expressed by many other cell types, including CNS resident cells (neurons and glia), leukocytes (lymphocytes and myeloid cells), epithelial cells (skin, lungs, breast, intestine, and prostate), endocrine cells (ovaries, pancreatic islets, parathyroid gland, placenta, testes, and thyroid gland), and several malignant cell types.[30,32,33] In these cells, vitamin D is believed to serve an autocrine or paracrine function, influencing the intracellular or microenvironmental vitamin D metabolism (Fig. 23.1). The absolute amounts of vitamin D processed in the extrarenal sites of vitamin D metabolism in health and disease are not well quantified, but could be substantial. For instance, low circulating levels of $1,25(OH)_2D$ could be measured in 46% of anephric subjects without vitamin D supplements, and in all anephric subjects with vitamin D supplements.[34]

It is assumed that vitamin D metabolism between subjects with MS and controls is similar. When MS patients without extensive disability or progressive disease were compared with controls, we could not find substantial difference in circulating 25(OH)D, $1,25(OH)_2D$, and DBP levels in our population.[8] When assessing common polymorphisms within vitamin D-related genes, a genome-wide association study found a differential prevalence in cases and controls of two common variants within the CYP27B1 and CYP24A1 genes with modest odds-ratios.[4] In frequently assessed common genetic

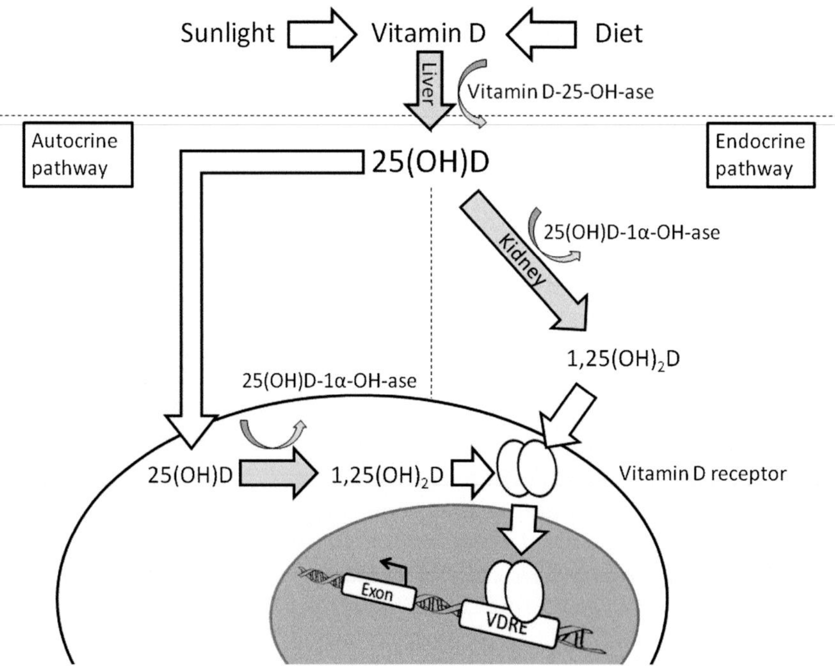

FIGURE 23.1 **Overview of vitamin D metabolism.** Vitamin D is acquired from both sun exposure and to a lesser extent from dietary intake. Vitamin D is hydroxylized into 25(OH)D, which can be hydroxylized to $1,25(OH)_2D$. The latter hydroxylation step can occur via an endocrine route (hydroxylation in the kidney) or via an autocrine route (hydroxylation within the target cell). $1,25(OH)_2D$ binds with high affinity to the intracellular vitamin D receptor, which can bind to a vitamin D response element in the promoter region of a vitamin D responsive gene and regulate expression. The gray arrows represent hydroxylation steps.

polymorphisms of the VDR, no consistent differences between subjects with MS and controls were shown.[35,36] There have been case-reports of families with cases of both MS and a homozygous rare mutation in the CYP27B1 gene causing an impaired capacity to form 1,25$(OH)_2$D and vitamin D-dependent rickets I.[37] Interestingly, in a whole exome sequencing study, an overtransmission of several rare mutations within the CYP27B1 gene was shown in affected individuals in families with several MS cases.[38] However, this finding could not be replicated in other cohorts.[39,40] These genetic findings support a role of vitamin D in the development of MS, but do not suggest a different vitamin D metabolism in the whole MS population.

23.3 VITAMIN D AND THE ADAPTIVE IMMUNE RESPONSE

As stated earlier, many distinct types of human leukocytes can express the VDR.[26] The functional implications of VDR-binding has been explored in several experimental settings. Modulation of the innate and adaptive immune response has been the subject of research.[41] We limit the scope of this chapter to the adaptive immune response, mainly because it is presumably most relevant for understanding potential interactions between vitamin D metabolism and the immune response in MS.

23.3.1 Vitamin D and the Adaptive Immune Response In Vitro

We and others extensively reviewed the effects of vitamin D metabolites on the adaptive immune response in vitro.[42,43] In short, upon activation by polyclonal receptor ligands, antigen-specific receptor ligands, or several pattern-recognition receptor ligands, lymphocytes ($CD4^+$ T cells, $CD8^+$ T cells, and B cells), and myeloid cells (dendritic cells, monocytes, macrophages) express the enzyme 1αOHase, catalyzing the local formation of 1,25$(OH)_2$D. Activated lymphocytes and myeloid cells express the VDR. Adding excessive amounts of exogenous 1,25$(OH)_2$D induces 24OHase expression in these cells, indicating the presence of a functional VDR. Upon addition of exogenous vitamin D, maturation and differentiation of dendritic cells is altered, resulting in tolerogenic dendritic cells and a reduced capacity to elicit an (autoreactive) T-cell response.[44] T-cell behavior, however, is not only modulated via the antigen-presenting cell compartment, but also directly. Exposure of $CD4^+$ T cells to 1,25$(OH)_2$D (but also to 25(OH)D) inhibits proliferation induced by antigen-specific or mitogenic stimuli.[45] Additionally, expression of several proinflammatory cytokines is inhibited (including interferon gamma (IFN-γ), interleukin (IL)-2, IL-6, IL-17, IL-21), while expression of several antiinflammatory cytokines (including IL-10) is promoted.[45,46] The proportion of $CD4^+$ T cells expressing FoxP3 and CTLA-4, markers for regulatory T cells (Tregs), is also increased.[45,46] Direct effects of 1,25$(OH)_2$D on B cells have also been demonstrated, with reduction of proliferation and plasma cell generation and increased expression of IL-10.[47,48] Obviously, effects of vitamin D on the B-cell response can also be a result of a modulation of the T-cell compartment and vice versa. In conclusion, vitamin D interacts with several key mechanisms of the adaptive immune response in vitro and functions as a negative feedback loop, suppressing proliferation and proinflammatory signals (Fig. 23.2).

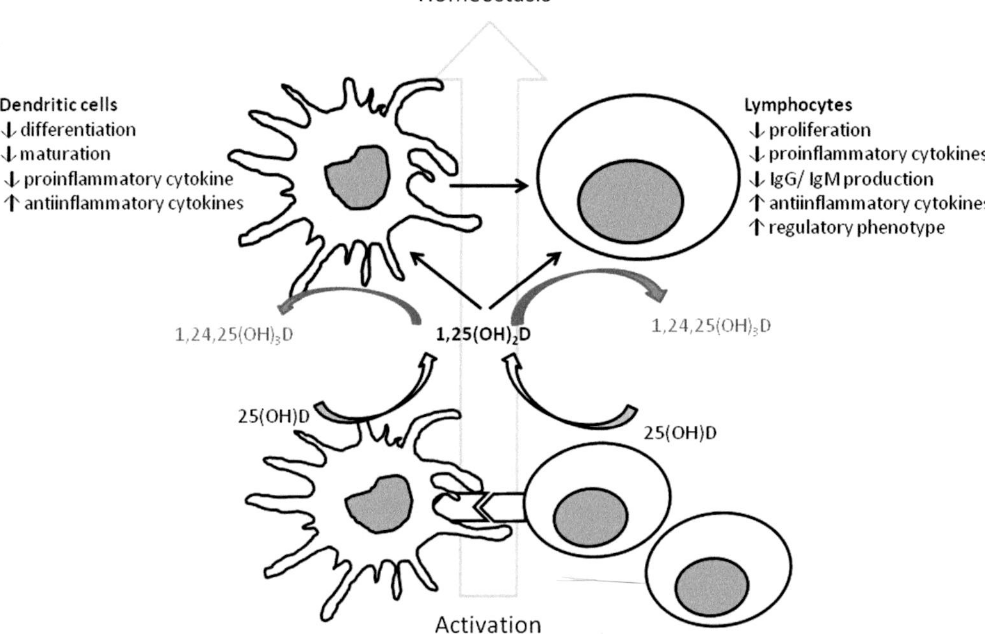

FIGURE 23.2 **Interaction between vitamin D and the immune response in vitro.** At sites of immune activation, expression of 1αOHase (CYP27B1) and subsequent formation of 1,25(OH)$_2$D out of 25(OH)D is induced in activated immune cells. This results in a microenvironment of high local levels of 1,25(OH)$_2$D. Binding of 1,25(OH)$_2$D to the vitamin D receptor, which is present in almost every activated immune cell, results in functional modulation of antigen-presenting cell/dendritic cell (APC/DC) and activated B and T cells. Vitamin D exerts its effects on lymphocytes both directly and via modulation of APC/DC. When exposed to 1,25(OH)$_2$D, lymphocytes and DC's express 24OHase (CYP24A1), catalyzing the inactivation of 1,25(OH)$_2$D. Altogether, activation of an immune response is accompanied by local activation of 25(OH)D, subsequently establishing increased levels of tolerance. *Adapted from Smolders J, Damoiseaux J. Vitamin D as a T cell modulator in multiple sclerosis.* Vitam Horm. *2011;86, Fig. 18.2, p. 410.*

23.3.2 Vitamin D and the Adaptive Immune Response in Experimental Autoimmune Encephalomyelitis

In vivo, these processes are believed to take place in the secondary lymphoid organs and within the CNS. Immunological studies exploring interactions between vitamin D and the immune response in vivo have mostly been performed in the experimental autoimmune encephalomyelitis (EAE) animal model for MS.

In animals with EAE, treatment with 1,25(OH)$_2$D before disease induction prevents disease, while treatment after disease induction cures or attenuates disease.[49,50] Supplementation with vitamin D, instead of the bioactive metabolite, resulted in a less impressive modulation of the disease course of EAE, and has been reported to be effective only in female mice.[51] The therapeutic effect of 1,25(OH)$_2$D treatment on the disease course of EAE was completely abrogated in rodent knockout models of VDR,[52] recombination activating gene (essential for mature B- and T-cell development),[53] and in a mice model with conditional targeting experiments with a dysfunctional VDR specifically in T cells.[54] Additionally, knockout of CD8$^+$

T cells in EAE had no impact on the effect of vitamin D on the disease course in these animals.[55] These studies point toward a direct effect of $1,25(OH)_2D$ on $CD4^+$ T cells in EAE.

In animals with EAE, cells derived from secondary lymphoid organs (spleen and draining lymph nodes) have been assessed. In mice with EAE, a study reported an expansion of $CD4^+$ T cells in the spleen after treatment with $1,25(OH)_2D$,[56] while another study reported a reduced fraction of splenic $CD4^+$ T cells.[57] When assessing cytokine profiles of splenocytes in EAE, it is important to note that in knockout experiments, the therapeutic effect of $1,25(OH)_2D$ in EAE was abrogated/attenuated by the knockout of IL-10 or IL-10 receptor,[58] and that IFN-γ was mandatory for the suppression of disease by vitamin D in female mice with EAE.[59] In secondary lymphoid organs of mice with EAE, transcription of IFN-γ was not affected by $1,25(OH)_2D$.[53,60] Accordingly, a proportion of cells staining positive for IFN-γ and IL-17 directly ex vivo was not affected, neither was the response of splenocytes (proliferation, IL-17, and IFN-γ production) to polyclonal or antigen-specific stimulation.[56] Contrastingly, a reduced production of IL-17A production by splenocytes and cells from draining lymph nodes after treatment of EAE animals with $1,25(OH)_2D$ has also been reported.[61] These differences may be time point- and animal model-dependent. A study assessing different time points in the EAE model found a reduced fraction of IL-17^+ and IFN-γ^+ splenic $CD4^+$ T cells 1 day after $1,25(OH)_2D$ treatment, which had disappeared 1 week later.[57] Both a significant increase as well as no increase of IL-4 transcripts in secondary lymphoid organs has been described.[53,60] An effect of $1,25(OH)_2D$ on Tregs has also been assessed in EAE. The proportion of splenic $CD25^+FoxP3^+$ T cells was not affected by $1,25(OH)_2D$ in one study in EAE mice,[56] yet was significantly increased in another study.[61] Likewise, treatment of EAE Lewis rats with $1,25(OH)_2D$ increased the proportions of $CD4^+$ $CD25^+FoxP3^+$ T cells in lymph nodes.[62] Again, a study on different time points found an expansion of splenic $Helios^+FoxP3^+$ Tregs 1 day post-$1,25(OH)_2D$, which was not found 1 week later.[57] Interestingly, a reduced expression of the chemokine receptor CXCR3 has been reported on T cells of EAE mice treated with $1,25(OH)_2D$.[56] This chemokine receptor mediates homing of T cells in inflamed CNS. The discrepancies between individual EAE-studies may be attributable to different rodent strains, different time points, different readout assays, and different treatment protocols.

These experiments suggest that treatment with $1,25(OH)_2D$ affects the microenvironment in the lymphoid organs in which immune cells are primed and activated, and hereby affects the disease course of EAE.

23.3.3 Vitamin D and the Adaptive Immune Response in Multiple Sclerosis

Whether the immunological findings in vitro and in EAE are directly applicable on the adaptive immune response in MS is not clear. First, although the pathophysiology of MS shares several characteristics with EAE, MS is not a disease driven by autoreactive cells against a single, well-defined epitope. Nevertheless, observations done in EAE may reflect general immune-modulating properties of vitamin D, which may be relevant for MS. Second, studies in vitro and in EAE have often been performed with (substantial doses of) $1,25(OH)_2D$, which may differentially affect immunological effector mechanisms when compared to vitamin D supplementation and subsequently elevated serum 25(OH)D levels. Third, human in vivo studies on vitamin D and leukocyte characteristics of cells acquired from secondary lymphoid or CNS tissue have not been performed.

Human in vivo studies on vitamin D status and lymphocyte characteristics mostly assessed correlations between vitamin D status and characteristics of circulating immune-markers and peripheral blood mononuclear cells (PBMCs). It should be mentioned that the circulation comprises a polyclonal pool of molecules and immune cells originating from primary lymphoid organs and circulating between secondary lymphoid organs and target tissue. Whether a specific vitamin D-related MS-modulating pathway reveals itself in the circulation is uncertain. However, correlations between MS disease activity status and circulating Treg functionality,[63] cytokine profile of T cells,[64] and B-cell differentiation profile have been reported.[65] Therefore, assessment of correlations between vitamin D status and these outcomes seems relevant. Furthermore, studying the polyclonal immune response in subjects with MS may reveal general immune-modulating properties of vitamin D, potentially relevant for MS. This view is supported by a whole-blood mRNA analysis of RRMS patients participating in the BENEFIT trial.[66] Expression of multiple genes was found to be regulated by 25(OH)D levels,[67] including the sphingosine-1-phosphate phosphatase 1 gene. This gene is involved in the egress of activated T cells from the lymph nodes, and its receptor is targeted by the second-line RRMS disease-modulating therapy fingolimod.[68]

T Cells

A study in a heterogeneous cohort of RRMS patients did show a positive correlation between the proportion of circulating Treg and vitamin D status.[69] Likewise, we assessed correlations between vitamin D status and several T-cell characteristics as measured with flow cytometry in RRMS.[70] We found no correlation between vitamin D status and circulating Treg proportions. However, the capacity of Tregs to suppress proliferation of polyclonally activated autologous T cells was impaired in subjects with a poor vitamin D status. It is not well defined whether this is attributable to a reduced functional capacity of the Treg, or a resistance to suppression of the responder cells. We performed a pilot experiment, in which 12-weeks supplementation of high-dose vitamin D_3 did not result in an improved Treg suppressive capacity in 15 subjects with RRMS.[71]

Furthermore, we assessed the cytokine profile of circulating $CD4^+$ T cells with flow cytometry. We found a negative correlation between vitamin D status and the ratio of IFN-γ^+ and IL-4^+ $CD4^+$ T cells in the circulation, with no significant correlations between vitamin D status and the individual proportions of IFN-γ, IL-4, IL-17, and IL-10^+ $CD4^+$ T cells.[70] The expression of these cytokines by $CD8^+$ T cells did also not correlate with 25(OH)D levels.[72] In the earlier mentioned pilot experiment, small changes in the $CD4^+$ T-cell cytokine profile were observed, trending toward a higher fraction of $CD4^+$ IL-10^+ T cells.[71] Interestingly, an impaired capacity of $CD4^+$ T cells of MS patients to express IL-10 upon stimulation with PHA and 1,25$(OH)_2$D in vitro has been reported.[73] In a pilot study in four healthy subjects, supplementation of vitamin D_3 resulted in increased IL-10 production by PBMC in vitro and a reduced proportion of IL-17^+ T cells.[74] Controlled studies should confirm an effect of vitamin D supplementation on these outcomes.

In addition to Treg and T-cell cytokine profiles, proliferative responses of PBMC to antigen-specific and polyclonal stimuli were assessed in RRMS patients.[75] In 25 RRMS patients supplemented with high doses of vitamin D_3, proliferative responses to the CNS-related antigens myelin basic protein (MBP) and the exon 2 epitope of MBP (Ex-2), declined. Likewise, in 31 RRMS patients, the patients with a positive T-cell proliferative response to a mix of seven myelin peptides displayed lower 25(OH)D levels.[76]

B Cells

We found no correlation between vitamin D status and the B-cell differentiation profile and IL-10 production in RRMS.[65] We explored in a pilot experiment the effect of 12-week vitamin D_3 supplementation on B cells in subjects with RRMS. We found no effect on the B-cell differentiation profile, total circulating immunoglobulins (IgG, IgA, and IgM), and circulating B-cell activating factor levels.[77] However, the circulating IgG load against the Epstein–Barr virus (EBV) nuclear antigen-1 (EBNA-1) was reduced in these subjects after supplementation.[78] Interestingly, higher serum concentrations of this molecule have been associated with a higher risk of developing MS,[79–81] although an association with relapse risk is uncertain.[82] Further controlled studies should confirm effects of vitamin D supplementation on anti-EBNA-1 antibody production.

Circulating Cytokines

In 85 RRMS patients not treated with immune-modulating drugs, vitamin D status correlated positively with circulating concentrations of IL-1 receptor antagonist and secreted frizzled-related protein 3.[83] In 25 MS patients supplemented with high doses of vitamin D_3, and 24 matched controls, concentrations of cytokines in serum and cell culture supernatants were reported to be too low to reliably assess an effect of supplementation.[75] In 17 MS patients supplemented with modest doses of vitamin D for 6 months, low circulating levels of TGF-beta1, as measured by ELISA, were increased when compared to 22 RRMS control patients, with no significant effect on TNF-α, IFN-γ, IL-2, and IL-13.[84] In 33 RRMS patients, supplementation of high-dose vitamin D for 12 months also induced an increase of low circulating latency-activated peptide (TGF-beta) levels as measured with a fluorescent bead immunoassay, but not in 29 placebo-supplemented control patients. In both groups, there was no significant change in circulating INF-γ, IL-17A, IL-2, IL-10, IL-9, IL-22, IL-6, IL-13, IL-4, IL-5, IL-1β, and TNF-α.[85]

23.4 VITAMIN D WITHIN THE CENTRAL NERVOUS SYSTEM

The ultimate target organ in MS is the CNS. By modulation of priming and T-cell activation in the secondary lymphoid organs, the disease process of MS and EAE in the CNS can be affected indirectly. However, vitamin D may also directly affect the behavior of leukocytes and resident cells within the CNS. Vitamin D gains access to the human CNS. In both MS patients and healthy controls, low levels of 25(OH)D were detected in cerebrospinal fluid (CSF).[86] Notably, these intrathecal levels of 25(OH)D correlated positively with the serum levels of 25(OH)D. In wild-type mice before EAE induction, spinal cord $1,25(OH)_2D$ levels were detectable, indicating the presence of vitamin D metabolites within the borders of the blood–brain barrier.[51] The carrier molecule for vitamin D metabolites, DBPs, is also frequently measured in CSF of MS patients and controls, which indicates active transport of this molecule through the blood–brain barrier.[87] We found no DBP mRNA transcription in normal-appearing white matter (NAWM) of donors with MS or controls.[33] It is not well defined whether active transport of DBP-bound vitamin D metabolites or passive passage of the circulating free-fraction is the main source of vitamin D within the undamaged blood–brain barrier.

23.4.1 CNS Infiltrating Leukocytes and Vitamin D in Experimental Autoimmune Encephalomyelitis and Multiple Sclerosis

In the EAE model of MS, the migration of autoreactive T cells to the CNS has been suggested to be inhibited by 1,25(OH)$_2$D. In a myelin oligodendrocyte glycoprotein (MOG)-mouse model of EAE, an expansion of MOG-specific T cells in blood and secondary lymphoid organs was found after 1,25(OH)$_2$D treatment, but cells did not cross the blood–brain barrier and no infiltrates with CD3$^+$ cells were found in the CNS tissue.[56] Alternatively, an increased rate of apoptosis of activated CD4$^+$ T cells in the CNS of 1,25(OH)$_2$D-treated animals has been suggested based on annexin V/propidium iodide stainings.[88] Although a reduced total load of CD4$^+$ T cells was found in CNS of 1,25(OH)$_2$D-treated animals, relative proportions of IFN-γ and IL-17 positive cells were unchanged.[57] Therefore, lower transcripts of IL-17 mRNA found in CNS homogenates of 1,25(OH)$_2$D-treated animals may be the result of a reduced T-cell load, rather than a specific effect on cytokine expression.[89] The proportion of CD4$^+$ T cells bearing a Helios$^+$FoxP3$^+$ regulatory phenotype has been reported to be increased in the CNS cell fraction.[57]

An impaired recruitment of carboxyfluorescein succinimidyl ester (CFSE)-labeled CD11b$^+$ monocytes into CNS of EAE animals treated with 1,25(OH)$_2$D has also been reported.[88] Earlier studies also reported reduced inflammatory infiltrates in the CNS of 1,25(OH)$_2$D-treated animals with EAE.[49] An effect on expression of several surface activation markers on predominantly myeloid cells has been studied in EAE. In semiquantitative studies, a decline of cells staining positive for CD11b, MHC-class II, inducible nitric oxide synthase (myeloid cells and astrocytes, confirmed with mRNA), and CD4 was observed.[88,90–92] CNS-derived cells staining for CD11b (activated macrophages) was reduced as assessed by flow cytometry, with a trend toward less CD4$^+$ cells.[93] Additionally, mRNA expression of the toll-like receptors (TLRs) 3, 4, 7, and 8 was suppressed, with also reduced immunostaining for TLR8 on CNS myeloid cells.[89] mRNA expression of the chemokines CXCL2, CXCL10, CCL2, CCL3, and CCL4 was reduced after 1,25(OH)$_2$D treatment.[88] It is unclear whether these outcomes are partly mediated by a direct effect of vitamin D on these cells, or via modulation of upstream components of the adaptive immune response locally or in the secondary lymphoid organs. Knockout experiments of lymphocytes and VDR on T cells suggest the latter scenario.[53,54]

In MS, infiltrating lymphocytes may also be directly responsive to vitamin D. In chronic active MS lesions, we observed intralesional T lymphocytes to stain positive for VDR.[33] Accordingly, 1,25(OH)$_2$D-responsive EBV-specific T-cell lines could be isolated from CSF of RRMS patients.[94] Regarding plasma cells, no association between CSF 25(OH)D and intrathecal total IgG-production in subjects with MS was found.[95]

23.4.2 Central Nervous System Resident Cells and Vitamin D in Experimental Studies and Multiple Sclerosis

In rat embryonal and adult rat CNS tissue, immunohistochemical staining, in situ-hybridization, and real-time polymerase chain reaction confirmed expression of VDR in different areas of brain and spinal cord.[96–98] In postmortem tissue of MS donors and donors without brain disease, VDR immunostaining colocalized with neuronal cell markers, confirming expression in human neurons.[33,99] Exposure of a neuronal cell line to 1,25(OH)$_2$D resulted in an upregulation of CYP24A1 mRNA, confirming the presence of a functional VDR.[33] The intensity of immunostaining is notably high in the substantia nigra, as well as the hypothalamic supraoptic and paraventricular nuclei, suggesting a specific biological role of vitamin D in these neurons.[99] Additionally, immunostaining for 1αOHase has been reported in human neurons.[32,99] The physiological role of vitamin D in neuronal homeostasis is not well defined. In cultures of rat cortical neurons, glutamate excitotoxicity was reduced by in vitro treatment with 1,25$(OH)_2$D.[100] A promoted expression of neurotrophic growth factor and neurothrophin-3 and downregulation of neurothrophin-4 in rat glial and neuronal cell cultures has been reported,[101–103] with coinciding inhibition of rat hippocampus neuronal cell proliferation and a promotion of neurite outgrowth.[101]

In both MS and control NAWM, we found nuclear staining of cells showing an oligodendrocyte morphology.[33] Accordingly, a functional VDR has been demonstrated in mouse Schwann cells and rat oligodendrocytes.[104,105] In two models of noninflammatory toxic demyelination, the cuprizone model and local stereotactic injection of ethidium bromide in the hippocampus, supplementation with high-dose vitamin D_3 protected against demyelination.[106,107] An effect of vitamin D_3 supplementation on remyelination was not shown in the cuprizone model.[106] However, in the ethidium bromide model, remyelination was promoted by vitamin D_3 supplements, coinciding with an increased expression of MBP.[107] Treatment of mice in the cuprizone model with 1,25$(OH)_2$D initially promoted demyelination, but during the remyelination phase much more myelin was restored when compared to the placebo group.[108] This improved repair coincided with a higher recruitment of NOGO-A positive cells (ie, oligodendrocytes).

In MS and control NAWM, hypothalamus and cortex, we and others showed colocalization of VDR immunostaining with major histocompatibility complex (MHC) class II and glial fibrillary acidic protein (GFAP), suggesting expression of VDR in microglia and astrocytes, respectively.[33,99] Upon exposure to 1,25$(OH)_2$D in vitro, both primary human astrocytes and microglia upregulated CYP24A1 expression, confirming the presence of a functional VDR.[33] Additionally, colocalization of immunostaining for 1αOHase and GFAP has been reported, suggesting expression of CYP27B1 in astrocytes.[99] Interestingly, the total VDR mRNA expression in NAWM of MS patients was higher when compared to control NAWM, while expression levels of CYP24A1 and CYP27B1 mRNA were similar.[33] In active MS lesions, immunostaining for VDR was shown in both human leukocyte antigen-positive microglia and GFAP-positive reactive astrocytes, while 24OHase staining was only observed in GFAP-positive reactive astrocytes.[33] Total VDR and CYP27B1 mRNA expression were higher in the rim of active lesions when compared to NAWM of MS patients.[33] In vitro, exposure of primary human astrocytes and microglia to inflammatory cytokines (TNF-α and IFN-γ) induced CYP27B1 but did not affect VDR mRNA expression.[33] Likewise, expression of CYP27B1 has been reported to be increased in inflamed CNS tissue in EAE with also corresponding increased local concentrations of 1,25$(OH)_2$D.[51] However, it remains to be seen whether this finding reflects an increased expression of VDR and CYP27B1 by resident cells or an influx of VDR- and 1αOHase-positive leukocytes, or both. Observations in animal studies suggest that 1,25$(OH)_2$D may induce an antiinflammatory phenotype in glial cells. Culture of LPS- or IFN-γ-challenged microglia with 1,25$(O)_2$D reduced production of TNF-α and IL-6.[109] In primary rat astrocytes and a glioma cell line challenged with LPS, culture with 1,25$(OH)_2$D reduced mRNA expression of TNF-α and M-CSF.[110]

In summary, several proteins and their encoding mRNA involved in vitamin D metabolism are found in human CNS tissue in resident CNS cells. Their regional distribution suggests a functional role of vitamin D metabolism within CNS homeostasis, which may be relevant for MS. The responsiveness of expression levels to inflammation in vitro and in situ in MS lesions suggests a potential role in the local disease process of MS. However, whether systemic vitamin D status is relevant for the vitamin D micromilieu in CNS tissue remains to be investigated. Further studies should unravel the functional role of vitamin D in the CNS.

23.5 CHALLENGES FOR CLINICAL RESEARCH

Although many studies have been performed, there is no conclusive evidence regarding the role of vitamin D in MS. Additionally, the relevance of systemic vitamin D status for the adaptive immune response in MS is also far from consolidated. Currently, there are several clinical trials on vitamin D supplementation being performed aimed on measuring clinical and radiological efficacy. These trials may provide an excellent framework to collect material that may shed more light on the role of vitamin D in the (auto)immune response. There are, however, several issues to be dealt with and several choices to be made.

The most basic question is which vitamin D compound should be chosen for supplementation. From a clinical perspective, the backbone of the clinical trials are the association studies on serum vitamin D status and clinical disease outcomes. Serum 25(OH)D levels comprise mostly 25(OH)D_3 and to a much lesser extent 25(OH)D_2.[28] In a study in a large cohort of men with osteoporosis, 25(OH)D_2 could not be measured with liquid-chromatography tandem mass spectrometry in 72.2%, while the mean level in the remaining subjects was 8.2 ng/mL, compared to a mean 25(OH)D_3 level of 24.9 ng/mL.[29] Therefore, the contribution of variation in 25(OH)D_3 to variation in total 25(OH)D levels as has been associated with clinical and radiological outcomes in MS exceeds 25(OH)D_2 by far. Furthermore, vitamin D_3 has been proposed to be more potent than D_2 in elevating serum 25(OH)D levels.[111] Therefore, supplementation of vitamin D_3 appears a more logical choice than vitamin D_2. Immunological studies in EAE, however, almost exclusively assessed effect of systemic treatment with 1,25$(OH)_2$D, with vitamin D supplementation resulting in more modest or less clear clinical and immunological effects.[51] Although hardly any correlation studies on circulating 1,25$(OH)_2$D levels and MS outcome have been published, a positive correlation between circulating levels of 25(OH)D and 1,25$(OH)_2$D has been published by us and others. Pilot studies and a small trial on 1,25$(OH)_2$D treatment in RRMS showed a good safety profile and trends toward clinical efficacy.[112,113] Therefore, although the relevance for clinical outcomes is less consolidated, trials with 1,25$(OH)_2$D may be very interesting from an immunological perspective. Another unexplored option is drugs that promote the bioavailability of 1,25$(OH)_2$D. In an EAE study, treatment with lovastatin promoted 1α-hydroxylation and inhibited 24-hydroxylation of vitamin D metabolites in the CNS, which also attenuated clinical disease activity.[114]

Another point of discussion is the dose of vitamin D to be supplemented. It is uncertain whether associations between vitamin D status and disease activity are predominantly driven by the lowest or the highest 25(OH)D levels. The implication of the first hypothesis is that only MS patients with the lowest 25(OH)D levels are likely to benefit from supplementation of vitamin D and should therefore be included in clinical trials. If the highest 25(OH)D levels are most important, supplements aiming to high physiological or supraphysiological ranges of 25(OH)D are most logical. Doses up to 10,000 IU/d have been argued to be safe.[115] However, cases of vitamin D toxicity in MS patients taking high doses of vitamin D have also been described.[116] Clinical trials on supplementation aiming at the supraphysiological range should extensively assess safety.

Since most clinical outcomes on MS disease activity and vitamin D status comprise markers of inflammation, inclusion of RRMS patients at the start of their disease into trials is most likely to yield positive results (as is the case in other trials in MS). Also from an immunological perspective, the chance of measuring an effect of an inflammation-modulating molecule appears largest when disease has the most inflammatory characteristics. Since vitamin D enters the CNS, modulation of inflammation inside the CNS, as is found in progressive MS, appears plausible.[117] Additionally, noninflammatory actions on CNS resident cells have been proposed, possibly affecting myelin- and neuroprotective mechanisms (vide supra). These observations provide at least a mechanistic rationale for clinical trials with vitamin D in progressive MS. Prospective data on disability progression and 25(OH)D status in progressive MS are at present notably lacking.

When designing trials, researchers should also think about combination therapies of vitamin D with disease-modulating drugs. Obviously, modulation of inflammation by vitamin D metabolites is not likely to be of added value to immune ablative strategies. Most observational studies included subjects treated with interferon beta or glatiramer acetate, and therefore appear to be likely candidates. Specific interactions between vitamin D status and interferon beta therapy have been reported.[67,118] However, also in subjects treated with natalizumab, low 25(OH)D levels have been associated with an increased relapse risk.[119] As already mentioned, serum 25(OH)D levels regulate the sphingosine-1-phosphate system, which is also targeted by fingolimod.[67] This finding may also provide a rationale for combination therapy.

Although appropriate clinical and radiological outcomes for trials on vitamin D supplementation in MS are quite obvious, immunological outcomes are not. For immunologists, assessment of the immune response within MS lesions in the CNS and in the CNS draining cervical lymph nodes would be most interesting. Yet, these studies are very challenging and appear impossible to include in the context of a clinical trial. Alternatively, CSF-derived leukocytes could be assessed, but the most easily accessible material and therefore most feasible for large-scale sampling remains the PBMC fraction. This material certainly has its limitations (vide supra, Fig. 23.3). In observational studies and pilot experiments several approaches have been used. These approaches range from hypothesis-free array experiments to hypothesis-driven functional cellular experiments addressing presumed components of MS pathophysiology. First immunological data from upcoming trials will likely provide useful directions for further studies to be designed.

23.6 CONCLUDING REMARKS

In this chapter, we described a consistent body of experimental data, showing that vitamin D may interfere with relevant pathophysiological mechanisms in MS. It is important to note that a causal role of vitamin D in the modulation of disease incidence and activity is

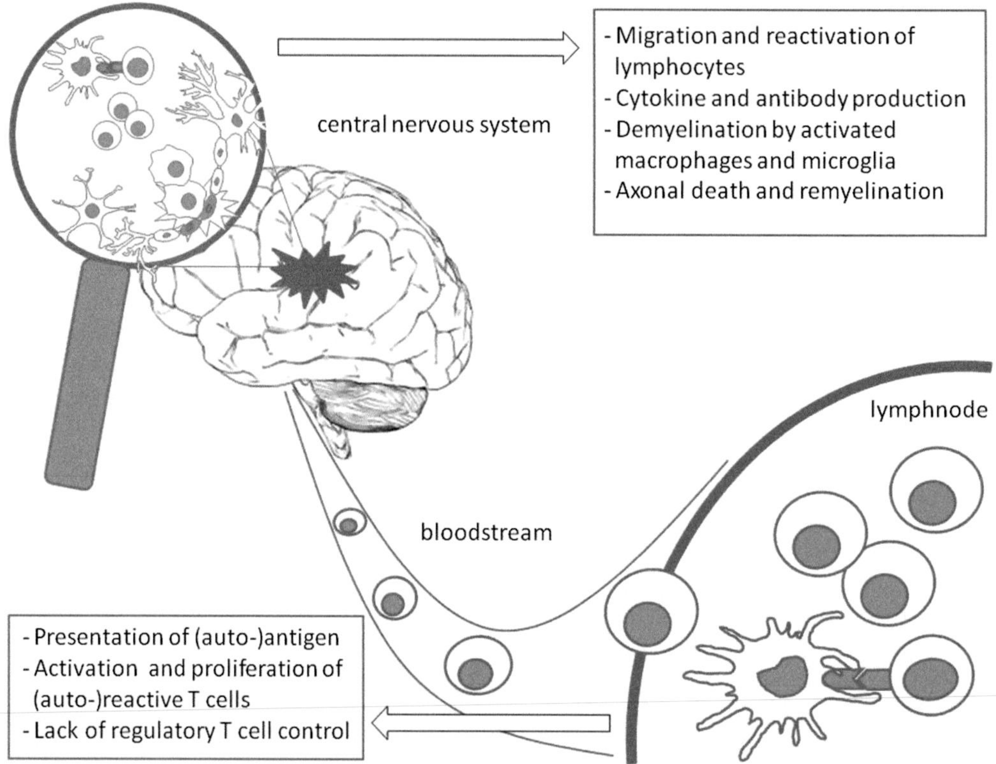

FIGURE 23.3 **Potential interactions between vitamin D and hallmarks of multiple sclerosis (MS) pathophysiology in vivo based on observations in vitro, in experimental autoimmune encephalomyelitis and in patients with relapsing-remitting MS.** Notably, the sites of interaction between vitamin D metabolites and cells involved in MS pathogenesis are presumably the draining lymph nodes and central nervous system (CNS) tissue. The peripheral blood mononuclear cell fraction contains, among other cells, cells trafficking from lymph nodes to CNS and vice versa.

still not fully consolidated. Plausible alternative hypotheses are readily available. The two main questions to be addressed are (1) whether lower 25(OH)D levels during active disease/inflammation are the consequence rather than the cause of more active disease, and (2) although $1,25(OH)_2D$ is a very exciting immune-modulating molecule in vitro and in EAE, it is not clear whether systemic 25(OH)D levels are relevant for immune homeostasis in vivo in health and disease. The ongoing and upcoming clinical trials in MS will provide extremely important data to answer these questions.

References

1. Goldberg D. Multiple sclerosis: vitamin D and calcium as environmental determinants of prevalence. *Int J Environ Stud*. 1974;6:19–27.
2. Munger KL, Zhang SM, O'Reilly E, et al. Vitamin D intake and incidence of multiple sclerosis. *Neurology*. 2004;62:60–65.
3. Munger K, Levin L, Hollis B, Howard N, Ascherio A. Serum 25-hydroxyvitamin D levels and risk of multiple sclerosis. *JAMA*. 2006;296:2832–2838.

4. Sawcer S, Hellenthal G, Pirinen M, et al. Genetic risk and a primary role for cell-mediated immune mechanisms in multiple sclerosis. *Nature*. 2011;476:214–219.
5. Soilu-Hanninen M, Airas L, Mononen I, Heikkila A, Viljanen M, Hanninen A. 25-Hydroxyvitamin D levels in serum at the onset of multiple sclerosis. *Mult Scler*. 2005;11:266–271.
6. Soilu-Hanninen M, Laaksonen M, Laitinen I, Eralinna JP, Lilius EM, Mononen I. A longitudinal study of serum 25-hydroxyvitamin D and intact PTH levels indicate the importance of vitamin D and calcium homeostasis regulation in multiple sclerosis. *J Neurol Neurosurg Psychiatry*. 2008;79:152–157.
7. Smolders J, Menheere P, Kessels A, Damoiseaux J, Hupperts R. Association of vitamin D metabolite levels with relapse rate and disability in multiple sclerosis. *Mult Scler*. 2008;14:1220–1224.
8. Smolders J, Peelen E, Thewissen M, Menheere P, Damoiseaux J, Hupperts R. Serum vitamin D binding protein levels are not associated with relapses or with vitamin D status in multiple sclerosis. *Mult Scler*. 2014;20:433–437.
9. Simpson S, Taylor B, Blizzard L, et al. Higher 25-hydroxyvitamin D is associated with lower relapse risk in multiple sclerosis. *Ann Neurol*. 2010;68:193–203.
10. Runia TF, Hop WC, de Rijke YB, Buljevac D, Hintzen RQ. Lower serum vitamin D levels are associated with a higher relapse risk in multiple sclerosis. *Neurology*. 2012;79:261–266.
11. Ascherio A, Munger KL, White R, et al. Vitamin D as an early predictor of multiple sclerosis activity and progression. *JAMA Neurol*. 2014;71:306–314.
12. Mowry EM, Krupp LB, Milazzo M, et al. Vitamin D status is associated with relapse rate in pediatric-onset multiple sclerosis. *Ann Neurol*. 2010;67:618–624.
13. van der Mei I, Ponsonby A, Dwyer T, et al. Vitamin D levels in people with multiple sclerosis and community controls in Tasmania, Australia. *J Neurol*. 2007;254:581–590.
14. Knippenberg S, Bol Y, Damoiseaux J, Hupperts R, Smolders J. A poor vitamin D status in patients with MS is negatively correlated with depression, but not with fatigue. *Acta Neurol Scand*. 2011;124:171–175.
15. Knippenberg S, Damoiseaux J, Bol Y, et al. Higher levels of reported sun exposure, and not vitamin D status, are associated with less depressive symptoms and fatigue in multiple sclerosis. *Acta Neurol Scand*. 2014;129:123–131.
16. Mowry EM, Waubant E, McCulloch CE, et al. Vitamin D status predicts new brain magnetic resonance imaging activity in multiple sclerosis. *Ann Neurol*. 2012;72:234–240.
17. Løken-Amsrud KI, Holmøy T, Bakke SJ, et al. Vitamin D and disease activity in multiple sclerosis before and during interferon-β treatment. *Neurology*. 2012;79:267–273.
18. Burton JM, Kimball S, Vieth R, et al. A phase I/II dose-escalation trial of vitamin D_3 and calcium in multiple sclerosis. *Neurology*. 2010;74:1852–1859.
19. Soliu-Hänninen M, Aivo J, Lindström BM, et al. A randomized, double-blind, placebo controlled trial with vitamin D_3 as an add on treatment to interferon β-1b in patients with multiple sclerosis. *J Neurol Neurosurg Psychiatry*. 2012;83:565–571.
20. Kampman MT, Steffensen LH, Mellgren SI, Jørgensen L. Effect of vitamin D_3 supplementation on relapses, disease progression, and measures of function in persons with multiple sclerosis: exploratory outcomes from a double-blind randomised controlled trial. *Mult Scler*. 2012;18:1144–1151.
21. Stein MS, Liu Y, Gray OM, et al. A randomized trial of high-dose vitamin D2 in relapsing-remitting multiple sclerosis. *Neurology*. 2011;77:1611–1618.
22. Holowaychuck MK, Birkenheuer AJ, Li J, Marr H, Boll A, Nordone SK. Hypocalcemia and hypovitaminosis D in dogs with induced endotoxemia. *J Vet Intern Med*. 2012;26:244–251.
23. Baughman RP, Janovcik J, Ray Et Al M. Calcium and vitamin D metabolism in sarcoidosis. *Sarcoidosis Vasc Diffuse Lung Dis*. 2013;30:113–120.
24. Wang Y, Marling SJ, Beaver EF, Severson KS, Deluca HF. UV light selectively inhibits spinal cord inflammation and demyelination in experimental autoimmune encephalomyelitis. *Arch Biochem Biophys*. 2014;567C:75–82.
25. Hayes CE. Vitamin D: a natural inhibitor of multiple sclerosis. *Proc Nutr Soc*. 2000;59:531–535.
26. Provvedini DM, Tsoukas CD, Deftos LJ, Manolagas SC. 1,25-dihydroxyvitamin D_3 receptors in human leukocytes. *Science*. 1983;221:1181–1183.
27. Holick MF. Vitamin D deficiency. *N Engl J Med*. 2007;357:266–281.
28. Heaney RP, Recker RR, Grote J, Horst RL, Armas LA. Vitamin D_3 is more potent than vitamin D2 in humans. *J Clin Endocrinol Metab*. 2011;96:E447–E452.
29. Swanson CM, Nielson CM, Shrestha S, et al. Higher 25(OH)D2 is associated with lower 25(OH)D_3 and 1,25(OH)2D_3. *J Clin Endocrinol Metab*. 2014;99:2736–2744.

30. Bikle DD. Vitamin D metabolism, mechanism of action, and clinical applications. *Chem Biol*. 2014;21:319–329.
31. Hollis BW. Assessment of vitamin D nutritional and hormonal status: what to measure and how to do it. *Calcif Tissue Int*. 1996;58:4–5.
32. Zehnder D, Bland R, Williams MC, et al. Extrarenal expression of 25-hydroxyvitamin D_3-1α-hydroxylase. *J Clin Endocrinol Metab*. 2001;86:888–894.
33. Smolders J, Schuurman K, van Strien M, et al. Expression of vitamin D receptor and metabolizing enzymes in multiple sclerosis-affected brain tissue. *J Neuropathol Exp Neurol*. 2013;72:91–105.
34. Jongen MJ, van der Vijgh WJ, Lips P, Netelenbos JC. Measurement of vitamin D metabolites in anephric subjects. *Nephron*. 1984;36:230–234.
35. Smolders J, Peelen E, Thewissen M, et al. The relevance of vitamin D receptor gene polymorphisms for vitamin D research in multiple sclerosis. *Autoimmun Rev*. 2009;8:621–626.
36. Orton SM, Ramagopalan SV, Para AE, et al. Vitamin D metabolic pathway genes and risk of multiple sclerosis in Canadians. *J Neurol Sci*. 2011;305:116–120.
37. Torkildsen Ø, Knappskog PM, Nyland HI, Myhr KM. Vitamin D-dependent rickets as a possible risk factor for multiple sclerosis. *Arch Neurol*. 2008;65:809–811.
38. Ramagopalan SV, Dyment DA, Cader MZ, et al. Rare variants in the CYP27B1 gene are associated with multiple sclerosis. *Ann Neurol*. 2011;70:881–886.
39. Barizzone N, Pauwels I, Luciano B, et al. No evidence for a role of rare CYP27B1 functional variations in multiple sclerosis. *Ann Neurol*. 2013;73:433–437.
40. Ban M, Caillier S, Mero IL, et al. No evidence of association between mutant alleles of the CYP27B1 gene and multiple sclerosis. *Ann Neurol*. 2013;73:430–432.
41. Bikle D. Vitamin D regulation of immune function. *Vitam Horm*. 2011;86:1–21.
42. Peelen E, Knippenberg S, Muris AH, et al. Effects of vitamin D on the peripheral adaptive immune system: a review. *Autoimmun Rev*. 2011;10:733–743.
43. Smolders J, Damoiseaux J. Vitamin D as a T cell modulator in multiple sclerosis. *Vitam Horm*. 2011;86:401–428.
44. van Halteren AG, van Etten E, de Jong EC, Bouillon R, Roep BO, Mathieu C. Redirection of human autoreactive T-cells upon interaction with dendritic cells modulated by TX527, an analog of 1, 25 dihydroxyvitamin D(3). *Diabetes*. 2002;51:2119–2125.
45. Correale J, Ysrraelit MC, Gaitán MI. Immunomodulatory effects of vitamin D in multiple sclerosis. *Brain*. 2009;132:1146–1160.
46. Jeffery LE, Burke F, Mura M, et al. 1, 25-Dihydroxyvitamin D_3 and IL-2 combine to inhibit T cell production of inflammatory cytokines and promote development of regulatory T cells expressing CTLA-4 and FoxP3. *J Immunol*. 2009;183:5458–5467.
47. Heine G, Niesner U, Chang HD, et al. 1, 25-dihydroxyvitamin D(3) promotes IL-10 production in human B cells. *Eur J Immunol*. 2008;38:2210–2218.
48. Chen S, Sims GP, Chen XX, Gu YY, Chen S, Lipsky PE. Modulatory effects of 1, 25-dihydroxyvitamin D_3 on human B cell differentiation. *J Immunol*. 2007;179:1634–1647.
49. Lemire JM, Archer DC. 1,25-dihydroxyvitamin D_3 prevents the in vivo induction of murine experimental autoimmune encephalomyelitis. *J Clin Invest*. 1991;87:1103–1107.
50. Cantorna MT, Hayes CE, DeLuca HF. 1, 25-Dihydroxyvitamin D_3 reversibly blocks the progression of relapsing encephalomyelitis, a model of multiple sclerosis. *Proc Natl Acad Sci USA*. 1996;93:7861–7864.
51. Spach KM, Hayes CE. Vitamin D_3 confers protection from autoimmune encephalomyelitis only in female mice. *J Immunol*. 2005;175:4119–4126.
52. Meehan TF, DeLuca HF. The vitamin D receptor is necessary for 1alpha,25-dihydroxyvitamin D(3) to suppress experimental autoimmune encephalomyelitis in mice. *Arch Biochem Biophys*. 2002;408:200–204.
53. Nashold FE, Hoag KA, Goverman J, Hayes CE. Rag-1-dependent cells are necessary for 1,25-dihydroxyvitamin D(3) prevention of experimental autoimmune encephalomyelitis. *J Neuroimmunol*. 2001;119:16–29.
54. Mayne CG, Spanier JA, Relland LM, Williams CB, Hayes CE. 1,25-Dihydroxyvitamin D_3 acts directly on the T lymphocyte vitamin D receptor to inhibit experimental autoimmune encephalomyelitis. *Eur J Immunol*. 2011;41:822–832.
55. Meehan TF, DeLuca HF. CD8(+) T cells are not necessary for 1 alpha,25-dihydroxyvitamin D(3) to suppress experimental autoimmune encephalomyelitis in mice. *Proc Natl Acad Sci USA*. 2002;99:5557–5560.
56. Grishkan IV, Fairchild AN, Calabresi PA, Gocke AR. 1,25-Dihydroxyvitamin D_3 selectively and reversibly impairs T helper-cell CNS localization. *Proc Natl Acad Sci USA*. 2013;110:21101–21106.

57. Nashold FE, Nelson CD, Brown LM, Hayes CE. One calcitriol dose transiently increases Helios$^+$ FoxP3$^+$ T cells and ameliorates autoimmune demyelinating disease. *J Neuroimmunol*. 2013;263:64–74.
58. Spach KM, Nashold FE, Dittel BN, Hayes CE. L-10 signaling is essential for 1,25-dihydroxyvitamin D_3-mediated inhibition of experimental autoimmune encephalomyelitis. *J Immunol*. 2006;177:6030–6037.
59. Spanier JA, Nashold FE, Olson JK, Hayes CE. The Ifng gene is essential for Vdr gene expression and vitamin D_3-mediated reduction of the pathogenic T cell burden in the central nervous system in experimental autoimmune encephalomyelitis, a multiple sclerosis model. *J Immunol*. 2012;189:3188–3197.
60. Cantorna MT, Woodward WD, Hayes CE, DeLuca HF. 1,25-dihydroxyvitamin D_3 is a positive regulator for the two anti-encephalitogenic cytokines TGF-beta 1 and IL-4. *J Immunol*. 1998;160:5314–5319.
61. Joshi S, Pantalena LC, Liu XK, et al. 1,25-dihydroxyvitamin D(3) ameliorates Th17 autoimmunity via transcriptional modulation of interleukin-17A. *Mol Cell Biol*. 2011;31:3653–3669.
62. Farias AS, Spagnol GS, Bordeaux-Rego P, et al. Vitamin D_3 induces IDO$^+$ tolerogenic DCs and enhances Treg, reducing the severity of EAE. *CNS Neurosci Ther*. 2013;19:269–277.
63. Viglietta V, Baecher-Allan C, Weiner H, Hafler D. Loss of functional suppression by CD4$^+$CD25$^+$ regulatory T cells in patients with multiple sclerosis. *J Exp Med*. 2004;199:971–979.
64. Edwards LJ, Robins RA, Constantinescu CS. Th17/Th1 phenotype in demyelinating disease. *Cytokine*. 2010;50:19–23.
65. Knippenberg S, Peelen E, Smolders J, et al. Reduction in IL-10 producing B cells (Breg) in multiple sclerosis is accompanied by a reduced naïve/memory Breg ratio during a relapse but not in remission. *J Neuroimmunol*. 2011;239:80–86.
66. Kappos L, Polman CH, Freedman MS, et al. Treatment with interferon beta-1b delays conversion to clinically definite and McDonald MS in patients with clinically isolated syndromes. *Neurology*. 2006;67:1242–1249.
67. Munger KL, Köchert K, Simon KC, et al. Molecular mechanism underlying the impact of vitamin D on disease activity of MS. *Ann Clin Transl Neurol*. 2014;1:605–617.
68. Kappos L, Antel J, Comi G, et al. FTY720 D2201 Study Group. Oral fingolimod (FTY720) for relapsing multiple sclerosis. *N Engl J Med*. 2006;355:1124–1140.
69. Royal III W, Mia Y, Li H, Naunton K. Peripheral blood regulatory T cell measurements correlate with serum vitamin D levels in patients with multiple sclerosis. *J Neuroimmunol*. 2009;213:135–141.
70. Smolders J, Thewissen M, Peelen E, et al. Vitamin D status is positively correlated with regulatory T cell function patients with multiple sclerosis. *PLoS One*. 2009;4:e6635.
71. Smolders J, Peelen E, Thewissen M, et al. Safety and T cell modulating effects of high dose vitamin D_3 supplementation in multiple sclerosis. *PLoS One*. 2010;5:e15235.
72. Peelen E, Thewissen M, Knippenberg S, et al. Fraction of IL-10$^+$ and IL-17$^+$ CD8 T cells is increased in MS patients in remission and during a relapse, but is not influenced by immune modulators. *J Neuroimmunol*. 2013;258:77–84.
73. Niino M, Fukazawa T, Miyazaki Y, et al. Suppression of IL-10 production by calcitriol in patients with multiple sclerosis. *J Neuroimmunol*. 2014;270:86–94.
74. Allen AC, Kelly S, Basdeo SA, et al. A pilot study of the immunological effects of high-dose vitamin D in healthy volunteers. *Mult Scler*. 2012;18:1797–1800.
75. Kimball S, Vieth R, Dosch HM, et al. Cholecalciferol plus calcium suppresses abnormal PBMC reactivity in patients with multiple sclerosis. *J Clin Endocrinol Metab*. 2011;96:2826–2834.
76. Grau-López L, Granada ML, Raïch-Regué D, et al. Regulatory role of vitamin D in T-cell reactivity against myelin peptides in relapsing-remitting multiple sclerosis patients. *BMC Neurol*. 2012;12:103.
77. Knippenberg S, Smolders J, Thewissen M, et al. Effect of vitamin D_3 supplementation on peripheral B cell differentiation and isotype switching in patients with multiple sclerosis. *Mult Scler*. 2011;17:1418–1423.
78. Disanto G, Handel AE, Damoiseaux J, et al. Vitamin D supplementation and antibodies against the Epstein-Barr virus in multiple sclerosis patients. *Mult Scler*. 2013;19:1679–1680.
79. Levin LI, Munger KL, Rubertone MV, et al. Temporal relationship between elevation of epstein-barr virus antibody titers and initial onset of neurological symptoms in multiple sclerosis. *JAMA*. 2005;293:2496–2500.
80. DeLorenze GN, Munger KL, Lennette ET, Orentreich N, Vogelman JH, Ascherio A. Epstein-Barr virus and multiple sclerosis: evidence of association from a prospective study with long-term follow-up. *Arch Neurol*. 2006;63:839–844.
81. Munger KL, Levin LI, O'Reilly EJ, Falk KI, Ascherio A. Anti-Epstein-Barr virus antibodies as serological markers of multiple sclerosis: a prospective study among United States military personnel. *Mult Scler*. 2011;17:1185–1193.
82. Horakova D, Zivadinov R, Weinstock-Guttman B, et al. Environmental factors associated with disease progression after the first demyelinating event: results from the multi-center SET study. *PLoS One*. 2013;8:e53996.

83. Røsjø E, Myhr KM, Løken-Amsrud KI, et al. Increasing serum levels of vitamin A, D and E are associated with alterations of different inflammation markers in patients with multiple sclerosis. *J Neuroimmunol*. 2014;271:60–65.
84. Mahon BD, Gordon SA, Cruz J, Cosman F, Cantorna MT. Cytokine profile in patients with multiple sclerosis following vitamin D supplementation. *J Neuroimmunol*. 2003;134:128–132.
85. Åivo J, HÄnninen A, Ilonen J, Soliu-Hänninen M. Vitamin D_3 administration to MS patients leads to increased serum levels of latency activated peptide (LAP) of TGF-beta. *J Neuroimmunol*. 2015;280:12–15.
86. Holmøy T, Moen SM, Gundersen TA, et al. 25-hydroxyvitamin D in cerebrospinal fluid during relapse and remission of multiple sclerosis. *Mult Scler*. 2009;15:1280–1285.
87. Dumont D, Noben JP, Raus J, Stinissen P, Robben J. Proteomic analysis of cerebrospinal fluid from multiple sclerosis patients. *Proteomics*. 2004;4:2117–2124.
88. Pedersen LB, Nashold FE, Spach KM, Hayes CE. 1,25-dihydroxyvitamin D_3 reverses experimental autoimmune encephalomyelitis by inhibiting chemokine synthesis and monocyte trafficking. *J Neurosci Res*. 2007;85:2480–2490.
89. Li B, Baylink DJ, Deb C, et al. 1,25-Dihydroxyvitamin D_3 suppresses TLR8 expression and TLR8-mediated inflammatory responses in monocytes in vitro and experimental autoimmune encephalomyelitis in vivo. *PLoS One*. 2013;8:e58808.
90. Garcion E, Sindji L, Nataf S, Brachet P, Darcy F, Montero-Menei CN. Treatment of experimental autoimmune encephalomyelitis in rat by 1,25-dihydroxyvitamin D_3 leads to early effects within the central nervous system. *Acta Neuropathol*. 2003;105:438–448.
91. Garcion E, Nataf S, Berod A, Darcy F, Brachet P. 1,25-Dihydroxyvitamin D_3 inhibits the expression of inducible nitric oxide synthase in rat central nervous system during experimental allergic encephalomyelitis. *Brain Res Mol Brain Res*. 1997;45:255–267.
92. Nataf S, Garcion E, Darcy F, Chabannes D, Muller JY, Brachet P. 1,25 Dihydroxyvitamin D_3 exerts regional effects in the central nervous system during experimental allergic encephalomyelitis. *J Neuropathol Exp Neurol*. 1996;55:904–914.
93. Nashold FE, Miller DJ, Hayes CE. 1,25-dihydroxyvitamin D_3 treatment decreases macrophage accumulation in the CNS of mice with experimental autoimmune encephalomyelitis. *J Neuroimmunol*. 2000;103:171–179.
94. Lossius A, Vartdal F, Holmøy T. Vitamin D sensitive EBNA-1 specific T cells in the cerebrospinal fluid of patients with multiple sclerosis. *J Neuroimmunol*. 2011;240–241:87–96.
95. Holmøy T, Lossius A, Gundersen TE, et al. Intrathecal levels of vitamin D and IgG in multiple sclerosis. *Acta Neurol Scand*. 2012;125:e28–31.
96. Veenstra TD, Prüfer K, Koenigsberger C, Brimjoin SW, Grande JP, Kumar R. 1,25-dihydroxyvitamin D receptors in the central nervous system of the rat embryo. *Brain Res*. 1998;804:193–205.
97. Prüfer K, Veenstra TD, Jirikowski GF, Kumar R. Distribution of 1,25-dihydroxyvitamin D_3 receptor immunoreactivity in the rat brain and spinal cord. *J Chem Neuroanat*. 1999;16:135–145.
98. Glaser SD, Prüfer K, Jirikowski GF. Vitamin D receptor is partly colocalized with oxytocin immunoreactivity in neurons of the male rat hypothalamus. *Cell Mol Biol*. 1997;43:543–548.
99. Eyles DW, Smith S, Kinobe R, Hewison M, McGrath JJ. Distribution of the vitamin D receptor and 1α-hydroxylase in human brain. *J Chem Neuroanat*. 2005;29:21–30.
100. Taniura H, Ito M, Sanasa N, et al. Chronic vitamin D_3 treatment protects against neurotoxicity by glutamate in association with upregulation of vitamin D receptor mRNA expression in cultured rat cortical neurons. *J Neurosci Res*. 2006;83:1179–1189.
101. Brown J, Bianco JI, McGrath JJ, Eyles DW. 1,25-dihydroxyvitamin D_3 induces nerve growth factor, promotes neurite outgrowth and inhibits mitosis in embryonic rat hippocampal neurons. *Neurosci Lett*. 2003;343:139–143.
102. Neveu I, Naveilhan P, Baudet C, Brachet P, Metsis M. 1,25-dihydroxyvitamin D_3 regulates NT-3, NT-4 but not BDNF mRNA in astrocytes. *Neuroreport*. 1994;6:124–126.
103. Neveu I, Naveilhan P, Jehan F, et al. 1,25-dihydroxyvitamin D regulates the synthesis of nerve growth factor in primary cultures of glial cells. *Mol Brain Res*. 1994;24:70–76.
104. Cornet A, Baudet C, Neveu I, Baron-van Evercooren A, Brachet P, Naveilhan P. 1,25-hydroxyvitamin D_3 regulates the expression of VDR and NGF gene in Schwann cells in vitro. *J Neurosci Res*. 1998;53:742–746.
105. Baas D, Prüfer K, Ittel ME, et al. Rat oligodendrocytes express the vitamin D_3 receptor and respond to 1,25-dihydroxyvitamin D. *Glia*. 2000;31:31–59.
106. Wergeland S, Torkildsen O, Myhr KM, Aksnes L, Mork S, Bo L. Dietary vitamin D_3 supplements reduces demyelination in the cuprizone model. *PLoS One*. 2011;6:e26262.

107. Goudarzvand M, Javan M, Mirnajafi-Zadeh J, Mozafari S, Tiraihi T. Vitamins E and D_3 attenuate demyelination and potentiate remyelination processes of hippocampal formation of rats following local injection of ethidium bromide. *Cell Mol Neurobiol*. 2010;30:289–299.
108. Nystad AE, Wergeland S, Aksnes L, Myhr KM, Bø L, Torkildsen O. Effect of high-dose 1.25 dihydroxyvitamin D_3 on remyelination in the cuprizone model. *APMIS*. 2014;122:1178–1186.
109. Lefebvre d' Hellencourt C, Montero-Menei CN, Bernard R, Couez D. Vitamin D_3 inhibits proinflammatory cytokines and nitric oxide production by the EOC13 microglial cell line. *J Neurosci Res*. 2003;71:575–582.
110. Furman I, Baudet C, Brachet P. Differential expression of M-CSF, LIF, and TNFalpha genes in normal and malignant rat glial cells: regulation by lipopolysaccharide and vitamin D. *J Neurosci Res*. 1996;46:360–366.
111. Tripkovic L, Lambert H, Hart K, et al. Comparison of vitamin D_2 and vitamin D_3 supplementation in raising serum 25-hydroxyvitamin D status: a systematic review and meta-analysis. *Am J Clin Nutr*. 2012;95:1357–1364.
112. Achiron A, Givon U, Magalashvili D, et al. Effect of Alfacalcidol on multiple sclerosis-related fatigue: a randomized, double-blind placebo-controlled study. *Mult Scler*. 2015;21:767–775.
113. Wingerchuk DM, Lesaux J, Rice GP, Kremenchutzky M, Ebers GC. A pilot study of oral calcitriol (1,25-dihydroxyvitamin D_3) for relapsing-remitting multiple sclerosis. *J Neurol Neurosurg Psychiatry*. 2005;76:1294–1296.
114. Paintlia AS, Paintlia MK, Hollis BW, Singh AK, Singh I. Interference with RhoA-ROCK signaling mechanism in autoreactive $CD4^+$ T cells enhances the bioavailability of 1,25-dihydroxyvitamin D_3 in experimental autoimmune encephalomyelitis. *Am J Pathol*. 2012;181:993–1006.
115. Hathcock JN, Shao A, Vieth R, Heaney R. Risk assessment for vitamin D. *Am J Clin Nutr*. 2007;85:6–18.
116. Fragoso YD, Adoni T, Damasceno A, et al. Unfavorable outcomes during treatment of multiple sclerosis with high doses of vitamin D. *J Neurol Sci*. 2014;346:341–342.
117. Lassmann H, Brück W, Lucchinetti CF. The immunopathology of multiple sclerosis: an overview. *Brain Pathol*. 2007;17:210–218.
118. Stewart N, Simpson Jr S, van der Mei I, et al. Interferon-β and serum 25-hydroxyvitamin D interact to modulate relapse risk in MS. *Neurology*. 2012;79:254–260.
119. Scott TF, Hackett CT, Dworek DC, Schramke CJ. Low vitamin D level is associated with higher relapse rate in natalizumab treated MS patients. *J Neurol Sci*. 2013;330:27–31.

CHAPTER

24

Symptomatic and Complementary Treatments

P.S. Sorensen, K. Schreiber, A.K. Andreasen
University Hospital Rigshospitalet, Copenhagen, Denmark

OUTLINE

24.1 INTRODUCTION

Although the efficacy of immunomodulatory therapies for multiple sclerosis (MS) have improved considerably during the last decade and to some extent seem to have altered the natural history of the disease, they still are only partially effective and do not ameliorate the accumulation of irreversible symptoms in the progressive phase of MS, for which no effective disease-modifying therapy is available. Hence, pharmacological treatment of MS-associated symptoms such as gait, spasticity, bladder dysfunction, pain, fatigue, and cognitive dysfunction remains an essential cornerstone of comprehensive care of patients with MS. In this chapter we review the evidence of efficacy of symptomatic therapies, of vitamin D as nutritional supplement, and also of complementary and alternative therapies.

Translational Neuroimmunology in Multiple Sclerosis
http://dx.doi.org/10.1016/B978-0-12-801914-6.00024-6

24.2 GAIT AND MOBILITY

Gait is rated as the most important function by MS patients,[1] and moderate or severe mobility problems are reported by more than 67% of the patients,[2] with up to two-thirds using a walking aid or wheelchair.[3] Today it is possible to improve gait function in approximately 35% of patients (responders) with mild to moderately severe gait disturbances (Expanded Disability Status Scale 2.5–7.0) with slow release fampridine (US: dalfampridine) tablets 10 mg twice daily.[4,5]

Fampridine (4-aminopyridine) is a blocker of voltage-dependent potassium channels.[6] In MS, demyelination leads to structural changes of the axon exposing potassium channels in the paranodal and internodal axonal membranes, which are normally covered by the myelin sheath.[7] Under these conditions, leakage of ionic current through the K^+ channel can contribute to the phenomenon of action potential conduction block. Fampridine blocks the function of exposed potassium channels and, thereby, improves action potential conduction in demyelinated motor axons.

In placebo-controlled phase III clinical trials of fampridine 10 mg twice daily fampridine-responders increased their walking speed in timed 25-foot walk (T25FW) by approximately 25%.[4,5]

In one trial ($N=239$) the number of patients who met the responder criterion (a faster walking speed for at least three of four visits during the double-blind treatment period as compared with the maximum speed for any of the five off-drug visits) was 51 of 119 (42.9%) in the fampridine-treated group, and 11 of 118 (9.3%) in the placebo-treated group ($p<0.0001$; Mantel–Haenszel odds ratio, 8.14; 95% confidence interval [CI], 3.73–17.74) and had an average change from baseline in the 12-item multiple sclerosis walking scale (MSWS-12) of −6.04 (95% CI, −9.57 to −2.52).[5] A change in MSWS-12 of −4 to −6 is considered clinically meaningful.[8]

In the other trial ($N=301$) the proportion of timed-walk responders was higher in the fampridine group (51/119 or 42.9%) compared to the placebo group (11/118 or 9.3%, $p<0.0001$).[4] The average improvement from baseline in walking speed among fampridine-treated responders was 24.7% (95% CI, 21.0–28.4%).

It was shown that fampridine may have a beneficial effect on upper limb function and in cognitive tests.[9]

Few adverse effects associated with fampridine use were as follows: hypotension, anxiety, abnormal coordination, blurred vision, chest discomfort, balance disorder, headache, paresthesia, and seizures were reported in a few patients.[4,5,10] Fampridine may elicit trigeminal neuralgia.[11]

24.3 SPASTICITY

Spasticity is a common disabling complication of MS and its prevalence increases as the disease progresses; after 10 years of MS, approximately 50% of patients have greater than mild spasticity and 30% have greater than moderate spasticity. Spasticity was defined by Lance in 1980 as a motor disorder characterized by velocity-dependent increase in tonic stretch reflexes (muscle tone) with exaggerated tendon jerks, resulting from hyperexcitability of the stretch reflex as one component of the upper motor syndrome. The velocity-dependent nature of spasticity distinguishes it from other forms of hypertonia, such as dystonia and rigidity. Other characteristics are exaggerated cutaneous reflexes (Babinski sign), flexor and extensor spasms, clonus, dystonia, and contractures. Accompanying negative symptoms include paresis, lack of dexterity, and fatigability. The neuropathophysiologic processes are complex and not fully understood, but there is a widely accepted hypothesis that spasticity represents a loss of balance in the central nervous system between inhibitory and excitatory inputs on the alpha motor neurons, leading to relative overexcitation of those neurons. It generally develops when suprasegmental control over spinal cord segmental reflexes is lost, due to the interruption of descending modulatory influences carried by the cortico-, vestibulo-, and reticulospinal tracts. Brain or spinal cord damage may prevent the release of gamma-aminobutyric acid (GABA) from interneurons, resulting in decreased inhibitory impulses. GABA is an inhibitory neurotransmitter that modulates excitatory input at the alpha motor neuron.[12,13]

Spasticity is typically described as muscle stiffness, tightening, or spasms with associated involuntary jerking, pain, and weakness and may result in secondary complications such as decubitus ulcers and contractures. Quality of life is impaired as patients with spasticity experience significant worse symptomatology in terms of spasms, urinary dysfunction, and sleep disturbances.[14–16]

A standardized evaluation facilitates the examination process, including a detailed history, assessments of pain, disability, functional impairment, and health-related quality of life of the patient. The manual passive stretch maneuver is used to assess resistance at different rates. A joint is passively moved while the corresponding muscles are lengthened and shortened. In mild spasticity, the muscles will only resist when stretched quickly (catch), whereas in moderate spasticity the resistance is noticed at a lower speed of movement, known as clasp-knife phenomenon. In severe spasticity it may be difficult or impossible to move the muscles (rigid). The most widely used clinical measure of spasticity is the Modified Ashworth Scale, ranging through six steps, where 0 is no increase in tone and up to 4 is a rigid extremity.[17] Advantages and disadvantages that patients may gain from their spasticity must be recognized, so that treatment strategies and goals may be defined. Disadvantages may include interference with activities of daily living, sleep disturbance, contractures, dislocations, skin breakdown, bowel and bladder dysfunction, impairment of respiratory function, pain with stretching, and the masking of residual voluntary movement. However, patients may rely on a certain amount of spasticity to maintain muscle tone to transfer and stand. Video cameras are often helpful during evaluation as the patient's movements can be recorded and compared to movements during and after treatment. Physiotherapy and exercise programs are always adjunctive to pharmacological treatment, just as pain must be relieved to alleviate spasticity.[18]

Pharmacotherapy is important, but is of little worth unless a therapeutic alliance exists between the provider and patient. All oral medications have limited effectiveness, and all have sedative side effects that may overwhelm the patient suffering from fatigue and cognitive impairment. As spasticity often worsens in the evening and night a larger nightly dose may be appreciated, thereafter small doses may be added as needed 2–4 times daily according to the drug. First-line oral agents for spasticity, baclofen and tizanidine, have different mechanisms of action and toxicities and can therefore be combined to improve therapy (Table 24.1). No solid evidence for the efficacy of these oral agents exists, so administration is mainly based on clinical experience.[19] The Δ-9 tetrahydrocannabinol and cannabidiol oromucosal Sativex spray (Nabiximols) has been approved in the European Union and Canada for symptom improvement in

TABLE 24.1 Oral Treatment of Spasticity

Drugs	Mode of Action	Comments
Baclofen (5–40 mg, every 3–6 h)	Gamma-aminobutyric acid antagonist	May cause weakness, sedation, confusion
Tizanidine (2–12 mg every 3–4 h)	Noradrenergic alpha-2 agonist	Sedation may help insomnia and painful nocturnal spasticity
THC:CBD (Sativex) spray	Cannabinoid	Dizziness, fatigue
Gabapentin (200–900 mg, 3–4 times daily)	Anticonvulsant	Useful for dystonic spasms and pain; no effect on hepatic enzymes
Carbamazepine (100–400 mg, thrice daily)	Anticonvulsant	Useful for dystonic spasms and pain; can induce severe weakness
Levetiracetam (250–1500 mg, twice daily)	Anticonvulsant	Less sedative, useful for pain
Diazepam (2–10 mg, 2 to 3 times daily)	Benzodiazepine	Sedation, may be helpful for nocturnal spasticity
Dantrolene (25–100 mg, thrice daily)	Prevents calcium release from muscle stores	Causes weakness; liver enzymes must be monitored every 3 months

adult patients with moderate to severe MS-related spasticity, who have not responded adequately to other spasmolytic medications, and who demonstrate clinically significant improvement during an initial 4-week trial of therapy. In the phase-III trial, approximately one-third of MS patients had a clinically relevant response to the oromucosal spray, measured on a subjective Neurological Rating Scale of 0–10 points.[20] In addition to a reduction in spasticity, responders experienced meaningful relief from associated symptoms as painful spasms and sleep disturbance. Adverse events were dizziness and fatigue, which were improved by slower titration and dose reduction. There was no evidence of euphoric effects and no withdrawal symptoms.

Botulinum toxin (BT) is a neurotoxin that paralyzes muscles by inhibiting release of acetylcholine from presynaptic vesicles at the neuromuscular junction. In 2010, Onabotulinum toxin type A was approved by the US Food and Drug Administration (FDA) to treat increased muscle stiffness in elbow, wrist, and fingers in adults with upper limb spasticity. In 2008, an evidence-based review found 14 class 1 randomized controlled studies in adults and concluded that BT therapy is safe and effective for reducing spasticity and is probably effective for improving active function.[21] Since the effect of BT injections is local, it is most beneficial for focal spasticity such as in elbow, wrist, fingers, and hip adductor spasticity, and safe when treated according to guidelines. The effect of BT on the muscles may take 2–4 weeks to reach its peak, generally lasts 2–6 months, and injections are usually repeated every 3 months. A serious side effect, albeit rare, is the spread of an effect beyond the muscles injected causing generalized weakness, dysphagia, and respiratory problems. Severe para- or tetraplegic spasticity is often medically intractable, whereas treatment with intrathecal baclofen can relieve spasticity and related pain in MS patients with advanced disability. Intrathecal baclofen therapy is administered through an implanted programmable pump with a reservoir and catheter that delivers baclofen directly into the cerebrospinal fluid, effective at about 1% of the daily oral dosage. Preferably patients who are nonambulatory profit from this treatment and have their oral medication tapered off, alleviating them from sedative side effects.[22–24] This is a potent treatment requiring specialized teams to care for patients. Main adverse effects include risk of infection, catheter/pump failure resulting in symptoms of baclofen withdrawal. Overdose is extremely rare and usually due to human error. BT may be an adjunctive therapy for residual focal spasticity in the arms, hands, and fingers.

24.4 BLADDER DYSFUNCTION

Urinary disorder with lower tract symptoms is very common in MS, where 50–90% of patients have disturbances already after disease duration of 6 years, and quality of life is poor.[25–30] It is a main cause of morbidity with complicating infections that may trigger relapses, and risk of urinary upper tract damage. In MS, loss of cortical inhibition of the detrusor reflex results in impaired control of the bladder emptying and is called overactive bladder or neurogenic detrusor overactivity. Overactive bladder is characterized by detrusor hyperreflexia associated with urgency, frequency, nocturia, and urge incontinence. Lesions in the spinal cord may cause loss of coordination between bladder and sphincter activity, and thereby cause obstruction. This is known as detrusor-sphincter dyssynergia and results in urinary hesitancy, incomplete emptying, and urinary retention. In more advanced disease when patients have paraplegic spasticity, the bladder may be areflexic. The patient cannot void properly and has urinary retention. Urinary dysfunction may lead to permanent urological changes such as hydronephrosis, reflux, recurrent infections, and stones. It is therefore important that neurologists and general practitioners can identify these problems, instigate appropriate examinations and first-line therapy, and refer patients to specialist neurourology units when "red flags" are spotted. Several national guidelines attempt to solve this logistical problem.[31–33] If ultrasound of the bladder and kidney is normal and if the postvoidal residual volumen (PVR) is a maximum of 100 mL, patients may be prescribed an anticholinergic (antimuscarinic) drug and a new evaluation is done 3 months later including PVR. The following are red flags that indicate the patient must be referred to the urologist: more than three urinary tract infections or severe infections with fever in the previous year, lumbar pain during voiding, and hematuria. In case of treatment failure or PVR increase, patients are referred. For voiding dysfunction without significant PVR, alpha-blockers can be prescribed and a revisit is done 3 months later. Urinary retention can be treated by clean intermittent catheterization (CIC), but referral to a urologist is recommended. Injection of onabotulinum toxin A in the detrusor muscle, in several well-designed trials, has proven to be a highly effective treatment for overactive bladder with incontinence, rendering the patients continent for an average of 10 months. Patients must be able to perform CIC as 31% sustain urinary retention, but even so this treatment is an improvement of quality of life and has resulted in a marked reduction in the need for indwelling catheters or surgical procedures.[34]

24.5 PAIN

Both chronic and acute pain have been reported among the most common symptoms in MS patients[35,36] with prevalence rates ranging as high as 83% in some studies.[37] Pain in MS encompasses both central neuropathic pain, spasticity-related pain, and musculoskeletal pain.

Pain occurs in all forms of MS although more frequent in patients with a progressive course of MS. Pooled prevalence of pain was 50.0% in relapsing-remitting MS and approximately 70% in primary- and secondary-progressive MS patients.[35] A proposed classification of pain conditions associated with MS included continuous central neuropathic pain (eg, dysesthetic extremity pain), intermittent central neuropathic pain (eg, trigeminal neuralgia), musculo-skeletal pain (eg, lower pain back), and mixed neuropathic and nonneuropathic pain (eg, headache).[38]

Major classes of pharmacological interventions include anticonvulsants, antidepressants, cannabinoids, dextromethorphan, and opioids/opioid antagonists.[36]

Central neuropathic pain, often described as being a constant, burning pain or a deep or muscular aching predominantly in the lower limbs, is seen in approximately 40% of patients with MS,[39] and the classical finding in patients with neuropathic pain is the combination of sensory loss and paradoxical presentation of hypersensitivity in the painful areas.[40]

In an animal model of MS, chronic relapsing experimental autoimmune encephalomyelitis, T lymphocytes, microglia, and macrophages, which are key mediators of development of pain-related hypersensitivity, are present in the superficial dorsal horn.[41–43]

For treatment of neuropathic pain, tricyclic antidepressants, amitriptyline, nortriptyline, and clomipramine, are by many considered the drugs of choice (Table 24.2).[44] Drowsiness, dry mouth, constipation, urinary retention, and hypotension are common adverse effects. Also selective serotonin re-uptake inhibitors and serotonin-norepinephrine re-uptake inhibitors have been tried although less successfully.[45]

A long range of antiepileptics have been tested for relief of central neuropathic pain (Table 24.2)[44]: gabapentin, pregabalin, valproic acid, topiramate, carbamazepine, oxcarbazepine, phenytoin, lamotrigine, and levetiracetam. Gabapentin (600 mg daily) and pregabalin (150–300 mg daily) have shown effect in open label trials.[46,47] Lamotrigine (400 mg daily) showed efficacy in open label studies, but not in a placebo-controlled,[48] while levetiracetem (3000 mg daily) markedly reduced pain in a placebo-controlled study,[49] but failed in another.[50] Common side effects with antiepileptics are somnolence, dizziness, and dyspepsia, and many patients do not tolerate these drugs in doses necessary for achieving pain relief.

Spasmolytics, baclofen (including intrathecal baclofen) and tizanidine, used for treatment of spasticity (see earlier), may also have a beneficial effect on central neuropathic pain.[51,52]

Opiods have not been proven effective in the treatment of central neuropathic pain.

Cannabinoids (dronabinol, oromucosal or sublingual Δ-9-tetrahydrocannabinol, and sublingual cannabidiol) seem to have a positive effect on pain in MS.[53–56] Nabiximols, a mixture of Δ-9 tetrahydrocannabinol and cannabidiol, has been approved for treatment of spasticity (see earlier), but may also have a beneficial effect on pain.[55,57] However, the efficacy of cannabinoids has never been compared to that of conventional pain therapies.

The prevalence of trigeminal neuralgia in patients with MS ranges from 1.0% to 6.3%.[58,59] There are two major hypotheses about the pathophysiology of trigeminal neuralgia: The ephaptic theory suggests that a short-circuiting between two demyelinated axons via an artificial synapse may lead to action potentials in both directions.[60] The compression theory postulates that the trigeminal nerve is compressed or irritated at the entry zone of its root by an MS lesion, brain-stem infarction, cerebellopontine angle tumor, or most commonly by the superior cerebellar artery.[60–63]

A joint American Academy of Neurology-European Federation of Neurological Societies consensus statement concluded that carbamazepine and oxcarbazepine are effective

TABLE 24.2 Evidence-Based Treatment Guidelines for Neuropathic Pain

Condition	Treatment	Evidence	NNT	NNH
Neuropathic pain[a] or central pain[b]	Tricyclic antidepressants	Level A	4.0[b]	14.7
	Gabapentin, pregabalin	Level A	4.7[a]	17.8
	Lamotrigine	Level B	4.9[a]	NS
	Opioids	Level B	2.5[a]	17.1
	Carbamazepine	Level B	2.0[a]	21.7
	Cannabinoids	Level B	3.4[#]	NS
	Topiramate	Level C	7.4[a]	6.3
Trigeminal neuralgia	Carbamazepine	Level A	1.7	21.7
	Oxcarbazepine	Level B	NA	NA
	Gabapentin		ND	ND
	Lamotrigine		2.1	NS
	Baclofen	Level C	NA	NA
	Topiramate		NA	NA
	Amitriptyline	Level U	ND	ND

NNT, combined numbers needed to treat (with 95% CI) to obtain one patient with more than 50% pain relief; *NNH*, combined numbers needed to harm (95% CI) to obtain one patient to withdraw because of side effects; *ND*, no studies done; *NA*, dichotomized data are not available; *NS*, relative risk not significant.

[a] *Heterogeneity across different pain conditions.*

[b] *Central pain.*

Modified from Finnerup NB, Otto M, McQuay HJ, Jensen TS, Sindrup SH. Algorithm for neuropathic pain treatment: an evidence based proposal. Pain. *2005;118(3):289–305; Nick ST, Roberts C, Billiodeaux S, et al. Multiple sclerosis and pain.* Neurol Res. *2012;34(9):829–841.*

first-line treatments for trigeminal neuralgia, with oxcarbazepine slightly better tolerated. If pain cannot be sufficiently controlled by carbamazepine or oxcarbazepine, adjuvant therapy with lamotrigine or baclofen may be tried.[64] The conclusion of another large review was that carbamazepine is a first-line treatment, fulfilling level A recommendation criteria. Gabapentin and oxcarbazepine are second-line treatments fulfilling level B recommendation criteria. Topiramate, baclofen, misoprostol, and valproic acid fulfill level C recommendation criteria.[65] Neurosurgical treatment such as microvascular decompression or Gamma Knife radiosurgery is only rarely indicated for trigeminal neuralgia in patients with MS.[65]

Painful tonic spasms affect 11% of patients with MS.[66] They are described as cramping, pulling pain that most commonly occurs in the lower extremities, often occurring during the night and triggered by movements or sensory stimuli and are thought to be caused by ectopic impulses resulting from demyelination and axonal damage.[38] Only limited evidence exists regarding treatment of painful tonic spasms with baclofen, benzodiazepines, gabapentin, tiagabin, and carbamazepine.[67,68]

Headache (migraine without aura, migraine with aura and tension-type headache) occurs in approximately 50% of patients with MS, which is not different from the occurrence in the background population, and headache in MS patients should be treated as generally recommended.[69]

Patients with MS frequently suffer from musculoskeletal pain, in particular those in more advanced stages of the disease when mobility is hampered due to muscle weakness and spasticity with abnormal posture and strain on muscles and skeleton. Physical activity, physiotherapy, and weight training are important for prevention of musculoskeletal pain. Treatment could consist of nonsteroidal antiinflammatory drugs within limited periods.[70]

24.6 COGNITIVE IMPAIRMENT AND INTERVENTIONAL STRATEGIES

Both cognitive disturbances and neuropsychiatric disorders are frequently observed in MS.

The prevalence of cognitive impairment among large groups of MS patients ranges from 40% to 70%.[71] The typical profile is impairments in information-processing speed, memory, and attention, but often executive skills are affected also, whereas language is relatively well preserved.[71]

Cognitive impairment is an important predictor of health-related quality of life at all stages of MS.[72] It reduces physical independence,[73] competence in daily activities,[74] coping,[75] medication adherence,[76] and rehabilitation potential.[77]

Diffuse brain damage and progressive central atrophy are among predictors of cognitive decline over time.[73,78,79]

Some MS patients are able to withstand considerable burden of disease as white matter lesions and cerebral atrophy without cognitive inefficiency and memory deterioration. Studies show that heritable and environmental factors attenuate the negative effect of disease burden on cognitive status. That is, persons with larger maximal lifetime brain growth, greater vocabulary knowledge, and/or greater early life participation in cognitive leisure activities (eg, reading, hobbies) are at less risk for early cognitive impairment.[80,81] However, this "cognitive reserve" protection may become ineffective with progression of brain pathology.[82]

A comprehensive neuropsychological evaluation is preferred to elucidate the nature of each person's skills and deficits. However full clinical cognitive assessment is expensive and requires expert staff and special equipment. Two cognitive batteries are particularly widely used in clinical and research settings. They have good psychometric properties and are relatively robust to the effects of other MS symptoms. They are Brief Repeatable Battery of Neuropsychological tests[83] and Minimal Assessment of Cognitive Function in MS which is a 90-min neuropsychological battery.[84] A much less time-demanding Brief International Cognitive Assessment for Multiple Sclerosis has been suggested when only 15 min or even 5 min can be allocated to neuropsychological testing.[85]

Disease-modifying drugs can protect against cognitive decline in patients with MS.[86–88] An increasing number of studies investigating the benefit of cognitive rehabilitation have been conducted, of which most have been performed in the remediation of learning and memory and there is evidence to support the use of story memory techniques, self-generation, and retrieval practice (quizzing themselves). A domain-specific approach may not necessarily generalize to other functional domains.[89–92]

Computerized home-based personalized cognitive training programs may also be a practical and valuable tool to improve cognitive skills.[93,94]

Exercise is shown to be neuroprotective in rodents and physical activity may diminish cognitive impairment in humans.[95,96] Recently one study demonstrated that aerobic exercise increases hippocampal volume and functional connectivity and improves memory in patients with MS.[97]

The lifetime prevalence of major depression in MS is close to 50%[98] and depression adversely affects cognitive processing capacity and, in particular, working memory.[99,100] Recently recommendations for assessment and management of psychiatric disorders in individuals with MS have been proposed.[101,102]

24.7 FATIGUE

In 1988, fatigue in MS (MS-fatigue) was defined by the Multiple Sclerosis Council on Clinical Practice Guidelines as a subjective lack of physical and/or mental energy that is perceived by the individual or caregiver to interfere with usual and desired activities. Fatigue is a subjective sensation, whereas fatigability refers to a measurable decrement in cognitive and motor performance.[103,104]

Structural and functional neuroimaging suggest a neurobiological substrate for primary fatigue. It has been demonstrated that regional atrophy and abnormalities within normal-appearing white matter contributes significantly to explain the presence of MS-fatigue. Among several areas involved are the inferior temporal gyrus and the anterior thalamic radiation. fMRI and PET studies provide additional evidence that MS-fatigue is related to impaired interactions between functionally related cortical and subcortical areas.[105–109]

The association between MS-related fatigue and elevated levels of proinflammatory cytokines has been inconsistent. Based on a review of the literature it has been proposed that fatigue is a feeling resulting from cytokine-mediated activity changes within brain areas involved in interoception and homeostasis including the insula, anterior cingulate, and hypothalamus.[110,111] MS-fatigue is assessed by questionnaires. Widely used are The Fatigue Severity Scale[112] and The Fatigue Scale for Motor and Cognitive Functions.[113]

Fatigue is among the most common symptoms, experienced by 75–95% of people with MS. When persisting for at least 6 weeks it is considered chronic as opposed to acute fatigue.[114,115]

It is beneficial to consider it as a phenomenon comprising cognitive and motor aspects. MS-fatigue has a multifactorial etiology. Primary fatigue is a direct consequence of MS, but it is often exaggerated by multiple MS-related and non-MS-related conditions that trigger secondary fatigue. Secondary contributing factors are adverse effects of drugs (eg, anticonvulsants, sedatives, or muscle relaxants), poor sleep, depression, and cognitive interpretation.[116] A cognitive–behavioral model for MS-fatigue encompassing a wide range of biological and psychological factors has been presented[116] (Fig. 24.1).

MS-fatigue management must be multidisciplinary and individualized, including pharmacological and nonpharmacological strategies.[117] Among the nonpharmacological options are energy management,[118] mindfulness[119] cognitive behavioral therapy,[120] and exercise.[121,122] Insomnia is frequent in MS, and often results from disease-related factors such as nocturia, pain, spasticity, depression, and medications. But also restless legs syndrome and sleep-disordered breathing is more frequent than in the general population. It has been demonstrated that treatment of sleep disorders has a positive effect on MS-fatigue.[123,124]

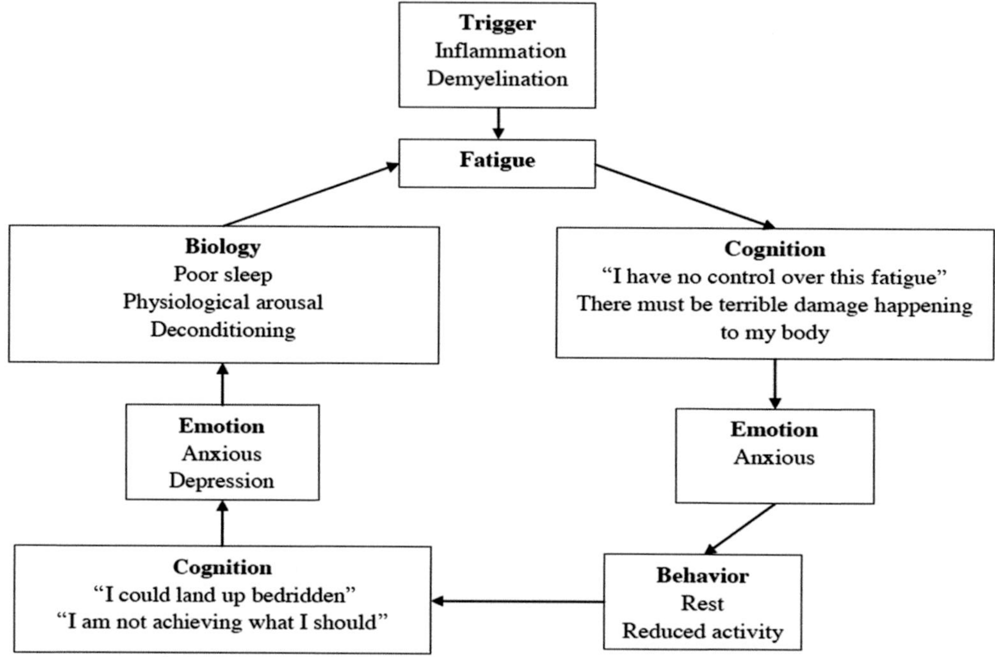

FIGURE 24.1 Cognitive–behavioral model of MS-fatigue. *From van Kessel K, Moss-Morris R. Understanding multiple sclerosis fatigue: a synthesis of biological and psychological factors.* J Psychosom Res. *2006;61(5):583–585.*

Particular attention must be given to depressive symptoms as there is a clear overlap with fatigue and it is unclear how to distinguish the two entities.[102] Treatment for depression is associated with reductions in the severity of fatigue due primarily to changes in mood.[125] MS-fatigue itself is not an indication for antidepressive treatment.[117]

Studies investigating the effect of natalizumab suggested improvement on MS-fatigue.[126]

No drugs have been approved by the medical authorities for treatment of MS-fatigue. Only an occasional treatment response can be expected and the effect must be monitored since the beneficial response attenuates with time. Amantadine, an antiviral agent and a dopamine agonist, is among the most widely used drugs. Recently a well-designed placebo-controlled double-blind study did not support modafinil as an effective treatment. However, modafinil and armodafinil are approved by the FDA to improve wakefulness in adult patients with excessive sleepiness associated with narcolepsy and obstructive sleep apnea.[127] Fampridine improves walking speed but it is still unclear whether it reduces fatigue.[128] A trial with aspirin 650 mg twice daily (morning and noon) improved fatigue in MS patients, perhaps due partly to relief from the Uhthoff's phenomenon.[129]

24.8 VITAMIN D

Most vitamin D in humans originates from skin exposure to sunlight ultraviolet-B radiation (wavelengths ranging from 280 to 315 nm) causing photolyses of 7-dehydrocholesterol in the skin to form previtamin D3, which then isomerizes to cholecalciferol. Cholecalciferol and ergocalciferol can also be acquired through the ingestion of fortified food products, fish, and vitamin supplements.[130] Cholecalciferol and ergocalciferol are hydroxylated in the liver to 25-hydroxyvitamin D (25(OH)D), which is the principal circulating form of vitamin D. This metabolite is then further hydroxylated in the kidney and other tissues to form 1,25-dihydroxyvitamin D (1,25$(OH)_2$D), the biologically active form of vitamin D, which in the body has a very short plasma half-life. 1,25$(OH)_2$D is actively and passively transported into the cell where it binds to the vitamin D receptor, which is present in most immune system cells.[131] 1,25$(OH)_2$D affects both the innate and adaptive immune system (humoral and cell-mediated immunity), promoting the innate immune response and inhibiting humoral and cell-mediated immunity, and the overall effect is antiinflammatory and antiproliferative.[132] The exact mechanisms by which vitamin D affects the pathogenesis of MS are uncertain, but it is well established that 1,25$(OH)_2$D has immunomodulatory effects that could be beneficial in autoimmune diseases.[133]

Low serum levels of 25-hydroxyvitamin D (25(OH)D), low intake of dietary vitamin D, and lack of sun exposure have been suggested as environmental risk factors for MS, indicating a potential role of vitamin D in the pathogenesis of MS.[134–136] MS patients in general have lower serum levels of 25(OH)D than do healthy controls, supporting a possible pathogenetic effect of vitamin D insufficiency in MS.[137] Moreover, protective effects of higher vitamin D levels have been shown in the disease course of MS.[138–141] Low serum levels of 25(OH)D in patients with clinically isolated syndromes (CIS) have been associated with increased risk of developing clinically definite MS.[142] This is supported by a study reporting that 25(OH)D is not only associated with the risk of conversion from CIS to clinically definite MS but also with disease activity and disability over a 5-year period.[143] Based on these findings it has been suggested that serum 25(OH)D levels at disease onset could predict the disease course in MS.

Interventional studies of vitamin D therapy, either as monotherapy or as add-on to interferon beta or glatiramer acetate, have shown conflicting, but mainly negative results. Out of five studies, four studies showed no beneficial effect on either relapses or MRI outcome measures,[144–147] while one study demonstrated a significantly lower number of T1-enhancing lesions.[148] However, all studies were small and had several methodological limitations, and, hence, larger randomized, double-blind studies are warranted.

24.9 COMPLEMENTARY AND ALTERNATIVE THERAPIES

Various forms of complementary or alternative therapies are used by a large proportion of people with MS. A study involving 6455 members of the five Nordic MS societies with data collected by an Internet-based questionnaire disclosed that the overall use of complementary and alternative medicine within the last 12 months varied from 46.0% in Sweden to 58.9% in Iceland. However, only between 9.5% in Finland and 18.4% in Norway reported use of complementary medicine as the sole therapy, while the rest used it as add-on to disease-modifying medicine for MS. The reason for using complementary medicine was usually non-specific or for preventative purposes such as strengthening the body in general, improving the body's muscle strength, and improving well-being.[149]

A European study comprising 1573 patients from Baden-Wuerttemberg, members of the German Multiple Sclerosis Society, who answered a mailed 53-item survey, reported 70%

lifetime use of at least one complementary or alternative medicine.[150] In comparison with conventional medicine, more patients displayed a positive attitude toward complementary and alternative medicine (44% vs. 38%, $p<0.05$). Use of complementary and alternative medicine was associated with religiosity, functional independence, female sex, white-collar job, and higher education. Among a wide variety of complementary and alternative medicine, diet modification (41%), omega-3 fatty acids (37%), removal of amalgam fillings (28%), vitamin E (28%), vitamin B (36%), vitamin C (28%), homeopathy (26%), and selenium (24%) were cited most frequently.

An American review concluded that up to 70% of people with MS have tried one or more complementary or alternative treatment for their MS and generally reported some benefit from the therapies. The authors concluded that low-fat diet, omega-3 fatty acids, lipoic acid, and vitamin D supplementation were the most promising antiinflammatory and neuroprotective agents in both relapsing and progressive forms of MS.[151]

An evidence-based guideline on complementary and alternative medicine in MS established by a subcommittee of the American Academy of Neurology concluded that there was Level A evidence of a beneficial effect of oral cannabis extract for spasticity symptoms and pain (excluding central neuropathic pain), level B evidence for effect of tetrahydrocannabinol for spasticity symptoms and pain (excluding central neuropathic pain), but level B evidence that these treatments were probably short-term ineffective for objective spasticity and tremor, and level C evidence that they were possibly long-term effective for spasticity and pain. Magnetic therapy was assessed probably effective for fatigue and probably ineffective for depression (Level B evidence); fish oil was categorized as probably ineffective for relapses, disability, fatigue, MRI lesions, and quality of life (Level B evidence); ginkgo biloba was classified as ineffective for cognition (Level A) and possibly effective for fatigue (Level C); reflexology was considered possibly effective for paresthesia (Level C); Cari Loder regimen was deemed possibly ineffective for disability, symptoms, depression, and fatigue (Level C); and bee sting therapy possibly ineffective for relapses, disability, fatigue, lesion burden/volume, and health-related quality of life (Level C). It was stressed that cannabinoids may cause adverse effects, hence clinicians should exercise caution regarding standardized versus nonstandardized cannabis extracts. Overall, the standard of quality control and nonregulation of complementary and alternative medicine should be seriously considered.[152]

References

1. Heesen C, Bohm J, Reich C, Kasper J, Goebel M, Gold SM. Patient perception of bodily functions in multiple sclerosis: gait and visual function are the most valuable. *Mult Scler*. 2008;14(7):988–991.
2. Gottberg K, Einarsson U, Ytterberg C, et al. Health-related quality of life in a population-based sample of people with multiple sclerosis in Stockholm County. *Mult Scler*. 2006;12(5):605–612.
3. Hobart J, Lamping D, Fitzpatrick R, Riazi A, Thompson A. The multiple sclerosis impact scale (MSIS-29): a new patient-based outcome measure. *Brain*. 2001;124(Pt 5):962–973.
4. Goodman AD, Brown TR, Edwards KR, et al. A phase 3 trial of extended release oral dalfampridine in multiple sclerosis. *Ann Neurol*. 2010;68(4):494–502.
5. Goodman AD, Brown TR, Krupp LB, et al. Sustained-release oral fampridine in multiple sclerosis: a randomised, double-blind, controlled trial. *Lancet*. 2009;373(9665):732–738.
6. Judge SI, Bever Jr CT. Potassium channel blockers in multiple sclerosis: neuronal Kv channels and effects of symptomatic treatment. *Pharmacol Ther*. 2006;111(1):224–259.

7. Bostock H, Sears TA, Sherratt RM. The effects of 4-aminopyridine and tetraethylammonium ions on normal and demyelinated mammalian nerve fibres. *J Physiol*. 1981;313:301–315.
8. Hobart JC, Riazi A, Lamping DL, Fitzpatrick R, Thompson AJ. Measuring the impact of MS on walking ability: the 12-Item MS Walking Scale (MSWS-12). *Neurology*. 2003;60(1):31–36.
9. Jensen H, Ravnborg M, Mamoei S, Dalgas U, Stenager E. Changes in cognition, arm function and lower body function after slow-release fampridine treatment. *Mult Scler*. 2014;20(14):1872–1880.
10. Goodman AD, Brown TR, Cohen JA, et al. Dose comparison trial of sustained-release fampridine in multiple sclerosis. *Neurology*. 2008;71(15):1134–1141.
11. Birnbaum G, Iverson J. Dalfampridine may activate latent trigeminal neuralgia in patients with multiple sclerosis. *Neurology*. 2014;83(18):1610–1612.
12. Balakrishnan S, Ward AB. The diagnosis and management of adults with spasticity. *Handb Clin Neurol*. 2013;110:145–160.
13. Pandyan AD, Gregoric M, Barnes MP, et al. Spasticity: clinical perceptions, neurological realities and meaningful measurement. *Disabil Rehabil*. 2005;27(1–2):2–6.
14. Barnes MP, Kent RM, Semlyen JK, McMullen KM. Spasticity in multiple sclerosis. *Neurorehabil Neural Repair*. 2003;17(1):66–70.
15. Haselkorn JK, Loomis S. Multiple sclerosis and spasticity. *Phys Med Rehabil Clin N Am*. 2005;16(2):467–481.
16. Kister I, Bacon TE, Chamot E, et al. Natural history of multiple sclerosis symptoms. *Int J MS Care*. 2013;15(3):146–158.
17. Bohannon RW, Smith MB. Interrater reliability of a modified Ashworth scale of muscle spasticity. *Phys Ther*. 1987;67(2):206–207.
18. Ward AB. A summary of spasticity management – a treatment algorithm. *Eur J Neurol*. 2002;9(suppl 1):48–52.
19. Shakespeare DT, Boggild M, Young C. Anti-spasticity agents for multiple sclerosis. *Cochrane Database Syst Rev*. 2003;4:CD001332.
20. Novotna A, Mares J, Ratcliffe S, et al. A randomized, double-blind, placebo-controlled, parallel-group, enriched-design study of nabiximols* (Sativex(®)), as add-on therapy, in subjects with refractory spasticity caused by multiple sclerosis. *Eur J Neurol*. 2011;18(9):1122–1131.
21. Simpson DM, Gracies JM, Graham HK, et al. Assessment: botulinum neurotoxin for the treatment of spasticity (an evidence-based review): report of the Therapeutics and Technology Assessment Subcommittee of the American Academy of Neurology. *Neurology*. 2008;70(19):1691–1698.
22. Ordia JI, Fischer E, Adamski E, Chagnon KG, Spatz EL. Continuous intrathecal baclofen infusion by a programmable pump in 131 consecutive patients with severe spasticity of spinal origin. *Neuromodulation*. 2002;5(1):16–24.
23. Dario A, Tomei G. Management of spasticity in multiple sclerosis by intrathecal baclofen. *Acta Neurochir Suppl*. 2007;97(Pt 1):189–192.
24. Ben SD, Peskine A, Roche N, Mailhan L, Thiebaut I, Bussel B. Intrathecal baclofen for treatment of spasticity of multiple sclerosis patients. *Mult Scler*. 2006;12(1):101–103.
25. de SM, Ruffion A, Denys P, Joseph PA, Perrouin-Verbe B. The neurogenic bladder in multiple sclerosis: review of the literature and proposal of management guidelines. *Mult Scler*. 2007;13(7):915–928.
26. Fowler CJ, Panicker JN, Drake M, et al. A UK consensus on the management of the bladder in multiple sclerosis. *J Neurol Neurosurg Psychiatry*. 2009;80(5):470–477.
27. Del PG, Panariello G, Del CF, De SG, Lombardi G. Diagnosis and therapy for neurogenic bladder dysfunctions in multiple sclerosis patients. *Neurol Sci*. 2008;29(suppl 4):S352–S355.
28. Nortvedt MW, Riise T, Frugard J, et al. Prevalence of bladder, bowel and sexual problems among multiple sclerosis patients two to five years after diagnosis. *Mult Scler*. 2007;13(1):106–112.
29. Nakipoglu GF, Kaya AZ, Orhan G, et al. Urinary dysfunction in multiple sclerosis. *J Clin Neurosci*. 2009;16(10):1321–1324.
30. Wiedemann A, Kaeder M, Greulich W, et al. Which clinical risk factors determine a pathological urodynamic evaluation in patients with multiple sclerosis? An analysis of 100 prospective cases. *World J Urol*. 2013;31(1):229–233.
31. Amarenco G, Chartier-Kastler E, Denys P, Jean JL, de SM, Lubetzski C. First-line urological evaluation in multiple sclerosis: validation of a specific decision-making algorithm. *Mult Scler*. 2013;19(14):1931–1937.
32. Stoffel JT. Contemporary management of the neurogenic bladder for multiple sclerosis patients. *Urol Clin North Am*. 2010;37(4):547–557.

33. Denys P, Phe V, Even A, Chartier-Kastler E. Therapeutic strategies of urinary disorders in MS. Practice and algorithms. *Ann Phys Rehabil Med*. 2014;57(5):297–301.
34. Ginsberg D, Gousse A, Keppenne V, et al. Phase 3 efficacy and tolerability study of onabotulinumtoxinA for urinary incontinence from neurogenic detrusor overactivity. *J Urol*. 2012;187(6):2131–2139.
35. Foley PL, Vesterinen HM, Laird BJ, et al. Prevalence and natural history of pain in adults with multiple sclerosis: systematic review and meta-analysis. *Pain*. 2013;154(5):632–642.
36. Jawahar R, Oh U, Yang S, Lapane KL. A systematic review of pharmacological pain management in multiple sclerosis. *Drugs*. 2013;73(15):1711–1722.
37. Truini A, Galeotti F, Cruccu G. Treating pain in multiple sclerosis. *Expert Opin Pharmacother*. 2011;12(15):2355–2368.
38. O'Connor AB, Schwid SR, Herrmann DN, Markman JD, Dworkin RH. Pain associated with multiple sclerosis: systematic review and proposed classification. *Pain*. 2008;137(1):96–111.
39. Beiske AG, Pedersen ED, Czujko B, Myhr KM. Pain and sensory complaints in multiple sclerosis. *Eur J Neurol*. 2004;11(7):479–482.
40. Treede RD, Jensen TS, Campbell JN, et al. Neuropathic pain: redefinition and a grading system for clinical and research purposes. *Neurology*. 2008;70(18):1630–1635.
41. Olechowski CJ, Truong JJ, Kerr BJ. Neuropathic pain behaviours in a chronic-relapsing model of experimental autoimmune encephalomyelitis (EAE). *Pain*. 2009;141(1–2):156–164.
42. Aicher SA, Silverman MB, Winkler CW, Bebo Jr BF. Hyperalgesia in an animal model of multiple sclerosis. *Pain*. 2004;110(3):560–570.
43. Lynch JL, Gallus NJ, Ericson ME, Beitz AJ. Analysis of nociception, sex and peripheral nerve innervation in the TMEV animal model of multiple sclerosis. *Pain*. 2008;136(3):293–304.
44. Finnerup NB, Otto M, McQuay HJ, Jensen TS, Sindrup SH. Algorithm for neuropathic pain treatment: an evidence based proposal. *Pain*. 2005;118(3):289–305.
45. Saarto T, Wiffen PJ. Antidepressants for neuropathic pain. *Cochrane Database Syst Rev*. 2007;4:CD005454.
46. Houtchens MK, Richert JR, Sami A, Rose JW. Open label gabapentin treatment for pain in multiple sclerosis. *Mult Scler*. 1997;3(4):250–253.
47. Solaro C, Boehmker M, Tanganelli P. Pregabalin for treating paroxysmal painful symptoms in multiple sclerosis: a pilot study. *J Neurol*. 2009;256(10):1773–1774.
48. Breuer B, Pappagallo M, Knotkova H, Guleyupoglu N, Wallenstein S, Portenoy RK. A randomized, double-blind, placebo-controlled, two-period, crossover, pilot trial of lamotrigine in patients with central pain due to multiple sclerosis. *Clin Ther*. 2007;29(9):2022–2030.
49. Rossi S, Mataluni G, Codeca C, et al. Effects of levetiracetam on chronic pain in multiple sclerosis: results of a pilot, randomized, placebo-controlled study. *Eur J Neurol*. 2009;16(3):360–366.
50. Finnerup NB, Grydehoj J, Bing J, et al. Levetiracetam in spinal cord injury pain: a randomized controlled trial. *Spinal Cord*. 2009;47(12):861–867.
51. Herman RM, D'Luzansky SC, Ippolito R. Intrathecal baclofen suppresses central pain in patients with spinal lesions. A pilot study. *Clin J Pain*. 1992;8(4):338–345.
52. Sadiq SA, Poopatana CA. Intrathecal baclofen and morphine in multiple sclerosis patients with severe pain and spasticity. *J Neurol*. 2007;254(10):1464–1465.
53. Rog DJ, Nurmikko TJ, Young CA. Oromucosal delta9-tetrahydrocannabinol/cannabidiol for neuropathic pain associated with multiple sclerosis: an uncontrolled, open-label, 2-year extension trial. *Clin Ther*. 2007;29(9):2068–2079.
54. Svendsen KB, Jensen TS, Bach FW. Does the cannabinoid dronabinol reduce central pain in multiple sclerosis? Randomised double blind placebo controlled crossover trial. *BMJ*. 2004;329(7460):253.
55. Wade DT, Makela P, Robson P, House H, Bateman C. Do cannabis-based medicinal extracts have general or specific effects on symptoms in multiple sclerosis? A double-blind, randomized, placebo-controlled study on 160 patients. *Mult Scler*. 2004;10(4):434–441.
56. Zajicek J, Fox P, Sanders H, et al. Cannabinoids for treatment of spasticity and other symptoms related to multiple sclerosis (CAMS study): multicentre randomised placebo-controlled trial. *Lancet*. 2003;362(9395):1517–1526.
57. Collin C, Davies P, Mutiboko IK, Ratcliffe S. Randomized controlled trial of cannabis-based medicine in spasticity caused by multiple sclerosis. *Eur J Neurol*. 2007;14(3):290–296.
58. Jensen TS, Rasmussen P, Reske-Nielsen E. Association of trigeminal neuralgia with multiple sclerosis: clinical and pathological features. *Acta Neurol Scand*. 1982;65(3):182–189.
59. Solaro C, Uccelli MM. Management of pain in multiple sclerosis: a pharmacological approach. *Nat Rev Neurol*. 2011;7(9):519–527.

60. Jannetta PJ. Arterial compression of the trigeminal nerve at the pons in patients with trigeminal neuralgia. 1967. *J Neurosurg*. 2007;107(1):216–219.
61. Cruccu G, Biasiotta A, Di RS, et al. Trigeminal neuralgia and pain related to multiple sclerosis. *Pain*. 2009;143(3):186–191.
62. Hilton DA, Love S, Gradidge T, Coakham HB. Pathological findings associated with trigeminal neuralgia caused by vascular compression. *Neurosurgery*. 1994;35(2):299–303.
63. Meaney JF, Eldridge PR, Dunn LT, Nixon TE, Whitehouse GH, Miles JB. Demonstration of neurovascular compression in trigeminal neuralgia with magnetic resonance imaging. Comparison with surgical findings in 52 consecutive operative cases. *J Neurosurg*. 1995;83(5):799–805.
64. Cruccu G, Gronseth G, Alksne J, et al. AAN-EFNS guidelines on trigeminal neuralgia management. *Eur J Neurol*. 2008;15(10):1013–1028.
65. Pollmann W, Feneberg W. Current management of pain associated with multiple sclerosis. *CNS Drugs*. 2008;22(4):291–324.
66. Solaro C, Brichetto G, Amato MP, et al. The prevalence of pain in multiple sclerosis: a multicenter cross-sectional study. *Neurology*. 2004;63(5):919–921.
67. Solaro C, Uccelli MM, Guglieri P, Uccelli A, Mancardi GL. Gabapentin is effective in treating nocturnal painful spasms in multiple sclerosis. *Mult Scler*. 2000;6(3):192–193.
68. Solaro C, Tanganelli P. Tiagabine for treating painful tonic spasms in multiple sclerosis: a pilot study. *J Neurol Neurosurg Psychiatry*. 2004;75(2):341.
69. Vacca G, Marano E, Brescia MV, et al. Multiple sclerosis and headache co-morbidity. A case-control study. *Neurol Sci*. 2007;28(3):133–135.
70. Thompson AJ. Symptomatic treatment in multiple sclerosis. *Curr Opin Neurol*. 1998;11(4):305–309.
71. Chiaravalloti ND, DeLuca J. Cognitive impairment in multiple sclerosis. *Lancet Neurol*. 2008;7(12):1139–1151.
72. Mitchell AJ, Benito-Leon J, Gonzalez JM, Rivera-Navarro J. Quality of life and its assessment in multiple sclerosis: integrating physical and psychological components of wellbeing. *Lancet Neurol*. 2005;4(9):556–566.
73. Rao SM, Leo GJ, Ellington L, Nauertz T, Bernardin L, Unverzagt F. Cognitive dysfunction in multiple sclerosis. II. Impact on employment and social functioning. *Neurology*. 1991;41:692–696.
74. Goverover Y, Genova HM, Hillary FG, DeLuca J. The relationship between neuropsychological measures and the Timed Instrumental Activities of Daily Living task in multiple sclerosis. *Mult Scler*. 2007;13(5):636–644.
75. Ehrensperger MM, Grether A, Romer G, et al. Neuropsychological dysfunction, depression, physical disability, and coping processes in families with a parent affected by multiple sclerosis. *Mult Scler*. 2008;14(8):1106–1112.
76. Bruce JM, Hancock LM, Arnett P, Lynch S. Treatment adherence in multiple sclerosis: association with emotional status, personality, and cognition. *J Behav Med*. 2010;33(3):219–227.
77. Langdon DW, Thompson AJ. Multiple sclerosis: a preliminary study of selected variables affecting rehabilitation outcome. *Mult Scler*. 1999;5(2):94–100.
78. Deloire MS, Ruet A, Hamel D, Bonnet M, Dousset V, Brochet B. MRI predictors of cognitive outcome in early multiple sclerosis. *Neurology*. 2011;76(13):1161–1167.
79. Potagas C, Giogkaraki E, Koutsis G, et al. Cognitive impairment in different MS subtypes and clinically isolated syndromes. *J Neurol Sci*. 2008;267(1–2):100–106.
80. Sumowski JF, Rocca MA, Leavitt VM, et al. Brain reserve and cognitive reserve in multiple sclerosis: what you've got and how you use it. *Neurology*. 2013;80(24):2186–2193.
81. Sumowski JF, Rocca MA, Leavitt VM, et al. Brain reserve and cognitive reserve protect against cognitive decline over 4.5 years in MS. *Neurology*. 2014;82(20):1776–1783.
82. Amato MP, Razzolini L, Goretti B, et al. Cognitive reserve and cortical atrophy in multiple sclerosis: a longitudinal study. *Neurology*. 2013;80(19):1728–1733.
83. Rao SM, Leo GJ, Bernardin L, Unverzagt F. Cognitive dysfunction in multiple sclerosis. I. Frequency, patterns, and prediction. *Neurology*. 1991;41:685–691.
84. Benedict RH, Fischer JS, Archibald CJ, et al. Minimal neuropsychological assessment of MS patients: a consensus approach. *Clin Neuropsychol*. 2002;16(3):381–397.
85. Langdon D, Amato M, Boringa J, et al. Recommendations for a brief international cognitive assessment for multiple sclerosis (BICAMS). *Mult Scler*. 2012;18(6):891–898.
86. Lang C, Reiss C, Maurer M. Natalizumab may improve cognition and mood in multiple sclerosis. *Eur Neurol*. 2012;67(3):162–166.
87. Mokhber N, Azarpazhooh A, Orouji E, et al. Cognitive dysfunction in patients with multiple sclerosis treated with different types of interferon beta: a randomized clinical trial. *J Neurol Sci*. 2014;342(1–2):16–20.

88. Patti F. Treatment of cognitive impairment in patients with multiple sclerosis. *Expert Opin Investig Drugs*. 2012;21(11):1679–1699.
89. Basso MR, Lowery N, Ghormley C, Combs D, Johnson J. Self-generated learning in people with multiple sclerosis. *J Int Neuropsychol Soc*. 2006;12(5):640–648.
90. Sumowski JF, Leavitt VM, Cohen A, Paxton J, Chiaravalloti ND, DeLuca J. Retrieval practice is a robust memory aid for memory-impaired patients with MS. *Mult Scler*. 2013;19(14):1943–1946.
91. Chiaravalloti ND, DeLuca J, Moore NB, Ricker JH. Treating learning impairments improves memory performance in multiple sclerosis: a randomized clinical trial. *Mult Scler*. 2005;11(1):58–68.
92. O'Brien AR, Chiaravalloti N, Goverover Y, DeLuca J. Evidenced-based cognitive rehabilitation for persons with multiple sclerosis: a review of the literature. *Arch Phys Med Rehabil*. 2008;89(4):761–769.
93. Mattioli F, Stampatori C, Zanotti D, Parrinello G, Capra R. Efficacy and specificity of intensive cognitive rehabilitation of attention and executive functions in multiple sclerosis. *J Neurol Sci*. 2010;288(1–2):101–105.
94. Shatil E, Metzer A, Horvitz O, Miller A. Home-based personalized cognitive training in MS patients: a study of adherence and cognitive performance. *Neurorehabilitation*. 2010;26(2):143–153.
95. Sandroff BM, Klaren RE, Pilutti LA, Dlugonski D, Benedict RH, Motl RW. Randomized controlled trial of physical activity, cognition, and walking in multiple sclerosis. *J Neurol*. 2014;261(2):363–372.
96. Briken S, Gold SM, Patra S, et al. Effects of exercise on fitness and cognition in progressive MS: a randomized, controlled pilot trial. *Mult Scler*. 2014;20(3):382–390.
97. Leavitt VM, Cirnigliaro C, Cohen A, et al. Aerobic exercise increases hippocampal volume and improves memory in multiple sclerosis: preliminary findings. *Neurocase*. 2014;20(6):695–697.
98. Feinstein A. Mood disorders in multiple sclerosis and the effects on cognition. *J Neurol Sci*. 2006;245(1–2):63–66.
99. Arnett PA, Higginson CI, Voss WD, Bender WI, Wurst JM, Tippin JM. Depression in multiple sclerosis: relationship to working memory capacity. *Neuropsychology*. 1999;13(4):546–556.
100. Arnett PA, Higginson CI, Voss WD, et al. Depressed mood in multiple sclerosis: relationship to capacity-demanding memory and attentional functioning. *Neuropsychology*. 1999;13(3):434–446.
101. Diaz-Olavarrieta C, Cummings JL, Velazquez J, Garcia de la C. Neuropsychiatric manifestations of multiple sclerosis. *J Neuropsychiatry Clin Neurosci*. 1999;11(1):51–57.
102. Minden SL, Feinstein A, Kalb RC, et al. Evidence-based guideline: assessment and management of psychiatric disorders in individuals with MS: report of the Guideline Development Subcommittee of the American Academy of Neurology. *Neurology*. 2014;82(2):174–181.
103. Kluger BM, Krupp LB, Enoka RM. Fatigue and fatigability in neurologic illnesses: proposal for a unified taxonomy. *Neurology*. 2013;80(4):409–416.
104. Multiple Sclerosis Council for Clinical Practice Guidelines. *Fatigue and Multiple Sclerosis: Evidence Based Management Strategies for Fatigue in Multiple Sclerosis*. Washington, DC.; 1998.
105. Andreasen AK, Jakobsen J, Soerensen L, et al. Regional brain atrophy in primary fatigued patients with multiple sclerosis. *Neuroimage*. 2010;50(2):608–615.
106. Calabrese M, Rinaldi F, Grossi P, et al. Basal ganglia and frontal/parietal cortical atrophy is associated with fatigue in relapsing-remitting multiple sclerosis. *Mult Scler*. 2010;16.
107. Filippi M, Rocca MA, Colombo B, et al. Functional magnetic resonance imaging correlates of fatigue in multiple sclerosis. *Neuroimage*. 2002;15(3):559–567.
108. Leocani L, Colombo B, Magnani G, et al. Fatigue in multiple sclerosis is associated with abnormal cortical activation to voluntary movement–EEG evidence. *Neuroimage*. 2001;13(6 Pt 1):1186–1192.
109. Rocca MA, Parisi L, Pagani E, et al. Regional but not global brain damage contributes to fatigue in multiple sclerosis. *Radiology*. 2014;273(2):511–520.
110. Hanken K, Eling P, Hildebrandt H. The representation of inflammatory signals in the brain – a model for subjective fatigue in multiple sclerosis. *Front Neurol*. 2014;5:264.
111. Krupp LB, Serafin DJ, Christodoulou C. Multiple sclerosis-associated fatigue. *Expert Rev Neurother*. 2010;10(9):1437–1447.
112. Krupp LB, LaRocca NG, Muir J, Steinberg AD. A study of fatigue in systemic lupus erythematosus. *J Rheumatol*. 1990;17(11):1450–1452.
113. Penner IK, Raselli C, Stocklin M, Opwis K, Kappos L, Calabrese P. The fatigue scale for motor and cognitive functions (FSMC): validation of a new instrument to assess multiple sclerosis-related fatigue. *Mult Scler*. 2009;15(12):1509–1517.
114. Bakshi R. Fatigue associated with multiple sclerosis: diagnosis, impact and management. *Mult Scler*. 2003;9(3):219–227.

115. Kos D, Kerckhofs E, Nagels G, D'Hooghe MB, Ilsbroukx S. Origin of fatigue in multiple sclerosis: review of the literature. *Neurorehabil Neural Repair*. 2008;22(1):91–100.
116. van Kessel K, Moss-Morris R. Understanding multiple sclerosis fatigue: a synthesis of biological and psychological factors. *J Psychosom Res*. 2006;61(5):583–585.
117. Amato MP, Portaccio E. Management options in multiple sclerosis-associated fatigue. *Expert Opin Pharmacother*. 2012;13(2):207–216.
118. Blikman LJ, Huisstede BM, Kooijmans H, Stam HJ, Bussmann JB, van MJ. Effectiveness of energy conservation treatment in reducing fatigue in multiple sclerosis: a systematic review and meta-analysis. *Arch Phys Med Rehabil*. 2013;94(7):1360–1376.
119. Grossman P, Kappos L, Gensicke H, et al. MS quality of life, depression, and fatigue improve after mindfulness training: a randomized trial. *Neurology*. 2010;75(13):1141–1149.
120. Knoop H, van KK, Moss-Morris R. Which cognitions and behaviours mediate the positive effect of cognitive behavioural therapy on fatigue in patients with multiple sclerosis? *Psychol Med*. 2012;42(1):205–213.
121. Andreasen AK, Stenager E, Dalgas U. The effect of exercise therapy on fatigue in multiple sclerosis. *Mult Scler*. 2011;17(9):1041–1054.
122. Latimer-Cheung AE, Martin Ginis KA, Hicks AL, et al. Development of evidence-informed physical activity guidelines for adults with multiple sclerosis. *Arch Phys Med Rehabil*. 2013;94(9):1829–1836.
123. Cote I, Trojan DA, Kaminska M, et al. Impact of sleep disorder treatment on fatigue in multiple sclerosis. *Mult Scler*. 2013;19(4):480–489.
124. Kaminska M, Kimoff RJ, Schwartzman K, Trojan DA. Sleep disorders and fatigue in multiple sclerosis: evidence for association and interaction. *J Neurol Sci*. 2011;302(1–2):7–13.
125. Mohr DC, Hart SL, Goldberg A. Effects of treatment for depression on fatigue in multiple sclerosis. *Psychosom Med*. 2003;65(4):542–547.
126. Svenningsson A, Falk E, Celius EG, et al. Natalizumab treatment reduces fatigue in multiple sclerosis. Results from the TYNERGY trial; a study in the real life setting. *PLoS One*. 2013;8(3):e58643.
127. Moller F, Poettgen J, Broemel F, Neuhaus A, Daumer M, Heesen C. HAGIL (Hamburg Vigil Study): a randomized placebo-controlled double-blind study with modafinil for treatment of fatigue in patients with multiple sclerosis. *Mult Scler*. 2011;17(8):1002–1009.
128. Goodman AD, Bethoux F, Brown TR, et al. Long-term safety and efficacy of dalfampridine for walking impairment in patients with multiple sclerosis: results of open-label extensions of two Phase 3 clinical trials. *Mult Scler*. 2015;21.
129. Wingerchuk DM, Benarroch EE, O'Brien PC, et al. A randomized controlled crossover trial of aspirin for fatigue in multiple sclerosis. *Neurology*. 2005;64(7):1267–1269.
130. Holick MF. Vitamin D deficiency. *N Engl J Med*. 2007;357(3):266–281.
131. Norman AW. Minireview: vitamin D receptor: new assignments for an already busy receptor. *Endocrinology*. 2006;147(12):5542–5548.
132. Smolders J, Peelen E, Thewissen M, et al. The relevance of vitamin D receptor gene polymorphisms for vitamin D research in multiple sclerosis. *Autoimmun Rev*. 2009;8(7):621–626.
133. Prietl B, Treiber G, Pieber TR, Amrein K. Vitamin D and immune function. *Nutrients*. 2013;5(7):2502–2521.
134. Munger KL, Levin LI, Hollis BW, Howard NS, Ascherio A. Serum 25-hydroxyvitamin D levels and risk of multiple sclerosis. *JAMA*. 2006;296(23):2832–2838.
135. Salzer J, Hallmans G, Nystrom M, Stenlund H, Wadell G, Sundstrom P. Vitamin D as a protective factor in multiple sclerosis. *Neurology*. 2012;79(21):2140–2145.
136. Munger KL, Zhang SM, O'Reilly E, et al. Vitamin D intake and incidence of multiple sclerosis. *Neurology*. 2004;62(1):60–65.
137. Soilu-Hanninen M, Airas L, Mononen I, Heikkila A, Viljanen M, Hanninen A. 25-Hydroxyvitamin D levels in serum at the onset of multiple sclerosis. *Mult Scler*. 2005;11(3):266–271.
138. Simpson Jr S, Taylor B, Blizzard L, et al. Higher 25-hydroxyvitamin D is associated with lower relapse risk in multiple sclerosis. *Ann Neurol*. 2010;68(2):193–203.
139. Pierrot-Deseilligny C, Rivaud-Pechoux S, Clerson P, de Paz R, Souberbielle JC. Relationship between 25-OH-D serum level and relapse rate in multiple sclerosis patients before and after vitamin D supplementation. *Ther Adv Neurol Disord*. 2012;5(4):187–198.
140. Mowry EM, Waubant E, McCulloch CE, et al. Vitamin D status predicts new brain magnetic resonance imaging activity in multiple sclerosis. *Ann Neurol*. 2012;72(2):234–240.
141. Scott TF, Hackett CT, Dworek DC, Schramke CJ. Low vitamin D level is associated with higher relapse rate in natalizumab treated MS patients. *J Neurol Sci*. 2013;330(1–2):27–31.

142. Martinelli V, Dalla CG, Colombo B, et al. Vitamin D levels and risk of multiple sclerosis in patients with clinically isolated syndromes. *Mult Scler*. 2014;20(2):147–155.
143. Ascherio A, Munger KL, White R, et al. Vitamin D as an early predictor of multiple sclerosis activity and progression. *JAMA Neurol*. 2014;71(3):306–314.
144. Burton JM, Kimball S, Vieth R, et al. A phase I/II dose-escalation trial of vitamin D3 and calcium in multiple sclerosis. *Neurology*. 2010;74(23):1852–1859.
145. Kampman MT, Steffensen LH, Mellgren SI, Jorgensen L. Effect of vitamin D3 supplementation on relapses, disease progression, and measures of function in persons with multiple sclerosis: exploratory outcomes from a double-blind randomised controlled trial. *Mult Scler*. 2012;18(8):1144–1151.
146. Mosayebi G, Ghazavi A, Ghasami K, Jand Y, Kokhaei P. Therapeutic effect of vitamin D3 in multiple sclerosis patients. *Immunol Investig*. 2011;40(6):627–639.
147. Stein MS, Liu Y, Gray OM, et al. A randomized trial of high-dose vitamin D2 in relapsing-remitting multiple sclerosis. *Neurology*. 2011;77(17):1611–1618.
148. Soilu-Hanninen M, Aivo J, Lindstrom BM, et al. A randomised, double blind, placebo controlled trial with vitamin D3 as an add on treatment to interferon beta-1b in patients with multiple sclerosis. *J Neurol Neurosurg Psychiatry*. 2012;83(5):565–571.
149. Skovgaard L, Nicolajsen PH, Pedersen E, et al. Use of complementary and alternative medicine among people with multiple sclerosis in the Nordic countries. *Autoimmune Dis*. 2012;2012:841085.
150. Schwarz S, Knorr C, Geiger H, Flachenecker P. Complementary and alternative medicine for multiple sclerosis. *Mult Scler*. 2008;14(8):1113–1119.
151. Yadav V, Shinto L, Bourdette D. Complementary and alternative medicine for the treatment of multiple sclerosis. *Expert Rev Clin Immunol*. 2010;6(3):381–395.
152. Yadav V, Bever Jr C, Bowen J, et al. Summary of evidence-based guideline: complementary and alternative medicine in multiple sclerosis: report of the guideline development subcommittee of the American Academy of Neurology. *Neurology*. 2014;82(12):1083–1092.
153. Nick ST, Roberts C, Billiodeaux S, et al. Multiple sclerosis and pain. *Neurol Res*. 2012;34(9):829–841.

CHAPTER

25

Cognitive Impairment in Multiple Sclerosis

M.P. Amato, B. Goretti
University of Florence, Florence, Italy

OUTLINE

Translational Neuroimmunology in Multiple Sclerosis
http://dx.doi.org/10.1016/B978-0-12-801914-6.00025-8

25.1 INTRODUCTION

For the greater part of the past century clinicians have grossly underestimated the relevance of cognitive impairment (CI) in multiple sclerosis (MS) patients, and only over the past decades have they become aware of its prevalence and profound impact on function.[1] The last decade, in particular, has seen a great deal of research addressing some of the key issues in this area: documenting CI in different disease stages and subtypes and its functional impact, also independent of the degree of physical impairment; developing and validating appropriate assessment tools; understanding the mechanisms underlying CI through neuroimaging studies; and addressing the potential for therapeutic interventions.

Cognitive dysfunction in MS presents a considerable burden to patients and to society, due to the negative impact on function, including maintaining employment, activities of daily living, social activity, coping strategies, adherence to therapies, and the capacity to benefit from in-patient rehabilitation.[1] Interventions to ameliorate or reduce CI may therefore benefit patient function and quality of life.

In this chapter, we summarize current understanding of MS-related CI and review available evidence on treatment strategies, based on both pharmacological approaches and cognitive rehabilitation, focusing on the most salient studies in the field.

25.2 PREVALENCE

Prevalence rates of CI in MS are estimated between 40% in more representative community-based studies up to ≥60% in clinic-based studies.[1,2] Mild dementia has been estimated in up to 20% of patients.[3–5]

CI has been detected in all the disease stages and clinical subtypes.[6] However, it tends to progress over time[2] and is usually prominent in secondary-progressive MS (SPMS).[6,7]

25.2.1 Early Relapsing-Remitting Multiple Sclerosis, Clinically Isolated Syndrome, and Radiologically Isolated Syndrome

Cognitive deficits are detected in approximately one-third of patients with early relapsing-remitting MS (RRMS).[8] In patients who have had only one attack suggestive of MS, so-called clinically isolated syndrome (CIS), frequency estimates of CI range between 20% up to more than 50%.[2] In general, the neuropsychological pattern of impairments is similar to that described in subjects with longstanding MS.[2] In one longitudinal study over 3.5 years[9] the presence of significant CI in CIS patients was an independent predictor of more rapid conversion to clinically definite MS, together with the presence of magnetic resonance (MR) criteria for dissemination in space at baseline scan. These findings, suggesting a prognostic role of early CI, are in line with those obtained in a cohort of early RRMS patients, where CI at baseline was predictive of increasing disability levels over a 7-year follow-up.[10]

The term radiologically isolated syndrome (RIS) refers to people with no symptoms or abnormalities on a neurological exam who have lesions typical for MS on MR.[11] In this situation, studies[12–14] have detected subtle cognitive changes in nearly 25% of subjects.[13] Whether the occurrence of CI in subjects with RIS may play a prognostic role remains an unanswered question that deserves long-term longitudinal studies.

25.2.2 Secondary-Progressive and Primary-Progressive Multiple Sclerosis

Studies comparing samples of patients with RRMS, SPMS, and primary-progressive MS (PPMS) have generally documented more frequent and severe CI in progressive compared with relapsing-remitting (RR) patients.[15,16] For SPMS, estimates of prevalence of CI range from 37%[17] up to 83%.[16] Interestingly, in a 10-year follow-up, the shift from the RR to the SP phase of the disease was predictive of a worse cognitive outcome together with increasing disability and older age.[18,19]

For PPMS, prevalence figures range from 7%[20] up to 57%.[16] In a collaborative study on cognitive functioning in 63 PPMS and transitional progressive the prevalence of CI was 28.6%.[21,22] In a cross-sectional study[23] comparing 41 PPMS, 60 RRMS, and a large group of healthy controls matched for age, sex, and education level, the prevalence of CI was 69.6% in PPMS patients and 21.7% in RRMS subjects.[23]

As for comparison between SPMS and PPMS patients, despite a few controversial findings, the majority of published studies show higher rates of impairment in SPMS.[16,17,20,24–26]

25.2.3 Other Disease Subtypes

Benign MS (BMS) is usually defined as a long-standing disease duration with minimal irreversible disability accrual.[27,28] Nevertheless, in one published study up to 45% of BMS subjects exhibited cognitive dysfunction with negative impact on work and social life.[27] The presence of CI in BMS was associated with more severe brain pathology detected at MR and worse clinical outcome over a 5-year follow-up.[29,30] The current expert consensus therefore includes also cognitive preservation among criteria for defining BMS.[31]

Patients with pediatric-onset MS may represent a particularly vulnerable population,[32] since in this group of subjects disease occurs while brain maturation is ongoing, during the learning curve and key formative years in the academic career. Estimates of prevalence of CI range from 30% up to 50% of the cases.[32] Moreover, in a large US cohort, 18% of pediatric subjects at their first clinical attack fulfilled criteria for CI.[33] The cognitive profile is generally consistent with that described in MS adults, although deficits in language and global intellectual efficiency have been reported in patients who are younger at the disease onset.[34,35]

25.3 NEUROPSYCHOLOGICAL PROFILE

MS-related CI is typically circumscribed (ie, confined to specific cognitive domains) rather than global. In general, the domains most frequently impaired are information processing speed (IPS) and working memory, memory (long-term, explicit, episodic) and executive functions.[36] Areas of cognition that are usually preserved are simple attention (eg, repeating digits), semantic memory (eg, memory for words), and linguistic skills (eg, word naming and comprehension).[36] Language is usually preserved[1,2] and most studies indicate that general intelligence remains intact.[37,38]

There is limited evidence on whether patients with different disease subtypes also have different cognitive profiles.[16,17,23,24,39] Pertinent research in each of these domains is summarized in the following.

25.3.1 Memory

Long-term memory—the ability to learn new information and to recall that information at a later time point—is one of the most consistently impaired cognitive functions in MS. Specifically, the deficit involves a form of explicit memory known as episodic memory (ie, memory for events and conversations). Early work on memory impairment in MS suggested that difficulty in retrieval from long-term storage was the primary memory problem,[40–42] whereas more recent research has suggested that the primary problem may be initial learning of information.[43,44] Thus, MS can disrupt both acquisition (ie, encoding and storage) and retrieval processes. Another form of explicit memory that involves remembering previously acquired knowledge, such as words (semantic memory), is relatively preserved in MS people. Implicit memory (learning and remembering without conscious awareness) is also substantially spared.[43,44]

25.3.2 Working Memory and Information Processing Speed

Information processing efficiency refers to the ability to maintain and manipulate information in the brain for a short time period (working memory) and to the speed with which we can process that information (processing speed).[45] These deficits are the most common cognitive deficits in MS, often documented from the earliest stages of the disease, so that tests tapping this domain, such as the Paced Auditory Serial Addition Test (PASAT)[46] and the Symbol Digit Modalities Test (SDMT),[46] have been included in short batteries[46,47] and quantitative scales of disability such as the Multiple Sclerosis Functional Composite (MSFC).[48] In general, basic attention tasks (eg, repeating digits) are unaffected in patients with MS, whereas impairments in complex attention, such as sustained, selective, divided, and alternating attention are more common.[4]

Confounding factors like fatigue and depression could influence IPS performances and therefore must be taken into account (see later).[49,50]

25.3.3 Executive Functions

Executive functions are higher order cognitive functions needed for goal-directed behavior and adaptation to environmental changes or demands. They include, among others, abstract and conceptual reasoning, planning, problem solving, and phonemic and semantic fluency.[36] Executive functioning is often impaired in MS subjects.[36] Measures of executive functioning are also particularly susceptible to the effects of depression in MS patients with MS.[49–52]

25.3.4 Visual Perceptual Functions

Visual perceptual functions include the recognition of a visual stimulus and the ability to perceive the characteristics of that stimulus accurately. Up to about 25% of people with MS may have deficits in visual perceptual functions, also after taking into account the possible negative impact of visual dysfunction on task performance.[53,54]

25.3.5 The Theory of Mind

Theory of mind (ToM) is the ability to deduce other people's thoughts and emotions on the basis of verbal and nonverbal cues.[55] It has been suggested that ToM may be frequently impaired in MS subjects.[55] Functional MRI (fMRI) studies suggest that underlying changes related to poor performance on ToM tasks include reorganization of the cortical networks controlling ToM.[56] Clearly this is a novel area of research that deservers further neuropsychological studies.

25.4 CLINICAL CORRELATES

25.4.1 Disease Duration and Disability

In general, in cross-sectional studies, the correlation between CI and disease duration is weak to moderate, whereas a stronger association has emerged in long-term longitudinal studies.[2]

The evidence of the relationship between CI and level of physical disability, which is generally measured on the Expanded Disability Status Scale (EDSS),[57] is sometimes conflicting. One large epidemiological survey[58] reported a slight but significant correlation, whereas other studies have failed to find any significant relationships.[2] In a study on a large sample of patients with RRMS or SPMS a significant correlation of CI with the EDSS was found, particularly with the cerebral and cerebellar systems.[59] Moreover, in a longitudinal study over 10 years[19] an increasing EDSS score and shift from RRMS to SPMS correlated with progressive cognitive deterioration over time, which suggests that, at least in the long term, CI and clinical parameters of disease severity can proceed in parallel.

25.4.2 Disease Activity

There are a few reports of transient fluctuations in cognitive performance during phases of clinical relapses and remissions.[60,61] Decreased performance on the SDMT was associated with clinical relapses,[62] whereas PASAT performance was shown to fluctuate in correspondence of MRI-detected phases of subclinical disease activity.[63] Recently, isolated cognitive relapses (ICRs) have been identified in patients who were unaware of any cognitive changes, in the absence of any neurological, fatigue, or depression changes.[64] ICRs were associated with a decrease of at least four points on the SDMT score and the presence of gadolinium-enhancing lesions at MR, particularly in the frontal and parietal lobes. ICRs in MS patients might contribute to the long-term burden of cognitive disability.[64]

Moreover, studies on animal models have shown the detrimental effects on cognition of several proinflammatory cytokines, and in particular the tumor necrosis factor-alpha, that can produce an impairment of synaptic transmission.[65]

25.4.3 Fatigue

Fatigue is reported in nearly 90% of MS patients and may have a physical and cognitive component.[66] The pathogenesis of fatigue, however, remains incompletely understood.

Imaging studies have suggested a physiopathological link between fatigue and cognitive disturbance. Studies with positron emission tomography[67] found a significant reduction of glucose metabolism in regions involved in cognitive functions in MS subjects. More recently, fMRI studies have suggested that fatigue is related to impaired interactions between functionally related cortical and subcortical areas.[68] No association has been reported between subjective reports of fatigue on the Fatigue Severity Scale[69] and objectively assessed cognitive performance.[70] However, studies using the Modified Fatigued Impact Scale[71] or more objective measure of fatigue—such as using a prolonged and continuous cognitive effort[72,73]—have documented a relationship between cognitive fatigue and decreased cognitive task performance.

25.4.4 Depression

Depression is common in MS and is thought to interfere significantly with cognitive and noncognitive activities.[74] In a critical review of the literature, Arnett et al.[74] found a positive link between depression and cognitive functioning. Depression can particularly affect working memory, processing speed, learning and memory functions, abstract reasoning, and executive functioning.[75] Patients with MS often take psychotropic drugs and corticosteroids. MS patients undergoing corticosteroid treatment for relapses may show transient worsening of cognitive functions.[76] Interestingly, hypothalamic–pituitary–adrenal axis dysregulation, observed in up to 50% of patients showing elevated disease activity,[77] correlated moderately with poorer performance in a test of processing speed and attention.[78]

As for the relationship between interferon beta (IFNB) therapy and depression, an extensive literature review[79] has shown that depression under treatment is mainly related to pretreatment depression. Providing regular monitoring of depressive symptoms and appropriate antidepressant treatment, depression rarely represents a contraindication to IFNB treatment in MS patients.[79]

25.5 IMAGING CORRELATES

Advancements in MR technology have dramatically enhanced our understanding of MS-related CI. A few authors have reviewed this interesting topic extensively.[80–83] In brief, available evidence shows that both focal and diffuse changes in white matter (WM) and gray matter (GM) are relevant to MS-related cognitive dysfunction. The presence of cerebral WM lesions may lead to a functional disconnection of different cortical and deep GM regions, and this may be responsible for development of CI in MS. In general, however, correlations improve when the specific location of the lesions is taken into account.[80] Overall brain atrophy and, in particular, cortical and subcortical GM atrophy are well correlated with CI in MS.[80] Moreover, cortical lesions detected through double-inversion recovery techniques show good relationships with CI in cross-sectional and longitudinal studies.[83] Abnormalities of so-called normal-appearing WM and GM of the brain are increasingly recognized as important factors in relation to CI in MS. Several quantitative MR-based techniques, including magnetization transfer[84] and diffusion tensor[85] MRI, can quantify the extent and improve the characterization of the nature of structural changes occurring within and outside focal MS lesions.

Proton MR spectroscopy[86] can add information on the biochemical nature of such changes. However, despite these multiparametric correlations, MR findings do not fully account for the severity of cognitive changes in MS subjects. This is at least in part explained by brain plasticity and ability to compensate for brain damage, as suggested by fMRI studies. Using different cognitive paradigms, fMRI changes have been detected in all the major clinical phenotypes of MS.[87–89] In cognitively preserved patients, such abnormalities usually consist of an increased recruitment of areas normally activated by healthy individuals when performing the same task, which is usually interpreted as an adaptive, compensatory phenomenon.[90] In contrast, cognitively impaired patients experience a poor pattern of cortical recruitment[26,90] or an abnormal pattern of activation of areas that are not usually activated by healthy individuals,[26] which is interpreted as the failure or dysfunction of compensatory mechanisms due to advanced degrees of brain damage. More recent fMRI studies have also dealt with the analysis of the resting state networks. These studies in people with MS have highlighted an imbalance in the function of two opposing cortical networks: the default network that is active at rest and the control network that is engaged during most cognitive tasks.[91]

25.5.1 The Cognitive Reserve Theory

The cognitive reserve (CR) theory, developed in research on aging and dementia, postulates that individual differences in the cognitive processes or neural networks underlying task performance allow some people to cope better than others with brain damage, so that patients with higher CR can withstand more advanced disease before suffering cognitive dysfunction.[92] Recently, the CR theory has been applied to MS patients. Overall, available evidence suggests that also in MS patients, higher CR can mediate between brain damage and cognitive performance, representing a protective factor against cognitive dysfunction. However, advanced brain atrophy can suppress this protective function, which points to the importance of early intervention strategies.[80,93,94] Developments in this novel research field may provide clues to identifying patients at higher risk of CI and treatment strategies based on intellectual enrichment.

25.6 ASSESSMENT TOOLS

To diagnose and quantify the extent of CI, appropriate assessments are essential. Patient report is unreliable and highly correlated with depressive symptoms.[95,96] Unfortunately, routine neurological examinations for MS are too insensitive to yield valid information on cognitive function. The development of the MSFC, which includes the PASAT[97] was a step forward toward incorporating a sensitive measure of cognition into a standardized MS assessment tool. The use of the SDMT instead of the PASAT has been suggested for screening purposes, given the practical advantage of this test and its good psychometric properties.[98] One assessment approach is to use brief or intermediate-length batteries that range from 30 to 90 min in duration and focus on the most frequently involved cognitive domains. Among these batteries, the most well validated are the Brief Repeatable Neuropsychological Battery[46] and the Minimal Assessment of Cognitive Function in MS.[99] These batteries differ in the specific verbal memory and visual–spatial memory tests employed, but assess similar domains, and are comparable in their overall sensitivity to disease status.[100] Recently, based on literature review and expert opinion, a Brief International Cognitive Assessment for MS has been proposed, which focuses on measures of processing speed, visual–spatial and verbal memory: validation studies of this instrument are currently ongoing in different countries.[47,101,102] Finally, computer-based assessment tools are increasingly developed in the literature and they are probably going to play a major role in the future for both screening and research purposes.[103–105]

25.7 APPROACHES TO TREATMENT

Approaches to the treatment of CI in MS patients are mainly based on the use of disease-modifying or symptomatic drugs and different types of cognitive rehabilitation techniques.

25.7.1 Pharmacologic Approaches: Disease-Modifying Drugs

The disease-modifying drugs (DMDs) have the potential to positively influence the cognitive outcome of the patients by acting on some key pathogenic mechanisms of MS-related CI. In particular, they can contain the inflammatory cascade in the brain, reduce the accumulation of irreversible nervous damage, as shown by the positive effects on T2 and T1 lesion loads, and some of them also have effects on the progression of brain atrophy.[106] It is also possible, based on preclinical data, that a few DMDs exert a direct neuroprotective/neurotrophic effect, via different mechanisms of action.[106]

However, available evidence in the field is limited. Interpretation of data is complicated by issues largely related to methodological problems of study design and execution. In clinical trials, cognitive assessment is often limited to just the PASAT, administered in the context of the MSFC.[107] In most of the published trials, the cognitive outcome represents a secondary end point and therefore patient inclusion criteria and sample size calculations may not be appropriate to assess cognitive outcomes. Patients recruited in the trials have most often preserved cognitive functioning, which renders it difficult to find out any significant changes in the short time window of a trial, limited to 2 or 3 years. Finally, when a positive result is documented, it is not linked to any clinically meaningful outcome measure. Despite these limitations, trial results have often reported at least some positive effects of DMDs on the cognitive outcome of the patients. The bulk of the evidence in the field has been collected with IFNB. The most widely cited study formed part of the IFNB-1a pivotal trial.[108] A total of 276 RRMS patients were recruited into this double-blind, placebo-controlled, randomized North American study. IFNB-1a intramuscularly was compared with placebo. Cognitive data on 166 patients was recorded at the end of the 2-year study. An extensive neuropsychological battery was administered, in two 3-h sessions, at the beginning and end of the 2 years. A subset of measures relating to information processing and memory was administered every 26 weeks in a 90-min session. The extensive neuropsychological battery was divided into Set A, comprising information processing and memory; Set B, comprising visuospatial and executive abilities; and Set C, verbal abilities and attention span. On direct comparison, both Set A (information processing and memory, $p=0.036$) and Set B (visuospatial and executive abilities, $p=0.005$) demonstrated an IFNB-1a treatment advantage. After correction for baseline performance, only Set A demonstrated an IFNB-1a treatment advantage ($p = -0.011$). However, data from only 60% of the baseline cohort was included in the analysis of change over 2 years. In addition, it is not specified how statistically significant practice effects were taken into account.[108] Results from the BENEFIT trial with IFNB-1b in CIS subjects[109] and its extension study at 5 years have suggested that early treatment can be the most effective strategy to prevent or slow down CI and maintain intact cognitive functioning.[109] The cognitive assessment in the BENEFIT trial was the PASAT, given every 6 months as part of the MSFC.[107] Of the 292 CIS subjects who were recruited into the early treatment group, 235 continued assessment for 5 years. Of the 176 recruited into the delayed treatment group, 123 continued assessment for 5 years. At 3 years, the early treatment group performed better on the PASAT than the late treatment group. The same advantage was in evidence at 5 years ($p = 0.005$).[109]

The potential of DMDs has also been addressed in a few observational, postmarketing studies. Among those including at least 100 patients, the Cognitive Impairment in Multiple Sclerosis study[110] was a prospective cohort study of 459 early RRMS patients treated with IFNB-1a s.c. 22 or 44 mcg in everyday clinical practice. Results after 3 years suggest a better cognitive outcome in patients treated with the higher dose.

The positive impact of natalizumab on cognitive functioning has been also suggested in a few large, postmarketing studies of RRMS patients.[111–114] In one study the positive effect of natalizumab on cognition was associated with improved fatigue scores,[112] although the absence of an appropriate control group does not take into account possible practice or learning effects. Further studies with fingolimod and new oral agents are under way.

Again, the intrinsic limitations of nonrandomized, observational studies and methodological shortcomings of published studies do not allow us to draw any firm inference. There is therefore an increasing consensus that, from now on, therapeutic trials on MS as well as methodologically rigorous, observational studies should include systematically brief and reliable cognitive measures to shed some light on the impact of DMDs on the subject's cognitive outcome.

25.7.2 Pharmacologic Approaches: Symptomatic Drugs

In contrast to the DMD clinical trials, studies using symptomatic drugs in MS have used specified inclusion criteria relative to cognitive performance and have focused on improving performance in specific, already impaired cognitive domains. The three main classes of drugs tested so far are represented by symptomatic treatments for fatigue, psychostimulants, and drugs used for the therapy of Alzheimer's disease. Symptomatic drugs for fatigue include pilot trials with potassium channel blockers and amantadine, yielding negative or inconsistent results, and a pilot trial with modafinil that showed positive results.[115] Studies with the new agent fampridine are underway. Studies using psychostimulants to enhance a specific cognitive function have included L-amphetamine, ginkgo biloba, and methylphenidate, obtaining in general negative or inconsistent findings.[115] A pilot double-blind, placebo-controlled study involving 19 MS patients[116] showed that single 45-mg doses of L-amphetamine sulfate in MS were associated with improved performance on neuropsychological tests measuring the speed of information processing. A subsequent L-amphetamine larger, multicentric trial testing 151 clinically definite MS patients did not reveal any significant improvement on measures of processing speed or on the subjective ratings of cognition. However, a reanalysis of the trial data set[117] focusing on patients with memory dysfunction at baseline showed a significant improvement of performance on memory tasks, suggesting that L-amphetamine might improve memory performance, possibly by enhancing hippocampal function.

Acethylcholine esterasy inhibitors, currently used in Alzheimer's disease, have been tested also in other cognitive disorders. In MS, it is hypothesized that cholinergic deficits might derive from disruption of cholinergic pathways and impaired axonal transport of acetylcholine due to demyelination and axonal transection. In MS, Alzheimer's drugs used have included donepezil, rivastigmine, and memantine, again with negative or inconsistent findings.[115] The pivotal studies in the field are two donepezil studies.[118,119] The first one[117] was a randomized, double-blind, placebo-controlled, single-center clinical trial of 69 patients with MS who were selected for initial memory difficulties. Donepezil improved memory performance on the selective reminding test (the primary outcome measure) when compared with placebo. A subsequent double-blind, parallel group design, placebo-controlled, multicenter randomized trial was performed by the same group of researchers and included 120 patients.[117] After 24 weeks, there were no improvements in memory performance in the donepezil group neither on the selective reminding test nor on other objective and subjective cognitive outcomes. Post hoc analyses, however, revealed significant improvements in the subgroup of patients exhibiting more severe degrees of cognitive dysfunction.[118]

In conclusion, the evidence for symptomatic drugs is mainly negative. When a positive finding has been shown in a pilot trial, this has not been confirmed in further larger-scale trials. A systematic review of pharmacologic symptomatic treatment for MS-associated memory disorder concluded that there is no evidence of efficacy. This is because most available trials have a limited quality and future large-scale randomized trials with higher methodological quality are needed.[120]

25.8 COGNITIVE REHABILITATION

Although available evidence in this field is still limited, the results obtained with cognitive rehabilitation approaches in MS appear to be promising, particularly due to improved trial methodology in the studies published over the past few years. The domain most frequently addressed in cognitive rehabilitation studies in MS is learning and memory. A Cochrane review concluded that there is some evidence that cognitive training can improve memory span and working memory.[121]

How cognitive rehabilitation may work is still debated: different potential mechanisms of action have been postulated. Cognitive rehabilitation may work as a preventive strategy, enhancing plasticity processes in the brain, as suggested by fMRI studies that show changes in the pattern of cortical activation and/or connectivity changes in subjects undergoing rehabilitation programs.[122–126] It can also have a restorative function, reducing cognitive deficits through cognitive retraining as best shown in the models of traumatic brain injury and stroke.[127] Compensation and maintenance can be achieved, improving coping strategies to deal with existing cognitive problems (eg, internal strategies for memory problems)[128] and teaching compensatory strategies (eg, external aids: notebooks, calendars, lists, computer-assisted aids, etc.).[128] Finally, a mainly palliative role is hypothesized, particularly in the most advanced stages of CI, based on neuropsychological support to help patients understand and accept their cognitive problems and increase psychological well-being.[129]

The intervention format used in the literature on MS has been heterogeneous, but mainly based either on computerized programs or behavioral techniques.

25.8.1 Behavioral Interventions

Behavioral interventions to improve learning and memory use techniques borrowed from cognitive psychology. The most important ones are the Modified Story Memory Technique (mSMT), which uses context and imagery to improve learning of new materials and, consequently, ability to recall; the Self-generation effect, based on the observation that items generated by subjects are remembered better than items simply presented; the Spacing effect, based on the observation that learning trials spaced over time result in significantly better performance than if trials are massed (eg, consecutive trials); and the Testing effect, based on the observation that testing (eg, quiz) improves subsequent recall more than restudying the material again.[44] A few studies have been published using behavioral techniques in MS.[130–138] The reference study in the field is the MEMRehab trial, a randomized controlled trial to rehabilitate learning and memory in MS subjects using the mSMT. This study has provided class-1 evidence for memory improvement after cognitive rehabilitation.[139] The trial included 86 MS patients (76% females, mostly with an RR course) randomized to a specific intervention, based on the mSTM, or nonspecific intervention. There were 10 sessions (each 45–60 min), two times per week for 5 weeks. Furthermore, the treatment group was randomized to monthly booster session or placebo booster session. Assessment was repeated at baseline, immediately after treatment and after 6 months. In the group undergoing to the specific intervention, the trial showed a significant improvement in the performance on the California Verbal Memory Test (the primary outcome measure) persisting after 6 months ($p < 0.05$). The specific intervention group also improved significantly on everyday functioning, based on patient self-report on the Functional Assessment of Multiple Sclerosis scale. Moreover, caregivers' assessment of patient apathy and dysexecutive functioning significantly favored the experimental group. There was no significant difference, however, based on the booster sessions.[139] These positive neuropsychological findings were corroborated by findings of fMRI performed in a subgroup of 16 patients (eight in the treatment and eight in the placebo group).[124] After treatment, greater activation was observed in the treatment group during performance of a memory task: activation involved a widespread memory network that included frontal, parietal, precuneus, and parahippocampal regions as well as the cerebellum.[124]

25.8.2 Computerized Programs

Table 25.1 synthesizes the most relevant studies using computer-assisted programs in the MS literature.

The office-based Rehacom program has been translated into several languages and used in several studies. However, home-based programs appear to be more friendly and cost/effective, particularly in a population of young adults. Shatil et al.[140] explored unprompted adherence to a personalized, home-based, computerized cognitive training program (CogniFIT)

TABLE 25.1 Summary of Computerized Cognitive Rehabilitation Studies in Multiple Sclerosis

Author	Program	Cognitive Functions	Pros/Cons	Number of Patients	Main Results
Plohman[143]	Aixtent-CogniPlus (new version)	Attention, memory, executive functions, neglect, visuospatial skills	Available in 14 languages/Sold exclusively to professionals Not specific to MS	22	Improvement of performance for the domains of alertness, divided attention, and all subsets of selective attention. Treatment effects were stable for 9 weeks. Daily functioning improved.
Mendozzi[144]	RehaCom	Attention, memory, executive functions, visuospatial functions	Available in 18 languages Periodically updated/Sold exclusively to professionals Not specific to MS	40	Improvements in 7 out of 11 memory and attention tests.
Solari[145]	RehaCom			77	Improvement occurred in 45% of study patients and 43% of control patients (not significant).
Tesar[146]	RehaCom			19	Improvement in patients with mild cognitive impairment.
Mattioli[147]	RehaCom			20	Improvement in tests of attention, information processing, and executive functions.
Cerasa[148]	RehaCom			23	Improvement in attention abilities.
Hildebrandt[149]	VILAT-G 1.0	Memory, working memory	Home-based/No longer available	42	Improvement in verbal learning.
Brenk[150]	Brain Gym	Attention, memory, executive functions	Home-based/No longer available	27	Improvements in short-term and working memory, complex attention functions, and visuoconstructive performances, generally maintained for at least 6 months. Improvements also observed in depression and quality of life.
Vogt[151]	Brain Stim	Working memory	Specific to MS/ Not yet available	45	Improvements in 7 out of 11 memory and attention tests.
Shatil[140]	MindFit-CogniFit Personal Coach (new version)	Attention, language, executive function, memory, processing speed, spatial perception	Home-based Distributed by pharmaceutical company, free for MS patients, allows customization/ Not specific to MS	107	Improvement in general memory, visual working memory, verbal working memory. Increased naming speed, speed of information recall, focussed attention, and visuomotor vigilance.

TABLE 25.1 Summary of Computerized Cognitive Rehabilitation Studies in Multiple Sclerosis—cont'd

Author	Program	Cognitive Functions	Pros/Cons	Number of Patients	Main Results
Brissart[152]	Procog-SEP	Memory, language, visuospatial skills	Specific to MS/ Few sessions	24	Improvement in verbal memory, visuospatial memory, verbal fluency, and response to conflicting orders.
Amato[141]	APT	Attention	Home-based/Not yet available/Not specific to MS	102	Improvement on tests of complex attention (PASAT-2 and -3).
Bonavita[153]	RehaCom	Attention, memory, executive functions, visuospatial functions	Available in 18 languages Periodically updated/Sold exclusively to professionals/ Not specific to MS	18/14	Improvement on SDMT, PASAT-3, PASAT-2, SRT-D, and 10/36 SPART.

MS, Multiple Sclerosis; *SRT*, Selective Reminding Test; *SDMT*, Symbol Digit Modalities Test; *PASAT-3*, Paced Auditory Serial Addition Test-3 seconds; *PASAT-2*, Paced Auditory Serial Addition Test-2 seconds; *SRT-D*, Selective Reminding Test-Delayed; *SPART*, Spatial Recall Test; *APT*, Attention Process Training.

in MS patients. Participants were assigned to a training ($n = 59$) or a control group ($n = 48$). Those in the training group were instructed to train three times a week for 12 weeks. The control group received no training. All participants were evaluated with a Neuropsychological Examination (N-CPC) at baseline and at the end of the study. In the training group, 42 (71.2%) participants adhered to the training schedule and 22 (37.3%) completed the entire training regimen. In the control group, 24 (50.0%) participants agreed to be retested on the N-CPC. The training group showed a significant improvement over that shown by the control group in three memory-based cognitive abilities (general memory, visual working memory and verbal working memory). Post hoc exploration of data from the N-CPC showed that cognitive training was also associated with increased naming speed, speed of information recall, focused attention, and visuomotor vigilance.

A randomized double-blind trial in MS has dealt with a home-based computerized program for the rehabilitation of attention, based on the Attention Processing Training program.[141] This is represented by hierarchically organized tasks that exercise different components of attention as indicated by the clinical model of attention[142]: sustained, selective, divided, and alternating attention. The study included 88 RRMS patients randomized to the specific computerized program and a nonspecific computerized program. Both interventions were performed at home, twice a week for 3 months (each session of nearly 1 h). Assessment was repeated at baseline, immediately after the intervention and after 6 months. The results showed, in the group undergoing the specific training, a significant improvement on measures of complex attention/working memory (PASAT 2 and 3, $p <$ or $= 0.002$). There was no improvement on other cognitive measures. Moreover, measures of handicap, fatigue, and depression, as well as self-assessment of attention on a visual analog scale significantly improved in both groups ($p < 0.05$), suggesting an aspecific/placebo effect of the intervention.[141]

25.9 COGNITIVE REHABILITATION: STATE OF THE ART

Evidence in the field of cognitive rehabilitation in MS is limited and many published studies suffer from important methodological limitations. There are still critical, unresolved questions: who would benefit most, what level or profile of impairment is best addressed, what should be the target, what form should training take, which is the optimal length and spacing of the sessions?

However, the methodological quality of the studies has improved over the past few years and a few well-designed studies have shown positive effects of rehabilitation, particularly in the field of learning and memory. Interventions based on robust psychological constructs and specific rehabilitation techniques (targeting specific functions) seem to work better. The efficacy of these interventions is also supported by findings of fMRI studies. In the future, home-based, computerized programs are likely to play an increasing role, favoring feasibility, adherence, and customization of the patients.

While cognitive rehabilitation should remain a key focus in research, rehabilitation interventions for CI in MS are available and should be offered to MS patients, when appropriate.

References

1. Amato MP, Zipoli V. Clinical management of cognitive impairment in multiple sclerosis: a review of current evidence. *Int MS J*. 2003;10(3):72–83.
2. Amato MP, Zipoli V, Portaccio E. Cognitive changes in multiple sclerosis. *Expert Rev Neurother*. 2008;8(10):1585–1596.
3. Rao SM. Neuropsychological aspects of multiple sclerosis. In: Raine CS, McFarland HF, Tourtellotte WW, eds. *Multiple sclerosis: clinical and pathogenetic basis*. London: Chapman & Hall; 1997:357–362.
4. Benedict RH, Cookfair D, Gavett R, et al. Validity of the minimal assessment of cognitive function in multiple sclerosis. *J Int Neuropsychol Soc*. 2006;12:549–558.
5. Benedict R, Bobholz J. Multiple sclerosis. *Semin Neurol*. 2007;27(1):78–85.
6. Amato MP, Portaccio E, Goretti B, et al. Cognitive impairment in early stages of multiple sclerosis. *Neurol Sci*. 2010;31(suppl 2):S211–S214.
7. Benedict RH, Bruce JM, Dwyer MG, et al. Neocortical atrophy, third ventricular width, and cognitive dysfunction in multiple sclerosis. *Arch Neurol*. 2006;63(9):1301–1306.
8. Amato MP, Portaccio E, Goretti B, et al. Relevance of cognitive deterioration in early relapsing-remitting MS: a 3-year follow-up study. *Mult Scler*. 2010;16:1474–1482.
9. Zipoli V, Goretti B, Hakiki B, et al. Cognitive impairment predicts conversion to multiple sclerosis in clinically isolated syndromes. *Mult Scler*. 2010;16(1):62–67.
10. Deloire M, Ruet A, Hamel D, Bonnet M, Brochet B. Early cognitive impairment in multiple sclerosis predicts disability outcome several years later. *Mult Scler*. 2010;16(5):581–587.
11. Okuda DT, Mowry EM, Cree BA, et al. Asymptomatic spinal cord lesions predict disease progression in radiologically isolated syndrome. *Neurology*. February 22, 2011;76(8):686–692.
12. Lebrun C, Blanc F, Brassat D, Zephir H, de Seze J, CFSEP. Cognitive function in radiologically isolated syndrome. *Mult Scler*. 2010;16(8):919–925.
13. Amato MP, Hakiki B, Goretti B, et al. Association of MRI metrics and cognitive impairment in radiologically isolated syndromes. *Neurology*. January 31, 2012;78(5):309–314.
14. Hakiki B, Goretti B, Portaccio E, Zipoli V, Amato MP. Subclinical MS': follow-up of four cases. *Eur J Neurol*. 2008;15(8):858–861.
15. Huijbregts SC, Kalkers NF, de Sonneville LM, de Groot V, Polman CH. Cognitive impairment and decline in different MS subtypes. *J Neurol Sci*. June 15, 2006;245(1–2):187–194.
16. Potagas C, Giogkaraki E, Koutsis G, et al. Cognitive impairment in different MS subtypes and clinically isolated syndromes. *J Neurol Sci*. April 15, 2008;267(1–2):100–106.

17. Wachowius U, Talley M, Silver N, Heinze HJ, Sailer M. Cognitive impairment in primary and secondary progressive multiple sclerosis. *J Clin Exp Neuropsychol*. 2005;27(1):65–77.
18. Amato MP, Ponziani G, Pracucci G, Bracco L, Siracusa G, Amaducci L. Cognitive impairment in early-onset multiple sc lerosis. Pattern, predictors, and impact on everyday life in a 4-year follow-up. *Arch Neurol*. 1995;52(2):168–172.
19. Amato MP, Ponziani G, Siracusa G, Sorbi S. Cognitive dysfunction in early-onset multiple sclerosis: a reappraisal after 10 years. *Arch Neurol*. 2001;58(10):1602–1606.
20. Comi G, Filippi M, Martinelli V, et al. Brain MRI correlates of cognitive impairment in primary and secondary progressive multiple sclerosis. *J Neurol Sci*. 1995;132(2):222–227.
21. Camp SJ, Stevenson VL, Thompson AJ, et al. Cognitive function in primary progressive and transitional progressive multiple sclerosis: a controlled study with MRI correlates. *Brain*. 1999;122(Pt 7):1341–1348.
22. Camp SJ, Stevenson VL, Thompson AJ, et al. A longitudinal study of cognition in primary progressive multiple sclerosis. *Brain*. 2005;128(Pt 12):2891–2898.
23. Ruet A, Deloire M, Charré-Morin J, Hamel D, Brochet B. Cognitive impairment differs between primary progressive and relapsing-remitting MS. *Neurology*. April 16, 2013;80(16):1501–1508.
24. Foong J, Rozewicz L, Davie CA, Thompson AJ, Miller DH, Ron MA. Correlates of executive function in multiple sclerosis: the use of magnetic resonance spectroscopy as an index of focal pathology. *J Neuropsychiatry Clin Neurosci*. 1999;11(1):45–50.
25. Riccitelli G, Rocca MA, Pagani E, et al. Cognitive impairment in multiple sclerosis is associated to different patterns of gray matter atrophy according to clinical phenotype. *Hum Brain Mapp*. 2011;32(10):1535–1543.
26. Rocca MA, Valsasina P, Absinta M, et al. Default-mode network dysfunction and cognitive impairment in progressive MS. *Neurology*. 2010;74(16):1252–1259.
27. Amato MP, Zipoli V, Goretti B, et al. Benign multiple sclerosis: cognitive, psychological and social aspects in a clinical cohort. *J Neurol*. 2006;253(8):1054–1059.
28. Amato MP, Portaccio E. Truly benign multiple sclerosis is rare: let's stop fooling ourselves–yes. *Mult Scler*. 2012;18(1):13–14.
29. Portaccio E, Stromillo ML, Goretti B, et al. Neuropsychological and MRI measures predict short-term evolution in benign multiple sclerosis. *Neurology*. August 18, 2009;73(7):498–503.
30. Amato MP, Portaccio E, Stromillo ML, et al. Cognitive assessment and quantitative magnetic resonance metrics can help to identify benign multiple sclerosis. *Neurology*. August 26, 2008;71(9):632–638.
31. Rovaris M, Barkhof F, Calabrese M, et al. MRI features of benign multiple sclerosis: toward a new definition of this disease phenotype. *Neurology*. May 12, 2009;72(19):1693–1701.
32. Amato MP. Cognitive and psychosocial issues in pediatric multiple sclerosis: where we are and where we need to go. *Neuropediatrics*. 2012;43(4):174–175.
33. Julian L, Serafin D, Charvet L, et al. Cognitive impairment occurs in children and adolescents with multiple sclerosis: results from a United States network. *J Child Neurol*. 2013;28(1):102–107.
34. Amato MP, Goretti B, Ghezzi A, et al. Cognitive and psychosocial features of childhood and juvenile MS. *Neurology*. May 13, 2008;70(20):1891–1897.
35. Amato MP, Goretti B, Ghezzi A, et al. Cognitive and psychosocial features in childhood and juvenile MS: two-year follow-up. *Neurology*. September 28, 2010;75(13):1134–1140.
36. Chiaravalloti ND, DeLuca J. Cognitive impairment in multiple sclerosis. *Lancet Neurol*. 2008;7(12):1139–1151.
37. Macniven JA, Davis C, Ho MY. Stroop performance in multiple sclerosis: information processing, selective attention, or executive functioning. *J Int Neuropsychol Soc*. 2008;14:805–814.
38. Peyser JM, Rao SM, LaRocca NG, Kaplan E. Guidelines for neuropsychological research in multiple sclerosis. *Arch Neurol*. 1990;47:94–97.
39. Huijbregts SC, Kalkers NF, de Sonneville LM, de Groot V, Reuling IE, Polman CH. Differences in cognitive impairment of relapsing remitting, secondary, and primary progressive MS. *Neurology*. July 27, 2004;63(2):335–339.
40. Caine ED, Bamford KA, Schiffer RB, Shoulson I, Levy S. A controlled neuropsychological comparison of Huntington's disease and multiple sclerosis. *Arch Neurol*. 1986;43:249–254.
41. Rao SM. Neuropsychology of multiple sclerosis:a critical review. *J Clin Exp Neuropsychol*. 1986;8:503–542.
42. Rao SM, Leo GJ, St Aubin-Faubert P. On the nature of memory disturbance in multiple sclerosis. *J Clin Exp Neuropsychol*. 1989;11:699–712.
43. DeLuca J, Barbieri-Berger S, Johnson SK. The nature of memory impairments in multiple sclerosis: acquisition versus retrieval. *J Clin Exp Neuropsychol*. 1994;16:183–189.

44. DeLuca J, Gaudino EA, Diamond BJ. Acquisition and storage deficits in multiple sclerosis. *J Clin Exp Neuropsychol*. 1998;20:376–390.
45. Baddeley A. Working memory. *Science*. 1992;255:556–559.
46. Rao SM. *A Manual for the Brief Repeatable Battery of Neuropsychological Tests in Multiple Sclerosis*. Milwaukee, WI; 1990.
47. Langdon DW, Amato MP, Boringa J, et al. Recommendations for a brief international cognitive assessment for multiple sclerosis (BICAMS). *Mult Scler*. 2012;18(6):891–898.
48. Strober L, Chiaravalloti N, Moore N, DeLuca J. Unemployment in multiple sclerosis (MS): utility of the MS Functional Composite and cognitive testing. *Mult Scler*. May; 2013;20.
49. Parmenter BA, Zivadinov R, Kerenyi L, et al. Validity of the Wisconsin card sorting and delis-kaplan executive function system (DKEFS) sorting tests in multiple sclerosis. *J Clin Exp Neuropsychol*. 2007;29:215–223.
50. Arnett PA, Higginson CI, Randolph JJ. Depression in multiple sclerosis: relationship to planning ability. *J Int Neuropsychol Soc*. 2001;7:665–674.
51. Denney DR, Lynch SG, Parmenter BA. Horne Cognitive impairment in relapsing and primary progressive multiple sclerosis: mostly a matter of speed. *J Int Neuropsychol Soc*. 2004;10:948–956.
52. Channon S, Baker J, Robertson M. Working memory in clinical depression: an experimental study. *Psychol Med*. 1993;23:87–91.
53. Vleugels L, Lafosse C, van Nunen A, et al. Visuospatial impairment in multiple sclerosis patients diagnosed with with neuropsychological tasks. *Mult Scler*. 2000;6:241–254.
54. Bruce JM, Bruce AS, Arnett PA. Mild visual acuity disturbances are associated with performance on tests of complex visual attention in MS. *J Int Neuropsychol Soc*. 2007;13:544–548.
55. Banati M, Sandor J, Mike A, et al. Social cognition and theory of mind in patients with relapsing–remitting multiple sclerosis. *Eur J Neurol*. 2009;17:426–433.
56. Jehna M, Langkammer C, Wallner-Blazek M, et al. Cognitively preserved MS patients demonstrate functional differences in processing neutral and emotional faces. *Brain Imaging Behav*. 2011;5:241–251.
57. Kurtzke JF. Rating neurologic impairment in multiple sclerosis: an expanded disability status scale (EDSS). *Neurology*. 1983;33(11):1444–1452.
58. Rao SM, Leo GJ, Bernardin L, Unverzagt F. Cognitive dysfunction in multiple sclerosis. I. Frequency, patterns, and prediction. *Neurology*. 1991;41(5):685–691.
59. Lynch SG, Parmenter BA, Denney DR. The association between cognitive impairment and physical disability in multiple sclerosis. *Mult Scler*. 2005;11(4):469–476.
60. Rozewicz L, Langdon DW, Davie CA, Thompson AJ, Ron M. Resolution of left hemisphere cognitive dysfunction in multiple sclerosis with magnetic resonance correlates: a case report. *Cogn Neuropsychiatry*. 1996;1(1):17–26.
61. Foong J, Rozewicz L, Quaghebeur G, Thompson AJ, Miller DH, Ron MA. Neuropsychological deficits in multiple sclerosis after acute relapse. *J Neurol Neurosurg Psychiatry*. 1998;64(4):529–532.
62. Morrow SA, Jurgensen S, Forrestal F, Munchauer FE, Benedict RH. Effects of acute relapses on neuropsychological status in multiple sclerosis patients. *J Neurol*. 2011;258(9):1603–1608.
63. Bellmann-Strobl J, Wuerfel J, Aktas O, et al. Poor PASAT performance correlates with MRI contrast enhancement in multiple sclerosis. *Neurology*. November 17, 2009;73(20):1624–1627.
64. Pardini M, Uccelli A, Grafman J, Yaldizli Ö, Mancardi G, Roccatagliata L. Isolated cognitive relapses in multiple sclerosis. *J Neurol Neurosurg Psychiatry*. 2014;85(9):1035–1037.
65. Centonze D, Muzio L, Rossi S, et al. Inflammation triggers synaptic alteration and degeneration in experimental autoimmune encephalomyelitis. *J Neurosci*. 2009;29:3442–3452.
66. Amato MP, Portaccio E. Management options in multiple sclerosis-associated fatigue. *Expert Opin Pharmacother*. 2012;13(2):207–216.
67. Roelcke U, Kappos L, Lechner-Scott J, et al. Reduced glucose metabolism in the frontal cortex and basal ganglia of multiple sclerosis patients with fatigue: a 18F fluorodeoxyglucose positron emission tomography study. *Neurology*. 1997;48(6):1566–1571.
68. Filippi M. Predictive value of MRI findings in multiple sclerosis. *Lancet Neurol*. 2002;1(1):9.
69. Krupp LB, LaRocca NG, Muir-Nash J. The fatigue severity scale. Application to patients with multiple sclerosis and systemic lupus erythematosus. *Arch Neurol*. 1989;46(10):1121–1123.
70. Beatty WW, Goretti B, Siracusa G. Changes in neuropsychological test performance over the workday in multiple sclerosis. *Clin Neuropsychol*. 2003;17(4):551–560.
71. Heesen C, Schulz KH, Fiehler J. Correlates of cognitive dysfunction in multiple sclerosis. *Brain Behav Immun*. 2010;24(7):1148–1155.

72. Krupp LB, Elkins LE. Fatigue and declines in cognitive functioning in multiple sclerosis. *Neurology*. October 10, 2000;55(7):934–939.
73. Rosti E, Hämäläinen P, Koivisto K, Hokkanen L. The PASAT performance among patients with multiple sclerosis: analyses of responding patterns using different scoring methods. *Mult Scler*. 2006;12(5):586–593.
74. Arnett PA, Barwick FH, Beeney JE. Depression in multiple sclerosis: review and theoretical proposal. *J Int Neuropsychol Soc*. 2008;14(5):691–724.
75. Julian LJ, Arnett PA. Relationships among anxiety, depression, and executive functioning in multiple sclerosis. *Clin Neuropsychol*. 2009;23(5):794–804.
76. Brunner R, Schaefer D, Hess K, Parzer P, Resch F, Schwab S. Effect of corticosteroids on short-term and long-term memory. *Neurology*. Janauary 25, 2005;64(2):335–337.
77. Heesen C, Gold SM, Huitinga I, Reul JM. Stress and hypothalamic-pituitary-adrenal axis function in experimental autoimmune encephalomyelitis and multiple sclerosis - a review. *Psychoneuroendocrinology*. 2007;32(6):604–618.
78. Heesen C, Gold SM, Raji A, Wiedemann K, Schulz KH. Cognitive impairment correlates with hypothalamo-pituitary-adrenal axis dysregulation in multiple sclerosis. *Psychoneuroendocrinology*. 2002;27(4):505–517.
79. Patten SB, Francis G, Metz LM, Lopez-Bresnahan M, Chang P, Curtin F. The relationship between depression and interferon beta-1a therapy in patients with multiple sclerosis. *Mult Scler*. 2005;11(2):175–181.
80. Benedict RH, Zivadinov R. Risk factors for and management of cognitive dysfunction in multiple sclerosis. *Nat Rev Neurol*. May 10, 2011;7(6):332–342.
81. Filippi M, Rocca MA. MRI and cognition in multiple sclerosis. *Neurol Sci*. 2010;31(suppl 2):S231–S234.
82. Giorgio A, De Stefano N. Cognition in multiple sclerosis: relevance of lesions, brain atrophy and proton MR spectroscopy. *Neurol Sci*. 2010;31(suppl 2):S245–S248.
83. Calabrese M, Agosta F, Rinaldi F, et al. Cortical lesions and atrophy associated with cognitive impairment in relapsing-remitting multiple sclerosis. *Arch Neurol*. 2009;66(9):1144–1150.
84. Filippi M, Agosta F. Magnetization transfer MRI in multiple sclerosis. *J Neuroimaging*. 2007;17(suppl 1):22S–26S.
85. Rovaris M, Agosta F, Pagani E, Filippi M. Diffusion tensor MR imaging. *Neuroimaging Clin N Am*. 2009;19(1):37–43.
86. Sajja BR, Wolinsky JS, Narayana PA. Proton magnetic resonance spectroscopy in multiple sclerosis. *Neuroimaging Clin N Am*. 2009;19(1):45–58.
87. Audoin B, Au Duong MV, Ranjeva JP, et al. Magnetic resonance study of the influence of tissue damage and cortical reorganization on PASAT performance at the earliest stage of multiple sclerosis. *Hum Brain Mapp*. 2005;24:216–228.
88. Mainero C, Caramia F, Pozzilli C, et al. MRI evidence of brain reorganization during attention and memory tasks in multiple sclerosis. *Neuroimage*. 2004;21:858–867.
89. Rocca MA, Valsasina P, Ceccarelli A, et al. Structural and functional MRI correlates of Stroop control in benign MS. *Hum Brain Mapp*. 2009;30:276–290.
90. Penner IK, Rausch M, Kappos L, Opwis K, Radü EW. Analysis of impairment related functional architecture in MS patients during performance of different attention tasks. *J Neurol*. 2003;250:461–472.
91. Huolman S, Hämäläinen P, Vorobyev V, et al. The effects of rivastigmine on processing speed and brain activation in patients with multiple sclerosis and subjective cognitive fatigue. *Mult Scler*. 2011;17:1351–1361.
92. Stern Y. What is cognitive reserve? Theory and research application of the reserve concept. *J Int Neuropsychol Soc*. 2002;8(3):448–460.
93. Sumowski JF, Chiaravalloti N, Leavitt VM, Deluca J. Cognitive reserve in secondary progressive multiple sclerosis. *Mult Scler*. 2012;18(10):1454–1458.
94. Amato MP, Razzolini L, Goretti B, et al. Cognitive reserve and cortical atrophy in multiple sclerosis: a longitudinal study. *Neurology*. May 7, 2013;80(19):1728–1733.
95. Benedict RH, Cox D, Thompson LL, Foley F, Weinstock-Guttman B, Munschauer F. Reliable screening for neuropsychological impairment in multiple sclerosis. *Mult Scler*. 2004;10(6):675–678.
96. Sonder JM, Bosma LV, van der Linden FA, Knol DL, Polman CH, Uitdehaag BM. Proxy measurements in multiple sclerosis: agreement on different patient-reported outcome scales. *Mult Scler*. 2012;18(2):196–201.
97. Rudick R, Antel J, Confavreux C, et al. Recommendations from the national multiple sclerosis society clinical outcomes assessment task force. *Ann Neurol*. 1997;42:379–382.
98. Drake AS, Weinstock-Guttman B, Morrow SA, Hojnacki D, Munschauer FE, Benedict RH. Psychometrics and normative data for the Multiple Sclerosis Functional Composite: replacing the PASAT with the Symbol Digit Modalities Test. *Mult Scler*. 2010;16(2):228–237.
99. Benedict RH, Fischer JS, Archibald CJ, et al. Minimal neuropsychological assessment of MS patients: a consensus approach. *Clin Neuropsychol*. 2002;16:381–397.

100. Strober L, Englert J, Munschauer F, Weinstock-Guttman B, Rao S, Benedict RH. Sensitivity of conventional memory tests in multiple sclerosis: comparing the Rao Brief Repeatable Neuropsychological Battery and the Minimal Assessment of Cognitive Function in MS. *Mult Scler*. 2009;15:1077–1084.
101. Benedict RH, Amato MP, Boringa J, et al. Brief International Cognitive Assessment for MS (BICAMS): international standards for validation. *BMCNeurol*. 2012:16;12:55.
102. Goretti B, Niccolai C, Hakiki B, et al. The brief international cognitive assessment for multiple sclerosis (BICAMS): normative values with gender, age and education corrections in the Italian population. *BMC Neurol*. September 10, 2014;14(1):171.
103. Utz KS, Hankeln TM, Jung L, et al. Visual search as a tool for a quick and reliable assessment of cognitive functions in patients with multiple sclerosis. *PLoS One*. November 25, 2013;8(11).
104. Lapshin H, Lanctôt KL, O'Connor P, Feinstein A. Assessing the validity of a computer-generated cognitive screening instrument for patients with multiple sclerosis. *Mult Scler*. December 2013;19(14):1905–1912.
105. Ruet A, Deloire MS, Charré-Morin J, Hamel D, Brochet B. A new computerized cognitive test for the detection of information processing speed impairment in multiple sclerosis. *Mult Scler*. October 2013;19(12):1665–1672.
106. Gold R, Wolinsky JS, Amato MP, Comi G. Evolving expectations around early management of multiple sclerosis. *Ther Adv Neurol Disord*. 2010;3(6):351–367.
107. Solari A, Radice D, Manneschi L, Motti L, Montanari E. The multiple sclerosis functional composite: different practice effects in the three test components. *J Neurol Sci*. January 15, 2005;228(1):71–74.
108. Fischer JS, Priore RL, Jacobs LD, et al. Neuropsychological effects of interferon beta-1a in relapsing multiple sclerosis. Multiple Sclerosis Collaborative Research Group. *Ann Neurol*. 2000;48(6):885–892.
109. Penner IK, Stemper B, Calabrese P, et al. Effects of interferon beta-1b on cognitive performance in patients with a first event suggestive of multiple sclerosis. *Mult Scler*. 2012;18(10):1466–1471. Epub 2012 April 4.
110. Patti F, Amato MP, Bastianello S, et al. Effects of immunomodulatory treatment with subcutaneous interferon beta-1a on cognitive decline in mildly disabled patients with relapsing-remitting multiple sclerosis. *Mult Scler*. January 2010;16(1):68–77.
111. Morrow SA, O'Connor PW, Polman CH, et al. Evaluation of the symbol digit modalities test (SDMT) and MS neuropsychological screening questionnaire (MSNQ) in natalizumab-treated MS patients over 48 weeks. *Mult Scler*. 2010;16(11):1385–1392.
112. Iaffaldano P, Viterbo RG, Paolicelli D, et al. Impact of natalizumab on cognitive performances and fatigue in relapsing multiple sclerosis:a prospective, open-label, two years observational study. *PLoS One*. 2012;7(4):e35843.
113. Portaccio E, Stromillo ML, Goretti B, et al. Natalizumab may reduce cognitive changes and brain atrophy rate in relapsing-remitting multiple sclerosis–a prospective, non-randomized pilot study. *Eur J Neurol*. June 2013;20(6):986–990.
114. Wilken J, Kane RL, Sullivan CL, et al. Changes in fatigue and cognition in patients with relapsing forms of multiple sclerosis treated with natalizumab: the ENER-G study. *Int J MS Care*. 2013;15(3):120–128.
115. Amato MP, Langdon D, Montalban X, et al. Treatment of cognitive impairment in multiple sclerosis: position paper. *J Neurol*. June 2013;260(6):1452–1468.
116. Benedict RH, Munschauer F, Zarevics P, et al. Effects of l-amphetamine sulfate on cognitive function in multiple sclerosis patients. *J Neurol*. June 2008;255(6):848–852.
117. Morrow SA, Kaushik T, Zarevics P, et al. The effects of L-amphetamine sulfate on cognition in MS patients: results of a randomized controlled trial. *J Neurol*. July 2009;256(7):1095–1102.
118. Krupp LB, Christodoulou C, Melville P, Scherl WF, MacAllister WS, Elkins LE. Donepezil improved memory in multiple sclerosis in a randomized clinical trial. *Neurology*. November 9, 2004;63(9):1579–1585.
119. Krupp LB, Christodoulou C, Melville P, et al. Multicenter randomized clinical trial of donepezil for memoryimpairment in multiple sclerosis. *Neurology*. November 9, 2011;63(9):1579–1585.
120. He D, Zhang Y, Dong S, Wang D, Gao X, Zhou H. Pharmacological treatment for memory disorder in multiple sclerosis. *Cochrane Database Syst Rev*. December 2013;12.
121. Rosti-Otajärvi EM, Hämäläinen PI. Neuropsychological rehabilitation for multiple sclerosis. *Cochrane Database Syst Rev*. 2014;11:2.
122. Penner IK, Opwis K, Kappos L. Relation between functional brain imaging, cognitive impairment and cognitive rehabilitation in patients with multiple sclerosis. *J Neurol*. 2007;254(suppl 2):II53–II57.
123. Sastre-Garriga J, Alonso J, Renom M, et al. A functional magnetic resonance proof of concept pilot trial of cognitive rehabilitation in multiple sclerosis. *Mult Scler*. 2011;17(4):457–467.

124. Chiaravalloti ND, Wylie G, Leavitt V, Deluca J. Increased cerebral activation after behavioraltreatment for memory deficits in MS. *J Neurol*. 2012;259(7):1337–1346.
125. Filippi M, Riccitelli G, Mattioli F, et al. Multiple sclerosis: effects of cognitive rehabilitation on structural and functional MR imaging measures–anexplorative study. *Radiology*. 2012;262(3):932–940.
126. Leavitt VM, Wylie GR, Girgis PA, DeLuca J, Chiaravalloti ND. Increased functional connectivity within memory networks following memory rehabilitation in multiple sclerosis. *Brain Imaging Behav*. 2014;8(3):394–402.
127. Cicerone KD, Dahlberg C, Kalmar K, et al. Evidence-based cognitive rehabilitation: recommendations for clinical practice. *Arch Phys Med Rehabil*. December 2000;81(12):1596–1615.
128. Wilson BA, Rous R, Sopena S. The current practice of neuropsychological rehabilitation in the United Kingdom. *Appl Neuropsychol*. 2008;15(4):229–240.
129. Mateer CA, Sira CS, O'Connell ME. Putting Humpty Dumpty together again: theimportance of integrating cognitive and emotional interventions. *J Head Trauma Rehabil*. 2005;20(1):62–75.
130. Jønsson A, Korfitzen EM, Heltberg A, Ravnborg MH, Byskov- Ottosen E. Effects of neuropsychological treatment in patients with multiple sclerosis. *Acta Neurol Scand*. 1993;88:394–400.
131. Allen DN, Goldstein G, Heyman RA, Rondinelli T. Teaching memory strategies to persons with multiple sclerosis. *J Rehabil Res Dev*. 1998;35:405–410.
132. Chiaravalloti ND, DeLuca J. Self-generation as a means of maximizing learning in multiple sclerosis: an application of the generation effect. *Arch Phys Med Rehabil*. 2002;83:1070–1079.
133. Chiaravalloti ND, DeLuca J, Moore NB, Ricker JH. Treating learning impairments improves memory performance in multiple sclerosis: a randomized clinical trial. *Mult Scler*. 2005;11:58–68.
134. Basso MR, Lowery N, Ghormley C, Combs D, Johnson J. Self-generated learning in people with multiple sclerosis. *J Int Neuropsychol Soc*. 2006;12:640–648.
135. Goverover Y, Chiaravalloti N, DeLuca J. Self-generation to improve learning and memory of functional activities in persons with multiple sclerosis: meal preparation and managing finances. *Arch Phys Med Rehabil*. 2008;89:1514–1521.
136. Goverover Y, Hillary FG, Chiaravalloti N, Arango-Lasprilla JC, DeLuca J. A functional application of the spacing effect to improve learning and memory in persons with multiple sclerosis. *J Clin Exp Neuropsychol*. 2009;31:513–522.
137. Sumowski JF, Wylie GR, Chiaravalloti ND, DeLuca J. Intellectual enrichment lessens the effect of brain atrophy on learning and memory in MS. *Neurology*. 2010;74:1942–1945.
138. Goverover Y, Basso MR, Wood H, Chiaravalloti N, DeLuca J. Examining the benefits of combining two learning strategies on recall of functional information in persons with multiple sclerosis. *Mult Scler*. 2011;17(12):1488–1497.
139. Chiaravalloti ND, Moore NB, Nikelshpur OM, DeLuca J. An RCT to treat learning impairment in multiple sclerosis: the MEMREHAB trial. *Neurology*. December 10, 2013;81(24):2066–2072.
140. Shatil E, Metzer A, Horvitz O, Miller A. Home-based personalized cognitive training in MS patients: a study of adherence and cognitive performance. *NeuroRehabilitation*. 2010;26(2):143–153.
141. Amato MP, Goretti B, Viterbo RG, et al. Computer-assisted rehabilitation of attention in patients withmultiple sclerosis: results of a randomized, double-blind trial. *Mult Scler*. January 2014;20(1):91–98.
142. Sohlberg MM, Mateer CA. Effectiveness of an attentiontraining program. *J Clin Exp Neuropsychol*. 1987;9:117–130.
143. Plohmann AM, Kappos L, Ammann W, et al. Computer assisted retraining of attentional impairments in patients with multiple sclerosis. *J Neurol Neurosurg Psychiatry*. 1998;64:455–462.
144. Mendozzi L, Pugnetti L, Motta A, Barbieri E, Gambini A, Cazzullo CL. Computer-assisted memory retraining of patients with multiple sclerosis. *Ital J Neurol Sci*. 1998;19:S431–S438.
145. Solari A, Motta A, Mendozzi L, et al. CRIMS Trial. Computer-aided retraining of memory and attention in people with multiple sclerosis: a randomized, double-blind controlled trial. *J Neurol Sci*. July 15, 2004;222(1–2):99–104. Erratum in: *J Neurol Sci*. September 15, 2004:224(1–2).
146. Tesar N, Bandion K, Baumhackl U. Efficacy of a neuropsychological training programme for patients with multiple sclerosis – a randomised controlled trial. *Wien Klin Wochenschr*. November 2005;117(21–22):747–754.
147. Mattioli F, Stampatori C, Zanotti D, Parrinello G, Capra R. Efficacy and specificity of intensive cognitive rehabilitation of attention and executive functions in multiple sclerosis. *J Neurol Sci*. January 15, 2010;288(1–2):101–105.
148. Cerasa A, Gioia MC, Valentino P, et al. Computer-assisted cognitive rehabilitation of attention deficits for multiple sclerosis: arandomized trial with fMRI correlates. *Neurorehabil Neural Repair*. May 2013;27(4):284–295.
149. Hildebrandt H, Lanz M, Hahn HK, et al. Cognitive training in MS: effects and relation to brain atrophy. *Restor Neurol Neurosci*. 2007;25(1):33–43.

150. Brenk A, Laun K, Haase CG. Short-term cognitive training improves mental efficiency and mood in patients with multiple sclerosis. *Eur Neurol*. 2008;60(6):304–309.
151. Vogt A, Kappos L, Calabrese P, et al. Working memory training in patients with multiple sclerosis - comparison of two different training schedules. *Restor Neurol Neurosci*. 2009;27(3):225–235.
152. Brissart H, Leroy M, Debouverie M. Cognitive rehabilitation in multiple sclerosis: preliminary results and presentation of a new program, PROCOG-SEP. *Rev Neurol Paris*. 2010;166:406–411.
153. Bonavita S, Sacco R, Della Corte M, et al. Computer-aided cognitive rehabilitation improves cognitive performances and induces brain functional connectivity changes in relapsing remitting multiple sclerosis patients: an exploratory study. *J Neurol*. October 2014;12.

SECTION VII

NOVEL & EMERGING STRATEGIES

CHAPTER

26

Personalized Medicine and Theranostics: Applications to Multiple Sclerosis

T. Paperna, E. Staun-Ram, N. Avidan, I. Lejbkowicz
Technion-Israel Institute of Technology, Haifa, Israel

A. Miller
Technion-Israel Institute of Technology, Haifa, Israel; Carmel Medical Center, Haifa, Israel

OUTLINE

Translational Neuroimmunology in Multiple Sclerosis
http://dx.doi.org/10.1016/B978-0-12-801914-6.00026-X

Care of the patient with multiple sclerosis (MS) has made considerable progress in the 2000s, including the increased accuracy in diagnosis of the disease, and the availability of new therapies for patients with relapsing MS (PwRMS). Currently, the key challenge in the application of optimized care for the patient with MS (PwMS) is the implementation of the personalized medicine-based approach, namely, choosing the right drug at the right dose, for the right patient at the right time and cost. This approach of tailored therapeutics based on diagnostic tests, called theranostics, consists of some cardinal conceptual changes in the mode of care in the field of MS: shifting paradigms from medical practice based on trial and error to medical practice based on predictive measures and biomarkers, moving from reactive to proactive medical care, and moving from treating the disease to treating the patient.

26.1 PERSONALIZED MEDICINE IN MULTIPLE SCLEROSIS: THE NEED

Approved therapies include the injectable interferon β (IFN-β) therapies and glatiramer acetate (GA), the oral therapies dimethyl fumarate (DMF), teriflunomide, fingolimod alemtuzumab for active disease, natalizumab, and other immunosuppressants such as mitoxantrone for refractory MS. Several potential therapies are currently being evaluated in clinical trials and will increase the armamentarium for MS in the future. The variety of therapy options is a benefit and a challenge for the PwRMS and the treating physician.

As for most drugs, response to MS therapies varies from individual to individual. Response variability has two aspects: efficiency and safety. Some PwRMS may show a clear beneficial effect for a specific therapy while for others the same therapy may fail to attenuate disease activity and disability accumulation. MS disease heterogeneity is well recognized and may reflect disease pathomechanisms in different patients, and may explain some of the therapy failures:

Therapies targeting specific pathological pathways may be effective solely for individuals with a specific MS subtype. Currently, as neither the underlying pathologies responsible for MS disease subtypes are fully understood, nor are criteria available to distinguish disease subtypes beyond the basic clinical definition of relapsing and progressive MS[1]; the targeting of MS subtypes with specific drug(s) (monotherapy or combination therapy) for the specific patient is not yet possible. Variability in efficiency of response is suggested to be caused by inherent differences between individuals that are determined by genetic and environmental exposures.

Adverse events (AEs) are experienced by some PwRMS and are a serious concern, as they can be severe and even fatal. AEs are usually drug-specific and not related to disease subtype. Three MS therapies, natalizumab, alemtuzumab, and mitoxantrone, were issued a black-box warning by the Food and Drug Administration (FDA)[2–4]: mitoxantrone for cardiotoxicity and risk of secondary leukemia; natalizumab for the risk of progressive multifocal leukoencephalopathy (PML); and alemtuzumab for risk of serious and fatal autoimmune reactions, infusion reactions, and malignancy. However, other MS therapies also have severe AEs associated with their use. For example, cardiovascular AEs can occur with fingolimod therapy, and elevation of liver enzymes is a well-recognized effect of the IFN-β therapies.[5] Beyond the devastating health complications of AEs for PwRMS, AEs present a significant economic burden to the health-care system as they incur additional hospitalizations and medical care.

Much effort over the years has been devoted to the identification of predictive and prognostic markers, with high sensitivity and specificity, to aid and guide MS diagnosis and therapy prescription. Ideally such markers should be detected in easily accessible bodily fluids, such as blood and its derived products—serum and plasma, urine, and saliva. Recently, the rising recognition of the role of the microbiome in disease suggests stool samples may be a source for biomarkers. Factors that determine the potential of a biomarker for translation from the basic science bench to the clinical lab include the stability of the biomarker and reproducibility across measurements and between laboratories, the level of expertise required for the bioassay, the invasiveness of the test, and last but not least, the cost of the assay.

In this chapter we review the use of biomarkers in MS diagnosis and describe recent progress on biomarkers for response to therapy for each of the currently approved MS therapies. In addition, we discuss how participatory medicine can contribute to the patient's health.

26.2 BIOMARKERS IN USE FOR THE DIAGNOSIS OF MULTIPLE SCLEROSIS

The set of criteria for the definite diagnosis of MS has evolved over time to include supportive laboratory and MRI-based data.[6–8] The criteria are continuously evaluated and modified, yet they are primarily based on clinical evidence and the expertise of the neurologist. The importance of early diagnosis is well recognized since early treatments can prevent or halt disability accumulation.[9] Prognosis is important to recognize very active disease (malignant MS),[10] which is likely to benefit from a more aggressive therapy approach than the standard first-line therapies. Prognosis is also a critical issue in considering the treatment of the patient with a clinically isolated syndrome (CIS), since approximately two-thirds progress to full-blown MS[11] and will benefit from early therapy,[9] while a third will experience only a single symptomatic episode and thus may avoid the MS therapies and risk for associated side effects.

26.2.1 Oligoclonal Bands and MRI

Since several variants of MS and central nervous system (CNS) demyelinating syndromes are known and treated differently, accurate diagnosis will allow tailoring of treatment to the individual patient.[12] The revised McDonald diagnostic criteria[8] integrate MRI assessments such as dissemination in space and time of gadolinium-enhancing T1 lesions and T2-hyperintense lesions with clinical observations, either monofocal or multifocal, as well as cerebrospinal fluid (CSF) findings. Additional MRI assessments such as T1-hypointense lesions (black holes) and whole brain, gray matter, and spinal cord atrophy seems to be prognostic biomarkers for disability and disease progression,[13] but have not yet been incorporated in the diagnostic criteria. A thorough description of the role of MRI in MS can be found in Chapter 13: Surrogate Markers in Multiple Sclerosis: The Role of Magnetic Resonance Imaging

IgG oligoclonal bands (OCB) in the CSF are present in ~60–70% of patients with an initial event of demyelination, defined as a CIS, and in up to 90% of clinically defined MS.[14–17] However, OCBs are not specific for MS and can be seen in other diseases such as CNS infections and disorders. In the 2010 revision of the McDonald criteria for MS diagnosis, positive IgG OCB is considered supportive evidence, but not a requirement for MS, and can be used instead of one of the two MRI criteria to ascertain diagnosis for primary-progressive MS (PPMS).[8]

26.2.2 Aquaporin-4 Autoantibodies for Differential Diagnosis in Multiple Sclerosis

Neuromyelitis optica (NMO), or Devic's disease, was originally considered a clinical subtype of MS, characterized by optic neuritis, myelitis, or both, and limited to signs or lesions in the spinal cord or optic nerves. The ability to clearly distinguish between MS and NMO is highly important as the treatments effective for these diseases are distinct.[18–22] Differential diagnosis of MS versus NMO is currently aided by testing for the serum IgG autoantibody targeting aquaporin-4 (AQP4), the most abundant water channel in the CNS, present almost exclusively in NMO patients with NMO, while absent in PwMS.[23,24] Results from 49 studies indicate a specificity of >96% for NMO versus MS.[22] NMO-IgG/AQP4-Ab has also been found in patients with Asian opticospinal MS[24–26] and in other neurological diseases.[22,27,28] Standardization and development of easy-to-use assays for detecting NMO IgG/AQP4-Ab is still warranted to allow for broader clinical use.

26.3 BIOMARKERS FOR RESPONSE TO THERAPY AND FOR TREATMENT SAFETY

Variability in response to therapy in MS is considerable. Because of the progressive nature of the disease, ineffective treatment will incur accumulation of disabilities, some of which may be irreversible. Pharmacogenomic research seeks to identify genetic factors and their products—RNA and protein—that affect drug efficacy and safety. Pharmacogenomic studies typically involve comparisons between extreme drug response phenotypes—a group of good responders and a group of poor or nonresponders, or with adverse responders—to identify predictor biomarkers for response. Most studies have been conducted in the MS clinic setting, presenting a limitation in distinguishing between associations with the natural course of the disease, or with the drug response. Intriguingly, although AE are probably the most serious issue in therapy decisions, only a minority of biomarker studies have directly addressed them. Candidate biomarkers were often selected for their involvement in the drug's mechanism of action, or in absorption and metabolism of the drug, or because they mark a disease subset. The definition of the response by the researcher has also been a source of variability across studies.

26.3.1 Interferon-β

IFN-β1b was the first therapy ever to be approved for MS therapy, and various IFN-β formulations have been in use for MS for over two decades. IFN-β is also the most studied drug in pharmacogenomic studies of MS. Up to 50% of PwRMS continue to experience relapses and disability progression when treated with IFN-β.[29] The pegylated form of IFN-β1a was only recently approved by the FDA (August 2014); no pharmacogenomic data are yet available for it.

Protein-Based Markers in Interferon-β Therapy

Protein biomarkers for prediction of response to IFN-β have occupied a large body of research. A comprehensive but not all-inclusive list is presented in Table 26.1. The development of neutralizing antibodies (NAbs) against MS drugs, most prevalently seen with IFN-β therapies, has a major impact on clinical response. NAbs develop in up to 45% of patients treated with IFN-β, usually within 6–18 months of therapy. Occurrence of NAbs is associated with reduced or loss of clinical efficacy, resulting in increased relapse rate and Expanded Disability Status Scale score progression.[30–33] The immunogenicity of IFN-β seems to depend upon chemical formulation, route of administration, and frequency of dosing, with intramuscular IFN-β1a less immunogenic than the subcutaneous IFN-β1a or IFN-β1b.[30] However, NAbs cross-react among different IFN-β formulations; thus switching to a different formulation is not beneficial for NAb-positive patients.[31,32,34] In light of the relative late appearance of NAbs after treatment initiation, NAbs should be monitored regularly and alternative therapies considered in patients with persistently high titers.[30] Noticeably, NAb status is not strictly correlated with response.[35–38] Since some NAb-positive patients are good responders, therapy decisions should be based upon clinical and radiological features, together with NAb status. Guidelines for monitoring NAbs and treatment decisions in patients treated with IFN-β were published by expert panels in 2010 and updated in 2014.[30,38]

Cytokine profiles before and during therapy have been suggested as potential biomarkers for monitoring clinical response to drugs.[39] Because IFN-β has been considered to be effective mainly in Th1-driven disease, high levels of Th17-related cytokines, such as IL17F and IL17A, were suggested as markers for disease subtypes less likely to respond to IFN-β. The findings, however, have been conflicting: In one study, high pretreatment IL17F serum levels were observed in a subset of nonresponders (6/14) compared to responders (12), suggesting that high IL17F may predict clinical nonresponsiveness to IFN-β.[40] However, two more powered, follow-up studies involving larger cohorts (54 good responders vs 64 poor responders, and 124 responders vs 27 nonresponders) were not able to replicate these findings, neither in pretreatment serum nor in posttreatment serum, and a more recent study reported even higher levels of IL17A in good responders.[41–43] IFN-β responder groups also did not differ in levels of IL17A and IL17F secretion

TABLE 26.1 Protein Biomarkers for Drug Response Prediction

Drug	Biomarker	Biological Sample	Main Findings	References
IFN-β				
IFN-β Antibodies	NAbs	Serum	NAbs against IFN-β develop in ~45% of IFN-β-treated patients after 6–18 months of therapy, and are associated with reduced or loss of response to therapy.	30–33,35,36,38,155
Cytokines	IL17	Serum	Proinflammatory cytokine reduced in IFN-β GR compared to PR, therefore a possible early marker of therapeutic effect. Conflicting results.	Supportive: 40,43, 156–158 Negative: 41,42,44,159
		CSF	IL17 CSF level is increased in PR after 6 months of therapy compared to GR.	39
	IL23	Serum	Proinflammatory cytokine reduced in response to IFN-β therapy. A possible early marker of therapeutic effect, in correlation with reduced radiology disease activity.	Supportive: 156,157
			Conflicting results.	Negative: 159
	IL12	Serum	Proinflammatory cytokine decreased after treatment with IFN-β, in correlation with reduced radiology disease activity.	156
	IL10	Serum	Antiinflammatory cytokine inversely correlated with MS severity and response to therapy. Predictor of responsiveness: IL10 is lower in GR compared to PR, and levels increase with treatment. Increased IL10/IFN-γ ratio after 3 months inversely correlates with number of relapses after 12 months.	39,160,161
		CSF	Significantly increased concentrations of CSF IL-10 after 2 years of IFN-β treatment correlates with a favorable therapeutic response.	162
	IL4	CSF	Th2 cytokine, increased after 6 months of therapy, inversely correlating with disability score after 12 months. IL4 and IL4/IFN-γ ratio are higher at baseline in PR compared to GR.	39
	IL7	Serum	Th1 cytokine, in which high pretreatment level, particularly when paired with low level of IL17F, is associated with good response to IFN-β.	46

Adhesion molecules	VLA-4, sVCAM-1	Serum/blood cells	VLA-4 is downregulated on T cells and VCAM-1 is upregulated in serum within 3–6 months of therapy in long-term (4 years) good responders. Increased serum sVCAM-I after 1 year of IFN-β treatment is associated with a favorable treatment response. Elevated sVCAM serum levels from 1 month inversely correlate with lower MRI activity.	39,48,49,51,52,163
	sICAM-1		sICAM-1 serum levels are significantly increased in 12 months of IFN-β treatment, paralleled with clinical and MRI improvement.	50
Costimulatory molecules	CD40, CD86, PD-L2	PBMCs	IFN-β treatment upregulates CD40, CD86, and PD-L2 expression on monocytes, in correlation with good response to therapy.	164
Others	Sema4A	Serum	Sema4A activates Th cells and promotes differentiation of Th1 and Th17 cells. High Sema4A levels cause Th17 skewing and are associated with unresponsiveness to IFN-β therapy.	165–167
	CXCL10	Serum	CXCL10 is increased after 3 months of IFN-β therapy in NAbs-free patients, while unchanged or reduced in NAbs-positive patients, with a sensitivity of 58% for prediction of development of NAbs. Potential biomarker for reduced IFN-β bioavailability and development of NAbs.	168,169
	TRAIL	Serum	Higher levels of TRAIL before treatment in GR than in PR to IFN-β (conflicting results).	Supportive: 170 Negative: 168,171
			TRAIL is increased after 3 months of IFN-β therapy in NAbs-free patients, while unchanged or reduced in NAbs-positive patients, with a sensitivity of 63% for prediction of development of NAbs.	
	MxA	Blood	MxA concentration is higher in NAbs-negative patients than in NAbs-positive patients.	172
	MMPs/TIMPs	Serum	Reduced levels of MMP-8, MMP-9, and MMP-9/TIMP-1 ratio after IFN-β therapy correlates with reduced radiology disease activity.	156,173
	BDNF	Serum	BDNF levels are significantly higher both at baseline and after 3 months of therapy in GR compared to PR.	39
	TLR4 and type I IFN pathways	PBMCs	Monocytes from PR and intermediate responders show increased baseline levels of endogenous IFN-β and elevated IFN receptor 1, and lack the capacity of IFN-β to induce its own expression, compared with GR. Baseline level of IRAK3, which negatively regulates TLR4 signaling, is decreased in GR compared with PR.	53

(Continued)

TABLE 26.1 Protein Biomarkers for Drug Response Prediction—*cont'd*

Drug	Biomarker	Biological Sample	Main Findings	References
GA				
Cytokines	IL2 IFN-γ IL13 IL15	Serum	A significant reduced percentage of proinflammatory IL2-producing and IFN-γ-producing $CD4^+$ and $CD8^+$ T. $CD4^+$ is seen after 12 months of therapy in GA-responders only. An increase in Th2 cytokines IL13 and IL5 is observed in GR after >6 months of therapy, positively correlating with the clinical response.	112,115
	Th1 to Th2 shift: (IL2 + IFN-γ)/(IL4 + IL10)	Serum	The ratio (IL2 + IFN-γ)/(IL4 + IL10) in serum is elevated in patients with relapses as compared to relapse-free patients after >12 months of therapy.	110
	IL4/IFN-γ	PBMCs	Increased IL4/IFN-γ ratio in PBMCs within 12 months of therapy is associated with a favorable clinical response.	111
Others	CD40	PBMCs	GA treatment decreases the expression of the activation marker CD40 on DCs, inversely correlating with clinical disease activity.	109
	BDNF	PBMCs/T cells	GA-responders show a significant increase in BDNF secretion from PBMCs after 6 months of therapy, while no change in GA-nonresponders.	115
NATALIZUMAB				
Antibodies	Anti-Nz Abs	Serum	Persistent anti-Nz antibodies develop in 3.5–6% of patients within the first few months of therapy and are associated with suboptimal response to treatment.	47,128,129
Neurofilaments	NfL NfH	CSF	NfL and to a lesser extent NfH are associated with neurodegeneration, are markedly reduced by Nz therapy, and are suggested to be surrogate markers of treatment response.	174,175
Cell markers	CD49d/α4 integrin	PBMCs	Nz treatment strongly reduces CD49d expression on PBMCs. Sustained CD49d expression is associated with the development of anti-Nz antibodies and reduced efficacy.	130,176,177
	L-selectin (CD62L)	CD4 T cells	Long-term Nz therapy significantly reduces the percentage of CD62L-expressing $CD4^+$ T cells. An unusually low percentage (9-fold lower) highly correlates with the risk of developing PML.	137,177

Cell count	Leukocyte count	CSF	Nz therapy reduces CSF cell count of CD4 and CD8 T cells, CD19 B cells, and CD138 plasma cells compared with OND patients and untreated MS patients. Effect is persistent even 6 months after cessation of Nz therapy. One patient had a clinical relapse, and showed the highest total leukocyte and CD4 and CD8 T-cell counts within treated patients, suggesting a possible inverse association with clinical response.	178
Others	Fetuin-A	CSF	Fetuin-A level is significantly reduced in Nz-treated patients after 6 months and 12 months of therapy. Reduction is strongly pronounced in GR, while absent in PR to Nz.	179
FINGOLIMOD				
	Leukocyte count	Serum	Fingolimod reduces the numbers of circulating CD4, Th17, and B cells and to a lesser extent CD8 T cells, reversing the CD4/CD8 T-cell ratio. Possible candidate biomarker.	123–125
		CSF	Fingolimod reduces CSF cell count of CD4 T cells, but not B cells, while increases the proportion of CD8 T cells, NK cells, and monocytes, reversing the CD4/CD8 T-cell ratio.	125

Abs, antibodies; *BDNF*, brain-derived neurotrophic factor; *CD*, cluster of differentiation; *CHI3L1*, chitinase-3-like protein 1; *CSF*, cerebrospinal fluid; *CXCL10*, C-X-C motif chemokine 10; *DCs*, dendritic cells; *GA*, glatiramer acetate; *GR*, good responders; *ICAM-1*, intercellular adhesion molecule; *IFN-β*, interferon beta; *IFN-γ*, interferon gamma; *IL*, interleukin; *IRAK3*, interleukin-1 receptor-associated kinase 3; *MMP*, matrix metalloproteinase; *MS*, multiple sclerosis; *MxA*, myxovirus A; *NAbs*, neutralizing antibodies; *NfL*, neurofilament light; *NfH*, neurofilament heavy; *Nz*, natalizumab; *OND*, other neurological disorders; *OPN*, osteopontin; *PBMC*, peripheral blood mononuclear cell; *PD-L2*, programmed death ligand 2; *PML*, progressive multifocal leukoencephalopathy; *PR*, poor responders; *s*, soluble; *Sema4A*, Semaphorin-4A; *Th*, T helper cells; *TIMP*, tissue inhibitors of metalloproteinases; *TLR*, toll-like receptor; *TRAIL*, tumor necrosis factor (TNF)-related apoptosis inducing ligand; *VCAM-1*, vascular cell adhesion protein; *VLA-4*, very late antigen 4.

from peripheral blood mononuclear cells (PBMCs) collected prior to therapy initiation.[44] Thus, the accumulating data do not support IL17 as a biomarker for IFN-β response.

The IL7 cytokine promotes the Th1 response, and genetic evidence implicates its receptor as a risk factor in MS.[45] Accordingly, in a small study of 26 RRMS patients, high levels of IL7, combined with low IL17F, which are indicative of Th1-based MS pathogenesis, correlated with full responsiveness to IFN-β.[46] However, this observation was made in a single small study and therefore mandates follow-up studies in larger cohorts.

Several adhesion molecules are affected by IFN-β therapy and have been suggested as potential biomarkers of treatment response. Downregulation of VLA-4 on T cells and increased levels of soluble VCAM-1 and ICAM (sVCAM and sICAM) in the serum were demonstrated in several studies within a few months of therapy, inversely correlating with MRI and disease activity[39,47–50] and positively correlating with favorable treatment response.[51,52] Since the effects are seen only after months of therapy, the value of adhesion molecules may be mainly as biomarkers to monitor treatment response, similar to NAbs.

Assessment of expression levels in PBMCs of proteins (and associated mRNAs, as described later) participating in the type-I IFN signaling pathways detected increased pretreatment levels of endogenous IFN-β and IFN receptor 1 (IFNAR1) in monocytes from nonresponders and intermediate responders, suggesting basal activation of IFN-β pathways in these patients.[53] Further studies are required to assess the utility of protein biomarkers of IFN-β pathways for prediction of the clinical response to IFN-β.

To summarize, despite much research effort, no validated protein-based biomarker for IFN-β treatment response is available. The optimal biomarker should be predictive at the pretreatment test, thus, the most promising markers mentioned in this chapter appear to be IL7/IL17 ratio, IL10, brain-derived neurotrophic factor (BDNF), endogenous IFN-β, and IFNAR1, all which require confirmatory replication in larger cohorts.

Pharmacogenetic Studies for Interferon-β Response

The majority of pharmacogenetic studies to identify genetic variants associated with the clinical response to IFN-β used a candidate-gene approach. Candidate genes were selected among genes involved in IFN-β signaling and response pathways, such as the IFN type I receptor genes *IFNRA1* and *IFNRA2*; *MXA*, the prototype IFN class I response gene; or genes with a genetic effect in MS such as HLA class I and class II genes.[54–61] Many of these studies failed to detect any association of response with variants within the candidate genes evaluated. A comprehensive candidate gene approach employed by Cunningham et al. detected association of several IFN-response genes, including *MXA*, *IFNRA1*, *LMP7*, and *CTSS* from among 100 genes containing the ISRE IFN-stimulated response element in their promoter.[62] However, these associations were not confirmed in subsequent studies.[60] An effect of the IFN-β response genes *IRF8* and *IRF5* on clinical response was observed in some studies, but not others.[61,63–65] A functional polymorphism within the *OAS1* gene was reported as associated with IFN-β clinical response, but this effect was not dissected from an effect of the single nucleotide polymorphism (SNP) on disease severity.[66] Recently, association of an SNP in the promoter of the IFN-β response gene *USP18* was reported.[67] In a Russian cohort of patients, several allele combinations were reported as associated with IFN-β beneficial response, each containing the *CCR5*d* + *IFNAR1*G* combination.[68,69]

The two genome-wide genetic association studies conducted for the IFN-β clinical response yielded nonidentical sets of associated genes, although both groups highlighted the glutamatergic neurotransmission pathways as associated with IFN-β clinical response.[70,71] The results of these studies suggest that variation in response to IFN-β therapy is not likely to be determined by large genetic effects of common variants. The *GPC5* (glypican 5) association with response to IFN-β therapy was replicated in one study but not in another.[65,72]

Association with IFN-β NAbs of *HLA DRB1*0401* and *DRB1*0408* alleles was reported and validated in over 200 patients in Germany, and in several follow-up studies, which identified additional effects of the *DRB1* alleles **1601, DRB1*07/DQA1*02* with *A*26* or *B*14*, and a protective effect for the *DRB1*03:01,*04:04,*11:04* alleles.[73–76] Non-*DRB1* SNPs with independent effects on NAbs have been reported as well.[75,77] Recently, the *DRB1*0401* effect was demonstrated also in the Swedish population, albeit only with the IFN-β1b formulation. In addition, a higher risk to develop NAbs was observed for *DRB1*15* carriers for patients treated with IFN-β1a which was not observed before.[78] Altogether, the consistent reports of association of *DRB1* alleles with the risk for developing NAbs suggest an effect that may be mediated through antigen presentation by the major histocompatibility complex (MHC) class II complex, and demonstrate a potential, if validated in additional cohorts, to select patients at low risk for NAbs for IFN-β therapy. The findings also highlight population-specific effects, which will need to be taken into account in future research.

In summary, as of March 2015, none of the putative genetic variants associated with response to IFN-β has been validated in more than two independent studies and in different populations. The *HLA-DRB1* association with NAbs requires extension of studies to other populations to evaluate their utility as predictive markers for NAb risk, and may be currently the most promising genetic variant related to the effect of IFN-β, keeping in mind that the correlation between NAb presence and lack of response is only partial.

RNA Expression-Based Markers for Interferon-β Response

Numerous studies have described the gene expression before and during IFN-β treatment. In this chapter we review mainly studies on gene expression profiles in therapy-naïve PwMS, for their potential to identify predictive patterns for treatment outcome, as relevant to application of personalized medicine. Similarly to pharmacogenetic studies, gene expression studies comparing between responders and poor responders to IFN-β initially focused on single candidate genes, selected for involvement in disease mechanisms or IFN-β downstream signaling pathways. To date, none of the candidate genes originally identified as associated with IFN-β response panned out in subsequent studies and crossed the line to clinical use as a predictive marker for IFN-β clinical response. Here we present information on a few selected genes, and review studies published since 2011 in Table 26.2. For a summary of earlier studies the reader is referred to the reviews from 2011 by Comabella et al. and Goertsches et al.[79,80]

The myxovirus resistance protein 1/MxA gene, a classic IFN type-1 response gene, is considered a marker for IFN-β bioactivity during a treatment course and has been suggested as a sensitive and specific indicator of the presence of NAbs to IFN-β.[32,81–83] However, data regarding the utility of MxA mRNA levels as a biomarker for IFN-β therapy response and relapse risk have been controversial.[82,84,85] Other IFN-β response genes also showed correlation of expression level with NAb.[86]

Whether the IFN-β pharmacological response, or part of it, is correlated with the IFN-β clinical response of the PwMS remains an open question. Several studies reported an

TABLE 26.2 Gene Expression Studies for Prediction of Drug Response Published Since 2011[a]

Drug	Study	Aims	Findings	Participants and Response Definitions	Study Details
IFN-β					
	Hecker et al.[180]	Evaluation of 110 DEGs previously identified in 18 studies, and IL17 related genes, in pretreatment RNA from clinical IFN-β response groups.	Five genes from previous studies identified as DEGs in the poor versus good responders: *CA11*,[181] *GPR3*,[181] *IL1RN*,[89] *PPFIA1*,[182] and *YEATS2*.[182] Lower expression of *GPR3* and IL17RC appeared to correlate with number of relapses and EDSS in 5 years.	48 PwMS: 30 responders: Relapse free and neurological relatively stable. 18 poor responders: ≥1 relapse in 2-year follow-up and/or strong worsening of EDSS. 4 poor responders were "very poor responders": EDSS change of >1 and average of 2 relapses in the 2-year follow-up. For validation-expression data from Gurevich et al.[182] was used: 43 good responders, 51 poor responders; and from Singh et al.[183]: 4 good responders and 1 nonresponder.	Sample type: PBMCs Analysis: HG-U133 A chip Affymetrix gene array.
	Bustamante et al.[53]	Analysis of expression of genes in the TLR4 and IFN-β pathways and clinical IFN-β response.	Decreased pretreatment gene expression levels of *IFNB1* (IFN-β), *IL1B*, *IRAK3*, and *IFNAR1* expression in PBMCs from responders was reported. Reduced protein expression was also demonstrated in monocytes. Reduced induction of IFN-β in monocytes by IFN-β was observed in nonresponders.	85 PwMS: 49 responders (57.6%): no relapses, no EDSS change/2 years. 18 nonresponders (21.2%): ≥1 relapse and ≥1 change in EDSS/2 years. 18 intermediate responders to IFN-β (21.2%).	Sample type: PBMCs analysis: RTPCR for *TLR2*, *TLR4*, *MYD88*, *TICAM1*, *IFN1B*, *IL1BCXCL10*, *IFNAR1*, *IRAK3*, *TYK2*, *PTPN6*, *PTPN11*, *SARM1*, *SIKE*, *SIGIRR*, and RIPK3. Flow cytometry for monocyte-specific protein expression.
	Rudick et al.[88]	Analysis of expression of 166 IFN-β response genes and the extent of induction by IFN-β in good and poor responders to IFN-β therapy.	The magnitude of the 12-hr biological response to IFN-β was increased in poor responders at therapy initiation (13 DEGs) and at 6 months (16 DEGs); however, the DEGs identified at the two time points were different (save one).	85 CIS and RRMS patients: 70 good responders based on MRI classification at 6 months after therapy initiation. 15 poor responders: Presenting with ≥3 new or enlarging lesions at 6-month time point.	Sample type: Whole blood PAXgene, collected before an IFN-β injection and after 12 h, at therapy initiation and at 6 months. Analysis: Custom macroarray for 166 IFN-regulated genes.
	Hundeshagen et al.[85]	Prospective comparison of the prognostic value of high or low *MXA* at treatment initiation.	Patients with *MXA* high or *MXA* low levels did not differ in 5-year EDSS or relapse rates. However, considering IFN-β formulation, relapse rates differed for the *MXA* high or low groups, albeit with opposing effects for IFN-β-1a or IFN-β-1b.	61 MS patients; 11 classified as *MXA* high-expressing, and 50 as *MXA* low-expressing based on levels of *MXA* measured before therapy initiation. Participants were followed up for 5 years.	Sample type: PBMC, obtained pretreatment and after 1 month on therapy. Analysis: HG-U133 A and B or plus 2.0 microarrays, Affymetrix, used to evaluate the expression of 56 IFN-β pathway-related genes.

Baranzini et al.[92]	Validation of gene triplet expression signatures for prediction of clinical response to IFN-β.	Gene triplets reported in 2005 by this group[91] were validated, albeit with a lower predictive accuracy, due to incorporation of MRI data in response definition. New triplet genes were identified, containing *CASP2*, the top classifier being *CASP2/IL10/IL12Rb1*	46 disease-free on treatment and 104 suboptimal responders classified by clinical (relapses, EDSS) and radiological (MRI) data. Subjects were NAb negative and participated in the IMPROVE study with 18 months of follow-up.	Sample type: Whole blood PAXgene Analysis: RTPCR (qPCR) Expression of 32 transcripts (based on Baranzini et al. study[91]), including IFN pathways, cell cycle, apoptosis, cytokines, and lymphocyte receptors.
Malhotra et al.[90]	Assessment of association between NLRP3 inflammasome genes expression and the clinical response to IFN-β.	Increased expression of *NLRP3* and *IL1B* observed in nonresponders at baseline. IFN-β increased the expression in responders but not in nonresponders for both genes.	I. 48 responders (49.5%): no relapses, and no EDSS change in 2 years. 22 nonresponders (22.7%): ≥1 relapse and ≥1 change in EDSS in 2 years. 27 intermediate responders (27.8%). II. With radiological data after follow-up of 1 year: 28 nonresponders (28.9%): Meeting 2 out of 3 criteria: 1. ≥1 relapse; 2. ≥1 change in EDSS; 3. ≥3 active lesions on the 1-year brain MRI scan. All others classified as responders (69, 71.1%).	Sample type: PBMCs, obtained before treatment, and for a subset also at 3 months after therapy initiation. Analysis: RTPCR (qPCR) analysis for the inflammasome-related genes *AIM2, NLRC4, NLRP1, NLRP3, IL10, IL1B*, and *IL18*.
GA				
Sellebjerg et al.[109]	To study the immunological response and immune-related gene expression in GA therapy and association with disease activity.	No association between relapse risk at 1 year of treatment and gene expression for the 23 genes tested. Increased levels of *GATA3* and *LTB* appeared to be associated with MRI activity in long-term treated patients (20 evaluated) and in 9 patients with MRI at 6 months from therapy initiation.	Stable disease (13, 45%): no relapse or progression of EDSS in 2-year follow-up. Breakthrough disease (15, 52%): Occurrence of a relapse or progression of ≥1 EDSS. MRI activity defined as presence of gadolinium-enhancing lesions.	Sample type: Whole blood PAXgene Analysis: RTPCR (qPCR) for 23 immune-related genes.

CIS, clinically isolated syndrome; *DEG*, differentially expressed genes; *EDSS*, Expanded Disability Status Scale; *GA*, glatiramer acetate; *HC*, healthy controls (non-MS individuals); *IFN-β*, *interferon beta*; *im*, intramuscular; *NAb*, neutralizing antibodies (to IFN-β); *PAXgene*, a whole blood RNA preparation method (Qiagen); *PBMC*, peripheral blood mononuclear cells; *PwMS*, patients with multiple sclerosis; *RRMS*, relapsing-remitting MS; *sc*, subcutane.

[a] *Studies listed are those published since 2011 and up to February 9, 2015, and are not included in the Goertsches et al.[80] table summarizing earlier studies on IFN-β and GA-related gene expression.*

increased capacity of response for IFN response genes that was already present at treatment initiation, in patients that proceeded to an exacerbated MS disease course, but other studies could not corroborate these findings.[35,86–88] Comabella and colleagues reported on a signature of increased expression of IFN-β pathway downstream genes in treatment-naïve patients that is correlated with nonresponder phenotype.[53,89] The best response predictor genes suggested in the first study included *IFIT1, IFIT2, IFIT3, RASGEF1B, OASL, IFI44, FADS1*, and *MARCKS*, and were reported to perform to a comparable extent in a validation cohort. The study suggested that the differential expression is enriched within the monocytic compartment. In the second study *IFN1β, IL1B, IFNRA1*, and *IRAK3* emerged. The same group reported that increased expression of two inflammasome-related genes—*NLRP3* and *IL1B*—prior to IFN-β therapy initiation, was correlated with nonresponsiveness to IFN-β.[90] The findings of all studies remain to be replicated in independent cohorts.

Baranzini and colleagues suggested that the predictive power of gene expression can be increased by using a triplet set of genes.[91] In a recent follow-up study, these investigators reported the validation of triplets of genes as correlated with clinical response in a large 155-PwMS cohort, albeit with a lower predictive power (up to 0.68) compared to the original 2005 study. Inclusion of the induction level by IFN-β and new triplet combinations resulted in improved predictive accuracies of up to 0.82.[92]

To summarize, despite the overwhelming amount of gene expression data for IFN-β therapy response that has accumulated over more than a decade, the evidence remains suggestive and has not been substantiated by sufficient validation studies to allow translation into the clinical practice. An important source of variation between studies using mRNA that needs to be accounted for is the RNA source. The choice between cell mixtures like PBMC or the total blood PAXgene format, or specific cell subsets, will determine the nature of signals detected. The expression patterns of single genes do not appear to suffice as predictors of disease activity under IFN-β therapy, and gene sets comprising triplets or more may be warranted for increased predictive power.

MicroRNAs as Biomarkers for Interferon-β Response

MicroRNAs (miRs) are abundant, small (21–25 nucleotides), single-stranded noncoding RNAs that regulate mRNA expression and impact a wide variety of cellular processes, such as cellular differentiation, proliferation, and apoptosis.[93] MiR levels have been suggested as drug response markers in cancer, and their properties make them attractive for clinical use.[94] MiRs can affect the levels of proteins relevant to drug response, including drug-metabolizing enzymes and immune-related proteins.[95] Only a handful of studies on the effect of IFN-β on miRs have been published, and of these, only one (De Felice et al.) assessed possible differences in miR expression between IFN-β responder groups.[96–98] De Felice and colleagues observed higher expression of the IFN-β response miR miR-26a-5p, even prior to therapy initiation in PwMS that later responded well to therapy, and this difference between response groups increased during therapy. In nonresponders, miR-26a-5p was not affected by IFN-β therapy. MiR-26a-5p targets many genes active within the nervous system, including genes within the glutamate signaling pathways. These results emphasize the need for additional studies on miRs as possible biomarkers for therapy response.

Biomarkers for Adverse Response to Interferon-β

The most common AE reported for the IFN-β group of therapies are the flu-like symptoms (FLS), including fever, muscle pain, malaise, and headaches. These symptoms usually resolve after the first 3 months. Occasionally, FLS severity requires therapy change.[99,100] Other severe, but not fatal AEs associated with IFN-β therapies include elevated liver function and thyroid-related autoimmunity. An in vitro assay based on the IL6 response to IFN-β in CD14+ cells was proposed to select patients likely to develop FLS, for whom low-dose oral steroid use at the beginning of therapy may be prescribed,[101] but no follow-up study has been published since. Using a network approach, Hecker and colleagues identified a set of six IFN-β response genes (*ESAM, ALOX12, ELOVL7, ANKRD9, GNAZ, SLC24A3*) that appear to have elevated mRNA expression levels prior to therapy in PwMS that develop FLS.[102] Only a small percentage of patients will actually cease IFN-β therapy due to FLS, therefore future studies should assess predictive power of these biomarkers to select for PwMS that are likely to suffer from the most severe presentation of this AE.

26.3.2 Glatiramer Acetate

Although GA became available as a first-line therapy almost in parallel to the IFN-β therapies, very few studies have been published on biomarkers for clinical response. A significant percent of patients respond poorly to GA (in some studies up to 50%, depending on the response definition),[54,103] with >22% experiencing two or more relapses during the clinical trials.[104]

Protein-Based Markers in Glatiramer Acetate Therapy

GA is immunogenic, and antibodies against GA can be detected in GA-treated patients, peaking at 3-months therapy; however, in contrast to IFN-β NAbs, the significance of the anti-GA antibody is not clear.[105–107] In vitro studies reported conflicting results, with reports of blockage of GA-stimulated proliferation of GA-specific T cells by anti-GA antibodies in one study, but lack of interference with, or even enhancement of the activities of GA in another.[106,108] This is in line with the results from three independent GA clinical trials, which showed that anti-GA antibodies affect neither the adverse reactions, nor the therapeutic efficacy of GA. Moreover, patients with higher anti-GA antibody titers tended to be relapse-free; suggesting that anti-GA antibodies may be part of the beneficial mechanism of action of GA.[107] Anti-GA antibodies appear to persist in long-term GA-treated patients, with no neutralizing activity on drug activity or correlation with clinical response.[105,109]

Several studies investigated cytokine profiles as possible biomarkers for GA clinical response, based on the known Th1 to Th2 shift induced by the therapy. Some cytokine expression differences were observed after several months of GA therapy period between relapse-free patients and relapsing patients, including the increased quotient (IL2 + IFN-γ)/(IL4 + IL10) in relapsing patients, an increased PBMC IL4/IFN-γ expression ratio, and elevated serum IL13 and IL5 levels in patients with a favorable clinical response.[110–112] CD40 and BDNF levels were affected by GA therapy, and associated with relapse rates or clinical response.[109,113–115] However, the cytokines or other immune-related proteins reported associations with clinical response were observed only after an extended period of therapy, consisting usually of several months, and therefore are not likely to be of any use for predictive personalized medicine.

Pharmacogenetic Studies for Glatiramer Acetate Response

Pharmacogenetic studies for GA in MS used a candidate gene design, based on the attributed mechanisms of action for GA as a competitor with the myelin basic protein for binding to the MHC class II complex, and other immunomodulatory and neuroprotective actions (see the chapter in this book *Glatiramer Acetate – From Bench to Bed and Back*). Conflicting results were obtained for the *HLA-DRB1*1501* and GA response: The first report of an increased frequency of the *HLA-DRB1*1501* allele in responders was corroborated by the larger study of Gross et al., demonstrating longer event-free survival for *HLA-DRB1*1501* homozygotes.[54,61] Dhib-Jalbut et al. also reported that DR15 and the haplotype DR15-DQ6 were significant predictors for good response to GA, whereas the DR17 allele was associated with poor response to GA.[116] The studies by Grossman et al. and Tsareva et al. failed to detect the *HLA-DRB1**1501 association with a beneficial response to GA.[103,117] Moreover, Tsareva et al. reported that allele combinations with the HLA-DR15 allele were associated with a poor response to GA. The studies by Grossman et al. and Tsareva and colleagues both tested a set of SNPs in immune-related candidate genes. Although some of the genes were tested in both studies, the results were divergent: Grossman and colleagues reported an association of GA clinical response with the *TRB@* (encoding a T-cell receptor subunit) and the *CTSS* gene, out of a set of 26 genes tested. Tsareva and colleagues were able to identify associations only in allele combinations, but not in single genes, reporting a significant effect on GA clinical response for the allele quad *DRB1*15+TGFB1*T+CCR5*d+IFNAR1*G*. An apparently drug-specific effect was observed for carriers of *DR16* and the GA response (but not IFN-β response).[69] The small study sizes and different populations assessed in the few GA pharmacogenomic studies likely underlie the heterogeneity of the reports.

Gene Expression Studies for Clinical Response to Glatiramer Acetate

The effects of GA on mRNA expression appear to be modest or not evident, at least in the first few months of therapy, depending on the statistical methodology used and the cell type (PBMC, monocytes) from which RNA is extracted.[118–120] Low levels of *GATA3* and *LTB* expression was suggested to be associated with MRI activity in GA-treated patients, but not with the clinical response defined by relapse rate and disability progression[109] (see Table 26.2). A study focusing on a few selected miRs reported that GA downregulates miR-142-3p and miR-146a, restoring expression values to levels observed in HC.[97] We are not aware of any other reports evaluating the pretreatment mRNA or miR expression profile correlation with the clinical response to GA.

Biomarkers for Adverse Response in Glatiramer Acetate Therapy

GA has a relatively good safety profile with few side effects, consisting mainly of injection-site and immediate systemic postinjection reactions[99,121] (see Chapter 13: Surrogate Markers in Multiple Sclerosis: The Role of Magnetic Resonance Imaging). There are no studies yet, to the best of our knowledge, that assess biomarkers for AE in GA therapy.

26.3.3 The Oral Multiple Sclerosis Drugs: Fingolimod, Teriflunomide, and Dimethyl Fumarate

The oral drugs fingolimod, teriflunomide, and DMF have an overall efficiency that is better, or at least as good as the injectable IFN-β and GA therapies, with the advantage of a more favorable mode of administration.[122] Each carries a distinct profile of associated AEs that requires attention and monitoring, including pretreatment screening tests and exclusion criteria, highlighting the importance of selecting the right drug for each patient (see the Chapter 19: Fingolimod (Gilenya®) and Chapter 20: Oral Dimethyl Fumarate (BG-12; Tecfidera®) for Multiple Sclerosis) As of March 2015, there are no publications for pharmacogenomic studies or any protein-based biomarkers for the oral drugs, and the percentages of good responders and nonresponders for these drugs are yet unreported.

Fingolimod has a strong lymphopenic effect, reducing the numbers of circulating CD4, Th17, and B cells, and to a lesser extent, CD8 T cells, reversing the CD4/CD8 T-cell ratio.[123–125] The CD4/CD8 T-cell ratio can serve as evidence of drug activity, but is not predictive of the clinical response. A model-predicted relationship between the annualized relapse rate and lymphocyte counts for different doses of fingolimod used in the phase-III trials suggests that the lymphocyte count can serve as a biomarker for clinical response; however, this has yet to be further evaluated.[126]

26.3.4 Natalizumab

Natalizumab was the first humanized monoclonal antibody approved for use in MS, designed to target the α4 subunit of the leukocyte VLA4 integrin to block migration across the endothelial layer. It is a highly efficient therapy for reduction of relapses, attenuation of disability progression, and development of new lesions.[127] Its main setback and reason for treatment cessation remain the elevated risk for PML. Currently, there are no reported studies assessing genetic markers associated with clinical response to natalizumab or with risk for AEs.

Protein-Based Markers in Natalizumab Therapy

Anti-natalizumab antibodies were found to develop in less than 10% of natalizumab-treated patients, as early as within 1 month of therapy, and in some patients are present only transiently.[47,128,129] In persistently positive patients natalizumab was less efficient as measured by disability progression, relapse rate, and MRI compared with antibody-negative patients, while transiently positive patients displayed beneficial response to the therapy, concomitantly with achievement of antibody negative status. Furthermore, in persistently positive patients the incidence of infusion-related AE was significantly higher. It was therefore suggested by Calabresi et al. that in patients with a suboptimal clinical response or persistent adverse event antibody testing should be considered to determine course of action.[47]

Recently it was suggested that the expression of CD49d (the target molecule of natalizumab, integrin α4) on PBMCs may be a surrogate biomarker for development of antinatalizumab antibodies and natalizumab efficiency.[130] Defer and colleagues observed a significant reduction of CD49d expression levels following the first infusion of natalizumab and during the therapy in most (84%) patients. In the other patients, CD49d expression levels returned to pretreatment levels concomitant with of the appearance of antinatalizumab antibodies. Sustained CD49d expression was suggested as an indication for testing for persistent antinatalizumab antibodies.[130]

Natalizumab therapy has been shown to affect the expression of several molecules (see Table 26.1); the association, however, of these expression changes with clinical efficacy remains to be further tested, and currently there are no established protein biomarkers of response to natalizumab therapy.

Biomarkers for Natalizumab-Related Adverse Response

PML continues to present the most serious and potentially fatal AE for natalizumab. PML is an opportunistic demyelinating disease caused by a pathogenic form of the John Cunningham virus (JCV). As of March 2015, the global incidence of PML among patients treated with natalizumab was reported as 3.87/1000 (https://medinfo.biogen.com/). At present, there is no available blood biomarker to diagnose PML. Safety programs have been set in place to screen and evaluate patients at risk prior to therapy and during therapy and a triad of major risk factors for PML has been identified: (1) Natalizumab treatment duration, with a significant increase after 24 months of therapy; (2) Previous exposure to immunosuppressants; (3) The presence of the JCV, indicated by presence of anti-JCV antibodies. JCV serology assays to detect anti-JCV antibodies have been developed and introduced into clinical use to screen patients prior to therapy initiation, as well as during therapy to detect seroconversion. Seroconversion in PwMS with PML has been reported as early as 6 months prior to appearance of PML symptoms and thus has the potential for early detection of increased risk to allow therapy switch.[131] A second-generation JCV enzyme-linked immunosorbent assay was demonstrated to differentiate PML risk in anti-JCV antibody-positive MS patients with no prior immunosuppressant use.[132] Using this assay of JCV antibody index, patients with serum/plasma JCV antibody index negative or below a threshold of 1.5 seem to be associated with a significantly lower risk for PML. Continued evaluation of anti-JCV antibody index and PML risk is warranted.

A JCV-PCR assay is available for PML diagnosis in CSF once PML-like symptoms or radiology is suspected.[127,132–135] Some biomarkers predictive of increased risk of PML have been suggested—a comprehensive review of these can be found in Ref. 136. Among these, a decrease in L-selectin (CD62L) expressing CD4 T cells has been associated with increased risk for PML.[137] This finding remains to be further validated in larger cohorts.

A single miR study identified three miRs, miR-320, miR-320b, and miR-629, as differentially expressed between patients that proceeded to develop PML and PML-free patients, suggesting these miRs might be biomarkers for individual PML risk assessment.[138] However, since only two PML patients were analyzed, follow-up studies with larger PML cohorts are required.

26.3.5 Immunosuppressant Therapy for Multiple Sclerosis: Mitoxantrone and Alemtuzumab

Immunosuppressant therapies for MS are recommended for PwRMS with active disease that is resistant to conventional disease-modifying therapies. Mitoxantrone has been in use for MS for more than a decade, and some potential clinical and genetic markers have been associated with clinical response as described later. There are no published studies yet for predictive markers for alemtuzumab, which are much needed in view of the severe and potentially fatal AEs that have been encountered in the clinical trials leading to its approval.[139]

Mitoxantrone

Although proven effective in reduction of the number of relapses and disability progression in randomized controlled trials, its potentially severe side effects restrict its use to a second- or third-line treatment.[140,141] Cardiotoxicity and secondary leukemia have been reported, and delayed side effects can occur, leading to a restriction on lifetime dose for mitoxantrone and a requirement for extended monitoring of patients for several years after treatment cessation.[122,140] Accordingly, only a small portion of PwMS are offered mitoxantrone therapy, and very few studies have evaluated markers for response and AE risk.

Based on the known pharmacokinetics of mitoxantrone, whose levels within cells are regulated by the activity of ABC transporters, Cotte and colleagues identified four SNPs in the genes *ABCB1* and *ABCG2* with a functional effect on mitoxantrone efflux from cells.[142] Lower rates of efflux were inversely correlated with cell death. Patients carrying the allele combination conferring the highest rate of efflux had the lowest rate of mitoxantrone responder phenotype (62.5%), and conversely, the low efflux genotype group had the highest response rate (84.8%, $p=0.039$). This genotype–phenotype effect was confined to mitoxantrone monotherapy and was not observed for a glucocorticoid/mitoxantrone combined therapy. A study from the same group evaluated whether the same ABC genotypes may serve as markers for response also in PPMS; however, no association with mitoxantrone therapy response was observed, and no overall benefit for PPMS therapy.[143]

26.4 PARTICIPATORY MEDICINE AND PERSONALIZED INFORMATION IN MULTIPLE SCLEROSIS

In recent years medicine is becoming gradually more participatory, where health-care providers and patients work in partnership, using modern information-based and communications tools, to increase the active participation of the patient in medical decisions related to his or her health. This approach of participatory medicine, based on increasing patient engagement, also holds promise to improve adherence to therapy as well as outcomes, reduce medical errors, increase patient satisfaction, and lower the cost of care.

Availability of health information technology (HIT) added a new dimension to personalized medicine: interactive information tailored to the individual patient. Studies of the information needs of PwMS found that interests and expectations change according to time from diagnosis and to disease activity and severity.[144] Research has shown that the majority of PwMS do not discuss the information found on the web with their physicians[144,145] and therefore they are exposed to information that has not undergone critical professional review. This is a problematic issue due to the variable quality of web sites for PwMS.[146] Beyond information, HIT offers PwMS e-tools for patient active involvement such as self-management programs for physical activity[147] and fatigue[148]; and personal health records.[149] Other mechanisms for increasing involvement of PwMS in their health care include implementation of patient-reported outcome using specific assessment questionnaires as the MSQoL-54 for quality of life,[150] the MSWS-12 for walking ability,[151] and the MSTEQ for adherence to therapy.[152]

Taken together, as personalized medicine in MS evolves by implementation of new pharmacogenetics and immune-marker-based approaches, awareness to adherence issues in the prolonged treatments of chronic diseases is increasing.[153,154] The importance of PwMS being more informed and actively involved in their own health care is being increasingly recognized and applied by patient-centered approaches of patient empowerment and participatory health care.

26.5 CONCLUSIONS

Significant efforts have been invested in the search for biomarkers that can aid in therapy choice. MS diagnosis currently is based on the clinical presentation, radiological findings, and aided by the long-time established marker of OCBs. Biomarkers for MS disease subtypes and prognosis have not yet been found. Although several markers and prognostic gene expression patterns have been suggested, none have been satisfactorily validated in independent groups, as required for transition to a clinical test. Several new assays have been introduced into the clinic and aid the neurologist in drug choice for the patients. These mostly consist of antibody-based tests that identify patients at risk for a specific drug-related AE (eg, JCV) or drug neutralizing activity (IFN-β NAbs). The expectation that pharmacogenomic-based markers will form the basis of personalized medicine and theranostics has not yet been realized.

Several significant challenges have shaped the arena of MS biomarker studies, and have likely contributed to lack of reproduction across studies. Heterogeneity across studies was derived from the inclusion of different populations, contributing different genetic backgrounds as well as environmental exposures; the use of different drug formulations as done in many studies for the IFN-β therapies; and different drug response definitions, including times of follow-up and measures used to define the clinical drug response. Distinction between the natural disease course and the effect of the drug on the disease course is another issue that requires attention. Is a person with an attenuated disease course simply a PwMS with a benign disease or a true drug-specific good responder? The answer to this question lies in the ability to include a placebo treatment group, or to apply a study design including several drugs that can assess the drug specificity of the biomarker tested. These options can be realized more easily within a clinical trial setting and should be a consideration in the design of the trial.

References

1. Lublin FD, Reingold SC, Cohen JA, et al. Defining the clinical course of multiple sclerosis: the 2013 revisions. *Neurology*. July 15, 2014;83(3):278–286.
2. *Novantrone (Mitoxantrone HCl) Injection – Medication Guide and Boxed Warning*; 2012. Accessed 26.10.14. http://www.fda.gov/safety/medwatch/safetyinformation/ucm219174.htm.
3. *Tysabri (Natalizumab) Injection – Medication Guide and Black Box Warning*; 2013. Accessed 26.10.14. http://www.fda.gov/safety/medwatch/safetyinformation/ucm355780.htm.
4. Label Approved on 11/14/2014 (PDF) for LEMTRADA, BLA no. 103948. http://www.accessdata.fda.gov/drugsatfda_docs/label/2014/103948s5139lbl.pdf; 2014 Accessed 01.12.14.
5. Buck D, Hemmer B. Treatment of multiple sclerosis: current concepts and future perspectives. *J Neurol*. October 2011;258(10):1747–1762.
6. Poser CM, Paty DW, Scheinberg L, et al. New diagnostic criteria for multiple sclerosis: guidelines for research protocols. *Ann Neurol*. March 1983;13(3):227–231.
7. McDonald WI, Compston A, Edan G, et al. Recommended diagnostic criteria for multiple sclerosis: guidelines from the International Panel on the diagnosis of multiple sclerosis. *Ann Neurol*. July 2001;50(1):121–127.
8. Polman CH, Reingold SC, Banwell B, et al. Diagnostic criteria for multiple sclerosis: 2010 revisions to the McDonald criteria. *Ann Neurol*. February 2011;69(2):292–302.
9. Noyes K, Weinstock-Guttman B. Impact of diagnosis and early treatment on the course of multiple sclerosis. *Am J Manag Care*. November 2013;19(17 suppl):s321–s331.
10. Gholipour T, Healy B, Baruch NF, Weiner HL, Chitnis T. Demographic and clinical characteristics of malignant multiple sclerosis. *Neurology*. June 7, 2011;76(23):1996–2001.

11. Brownlee WJ, Miller DH. Clinically isolated syndromes and the relationship to multiple sclerosis. *J Clin Neurosci.* July 11, 2014;21(12):2065–2071.
12. Zettl UK, Stuve O, Patejdl R. Immune-mediated CNS diseases: a review on nosological classification and clinical features. *Autoimmun Rev.* January 2012;11(3):167–173.
13. Katsavos S, Anagnostouli M. Biomarkers in multiple sclerosis: an up-to-date overview. *Mult Scler Int.* 2013;2013:340508.
14. Avasarala JR, Cross AH, Trotter JL. Oligoclonal band number as a marker for prognosis in multiple sclerosis. *Arch Neurol.* December 2001;58(12):2044–2045.
15. Awad A, Hemmer B, Hartung HP, Kieseier B, Bennett JL, Stuve O. Analyses of cerebrospinal fluid in the diagnosis and monitoring of multiple sclerosis. *J Neuroimmunol.* February 26, 2010;219(1–2):1–7.
16. Miller D, Barkhof F, Montalban X, Thompson A, Filippi M. Clinically isolated syndromes suggestive of multiple sclerosis, part 2: non-conventional MRI, recovery processes, and management. *Lancet Neurol.* June 2005;4(6):341–348.
17. Zipoli V, Hakiki B, Portaccio E, et al. The contribution of cerebrospinal fluid oligoclonal bands to the early diagnosis of multiple sclerosis. *Mult Scler.* April 2009;15(4):472–478.
18. Kleiter I, Hellwig K, Berthele A, et al. Failure of natalizumab to prevent relapses in neuromyelitis optica. *Arch Neurol.* February 2012;69(2):239–245.
19. Palace J, Leite MI, Nairne A, Vincent A. Interferon Beta treatment in neuromyelitis optica: increase in relapses and aquaporin 4 antibody titers. *Arch Neurol.* August 2010;67(8):1016–1017.
20. Shimizu J, Hatanaka Y, Hasegawa M, et al. IFNbeta-1b may severely exacerbate Japanese optic-spinal MS in neuromyelitis optica spectrum. *Neurology.* October 19, 2010;75(16):1423–1427.
21. Uzawa A, Mori M, Hayakawa S, Masuda S, Kuwabara S. Different responses to interferon beta-1b treatment in patients with neuromyelitis optica and multiple sclerosis. *Eur J Neurol.* May 2010;17(5):672–676.
22. Jarius S, Wildemann B. Aquaporin-4 antibodies (NMO-IgG) as a serological marker of neuromyelitis optica: a critical review of the literature. *Brain Pathol.* November 2013;23(6):661–683.
23. Lennon VA, Kryzer TJ, Pittock SJ, Verkman AS, Hinson SR. IgG marker of optic-spinal multiple sclerosis binds to the aquaporin-4 water channel. *J Exp Med.* August 15, 2005;202(4):473–477.
24. Lennon VA, Wingerchuk DM, Kryzer TJ, et al. A serum autoantibody marker of neuromyelitis optica: distinction from multiple sclerosis. *Lancet.* December 11–17, 2004;364(9451):2106–2112.
25. Matsuoka T, Matsushita T, Kawano Y, et al. Heterogeneity of aquaporin-4 autoimmunity and spinal cord lesions in multiple sclerosis in Japanese. *Brain.* May 2007;130(Pt 5):1206–1223.
26. Matsushita T, Isobe N, Matsuoka T, et al. Aquaporin-4 autoimmune syndrome and anti-aquaporin-4 antibody-negative opticospinal multiple sclerosis in Japanese. *Mult Scler.* July 2009;15(7):834–847.
27. Matiello M, Lennon VA, Jacob A, et al. NMO-IgG predicts the outcome of recurrent optic neuritis. *Neurology.* June 3, 2008;70(23):2197–2200.
28. Weinshenker BG, Wingerchuk DM, Vukusic S, et al. Neuromyelitis optica IgG predicts relapse after longitudinally extensive transverse myelitis. *Ann Neurol.* March 2006;59(3):566–569.
29. Mahurkar S, Suppiah V, O'Doherty C. Pharmacogenomics of interferon beta and glatiramer acetate response: a review of the literature. *Autoimmun Rev.* February 2014;13(2):178–186.
30. Polman CH, Bertolotto A, Deisenhammer F, et al. Recommendations for clinical use of data on neutralising antibodies to interferon-beta therapy in multiple sclerosis. *Lancet Neurol.* July 2010;9(7):740–750.
31. Bertolotto A, Malucchi S, Sala A, et al. Differential effects of three interferon betas on neutralising antibodies in patients with multiple sclerosis: a follow up study in an independent laboratory. *J Neurol Neurosurg Psychiatry.* August 2002;73(2):148–153.
32. Sorensen PS, Deisenhammer F, Duda P, et al. Guidelines on use of anti-IFN-beta antibody measurements in multiple sclerosis: report of an EFNS Task Force on IFN-beta antibodies in multiple sclerosis. *Eur J Neurol.* November 2005;12(11):817–827.
33. Paolicelli D, D'Onghia M, Pellegrini F, et al. The impact of neutralizing antibodies on the risk of disease worsening in interferon beta-treated relapsing multiple sclerosis: a 5year post-marketing study. *J Neurol.* June 2013;260(6):1562–1568.
34. Farrell RA, Marta M, Gaeguta AJ, Souslova V, Giovannoni G, Creeke PI. Development of resistance to biologic therapies with reference to IFN-beta. *Rheumatology.* April 2012;51(4):590–599.
35. Hesse D, Krakauer M, Lund H, et al. Breakthrough disease during interferon-[beta] therapy in MS: no signs of impaired biologic response. *Neurology.* May 4, 2010;74(18):1455–1462.

36. Sbardella E, Tomassini V, Gasperini C, et al. Neutralizing antibodies explain the poor clinical response to interferon beta in a small proportion of patients with multiple sclerosis: a retrospective study. *BMC Neurol*. October 13, 2009;9:54.
37. Steinberg SC, Faris RJ, Chang CF, Chan A, Tankersley MA. Impact of adherence to interferons in the treatment of multiple sclerosis: a non-experimental, retrospective, cohort study. *Clin Drug Investig*. 2010;30(2):89–100.
38. Bertolotto A, Capobianco M, Amato MP, et al. Guidelines on the clinical use for the detection of neutralizing antibodies (NAbs) to IFN beta in multiple sclerosis therapy: report from the Italian Multiple Sclerosis Study group. *Neurol Sci*. February 2014;35(2):307–316.
39. Dhib-Jalbut S, Sumandeep S, Valenzuela R, Ito K, Patel P, Rametta M. Immune response during interferon beta-1b treatment in patients with multiple sclerosis who experienced relapses and those who were relapse-free in the START study. *J Neuroimmunol*. January 15, 2013;254(1–2):131–140.
40. Axtell RC, de Jong BA, Boniface K, et al. T helper type 1 and 17 cells determine efficacy of interferon-beta in multiple sclerosis and experimental encephalomyelitis. *Nat Med*. April 2010;16(4):406–412.
41. Bushnell SE, Zhao Z, Stebbins CC, et al. Serum IL-17F does not predict poor response to IM IFNbeta-1a in relapsing-remitting MS. *Neurology*. August 7, 2012;79(6):531–537.
42. Hartung HP, Steinman L, Goodin DS, et al. Interleukin 17F level and interferon beta response in patients with multiple sclerosis. *JAMA Neurol*. August 2013;70(8):1017–1021.
43. Dimisianos N, Rodi M, Kalavrizioti D, Georgiou V, Papathanasopoulos P, Mouzaki A. Cytokines as biomarkers of treatment response to IFN beta in relapsing-remitting multiple sclerosis. *Mult Scler Int*. 2014;2014:436764.
44. Bustamante MF, Rio J, Castro Z, Sanchez A, Montalban X, Comabella M. Cellular immune responses in multiple sclerosis patients treated with interferon-beta. *Clin Exp Immunol*. March 2013;171(3):243–246.
45. Lundmark F, Duvefelt K, Iacobaeus E, et al. Variation in interleukin 7 receptor alpha chain (IL7R) influences risk of multiple sclerosis. *Nat Genet*. September 2007;39(9):1108–1113.
46. Lee LF, Axtell R, Tu GH, et al. IL-7 promotes T(H)1 development and serum IL-7 predicts clinical response to interferon-beta in multiple sclerosis. *Sci Transl Med*. July 27, 2011;3(93):93ra68.
47. Calabresi PA, Giovannoni G, Confavreux C, et al. The incidence and significance of anti-natalizumab antibodies: results from AFFIRM and SENTINEL. *Neurology*. October 2, 2007;69(14):1391–1403.
48. Graber J, Zhan M, Ford D, et al. Interferon-beta-1a induces increases in vascular cell adhesion molecule: implications for its mode of action in multiple sclerosis. *J Neuroimmunol*. April 2005;161(1–2):169–176.
49. Rieckmann P, Kruse N, Nagelkerken L, et al. Soluble vascular cell adhesion molecule (VCAM) is associated with treatment effects of interferon beta-1b in patients with secondary progressive multiple sclerosis. *J Neurol*. May 2005;252(5):526–533.
50. Trojano M, Defazio G, Avolio C, et al. Effects of rIFN-beta-1b on serum circulating ICAM-1 in relapsing remitting multiple sclerosis and on the membrane-bound ICAM-1 expression on brain microvascular endothelial cells. *J Neurovirol*. May 2000;6(suppl 2):S47–S51.
51. Rieckmann P, Altenhofen B, Riegel A, Kallmann B, Felgenhauer K. Correlation of soluble adhesion molecules in blood and cerebrospinal fluid with magnetic resonance imaging activity in patients with multiple sclerosis. *Mult Scler*. June 1998;4(3):178–182.
52. Soilu-Hanninen M, Laaksonen M, Hanninen A, Eralinna JP, Panelius M. Downregulation of VLA-4 on T cells as a marker of long term treatment response to interferon beta-1a in MS. *J Neuroimmunol*. October 2005;167(1–2):175–182.
53. Bustamante MF, Fissolo N, Rio J, et al. Implication of the Toll-like receptor 4 pathway in the response to interferon-beta in multiple sclerosis. *Ann Neurol*. October 2011;70(4):634–645.
54. Fusco C, Andreone V, Coppola G, et al. HLA-DRB1*1501 and response to copolymer-1 therapy in relapsing-remitting multiple sclerosis. *Neurology*. December 11, 2001;57(11):1976–1979.
55. Villoslada P, Barcellos LF, Rio J, et al. The HLA locus and multiple sclerosis in Spain. Role in disease susceptibility, clinical course and response to interferon-beta. *J Neuroimmunol*. September 2002;130(1–2):194–201.
56. Sriram U, Barcellos LF, Villoslada P, et al. Pharmacogenomic analysis of interferon receptor polymorphisms in multiple sclerosis. *Genes Immun*. March 2003;4(2):147–152.
57. Leyva L, Fernandez O, Fedetz M, et al. IFNAR1 and IFNAR2 polymorphisms confer susceptibility to multiple sclerosis but not to interferon-beta treatment response. *J Neuroimmunol*. June 2005;163(1–2):165–171.
58. Fernandez O, Fernandez V, Mayorga C, et al. HLA class II and response to interferon-beta in multiple sclerosis. *Acta Neurol Scand*. December 2005;112(6):391–394.

59. Comabella M, Fernandez-Arquero M, Rio J, et al. HLA class I and II alleles and response to treatment with interferon-beta in relapsing-remitting multiple sclerosis. *J Neuroimmunol*. May 29, 2009;210(1–2):116–119.
60. Weinstock-Guttman B, Tamano-Blanco M, Bhasi K, Zivadinov R, Ramanathan M. Pharmacogenetics of MXA SNPs in interferon-beta treated multiple sclerosis patients. *J Neuroimmunol*. January 2007;182(1–2):236–239.
61. Gross R, Healy BC, Cepok S, et al. Population structure and HLA DRB1 1501 in the response of subjects with multiple sclerosis to first-line treatments. *J Neuroimmunol*. April 2011;233(1–2):168–174.
62. Cunningham S, Graham C, Hutchinson M, et al. Pharmacogenomics of responsiveness to interferon IFN-beta treatment in multiple sclerosis: a genetic screen of 100 type I interferon-inducible genes. *Clin Pharmacol Ther*. December 2005;78(6):635–646.
63. Vosslamber S, van der Voort LF, van den Elskamp IJ, et al. Interferon regulatory factor 5 gene variants and pharmacological and clinical outcome of Interferonbeta therapy in multiple sclerosis. *Genes Immun*. September 2011;12(6):466–472.
64. Vandenbroeck K, Alloza I, Swaminathan B, et al. Validation of IRF5 as multiple sclerosis risk gene: putative role in interferon beta therapy and human herpes virus-6 infection. *Genes Immun*. January 2011;12(1):40–45.
65. Sellebjerg F, Sondergaard HB, Koch-Henriksen N, Sorensen PS, Oturai AB. Prediction of response to interferon therapy in multiple sclerosis. *Acta Neurol Scand*. October 2014;130(4):268–275.
66. O'Brien M, Lonergan R, Costelloe L, et al. OAS1: a multiple sclerosis susceptibility gene that influences disease severity. *Neurology*. August 3, 2010;75(5):411–418.
67. Malhotra S, Morcillo-Suarez C, Nurtdinov R, et al. Roles of the ubiquitin peptidase USP18 in multiple sclerosis and the response to interferon-beta treatment. *Eur J Neurol*. October 2013;20(10):1390–1397.
68. Kulakova OG, Tsareva EY, Boyko AN, et al. Allelic combinations of immune-response genes as possible composite markers of IFN-beta efficacy in multiple sclerosis patients. *Pharmacogenomics*. November 2012;13(15):1689–1700.
69. Kulakova OG, Tsareva EY, Lvovs D, Favorov AV, Boyko AN, Favorova OO. Comparative pharmacogenetics of multiple sclerosis: IFN-beta versus glatiramer acetate. *Pharmacogenomics*. April 2014;15(5):679–685.
70. Byun E, Caillier SJ, Montalban X, et al. Genome-wide pharmacogenomic analysis of the response to interferon beta therapy in multiple sclerosis. *Arch Neurol*. March 2008;65(3):337–344.
71. Comabella M, Craig DW, Morcillo-Suarez C, et al. Genome-wide scan of 500,000 single-nucleotide polymorphisms among responders and nonresponders to interferon beta therapy in multiple sclerosis. *Arch Neurol*. August 2009;66(8):972–978.
72. Cénit MD, Blanco-Kelly F, de las Heras V, et al. Glypican 5 is an interferon-beta response gene: a replication study. *Mult Scler*. August 2009;15(8):913–917.
73. Hoffmann S, Cepok S, Grummel V, et al. HLA-DRB1*0401 and HLA-DRB1*0408 are strongly associated with the development of antibodies against interferon-beta therapy in multiple sclerosis. *Am J Hum Genet*. August 2008;83(2):219–227.
74. Buck D, Cepok S, Hoffmann S, et al. Influence of the HLA-DRB1 genotype on antibody development to interferon beta in multiple sclerosis. *Arch Neurol*. April 2011;68(4):480–487.
75. Weber F, Cepok S, Wolf C, et al. Single-nucleotide polymorphisms in HLA- and non-HLA genes associated with the development of antibodies to interferon-beta therapy in multiple sclerosis patients. *Pharmacogenomics J*. June 2012;12(3):238–245.
76. Nunez C, Cenit MC, Alvarez-Lafuente R, et al. HLA alleles as biomarkers of high-titre neutralising antibodies to interferon-beta therapy in multiple sclerosis. *J Med Genet*. June 2014;51(6):395–400.
77. Enevold C, Oturai AB, Sorensen PS, Ryder LP, Koch-Henriksen N, Bendtzen K. Polymorphisms of innate pattern recognition receptors, response to interferon-beta and development of neutralizing antibodies in multiple sclerosis patients. *Mult Scler*. August 2010;16(8):942–949.
78. Link J, Lundkvist Ryner M, Fink K, et al. Human leukocyte antigen genes and interferon beta preparations influence risk of developing neutralizing anti-drug antibodies in multiple sclerosis. *PLoS One*. 2014;9(3):e90479.
79. Comabella M, Vandenbroeck K. Pharmacogenomics and multiple sclerosis: moving toward individualized medicine. *Curr Neurol Neurosci Rep*. October 2011;11(5):484–491.
80. Goertsches RH, Zettl UK, Hecker M. Sieving treatment biomarkers from blood gene-expression profiles: a pharmacogenomic update on two types of multiple sclerosis therapy. *Pharmacogenomics*. March 2011;12(3):423–432.
81. Hesse D, Sellebjerg F, Sorensen PS. Absence of MxA induction by interferon beta in patients with MS reflects complete loss of bioactivity. *Neurology*. August 4, 2009;73(5):372–377.
82. Malucchi S, Gilli F, Caldano M, et al. Predictive markers for response to interferon therapy in patients with multiple sclerosis. *Neurology*. March 25, 2008;70(13 Pt 2):1119–1127.

83. van der Voort LF, Kok A, Visser A, et al. Interferon-beta bioactivity measurement in multiple sclerosis: feasibility for routine clinical practice. *Mult Scler*. February 2009;15(2):212–218.
84. Weinstock-Guttman B, Bhasi K, Badgett D, et al. Genomic effects of once-weekly, intramuscular interferon-beta1a treatment after the first dose and on chronic dosing: relationships to 5-year clinical outcomes in multiple sclerosis patients. *J Neuroimmunol*. December 15, 2008;205(1–2):113–125.
85. Hundeshagen A, Hecker M, Paap BK, et al. Elevated type I interferon-like activity in a subset of multiple sclerosis patients: molecular basis and clinical relevance. *J Neuroinflammation*. 2012;9:140.
86. Malhotra S, Bustamante MF, Perez-Miralles F, et al. Search for specific biomarkers of IFNbeta bioactivity in patients with multiple sclerosis. *PLoS One*. 2011;6(8):e23634.
87. van Baarsen LG, Vosslamber S, Tijssen M, et al. Pharmacogenomics of interferon-beta therapy in multiple sclerosis: baseline IFN signature determines pharmacological differences between patients. *PLoS One*. 2008;3(4):e1927.
88. Rudick RA, Rani MR, Xu Y, et al. Excessive biologic response to IFNbeta is associated with poor treatment response in patients with multiple sclerosis. *PLoS One*. 2011;6(5):e19262.
89. Comabella M, Lunemann JD, Rio J, et al. A type I interferon signature in monocytes is associated with poor response to interferon-beta in multiple sclerosis. *Brain*. December 2009;132(Pt 12):3353–3365.
90. Malhotra S, Rio J, Urcelay E, et al. NLRP3 inflammasome is associated with the response to IFN-beta in patients with multiple sclerosis. *Brain*. January 12, 2015;138.
91. Baranzini SE, Mousavi P, Rio J, et al. Transcription-based prediction of response to IFNbeta using supervised computational methods. *PLoS Biol*. January 2005;3(1):e2.
92. Baranzini SE, Madireddy LR, Cromer A, et al. Prognostic biomarkers of IFNb therapy in multiple sclerosis patients. *Mult Scler*. November 12, 2014;21.
93. Singh RP, Massachi I, Manickavel S, et al. The role of miRNA in inflammation and autoimmunity. *Autoimmun Rev*. October 2013;12(12):1160–1165.
94. Sempere LF. Integrating contextual miRNA and protein signatures for diagnostic and treatment decisions in cancer. *Expert Rev Mol Diagn*. November 2011;11(8):813–827.
95. Yokoi T, Nakajima M. microRNAs as mediators of drug toxicity. *Annu Rev Pharmacol Toxicol*. 2013;53:377–400.
96. Hecker M, Thamilarasan M, Koczan D, et al. MicroRNA expression changes during interferon-beta treatment in the peripheral blood of multiple sclerosis patients. *Int J Mol Sci*. August 5, 2013;14(8):16087–16110.
97. Waschbisch A, Atiya M, Linker RA, Potapov S, Schwab S, Derfuss T. Glatiramer acetate treatment normalizes deregulated microRNA expression in relapsing remitting multiple sclerosis. *PLoS One*. 2011;6(9):e24604.
98. De Felice B, Mondola P, Sasso A, et al. Small non-coding RNA signature in multiple sclerosis patients after treatment with interferon-beta. *BMC Med Genomics*. May 17, 2014;7:26.
99. Broadley SA, Barnett MH, Boggild M, et al. Therapeutic approaches to disease modifying therapy for multiple sclerosis in adults: an Australian and New Zealand perspective Part 3 treatment practicalities and recommendations. *J Clin Neurosci*. June 30, 2014;21(11).
100. Wiese MD, Suppiah V, O'Doherty C. Metabolic and safety issues for multiple sclerosis pharmacotherapy–opportunities for personalised medicine. *Expert Opin Drug Metab Toxicol*. August 2014;10(8):1145–1159.
101. Montalban X, Duran I, Rio J, Saez-Torres I, Tintore M, Martinez-Caceres EM. Can we predict flu-like symptoms in patients with multiple sclerosis treated with interferon-beta? *J Neurol*. April 2000;247(4):259–262.
102. Hecker M, Goertsches RH, Fatum C, et al. Network analysis of transcriptional regulation in response to intramuscular interferon-beta-1a multiple sclerosis treatment. *Pharmacogenomics J*. August 2012;12(4):360.
103. Grossman I, Avidan N, Singer C, et al. Pharmacogenetics of glatiramer acetate therapy for multiple sclerosis reveals drug-response markers. *Pharmacogenet Genomics*. August 2007;17(8):657–666.
104. Martinelli Boneschi F, Rovaris M, Johnson KP, et al. Effects of glatiramer acetate on relapse rate and accumulated disability in multiple sclerosis: meta-analysis of three double-blind, randomized, placebo-controlled clinical trials. *Mult Scler*. August 2003;9(4):349–355.
105. Karussis D, Teitelbaum D, Sicsic C, Brenner T, group AC001 Multi-Center Israeli Study Group. Long-term treatment of multiple sclerosis with glatiramer acetate: natural history of the subtypes of anti-glatiramer acetate antibodies and their correlation with clinical efficacy. *J Neuroimmunol*. March 30, 2010;220(1–2):125–130.
106. Teitelbaum D, Brenner T, Abramsky O, Aharoni R, Sela M, Arnon R. Antibodies to glatiramer acetate do not interfere with its biological functions and therapeutic efficacy. *Mult Scler*. December 2003;9(6):592–599.
107. Brenner T, Arnon R, Sela M, et al. Humoral and cellular immune responses to Copolymer 1 in multiple sclerosis patients treated with Copaxone. *J Neuroimmunol*. April 2, 2001;115(1–2):152–160.

108. Salama HH, Hong J, Zang YC, El-Mongui A, Zhang J. Blocking effects of serum reactive antibodies induced by glatiramer acetate treatment in multiple sclerosis. *Brain*. December 2003;126(Pt 12):2638–2647.
109. Sellebjerg F, Hedegaard CJ, Krakauer M, et al. Glatiramer acetate antibodies, gene expression and disease activity in multiple sclerosis. *Mult Scler*. March 2012;18(3):305–313.
110. Tumani H, Kassubek J, Hijazi M, et al. Patterns of TH1/TH2 cytokines predict clinical response in multiple sclerosis patients treated with glatiramer acetate. *Eur Neurol*. 2011;65(3):164–169.
111. Valenzuela RM, Costello K, Chen M, Said A, Johnson KP, Dhib-Jalbut S. Clinical response to glatiramer acetate correlates with modulation of IFN-gamma and IL-4 expression in multiple sclerosis. *Mult Scler*. July 2007;13(6):754–762.
112. Wiesemann E, Klatt J, Wenzel C, Heidenreich F, Windhagen A. Correlation of serum IL-13 and IL-5 levels with clinical response to Glatiramer acetate in patients with multiple sclerosis. *Clin Exp Immunol*. September 2003;133(3):454–460.
113. Sarchielli P, Zaffaroni M, Floridi A, et al. Production of brain-derived neurotrophic factor by mononuclear cells of patients with multiple sclerosis treated with glatiramer acetate, interferon-beta 1a, and high doses of immunoglobulins. *Mult Scler*. April 2007;13(3):313–331.
114. Ziemssen T, Kumpfel T, Schneider H, Klinkert WE, Neuhaus O, Hohlfeld R. Secretion of brain-derived neurotrophic factor by glatiramer acetate-reactive T-helper cell lines: Implications for multiple sclerosis therapy. *J Neurol Sci*. June 15, 2005;233(1–2):109–112.
115. Blanco Y, Moral EA, Costa M, et al. Effect of glatiramer acetate (Copaxone) on the immunophenotypic and cytokine profile and BDNF production in multiple sclerosis: a longitudinal study. *Neurosci Lett*. October 9, 2006;406(3):270–275.
116. Dhib-Jalbut S, Valenzuela RM, Ito K, Kaufman M, Ann Picone M, Buyske S. HLA DR and DQ alleles and haplotypes associated with clinical response to glatiramer acetate in multiple sclerosis. *Mult Scler Relat Disord*. October 2013;2(4):340–348.
117. Tsareva EY, Kulakova OG, Boyko AN, et al. Allelic combinations of immune-response genes associated with glatiramer acetate treatment response in Russian multiple sclerosis patients. *Pharmacogenomics*. January 2012;13(1):43–53.
118. Achiron A, Feldman A, Gurevich M. Molecular profiling of glatiramer acetate early treatment effects in multiple sclerosis. *Dis Mark*. 2009;27(2):63–73.
119. Thamilarasan M, Hecker M, Goertsches RH, et al. Glatiramer acetate treatment effects on gene expression in monocytes of multiple sclerosis patients. *J Neuroinflammation*. October 17, 2013;10:126.
120. Ottoboni L, Keenan BT, Tamayo P, et al. An RNA profile identifies two subsets of multiple sclerosis patients differing in disease activity. *Sci Transl Med*. September 26, 2012;4(153):153ra131.
121. Rommer PS, Zettl UK, Kieseier B, et al. Requirement for safety monitoring for approved multiple sclerosis therapies: an overview. *Clin Exp Immunol*. March 2014;175(3):397–407.
122. Wingerchuk DM, Carter JL. Multiple sclerosis: current and emerging disease-modifying therapies and treatment strategies. *Mayo Clin Proc*. February 2014;89(2):225–240.
123. Mehling M, Brinkmann V, Antel J, et al. FTY720 therapy exerts differential effects on T cell subsets in multiple sclerosis. *Neurology*. October 14, 2008;71(16):1261–1267.
124. Mehling M, Lindberg R, Raulf F, et al. Th17 central memory T cells are reduced by FTY720 in patients with multiple sclerosis. *Neurology*. August 3, 2010;75(5):403–410.
125. Kowarik MC, Pellkofer HL, Cepok S, et al. Differential effects of fingolimod (FTY720) on immune cells in the CSF and blood of patients with MS. *Neurology*. April 5, 2011;76(14):1214–1221.
126. Lee JY, Wang Y. Use of a biomarker in exposure-response analysis to support dose selection for fingolimod. *CPT Pharmacometrics Syst Pharmacol*. August 21, 2013;2:e67.
127. Kappos L, Bates D, Edan G, et al. Natalizumab treatment for multiple sclerosis: updated recommendations for patient selection and monitoring. *Lancet Neurol*. August 2011;10(8):745–758.
128. Sorensen PS, Jensen PE, Haghikia A, et al. Occurrence of antibodies against natalizumab in relapsing multiple sclerosis patients treated with natalizumab. *Mult Scler*. September 2011;17(9):1074–1078.
129. Oliver-Martos B, Orpez-Zafra T, Urbaneja P, Maldonado-Sanchez R, Leyva L, Fernandez O. Early development of anti-natalizumab antibodies in MS patients. *J Neurol*. September 2013;260(9):2343–2347.
130. Defer G, Mariotte D, Derache N, et al. CD49d expression as a promising biomarker to monitor natalizumab efficacy. *J Neurol Sci*. March 15, 2012;314(1–2):138–142.

131. Sorensen PS, Bertolotto A, Edan G, et al. Risk stratification for progressive multifocal leukoencephalopathy in patients treated with natalizumab. *Mult Scler*. February 2012;18(2):143–152.
132. Plavina T, Subramanyam M, Bloomgren G, et al. Anti-JC virus antibody levels in serum or plasma further define risk of natalizumab-associated progressive multifocal leukoencephalopathy. *Ann Neurol*. December 2014;76(6):802–812.
133. Hoepner R, Faissner S, Salmen A, Gold R, Chan A. Efficacy and side effects of natalizumab therapy in patients with multiple sclerosis. *J Cent Nerv Syst Dis*. April 28, 2014;6:41–49.
134. Bloomgren G, Richman S, Hotermans C, et al. Risk of natalizumab-associated progressive multifocal leukoencephalopathy. *N Engl J Med*. May 17, 2012;366(20):1870–1880.
135. Rossi F, Newsome SD, Viscidi R. Molecular diagnostic tests to predict the risk of progressive multifocal leukoencephalopathy in natalizumab-treated multiple sclerosis patients. *Mol Cell Probes*. February 2015;29(1):54–62.
136. Serana F, Chiarini M, Sottini A, et al. Immunological biomarkers identifying natalizumab-treated multiple sclerosis patients at risk of progressive multifocal leukoencephalopathy. *J Neuroimmunol*. December 15, 2014;277(1–2):6–12.
137. Schwab N, Schneider-Hohendorf T, Posevitz V, et al. L-selectin is a possible biomarker for individual PML risk in natalizumab-treated MS patients. *Neurology*. September 3, 2013;81(10):865–871.
138. Munoz-Culla M, Irizar H, Castillo-Trivino T, et al. Blood miRNA expression pattern is a possible risk marker for natalizumab-associated progressive multifocal leukoencephalopathy in multiple sclerosis patients. *Mult Scler*. May 22, 2014;20.
139. Hartung HP, Aktas O, Boyko AN. Alemtuzumab: a new therapy for active relapsing-remitting multiple sclerosis. *Mult Scler*. October 24, 2014;21.
140. Morrissey SP, Le Page E, Edan G. Mitoxantrone in the treatment of multiple sclerosis. *Int MS J*. November 2005;12(3):74–87.
141. Le Page E, Leray E, Taurin G, et al. Mitoxantrone as induction treatment in aggressive relapsing remitting multiple sclerosis: treatment response factors in a 5 year follow-up observational study of 100 consecutive patients. *J Neurol Neurosurg Psychiatry*. January 2008;79(1):52–56.
142. Cotte S, von Ahsen N, Kruse N, et al. ABC-transporter gene-polymorphisms are potential pharmacogenetic markers for mitoxantrone response in multiple sclerosis. *Brain*. July 15, 2009;132.
143. Grey Nee Cotte S, Salmen Nee Stroet A, von Ahsen N, et al. Lack of efficacy of mitoxantrone in primary progressive Multiple Sclerosis irrespective of pharmacogenetic factors: a multi-center, retrospective analysis. *J Neuroimmunol*. January 15, 2015;278:277–279.
144. Lejbkowicz I, Paperna T, Stein N, Dishon S, Miller A. Internet usage by patients with multiple sclerosis: implications to participatory medicine and personalized healthcare. *Mult Scler Int*. 2010;2010:640749.
145. Hay MC, Strathmann C, Lieber E, Wick K, Giesser B. Why patients go online: multiple sclerosis, the internet, and physician-patient communication. *Neurologist*. November 2008;14(6):374–381.
146. Harland J, Bath P. Assessing the quality of websites providing information on multiple sclerosis: evaluating tools and comparing sites. *Health Inf J*. September 2007;13(3):207–221.
147. Motl RW, Dlugonski D, Wojcicki TR, McAuley E, Mohr DC. Internet intervention for increasing physical activity in persons with multiple sclerosis. *Mult Scler*. January 2011;17(1):116–128.
148. Twomey F, Robinson K. Pilot study of participating in a fatigue management programme for clients with multiple sclerosis. *Disabil Rehabil*. 2010;32(10):791–800.
149. Miller DM, Moore SM, Fox RJ, et al. Web-based self-management for patients with multiple sclerosis: a practical, randomized trial. *Telemed J E Health*. January–February 2011;17(1):5–13.
150. Riazi A. Patient-reported outcome measures in multiple sclerosis. *Int MS J*. November 2006;13(3):92–99.
151. Hobart JC, Riazi A, Lamping DL, Fitzpatrick R, Thompson AJ. Measuring the impact of MS on walking ability: the 12-Item MS Walking Scale (MSWS-12). *Neurology*. January 14, 2003;60(1):31–36.
152. Ozura A, Kovac L, Sega S. Adherence to disease-modifying therapies and attitudes regarding disease in patients with multiple sclerosis. *Clin Neurol Neurosurg*. December 2013;115(suppl 1):S6–S11.
153. Bruce JM, Hancock LM, Lynch SG. Objective adherence monitoring in multiple sclerosis: initial validation and association with self-report. *Mult Scler*. January 2010;16(1):112–120.
154. Bruce JM, Lynch SG. Multiple sclerosis: MS treatment adherence–how to keep patients on medication? *Nat Rev Neurol*. August 2011;7(8):421–422.
155. Sorensen PS, Koch-Henriksen N, Ross C, Clemmesen KM, Bendtzen K. Danish Multiple Sclerosis Study G. Appearance and disappearance of neutralizing antibodies during interferon-beta therapy. *Neurology*. July 12, 2005;65(1):33–39.

156. Alexander JS, Harris MK, Wells SR, et al. Alterations in serum MMP-8, MMP-9, IL-12p40 and IL-23 in multiple sclerosis patients treated with interferon-beta1b. *Mult Scler*. July 2010;16(7):801–809.
157. Kurtuncu M, Tuzun E, Turkoglu R, et al. Effect of short-term interferon-beta treatment on cytokines in multiple sclerosis: significant modulation of IL-17 and IL-23. *Cytokine*. August 2012;59(2):400–402.
158. Axtell RC, Raman C, Steinman L. Interferon-beta exacerbates Th17-mediated inflammatory disease. *Trends Immunol*. June 2011;32(6):272–277.
159. Petek-Balci B, Coban A, Shugaiv E, et al. Predictive value of early serum cytokine changes on long-term interferon beta-1a efficacy in multiple sclerosis. *Int J Neurosci*. July 17, 2014;125.
160. Graber JJ, Ford D, Zhan M, Francis G, Panitch H, Dhib-Jalbut S. Cytokine changes during interferon-beta therapy in multiple sclerosis: correlations with interferon dose and MRI response. *J Neuroimmunol*. April 2007;185(1–2):168–174.
161. Bartosik-Psujek H, Stelmasiak Z. The interleukin-10 levels as a potential indicator of positive response to interferon beta treatment of multiple sclerosis patients. *Clin Neurol Neurosurg*. October 2006;108(7):644–647.
162. Rudick RA, Ransohoff RM, Lee JC, et al. In vivo effects of interferon beta-1a on immunosuppressive cytokines in multiple sclerosis. *Neurology*. May 1998;50(5):1294–1300.
163. Calabresi PA, Tranquill LR, Dambrosia JM, et al. Increases in soluble VCAM-1 correlate with a decrease in MRI lesions in multiple sclerosis treated with interferon beta-1b. *Ann Neurol*. May 1997;41(5):669–674.
164. Wiesemann E, Deb M, Trebst C, Hemmer B, Stangel M, Windhagen A. Effects of interferon-beta on co-signaling molecules: upregulation of CD40, CD86 and PD-L2 on monocytes in relation to clinical response to interferon-beta treatment in patients with multiple sclerosis. *Mult Scler*. March 2008;14(2):166–176.
165. Koda T, Okuno T, Takata K, et al. Sema4A inhibits the therapeutic effect of IFN-beta in EAE. *J Neuroimmunol*. March 15, 2014;268(1–2):43–49.
166. Nakatsuji Y, Okuno T, Moriya M, et al. Elevation of Sema4A implicates Th cell skewing and the efficacy of IFN-beta therapy in multiple sclerosis. *J Immunol*. May 15, 2012;188(10):4858–4865.
167. Nakatsuji Y. Sema4A as a biomarker predicting responsiveness to IFN beta treatment. *Rinsho Shinkeigaku*. 2014;54(12):972–974.
168. Hegen H, Millonig A, Bertolotto A, et al. Early detection of neutralizing antibodies to interferon-beta in multiple sclerosis patients: binding antibodies predict neutralizing antibody development. *Mult Scler*. April 2014;20(5):577–587.
169. Sellebjerg F, Krakauer M, Hesse D, et al. Identification of new sensitive biomarkers for the in vivo response to interferon-beta treatment in multiple sclerosis using DNA-array evaluation. *Eur J Neurol*. December 2009;16(12):1291–1298.
170. Wandinger KP, Lunemann JD, Wengert O, et al. TNF-related apoptosis inducing ligand (TRAIL) as a potential response marker for interferon-beta treatment in multiple sclerosis. *Lancet*. June 14, 2003;361(9374):2036–2043.
171. Buttmann M, Merzyn C, Hofstetter HH, Rieckmann P. TRAIL, CXCL10 and CCL2 plasma levels during long-term Interferon-beta treatment of patients with multiple sclerosis correlate with flu-like adverse effects but do not predict therapeutic response. *J Neuroimmunol*. October 2007;190(1–2):170–176.
172. Deisenhammer F, Reindl M, Harvey J, Gasse T, Dilitz E, Berger T. Bioavailability of interferon beta 1b in MS patients with and without neutralizing antibodies. *Neurology*. April 12, 1999;52(6):1239–1243.
173. Avolio C, Filippi M, Tortorella C, et al. Serum MMP-9/TIMP-1 and MMP-2/TIMP-2 ratios in multiple sclerosis: relationships with different magnetic resonance imaging measures of disease activity during IFN-beta-1a treatment. *Mult Scler*. August 2005;11(4):441–446.
174. Kuhle J, Malmestrom C, Axelsson M, et al. Neurofilament light and heavy subunits compared as therapeutic biomarkers in multiple sclerosis. *Acta Neurol Scand*. December 2013;128(6):e33–e36.
175. Gunnarsson M, Malmestrom C, Axelsson M, et al. Axonal damage in relapsing multiple sclerosis is markedly reduced by natalizumab. *Ann Neurol*. January 2011;69(1):83–89.
176. Wipfler P, Oppermann K, Pilz G, et al. Adhesion molecules are promising candidates to establish surrogate markers for natalizumab treatment. *Mult Scler*. January 2011;17(1):16–23.
177. Jilek S, Mathias A, Canales M, et al. Natalizumab treatment alters the expression of T-cell trafficking marker LFA-1 alpha-chain (CD11a) in MS patients. *Mult Scler*. November 20, 2013;20.
178. Stuve O, Marra CM, Jerome KR, et al. Immune surveillance in multiple sclerosis patients treated with natalizumab. *Ann Neurol*. May 2006;59(5):743–747.
179. Harris VK, Donelan N, Yan QJ, et al. Cerebrospinal fluid fetuin-A is a biomarker of active multiple sclerosis. *Mult Scler*. October 2013;19(11):1462–1472.

180. Hecker M, Paap BK, Goertsches RH, et al. Reassessment of blood gene expression markers for the prognosis of relapsing-remitting multiple sclerosis. *PLoS One*. 2011;6(12):e29648.
181. Achiron A, Gurevich M, Snir Y, Segal E, Mandel M. Zinc-ion binding and cytokine activity regulation pathways predicts outcome in relapsing-remitting multiple sclerosis. *Clin Exp Immunol*. August 2007;149(2):235–242.
182. Gurevich M, Tuller T, Rubinstein U, Or-Bach R, Achiron A. Prediction of acute multiple sclerosis relapses by transcription levels of peripheral blood cells. *BMC Med Genomics*. July 22, 2009;2:46.
183. Singh MK, Scott TF, LaFramboise WA, Hu FZ, Post JC, Ehrlich GD. Gene expression changes in peripheral blood mononuclear cells from multiple sclerosis patients undergoing beta-interferon therapy. *J Neurol Sci*. July 15, 2007;258(1–2):52–59.

CHAPTER

27

Stem Cell-Based Therapies, Remyelination, and Repair Promotion in the Treatment of Multiple Sclerosis

A. Merlini, D. De Feo, M. Radaelli, C. Laterza, F. Ciceri, G. Martino

San Raffaele Scientific Institute, Milan, Italy

OUTLINE

Translational Neuroimmunology in Multiple Sclerosis
http://dx.doi.org/10.1016/B978-0-12-801914-6.00027-1

27.1 INTRODUCTION

Multiple sclerosis (MS) arises from an autoimmune response against central nervous system (CNS) antigens, which leads to irreversible neural and myelin damage, thus leading to neurodegeneration.

Current treatments for MS encompass immunomodulatory and immunosuppressive drugs that are, however, only partially effective in controlling peripheral inflammation and do not have a clear direct effect on CNS pathology, showing limited efficacy in patients with either a progressive or an aggressive disease course.[1]

Stem cell-based therapeutic approaches have received great attention as a potential therapeutic strategy for MS, given their ability to replace damaged cells and to dramatically modify their behavior in response to microenvironment (Fig. 27.1).

In this chapter we will address three main classes of adult stem cells for the treatment of MS: hematopoietic stem cells (HSCs), mesenchymal stem cells (MSCs), and neural stem/precursor cells (NPCs).

HSC transplantation (HSCT), which has been proposed and used in the past 15 years to control or even cure refractory cases of MS,[2,3] has the aim to renew and reset the immune system with high-dose chemotherapy, lymphodepletion, and secondary induction of tolerogenic cell subtypes.

MSCs on the other hand, rather than replacing the immune cell repertoire, as in HSC transplantation, interact directly with immune cells, driving peripheral tolerance and inhibiting the autoinflammatory cascade.

NPCs possess the ability to repair to some extent CNS tissue damage via cell replacement and at the same time promote neuroprotection and immunomodulation not only in the

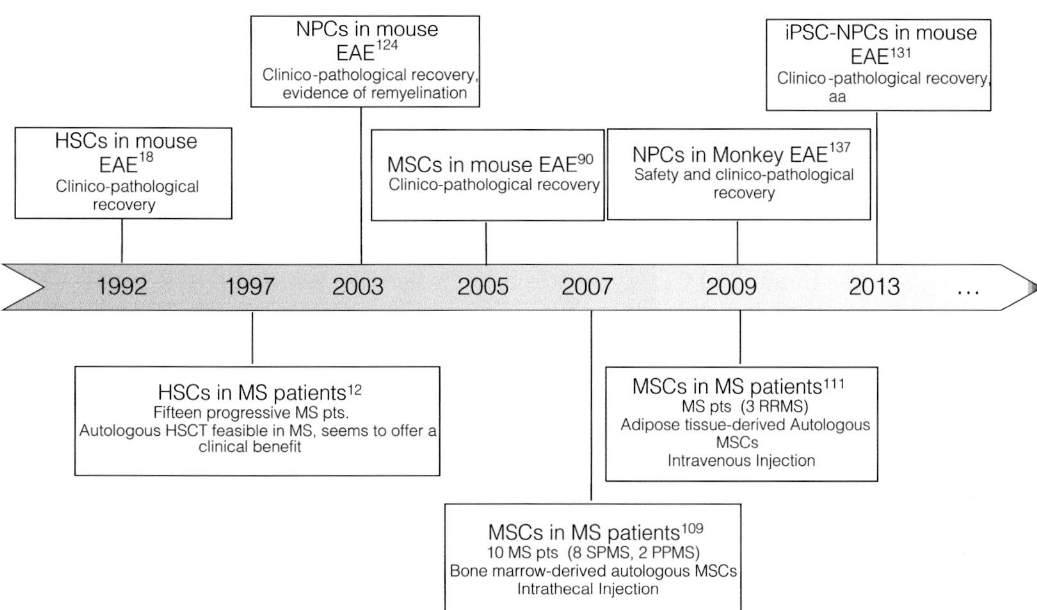

FIGURE 27.1 **Preclinical and clinical milestones in the advancement of stem cell therapy in multiple sclerosis (MS).** The first experimental evidence that stem cells could be a useful immunomodulatory therapeutic approach in MS comes from transplantation studies in experimental autoimmune encephalomyelitis (EAE) mice of intravenous infusion of bone marrow-derived stem cells after intense immunosuppression.[17] Soon after, the first trial of autologous hematopoietic stem cell transplantation in progressive MS was published.[11] A few years later, in 2003, stem cell-based regenerative therapies were initiated and tested in preclinical models of MS. The first stem cells to be used were neural stem/precursor cells (NPCs).[123] Such cells, transplanted intrathecally and intravenously in mice with EAE after disease onset, promoted both clinical and pathological disease amelioration. Years later human fetal brain-derived NPCs have been successfully injected, both intrathecally and intravenously, in a nonhuman primate model of MS, thus confirming the regenerative potential of NPCs in autoimmune demyelination as demonstrated in rodent studies.[136] Inspired by the positive results obtained using NPCs, in 2005 MSCs have been intravenously injected into EAE mice.[89] Results have been extremely positive and this was considered sufficient to start safety trials in MS patients using autologous bone marrow-derived MSCs.[108,110]

periphery but also within the CNS, where they can persist in ectopic stem cell niches producing neurotrophic and immunomodulatory factors.

While the therapeutic aim of these three stem cell subtypes markedly varies, a common leitmotif is that HSC, MSC, and NPC therapeutic effect depends on the presence of active inflammation, thus suggesting that stem cells also need a permissive microenvironment to exert their therapeutic function.

27.2 HEMATOPOIETIC STEM CELLS

27.2.1 Rationale and Procedure of Hematopoietic Stem Cell Transplantation

HSCT exerts its therapeutic effect in MS through various mechanisms: on one hand, the immunosuppressive conditioning regimen prior to HSCT is able to temporarily eradicate the autoreactive cells; on the other hand, the regeneration/renewal of the immune system resets the aberrant immune response to self-antigens.[4] Additional biological mechanisms are considered as contributors to the therapeutic effect of HSCT: the induction of immune tolerance, the halting of the inflammatory activity, and the prevention of further relapses of the disease are considered the most important.[5]

In the first phase of HSCT, autologous HSCs can be collected directly from bone marrow or by leukapheresis from the peripheral blood. Nowadays, there is general agreement on the mobilization procedure: it is recommended to perform a peripheral blood mobilization with cyclophosphamide (Cy) 2–4 gr/m^2 plus granulocyte-colony stimulating factor (G-CSF) 5–10 g/kμg and the number of CD34$^+$ cells to be collected is a minimum of 2–3 × 10^6 CD34$^+$ cells/kg.[2,6] Cy was introduced to reduce the occurrence of relapses after G-CSF administration.[7–9] Cryopreservation techniques allow long-term storage of HSCs without significant loss of viability. Afterward, HSCs are infused into the patient after a conditioning regimen with high-dose chemotherapy. Lymphocyte depletion of autologous HSCs can be achieved by CD34$^+$ positive selection of the graft or by in vivo T-cell depletion with antithymocyte globulin (ATG). Different conditioning regimens may be administered before the reinfusion of CD34$^+$ autologous cells. They can be categorized according to their myeloablative potentials: (1) high-intensity conditioning regimens include total body irradiation (TBI) or high-dose busulphan; (2) low-intensity (ie, nonmyeloablative) conditioning regimens include Cy alone, melphalan alone, and fludarabine-based regimens; and (3) intermediate-intensity conditioning regimens include other combinations, such as in most patients, BEAM (carmustine 300 mg/m^2 on day 6, cytosine arabinoside, 200 mg/m^2 and etoposide 200 mg/m^2 day 5 to day 2, melphalan 140 mg/m^2 day 1), or the combined use of ATG with high-dose Cy.

In the first studies the use of high-intensity regimens such as TBI or busulfan was preferred, whereas in more recent studies low-intensity, immune-suppressive, nonmyeloablative treatments were introduced.[10] The different procedures used during the last years reflect the tendency by hematologists and neurologists toward patient-tailored approaches. In general, the most frequent conditioning regiment for HSCT in MS is BEAM-ATG. In the 2006 revision of the European Group for Blood and Marrow Transplantation (EBMT) database it was used in 40% (74/178) of MS-treated patients. This conditioning regimen is also suggested in the HSCT guidelines for MS patients.[11,12]

27.2.2 Effect of Autologous Hematopoietic Stem Cell Transplantation on Peripheral Immunity

Remission of disease activity in patients with severe active MS upon autologous HSCT relies not only on lymphodepletion but also on the renewal of the immune compartment.[13]

Increased thymopoiesis likely contributes to T-cell renewal after autologous HSCT.[13,14] A thymic origin of the expanded naive T-cell subsets was confirmed by the long-term analysis of immune reconstitution after autologous HSCT in MS patients.[13] Myeloablative conditioning followed by autologous HSCT extensively renewed CD4$^+$ T-cell clone subsets (>90% in most cases). Importantly, patients who failed to respond to treatment had less diversity in their T-cell repertoire early during the reconstitution process, suggesting that repertoire complexity is critical for the reestablishment of immune tolerance.[15]

Interestingly, on the other hand, after nonmyeloablative conditioning, both the patient's bone marrow HSC and graft repopulate the T-cell repertoire.[16]

Darlington and colleagues showed that, while T helper (T_H) cells type 1 were unchanged upon autologous HSCT, encephalitogenic T_H17 cells were selectively reduced. Moreover, reduction of IL-1β and IL-6 after autologous HSCT might influence the functional reconstitution of antigen-presenting cells, which in turn could impair T_H17 polarization.[14]

Beside changes in the T_H repertoire, autologous HSCT induces profound modifications also in the immunoregulatory compartment with a transient increase in regulatory FoxP3$^+$ T cells and CD56^{high} natural killer (NK) cells.[14,16] After nonmyeloablative conditioning, CD8$^+$CD57$^+$ cytotoxic T cells, a suggested immunoregulatory population, were persistently increased. The detailed characterization of CD8$^+$ subsets revealed that a proinflammatory CD161^{high}CD8$^+$ T cell population, readily detectable in all patients before treatment, was depleted after autologous HSCT. Such T cells were mucosal-associated invariant T cells (MAIT cells), which originate in the gut mucosa, produce IFN-γ, TNF-α, and IL-17 at high levels and express the CNS-homing receptor CCR6. MAIT cells were also found in MS active lesions, suggesting their involvement in MS pathogenesis.

The depletion of CD8$^+$ cells that produce IL-17 (Tc17) after autologous HSCT complements the finding that the capacity to mount T_H17 responses is diminished posttherapy.[14] Together, these studies indicate that alterations of the T_H17 and Tc17 might represent a crucial mechanism in the therapeutic effect of autologous HSCT in MS.

In conclusion, autologous HSCT in MS renews the CD4$^+$ repertoire, selectively blunts the encephalitogenic effector response by reducing Tc17 and T_H17, impairs antigen presentation, and increases immunoregulatory cell subsets.

27.2.3 Effect of Autologous Hematopoietic Stem Cell Transplantation on Central Nervous System Pathology

Preclinical studies of autologous HSCT in experimental autoimmune encephalomyelitis (EAE, the animal model of MS) contributed to the optimization of autologous HSCT and to the understanding of its effect on CNS pathology. While transplantation at the peak of EAE neuroinflammation resulted in accelerated recovery,[17] protection from disability was not seen when autologous HSCT was carried out in the chronic stages of EAE or when EAE was especially severe, with chronic activation of microglial cells and maturation of reactive astrocytes.[18]

Indeed, histological analysis of the brain of MS patients who had received autologous HSCT showed that, despite a profound effect on CNS inflammatory infiltrate, with very few CD8$^+$ cells, no B and plasma cells, numerous lesions showed active demyelination and axonal loss, indicative of ongoing parenchymal damage. Areas of active tissue loss were accompanied by reactive inflammation with macrophages and resident activated microglia, suggesting that autologous HSCT might only partially affect CNS-confined inflammation.[19]

27.2.4 Clinical Outcome: Safety and Efficacy

According to the EBMT Registry, so far, 572 MS patients have undergone HSCT.[2]

Overall, data from four different studies are available about long-term follow-up of MS patients undergoing HSCT and enrolled in trials conducted from 1997.[11,12] Mobilization was successfully obtained in almost all MS patients without severe adverse events. The most frequent side effect observed during this treatment phase was neurological deterioration due to disease reactivation. This risk was reduced with the concomitant use of Cy and steroids in association with G-CSF although even in this case a transient worsening of MS symptoms was reported by some authors. In a recent review of the EBMT database, the transplant related mortality of autologous HSCT in MS patients dropped down to 1.3%.[2] These positive results can be attributed to both a better supportive care and a better selection of patients, as more recent studies included patients with lower Expanded Disability Status Scale (EDSS) scores and in a relapsing-remitting phase of the disease. They also support the hypothesis of the negative effect of a too intensive conditioning regimen, as none of the studies beginning after 2000 used TBI or busulfan or an ex vivo purging of the graft.

In the majority of reports, efficacy was expressed as progression-free survival, defined as the absence of a confirmed increase of EDSS by at least one point. In the studies with a follow-up of at least 2 years it ranged from 36% to 100%; only a minority of patients showed a real improvement of EDSS.[11,20–22]

A better outcome seems to be more evident in MS patients with an active disease,[21–24] a short disease duration,[21,25] and a lower EDSS score,[10,25,26] and in studies including relapsing-remitting MS rather than primary-progressive MS or secondary-progressive MS patients,[26,27] thus suggesting that the prevalent role of HSCT in arresting disability progression is mainly related to the control of the peripheral immunity rather than to a direct effect on CNS-confined pathology. A dramatic effect on the reduction in the relapse rate was observed in almost all studies.[23] When available, MRI data confirmed the same positive effect of HSCT in patients with active disease.[22,23,28] The important effects on the inflammatory phases of the disease was further confirmed in patients affected by malignant forms of MS, where a marked and sustained improvement of the EDSS was observed even with a relatively long-term follow-up.[29]

27.2.5 Beyond Autologous Hematopoietic Stem Cell Transplantation in Multiple Sclerosis: Hematopoietic Stem Cell Transplantation in Other Autoimmune Central Nervous System Disorders and Allogeneic Hematopoietic Stem Cell Transplantation

Concerning other inflammatory disorders of the CNS, autologous HSCT has been reported as a possible therapeutic approach in cases of systemic lupus erythematosus with CNS involvement,[30,31] and brain histiocytosis.[32] However, the number of treated patients is too few to draw any meaningful conclusion about the therapeutic efficacy of HCST in the aforementioned diseases.

On the other hand, a retrospective survey of the EBMT working group on autoimmune disorders reported the results of autologous HSCT treatment in 16 cases of neuromyelitis optica (NMO)[33]: after a median follow-up of 47 months, 3/16 (18%) patients remained progression- and treatment-free, 9/16 patients (56%) progressed, and 13/16 (81%) patients experienced relapses within 7 months of follow-up. Importantly, the pathogenic antiaquaporin 4 (AQP4) antibodies persisted at follow-up in 8/8 patients (ie, all the patients for which this information was available).

Concerning allogeneic HSCT (allo-HSCT), it is certainly plausible to consider this approach to be more effective than autologous HSCT in malignant cases of CNS autoimmunity due to a stronger conditioning regimen, donor-versus-host alloreactivity (graft vs. autoimmunity), and generation of a healthy immune system, tolerant to autoantigens.[34] Indeed, two NMO patients who failed to benefit from autologous HSCT have been reported to have been treated with allo-HSCT. Both patients showed at the latest follow-up, >3 years after allo-HSCT, clinical and radiological stability with actively improving motor abilities. The impressive amelioration correlated with the disappearance of anti-AQP4 antibodies.[35] On the other hand, five MS patients who underwent allo-HSCT because of concomitant hematological malignancy showed inconsistent effects of allo-HSCT on the course of the disease. Three patients showed improvement or stabilization,[36–38] while two patients had persistent clinical and MRI activity at follow-up.[39,40] Indeed, postmortem findings showed persistent CNS-confined inflammation and active demyelination,[41] with most $CD45^+$ and $CD68^+$ cells within the brain still of the recipient's origin.[39]

27.2.6 Conclusions

Autologous HSCT has been recognized as a clinical option for refractory MS. HSCT seems particularly fit for young patients, with short disease duration, no or minor irreversible disability, and a malignant presentation. Those patients could benefit from considerable persistent suppression of disease activity without the need for further therapies. However, the ever-growing spectrum of very effective approved drugs for MS, such as natalizumab and, more recently, alemtuzumab, imposes a reevaluation of autologous HSCT indications. On that note, an important step toward the positioning of autologous HSCT within the treatment algorithm of MS patients has been performed. A multicenter, randomized, phase II study (ie, Autologous Hematopoietic Stem Cell Transplantation trial in MS) has been performed to assess the effect on the disease activity of autologous HSCT compared to mitoxantrone (MTX), an approved treatment for aggressive MS.[42] Autologous HSCT showed to be superior to MTX in reducing the number of new and active MRI lesions as well as the annualized relapse rate but not disease progression.

27.3 MESENCHYMAL STEM CELLS

27.3.1 Mesenchymal Stem Cells and Their Role in Bone Marrow

MSCs were originally identified as colony-forming unit fibroblasts in bone marrow stroma capable of transferring the microenvironment of hematopoietic tissues upon transplantation to an ectopic site.[43] They can be isolated from almost any connective tissue but this chapter will focus on bone marrow-derived MSCs, which are the most well-characterized subtype.

MSCs are defined in vitro by several features: (1) growth as adherent cells; (2) fibroblast-like morphology; (3) formation of colonies in vitro that support hematopoiesis; (4) capacity to differentiate into cells of the three mesodermal lineage; and (5) maintenance of a stable phenotype characterized by stromal markers such as CD29, CD51, CD73, CD90, and CD105, but not hematopoietic lineage markers as CD31, CD34, or CD45.[44]

Conversely in vivo there are no solid markers for tracking MSC. Perivascular CD45-$CD146^+$ pericytes were proposed to be the progenitors of MSCs.[45,46] However, Mendez-Ferrer and colleagues identified MSCs as a rare subset of perivascular nestin-positive cells that may be distinct from pericytes and are an essential component of the niche of HSCs.[47] Inside the HSC niche, MSCs support the maintenance and the self-renewal of HSCs by shielding them from differentiation and apoptotic stimuli and controlling their recruitment in the vascular niche.[48] A similar role for MSCs was recently suggested in immunological memory, in which bone marrow stromal cells are able to maintain memory T cells and plasma cells in a nonproliferative state.[49]

It is remarkable that these niche functions anticipate most of the therapeutic properties of the MSC observed in experimental models.

27.3.2 Mechanisms of Action

Systemically administered MSCs, despite being trapped in large numbers in the lungs,[50] preferentially home to sites of injury.[51] As do immune cells and other stem cells,[52] MSCs show a coordinated rolling and adhesion behavior on endothelial cells[53] and migrate in response to several chemokines[54] that bind to cognate receptors expressed on their cell surface and lead to the activation of matrix metalloproteinases that degrade the basement membrane and allow subsequent extravasation.[55] Many studies have attempted to exploit the potential of MSCs to replace damaged resident cells; however, evidence of MSCs' putative ability to transdifferentiate into neuronal cell lines is limited.[56–58] However, in most cases of systemic injection, MSC engraftment is poor, and engrafted MSCs tend to be short-lived.[59,60] Also, in experimental animal models, the success of MSC therapy does not correlate with the efficiency of cell engraftment and replacement.[50,61] Accordingly, inflammatory diseases have been effectively treated with only the culture supernatants of MSCs (the MSC secretome) containing growth factors, such as hepatocyte growth factor (HGF)[61] or TSG6 (tumor-necrosis factor (TNF)-stimulated gene 6).[50] It has been shown that MSCs are the target of NK[62] and γδ T cell[63] killing in vitro and that are rejected in vivo under mismatched conditions. These findings suggest that extended engraftment of MSCs is unnecessary for their therapeutic effect.[64,65]

The therapeutic effects of MSCs largely depend on their capacity to regulate inflammation and tissue homeostasis via an array of immunosuppressive factors, cytokines, growth factors, and differentiation factors.[44] These include interleukin 6 (IL-6), transforming growth factor-β (TGF-β), prostaglandin E2, hepatocyte growth factor (HGF), epidermal growth factor (EGF), fibroblast growth factor (FGF), platelet-derived growth factor (PDGF), vascular endothelial growth factor, insulin growth factor, stromal cell-derived factor 1, the tryptophan-catabolic enzyme IDO, and nitric oxide.[66] Together, these secreted factors may inhibit inflammatory responses, promote endothelial and fibroblast activities, and facilitate the proliferation and differentiation of in situ progenitor cells of the tissues.

Several studies have demonstrated that MSCs can modulate the effector functions of macrophages, NK cells, dendritic cells (DCs), T and B lymphocytes, which have a role in pathogenesis of autoimmune diseases.[44] Beside the initial observation of the effect on T-cell proliferation,[67] MSCs are known to inhibit T_H1[68] and T_H17 differentiation, thus promoting the generation of regulatory T cells (Tregs),[69] and to foster the production of TGF-β by macrophage, which in turn amplifies Treg population.[70] MSCs can also interfere with the humoral arm of adaptive immunity by inhibiting B-cell differentiation and the constitutive expression of chemokine receptors[71] through the release of soluble factors and by cell–cell contact.[72]

B-cell responses are mainly T cell-dependent and therefore the final outcome of the interaction between MSCs and B cells in vivo might be significantly influenced by the MSC-mediated inhibition of T-cell functions. This is the case of a study of proteolipid protein peptide (PLP)-induced EAE, in which the production of antigen-specific antibodies in vivo was inhibited by the infusion of MSCs, in parallel to a significant downregulation of PLP-specific T-cell responses, which indicates that the two events were closely linked.[73]

Moreover, MSCs can modulate also the innate immune response, by inhibiting DC differentiation and maturation in vitro,[74,75] inducing a tolerogenic phenotype[76,77] and promoting the switch from a proinflammatory type 1 to an antiinflammatory type 2 phenotype[78,79] in macrophages.

Studies highlighted that MSCs need to be activated to exert their immunomodulatory effect by the stimulation with IFNγ[80,81] and other inflammatory cytokines such as TNF-α, IL-1β,[80] or IL-17.[82] On the contrary, the exposure to toloregenic cytokines as TGF-β and IL-10 impairs MSC immunoregulatory functions. Considering the interactions of MSCs within an inflamed tissue, these observations lead a series of consequences: (1) given that MSCs also secrete a number of chemokines able to attract T cells in close proximity to the stem cells,[80] if MSCs are not properly licensed by the microenvironment they can potently foster inflammation instead of controlling it; (2) since upon activation MSCs secrete and induce the production in the tissue of resolution cytokines as TGF-β,[83] the described effect of TGF-β[83] and IL-10 on MSCs represent a negative feedback loop that controls the immunomodulatory properties of MSCs[84]; (3) the coadministration of immunosuppressant agents might ablate the MSC immunomodulatory effect.[85,86]

Finally, together with the microenvironment also the tissue type and the species from which MSCs were derived can affect their immunomodulatory properties.[66,87]

27.3.3 In Vivo Effects in Preclinical Model of Central Nervous System Autoimmunity

The initial evidence that MSCs can affect T-cell proliferation together with the idea that systemically injected MSCs home to different organs including the CNS introduced the possibility that MSCs might be effective in halting inflammation and promoting repair in autoimmune disorders such as MS.

Zappia and coworkers demonstrated that upon intravenous (i.v.) infusion in EAE mice, syngeneic MSCs homed in secondary lymphoid organs, inducing peripheral T-cell tolerance to the antigen used for the immunization (ie, myelin oligodendrocyte glycoprotein (MOG_{35-55})), leading to a large improvement in clinical course, to the reduction of demyelination and CNS infiltration by T cells and macrophages.[88] Consistently with the idea that MSCs have to be licensed by the inflammatory milieu to exert their immunomodulation,[66] this effect was reported exclusively when MSCs were infused at disease onset, but could not improve neurological deficits when the disease was already in the chronic progressive phase.[88]

A subsequent report demonstrated that i.v. administration of MSCs to EAE-affected SJL mice also inhibited the pathogenic B-cell response.[73] Many other groups have subsequently confirmed that MSCs, isolated from either human or mouse tissues, can improve the clinical course in EAE, decreasing demyelination and promoting tissue repair.[64,73,89–91]

Interestingly, the effect of MSCs on encephalitogenic IL-17-releasing CD4 T cells was achieved through a metalloproteinase-mediated paracrine proteolysis of CCL2, leading to an increase in PDL1.[64] Other investigators reported that interactions between PDL1 on MSCs and PD1 on T cells are involved in the inhibition of T-cell[72] and B-cell proliferation.[92] The evidence suggests that there is an important cross-talk between MSC and lymphocytes that requires both cell-to-cell and paracrine interactions.

However, only few studies provided some evidence that a small number of MSCs can engraft in the CNS and possibly acquire a neural phenotype,[93] particularly if injected directly in the CNS.[94] However we cannot exclude the possibility that MSCs may have undergone cell fusion[95,96] or, upon cell death, may have phagocytized by macrophage or microglia.[97]

Irrespective of little or no integration in the CNS, MSCs not only act in the periphery to inhibit autoimmune attack of the CNS, but they can also support some degree of tissue repair inside the CNS. Although halting of immune-mediated damage to the CNS certainly can lead to sparing of axons,[73,93,94] several lines of evidence suggest that MSCs can directly protect neural tissues through paracrine mechanisms.[44] In animal models of neurological diseases, a direct in vivo neuroprotective effect mediated by MSCs is sustained by their ability to recruit local oligodendrocyte precursors, thus inducing oligodendrogenesis,[89–91,98] rescue of neural cells from apoptosis,[99–101] secretion of neurotrophic factors,[91] inhibition of inflammation-associated oxidative stress,[102] and modulation of microglia activation, possibly inducing a switch toward a protective phenotype.[100,103]

27.3.4 Mesenchymal Stem Cells for Multiple Sclerosis Treatment: Clinical Studies

Until now, four clinical reports have been published[104–107] and 17 clinical trials are still ongoing (clinicaltrial.gov) the treatment of MS with MSCs. All studies had an open-label design and employed autologous MSCs.

In these pioneer clinical studies, MSCs were administered intrathecally,[105,106] intrathecally plus intravenously,[107] or exclusively intravenously.[104] The rationale for intrathecal administration was to deliver the cells directly into the CNS, thereby overcoming the limited amount of cells engrafting upon i.v. administration[60] and enhancing MSC ability to promote repair by secreting neurotrophic factors, such as brain-derived neurotrophic factor (BDNF)[108] and antioxidant molecules.[109] However, it should be emphasized that most beneficial results observed in EAE were achieved following i.v. administration, which was sufficient not only to induce peripheral tolerance to myelin antigens[88] but also to preserve axons[73,93,94] and foster remyelination.[90,91] Furthermore, the recent demonstration that soluble molecules released by MSC recapitulate all the beneficial effects observed when MSC are injected in mice with EAE further minimizes the importance of CNS engraftment.[61] Accordingly, a study comparing i.v. versus intrathecal administration in mice with EAE demonstrated similar effects on clinical and histological parameters.[110]

The dose of MSCs injected in clinical trials with MS patients vary among studies and oscillates in a range of 1–2 × 106 per Kg of body weight, as previously established in pivotal trials with hematologic diseases[111] and in patients with other immune-mediated diseases such as Crohn's disease.[112] Due to the lack of effective treatments for the progressive phase of MS, when inflammation fades and neurodegeneration is the major determinant of disability, subjects with progressive MS,[104–106] or with active disease not responding to approved treatments,[107] represent the ideal candidates for adult stem cell administration. In fact, the recent trial published by Connick et al. demonstrated the effectiveness of i.v.-administered MSC on visual parameters in progressive patients.[104]

No serious adverse events have been reported in clinical studies of MSC for MS, excluding a case of transient encephalopathy with seizures in a subject receiving a dose equal to 100 × 106 MSCs intrathecally,[105] or of meningeal irritation after intrathecal MSC injection.[105–107] These data, together with the current experience with MSC in other diseases, suggests that they are safe and well tolerated.

However, the uncontrolled nature of the trials and the limited number of patients enrolled make these encouraging results yet anecdotal. On the basis of these considerations, scientists and clinicians reached a consensus in 2010 to guide the use of MSCs for the treatment of MS.[113] This consensus has been used to develop a clinical trial as a coordinated international effort to address the safety and the efficacy of autologous MSCs on MRI-based outcomes in inflammatory forms of MS.

27.3.5 Considering Immunological Plasticity for Long-Term Safety and Efficacy

The therapeutic plasticity of MSCs, based on the release of antiinflammatory, antiapoptotic, and trophic cytokines, modify the microenvironment of injured tissue, recapitulating the physiological activity of this cells within the HSC niche. However, also MSC immunomodulatory activity seems to be tightly modulated by pro- and antiinflammatory cues as confirmed by the evidence of efficacy in preclinical studies only in the acute phase of the disease. The limited evidence of persistent engraftment in damaged tissue upon transplantation, despite the proved therapeutic efficacy in several animal models of autoimmune diseases, fosters the idea of a touch-and-go mechanism for MSCs that are rapidly recruited to the site of tissue injury and rapidly cleared after the release of their healing molecules.

27.4 NEURAL/PRECURSOR STEM CELLS

27.4.1 Sources and Definition of Neural/Precursor Stem Cells

NPCs are proliferating, self-renewing, multipotent cells that can be isolated from embryonic, fetal, neonatal, or adult CNS tissues.[114] NPCs are cultivated in a serum-free selective medium enriched with EGF, and FGF-2, which maintain NPCs in a proliferative undifferentiated state as multicellular floating spheres (neurospheres) or as an adherent monolayer. Upon growth factor withdrawal, NPCs spontaneously differentiate into postmitotic neuroectodermal cells (ie, neurons, astrocytes, or oligodendrocytes).[115] In the adult CNS, at least two distinct germinal neurogenic niches have been detected, the subventricular zone of the lateral ventricles and the subgranular zone of the hippocampal dentate gyrus.

27.4.2 Rationale of Neural Stem/Precursor Cell Therapy in Multiple Sclerosis

In the unperturbed adult brain, the main function of NPCs is supplying previously formed circuits with new neurons. Disease and injury states in the brain induce NPCs to start gliogenesis or neurogenesis in variable proportions according to the pathogenic process.[116]

In animal models, focal demyelination induced by toxic agents (ie, cuprizone and lysolecithin) triggers oligodendrogenesis by subventricular zone neural stem cells, which substantially contribute to the remyelination process.[117,118]

Similarly, neuronal loss, induced by acute ischemia or excitotoxicity, is followed by neurogenesis from NPCs.[119]

Therefore, given NPCs' ability to mount an appropriate repair response upon environmental cues, NPC transplantation has been advocated as a potential therapeutic strategy for numerous neurological diseases, including MS, characterized by irreversible glial and/or neuronal loss.[120]

27.4.3 Preclinical Evidence of Transplanted Neural Stem/Precursor Cell Efficacy in Animal Models of Multiple Sclerosis

Upon intravenous, intraparenchymal, or intrathecal administration, transplanted NPCs induce clinical amelioration in EAE. The clinical amelioration is substantiated by reduced axonal loss and demyelination at neuropathology analysis.[121]

A small population of NPCs is able to integrate within the EAE parenchyma upon either systemic or local transplantation. Two-photon and electron microscopy revealed that transplanted NPCs contribute to remyelination and limit axonal damage by differentiating into oligodendroglial cells and interacting with host axons.[121,122] However, the majority of transplanted NPCs in EAE remains undifferentiated and localizes preferentially within inflammatory perivascular infiltrates.[52]

This phenomenon has been ascribed to the shared signaling pool between the inflammatory cascade and neural stem cell niche. In particular, during EAE, inflammatory infiltrating cells and resident CNS cells produce inflammatory mediators (CXCL12, IL1beta, IL6), growth factors (VEGF, TGF-β, EPO, LIF, CNTF, PDGF), and stem cell regulators (BMP4, noggin, SHH, tenascin C) that are known to have a direct effect on NPC differentiation and proliferation.[123] In turn, NPCs constitutively express integrins ($\alpha_4\beta_1$), adhesion molecules (CD44, PSA-NCAM), and chemokine receptors (CCR1-2-3-5, CXCR3-4, CX3CR1, c-Met), which make them competent to sense inflammatory signals and migrate toward sites of active damage or inflammation.[52,124–126]

There, undifferentiated NPCs exert a neuroprotective effect by upregulating neurotrophin levels within EAE tissue and preventing glial scar formation via the secretion of TGF-β and FGF-2.[121] As a partial confirmation of this mechanism, in toxic-induced demyelination, NPC transplantation induces proliferation and differentiation of oligodendrocyte progenitor cells (OPCs) via the secretion of PDGF and FGF-2.[127] In the presence of inflammation, transplanted NPCs also secrete leukemia inhibitory factor (LIF), which enhances oligodendrocyte survival and oligodendrocyte precursor differentiation.[128]

This bystander effect of NPCs in EAE is not limited to resident CNS cells and affects the autoreactive inflammatory response as well. In particular, in vitro, neurospheres inhibit the proliferation and activation of T cells, partly through the inhibition of interleukin-2 and interleukin-6 signaling,[129,130] and T cells cocultured with neurospheres produce less IFNγ and TNF-α, while increasing TGF-β expression. NPCs selectively increase the apoptosis of antigen-specific Th1 proinflammatory cells both through death receptors, including FasL, TRAIL, and APO3L, on the surface of NPCs.[52]

NPCs' profound effect on T-cell autoreactivity has been confirmed by the observation that lymph node cells obtained from i.v. NPC-treated mice exhibited poor encephalitogenicity on transfer to naive mice and caused a milder EAE compared with those obtained from non-treated mice.[131]

Transplanted NPCs influence also Th17 differentiation via the secretion of LIF, which antagonizes IL-6-mediated phosphorylation of signal transducer and activator of transcription 3 (STAT3), a fundamental step for Th17 cell differentiation in peripheral lymphoid organs.[132]

NPC transplantation also impairs antigen presentation by DCs. After subcutaneous injection in EAE, NPCs accumulate in draining lymph nodes, where they hinder DC maturation to professional antigen-presenting cells, presumably via a BMP-4-dependent mechanism.[133] Most of the presented findings have been discovered in rodent NPCs; importantly, human-derived NPCs have also been demonstrated to possess immunomodulatory properties. First, fetal human NPCs inhibit T lymphocytes proliferation as well as DC maturation in vitro.[134] Second, human embryonic stem cell-derived NPCs, transplanted within the rodent spinal cord parenchyma, although short-lived, restrict T-cell and macrophage infiltration into the CNS and increase Treg recruitment, via the production of TGF-β1 and TGF-β2.[135]

In conclusion, while a small quota of NPCs integrates into the EAE parenchyma, NPCs' therapeutic effect in EAE should be ascribed mainly to their bystander paracrine effect, promoting neuroprotection and at the same time inhibiting the autoreactive inflammatory response.

27.4.4 Clinical Translation of Neural Stem/Precursor Cell Transplantation in Multiple Sclerosis

The most important steps toward clinical translation of NPC transplantation in MS have been the demonstration of NPC efficacy in nonhuman primates[134] and the publication in 2010, a consensus statement concerning stem cell transplantation in MS,[136] which advocated proof-of-principle studies of NPC transplantation in patients with early secondary-progressive MS that is refractory to conventional therapy. Intrathecal delivery was the preferred route of administration since central immunomodulatory and, to a lesser extent, remyelinating effects have been observed after intrathecal delivery of NPCs.

As recent as September 2014, there has been the communication of an interim analysis of a phase I clinical trial on intrathecal administration of MSC-neural progenitors (MSC-NPs) in MS.[112] MSC-NPs are an autologous bone marrow-derived population of regenerative cells currently under investigation as a novel MS treatment. In preclinical studies in mouse EAE, intrathecal delivery of MSC-NPs was associated with cell migration to lesion areas, suppression of local inflammatory response, and trophic support for damaged cells at the lesion site.[137] The initial clinical experience with IT administration of autologous MSC-NPs in seven MS patients also supported the dosing, safety, feasibility, and potential efficacy of this therapeutic approach, which led to the FDA approval of a phase I clinical trial investigating autologous MSC-NPs in 20 patients with MS. Preliminary data in the first five patients indicate safety and tolerability of the treatment.[138]

27.4.5 Open Issues in Neural Stem/Precursor Cell Transplantation in Multiple Sclerosis and Future Developments

The main limitation in clinical translation of NPC-based therapies in MS is their unexpected therapeutic mechanism, based on bystander effects rather than cell replacement. So far, most NPC transplants have been carried out in the very early phases of EAE, from the preclinical phase to the peak of disease severity, when inflammation is florid. It still needs to be addressed if such bystander effect is long-lasting in the course of a chronic disease such as MS and if it adds any further benefit compared to standard approved treatments. Preclinical data suggest that timing of NPC transplantation might play a crucial role in determining its effect: transplantation in a chronic phase of the disease, when inflammation has waned out, seems to result in enhanced differentiation of NPCs,[139] while limiting their survival and therapeutic efficacy.[139,140]

Another area of uncertainty is the extent of NPC persistence upon transplantation. Bioluminescence and magnetic resonance experiments suggest that NPCs are short-lived in EAE,[135,141–144] thus raising the question whether multiple rounds of NPC transplantation would be needed to sustain the therapeutic effect over time.

27.5 FUTURE PERSPECTIVES IN THE STEM CELL TRANSPLANTATION FIELD

27.5.1 New Sources of Stem Cells: Induced Pluripotent Stem Cells, Direct Reprogramming, and In Vivo Reprogramming

The discovery that somatic cells can travel back in development and become pluripotent through the forced expression of specific factors (ie, Oct4, Sox2, Klf4, and cMyc),[145] and the possibility to direct these induced pluripotent stem cells (iPSCs) to differentiate ideally in any cell type of our body, has ensured the 2012 Nobel Prize in Medicine to its discoverer, Shinya Yamanaka, and has triggered an enormous global research effort to realize autologous cell-based treatments for a wide spectrum of neurodegenerative diseases. Indeed, autologous somatic cells of different origin can be reprogrammed into iPSCs and then differentiated in vitro into large quantities of autologous NPCs suitable for transplantation. iPSC-derived NPCs could potentially overcome the limitations of allogeneic stem cell sources and the ethical issue related to the use of embryonic stem cells.

Very few preclinical data are so far available on the use of iPSC-derived NPCs in MS. While a highly enriched population of human iPSC-derived OPCs was able to robustly myelinate the brain of myelin-deficient shiverer mice,[146] encouraging results have been recently reported in an autoimmune demyelinating setting, showing that mouse iPSC-NPCs, transplanted into the mouse model of MS, might ameliorate the disease course through a bystander effect rather than a cell replacement. Such cells were capable of fostering the endogenous mechanism of repair and promote remyelination via the in situ production of LIF.[128]

The next step would be the translation of this work in an experimental setting closer to clinic. Only two reports have been published so far,[147,148] showing the generation of iPSCs from MS patients and their differentiation into mature astrocytes, oligodendrocytes, and neurons. These cells did not show any evident defect compared to healthy control lines, confirming the possibility of using them as a cell source for regenerative medicine.

Despite the promising applications of iPSC-based therapies, several fundamental aspects remain to be thoroughly investigated before the generation of clinical-graded iPSCs will be a reality: the choice of the reprogramming technique, which should be a balance between safety and efficacy; the epigenetic memory of iPSCs, which can restrain the differentiation capability of cells; the residual presence of undifferentiated cells that can cause teratoma formation upon iPSC-derived cell transplantation; and the genomic instability of iPSCs caused by reprogramming process.[149–152] Furthermore, the generation of clinical-graded iPSCs from each single patient would be very challenging, both in terms of economical resources and logistical requirements. For these reasons, the establishment of banks of allogeneic iPSCs has been proposed. Indeed, it has been calculated that 75 iPSC lines derived from homozygous human leukocyte antigen donors would be enough to match the 80% of the population in Japan[153] and a pool of 150 cell lines would be enough to match the 93% of the UK population.[154]

Nevertheless, considering the brief story of reprogramming, the field is moving rapidly toward clinical applications, more than expected. The first clinical study involving the transplantation of iPSC-derived retinal pigment epithelium in retina of patients suffering from age-related macular degeneration[155,156] has been recently approved by the Japanese government and the first patient has been already treated. Moreover, another Japanese group intends to apply in the near future for the authorization of a clinical trial for the treatment of Parkinson's disease patients with iPSC-derived dopaminergic neurons.[157]

Many efforts have been done to refine reprogramming technique. However iPSCs, while representing a promising source for cell transplantation, retain a very high tumorigenic capacity[158] that needs to be overcome before their clinical application can be envisaged.

After the seminal reprogramming work of Yamanaka and coworkers, many researchers started testing the hypothesis that overexpression of neural-specific transcriptional factors may directly convert nonneural somatic cells into neural cells. These directly reprogrammed or induced neural cells can be produced within a relatively short period of time (couple of weeks) and may be safer than iPSCs, since they do not pass through the pluripotent stem cell stage.[159] Induced neural stem cells (iNSCs) are now gaining very much interest. Indeed, it has been reported that fibroblasts can be converted into iNSCs by overexpression of Sox2, with or without other factors, and their transplantation into mouse brain did not generate tumors.[159–161] However much work is still required to investigate their therapeutic potential in different experimental settings.

Another promising evolution of reprogramming technique involves the possibility of in vivo directly converting one cell type into another. In this strategy, endogenous parenchymal nonneuronal cells (ie, astrocytes) can be changed to a required neuronal or oligodendroglial cell fate.[162] Although the concept of in vivo reprogramming is still in its infancy, it can potentially overcome some of the limitations related to in vitro reprogrammed cells. Indeed, in vivo reprogrammed cells, being susceptible to tissue microenvironment, can redifferentiate into normally functional cells bypassing ex vivo and in vitro procedures and cell manipulations.

27.5.2 Stem Cells as Sources of Novel Paracrine Mediators

As detailed in the previous chapters, one of the main advantages of stem cells in regenerative medicine relies on their intrinsic capability to secrete, in response to environmental cues, a vast array of molecules including growth factors, cytokines, chemokines, metabolites, and bioactive lipids, which results in a trophic effect on endogenous cells and a beneficial immunomodulatory action.[163–167]

The idea that stem cell produced factors could be sufficient to mediate the therapeutic effect of stem cells without implanting stem cells themselves, gives many advantages with regard to reproducibility, scalability, quality control, safety, and regulatory considerations.[168–170]

Different approaches have been developed to enhance the stem cell paracrine effect increasing the production of desired factors (both trophic and/or immune regulatory factors). These strategies include the preconditioning of stem cells and stem cell engineering via gene expression approaches. Concerning stem cell preconditioning, various stimuli able to activate stem cell-dependent functional effects have been taken into consideration. Hypoxic exposure is able to enhance MSC secretion of prosurvival and proangiogenic autocrine, paracrine of juxtacrine factors,[171] and preconditioned MSCs performed better once transplanted into a rat model of traumatic brain injury than nontreated MSCs.[172] Proinflammatory cytokines have been shown to play a fundamental role in the activation of stem cell-specific therapeutic functions, and NPCs pretreated with IL-6 have shown enhanced therapeutic effect in experimental ischemic stroke in mice.[173] Another option consists in the mechanical preconditioning, in which cell-to-cell interaction is promoted via in vitro 3D culture methods. In this condition, cells are forced to adapt their native morphology and consequently their signaling activity,[174] and MSCs cultured in 3D spheroids have shown an enhanced antiinflammatory effect.[175]

All these preconditioning methods could have only a transient impact on stem cell secretome, which can be in some situation an advantage. However, once a sustained response is needed, stem cell engineering could represent the best option. Different viral and nonviral approaches have been applied to enhance or force the expression of defined factors in stem cells.[164] The delivery of BDNF into the CNS through genetically modified MSCs, overcoming some of the limitations related to BDNF, use as the short half-life and the low permeability to the blood–brain barrier, allowed the exploration of the effect of BDNF in EAE mice, showing a significant delay in EAE onset and an amelioration of the clinical outcome.[176] Similar results have been obtained with IL-10 overexpression in NPCs, which showed enhanced therapeutic activity promoting clinical recovery, remyelination, and induction of apoptosis of inflammatory cells both in vitro and in vivo in the EAE CNS.[177] Other strategies include the overexpression of chemokine receptors, such as CCR5, and adhesion molecules, such as CD44, which resulted in enhanced migration toward sites of CNS inflammation and eventually in a better clinical recovery,[126,178] or of transcription factors, such as Olig2, that increased, upon transplantation in EAE, the differentiation of NPCs in oligodendrocyte precursors and their survival.[141]

In addition to soluble secreted factors, stem cells release extracellular vesicles (EVs), which might be considered as paracrine or endocrine signaling vehicles.[179,180] EVs are lipid membrane vesicles containing a variety of RNA species (mRNA, miRNA), soluble proteins, and transmembrane proteins, and consist of exosomes; generated inside the cytosol of the cells from multivesicular bodies and released by exocytosis from the plasma membrane (40–150 nm in size), microvesicles that shed directly from the plasma membrane and are much larger than exosomes (50–1000 nm) and apoptotic bodies.[181–185]

Most cell types within the nervous system communicate between each other through EV release. Indeed, EVs play fundamental roles in nervous system development, normal function, and pathophysiology,[186] as observed in neuron–oligodendrocyte communication where EVs play a critical role in promoting axonal integrity.[187] Moreover, several reports revealed the therapeutic potential of EVs, showing that Schwann cell-derived EVs are able to induce axonal regeneration after injury,[188] and MSC-derived EVs induce neurite outgrowth and enhance functional recovery from stroke.[189,190]

Apart from the great potential in regenerative medicine through the transfer of neurotrophic and proregenerative signals, EVs possess a modulatory role of immune response. Accordingly, MSC-derived EVs are able to induce an M2-like phenotype in monocytes in vitro,[191] promote the secretion of antiinflammatory cytokines, and contain an array of tolerogenic molecules,[192] which ensured them the capacity to reduce inflammation upon administration in a myocardial ischemia/reperfusion model.[193]

The most straightforward utilization of EVs as therapeutic delivery tool involves the use of unmodified EVs containing their natural cargo as surrogate of the stem cell source of origin.[168] However, many limitations are associated to native EVs, in particular concerning their content and targeting capacity. The idea of engineering EVs, modifying their content or developing synthetic alternatives, would allow a tighter control of this drug delivery device, facilitating its clinical translation. On one hand, the in vitro manipulation of stem cell source would enhance the production of EVs charged with proregenerative factors or therapeutic molecules. On the other, the complete control over EVs could be achieved by synthetic strategies, like liposomes. Synthetic EVs have been investigated as therapeutic delivery systems over the last 40 years but are not able to mimic completely the lipid and protein composition of EVs yet, which may be crucial for their intercellular interactions.[194]

27.6 CONCLUSIONS

In conclusion, given the therapeutic plasticity of stem/precursors cells, stem cell-based approaches for neurological disorders should be seriously considered. The different therapeutic properties of the different stem cell types available could be the major advantage of such an approach to foster tissue repair in a very complex and pathologically heterogeneous disease such as MS. However, much work is still need to shed light on some unsolved questions regarding the best way to control the in vivo fate of transplanted cells and how to enhance one specific pathway compared to the others, the proper choice of candidate patient, the type of stem/precursor cells, and the route and timing of cell delivery. Moreover, defining the factors responsible for the stem cell-mediated therapeutic effects would be crucial to better exploit stem cell potentiality both via stem cell engineering to push their efficacy and through the use of stem cell surrogates, such as EVs or trophic factors, instead of living cells with obvious advantages in terms of safety.

References

1. Comi G. Induction vs. escalating therapy in multiple sclerosis: practical implications. *Neurol Sci*. 2008;29(suppl 2): S253–S255.
2. Snowden JA, Saccardi R, Allez M, et al. Haematopoietic SCT in severe autoimmune diseases: updated guidelines of the European Group for Blood and Marrow Transplantation. *Bone Marrow Transplant*. 2012;47(6):770–790.
3. Saccardi R, Di Gioia M, Bosi A. Haematopoietic stem cell transplantation for autoimmune disorders. *Curr Opin Hematol*. 2008;15(6):594–600.
4. Radaelli M, Merlini A, Greco R, et al. Autologous bone marrow transplantation for the treatment of multiple sclerosis. *Curr Neurol Neurosci Rep*. 2014;14(9):478.
5. Pfender N, Saccardi R, Martin R. Autologous hematopoietic stem cell transplantation as a treatment option for aggressive multiple sclerosis. *Curr Treat Options Neurol*. 2013;15(3):270–280.
6. Saccardi R, Freedman MS, Sormani MP, et al. A prospective, randomized, controlled trial of autologous haematopoietic stem cell transplantation for aggressive multiple sclerosis: a position paper. *Mult Scler*. 2012;18(6):825–834.
7. Burt RK, Cohen BA, Russell E, et al. Hematopoietic stem cell transplantation for progressive multiple sclerosis: failure of a total body irradiation-based conditioning regimen to prevent disease progression in patients with high disability scores. *Blood*. 2003;102(7):2373–2378.
8. Nash RA, Bowen JD, McSweeney PA, et al. High-dose immunosuppressive therapy and autologous peripheral blood stem cell transplantation for severe multiple sclerosis. *Blood*. 2003;102(7):2364–2372.
9. Openshaw H, Stuve O, Antel JP, et al. Multiple sclerosis flares associated with recombinant granulocyte colony-stimulating factor. *Neurology*. 2000;54(11):2147–2150.
10. Burt RK, Balabanov R, Han X, et al. Association of nonmyeloablative hematopoietic stem cell transplantation with neurological disability in patients with relapsing-remitting multiple sclerosis. *JAMA*. 2015;313(3):275–284.
11. Fassas A, Kimiskidis VK, Sakellari I, et al. Long-term results of stem cell transplantation for MS: a single-center experience. *Neurology*. 2011;76(12):1066–1070.
12. Bowen JD, Kraft GH, Wundes A, et al. Autologous hematopoietic cell transplantation following high-dose immunosuppressive therapy for advanced multiple sclerosis: long-term results. *Bone Marrow Transplant*. 2012;47(7):946–951.
13. Muraro PA, Douek DC, Packer A, et al. Thymic output generates a new and diverse TCR repertoire after autologous stem cell transplantation in multiple sclerosis patients. *J Exp Med*. 2005;201(5):805–816.
14. Darlington PJ, Touil T, Doucet JS, et al. Diminished Th17 (not Th1) responses underlie multiple sclerosis disease abrogation after hematopoietic stem cell transplantation. *Ann Neurol*. 2013;73(3):341–354.
15. Muraro PA, Robins H, Malhotra S, et al. T cell repertoire following autologous stem cell transplantation for multiple sclerosis. *J Clin Invest*. 2014;124(3):1168–1172.
16. Abrahamsson SV, Angelini DF, Dubinsky AN, et al. Non-myeloablative autologous haematopoietic stem cell transplantation expands regulatory cells and depletes IL-17 producing mucosal-associated invariant T cells in multiple sclerosis. *Brain*. 2013;136(Pt 9):2888–2903.
17. Burt RK, Padilla J, Begolka WS, Canto MC, Miller SD. Effect of disease stage on clinical outcome after syngeneic bone marrow transplantation for relapsing experimental autoimmune encephalomyelitis. *Blood*. 1998;91(7):2609–2616.
18. Cassiani-Ingoni R, Muraro PA, Magnus T, et al. Disease progression after bone marrow transplantation in a model of multiple sclerosis is associated with chronic microglial and glial progenitor response. *J Neuropathol Exp Neurol*. 2007;66(7):637–649.
19. Metz I, Lucchinetti CF, Openshaw H, et al. Autologous haematopoietic stem cell transplantation fails to stop demyelination and neurodegeneration in multiple sclerosis. *Brain*. 2007;130(Pt 5):1254–1262.
20. Chen B, Zhou M, Ouyang J, et al. Long-term efficacy of autologous haematopoietic stem cell transplantation in multiple sclerosis at a single institution in China. *Neurol Sci*. 2012;33(4):881–886.
21. Krasulova E, Trneny M, Kozak T, et al. High-dose immunoablation with autologous haematopoietic stem cell transplantation in aggressive multiple sclerosis: a single centre 10-year experience. *Mult Scler*. 2010;16(6):685–693.
22. Mancardi GL, Sormani MP, Di Gioia M, et al. Autologous haematopoietic stem cell transplantation with an intermediate intensity conditioning regimen in multiple sclerosis: the Italian multi-centre experience. *Mult Scler*. 2012;18(6):835–842.

23. Burman J, Iacobaeus E, Svenningsson A, et al. Autologous haematopoietic stem cell transplantation for aggressive multiple sclerosis: the Swedish experience. *J Neurol Neurosurg Psychiatry*. 2014;85:1116–1121.
24. Fassas A, Anagnostopoulos A, Kazis A, et al. Autologous stem cell transplantation in progressive multiple sclerosis–an interim analysis of efficacy. *J Clin Immunol*. 2000;20(1):24–30.
25. Shevchenko JL, Kuznetsov AN, Ionova TI, et al. Autologous hematopoietic stem cell transplantation with reduced-intensity conditioning in multiple sclerosis. *Exp Hematol*. 2012;40(11):892–898.
26. Burt RK, Loh Y, Cohen B, et al. Autologous non-myeloablative haemopoietic stem cell transplantation in relapsing-remitting multiple sclerosis: a phase I/II study. *Lancet Neurol*. 2009;8(3):244–253.
27. Fagius J, Lundgren J, Oberg G. Early highly aggressive MS successfully treated by hematopoietic stem cell transplantation. *Mult Scler*. 2009;15(2):229–237.
28. Saccardi R, Mancardi GL, Solari A, et al. Autologous HSCT for severe progressive multiple sclerosis in a multicenter trial: impact on disease activity and quality of life. *Blood*. 2005;105(6):2601–2607.
29. Mancardi GL, Murialdo A, Rossi P, et al. Autologous stem cell transplantation as rescue therapy in malignant forms of multiple sclerosis. *Mult Scler*. 2005;11(3):367–371.
30. Trysberg E, Lindgren I, Tarkowski A. Autologous stem cell transplantation in a case of treatment resistant central nervous system lupus. *Ann Rheum Dis*. 2000;59(3):236–238.
31. Lehnhardt FG, Scheid C, Holtik U, et al. Autologous blood stem cell transplantation in refractory systemic lupus erythematodes with recurrent longitudinal myelitis and cerebral infarction. *Lupus*. 2006;15(4):240–243.
32. Gaspar N, Boudou P, Haroche J, et al. High-dose chemotherapy followed by autologous hematopoietic stem cell transplantation for adult histiocytic disorders with central nervous system involvement. *Haematologica*. 2006;91(8):1121–1125.
33. Greco R, Bondanza A, Oliveira MC, et al. Autologous hematopoietic stem cell transplantation in neuromyelitis optica: a registry study of the EBMT Autoimmune Diseases Working Party. *Mult Scler*. 2015;21(2):189–197.
34. Van Wijmeersch B, Sprangers B, Rutgeerts O, et al. Allogeneic bone marrow transplantation in models of experimental autoimmune encephalomyelitis: evidence for a graft-versus-autoimmunity effect. *Biol Blood Marrow Transplant*. 2007;13(6):627–637.
35. Greco R, Bondanza A, Vago L, et al. Allogeneic hematopoietic stem cell transplantation for neuromyelitis optica. *Ann Neurol*. 2014;75(3):447–453.
36. McAllister LD, Beatty PG, Rose J. Allogeneic bone marrow transplant for chronic myelogenous leukemia in a patient with multiple sclerosis. *Bone Marrow Transplant*. 1997;19(4):395–397.
37. Mandalfino P, Rice G, Smith A, Klein JL, Rystedt L, Ebers GC. Bone marrow transplantation in multiple sclerosis. *J Neurol*. 2000;247(9):691–695.
38. La Nasa G, Littera R, Cocco E, Battistini L, Marrosu MG, Contu L. Allogeneic hematopoietic stem cell transplantation in a patient affected by large granular lymphocyte leukemia and multiple sclerosis. *Ann Hematol*. 2004;83(6):403–405.
39. Lu JQ, Storek J, Metz L, et al. Continued disease activity in a patient with multiple sclerosis after allogeneic hematopoietic cell transplantation. *Arch Neurol*. 2009;66(1):116–120.
40. Jeffery DR. Failure of allogeneic bone marrow transplantation to arrest disease activity in multiple sclerosis. *Mult Scler*. 2007;13(8):1071–1075.
41. Lu JQ, Joseph JT, Nash RA, et al. Neuroinflammation and demyelination in multiple sclerosis after allogeneic hematopoietic stem cell transplantation. *Arch Neurol*. 2010;67(6):716–722.
42. Mancardi GL, Sormani MP, Gualandi F, et al. Autologous hematopoietic stem cell transplantation in multiple sclerosis: a phase II trial. *Neurology*. 2015;84(10):981–988.
43. Friedenstein AJ, Chailakhjan RK, Lalykina KS. The development of fibroblast colonies in monolayer cultures of guinea-pig bone marrow and spleen cells. *Cell Tissue Kinet*. 1970;3(4):393–403.
44. Uccelli A, Moretta L, Pistoia V. Mesenchymal stem cells in health and disease. *Nat Rev Immunol*. 2008;8(9):726–736.
45. Sacchetti B, Funari A, Michienzi S, et al. Self-renewing osteoprogenitors in bone marrow sinusoids can organize a hematopoietic microenvironment. *Cell*. 2007;131(2):324–336.
46. Crisan M, Yap S, Casteilla L, et al. A perivascular origin for mesenchymal stem cells in multiple human organs. *Cell Stem Cell*. 2008;3(3):301–313.
47. Mendez-Ferrer S, Michurina TV, Ferraro F, et al. Mesenchymal and haematopoietic stem cells form a unique bone marrow niche. *Nature*. 2010;466(7308):829–834.
48. Pedemonte E, Benvenuto F, Casazza S, et al. The molecular signature of therapeutic mesenchymal stem cells exposes the architecture of the hematopoietic stem cell niche synapse. *BMC Genomics*. 2007;8:65.

49. Tokoyoda K, Hauser AE, Nakayama T, Radbruch A. Organization of immunological memory by bone marrow stroma. *Nat Rev Immunol*. 2010;10(3):193–200.
50. Lee RH, Pulin AA, Seo MJ, et al. Intravenous hMSCs improve myocardial infarction in mice because cells embolized in lung are activated to secrete the anti-inflammatory protein TSG-6. *Cell Stem Cell*. 2009;5(1):54–63.
51. Li Y, Chen J, Chen XG, et al. Human marrow stromal cell therapy for stroke in rat: neurotrophins and functional recovery. *Neurology*. 2002;59(4):514–523.
52. Pluchino S, Zanotti L, Rossi B, et al. Neurosphere-derived multipotent precursors promote neuroprotection by an immunomodulatory mechanism. *Nature*. 2005;436(7048):266–271.
53. Ruster B, Gottig S, Ludwig RJ, et al. Mesenchymal stem cells display coordinated rolling and adhesion behavior on endothelial cells. *Blood*. 2006;108(12):3938–3944.
54. Sordi V, Malosio ML, Marchesi F, et al. Bone marrow mesenchymal stem cells express a restricted set of functionally active chemokine receptors capable of promoting migration to pancreatic islets. *Blood*. 2005;106(2):419–427.
55. Son BR, Marquez-Curtis LA, Kucia M, et al. Migration of bone marrow and cord blood mesenchymal stem cells in vitro is regulated by stromal-derived factor-1-CXCR4 and hepatocyte growth factor-c-met axes and involves matrix metalloproteinases. *Stem Cells*. 2006;24(5):1254–1264.
56. Cho KA, Ju SY, Cho SJ, et al. Mesenchymal stem cells showed the highest potential for the regeneration of injured liver tissue compared with other subpopulations of the bone marrow. *Cell Biol Int*. 2009;33(7):772–777.
57. Rose RA, Jiang H, Wang X, et al. Bone marrow-derived mesenchymal stromal cells express cardiac-specific markers, retain the stromal phenotype, and do not become functional cardiomyocytes in vitro. *Stem Cells*. 2008;26(11):2884–2892.
58. Qian H, Yang H, Xu W, et al. Bone marrow mesenchymal stem cells ameliorate rat acute renal failure by differentiation into renal tubular epithelial-like cells. *Int J Mol Med*. 2008;22(3):325–332.
59. Prockop DJ, Kota DJ, Bazhanov N, Reger RL. Evolving paradigms for repair of tissues by adult stem/progenitor cells (MSCs). *J Cell Mol Med*. 2010;14(9):2190–2199.
60. von Bahr L, Batsis I, Moll G, et al. Analysis of tissues following mesenchymal stromal cell therapy in humans indicates limited long-term engraftment and no ectopic tissue formation. *Stem Cells*. 2012;30(7):1575–1578.
61. Bai L, Lennon DP, Caplan AI, et al. Hepatocyte growth factor mediates mesenchymal stem cell-induced recovery in multiple sclerosis models. *Nat Neurosci*. 2012;15(6):862–870.
62. Spaggiari GM, Capobianco A, Becchetti S, Mingari MC, Moretta L. Mesenchymal stem cell-natural killer cell interactions: evidence that activated NK cells are capable of killing MSCs, whereas MSCs can inhibit IL-2-induced NK-cell proliferation. *Blood*. 2006;107(4):1484–1490.
63. Prigione I, Benvenuto F, Bocca P, Battistini L, Uccelli A, Pistoia V. Reciprocal interactions between human mesenchymal stem cells and gammadelta T cells or invariant natural killer T cells. *Stem Cells*. 2009;27(3):693–702.
64. Rafei M, Campeau PM, Aguilar-Mahecha A, et al. Mesenchymal stromal cells ameliorate experimental autoimmune encephalomyelitis by inhibiting CD4 Th17 T cells in a CC chemokine ligand 2-dependent manner. *J Immunol*. 2009;182(10):5994–6002.
65. Eliopoulos N, Stagg J, Lejeune L, Pommey S, Galipeau J. Allogeneic marrow stromal cells are immune rejected by MHC class I- and class II-mismatched recipient mice. *Blood*. 2005;106(13):4057–4065.
66. Wang Y, Chen X, Cao W, Shi Y. Plasticity of mesenchymal stem cells in immunomodulation: pathological and therapeutic implications. *Nat Immunol*. 2014;15(11):1009–1016.
67. Glennie S, Soeiro I, Dyson PJ, Lam EW, Dazzi F. Bone marrow mesenchymal stem cells induce division arrest anergy of activated T cells. *Blood*. 2005;105(7):2821–2827.
68. Aggarwal S, Pittenger MF. Human mesenchymal stem cells modulate allogeneic immune cell responses. *Blood*. 2005;105(4):1815–1822.
69. Luz-Crawford P, Kurte M, Bravo-Alegria J, et al. Mesenchymal stem cells generate a $CD4^+CD25^+Foxp3^+$ regulatory T cell population during the differentiation process of Th1 and Th17 cells. *Stem Cell Res Ther*. 2013;4(3):65.
70. Akiyama K, Chen C, Wang D, et al. Mesenchymal-stem-cell-induced immunoregulation involves FAS-ligand-/FAS-mediated T cell apoptosis. *Cell Stem Cell*. 2012;10(5):544–555.
71. Corcione A, Benvenuto F, Ferretti E, et al. Human mesenchymal stem cells modulate B-cell functions. *Blood*. 2006;107(1):367–372.
72. Augello A, Tasso R, Negrini SM, et al. Bone marrow mesenchymal progenitor cells inhibit lymphocyte proliferation by activation of the programmed death 1 pathway. *Eur J Immunol*. 2005;35(5):1482–1490.
73. Gerdoni E, Gallo B, Casazza S, et al. Mesenchymal stem cells effectively modulate pathogenic immune response in experimental autoimmune encephalomyelitis. *Ann Neurol*. 2007;61(3):219–227.

74. Jiang XX, Zhang Y, Liu B, et al. Human mesenchymal stem cells inhibit differentiation and function of monocyte-derived dendritic cells. *Blood*. 2005;105(10):4120–4126.
75. Nauta AJ, Westerhuis G, Kruisselbrink AB, Lurvink EG, Willemze R, Fibbe WE. Donor-derived mesenchymal stem cells are immunogenic in an allogeneic host and stimulate donor graft rejection in a nonmyeloablative setting. *Blood*. 2006;108(6):2114–2120.
76. Li YP, Paczesny S, Lauret E, et al. Human mesenchymal stem cells license adult $CD34^+$ hemopoietic progenitor cells to differentiate into regulatory dendritic cells through activation of the Notch pathway. *J Immunol*. 2008;180(3):1598–1608.
77. Deng Y, Yi S, Wang G, et al. Umbilical cord-derived mesenchymal stem cells instruct dendritic cells to acquire tolerogenic phenotypes through the IL-6-mediated upregulation of SOCS1. *Stem Cells Dev*. 2014;23(17):2080–2092.
78. Nemeth K, Leelahavanichkul A, Yuen PS, et al. Bone marrow stromal cells attenuate sepsis via prostaglandin E(2)-dependent reprogramming of host macrophages to increase their interleukin-10 production. *Nat Med*. 2009;15(1):42–49.
79. Abumaree MH, Al Jumah MA, Kalionis B, et al. Human placental mesenchymal stem cells (pMSCs) play a role as immune suppressive cells by shifting macrophage differentiation from inflammatory M1 to anti-inflammatory M2 macrophages. *Stem Cell Rev*. 2013;9(5):620–641.
80. Ren G, Zhang L, Zhao X, et al. Mesenchymal stem cell-mediated immunosuppression occurs via concerted action of chemokines and nitric oxide. *Cell Stem Cell*. 2008;2(2):141–150.
81. Krampera M, Cosmi L, Angeli R, et al. Role for interferon-gamma in the immunomodulatory activity of human bone marrow mesenchymal stem cells. *Stem Cells*. 2006;24(2):386–398.
82. Han X, Yang Q, Lin L, et al. Interleukin-17 enhances immunosuppression by mesenchymal stem cells. *Cell Death Differ*. 2014;21(11):1758–1768.
83. Xu C, Yu P, Han X, et al. TGF-β promotes immune responses in the presence of mesenchymal stem cells. *J Immunol*. 2014;192(1):103–109.
84. Renner P, Eggenhofer E, Rosenauer A, et al. Mesenchymal stem cells require a sufficient, ongoing immune response to exert their immunosuppressive function. *Transpl Proc*. 2009;41(6):2607–2611.
85. Inoue S, Popp FC, Koehl GE, et al. Immunomodulatory effects of mesenchymal stem cells in a rat organ transplant model. *Transplantation*. 2006;81(11):1589–1595.
86. Chen X, Gan Y, Li W, et al. The interaction between mesenchymal stem cells and steroids during inflammation. *Cell Death Dis*. 2014;5:e1009.
87. Su J, Chen X, Huang Y, et al. Phylogenetic distinction of iNOS and IDO function in mesenchymal stem cell-mediated immunosuppression in mammalian species. *Cell Death Differ*. 2014;21(3):388–396.
88. Zappia E, Casazza S, Pedemonte E, et al. Mesenchymal stem cells ameliorate experimental autoimmune encephalomyelitis inducing T-cell anergy. *Blood*. 2005;106(5):1755–1761.
89. Zhang J, Li Y, Chen J, et al. Human bone marrow stromal cell treatment improves neurological functional recovery in EAE mice. *Exp Neurol*. 2005;195(1):16–26.
90. Bai L, Lennon DP, Eaton V, et al. Human bone marrow-derived mesenchymal stem cells induce Th2-polarized immune response and promote endogenous repair in animal models of multiple sclerosis. *Glia*. 2009;57(11):1192–1203.
91. Constantin G, Marconi S, Rossi B, et al. Adipose-derived mesenchymal stem cells ameliorate chronic experimental autoimmune encephalomyelitis. *Stem Cells*. 2009;27(10):2624–2635.
92. Schena F, Gambini C, Gregorio A, et al. Interferon-γ-dependent inhibition of B cell activation by bone marrow-derived mesenchymal stem cells in a murine model of systemic lupus erythematosus. *Arthritis Rheum*. 2010;62(9):2776–2786.
93. Zhang J, Li Y, Lu M, et al. Bone marrow stromal cells reduce axonal loss in experimental autoimmune encephalomyelitis mice. *J Neurosci Res*. 2006;84(3):587–595.
94. Kassis I, Grigoriadis N, Gowda-Kurkalli B, et al. Neuroprotection and immunomodulation with mesenchymal stem cells in chronic experimental autoimmune encephalomyelitis. *Arch Neurol*. 2008;65(6):753–761.
95. Wang X, Willenbring H, Akkari Y, et al. Cell fusion is the principal source of bone-marrow-derived hepatocytes. *Nature*. 2003;422(6934):897–901.
96. Terada N, Hamazaki T, Oka M, et al. Bone marrow cells adopt the phenotype of other cells by spontaneous cell fusion. *Nature*. 2002;416(6880):542–545.
97. Pawelczyk E, Jordan EK, Balakumaran A, et al. In vivo transfer of intracellular labels from locally implanted bone marrow stromal cells to resident tissue macrophages. *PLoS One*. 2009;4(8):e6712.

98. Rivera FJ, Couillard-Despres S, Pedre X, et al. Mesenchymal stem cells instruct oligodendrogenic fate decision on adult neural stem cells. *Stem Cells*. 2006;24(10):2209–2219.
99. Chen J, Li Y, Katakowski M, et al. Intravenous bone marrow stromal cell therapy reduces apoptosis and promotes endogenous cell proliferation after stroke in female rat. *J Neurosci Res*. 2003;73(6):778–786.
100. Ohtaki H, Ylostalo JH, Foraker JE, et al. Stem/progenitor cells from bone marrow decrease neuronal death in global ischemia by modulation of inflammatory/immune responses. *Proc Natl Acad Sci USA*. 2008;105(38):14638–14643.
101. Zhang J, Brodie C, Li Y, et al. Bone marrow stromal cell therapy reduces proNGF and p75 expression in mice with experimental autoimmune encephalomyelitis. *J Neurol Sci*. 2009;279(1–2):30–38.
102. Lanza C, Morando S, Voci A, et al. Neuroprotective mesenchymal stem cells are endowed with a potent antioxidant effect in vivo. *J Neurochem*. 2009;110(5):1674–1684.
103. Kim YJ, Park HJ, Lee G, et al. Neuroprotective effects of human mesenchymal stem cells on dopaminergic neurons through anti-inflammatory action. *Glia*. 2009;57(1):13–23.
104. Connick P, Kolappan M, Crawley C, et al. Autologous mesenchymal stem cells for the treatment of secondary progressive multiple sclerosis: an open-label phase 2a proof-of-concept study. *Lancet Neurol*. 2012;11(2):150–156.
105. Yamout B, Hourani R, Salti H, et al. Bone marrow mesenchymal stem cell transplantation in patients with multiple sclerosis: a pilot study. *J Neuroimmunol*. 2010;227(1–2):185–189.
106. Mohyeddin Bonab M, Yazdanbakhsh S, Lotfi J, et al. Does mesenchymal stem cell therapy help multiple sclerosis patients? Report of a pilot study. *Iran J Immunol*. 2007;4(1):50–57.
107. Karussis D, Karageorgiou C, Vaknin-Dembinsky A, et al. Safety and immunological effects of mesenchymal stem cell transplantation in patients with multiple sclerosis and amyotrophic lateral sclerosis. *Arch Neurol*. 2010;67(10):1187–1194.
108. Wilkins A, Kemp K, Ginty M, Hares K, Mallam E, Scolding N. Human bone marrow-derived mesenchymal stem cells secrete brain-derived neurotrophic factor which promotes neuronal survival in vitro. *Stem Cell Res*. 2009;3(1):63–70.
109. Kemp K, Gray E, Mallam E, Scolding N, Wilkins A. Inflammatory cytokine induced regulation of superoxide dismutase 3 expression by human mesenchymal stem cells. *Stem Cell Rev*. 2010;6(4):548–559.
110. Morando S, Vigo T, Esposito M, et al. The therapeutic effect of mesenchymal stem cell transplantation in experimental autoimmune encephalomyelitis is mediated by peripheral and central mechanisms. *Stem Cell Res Ther*. 2012;3(1):3.
111. Le Blanc K, Frassoni F, Ball L, et al. Mesenchymal stem cells for treatment of steroid-resistant, severe, acute graft-versus-host disease: a phase II study. *Lancet*. 2008;371(9624):1579–1586.
112. Duijvestein M, Vos AC, Roelofs H, et al. Autologous bone marrow-derived mesenchymal stromal cell treatment for refractory luminal Crohn's disease: results of a phase I study. *Gut*. 2010;59(12):1662–1669.
113. Freedman MS, Bar-Or A, Atkins HL, et al. The therapeutic potential of mesenchymal stem cell transplantation as a treatment for multiple sclerosis: consensus report of the International MSCT Study Group. *Mult Scler*. 2010;16(4):503–510.
114. Reynolds BA, Weiss S. Generation of neurons and astrocytes from isolated cells of the adult mammalian central nervous system. *Science*. 1992;255(5052):1707–1710.
115. Reynolds BA, Rietze RL. Neural stem cells and neurospheres–re-evaluating the relationship. *Nat Methods*. 2005;2(5):333–336.
116. Gage FH, Temple S. Neural stem cells: generating and regenerating the brain. *Neuron*. 2013;80(3):588–601.
117. Xing YL, Roth PT, Stratton JA, et al. Adult neural precursor cells from the subventricular zone contribute significantly to oligodendrocyte regeneration and remyelination. *J Neurosci*. 2014;34(42):14128–14146.
118. Nait-Oumesmar B, Decker L, Lachapelle F, Avellana-Adalid V, Bachelin C, Baron-Van Evercooren A. Progenitor cells of the adult mouse subventricular zone proliferate, migrate and differentiate into oligodendrocytes after demyelination. *Eur J Neurosci*. 1999;11(12):4357–4366.
119. Butti E, Cusimano M, Bacigaluppi M, Martino G. Neurogenic and non-neurogenic functions of endogenous neural stem cells. *Front Neurosci*. 2014;8:92.
120. Aboody K, Capela A, Niazi N, Stern JH, Temple S. Translating stem cell studies to the clinic for CNS repair: current state of the art and the need for a Rosetta stone. *Neuron*. 2011;70(4):597–613.
121. Pluchino S, Quattrini A, Brambilla E, et al. Injection of adult neurospheres induces recovery in a chronic model of multiple sclerosis. *Nature*. 2003;422(6933):688–694.

122. Greenberg ML, Weinger JG, Matheu MP, et al. Two-photon imaging of remyelination of spinal cord axons by engrafted neural precursor cells in a viral model of multiple sclerosis. *Proc Natl Acad Sci USA*. 2014;111(22):E2349–E2355.
123. Pluchino S, Muzio L, Imitola J, et al. Persistent inflammation alters the function of the endogenous brain stem cell compartment. *Brain*. 2008;131(Pt 10):2564–2578.
124. Ben-Hur T, Einstein O, Mizrachi-Kol R, et al. Transplanted multipotential neural precursor cells migrate into the inflamed white matter in response to experimental autoimmune encephalomyelitis. *Glia*. 2003;41(1):73–80.
125. Cohen ME, Fainstein N, Lavon I, Ben-Hur T. Signaling through three chemokine receptors triggers the migration of transplanted neural precursor cells in a model of multiple sclerosis. *Stem Cell Res*. 2014;13(2):227–239.
126. Deboux C, Ladraa S, Cazaubon S, et al. Overexpression of CD44 in neural precursor cells improves trans-endothelial migration and facilitates their invasion of perivascular tissues in vivo. *PLoS One*. 2013;8(2):e57430.
127. Einstein O, Friedman-Levi Y, Grigoriadis N, Ben-Hur T. Transplanted neural precursors enhance host brain-derived myelin regeneration. *J Neurosci*. 2009;29(50):15694–15702.
128. Laterza C, Merlini A, De Feo D, et al. iPSC-derived neural precursors exert a neuroprotective role in immune-mediated demyelination via the secretion of LIF. *Nat Commun*. 2013;4:2597.
129. Einstein O, Karussis D, Grigoriadis N, et al. Intraventricular transplantation of neural precursor cell spheres attenuates acute experimental allergic encephalomyelitis. *Mol Cell Neurosci*. 2003;24(4):1074–1082.
130. Fainstein N, Vaknin I, Einstein O, et al. Neural precursor cells inhibit multiple inflammatory signals. *Mol Cell Neurosci*. 2008;39(3):335–341.
131. Einstein O, Fainstein N, Vaknin I, et al. Neural precursors attenuate autoimmune encephalomyelitis by peripheral immunosuppression. *Ann Neurol*. 2007;61(3):209–218.
132. Cao W, Yang Y, Wang Z, et al. Leukemia inhibitory factor inhibits T helper 17 cell differentiation and confers treatment effects of neural progenitor cell therapy in autoimmune disease. *Immunity*. 2011;35(2):273–284.
133. Pluchino S, Zanotti L, Brambilla E, et al. Immune regulatory neural stem/precursor cells protect from central nervous system autoimmunity by restraining dendritic cell function. *PLoS One*. 2009;4(6):e5959.
134. Pluchino S, Gritti A, Blezer E, et al. Human neural stem cells ameliorate autoimmune encephalomyelitis in non-human primates. *Ann Neurol*. 2009;66(3):343–354.
135. Chen L, Coleman R, Leang R, et al. Human neural precursor cells promote neurologic recovery in a viral model of multiple sclerosis. *Stem Cell Rep*. 2014;2(6):825–837.
136. Martino G, Franklin RJ, Baron Van Evercooren A, Kerr DA, Stem Cells in Multiple Sclerosis Consensus Group. Stem cell transplantation in multiple sclerosis: current status and future prospects. *Nat Rev Neurol*. 2010;6(5):247–255.
137. Harris VK, Faroqui R, Vyshkina T, Sadiq SA. Characterization of autologous mesenchymal stem cell-derived neural progenitors as a feasible source of stem cells for central nervous system applications in multiple sclerosis. *Stem Cells Transl Med*. 2012;1(7):536–547.
138. Sadiq SA, Harris VK, Vyshkina T, Chirls S. Intrathecal administration of mesenchymal stem cell- neural progenitors in multiple sclerosis: an interim analysis of a phase I clinical trial. In: *Paper Presented at: 2014 Joint ACTRIMS-ECTRIMS Meeting 2014*. 2014. Boston.
139. Giannakopoulou A, Grigoriadis N, Polyzoidou E, Lourbopoulos A, Michaloudi E, Papadopoulos GC. Time-dependent fate of transplanted neural precursor cells in experimental autoimmune encephalomyelitis mice. *Exp Neurol*. 2011;230(1):16–26.
140. Fainstein N, Einstein O, Cohen ME, Brill L, Lavon I, Ben-Hur T. Time limited immunomodulatory functions of transplanted neural precursor cells. *Glia*. 2013;61(2):140–149.
141. Sher F, Amor S, Gerritsen W, et al. Intraventricularly injected Olig2-NSCs attenuate established relapsing-remitting EAE in mice. *Cell Transplant*. 2012;21(9):1883–1897.
142. Muja N, Cohen ME, Zhang J, et al. Neural precursors exhibit distinctly different patterns of cell migration upon transplantation during either the acute or chronic phase of EAE: a serial MR imaging study. *Magnetic Reson Med*. 2011;65(6):1738–1749.
143. Ben-Hur T, van Heeswijk RB, Einstein O, et al. Serial in vivo MR tracking of magnetically labeled neural spheres transplanted in chronic EAE mice. *Magnetic Reson Med*. 2007;57(1):164–171.
144. Politi LS, Bacigaluppi M, Brambilla E, et al. Magnetic-resonance-based tracking and quantification of intravenously injected neural stem cell accumulation in the brains of mice with experimental multiple sclerosis. *Stem Cells*. 2007;25(10):2583–2592.

145. Takahashi K, Yamanaka S. Induction of pluripotent stem cells from mouse embryonic and adult fibroblast cultures by defined factors. *Cell*. 2006;126(4):663–676.
146. Wang S, Bates J, Li X, et al. Human iPSC-derived oligodendrocyte progenitor cells can myelinate and rescue a mouse model of congenital hypomyelination. *Cell Stem Cell*. 2013;12(2):252–264.
147. Song B, Sun G, Herszfeld D, et al. Neural differentiation of patient specific iPS cells as a novel approach to study the pathophysiology of multiple sclerosis. *Stem Cell Res*. 2012;8(2):259–273.
148. Douvaras P, Wang J, Zimmer M, et al. Efficient generation of myelinating oligodendrocytes from primary progressive multiple sclerosis patients by induced pluripotent stem cells. *Stem Cell Rep*. 2014;3(2):250–259.
149. Gore A, Li Z, Fung HL, et al. Somatic coding mutations in human induced pluripotent stem cells. *Nature*. 2011;471(7336):63–67.
150. Hussein SM, Batada NN, Vuoristo S, et al. Copy number variation and selection during reprogramming to pluripotency. *Nature*. 2011;471(7336):58–62.
151. Yamanaka S. A fresh look at iPS cells. *Cell*. 2009;137(1):13–17.
152. Pasi CE, Dereli-Oz A, Negrini S, et al. Genomic instability in induced stem cells. *Cell Death Differ*. 2011;18(5):745–753.
153. Nakajima F, Tokunaga K, Nakatsuji N. Human leukocyte antigen matching estimations in a hypothetical bank of human embryonic stem cell lines in the Japanese population for use in cell transplantation therapy. *Stem Cells*. 2007;25(4):983–985.
154. Taylor CJ, Peacock S, Chaudhry AN, Bradley JA, Bolton EM. Generating an iPSC bank for HLA-matched tissue transplantation based on known donor and recipient HLA types. *Cell Stem Cell*. 2012;11(2):147–152.
155. Kamao H, Mandai M, Okamoto S, et al. Characterization of human induced pluripotent stem cell-derived retinal pigment epithelium cell sheets aiming for clinical application. *Stem Cell Rep*. 2014;2(2):205–218.
156. Okamoto S, Takahashi M. Induction of retinal pigment epithelial cells from monkey iPS cells. *Invest Ophthalmol Vis Sci*. 2011;52(12):8785–8790.
157. Kikuchi T, Morizane A, Doi D, et al. Survival of human induced pluripotent stem cell-derived midbrain dopaminergic neurons in the brain of a primate model of Parkinson's disease. *J Park Dis*. 2011;1(4):395–412.
158. Tong M, Lv Z, Liu L, et al. Mice generated from tetraploid complementation competent iPS cells show similar developmental features as those from ES cells but are prone to tumorigenesis. *Cell Res*. 2011;21(11):1634–1637.
159. Ring KL, Tong LM, Balestra ME, et al. Direct reprogramming of mouse and human fibroblasts into multipotent neural stem cells with a single factor. *Cell Stem Cell*. 2012;11(1):100–109.
160. Han DW, Tapia N, Hermann A, et al. Direct reprogramming of fibroblasts into neural stem cells by defined factors. *Cell Stem Cell*. 2012;10(4):465–472.
161. Lujan E, Chanda S, Ahlenius H, Sudhof TC, Wernig M. Direct conversion of mouse fibroblasts to self-renewing, tripotent neural precursor cells. *Proc Natl Acad Sci USA*. 2012;109(7):2527–2532.
162. Torper O, Pfisterer U, Wolf DA, et al. Generation of induced neurons via direct conversion in vivo. *Proc Natl Acad Sci USA*. 2013;110(17):7038–7043.
163. Kokaia Z, Martino G, Schwartz M, Lindvall O. Cross-talk between neural stem cells and immune cells: the key to better brain repair? *Nat Neurosci*. 2012;15(8):1078–1087.
164. Baraniak PR, McDevitt TC. Stem cell paracrine actions and tissue regeneration. *Regen Med*. 2010;5(1):121–143.
165. Shi Y, Su J, Roberts AI, Shou P, Rabson AB, Ren G. How mesenchymal stem cells interact with tissue immune responses. *Trends Immunol*. 2012;33(3):136–143.
166. De Feo D, Merlini A, Laterza C, Martino G. Neural stem cell transplantation in central nervous system disorders: from cell replacement to neuroprotection. *Curr Opin Neurol*. 2012;25(3):322–333.
167. Pluchino S, Martino G. The therapeutic plasticity of neural stem/precursor cells in multiple sclerosis. *J Neurol Sci*. 2008;265(1–2):105–110.
168. Lamichhane TN, Sokic S, Schardt JS, Raiker RS, Lin JW, Jay SM. Emerging roles for extracellular vesicles in tissue engineering and regenerative medicine. *Tissue Eng Part B Rev*. 2015;21(1):45–54.
169. Drago D, Cossetti C, Iraci N, et al. The stem cell secretome and its role in brain repair. *Biochimie*. 2013;95(12):2271–2285.
170. Cossetti C, Alfaro-Cervello C, Donega M, Tyzack G, Pluchino S. New perspectives of tissue remodelling with neural stem and progenitor cell-based therapies. *Cell Tissue Res*. 2012;349(1):321–329.
171. Oskowitz A, McFerrin H, Gutschow M, Carter ML, Pochampally R. Serum-deprived human multipotent mesenchymal stromal cells (MSCs) are highly angiogenic. *Stem Cell Res*. 2011;6(3):215–225.

172. Chang CP, Chio CC, Cheong CU, Chao CM, Cheng BC, Lin MT. Hypoxic preconditioning enhances the therapeutic potential of the secretome from cultured human mesenchymal stem cells in experimental traumatic brain injury. *Clin Sci (Lond)*. 2013;124(3):165–176.
173. Sakata H, Narasimhan P, Niizuma K, Maier CM, Wakai T, Chan PH. Interleukin 6-preconditioned neural stem cells reduce ischaemic injury in stroke mice. *Brain*. 2012;135(Pt 11):3298–3310.
174. Frith JE, Thomson B, Genever PG. Dynamic three-dimensional culture methods enhance mesenchymal stem cell properties and increase therapeutic potential. *Tissue Eng Part C Methods*. 2010;16(4):735–749.
175. Bartosh TJ, Ylostalo JH, Mohammadipoor A, et al. Aggregation of human mesenchymal stromal cells (MSCs) into 3D spheroids enhances their antiinflammatory properties. *Proc Natl Acad Sci USA*. 2010;107(31):13724–13729.
176. Makar TK, Bever CT, Singh IS, et al. Brain-derived neurotrophic factor gene delivery in an animal model of multiple sclerosis using bone marrow stem cells as a vehicle. *J Neuroimmunol*. 2009;210(1–2):40–51.
177. Yang J, Jiang Z, Fitzgerald DC, et al. Adult neural stem cells expressing IL-10 confer potent immunomodulation and remyelination in experimental autoimmune encephalitis. *J Clin Invest*. 2009;119(12):3678–3691.
178. Yang J, Yan Y, Ma CG, et al. Accelerated and enhanced effect of CCR5-transduced bone marrow neural stem cells on autoimmune encephalomyelitis. *Acta Neuropathol*. 2012;124(4):491–503.
179. Camussi G, Deregibus MC, Cantaluppi V. Role of stem-cell-derived microvesicles in the paracrine action of stem cells. *Biochem Soc Trans*. 2013;41(1):283–287.
180. Biancone L, Bruno S, Deregibus MC, Tetta C, Camussi G. Therapeutic potential of mesenchymal stem cell-derived microvesicles. *Nephrol Dial Transpl*. 2012;27(8):3037–3042.
181. Gould SJ, Raposo G. As we wait: coping with an imperfect nomenclature for extracellular vesicles. *J Extracell Vesicles*. 2013;2.
182. Cocucci E, Racchetti G, Meldolesi J. Shedding microvesicles: artefacts no more. *Trends Cell Biol*. 2009;19(2):43–51.
183. Stoorvogel W, Kleijmeer MJ, Geuze HJ, Raposo G. The biogenesis and functions of exosomes. *Traffic*. 2002;3(5):321–330.
184. Thery C, Zitvogel L, Amigorena S. Exosomes: composition, biogenesis and function. *Nat Rev Immunol*. 2002;2(8):569–579.
185. Valadi H, Ekstrom K, Bossios A, Sjostrand M, Lee JJ, Lotvall JO. Exosome-mediated transfer of mRNAs and microRNAs is a novel mechanism of genetic exchange between cells. *Nat Cell Biol*. 2007;9(6):654–659.
186. Lai CP, Breakefield XO. Role of exosomes/microvesicles in the nervous system and use in emerging therapies. *Front Physiol*. 2012;3:228.
187. Fruhbeis C, Frohlich D, Kuo WP, Kramer-Albers EM. Extracellular vesicles as mediators of neuron-glia communication. *Front Cell Neurosci*. 2013;7:182.
188. Lopez-Verrilli MA, Picou F, Court FA. Schwann cell-derived exosomes enhance axonal regeneration in the peripheral nervous system. *Glia*. 2013;61(11):1795–1806.
189. Xin H, Li Y, Cui Y, Yang JJ, Zhang ZG, Chopp M. Systemic administration of exosomes released from mesenchymal stromal cells promote functional recovery and neurovascular plasticity after stroke in rats. *J Cereb Blood Flow Metab*. 2013;33(11):1711–1715.
190. Xin H, Li Y, Buller B, et al. Exosome-mediated transfer of miR-133b from multipotent mesenchymal stromal cells to neural cells contributes to neurite outgrowth. *Stem Cells*. 2012;30(7):1556–1564.
191. Zhang B, Yin Y, Lai RC, Tan SS, Choo AB, Lim SK. Mesenchymal stem cells secrete immunologically active exosomes. *Stem Cells Dev*. 2014;23(11):1233–1244.
192. Mokarizadeh A, Delirezh N, Morshedi A, Mosayebi G, Farshid AA, Mardani K. Microvesicles derived from mesenchymal stem cells: potent organelles for induction of tolerogenic signaling. *Immunol Lett*. 2012;147(1–2):47–54.
193. Arslan F, Lai RC, Smeets MB, et al. Mesenchymal stem cell-derived exosomes increase ATP levels, decrease oxidative stress and activate PI3K/Akt pathway to enhance myocardial viability and prevent adverse remodeling after myocardial ischemia/reperfusion injury. *Stem Cell Res*. 2013;10(3):301–312.
194. van der Meel R, Fens MH, Vader P, van Solinge WW, Eniola-Adefeso O, Schiffelers RM. Extracellular vesicles as drug delivery systems: lessons from the liposome field. *J Control Release*. 2014;195:72–85.

CHAPTER

28

Stem Cells in Multiple Sclerosis

I. Kassis, P. Petrou, D. Karussis

Hebrew University, Ein–Karem, Jerusalem, Israel

OUTLINE

28.1 BACKGROUND AND RATIONALE FOR STEM CELL THERAPIES IN MULTIPLE SCLEROSIS

28.1.1 Multiple Sclerosis

Multiple sclerosis (MS) is a chronic inflammatory multifocal demyelinating disease of the central nervous system (CNS) that affects predominantly young adults. MS is the leading cause of chronic neurological disability in young age. While its pathogenesis is still obscure, and multiple (genetic, environmental, and infectious) factors seem to be involved, it is widely accepted that the final pathogenetic pathway is that of an autoimmune attack against myelin

Translational Neuroimmunology in Multiple Sclerosis
http://dx.doi.org/10.1016/B978-0-12-801914-6.00028-3

components (the outer layer/lining of all the axons in the CNS, which greatly facilitates the transmission of the electrical stimuli of the brain), which causes damage of myelin (demyelination) in multiple discrete areas of the CNS (the plaques of MS, which are the pathological hallmark of the disease), dysfunction of multiple neuronal circuits in the CNS, therefore leading to neurological disability. Additional mechanisms have been uncovered, including damage of the axons in the CNS and a degenerative process, which is probably the result of inflammation, and which causes accumulating and irreversible damage with time.[1]

Extensive studies have provided strong evidence for neurodegeneration in MS, including the finding of amyloid precursor protein accumulation in neurons[2]; a reduction in the n-acetylaspartate/creatine ratio in magnetic resonance spectroscopy, which correlates well with the degree of disability[3]; the finding of axonal ovoids/transected axons at the edge and the core of active lesions[4] and of oxidative damage in mitochondrial DNA and impaired activity of mitochondrial enzyme complexes[5]; the reduction in axonal density in normally appearing white matter (NAWM) early in MS; and a more prominent reduction of axonal density in spinal cord NAWM in progressive MS patients.[6] It is not clear which factors are responsible for the variability in the course of MS in different patients and the heterogeneity of the morphological alterations of the CNS found by magnetic resonance imaging (MRI)[7] or by histopathological evaluation,[8] as well as a wide variability of the response to the immunomodulatory treatments. Possible explanations may include the complex genetic trait that translates into different immune abnormalities and/or increased vulnerability of CNS tissue to inflammatory insult or reduced ability to repair damage.

The precise reasons for the progression of disability in MS, despite the increasingly effective immunomodulating therapies, also remain largely obscure and puzzling. Recent findings indicate that progressive disease may be caused by the previously mentioned axonal damage and atrophy, but can be also related to compartmentalized inflammation, which is purely amendable to the treatments and may be mediated by humoral mechanisms and B cells, and cortical (submeningeal) and deep gray matter damage.[9–11] All the currently approved treatments for MS target the immune system, aiming to suppress the inflammatory components of the disease in a nonspecific manner (generalized immunosuppression) and/or in a more restricted way (immunomodulation). In both cases, these treatments are only preventing to some extent the appearance of new relapses and the progression of the disease, but cannot reverse existing disability.

It seems therefore that in order to affect the progressive phases of the disease, different immunomodulating modalities (ie, more effective in suppression of the localized deep inflammation and in downregulation of humoral mechanisms) and/or neuroprotective approaches are essential. These may include cellular therapies, including stem cell treatments that may be more effective in suppression of compartmentalized inflammation and in induction of neurotrophic effects, neuroprotection, and possibly even enhanced regeneration.

28.1.2 Stem Cell Therapy for Multiple Sclerosis

Stem cells are a diverse group of multipotent cells. In general, these cells are relatively undifferentiated and unspecialized, and can give rise to the differentiated and specialized cells of the body. All stem cells exert two characteristic features: (1) the capacity for self-renewal, which preserves a pool of undifferentiated stem cells, and (2) the potential for transdifferentiation (the ability to produce various differentiated cell types). There are different kinds of stem cells that can be isolated from embryonic and adult tissues. Embryonic stem cells (ESCs), which are the prototype of all stem cells, are cells derived from the inner cell mass of embryos[12] at the blastocyte stage (5–9 days after fertilization). The only source for human stem cells is from embryos obtained from in vitro fertilization.

A significant breakthrough in stem cell research was the discovery of stem cells that reside in various body tissues (including the brain) during the adult life. These are defined as adult stem cells and represent a more differentiated cell population than ESCs. They can be isolated from various tissues, including muscle,[13] adipose tissue,[14] CNS (neural stem cells (NSCs)),[15] and bone marrow (mesenchymal stem cells (MSCs), and hematopoietic stem cells (HSCs)).[16,17] All of the previously described stem cells carry a potential for specific and nonspecific tissue repair.

28.2 DIFFERENT TYPES OF STEM CELLS

28.2.1 Embryonic Stem Cells

Human embryonic stem cells (hESC) are pluripotent cells derived from the inner cell mass of the blastocyst[18] that represent the prototype of all stem cells. hESCs have unlimited, proliferation potential in vitro and carry the potential to generate cell types from all germ layers.[19] They can also serve as a source of neural cells for transplantation in neurological disorders, such as MS.[20] Different studies have proposed that the actions of these ESC-derived neural progenitors may be at least partially exerted via mechanisms of immunomodualtion[21] and/or neuroprotection[22] rather than tissue repair and neuroregeneration. In the study by Kim et al. transplantation of oligodendroglial progenitors (OPs) derived from hESCs lead to amelioration of the disease symptoms of experimental autoimmune encephalomyelitis (EAE), an animal model for MS.[21] EAE mice that received hESC-OPs showed a significant improvement in neurological disability scores compared to that of control animals at day 15 posttransplantation. Histopathologically, transplanted hESC-OPs generated TREM2-positive CD45 cells, increased TIMP-1 expression, confined inflammatory cells within the subarachnoid space, and gave rise to higher numbers of Foxp3-positive regulatory T cells in the spinal cord and spleen.[21] These results suggest that transplantation of hESC-OPs can alter the pathogenesis of EAE through immunomodulation.

In the study by Ahronowiz et al. hESC-derived early multipotent neural precursors (NPs) were transplanted into the brain ventricles of mice induced with EAE[22] and significantly reduced the clinical signs of EAE. Histopathological evaluation showed migration of the transplanted NPs to the host white matter. However, differentiation to mature oligodendrocytes and remyelination were negligible.[22] Time course analysis of the evolution and progression of CNS inflammation and tissue injury showed an attenuation of the inflammatory process in transplanted animals, which was correlated with the reduction of both axonal damage and demyelination.[22] ESCs, which have a tremendous transdifferentiation ability, therefore may represent the ideal cell population for neural cell replacement. However, the transplantation of undifferentiated ESC does not seem to provide a first-line option for clinical applications in the meantime, as it may be associated with a high risk of potential carcinomatous transformation.[23]

28.2.2 Adult Neural Stem Cells

Adult NSCs, or CNS neural stem cells, are cells that are cultured as sparse adherent cells[24] or as aggregates of floating cells called neurospheres in serum-free medium on a nonadherent surface in the presence of epidermal growth factor and fibroblast growth factor 2.[25] NSCs express markers as Nestin PSA-NCAM and Sox2. As their name implies, these cells can give rise to neural, astrocytic, and oligodendrocytic precursors that in turn may differentiate into neurons, astrocytes, and Oligodendrocytes.[26,27] Based on their ability for neural and glial cell generation, NSCs were tested in the animal model of MS and EAE. In a study by Pluchino et al.[28] it was shown that adult NSCs cultured and injected into EAE-mice—intravenously (iv) or intracerebroventricularly (icv)—could migrate into the demyelinating CNS area and differentiate into mature brain cells. It was noticed in this study that intrinsic oligodendrocyte progenitors were significantly increased in the CNS of the NSC-injected animals. Clinically, EAE symptoms were strongly downregulated in the transplanted animals. In additional studies by Einstein et al. and Ben-Hur et al.[29–31] from our medical center and laboratory at the Hadassah neurology department, it was found that transplanted NP neurospheres (icv) migrated into the inflamed white matter in EAE, attenuated the severity of clinical signs, and reduced brain inflammation. In these studies, it was shown that when NPs were administered intravenously, EAE was suppressed by a peripheral immunosuppressive effect, which inhibited T-cell activation and proliferation in the lymph nodes.[32] Moreover, the transplanted NP cells could downregulate the inflammatory brain process in situ, as indicated by the reduction in the number of perivascular infiltrates and of brain $CD3^+$ T cells, a reduction in the expression of ICAM-1 and LFA-1 in the brain, and an increase in the number and proportion of regulatory T cells in the CNS.[32] The latter may indicate an active immunoregulation induced by the NPC.

Despite these promising results, which place the NSC as the optimal adult stem cell population for cell replacement therapy in CNS diseases, these cells are associated with significant drawbacks. First is the difficulty to culture neurospheres from regions of the adult brain that do not normally undergo self-renewal.[33] Second, although NSC can be propagated for extended periods of time and differentiated into both neuronal and glial cells, studies suggest that this behavior is induced by the culture conditions of the progenitor cells, and seems to be restricted to a limited number of replication cycles in vivo.[34] Furthermore, neurosphere-derived cells do not necessarily behave as stem cells when transplanted back into the brain.[35] Additional concerns include the potential for immune rejection of NSC, the danger of tumor development in the host brain, and various ethical aspects related to the donor tissue origin.

28.2.3 Induced Pluripotent Stem Cells

Induced pluripotent stem cells (iPSCs) are adult cells that have been genetically reprogrammed to an ESC-like state by making these cells express genes and factors important for maintaining the defining properties of ESCs. In 2006, Yamanaka and colleagues showed that iPSCs can be generated from mouse fibroblasts by the retrovirus-mediated transfection of four transcription factors, namely Oct3/4, Sox2, c-Myc, and Klf4.[36] These iPSCs were found to have the ESCs morphology and several biological features including gene expression and teratoma formation. In 2007, the same group presented similar results for human iPSCs.[37]

They demonstrated the generation of iPSCs from adult human dermal fibroblasts with the same four factors: Oct3/4, Sox2, Klf4, and c-Myc. Human iPSCs were similar to hESCs in morphology, proliferation, cell surface antigens, gene expression profile, and telomerase activity. These cells were shown capable of differentiating into cell types of the three germ layers in vitro and in teratomas.

Recently, the generation of patient-specific cells from iPSCs has emerged as a promising strategy for the development of autologous cell therapies.[38] In a hypomyelinated mouse model it was shown that oligodendrocyte progenitor cells (OPCs) derived from iPSC lines induced a high degree of remyelination and in this myelin-defective mouse.[39] However, the differentiation protocols are still inefficient and not easily reproducible and require over 120 days in culture. In 2014, Douvaras et al. reported a highly reproducible protocol to produce OPCs and mature oligodendrocytes from iPSCs.[40] The major elements of the protocol include adherent cultures, dual SMAD inhibition, and addition of retinoids from the beginning of differentiation, which lead to increased yields of OLIG2 progenitors and high numbers of OPCs within 75 days.[40] Moreover, the authors reported the generation of integration-free iPSCs from primary-progressive MS (PPMS) patients and their efficient differentiation to oligodendrocytes. PPMS OPCs were found to be functional, as demonstrated by in vivo myelination in the shiverer mouse model.[40] These results represent encouraging advances toward the future development of autologous cell therapies using iPSCs.

28.2.4 Hematopoietic Stem Cells

HSCs are the main stem cell population of the bone marrow. These cells typically express the surface marker phenotypes of $CD34^+$, $CD133^+$, $CD45^+$, and CD38−. HSCs are the precursor cells that give rise to all types of blood cells, including T cells, B cells, natural killer (NK) cells, macrophages, red blood cells, granulocytes, and other monocytes.[41,42] Several studies have shown the ability of HSCs to transdifferentiate into CNS cells including neurons, astrocytes, and oligodendrocytes.[43–45] HSC transplantation (HSCT) or bone marrow transplantation (BMT) is widely used in hematological malignancies. During the last few years HSCT or BMT also has been applied for the treatment of autoimmune diseases such as MS.[46–53] The rationale for this application has been provided by studies from our group,[46–51] which have shown that high-dose cyclophosphamide for the elimination of immunocompetent lymphocytes, followed by syngeneic BMT rescue, could suppress chronic EAE and induce tolerance to the immunizing antigens.

Several small open trials over the last two decades (reviewed by us[54]) have shown the efficacy of autologous HSCT in suppressing the inflammatory activity in patients with severe MS. Younger patients with relapsing forms of MS are most likely to respond to the treatment. Some open trials in 26 patients with active relapsing MS and a long median follow-up period (186 weeks) showed an overall event-free survival of 78.4% at 3 years.[55] Progression-free survival and clinical relapse-free survival were 90.9% and 86.3%, respectively. Adverse events were consistent with expected toxic effects associated with HDIT/HCT, and no acute treatment-related neurologic adverse events were observed. Improvements were noted in neurologic disability, quality of life, and functional scores showed.

The apparent problem of the use of HSCT or BMT in autoimmunity is the need for strong (lethal) immunosuppression, which is associated with significant morbidity and mortality.

The use of lower doses of immunosuppression is not only less efficient but may actually provoke relapses of EAE.[56–58] In addition, despite the significant effects of such protocols in suppressing the inflammatory activity, still the clinical effects were not equally impressive, especially in the progressed stages of MS.[59–61] In reports from two groups,[59,62] it was shown that despite the efficacy of HSCT in MS patients in terms of elimination of the inflammatory lesions in the MRI and clinical stabilization (progress free patients >60% in 3 years), this treatment still did not prevent continuation of brain atrophy[63] and did not induce any functional improvement in most of the patients, especially in those with high disability scores.[64] Moreover, the procedure-related mortality (solely attributed to the cytotoxic conditioning protocol) in these studies may be in some cases as high as 5%, which seems to be an unacceptable risk for the majority of MS patients. A randomized controlled clinical trial with HSCT versus immunosuppressive modalities in MS is underway.[61] A report of a small group (21 patients; 9 were randomized in the autologous HSCT (AHSCT) and 12 in the mitoxantrone arm) from this randomized study showed that AHSCT reduced by 79% the number of new T2 lesions as compared to mitoxantrone 20 mg/month (rate ratio 0.21, $p=0.00016$).[65] It also reduced Gd^+ lesions as well as the annualized relapse rate. No difference was found in the progression of disability. The authors concluded that AHSCT is significantly superior to cytotoxic treatment (mitoxantrone) alone in reducing MRI activity in severe cases of MS.

28.2.5 Mesenchymal Stem Cells

MSCs are another important member of the bone marrow stem cell repertoire. These cells are described as nonhematopoietic stromal cells and their classical role is to support the process of hematopoiesis and HSC engraftment and to give rise to cells of mesodermal origin, such as osteoblasts, adipocytes, and chondrocytes.[66] These cells do not have a specific surface marker profile, but it is widely accepted that they are negative for CD34, CD45, and CD14, and positive for CD29, CD73, CD90, CD105, and CD166.[67] During the 2010s, significant interactions between MSCs and cells from the immune system have been demonstrated in vitro: MSCs were found to downregulate T and B lymphocytes, NK cells, and antigen-presenting cells through various mechanisms, including cell-to-cell interaction and soluble factor production.[68] Besides these immunomodulatory effects, studies in vitro and in vivo have indicated neuroprotective effects.[68] MSCs are believed to promote functional recovery following CNS injury or inflammation by producing trophic factors that may facilitate the mobilization of endogenous NSCs and promote the regeneration or the survival of the affected neurons.

The immunomodulatory and neuroprotective properties of MSCs in vitro were confirmed by us and other groups mainly in the EAE model. Zappia and colleagues demonstrated that the injection of syngeneic MSC indeed ameliorated the clinical severity of the disease in a mouse model of acute monophasic EAE and reduced demyelination and leukocyte infiltration of the CNS.[69] The findings were explained by the induction of T-cell anergy by MSC-treatment. In the study by Zhang et al. it was shown that intravenous administration of MSCs could suppress the disease in a relapsing-remitting model of EAE induced in SJL mice.[70] MSCs migrated into the CNS where they promoted brain-derived neurotrophic factor (BDNF) production and induced proliferation of a limited number of oligodendrocyte progenitors. Evidence of neuroprotection in EAE following MSC treatment was also shown by Chops et al.[71] accompanied by indications of in vivo neural differentiation of the transplanted cells.[71]

Gerdoni et al. used in his study the relapsing-remitting model of EAE that was induced with the proteolipid protein in SJL mice.[72] Intravenously treated mice with MSCs had a milder disease and developed fewer relapses than the untreated control animals.[72]In studies from our group, a model of chronic EAE (more reminiscent of human MS) was used and the effect of MSC transplantation via additional routes (both intravenous and intraventricularly, directly into the brain and CSF) was evaluated. Although in previous studies the suggested mechanism of suppression of EAE following intravenous injection of MSCs was suggested to be that of induction of peripheral immunomodulation/anergy,[69,72] in our experimental setting, we verified the advantages of direct injection of MSCs into the ventricles of the brain, where they induced a more prominent reduction in infiltrating lesions, indicating an additional in situ immunomodulation. The peripheral immunomodulatory effects of MSCs are probably equally important and the migratory ability of these cells to the lymph nodes and other lymphatic organs (when injected intravenously), shown in this work, argue in favor of such additional peripheral mechanisms. It is therefore logical to assume that the main immunomodulatory activity of MSCs is exerted in the peripheral lymphoid organs where MSCs migrate following i.v. administration, inhibiting T cells homing in the CNS.[69,73] In addition to these peripheral effects, MSCs migrating to the CNS following i.v. and i.c.v injection may also locally further modulate the CNS autoimmune process, stimulate endogenous neurogenesis, and protect neurons and oligodendrocytes by similar paracrinic and neurotrophic mechanisms.[73]

These in vivo experiments in EAE utilized the model of autologous MSC transplantation. This setting is logically considered more convenient for clinical transplantation since it does not hold any risks of rejection of the transplanted cells. However, in clinical reality, it is not always feasible that the patient can serve as a donor, due to his or her progressed medical condition. Moreover, if genetic factors are involved in the pathogenesis of MS, it would be preferable to avoid transplantation of stem cells carrying a putatively defective genome. Therefore, the possibility of allogenic MSC transplantation (using MSCs obtained from healthy donors) might be considered, especially since MSCs were shown to escape rejection by masking parts of the immune response, such as the complement system.[74]

Three main mechanisms contribute to this immune-privileged status of MSCs: (1) MSCs are hypoimmunogenic, often lacking MHC-II and costimulatory molecules expression[75]; (2) MSCs prevent T-cell responses indirectly through modulation of the dendritic cells and directly through downregulation of the NK, $CD8^+$, and $CD4^+$ T-cell functions[76]; (3) MSCs induce a suppressive local microenvironment through the production of prostaglandins and interleukin-10 as well as by the expression of indoleamine 2,3,-dioxygenase, which depletes the local milieu of tryptophan.[76] A possible attractive explanation of the reported efficacy of allogeneic MSC-transplantation may involve a mechanism of a single hit (probably immunomodulatory or neurotrophic in its nature), directly following the injection of MSCs, and before any putative rejection process may take place. Supporting such a possibility studies consistently reported that MSCs induce significant beneficial clinical effects and potent immunomodulation and neuroprotection mediated by the production of neurotrophic factors and/or through the recruitment of local/intrinsic CNS precursor cells[77–80] or paracrinic mechanisms.

These immunomodulatory and neuroprotective features could make MSCs potential candidates for future therapeutic modalities in immune-mediated and neurodegenerative diseases.

28.3 CLINICAL EXPERIENCE WITH STEM CELLS IN MULTIPLE SCLEROSIS

28.3.1 General Considerations and Rationale

Due to the aforementioned practical advantages of MSCs, bone marrow MSCs are, to date, the most commonly used stem cell population in clinical trials, with the exception of HSCs, especially regarding the treatment of neurological diseases in general and MS specifically. As these cells seem to be able to cross the blood–brain barrier, the need for invasive intracerebral surgery can be avoidable neurological diseases and at least the peripheral systemic administration has been proven a rather safe and efficient way for cell delivery in humans.[81] In a recent metaanalysis of clinical trials utilizing intravascular delivery of MSC (intravenously or intraarterially) testing immediate events (toxicity, fever), organ system complications, infection, and long-term adverse events (death, malignancy), it was found that peripheral MSC administration is safe. The data revealed from randomized control trials did not detect any association between acute infusional toxicity, organ system complications, infections, or deaths.[81] However, the extent to which MSC can be directed to a neural or other than mesodermal cellular fate either ex vivo or in vivo following transplantation is still a point of controversy.

For clinical trials, isolated MSC should be produced according to Good Manufacturing Practice. The culture process should be reproducible and efficient. According to guidelines of the International Society for Cell Therapy, the minimal criteria to define human MSC are (1) MSC must be plastic-adherent when maintained in standard culture conditions using tissue culture flasks; and (2) more than 95% of the MSC population must express CD105, CD73, and CD90, as measured by flow cytometry. Additionally, these cells should be negative for the CD45, CD34, CD14, CD11b, CD79a, or CD19 and HLA class II markers. (3) The cells must be able to differentiate into osteoblasts, adipocytes, and chondroblasts under in vitro differentiating conditions.[67] Safety is the major concern during the culture process as well as quality control of these cells. Several levels of quality control during the production of MSCs are necessary. These should include various microbiological tests (including bacterial, viral, and mycoplasma detection and LPS levels) and genetic-karyotype testing to exclude contaminations and genetic transformations or instability. The main source for MSCs used for clinical trials is the bone marrow compartment,[82] but MSCs can be also harvested by other tissues such as the adipose tissue and the umbilical cord.[14]

The culture process is also an issue of major consideration, since the culture can be started from either unfractioned (whole BM) or fractioned cells (mononuclear fraction of BM after gradient density). The medium of choice for culturing is of equally high importance for the efficacy and safety of MSCs. Generally, Dulbecco's Modified Eagle's medium or alpha-minimal essential medium are used for culturing with the addition of fetal bovine serum, fetal calf serum, human serum, plasma, and platelets lysates, with the addition of growth factors as fibroblast growth factor. The use of serum is one of the controversial parameters of the culture having an impact on the batch-to-batch variability and the risks of contamination. The use of chemical-defined, xeno-free, serum-free medium may provide a preferable solution. The final product of MSCs that will be used for transplantation should be tested microbiologically to detect the presence of aerobic and anaerobic microbes, mycoplasma, and endotoxin levels and genetically (karyotypic profile and stability) before the administration to patients. Viability of the cells should be also checked and be more than 80%.

The lack of standardization for MSC isolation and culturing has delayed the progress in the field of MSC use in human diseases since the comparison of results from different laboratories was sometimes impossible. Any differences in the culture conditions might selectively favor the expansion of different subpopulations. Based on morphology, several distinct cell types can be distinguished: spindle-shaped fibroblast-like cells, large flat cells, and small round-cell subpopulations.[83] The quality of preparations from different protocols are various and the cell products therefore heterogeneous. The source and quality of the starting material, culture media used, use of animal serum, cytokine supplements, initial seeding cell density, number of passages upon culture, and even type of cell culture dishes all have a significant influence on the cell populations that are finally produced. Therefore, there is an urgent need for the development of standardized cell culture reagents and products, common guidelines and standards for MSC preparations and of molecular and cellular markers to define subpopulations with different potentials. Only by these standardizations, the potential of MSCs in the treatment of different human diseases will be effectively evaluated.

28.3.2 Modified Mesenchymal Stem Cells

MSCs can be manipulated to secrete higher quantities of neuronal growth factors and therefore pushed toward a neuronal transdifferentiation (neuralization) and gain higher neurotrophic and neuroprotective effects. Such a method has been developed by Brainstorm Co., and used by our group in a recently completed trial in ALS with very promising results (Petrou et al. manuscript submitted). The MSCs are transformed to neurotrophic factor-secreting cells (MSC-NTF), following a medium-based differentiation process. These cells not only secrete NTFs such as glial-derived neurotrophic factor and BDNF but also express astrocytic markers. MSC-NTFs can migrate better toward brain lesions and were shown effective in several models of neuronal damage, such as the 6-hydroxy dopamine model for Parkinson's disease,[84] models for Huntington's disease,[85,86] models of optic nerve damage,[87] and in the model of MS, EAE,[88] and after sciatic nerve injury.[89] A trial in MS with these modified MSCs is under preparation in our center.

Another group has used a different method to produce bone marrow MSC-derived neural progenitors (MSC-NPs) as an autologous source of stem cells. MSC-NPs have stronger neural progenitor and immunoregulatory properties, and a reduced capacity for mesodermal differentiation and showed efficacy in the animal model of EAE.[90]

28.3.3 Pilot Clinical Trials with Stem Cells in Multiple Sclerosis

Phase I/II safety studies with MSC or bone marrow-derived cells have been performed in MS.[91–93] Overall, MSCs given intravenously or intrathecally were well tolerated, with some preliminary evidence of efficacy.[91]

On the basis of the preclinical data from our studies and the cumulative data from other centers, an exploratory clinical trial with autologous bone marrow-derived MSCs in 15 patients with intractable MS was initiated at Hadassah.[91,94] In this trial, based on the data in EAE models (indicating probably two distinct mechanisms of action by the two different routes of

MSC-administration), a combined intrathecal and intravenous administration was used to maximize the potential therapeutic benefit by accessing the CNS through the cerebrospinal fluid and the systemic circulation. In nine patients, MSCs were labeled with the superparamagnetic iron oxide MRI contrast agent ferumoxides (Feridex™) to track cell migration after local grafting.

Follow-up of the patients for 6 months showed that the mean Expanded Disability Status Scale (EDSS) (disability) score of the transplanted MS patients improved from 6.7 ± 1.0 to 5.9 ± 1.6. MRI visualized the MSCs in the occipital horns of the ventricles, indicative of the possible migration of the labeled cells in the meninges, subarachnoid space, and spinal cord. Immunological analysis at 24 h posttransplantation revealed an increase in the proportion of $CD4^{+}CD25^{+}$ regulatory T cells, a decrease in the proliferative responses of lymphocytes, and the expression of $CD40^{+}$, $CD83^{+}$, $CD86^{+}$, and HLA-DR on myeloid dendritic cells.

Since this was a pilot feasibility study, the most important finding was the acceptable safety profile of transplantation of autologous stem cells from the bone marrow in patients with MS. None of the patients experienced significant adverse effects during the 6- to 25-month observation. The follow-up MRI 1 year after transplantation did not reveal any unexpected pathology or significant new activity of the disease. Several clinical trials in nonneurological diseases have also indicated that intravenous administration of MSCs is a safe procedure.[95] The study in Hadassah Center additionally showed an acceptable short-term safety profile of the intrathecal route of administration of stem cells at doses of up to 70 million cells per injection per patient. The intrathecal approach (which was supported by the preclinical data from our group showing that this route of administration could induce superior neurotrophic and neuroprotective effects)[73] may be more advantageous for cell-based therapies in neurological diseases, in which the areas of tissue damage are widespread throughout the neuroaxis, since it may increase the possibility of migration of the injected cells to the proximity of the CNS lesions. The injected cells may circulate with the flow of the cerebrospinal fluid and have a better chance of reaching the affected CNS areas. However, the optimal route of stem cell administration in general and particularly MSC administration in patients with neurological diseases remains debatable and is currently tested in a large double-blind trial that is running in our center at Hadassah.

In the trial described earlier, MSCs were labeled with iron particles for MRI analysis. Such labeling of MSCs with the commercially used paramagnetic material, Feridex, was shown to be safe and had no negative effect on the functional (immunomodulatory and neurotrophic) properties of MSCs.[96] It seems therefore that Feridex may be used for the tracking of this type of stem cell in clinical applications, without compromising their major functional properties.

Additional clinical trials explored the safety and therapeutic benefit of intrathecal injection of ex vivo expanded autologous bone marrow-derived MSCs in patients with advanced MS.[93] In the later study, assessment of the patients at 3–6 months revealed an improvement in EDSS score in 5/7, stabilization in 1/7, and worsening only in 1/7 patients. Vision and low-contrast sensitivity testing at 3 months showed improvement in 5/6 and worsening in 1/6 patients. These preliminary results indicate additional (to the Hadassah trial) hints of clinical but not radiological efficacy and evidence of safety with no serious adverse events. A phase 2a study[97] in 10 patients with secondary-progressive MS showed an improvement in visual acuity and visual-evoked response latency, accompanied by an increase in optic nerve area, following intravenous transplantation of autologous MSCs. Although no significant effects on other visual parameters, retinal nerve fiber layer thickness, or optic nerve magnetization transfer ratio, were observed, this study provides a strong indication for induction of tissue repair with MSC transplantation in humans.

Another small pilot trial recruited 25 patients with progressive MS with an EDSS score of up to 6.5, unresponsive to conventional treatments.[98] The patients received a single intrathecal injection of ex vivo expanded MSCs (mean dose: 29.5 × 106 cells). Short-term adverse events of the injection included transient low-grade fever, nausea/vomiting, weakness in the lower limbs, and headache. No major delayed adverse effect was reported. The clinical course of the disease improved in 4 patients, deteriorated in 6, and did not change in 12 patients.

28.3.4 Ongoing or Recently Completed Clinical Trials with Mesenchymal Stem Cells in Multiple Sclerosis

A small, open-label, phase-1 clinical trial led by Dr. Jeffrey A. Cohen at Cleveland Clinic tested the ability of an individual's own MSCs to downregulate inflammatory mechanisms and to augment intrinsic tissue repair processes in people with relapsing forms of MS. They were given intravenously (infused into the vein). This trial, which was designed to evaluate safety and not designed to determine benefits, was completed and preliminary results were presented at the ECTRIMS meeting (Sep. 2014), suggesting that this approach was safe and warrants a phase-2 trial, which is now in planning stages.

A small, open-label, phase-1 stem cell trial has begun at the Tisch MS Research Center of New York using individuals' own MSCs to derive more specific stem cells called neural progenitor cells. The cells are expanded in the laboratory and then injected into the space around the spinal cord (intrathecal). The goal is to inhibit immune mechanisms and to augment tissue repair.

Another small, open-label, phase-1 trial of stem cells derived from placenta (known as PDA-001, manufactured by Celgene Cellular Therapeutics) was completed in 2014, and results suggested this approach was safe. The study involved 16 people with relapsing-remitting or secondary-progressive MS at sites in the United States and Canada. This study was designed to evaluate safety and not designed to show effectiveness. In the published paper, the researchers comment that the next step, a proof-of-concept clinical trial, is planned.

A placebo-controlled, phase-2 stem cell trial involving patients with secondary-progressive MS and primary-progressive MS has begun at Frenchay Hospital in Bristol, United Kingdom, testing the benefits and safety of using individuals' own bone marrow cells. The cells are extracted and then given by intravenous infusion immediately or 1 year after the extraction. The goal is to inhibit immune mechanisms and to augment tissue repair.

The largest and only randomized double-blind control trial in 48 MS patients is currently running at Hadassah Medical Center (PI: Dimitrios Karussis) and is aimed to answer the critical questions of efficacy, neuroregeneration potential, and optimal route of administration of the MSCs and the added value of repeated injections.

28.4 CONCLUSION

In conclusion, a decade of intensive preclinical and clinical research has greatly advanced our understanding of the role of stem cells in neurological diseases in general and specifically in MS, but has not yet fully clarified the picture. A great deal of positive data support the possibility of neuronal regeneration and neuroprotection. The mode of administration of the stem cells (and especially the intrathecal route that brings the cells in higher proximity to extensive areas of the CNS and has been long advocated and applied by our group) may be also crucial for the response of MS patients.[14,25] Long-term safety data are still missing along with substantially proven efficacy. Stem cells do not yet represent a panacea for all MS cases (or neurological conditions in general) but also should neither be the red flag in neurological research. Controlled studies using suitable clinical and surrogate markers (novel MRI and electrophysiological techniques) to substantiate regeneration and restoration of neurological function may provide the missing information.

References

1. Steinman L. Multiple sclerosis: a two-stage disease. *Nat Immunol*. 2001;2:762–764.
2. Gehrmann J, Banati RB, Cuzner ML, et al. Amyloid precursor protein (APP) expression in multiple sclerosis lesions. *Glia*. 1995;15:141–151.
3. Foong J, Rozewicz L, Davie CA, et al. Correlates of executive function in multiple sclerosis: the use of magnetic resonance spectroscopy as an index of focal pathology. *J Neuropsychiatry Clin Neurosci*. 1999;11:45–50.
4. Trapp BD, Peterson J, Ransohoff RM, et al. Axonal transection in the lesions of multiple sclerosis. *N Engl J Med*. 1998;338:278–285.
5. Vladimirova O, O'Connor J, Cahill A, et al. Oxidative damage to DNA in plaques of MS brains. *Mult Scler*. 1998;4:413–418.
6. Evangelou N, Esiri MM, Smith S, et al. Quantitative pathological evidence for axonal loss in normal appearing white matter in multiple sclerosis. *Ann Neurol*. 2000;47:391–395.
7. McFarland HF. Correlation between MR and clinical findings of disease activity in multiple sclerosis. *AJNR Am J Neuroradiol*. 1999;20:1777–1778.
8. Raine CS, Scheinberg LC. On the immunopathology of plaque development and repair in multiple sclerosis. *J Neuroimmunol*. 1988;20:189–201.
9. Popescu BF, Lucchinetti CF. Meningeal and cortical grey matter pathology in multiple sclerosis. *BMC Neurol*. 2012;12:11.
10. Magliozzi R, Howell O, Vora A, et al. Meningeal B-cell follicles in secondary progressive multiple sclerosis associate with early onset of disease and severe cortical pathology. *Brain J Neurol*. 2007;130:1089–1104.
11. Lucchinetti CF, Popescu BF, Bunyan RF, et al. Inflammatory cortical demyelination in early multiple sclerosis. *N Engl J Med*. 2011;365:2188–2197.
12. Conley BJ, Young JC, Trounson AO, et al. Derivation, propagation and differentiation of human embryonic stem cells. *Int J Biochem Cell Biol*. 2004;36:555–567.
13. Alessandri G, Pagano S, Bez A, et al. Isolation and culture of human muscle-derived stem cells able to differentiate into myogenic and neurogenic cell lineages. *Lancet*. 2004;364:1872–1883.
14. Gimble J, Guilak F. Adipose-derived adult stem cells: isolation, characterization, and differentiation potential. *Cytotherapy*. 2003;5:362–369.
15. Bottai D, Fiocco R, Gelain F, et al. Neural stem cells in the adult nervous system. *J Hematother Stem Cell Res*. 2003;12:655–670.
16. Bernstein ID, Andrews RG, Rowley S. Isolation of human hematopoietic stem cells. *Blood Cells*. 1994;20:15–23. discussion 24.
17. Lennon DP, Caplan AI. Isolation of human marrow-derived mesenchymal stem cells. *Exp Hematol*. 2006; 34:1604–1605.
18. Thomson JA, Itskovitz-Eldor J, Shapiro SS, et al. Embryonic stem cell lines derived from human blastocysts. *Science*. 1998;282:1145–1147.
19. Vazin T, Freed WJ. Human embryonic stem cells: derivation, culture, and differentiation: a review. *Restor Neurol Neurosci*. 2010;28:589–603.
20. Zhang SC. Embryonic stem cells for neural replacement therapy: prospects and challenges. *J Hematother Stem Cell Res*. 2003;12:625–634.

21. Kim H, Walczak P, Kerr C, et al. Immunomodulation by transplanted human embryonic stem cell-derived oligodendroglial progenitors in experimental autoimmune encephalomyelitis. *Stem Cells*. 2012;30:2820–2829.
22. Aharonowiz M, Einstein O, Fainstein N, et al. Neuroprotective effect of transplanted human embryonic stem cell-derived neural precursors in an animal model of multiple sclerosis. *PLoS One*. 2008;3:e3145.
23. Ben-David U, Benvenisty N. The tumorigenicity of human embryonic and induced pluripotent stem cells. Nature reviews. *Cancer*. 2011;11:268–277.
24. Hsu YC, Lee DC, Chiu IM. Neural stem cells, neural progenitors, and neurotrophic factors. *Cell Transplant*. 2007;16:133–150.
25. Campos LS. Neurospheres: insights into neural stem cell biology. *J Neurosci Res*. 2004;78:761–769.
26. Doetsch F, Caille I, Lim DA, et al. Subventricular zone astrocytes are neural stem cells in the adult mammalian brain. *Cell*. 1999;97:703–716.
27. Eriksson PS, Perfilieva E, Bjork-Eriksson T, et al. Neurogenesis in the adult human hippocampus. *Nat Med*. 1998;4:1313–1317.
28. Pluchino S, Quattrini A, Brambilla E, et al. Injection of adult neurospheres induces recovery in a chronic model of multiple sclerosis. *Nature*. 2003;422:688–694.
29. Ben-Hur T, Einstein O, Mizrachi-Kol R, et al. Transplanted multipotential neural precursor cells migrate into the inflamed white matter in response to experimental autoimmune encephalomyelitis. *Glia*. 2003;41:73–80.
30. Einstein O, Grigoriadis N, Mizrachi-Kol R, et al. Transplanted neural precursor cells reduce brain inflammation to attenuate chronic experimental autoimmune encephalomyelitis. *Exp Neurol*. 2006;198:275–284.
31. Einstein O, Karussis D, Grigoriadis N, et al. Intraventricular transplantation of neural precursor cell spheres attenuates acute experimental allergic encephalomyelitis. *Mol Cell Neurosci*. 2003;24:1074–1082.
32. Einstein O, Fainstein N, Vaknin I, et al. Neural precursors attenuate autoimmune encephalomyelitis by peripheral immunosuppression. *Ann Neurol*. 2006;61.
33. Shihabuddin LS, Horner PJ, Ray J, et al. Adult spinal cord stem cells generate neurons after transplantation in the adult dentate gyrus. *J Neurosci*. 2000;20:8727–8735.
34. Doetsch F, Verdugo JM, Caille I, et al. Lack of the cell-cycle inhibitor p27Kip1 results in selective increase of transit-amplifying cells for adult neurogenesis. *J Neurosci*. 2002;22:2255–2264.
35. Marshall 2nd GP, Laywell ED, Zheng T, et al. In vitro-derived "neural stem cells" function as neural progenitors without the capacity for self-renewal. *Stem Cells*. 2006;24:731–738.
36. Takahashi K, Yamanaka S. Induction of pluripotent stem cells from mouse embryonic and adult fibroblast cultures by defined factors. *Cell*. 2006;126:663–676.
37. Takahashi K, Tanabe K, Ohnuki M, et al. Induction of pluripotent stem cells from adult human fibroblasts by defined factors. *Cell*. 2007;131:861–872.
38. Goldman SA, Nedergaard M, Windrem MS. Glial progenitor cell-based treatment and modeling of neurological disease. *Science*. 2012;338:491–495.
39. Wang S, Bates J, Li X, et al. Human iPSC-derived oligodendrocyte progenitor cells can myelinate and rescue a mouse model of congenital hypomyelination. *Cell Stem Cell*. 2013;12:252–264.
40. Douvaras P, Wang J, Zimmer M, et al. Efficient generation of myelinating oligodendrocytes from primary progressive multiple sclerosis patients by induced pluripotent stem cells. *Stem Cell Rep*. 2014;3:250–259.
41. Gallacher L, Murdoch B, Wu DM, et al. Isolation and characterization of human CD34(−)Lin(−) and CD34(+) Lin(−) hematopoietic stem cells using cell surface markers AC133 and CD7. *Blood*. 2000;95:2813–2820.
42. Wognum AW, Eaves AC, Thomas TE. Identification and isolation of hematopoietic stem cells. *Arch Med Res*. 2003;34:461–475.
43. Brazelton TR, Rossi FM, Keshet GI, et al. From marrow to brain: expression of neuronal phenotypes in adult mice. *Science*. 2000;290:1775–1779.
44. Locatelli F, Corti S, Donadoni C, et al. Neuronal differentiation of murine bone marrow Thy-1- and Sca-1-positive cells. *J Hematother Stem Cell Res*. 2003;12:727–734.
45. Mezey E, Chandross KJ, Harta G, et al. Turning blood into brain: cells bearing neuronal antigens generated in vivo from bone marrow. *Science*. 2000;290:1779–1782.
46. Karussis D, Slavin S. Hematopoietic stem cell transplantation in multiple sclerosis: experimental evidence to rethink the procedures. *J Neurol Sci*. 2004;223:59–64.
47. Karussis D, Vaknin-Dembinsky A. Hematopoietic stem cell transplantation in multiple sclerosis: a review of the clinical experience and a report of an international meeting. *Expert Rev Clin Immunol*. 2010;6:347–352.

48. Karussis D, Vourka-Karussis U, Mizrachi-Koll R, et al. Acute/relapsing experimental autoimmune encephalomyelitis: induction of long lasting, antigen-specific tolerance by syngeneic bone marrow transplantation. *Mult Scler*. 1999;5:17–21.
49. Karussis DM, Slavin S, Ben-Nun A, et al. Chronic-relapsing experimental autoimmune encephalomyelitis (CR-EAE): treatment and induction of tolerance, with high dose cyclophosphamide followed by syngeneic bone marrow transplantation. *J Neuroimmunol*. 1992;39:201–210.
50. Karussis DM, Slavin S, Lehmann D, et al. Prevention of experimental autoimmune encephalomyelitis and induction of tolerance with acute immunosuppression followed by syngeneic bone marrow transplantation. *J Immunol*. 1992;148:1693–1698.
51. Karussis DM, Vourka-Karussis U, Lehmann D, et al. Immunomodulation of autoimmunity in MRL/lpr mice with syngeneic bone marrow transplantation (SBMT). *Clin Exp Immunol*. 1995;100:111–117.
52. van Gelder M, van Bekkum DW. Treatment of relapsing experimental autoimmune encephalomyelitis in rats with allogeneic bone marrow transplantation from a resistant strain. *Bone Marrow Transplant*. 1995;16:343–351.
53. van Gelder M, van Bekkum DW. Effective treatment of relapsing experimental autoimmune encephalomyelitis with pseudoautologous bone marrow transplantation. *Bone Marrow Transplant*. 1996;18:1029–1034.
54. Karussis D, Petrou P, Vourka-Karussis U, et al. Hematopoietic stem cell transplantation in multiple sclerosis. *Expert Rev Neurother*. 2013;13:567–578.
55. Nash RA, Hutton GJ, Racke MK, et al. High-dose immunosuppressive therapy and autologous hematopoietic cell transplantation for relapsing-remitting multiple sclerosis (HALT-MS): a 3-year interim report. *JAMA Neurol*. 2015;72:159–169.
56. Minagawa H, Takenaka A, Itoyama Y, et al. Experimental allergic encephalomyelitis in the Lewis rat. A model of predictable relapse by cyclophosphamide. *J Neurol Sci*. 1987;78:225–235.
57. Miyazaki C, Nakamura T, Kaneko K, et al. Reinduction of experimental allergic encephalomyelitis in convalescent Lewis rats with cyclophosphamide. *J Neurol Sci*. 1985;67:277–284.
58. Polman CH, Matthaei I, de Groot CJ, et al. Low-dose cyclosporin A induces relapsing remitting experimental allergic encephalomyelitis in the Lewis rat. *J Neuroimmunol*. 1988;17:209–216.
59. Fassas A, Passweg JR, Anagnostopoulos A, et al. Hematopoietic stem cell transplantation for multiple sclerosis. A retrospective multicenter study. *J Neurol*. 2002;249:1088–1097.
60. Mancardi GL, Saccardi R, Filippi M, et al. Autologous hematopoietic stem cell transplantation suppresses Gd-enhanced MRI activity in MS. *Neurology*. 2001;57:62–68.
61. Saccardi R, Kozak T, Bocelli-Tyndall C, et al. Autologous stem cell transplantation for progressive multiple sclerosis: update of the European Group for Blood and Marrow Transplantation autoimmune diseases working party database. *Mult Scler*. 2006;12:814–823.
62. Kotter I, Daikeler T, Amberger C, et al. Autologous stem cell transplantation of treatment-resistant systemic vasculitis–a single center experience and review of the literature. *Clin Nephrol*. 2005;64:485–489.
63. Chen JT, Collins DL, Atkins HL, et al. Brain atrophy after immunoablation and stem cell transplantation in multiple sclerosis. *Neurology*. 2006;66:1935–1937.
64. Burt RK, Cohen BA, Russell E, et al. Hematopoietic stem cell transplantation for progressive multiple sclerosis: failure of a total body irradiation-based conditioning regimen to prevent disease progression in patients with high disability scores. *Blood*. 2003;102:2373–2378.
65. Mancardi GL, Sormani MP, Gualandi F, et al. Autologous hematopoietic stem cell transplantation in multiple sclerosis: a phase II trial. *Neurology*. 2015;84:981–988.
66. Pittenger MF, Mackay AM, Beck SC, et al. Multilineage potential of adult human mesenchymal stem cells. *Science*. 1999;284:143–147.
67. Dominici M, Le Blanc K, Mueller I, et al. Minimal criteria for defining multipotent mesenchymal stromal cells. The International Society for Cellular Therapy position statement. *Cytotherapy*. 2006;8:315–317.
68. Kassis I, Vaknin-Dembinsky A, Karussis D. Bone marrow mesenchymal stem cells: agents of immunomodulation and neuroprotection. *Curr Stem Cell Res Ther*. 2011;6:63–68.
69. Zappia E, Casazza S, Pedemonte E, et al. Mesenchymal stem cells ameliorate experimental autoimmune encephalomyelitis inducing T-cell anergy. *Blood*. 2005;106:1755–1761.
70. Zhang J, Li Y, Chen J, et al. Human bone marrow stromal cell treatment improves neurological functional recovery in EAE mice. *Exp Neurol*. 2005;195:16–26.
71. Zhang J, Li Y, Lu M, et al. Bone marrow stromal cells reduce axonal loss in experimental autoimmune encephalomyelitis mice. *J Neurosci Res*. 2006;84:587–595.

72. Gerdoni E, Gallo B, Casazza S, et al. Mesenchymal stem cells effectively modulate pathogenic immune response in experimental autoimmune encephalomyelitis. *Ann Neurol*. 2007;61:219–227.
73. Kassis I, Grigoriadis N, Gowda-Kurkalli B, et al. Neuroprotection and immunomodulation with mesenchymal stem cells in chronic experimental autoimmune encephalomyelitis. *Arch Neurol*. 2008;65:753–761.
74. Ryan JM, Barry FP, Murphy JM, et al. Mesenchymal stem cells avoid allogeneic rejection. *J Inflamm (Lond)*. 2005;2:8.
75. Uccelli A, Moretta L, Pistoia V. Immunoregulatory function of mesenchymal stem cells. *Eur J Immunol*. 2006;36:2566–2573.
76. Kassis I, Vaknin-Dembinsky A, Karussis D. Bone marrow mesenchymal stem cells: agents of immunomodulation and neuroprotection. *Curr Stem Cell Res Ther*. 2010;6.
77. Gordon D, Pavlovska G, Glover CP, et al. Human mesenchymal stem cells abrogate experimental allergic encephalomyelitis after intraperitoneal injection, and with sparse CNS infiltration. *Neurosci Lett*. 2008;448:71–73.
78. Rafei M, Campeau PM, Aguilar-Mahecha A, et al. Mesenchymal stromal cells ameliorate experimental autoimmune encephalomyelitis by inhibiting CD4 Th17 T cells in a CC chemokine ligand 2-dependent manner. *J Immunol*. 2009;182:5994–6002.
79. Constantin G, Marconi S, Rossi B, et al. Adipose-derived mesenchymal stem cells ameliorate chronic experimental autoimmune encephalomyelitis. *Stem Cells*. 2009;27:2624–2635.
80. Bai L, Lennon DP, Eaton V, et al. Human bone marrow-derived mesenchymal stem cells induce Th2-polarized immune response and promote endogenous repair in animal models of multiple sclerosis. *Glia*. 2009;57:1192–1203.
81. Lalu MM, McIntyre L, Pugliese C, et al. Safety of cell therapy with mesenchymal stromal cells (SafeCell): a Systematic review and meta-analysis of clinical trials. *PLoS One*. 2012;7:e47559.
82. Prockop DJ, Sekiya I, Colter DC. Isolation and characterization of rapidly self-renewing stem cells from cultures of human marrow stromal cells. *Cytotherapy*. 2001;3:393–396.
83. Colter DC, Sekiya I, Prockop DJ. Identification of a subpopulation of rapidly self-renewing and multipotential adult stem cells in colonies of human marrow stromal cells. *Proc Natl Acad Sci USA*. 2001;98:7841–7845.
84. Sadan O, Bahat-Stromza M, Barhum Y, et al. Protective effects of neurotrophic factor-secreting cells in a 6-OHDA rat model of Parkinson disease. *Stem Cells Dev*. 2009;18:1179–1190.
85. Sadan O, Shemesh N, Barzilay R, et al. Mesenchymal stem cells induced to secrete neurotrophic factors attenuate quinolinic acid toxicity: a potential therapy for Huntington's disease. *Exp Neurol*. 2012;234:417–427.
86. Sadan O, Melamed E, Offen D. Intrastriatal transplantation of neurotrophic factor-secreting human mesenchymal stem cells improves motor function and extends survival in R6/2 transgenic mouse model for Huntington's disease. *PLoS Curr*. 2012;4. e4f7f6dc013d014e.
87. Levkovitch-Verbin H, Sadan O, Vander S, et al. Intravitreal injections of neurotrophic factors secreting mesenchymal stem cells are neuroprotective in rat eyes following optic nerve transection. *Invest Ophthalmol Vis Sci*. 2010;51:6394–6400.
88. Barhum Y, Gai-Castro S, Bahat-Stromza M, et al. Intracerebroventricular transplantation of human mesenchymal stem cells induced to secrete neurotrophic factors attenuates clinical symptoms in a mouse model of multiple sclerosis. *J Mol Neurosci*. 2010;41:129–137.
89. Dadon-Nachum M, Sadan O, Srugo I, et al. Differentiated mesenchymal stem cells for sciatic nerve injury. *Stem Cell Rev*. 2011;7:664–671.
90. Harris VK, Yan QJ, Vyshkina T, et al. Clinical and pathological effects of intrathecal injection of mesenchymal stem cell-derived neural progenitors in an experimental model of multiple sclerosis. *J Neurol Sci*. 2012;313:167–177.
91. Karussis D, Karageorgiou C, Vaknin-Dembinsky A, et al. Safety and immunological effects of mesenchymal stem cell transplantation in patients with multiple sclerosis and amyotrophic lateral sclerosis. *Arch Neurol*. 2010;67:1187–1194.
92. Mohyeddin Bonab M, Yazdanbakhsh S, Lotfi J, et al. Does mesenchymal stem cell therapy help multiple sclerosis patients? Report of a pilot study. *Iran J Immunol*. 2007;4:50–57.
93. Yamout B, Hourani R, Salti H, et al. Bone marrow mesenchymal stem cell transplantation in patients with multiple sclerosis: a pilot study. *J Neuroimmunol*. 2010;227:185–189.
94. Karussis D, Kassis I, Kurkalli BG, et al. Immunomodulation and neuroprotection with mesenchymal bone marrow stem cells (MSCs): a proposed treatment for multiple sclerosis and other neuroimmunological/neurodegenerative diseases. *J Neurol Sci*. 2008;265:131–135.

95. Giordano A, Galderisi U, Marino IR. From the laboratory bench to the patient's bedside: an update on clinical trials with mesenchymal stem cells. *J Cell Physiol*. 2007;211:27–35.
96. Kassis I, Vaknin-Dembinsky A, Bulte JW, et al. Effects of supermagnetic iron oxide labelling on the major functional properties of human mesenchymal stem cells from multiple sclerosis patients. *Int J Stem Cells*. 2010;3:144–153.
97. Connick P, Kolappan M, Crawley C, et al. Autologous mesenchymal stem cells for the treatment of secondary progressive multiple sclerosis: an open-label phase 2a proof-of-concept study. *Lancet Neurol*. 2012;11:150–156.
98. Bonab MM, Sahraian MA, Aghsaie A, et al. Autologous mesenchymal stem cell therapy in progressive multiple sclerosis: an open label study. *Curr Stem Cell Res Ther*. 2012;7:407–414.

CHAPTER

29

T-Cell Vaccination: An Insight Into T-Cell Regulation

I.R. Cohen, N. Friedman
The Weizmann Institute of Science, Rehovot, Israel

F.J. Quintana
Harvard Medical School, Boston, MA, United States

OUTLINE

Translational Neuroimmunology in Multiple Sclerosis
http://dx.doi.org/10.1016/B978-0-12-801914-6.00029-5

29.1 INTRODUCTION TO T-CELL VACCINATION

T-cell vaccination (TCV) was discovered in 1981 as an unexpected outcome of our study of experimental autoimmune encephalomyelitis (EAE) in the Lewis rat, a model of multiple sclerosis (MS). Our initial aim was to isolate lines and clones of T cells specifically reactive to myelin basic protein (MBP) that could adoptively produce EAE in otherwise healthy rats. Indeed, we succeeded in isolating encephalitogenic T cells that could cause clinical EAE within 4 days following intravenous inoculation of a million or more cells. Adoptive transfer of disease required that the transferred T cells be in a state of activation triggered in vitro by culture with a specific antigen or T-cell mitogen; inoculation of 50-fold greater numbers of nonactivated autoimmune T cells could not produce EAE[1]; EAE was a function of the state of the autoimmune T cells, not of their mere presence in the body. Activated anti-MBP T cells were able to satisfy Koch's Postulates for defining specifically autoimmune T cells as the etiologic agents responsible for EAE.[2] This pure culture methodology has made it possible to study T-cell behavior generally—their roles, for example, in immune pathophysiology,[3] in detecting new target antigens,[4] in characterizing T-cell migrations,[5,6] and in probing immune system behavior generally.[7] TCV is only one outcome of this platform technology for isolating functional cultures of T cells.

A century before our T-cell investigations, Pasteur, Koch, and their colleagues had discovered that the etiologic agents of various infectious diseases, upon attenuation, could be used as vaccines to induce resistance to the disease caused by the virulent agent of that disease.[8] The conceptual connection between the pathogenic agents of infection—bacteria or viruses—and the T-cell agents of autoimmune disease suggested the fanciful idea that, just as attenuated pathogens could vaccinate individuals against infection, attenuated autoimmune T cells might serve to vaccinate individuals against an autoimmune disease.

Experimenting with this idea, we found that resistance to EAE could be induced by injecting rats with encephalitogenic T-cell lines or clones whose virulence had been attenuated by irradiation or chemical cross-linkers.[9] We termed this vaccination procedure TCV. As a result of TCV, experimental animals acquired resistance to EAE produced by administering the virulent T cells, but TCV also induced resistance to EAE actively induced by immunization to MBP. Later, we found that resistance to EAE could also be induced by injecting rats with activated, intact anti-MBP T cells; resistance without clinical EAE was obtained by injecting numbers of unattenuated T cells lower than the number needed to cause clinically overt EAE.[10] Thus, resistance or disease was an outcome of the numbers as well as the state of the autoimmune T cells.

We found that TCV was effective in preventing or treating various other model autoimmune diseases: adjuvant arthritis in rats,[11] and thyroiditis,[12] lupus,[13,14] and spontaneous type 1 diabetes[15] in mice. We also found that TCV using specific alloreactive T cells could inhibit allograft rejection.[16] Further studies emerging from the initial discovery of TCV have proceeded in three directions: clinical trials of TCV in human autoimmune diseases, characterization of the regulatory mechanisms induced by TCV, and identification of the target T-cell molecules that induce TCV. We shall summarize here some of the findings generated by these studies, present new observations, and draw some conclusions about immune regulation in the light of TCV.

29.2 T-CELL VACCINATION CLINICAL TRIALS

The first trials of TCV to treat human autoimmune disease were undertaken within a decade after the first reports of TCV in experimental animals,[17,18] but only now, after two more decades of encouraging clinical reports, is TCV being developed by a pharmaceutical entity for the treatment of MS[19]; the initial results have earned TCV fast-track status by the US Food and Drug Administration.[20] We might wonder how it is that a promising, effective, remarkably safe, and relatively low-cost form of treatment for serious illness has tarried so long in its clinical translation—the development of new therapies is a complicated business indeed.

In any case, clinical trials of TCV to treat MS have been undertaken in the United States, Europe, and Israel; these trials have used autologous T cells isolated from the patient's peripheral blood and selected by their reactivity to myelin antigens. The TCV trials have demonstrated significant safety, in marked contrast to the undesirable side effects that may accompany the use of biologic agents and immune suppression.[21] The effects of TCV on the manifestations of MS have been generally beneficial; these trials have included double-blinded, placebo-controlled studies[22] and the arrest of disease progression in subjects who had failed earlier to respond to standard immune suppression therapy.[23] The clinical studies of TCV in MS have been reviewed over the years,[24–27] and interested persons may consult these reviews for specific information.

Some clinical results of TCV in rheumatoid arthritis[18,28–31] and in lupus[32] have been reported; the trials have been relatively small and were not placebo controlled; but the results in these diseases too are encouraging. Cell-based, personalized therapies for complex diseases are gaining acceptance, and TCV, as we learn more about how to use it clinically, might well be of benefit to many people.

29.3 GENERAL IMMUNE REGULATORY MECHANISMS UNCOVERED BY T-CELL VACCINATION

TCV provides an insight into the general physiology of immune regulation, we believe, because TCV activates the immune system to regulate itself; in contrast to injecting the subject with a preformed, pharmaceutically designed monoclonal antibody to this or that cytokine or receptor molecule, TCV queries the immune system itself: what do you do, immune system, when you are confronted with a population of syngeneic, activated T cells?

Historically, we explored TCV in the context of autoimmune conditions, but we now know that TCV can activate immune regulation even when using T-cell vaccines directed to foreign molecules.[33,34] This implies that T-cell activation, resulting from either self-antigens or foreign antigens, can induce endogenous regulatory mechanisms. Autoimmune disease is only a special case of a general phenomenon—any perturbation of a complex system like the immune system requires resolution: what goes up must come down. Immune reactions to any antigen need to be regulated—unnecessary or persisting immune inflammation in itself constitutes a disease. TCV shows us that the immune system can regulate inflammation generally.

29.4 THE T-CELL RECEPTOR IS A MAJOR INDUCER OF T-CELL VACCINATION REGULATION

TCV activates four types of regulatory mediators: antiidiotype, antiergotype, Qa1-restricted $CD8^+$ suppressor T cells, and anti-TCR antibodies (Fig. 29.1), at least as defined by variously different experimental protocols. These four mechanisms are neither mutually exclusive nor exhaustive: some of these reports may merely reflect different expressions of a shared mechanism and additional regulatory mechanisms may yet be discovered. But these four mechanisms do provide useful insights into immune system regulation. Note that all four mechanisms recognize the T-cell receptor (TCR), among other molecules expressed by activated T cells (see Fig. 29.1). We shall begin our discussion with the TCR as a target of TCV.

The TCR is composed of three different functional domains encoded by separate genetic elements: the C (constant) domain, which is critical to transducing the antigen-recognition signal into the cell; the V (variable) domain, which contains complementarity determining regions (CDR1 and CDR2) that mainly interact with the major histocompatibility complex (MHC) molecules of the antigen-presenting cell (APC), and so determine MHC restriction; and the hypervariable CDR3 region formed by variable–diversity–joining (V–D–J) recombination, which interacts directly with the antigen epitope presented by the MHC. Investigation of the regulatory response to TCV has uncovered T cell regulatory mechanisms targeted to each of the three TCR domains, along with anti-TCR antibodies (Fig. 29.2).

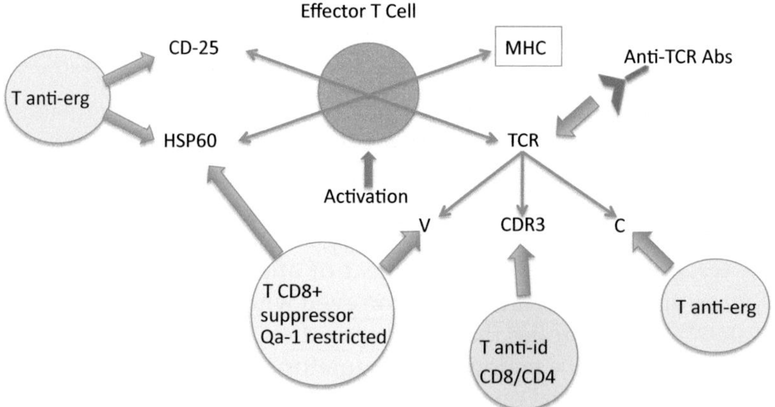

FIGURE 29.1 T-cell vaccination regulatory mechanisms. Activation of autoimmune (and other) effector T cells endows them with the capacity to mount proinflammatory immune reactions, but activation also induces the effector T cells to upregulate their major histocompatibility complex molecules and present signals to regulatory T cells such as T-cell receptor idiotopes and ergotopes that include HSP60 and CD-25 epitopes; these signals invite regulation by a variety of anti-id, anti-erg, and antibody regulators.

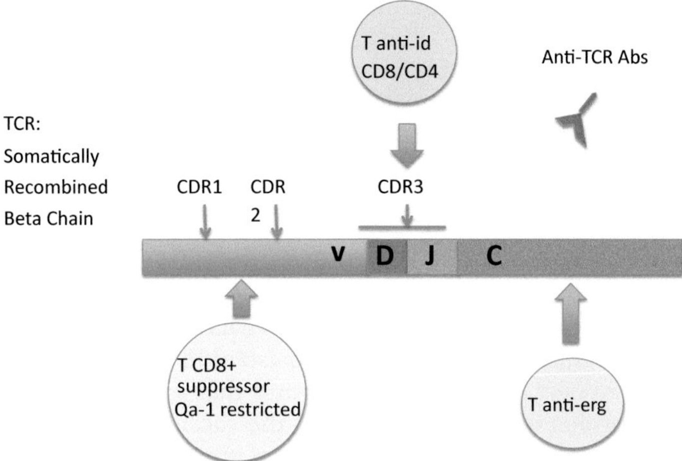

FIGURE 29.2 The beta chain of the T-cell receptor (TCR) provides signals for regulation by anti-TCR antibodies and T cells: anti-CDR3 region (anti-id), anti-C region (anti-erg), and anti-V region (CD8 suppressors).

29.4.1 T Antiidiotypic Regulation

T cell clones differ from one another by their somatically rearranged antigen-receptor V (variable), D (diversity), and J (joining) minigenes that encode its TCR and beta chains; TCR alpha chains do not include D segments.[35,36] The CDR3 regions of the TCR, which interact with the antigen epitope presented in the MHC cleft, comprise the T cell's idiotype; we might expect that a specific T-cell vaccine should induce a regulatory immune response to its CDR3 idiotopes. Indeed, evidence that TCV induces antiidiotypic (anti-id) T-cell responses was obtained early on. TCV was found to activate T cells, both CD4 and CD8, that could distinguish between different T-cell clones and their TCRs[37]; moreover, anti-id T cells induced by TCV could respond to peptides of TCR variable regions[38]; and such peptides were reported to be presented by activated T cells in association with MHC molecules.[39] Moreover, cytotoxic CD8 T cells induced by TCV could lyse autoimmune effector T cells,[40] and TCV led to reduced numbers of specific autoimmune T cells in vaccinated subjects.[41,42]

Anti-id T cells alone, however, cannot explain all of the observations associated with effective TCV. First, TCV using a defined T-cell clone or line responsive to a particular autoantigen is able to downregulate autoimmune diseases known to target a number of different autoantigen epitopes on one or more self-antigens: for example, type 1 diabetes, known to involve effector autoimmunity to several different self-antigens, can be arrested by TCV with a single clone of T cells responsive to peptide p277, only one of the epitopes on HSP60, which is only one of the different target molecules.[15] Moreover, TCV may reduce, but not abolish T-cell autoreactivity to the target self-antigens. Although a vaccinated subject may resist the particular autoimmune disease, that subject can still maintain T cells reactive to the relevant self-antigens.[43] TCV affects the phenotype of the response to the antigen, not the ability of the subject's T-cell repertoire to recognize that self-antigen.[44] In other words, TCV regulates the inflammatory nature of the autoimmune response, and not only the antigen recognition repertoire (Figs. 29.1 and 29.2). An autoimmune disease reflects the proinflammatory phenotype of activated autoimmune T cells—the presence in the subject's repertoire of T cells that can recognize self-antigens is necessary, but not sufficient to cause a disease.[45] This implies that an autoimmune disease might be treated successfully by downregulating the proinflammatory state of existing, activated T cells.

29.4.2 T Antiergotypic Regulation

Shortly after the observation of anti-id regulation, we discovered that TCV also activated a T-cell regulatory mechanism that was not directed to a TCR idiotype, but to molecules expressed by activated T cells generally—a process we have termed antiergotypic (anti-erg) regulation.[46] In practice, rats could be rendered resistant to EAE by vaccinating them with T cells that did not respond to myelin antigens and that were not encephalitogenic. The critical requirement for this type of TCV was that the vaccinating T cells had to have been activated. The Greek word for activation or work is *ergon* (εϱγον)—hence we coined the terms *ergotope* (a molecule that marks a state of activation) and *antiergotypic* T cell (a T cell that responds to an ergotope and regulates an immune response). Functionally, we define an anti-erg T cell as any T cell that kills, proliferates, or secretes cytokines when stimulated by syngeneic T cells that are in a state of activation; the anti-erg T cell does not respond to the same syngeneic T cells when they are not in an activated state. Anti-erg T cells may also be detected by their response to isolated ergotope antigens.[47] Anti-erg regulation focuses on the state of the system's T cells[48]; anti-id regulation, in contrast, focuses on the specificity of the system's T cells. The two targets of regulation would seem to complement each other (Figs. 29.1 and 29.2).

We have observed (F.J. Quintana and I.R. Cohen; in preparation) that each of the two C-region variants in the rat TCR includes ergotopes that are presented in the context of MHC II by rat T cells that have been activated by specific antigen—the same clones of T cells in a resting state do not present their C-region ergotopes to anti-erg T cells. We found that a model autoimmune disease such as adjuvant arthritis[4] can be downregulated significantly by the anti-C region anti-erg T cell regulatory response; inhibition of disease was associated with changes in the cytokine responses of the effector T cells to their target antigens. The protection against arthritis could be adoptively transferred using isolated anti-erg T cells. The details are being prepared for publication, but important for the present discussion is the fact that the C region of the TCR provides functional ergotopes, shown to be effective in downregulating the inflammatory behavior of activated effector T cells. Note that the TCR C region is common to all effector T cells, irrespective of their specificity for self-antigens or foreign antigens; thus TCV appears to have uncovered a regulatory mechanism that operates generally to downregulate inflammation induced by any T-cell immune response.

Anti-erg T cells are heterogeneous; they are present in the thymuses of newborn rats and are found in the spleen some days later.[49] They include αβ TCR CD4 and CD8 T cells restricted respectively by MHC II and MHC I molecules expressed on the stimulating, activated T cell or on other types of APC. Anti-erg T cells also include γδ TCR T cells that may not be MHC restricted.[50] Some anti-erg T cells proliferate in response to ergotopes without secreting detectable cytokines; some secrete different sets of regulatory cytokines including IL-10, IL-4, IL-2, TGFβ, TNFα, and IFNγ.[51] Anti-erg T cells were first demonstrated following TCV, but they are detectable in naïve humans and rodents; they expand in response to strong immunization to foreign antigens such as complete Freund's adjuvant[51]; they appear to be decreased in subjects suffering from MS, but expand after TCV.[51]

CD-25 Ergotope

In addition to the TCR C region, other peptide ergotopes are presented by activated T cells, including peptides of the CD-25 molecular component of the alpha chain of the IL-2 receptor (Fig. 29.1), which is upregulated upon T-cell activation.[44,52] The CD-132 γ chain component of the IL-2 receptor, which is not upregulated in activated T cells, does not feature ergotope peptides. The TNF receptor also provides ergotopes presented by activated T cells.[53]

HSP60 ERGOTOPES

A notable source of ergotope peptides is the HSP60 molecule (Fig. 29.1); activated T cells present HSP60 peptide epitopes in the clasp of their MHC II molecules, which are also upregulated, to anti-erg regulatory T cells (Tregs).[47] HSP60 and other stress protein molecules function within the cell as chaperones to assist in the proper folding of newly synthesized proteins and to protect the cell from denatured proteins, among other roles.[47,54] The essential functions of heat shock protein (HSP) molecules for the health of cells and organisms make them ideal signals for the immune system—the expression of HSP molecules reflects the states of cells and tissues. These molecules provide the immune system with reliable biomarkers, which assist the immune system in regulating inflammation.[55] The HSP60 molecule is central to a network of immune cells, in which its molecular concentration, its particular peptide moiety, and the cell types that respond to it are integrated to determine a proinflammatory or antiinflammatory outcome.[54]

Moreover, we have discovered that HSP60 is immunologically related to HSP70 and HSP90 as immune system regulatory signals: intramuscular administration of DNA encoding HSP70 or HSP90, like administration of DNA encoding HSP60,[56,57] inhibits autoimmune disease.[58] We found that the three HSP molecules are mutually connected in a network—administration of HSP70 or HSP90 induces Tregs that respond to epitopes of HSP60 as well as Tregs that respond to HSP70 and HSP90 epitopes.[58]

A peptide derived from human HSP60, p277, is a T-cell ergotope recognized by anti-erg Tregs.[47] This HSP60 ergotope peptide[59] is presently completing a phase-III clinical trial to arrest the autoimmune destruction of beta cells in type 1 diabetes. In addition to serving as a T-cell ergotope, peptide p277 is an agonist by way of toll-like receptor (TLR) 2 (TLR2) signaling in $CD4^+CD25^+$ Tregs.[60] Later, we shall note that HSP60 epitopes are also presented by the Qa1 antigen-presenting molecule to $CD8^+$ suppressor T cells; HSP60 is truly a multipotent, key regulator of inflammation.

CD4+ Regulatory T Cells

Classical Tregs of the $CD4^+$ type,[61] now defined by various markers ($CD4^+CD25^+$ FoxP3, etc.) had not yet been discovered when TCV first came to the attention of immunologists, so work on $CD4^+$ Tregs in the context of TCV was late and has been scanty. However, there has been at least one report of standard Tregs induced by TCV.[44] In a clinical trial of TCV in rheumatoid arthritis patients, Jingwu Zhang and associates constructed T-cell vaccines using autologous synovial fluid T cells cultured in low concentrations of IL-2 to preferentially expand T cells that had already been activated in the synovial exudate in vivo. They reported clinical and immunologic improvement in two-thirds of the treated subjects associated with $CD8^+$ cytotoxic T cells specific for the T-cell vaccine along with $CD4^+CD25^+$ Foxp3hi, IL-10 secreting Tregs reactive to peptides of CD25, the IL-2 receptor α-chain.[44] It is likely that the vaccine-specific $CD8^+$ cytotoxic T cells were anti-id and that the Tregs were anti-erg, as we have defined them here. Thus, human subjects respond to TCV with a combination of anti-id and anti-erg regulators that include classical $CD4^+$ Tregs.

29.4.3 Qa1-Restricted $CD8^+$ Suppressor T Cells

$CD8^+$ suppressor T cells (Figs. 29.1 and 29.2) were first discovered in the 1970s by Gershon and colleagues,[62] but molecular-scale investigation of this class of cells was delayed for decades until the emergence of new molecular technologies made it possible to revisit the subject.[63] Even now, however, the relatively newer $CD4^+$ Tregs have gained most of the attention; but, irrespective of research fashions, these $CD8^+$ regulatory cells are relevant to TCV. The important points for our present discussion are that (1) these suppressors are targeted to epitopes presented by the nonclassical Qa-1 antigen-presenting molecule (so named in mice; the human equivalent is termed HLA-E)[64]; (2) Qa-1 restricted epitopes include peptides of the V regions of the TCR[65] and peptides of the HSP60 molecule[63]; (3) a range of epitopes is presented by Qa-1 depending on the state of the presenting cell (activated T cells, but not resting T cells present their V region peptides[66,67]); (4) Qa-1-restricted epitopes can be cross-presented by non-T cell APC to the $CD8^+$ T regulators[63–67]; and (5) most importantly for us here, these $CD8^+$ regulators are activated by TCV and contribute significantly to TCV-induced regulation of inflammation.[63–67]

29.4.4 Anti-T-Cell Receptor Antibodies

TCV has been reported to induce antibodies with regulatory properties (Figs. 29.1 and 29.2).[34,68] Much remains to be learned about these antibodies, their target epitopes on the TCR, and their regulatory functions, but we can list several of their relevant features: (1) anti-TCR binding antibodies can be detected following TCV, but also following spontaneous recovery from an autoimmune disease and following intense immunization to a foreign antigen[34]; (2) anti-TCR antibodies may be anti-id,[69] but TCV can also induce antibodies to many other epitopes expressed by activated T cells[33]; (3) post-TCV antibodies bind to activated T cells much more than they do to resting T cells[34] (the antibodies, like TCV-induced anti-erg T cells, recognize the activated state of the T cell they regulate); (4) post-TCV antibodies can downregulate T-cell responses to self-antigens or foreign antigens[34,68]; and (5) anti-T cell antibodies stimulated by a foreign T-cell vaccine can downregulate an autoimmune disease response[34] (these antibodies too are anti-erg). Thus we can conclude that regulatory antibodies, and not only Tregs, are functionally activated by TCV. We have yet to learn how these antibodies downregulate inflammation, but it is possible that anti-TCR antibodies induced by TCV trigger T-cell activation in a way that results in T-cell deletion and/or anergy.

29.5 PUBLIC CDR3 DOMAINS AND IMMUNE REGULATION

The discovery of shared, public CDR3 domains of the TCR[70] casts a new light on TCV anti-id regulation. Research on public CDR3 domains is in an early stage, and we cannot yet draw firm conclusions; nevertheless, the findings invite us to consider the possible roles of public TCR segments in antigen-specific T cell regulation.

What is a public CDR3 domain? The CDR3 segment of the TCR is formed by recombination of V–D–J mini-gene segments (Fig. 29.2); this recombination event has been assumed to be diversely random, and the diversity of the CDR3 segment is further amplified by insertions and deletions of DNA nucleotide sequences not encoded in the genome. Diversification of the TCR is thought to fashion about 10 billion possible CDR3 amino acid sequences[35,36]; this is larger than the estimated number of T-cell specificities in an individual mouse or human (10^7–10^9). Thus, on statistical grounds, the expression of identical CDR3 sequences in different subjects would be unexpected.

Despite expectations, we have reported that the naïve mouse T-cell repertoire contains CDR3 TCR beta chain amino acid sequences shared by most mice; such highly shared sequences can be termed public. We sequenced hundreds of thousands of TCR molecules present in the $CD4^+$ T cells obtained from the spleens of 28 young and healthy C57BL/6 mice; about 300 public CDR3 sequences were found to be expressed in all 28 mice, and over 1000 sequences were shared by 70% or more of the mice.[70]

These highly shared CDR3 sequences differed in two features from the more private sequences found in only one or a few mice: the more private sequences exhibited a range of frequencies from very low (small clone size) to very high (large clone size); the more public sequences, in contrast, were all found at relatively high frequency—low frequency sequences were missing. Since a high-frequency T cell is assumed to have been expanded in response to antigen stimulation,[70] we can surmise that all the T cells bearing public CDR3 sequences had undergone positive selection, probably in response to antigen stimulation; in contrast, only some T cells among the more private sequences were likely to have been expanded by contact with their specific antigens.

Another feature of public CDR3 sequences was their high degree of convergent recombination.[70] Due to codon degeneracy, a single CDR3 amino acid sequence can be encoded by more than one nucleic acid recombination event. We found that private CDR3 amino acid sequences were derived from an average of only one nucleic acid recombination event. Public sequences, in contrast, were each encoded by many different nucleic acid recombinations. The average number of nucleic acid recombinations that converged to encode a single public amino acid sequence was 35; some individual public amino acid CDR3 sequences were the products of up to 100 different nucleic acid recombinations in the mouse population. A high degree of convergent recombination could be most easily explained by the positive selection of different nucleic acid recombinations all encoding the same amino acid sequence. What can explain this selective expansion?

It has been proposed that public TCR sequences could emerge from biases in the process of V–D–J recombination.[71,72] Indeed, the results of our computer simulations suggested that there does exist a fundamental biochemical–mechanical bias in particular V–D–J nucleic acid recombination partners, probably due to favorable positioning of certain V, D and J genes on the chromosome.[70] Our experimental observations suggested that recombination bias, as estimated from TCR repertoire data, can explain the higher frequencies of many public CDR3 sequences, as well as the high level of convergent recombination for these sequences.[70] This biophysical bias in particular V–D–J nucleic acid recombination partners would appear to be further fine-tuned by antigen selection. Public sequences thus result both from recombination bias and from subsequent antigen selection in different individuals.[70]

Obviously, the T-cell repertoires of different subjects cannot be coordinated at the somatic level—my T cells can know nothing about the specificities of your T cells. Although the TCR is created by somatic V–D–J recombination in each individual, CDR3 publicness must be a product of the evolutionary experience of the species encoded in germline DNA inherited in common by different individuals. Thus the germline would have to encode both favorable V–D–J positioning on the chromosome and TCR selection by certain self-antigens common to different individuals. In contrast to public CDR3 sequences, private sequences would entail somatic selection restricted to the level of the individual. This hypothesis would gain credibility if indeed public and private CDR3 sequences would be found to recognize different classes of antigens. How can we derive a likely target antigen from a CDR3 TCR sequence?

At the present state of immunology, knowledge of a CDR3 amino acid sequence alone cannot tell us the sequence of the antigen epitope recognized by that TCR; this uncertainty is compounded by the fact that the functioning TCR is formed by the association of both alpha and beta chains. So the sequences of CDR3 beta chains, public or private, cannot tell us which antigens the TCR sequences might recognize. However, we were able to circumvent this limitation to some degree and annotate our CDR3 sequences by searching for identical sequences present in T-cell clones of known specificities published in the past.

Over the years, various laboratories have published the TCR sequences of T cells associated with variously specific immune reactivities; any of our CDR3 TCRβ sequences identical to any of these annotated TCRβ sequences would suggest that our naïve C57BL/6 sequences might be associated with a similar immune function. We culled from the literature some 250 TCR beta chain sequences obtained from various strains of mice (with different MHC haplotypes) associated with known immune reactivities; amazingly, 124 of the 250 annotated CDR3 TCR beta chain sequences were present in our CDR3 database.

The use of identical CDR3 sequences by mice of differing MHC genotypes can be explained by the fact that the CDR1 and CDR2 regions of the TCR, which bind to the antigen-presenting MHC molecules,[73] are present in the V region and not in the V–D–J region (Fig. 29.2); thus MHC restrictions are more linked to the V region than they are to the CDR3 joining region. Different MHC molecules that can present the same peptide epitope in their clasp could select a common CDR3 region appended to different TCR V segments.[70] Indeed, we find that the same CDR3 region can be encoded by a number of different V genes, and the Vβ genes used to encode a given CDR3 sequence vary between mice of different MHC haplotype.[70]

We sorted the 124 public CDR3 sequences into four annotated T-cell categories: TCR responses to foreign antigens associated with pathogens; TCR responses to known self-antigens or associated with autoimmune conditions; TCRs from T cells infiltrating tumors or responding to known tumor antigens; and TCRs associated with allograft reactions. Most tumor antigens are self-antigens or modified self-antigens, and allograft responses include antiself reactions.[74,75] Thus, the four categories of annotated T-cell reactivity could be reduced to two: antiforeign and antiself associations. These two categories nicely separated the private and public CDR3 sequences: the more private CDR3 sequences were prominent in the antiforeign category and deficient in the antiself category; the more public CDR3 sequences were present in the antiforeign category, but they were relatively enriched in the antiself category.[70]

Note that publicness constitutes an unanticipated distinction between self- and notself-antigens: both types of antigen are recognized by the T cell repertoire, but the TCRs for self tend to be more public and those for foreign tend to be more private. This distinction, of course, is quantitatively relative, not absolute.

The association of a set of public CDR3 segments with self and self-like reactivity in healthy mice is compatible with the Immunological Homunculus theory—autoimmune repertoires of T cells and B cells in healthy individuals serve an ongoing dialog between the immune system and the body.[76,77] Be that as it may, does the set of public CDR3 sequences serve TCV and immune system regulation? The story of the C9 T-cell clone suggests a positive answer, at least for one member of the public set.

29.5.1 C9 CDR3 Regulatory Functions

We isolated the C9 T cell clone in 1991 from NOD strain mice on their way to spontaneously develop type 1 diabetes (T1D)[78] and reported its TCR alpha and beta chain sequence in 1999.[79] The C9 clone responded to human or mouse HSP60 and to a defined peptide of the human HSP60 molecule, peptide p277.[78,80] Clone C9 was able to adoptively transfer insulitis and hyperglycemia to otherwise healthy recipient mice; albeit, the transferred diabetes was mild and transient.[78] After spontaneous recovery, the mice were no longer susceptible to T1D. The functional relevance of clone C9 to autoimmune regulation was further confirmed by its ability to vaccinate NOD mice against the spontaneous development of autoimmune T1D.[78] T1D is known to be associated with autoimmunity to numerous self-antigens[81] including insulin and glutamic acid decarboxylase; despite the multiple autoimmune targets, C9 was an effective T-cell vaccine.

We found that the CDR3 beta chain peptide segment of the C9 TCR was public in different NOD mice,[79] and the peptide itself could be used to vaccinate against T1D.[78,82] The outcome of this TCV-induced inhibition of T1D appeared to involve a cytokine shift to an antiinflammatory phenotype in the diabetogenic T cells, rather than to their elimination.[83]

Interestingly, inhibition of T1D in NOD mice could be affected by a single subcutaneous administration of 100 μg of the target peptide of C9, p277.[82] The regulatory effects of p277 appear to be relevant to humans too; administration several times a year of 1 mg of p277 subcutaneously preserved beta-cell function in humans with newly diagnosed T1D,[59,84] and p277 peptide therapy is presently completing a phase-3 clinical trial. The half-life of peptide p277 in human plasma or serum is about 3 min; hence, resistance to T1D is mediated by an endogenous regulatory mechanism set into motion by a transient encounter with the peptide.

Indeed, downregulation of T1D by peptide p277 in NOD mice is marked by the activation of anti-C9 anti-id T cells specific for the CDR3 peptide of C9[15]; likewise, vaccination with the C9 CDR3 peptide upregulates the anti-C9 anti-id response and downregulates the inflammatory phenotype of the spontaneous T-cell response to p277.[15] Moreover, adoptive transfer of anti-C9 anti-id T cells also inhibits T1D.[15] Thus an immune regulatory network links the public C9 CDR3 domain of the TCR repertoire with anti-id T cells and with the p277 peptide of HSP60. Fig. 29.3 shows that the anti-C9 anti-id is the agent that downregulates the autoimmune T1D, and this anti-id is activated by the public CDR3 segment of the C9 TCR.

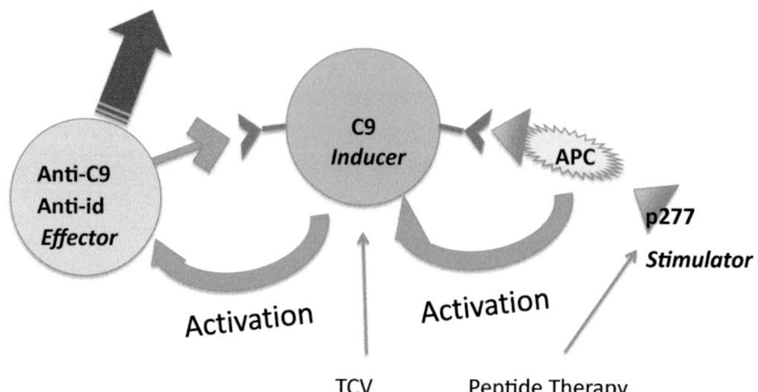

FIGURE 29.3 Autoimmune and inflammatory disease inhibition. The C9 T cell illustrates regulatory network connectivity: the effector of regulation is the anti-C9 anti-id response; C9 is the inducer of this effector of regulation; and peptide p277, presented to C9 by APC, is a stimulator of the C9 inducer; T-cell vaccination (TCV) using C9 T cells directly induces the anti-C9 regulatory response. C9 thus connects peptide therapy with TCV.

Thus, T1D can be downregulated alternatively by vaccinating with C9, by administering its p277 target self-antigen, or by adoptively transferring anti-C9 anti-id T cells. In other words, C9 vaccination activates anti-C9 anti-id downregulators of inflammation; peptide p277, via presentation by APC, activates C9 to activate, and in turn, anti-C9 anti-id downregulators.

This sounds dauntingly complicated, but we might simplify C9-based regulation thusly: take another look at Fig. 29.3; the actual effector of regulation is the anti-C9 response; C9 is the inducer of regulation; and peptide p277 is the stimulator of the inducer. Peptide p277 is also a coactivator of $CD4^+$ Tregs via TLR2 of the Tregs.[60] The CDR3 idiotope of the C9 inducer is at the center of this regulatory network. Do other public CDR3 idiotopes play similar roles as inducers of regulation? If so, anti-id responses directed to public CDR3 segments might by exploited generically to facilitate the enforcement of immune regulation. Indeed, public CDR3 segments might provide anti-id regulation with an anti-erg character widely shared among different individuals. Obviously, these speculations are only working hypotheses. The fact that naïve, healthy C57BL/6 mice harbor public TCR sequences such as C9,[70] which are also expressed in the autoimmune T1D process in NOD mice, suggests that such regulatory networks are built into the physiology of the healthy immune system.[76,77]

29.6 REGULATION DEPENDS ON TISSUE CONTEXT

Most autoimmune diseases result from the activities of a large number of activated, autoimmune T cells—no autoimmune T cells, no autoimmune disease. How then does TCV downregulate an inflammatory autoimmune disease if it merely adds to the body more activated, autoimmune T cells? Obviously, the TCV cells are attenuated, so they are harmless, but even so, how does TCV downregulate the disease process? What does the vaccine do to the host immune system that the endogenous pathogenic T cells don't do? This question brings us to the regulation of the regulators.

The regulators of activated T cells induced by TCV downregulate the effectors, but the regulators too need to be downregulated. The reason is obvious: unless these regulators get removed, the immune system would become paralyzed by regulator-cell overload; after all, immunity to pathogens and tumor surveillance requires vigorous effector responses. There are probably a number of different mechanisms that work to restrain overactive immune suppression. We have uncovered one way that helps keep regulators and effectors balanced.

Fig. 29.4 illustrates our finding that anti-erg regulators are subject to loss or renewal, depending on who presents them with their target ergotope. Anti-erg regulators become inactivated following a direct interaction with effector T cells presenting ergotopes, but the anti-erg regulators proliferate when they interact with APC (non-T cell) that present the same ergotope antigens.[47,85] Thus the cellular context can direct the outcome; at sites of inflammation, the anti-erg T cells are lost when they downregulate activated effector T cells—the effectors are inactivated, but so are the regulators. But at noninflamed sites outside the diseased tissue, the regulators proliferate when they interact with APC (non-T cell) that cross-present T-cell ergotopes. Inflammatory sites are marked by concentrations of cytokines and other innate receptor agonists that are absent at noninflamed sites. Outside a site of inflammation, the ergotope protein, its peptide ergotope, or even DNA encoding the regulatory ergotope can all induce expansion of the regulators when mediated by APC (non-T cell) (Quintana, FJ and Cohen IR; in preparation).

Likewise, the anatomic context of T-cell activation would also seem to be important for generating autoimmune effector T cells; it has been proposed that the autoimmune effectors of MS are generated outside the nervous system, possibly in the gut.[86]

In any case, we might reason that TCV using attenuated effector cells, or even a small number of intact effector cells administered into a peaceful body site such as the subcutaneous

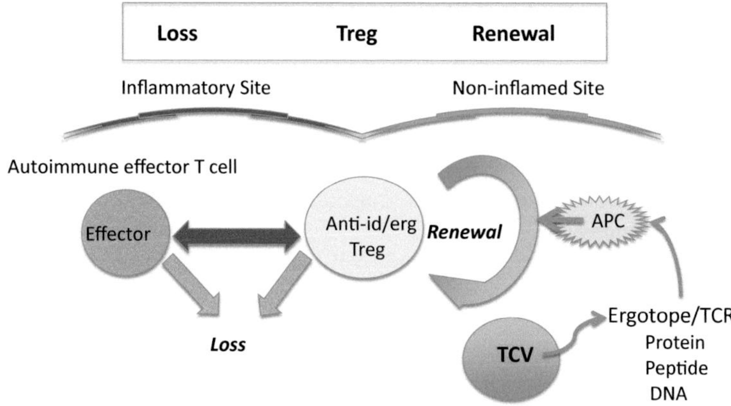

FIGURE 29.4 The regulation of the regulators. Regulatory T cells (Tregs) (anti-id/anti-erg) undergo loss or renewal depending on the context of inflammation and the type of antigen-presenting cell (APC) that presents the regulatory simulator signal. At the site of inflammation, the Tregs are lost when they downregulate the effector T cells. The same Tregs can be stimulated to proliferate and renew regulation when they encounter their target id or erg presented by non-T cell APCs at noninflamed sites. The target id or erg can be presented directly via T-cell vaccination or cross-presented by APCs that have taken up the relevant id/erg protein, peptide, or DNA encoding the protein.

tissue, enables APC cross-presentation of T-cell ergotopes and idiotopes or even of T-cell DNA that leads to the activation and expanded renewal of anti-erg and anti-id T regulators, free of interference by inflammatory effector T cells. Indeed, we have found that naked DNA encoding ergotopes such as CD-25,[47] HSP60/70/90,[56–58] and the TCR C region (in preparation) can each induce downregulation of model autoimmune diseases. Once stimulated by APC presenting the ergotope, the amplified regulators than can migrate to the site of inflammation where they downregulate the effectors, while also getting themselves downregulated in the process of regulation. Note that the generation of $CD8^+$ T suppressors has been shown to be mediated by APC (non-T cell) that present the TCR idiotope to the Tregs.[67] Thus, the exhaustion or renewal of regulators depends on the tissue site and cellular context of the reaction.

29.7 EPILOGUE

TCV can teach a number of key points about immune regulation. First, the immune system is capable of regulating its own inflammatory reactivity; we can reason, in fact, that endogenous self-regulation is going on all the time.[76,77] The relapsing-remitting phase of MS is probably a clinical expression of the dynamic struggle between effectors and regulators. TCV can be seen in this light as a way to enhance the collective strength of the regulators. The safety and effectiveness of TCV can be attributed to its activation of endogenous agents of regulation; TCV talks to the system (Fig. 29.3), and does not punish it with pharmaceutical suppression.

Second, the TCR and other molecules involved in T-cell activation are prime targets for regulators (Figs. 29.1 and 29.2). Regulation is expressed by networks (Fig. 29.3).

Third, TCV shows us that regulation is mediated by variously different populations of T cells and antibodies directed to diverse biomarker molecules; regulation is not the exclusive domain of $CD4^+$ Tregs (Fig. 29.1). Regulation can be antigen-specific (anti-id T cells and antibodies) or antigen-nonspecific (anti-erg responses). Regulation is an expression of systems immunology.

Fourth, some mechanisms of regulation are sensitive to the contexts of inflammation and the anatomy of interactions of different cell types (Fig. 29.4).

Fifth, regulation is structured by evolutionary experience that has generated regulatory repertoires shared by different individuals, manifested as an Immunological Homunculus[87] and public CDR3 sequences.[70]

Perhaps the time is ripe for individualized, cell-based, physiological treatments.

References

1. Ben-Nun A, Wekerle H, Cohen IR. The rapid isolation of clonable antigen-specific T lymphocyte lines capable of mediating autoimmune encephalomyelitis. *Eur J Immunol*. 1981;11(3):195–199.
2. Cohen IR. Regulation of autoimmune disease physiological and therapeutic. *Immunol Rev*. 1986;94:5–21.
3. Yarom Y, et al. Immunospecific inhibition of nerve conduction by T lymphocytes reactive to basic protein of myelin. *Nature*. 1983;303(5914):246–247.
4. van Eden W, et al. Cloning of the mycobacterial epitope recognized by T lymphocytes in adjuvant arthritis. *Nature*. 1988;331(6152):171–173.
5. Kawakami N, et al. Live imaging of effector cell trafficking and autoantigen recognition within the unfolding autoimmune encephalomyelitis lesion. *J Exp Med*. 2005;201(11):1805–1814.

6. Naparstek Y, et al. Effector T lymphocyte line cells migrate to the thymus and persist there. *Nature*. 1982; 300(5889):262–264.
7. Liblau RS, Singer SM, McDevitt HO. Th1 and Th2 CD4+ T cells in the pathogenesis of organ-specific autoimmune diseases. *Immunol Today*. 1995;16(1):34–38.
8. Pasteur L, Chamberland, Roux. Summary report of the experiments conducted at Pouilly-le-Fort, near Melun, on the anthrax vaccination, 1881. *Yale J Biol Med*. 2002;75(1):59–62.
9. Ben-Nun A, Wekerle H, Cohen IR. Vaccination against autoimmune encephalomyelitis with T-lymphocyte line cells reactive against myelin basic protein. *Nature*. 1981;292(5818):60–61.
10. Lider O, et al. Vaccination against experimental autoimmune encephalomyelitis using a subencephalitogenic dose of autoimmune effector T cells. (2). Induction of a protective anti-idiotypic response. *J Autoimmun*. 1989;2(1):87–99.
11. Holoshitz J, et al. Lines of T lymphocytes induce or vaccinate against autoimmune arthritis. *Science*. 1983; 219(4580):56–58.
12. Maron R, et al. T lymphocyte line specific for thyroglobulin produces or vaccinates against autoimmune thyroiditis in mice. *J Immunol*. 1983;131(5):2316–2322.
13. Ben-Yehuda A, et al. Lymph node cell vaccination against the lupus syndrome of MRL/lpr/lpr mice. *Lupus*. 1996;5(3):232–236.
14. De Alboran IM, et al. lpr T cells vaccinate against lupus in MRL/lpr mice. *Eur J Immunol*. 1992;22(4):1089–1093.
15. Elias D, et al. Regulation of NOD mouse autoimmune diabetes by T cells that recognize a TCR CDR3 peptide. *Int Immunol*. 1999;11(6):957–966.
16. Shapira OM, et al. Prolongation of survival of rat cardiac allografts by T cell vaccination. *J Clin Invest*. 1993;91(2):388–390.
17. Hafler DA, et al. T cell vaccination in multiple sclerosis: a preliminary report. *Clin Immunol Immunopathol*. 1992;62(3):307–313.
18. van Laar JM, et al. Effects of inoculation with attenuated autologous T cells in patients with rheumatoid arthritis. *J Autoimmun*. 1993;6(2):159–167.
19. Rivera VM. Tovaxin for multiple sclerosis. *Expert Opin Biol Ther*. 2011;11(7):961–967.
20. http://www.nationalmssociety.org/About-the-Society/News/Tovaxin%C2%AE-%28T-cell-vaccination%29-granted-fast-track-d.
21. Nam JL, et al. Current evidence for the management of rheumatoid arthritis with biological disease-modifying antirheumatic drugs: a systematic literature review informing the EULAR recommendations for the management of RA. *Ann Rheum Dis*. 2010;69(6):976–986.
22. Karussis D, et al. T cell vaccination benefits relapsing progressive multiple sclerosis patients: a randomized, double-blind clinical trial. *PLoS One*. 2012;7(12):e50478.
23. Achiron A, et al. T cell vaccination in multiple sclerosis relapsing-remitting nonresponders patients. *Clin Immunol*. 2004;113(2):155–160.
24. Achiron A, Mandel M. T-cell vaccination in multiple sclerosis. *Autoimmun Rev*. 2004;3(1):25–32.
25. Vandenbark AA, Abulafia-Lapid R. Autologous T-cell vaccination for multiple sclerosis: a perspective on progress. *BioDrugs*. 2008;22(4):265–273.
26. Zhang J, et al. T cell vaccination: clinical application in autoimmune diseases. *J Mol Med Berl*. 1996;74(11):653–662.
27. Huang X, Wu H, Lu Q. The mechanisms and applications of T cell vaccination for autoimmune diseases: a comprehensive review. *Clin Rev Allergy Immunol*. 2014;47(2):219–233.
28. Bridges Jr SL, Moreland LW. T-cell receptor peptide vaccination in the treatment of rheumatoid arthritis. *Rheum Dis Clin North Am*. 1998;24(3):641–650.
29. Chen G, et al. Vaccination with selected synovial T cells in rheumatoid arthritis. *Arthritis Rheum*. 2007;56(2): 453–463.
30. Moreland LW, et al. V beta 17 T cell receptor peptide vaccination in rheumatoid arthritis: results of phase I dose escalation study. *J Rheumatol*. 1996;23(8):1353–1362.
31. Moreland LW, et al. T cell receptor peptide vaccination in rheumatoid arthritis: a placebo-controlled trial using a combination of Vbeta3, Vbeta14, and Vbeta17 peptides. *Arthritis Rheum*. 1998;41(11):1919–1929.
32. Li ZG, et al. T cell vaccination in systemic lupus erythematosus with autologous activated T cells. *Lupus*. 2005;14(11):884–889.
33. Xu W, Yuan Z, Gao X. Immunoproteomic analysis of the antibody response obtained in mouse following vaccination with a T-cell vaccine. *Proteomics*. 2011;11(22):4368–4375.

34. Zhang XY, et al. Anti-T-cell humoral and cellular responses in healthy BALB/c mice following immunization with ovalbumin or ovalbumin-specific T cells. *Immunology*. 2003;108(4):465–473.
35. Alt FW, et al. Mechanisms of programmed DNA lesions and genomic instability in the immune system. *Cell*. 2013;152(3):417–429.
36. Schatz DG, Swanson PC. V(D)J recombination: mechanisms of initiation. *Annu Rev Genet*. 2011;45:167–202.
37. Lider O, et al. Anti-idiotypic network induced by T cell vaccination against experimental autoimmune encephalomyelitis. *Science*. 1988;239(4836):181–183.
38. Howell MD, et al. Vaccination against experimental allergic encephalomyelitis with T cell receptor peptides. *Science*. 1989;246(4930):668–670.
39. Lal G, Shaila MS, Nayak R. Activated mouse T cells downregulate, process and present their surface TCR to cognate anti-idiotypic CD4$^+$ T cells. *Immunol Cell Biol*. 2006;84(2):145–153.
40. Sun D, et al. Suppression of experimentally induced autoimmune encephalomyelitis by cytolytic T-T cell interactions. *Nature*. 1988;332(6167):843–845.
41. Ivanova IP, et al. Induction of antiidiotypic immune response with autologous T-cell vaccine in patients with multiple sclerosis. *Bull Exp Biol Med*. 2008;146(1):133–138.
42. Volovitz I, et al. T cell vaccination induces the elimination of EAE effector T cells: analysis using GFP-transduced, encephalitogenic T cells. *J Autoimmun*. 2010;35(2):135–144.
43. Bouwer HG, Hinrichs DJ. T-cell vaccination prevents EAE effector cell development but does not inhibit priming of MBP responsive cells. *J Neurosci Res*. 1996;45(4):455–462.
44. Hong J, et al. CD4$^+$ regulatory T cell responses induced by T cell vaccination in patients with multiple sclerosis. *Proc Natl Acad Sci USA*. 2006;103(13):5024–5029.
45. Cohen IR. The cognitive paradigm and the immunological homunculus. *Immunol Today*. 1992;13(12):490–494.
46. Lohse AW, et al. Control of experimental autoimmune encephalomyelitis by T cells responding to activated T cells. *Science*. 1989;244(4906):820–822.
47. Quintana FJ, et al. HSP60 as a target of anti-ergotypic regulatory T cells. *PLoS One*. 2008;3(12):e4026.
48. Cohen IR. Real and artificial immune systems: computing the state of the body. *Nat Rev Immunol*. 2007;7(7): 569–574.
49. Mimran A, Cohen IR. Regulatory T cells in autoimmune diseases: anti-ergotypic T cells. *Int Rev Immunol*. 2005;24(3–4):159–179.
50. Mimran A, et al. Anti-ergotypic T cells in naive rats. *J Autoimmun*. 2005;24(3):191–201.
51. Quintana FJ, Cohen IR. Anti-ergotypic immunoregulation. *Scand J Immunol*. 2006;64(3):205–210.
52. Mimran A, et al. DNA vaccination with CD25 protects rats from adjuvant arthritis and induces an antiergotypic response. *J Clin Invest*. 2004;113(6):924–932.
53. Mor F, et al. IL-2 and TNF receptors as targets of regulatory T-T interactions: isolation and characterization of cytokine receptor-reactive T cell lines in the Lewis rat. *J Immunol*. 1996;157(11):4855–4861.
54. Quintana FJ, Cohen IR. The HSP60 immune system network. *Trends Immunol*. 2011;32(2):89–95.
55. Cohen IR. Autoantibody repertoires, natural biomarkers, and system controllers. *Trends Immunol*. 2013;34(12): 620–625.
56. Quintana FJ, Carmi P, Cohen IR. DNA vaccination with heat shock protein 60 inhibits cyclophosphamide-accelerated diabetes. *J Immunol*. 2002;169(10):6030–6035.
57. Quintana FJ, et al. Inhibition of adjuvant arthritis by a DNA vaccine encoding human heat shock protein 60. *J Immunol*. 2002;169(6):3422–3428.
58. Quintana FJ, et al. Inhibition of adjuvant-induced arthritis by DNA vaccination with the 70-kd or the 90-kd human heat-shock protein: immune cross-regulation with the 60-kd heat-shock protein. *Arthritis Rheum*. 2004;50(11):3712–3720.
59. Raz I, et al. Beta-cell function in new-onset type 1 diabetes and immunomodulation with a heat-shock protein peptide (DiaPep277): a randomised, double-blind, phase II trial. *Lancet*. 2001;358(9295):1749–1753.
60. Zanin-Zhorov A, et al. Heat shock protein 60 enhances CD4$^+$ CD25$^+$ regulatory T cell function via innate TLR2 signaling. *J Clin Invest*. 2006;116(7):2022–2032.
61. Kuniyasu Y, et al. Naturally anergic and suppressive CD25(+)CD4(+) T cells as a functionally and phenotypically distinct immunoregulatory T cell subpopulation. *Int Immunol*. 2000;12(8):1145–1155.
62. Gershon RK, Kondo K. Cell interactions in the induction of tolerance: the role of thymic lymphocytes. *Immunology*. 1970;18(5):723–737.
63. Lu L, Cantor H. Generation and regulation of CD8(+) regulatory T cells. *Cell Mol Immunol*. 2008;5(6):401–406.

64. Jiang H, et al. T cell vaccination induces T cell receptor Vbeta-specific Qa-1-restricted regulatory CD8(+) T cells. *Proc Natl Acad Sci USA*. 1998;95(8):4533–4537.
65. Varthaman A, et al. Physiological induction of regulatory Qa-1-restricted $CD8^+$ T cells triggered by endogenous $CD4^+$ T cell responses. *PLoS One*. 2011;6(6):e21628.
66. Sarantopoulos S, Lu L, Cantor H. Qa-1 restriction of $CD8^+$ suppressor T cells. *J Clin Invest*. 2004;114(9):1218–1221.
67. Varthaman A, et al. Control of T cell reactivation by regulatory Qa-1-restricted $CD8^+$ T cells. *J Immunol*. 2010;184(12):6585–6591.
68. Herkel J, et al. Humoral mechanisms in T cell vaccination: induction and functional characterization of anti-lymphocytic autoantibodies. *J Autoimmun*. 1997;10(2):137–146.
69. Hong J, et al. Reactivity and regulatory properties of human anti-idiotypic antibodies induced by T cell vaccination. *J Immunol*. 2000;165(12):6858–6864.
70. Madi A, et al. T-cell receptor repertoires share a restricted set of public and abundant CDR3 sequences that are associated with self-related immunity. *Genome Res*. 2014;24(10):1603–1612.
71. Miles JJ, Douek DC, Price DA. Bias in the alphabeta T-cell repertoire: implications for disease pathogenesis and vaccination. *Immunol Cell Biol*. 2011;89(3):375–387.
72. Venturi V, et al. The molecular basis for public T-cell responses? *Nat Rev Immunol*. 2008;8(3):231–238.
73. Garcia KC, Teyton L, Wilson IA. Structural basis of T cell recognition. *Annu Rev Immunol*. 1999;17:369–397.
74. Hagedorn PH, et al. Chronic rejection of a lung transplant is characterized by a profile of specific autoantibodies. *Immunology*. 2010;130(3):427–435.
75. Merbl Y, et al. A systems immunology approach to the host-tumor interaction: large-scale patterns of natural autoantibodies distinguish healthy and tumor-bearing mice. *PLoS One*. 2009;4(6):e6053.
76. Cohen IR. Discrimination and dialogue in the immune system. *Semin Immunol*. 2000;12(3):215–219. discussion 257–344.
77. Cohen IR. Tending Adam's garden: evolving the cognitive immune self. In: Cohen IR, ed. *Tending Adam's Garden*. London: Academic Press; 2000.
78. Elias D, et al. Vaccination against autoimmune mouse diabetes with a T-cell epitope of the human 65-kDa heat shock protein. *Proc Natl Acad Sci USA*. 1991;88(8):3088–3091.
79. Tikochinski Y, et al. A shared TCR CDR3 sequence in NOD mouse autoimmune diabetes. *Int Immunol*. 1999;11(6):951–956.
80. Birk OS, et al. NOD mouse diabetes: the ubiquitous mouse hsp60 is a beta-cell target antigen of autoimmune T cells. *J Autoimmun*. 1996;9(2):159–166.
81. Wallberg M, Cooke A. Immune mechanisms in type 1 diabetes. *Trends Immunol*. 2013;34(12):583–591.
82. Elias D, Cohen IR. Peptide therapy for diabetes in NOD mice. *Lancet*. 1994;343(8899):704–706.
83. Elias D, et al. Hsp60 peptide therapy of NOD mouse diabetes induces a Th2 cytokine burst and downregulates autoimmunity to various beta-cell antigens. *Diabetes*. 1997;46(5):758–764.
84. Schloot NC, Cohen IR. DiaPep277(R) and immune intervention for treatment of type 1 diabetes. *Clin Immunol*. 2013;149(3):307–316.
85. Cohen IR, Quintana FJ, Mimran A. Tregs in T cell vaccination: exploring the regulation of regulation. *J Clin Invest*. 2004;114(9):1227–1232.
86. Wekerle H, Berer K, Krishnamoorthy G. Remote control-triggering of brain autoimmune disease in the gut. *Curr Opin Immunol*. 2013;25(6):683–689.
87. Madi A, et al. Tumor-associated and disease-associated autoantibody repertoires in healthy colostrum and maternal and newborn cord sera. *J Immunol*. 2015;194(11):5272–5281.

CHAPTER

30

Reversal of Misfortune: Therapeutic Strategies on the Horizon

L. Steinman

Stanford University, Stanford, CA, United States

OUTLINE

There are 10 approved drugs for relapsing-remitting multiple sclerosis (RRMS). The major challenge in RRMS is for predictive biomarkers that would enable us to optimize the already approved therapies and any future therapies. We need to know the best intervention, or combination of therapeutics, for a given individual. So far, there is one major biomarker that is validated for multiple sclerosis (MS). This biomarker is a risk-mitigation assay, approved by regulatory authorities in North America and Europe, to determine who is at risk for the development of progressive multifocal leukoencephalopathy after treatment with natalizumab.[1–3] The major challenge is the development of new therapies for the progressive forms of disease, both primary-progressive MS (PPMS) and secondary-progressive MS (SPMS). On the horizon for progressive forms of MS, I shall describe three approaches with promise: First, approaches targeting B cells and even plasma cells are logical targets for the progressive phases of disease given our understanding of the neuropathology of these forms of disease.[4,5] Second, therapies

Translational Neuroimmunology in Multiple Sclerosis
http://dx.doi.org/10.1016/B978-0-12-801914-6.00030-1

with small molecules, especially those that either cross the blood–brain barrier or those that at least meet the blood–brain barrier, may be attractive new therapeutic entities for progressive disease. A third approach that is worth exploring is the arming of monocytic cells with protective cytokines, while disarming them from release of pathogenic mediators. After all, monocytic cells have elaborate machinery for traversing from the blood to the brain. This machinery could be harvested for benefit. Much worthy attention has been given to implantation of stem cells in the brain. I propose here to exploit mechanisms whereby protective immune cells gain entry into brain, in order to provide a new opportunity in regenerative medicine. This third approach would be an ironic intervention for progressive forms of MS. A central aspect of the pathogenesis of MS is the transmigration of pathogenic immune cells. This approach of reprogramming cells with guardian molecules would represent a true reversal of misfortune.

30.1 TARGETING B CELLS FOR PROGRESSIVE ASPECTS OF MULTIPLE SCLEROSIS

The B cell and the plasmablast comprise an important component of lesions analyzed in both PPMS and SPMS.[4] Treatment with anti-CD20 and anti-CD19 is currently being attempted in RRMS and PPMS. However, it is unclear whether sufficient amounts of these antibodies will enter the central nervous system (CNS) during progressive phases of MS. It is also unclear whether anti-CD20 or anti-CD19 can target these nests of B cells. Hauser writes in his Charcot Prize lecture, "If established B cell nests residing in lymphoid follicle-like structures in the meninges are drivers of a chronic neurodegenerative process that ultimately results in progressive MS, anti-CD20 therapy would likely fail to deplete B cells from these sites. This resistance of B cells in protective niches could explain the relatively meager response of Rituxan (RTX) in primary progressive MS, and our observation that individuals with relapsing MS can evolve to secondary progressive MS despite ongoing RTX treatment."[5]

30.2 TARGETING MYELOID CELLS IN THE CENTRAL NERVOUS SYSTEM WITH SMALL MOLECULES THAT CROSS THE BLOOD–BRAIN BARRIER FOR PROGRESSIVE MULTIPLE SCLEROSIS

To target B cells in their protective niches in brain, approaches with small molecules may be best. Ibruntinib, the Bruton tyrosine kinase inhibitor, might be a productive approach to targeting such B cells. Mice with a deficiency of the Bruton tyrosine kinase in myeloid lineages develop reduced severity of experimental autoimmune encephalomyelitis (EAE).[6]

Other tyrosine kinase inhibitors like imatinib are effective at reducing clinical and pathologic disease in EAE, and might be tried in progressive MS.[7] Clinical trials with small molecules including modulators of sphingosine phosphate receptors may be promising for progressive forms of MS, as they cross the blood–brain barrier.[8,9] However, a trial of fingolimod in PPMS was unsuccessful.[10] Whether fingolimod may be beneficial in SPMS is an open matter.

Other potential approaches to treatment of progressive forms of MS include dimethylfumarate and cysteamine. Both drugs are effective in EAE, and cross the blood–brain barrier.

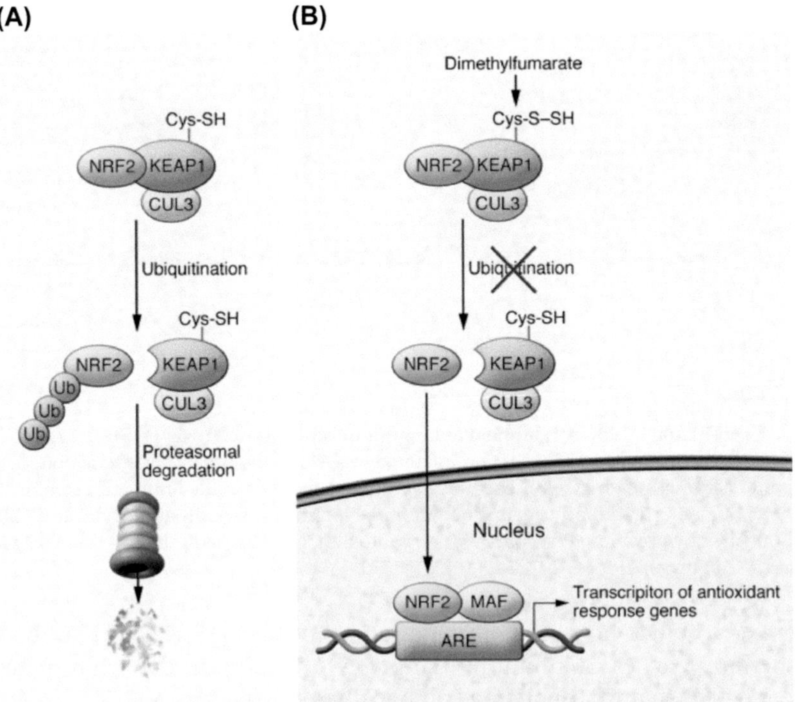

FIGURE 30.1 **Mechanism of action of dimethylfumarate in the NRF2 pathway.** Inflammation and oxidative stress are thought to promote tissue damage and multiple sclerosis (MS), and recent data point to a protective role for antioxidant pathways, including the transcription factor NRF2, in MS. The drug dimethylfumarate, which targets the NRF2 pathway, is approved for the treatment of relapsing-remitting MS and is under investigation in clinical trials for progressive forms of MS. Dimethylfumarate has pleiotropic mechanisms of action, but activation of NRF2 accounts for some of its antiinflammatory activity. (A) KEAP1, which is part of a cullin family E3 ubiquitin ligase complex, normally mediates ubiquitination and proteasomal degradation of NRF2. (B) When KEAP1 undergoes sulfhydration of cysteine 151 via binding of fumarate, it becomes unable to interact with NRF2, which accumulates in the nucleus and induces the expression of NRF2-dependent antioxidant and cytoprotective genes. Activation of NRF2 leads to increased production of a spectrum of antioxidants, including glutathione and cystathionine. *Reprinted with author's permission from Steinman L. No quiet surrender.* J Clin Invest. *April 1, 2015;125(4):1371–1378.*

They may be effective in downregulating overactive oxidative pathophysiology in neurodegenerative conditions. Both dimethylfumarate and cysteamine interact with kelch-like ECH-associated protein (KEAP1), and thereby activate the nuclear factor (erythroid-derived 2)–like 2 (NRF2) antioxidant pathway[8] (Fig. 30.1).

30.3 ARMING PENETRATING MYELOID CELLS WITH SUPPRESSIVE AND RESTORATIVE MOLECULES

MS is in part a disease directed at the CNS from the outside. We surmise from the success of blocking lymphocyte homing to the brain that inhibiting immune cell infiltration reduces the inflammatory response in MS to a large degree. Fig. 30.2 is a schematic demonstrating

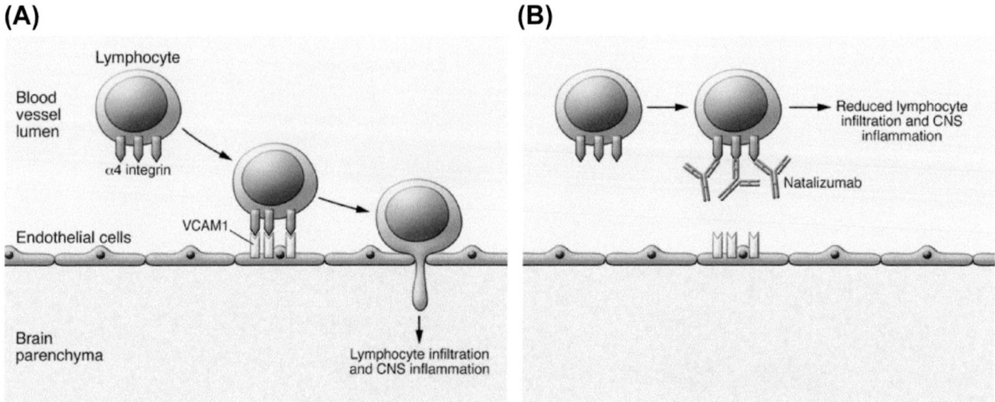

FIGURE 30.2 **Natalizumab blocks lymphocyte homing in multiple sclerosis (MS).** (A) Alpha4 integrin binds to vascular cell adhesion molecule 1 (VCAM1) on inflamed brain endothelium. This interaction gives lymphocytes access to the central nervous system (CNS). The presence of immune cells in the brain is a prominent feature of MS. (B) Natalizumab, a humanized antibody against α4 integrin, blocks binding of lymphocytes to VCAM on inflamed brain endothelium, thereby preventing lymphocyte entry into the CNS. *Used with author's permission from Steinman L. No quiet surrender.* J Clin Invest. *April 1, 2015;125(4):1371–1378.*

how we are able to block homing and penetration to the CNS with an antibody to alpha4 integrin. Other approaches to inhibiting lymphocyte homing to the brain include blockade with small molecules that modulate the sphingosine phosphate receptor.

On the horizon with our increasing capacity to transduce the genetic program of cells, I predict that we can update an approach published 18 years ago. By transfecting infiltrating cells with molecules like IL-4, our group was able to attenuate EAE via local delivery of the cytokine to areas of inflammation.[8,11] Thus capitalizing on the well-known, and elaborately effective mechanism whereby myeloid cells penetrate into the CNS parenchyma, we can engage in a classic example of immunologic jujutsu. We might reverse pathology in MS, by arming myeloid cells with suppressive cytokines like IL-4, IL-10, and IL-27, regulatory transcription factors like FoxP3, or even with growth factors like brain-derived neurotrophic factor.[8] Immunologic jujutsu for a reversal of neuroimmunologic misfortune is a wonderful aspiration for future therapies in MS!

References

1. Rudick R, Polman C, Clifford D, Miller D, Steinman L. Natalizumab bench to bedside and beyond. *JAMA Neurol.* 2013;70(2):172–182.
2. US Food and Drug Administration. FDA Permits Marketing of First Test for Risk of Rare Brain Infection in Some People Treated with Tysabri. http://www.fda.gov/NewsEvents/Newsroom/PressAnnouncements/ucm288471.htm; Accessed 17.04.15.
3. Gorelik L, Lerne RM, Bixler S, et al. Anti-JC virus antibodies: implications for PML risk stratification. *Ann Neurol.* 2010;68(3):295–303.
4. Frischer JM, Bramow S, Dal-Bianco A, et al. The relation between inflammation and neurodegeneration in multiple sclerosis brains. *Brain.* 2009;132:1175–1189.
5. Hauser SL. The Charcot Lecture | beating MS: a story of B cells, with twists and turns. *Mult Scler.* 2015;21(1):8–21.

6. Mangla A, Khare A, Vineeth V, et al. Pleiotropic consequences of Bruton tyrosine kinase deficiency in myeloid lineages lead to poor inflammatory responses. *Blood*. August 15, 2004;104(4):1191–1197.
7. Crespo O, Kang SC, Daneman R, et al. Tyrosine kinase inhibitors ameliorate autoimmune encephalomyelitis in a mouse model of multiple sclerosis. *J Clin Immunol*. 2011;231(6):1010–1020.
8. Steinman L. No quiet surrender. *J Clin Invest*. April 1, 2015;125(4):1371–1378.
9. Proia R, Hla T. Emerging biology of sphingosine-1-phosphate: its role in pathogenesis and therapy. *J Clin Invest*. April 1, 2015;125(4):1379–1387.
10. http://www.novartis.com/newsroom/media-releases/en/2014/1875463.shtml; Accessed 18.04.15.
11. Shaw MK, Lorens JB, Dhawan A, et al. Local delivery of interleukin 4 by retrovirus-transduced T lymphocytes ameliorates experimental autoimmune encephalomyelitis. *J Exp Med*. May 5, 1997;185(9):1711–1714.

Index

'*Note*: Page numbers followed by "f" indicate figures, "t" indicate tables and "b" indicate boxes.'

D

Printed in the United States
By Bookmasters